The Organic Chemistry of Iron

Volume 1

ORGANOMETALLIC CHEMISTRY

A Series of Monographs

BRIAN G. RAMSEY: Electronic Transitions in Organometalloids, 1969.
R. C. POLLER: The Chemistry of Organotin Compounds, 1970.
RUSSELL N. GRIMES: Carboranes, 1970.
PETER M. MAITLIS: The Organic Chemistry of Palladium, Volumes I and II, 1971.
DONALD S. MATTESON: Organometallic Reaction Mechanisms of the Nontransition Elements, 1974.
RICHARD F. HECK: Organotransition Metal Chemistry: A Mechanistic Approach, 1974.
P. W. JOLLY AND G. WILKE: The Organic Chemistry of Nickel, Volume I, Organonickel Complexes, 1974. Volume II, Organic Synthesis, 1975.
P. C. WAILES, R. S. P. COUTTS, AND H. WEIGOLD: Organometallic Chemistry of Titanium, Zirconium, and Hafnium, 1974.
U. BELLUCO: Organometallic and Coordination Chemistry of Platinum, 1974.
P. S. BRATERMAN: Metal Carbonyl Spectra, 1974.
L. MALATESTA AND S. CENINI: Zerovalent Compounds of Metals, 1974.
THOMAS ONAK: Organoborane Chemistry, 1975.
R. P. A. SNEEDEN: Organochromium Compounds, 1975.
A. G. SHARPE: The Chemistry of Cyano Complexes of the Transition Metals, 1976.
ERNST A. KOERNER VON GUSTORF, FRIEDRICH-WILHELM GREVELS, AND INGRID FISCHLER: The Organic Chemistry of Iron, Vol. 1, 1978.

The Organic Chemistry of Iron

Volume 1

edited by

ERNST A. KOERNER VON GUSTORF

FRIEDRICH-WILHELM GREVELS

INGRID FISCHLER

Institut für Strahlenchemie
Max-Planck-Institut für Kohlenforschung
Mülheim a. d. Ruhr, Germany

Academic Press
New York San Francisco London 1978
A Subsidiary of Harcourt Brace Jovanovich, Publishers

ACADEMIC PRESS, INC.
111 Fifth Avenue, New York, New York 10003

United Kingdom Edition published by
ACADEMIC PRESS, INC. (LONDON) LTD.
24/28 Oval Road, London NW1 7DX

Library of Congress Cataloging in Publication Data

Main entry under title:

The Organic chemistry of iron.

(Organometallic chemistry series)
Includes index.
1. Organoiron compounds. I. Koerner von Gustorf, 1932–1975. II. Fischler, Ingrid. III. Grevels, Friedrich Wilhelm.
QD412.F4073 547'.05'621 77-16071
ISBN 0-12-417101-X (v. 1)

PRINTED IN THE UNITED STATES OF AMERICA

To the memory of

Ernst A. Koerner von Gustorf

(1932–1975)

Contents

List of Contributors

Numbers in parentheses indicate the pages on which authors' contributions begin.

B. L. Barnett (1), Max-Planck-Institut für Kohlenforschung, Lembkestrasse 5, Mülheim a. d. Ruhr, Germany

F. L. Bowden (345), Chemistry Department, The University of Manchester, Institute of Science and Technology, Manchester M6O 1QD, England

D. Brauer (1), Max-Planck-Institut für Kohlenforschung, Lembkestrasse 5, Mülheim a. d. Ruhr, Germany

Henri Brunner (299), Chemisches Institut der Universität, Regensburg, Germany

R. B. King (397, 463, 525), Department of Chemistry, University of Georgia, Athens, Georgia 30602

Edgar König (213, 257), Institut für Physikalische Chemie II, Universität Erlangen-Nürnberg, 8520 Erlangen, Germany

Carl Krüger (1), Max-Planck-Institut für Kohlenforschung, Lembkestrasse 5, Mülheim a. d. Ruhr, Germany

Joseph M. Landesberg (627), Department of Chemistry, Adelphi University, Garden City, New York 11530

Tobin J. Marks (113), Department of Chemistry, Northwestern University, Evanston, Illinois 60201

Jörn Müller (145), Institut für Anorganische und Analytische Chemie der Technischen Universität, Berlin, Germany

R. V. Parish (175), The University of Manchester, Institute of Science and Technology, Manchester M6O 1QD, England

L. H. Wood (345), Chemistry Department, The University of Manchester, Institute of Science and Technology, Manchester M6O 1QD, England

Foreword

The large body of information that today forms organo–transition metal chemistry can be classified and discussed in two ways. The properties and reactivities of given ligands when attached to different metals can be emphasized; alternatively, the chemistry of one metal can be defined in terms of the effect it has on differing ligands. A complete understanding requires both approaches and one long-term aim of the Organometallic Chemistry Monograph Series is to provide this.

In addition to books on specific ligands and on more general topics, some years ago we inaugurated the survey of the organic chemistry of individual transition metals. Volumes on organo–titanium, –zirconium, –hafnium, –chromium, –nickel, –palladium, and –platinum chemistry have already appeared, and monographs on organo–molybdenum, –tungsten, –cobalt, –rhodium, –iridium, –ruthenium, and –osmium chemistry are in preparation.

We were particularly pleased, at the time that these volumes were being planned, that Dr. Ernst A. Koerner von Gustorf of the Institut für Strahlenchemie im Max-Planck-Institut in Mülheim agreed to undertake *The Organic Chemistry of Iron.* Koerner von Gustorf was an ideal choice; he had established himself as a very innovative organometallic chemist, particularly in respect to his work on the photochemical syntheses of organo–iron complexes, and he also had access to the archives of the Max-Planck-Institut in Mülheim with their reference files on organometallic chemistry.

Even with this background, the task of organizing and compiling all the knowledge of organo–iron chemistry was impossible for one man to accomplish within a reasonable space of time, and Koerner von Gustorf wisely enlisted the help of a number of other distinguished experts to author specific chapters of the work.

The organization of all this material was just beginning at the time of Koerner von Gustorf's untimely and tragic death in September 1975 at the age of 43. It accordingly fell to his collaborators, Dr. F. W. Grevels and Dr. Ingrid Fischler, to actually undertake the onerous task of collating and editing the final manuscripts. This task was greater than it had been for other volumes in the series since it had been decided to print directly from camera-ready typescript and the work of editing as well as of organizing both the typing and the drawing of the diagrams and the formulas was a very extensive one. We are very grateful to Dr. Grevels and Dr. Fischler for having succeeded so well.

From our knowledge of the scope of the topic we had expected a large volume; in the event the project has grown beyond our original estimate and has, owing to Koerner von Gustorf's death, taken longer. We hope that this first of two volumes will be well received and that the extra material that has gone into it will make it even more useful. We would also like to thank all the individual authors for their carefully and comprehensively organized contributions and also for their patience with the delays that have occurred. Doctors Grevels and Fischler have been able to ensure that all the contributions have been updated and in most cases the literature has been covered up to and including the year 1975.

It was not possible to adhere entirely to the order that Koerner von Gustorf had planned for the material, but the reader will see that a logical arrangement has been followed. Volume 1 covers the structures and bonding and the applications of a variety of physical techniques to organo–iron compounds, optically active compounds, as well as chapters on σ-bonded, η^2-, η^3-, and η^4-organo–iron compounds. Volume 2 will be developed in a similar manner and will include further chapters on spectroscopy, both electronic and vibrational (Poliakoff and Turner), metal–metal bonded compounds and iron cluster chemistry (Dahl and Sinclair; Chini), complexes of polyenes (Kerber), arenes (King), and those derived from acetylenes and cumulenes (Müller), complexes with N–bonded ligands (tom Dieck and De Paoli), as well as a short discussion of ferrocene chemistry (Eagar and Richards).

P. M. MAITLIS
F. G. A. STONE
ROBERT WEST

Structure and Bonding in Organic Iron Compounds

By CARL KRÜGER

and

B.L. BARNETT and D. BRAUER

Max-Planck-Institut für Kohlenforschung
Lembkestr. 5, Mülheim a.d.Ruhr, Germany

TABLE OF CONTENTS

ISBN 0-12-417101-X (v.1)

I. INTRODUCTION

SCOPE OF THE SURVEY.

This article is concerned with the results of structural analyses, and to a lesser extent, with the bonding principles of organometallic iron compounds. The aim is to survey as completely as possible structural details presently available on these compounds, and to indicate problems which are still under discussion. Literature has been searched up to 1975, and all relevant structural results are summarized by tables referred to by each section of the article. However, due to delays in the editing, the discussion part of this review only covers those reports published up to 1972. Data quoted in these tables represent published values, including non-significant digits. The outline of this review follows the editorial arrangement of this volume.

For each class of compounds discussed, one or more typical examples were selected for computer drawings. The selected illustrations are not intended to reflect the accuracy of the individual structural determinations.

The author wishes to caution the uninitiated reader of structural publications from uncritical acceptance of accuracy of crystallographic data. Neither a low R-value (residual index, defined as

$$R = \frac{\Sigma \| F_o | - | F_c \|}{\Sigma | F_o |}$$

thus describing the fit of a molecular model (F_c) to a given data set (F_o)) nor low standard deviations are *alone* an absolute measure for the accuracy of an X-ray structural determination. A low R-value, which represents good precision and not necessarily good accuracy, could describe an excellent fit of a distorted structural model to a poor data set, hence giving rise to erroneous interpretations. Systematic, but not obvious errors in intensity measurements, can give rise to underestimates of standard deviations. These values are obtained from the inverse error matrices of the usual least-squares refinement, and decrease with the increasing symmetry of the crystal system. Therefore one should avoid overly interpreting crystallographic data. If these data are the basis for theoretical computations, the advice of a specialist in the field should be sought.

However, the precision of structural work increased tremendously during the last decade with the introduction of diffractometers and better mathematical formalisms and computational techniques. The estimated errors in bonding distances between the non-hydrogen atoms may now be in the range of

0.005 to 0.01 Å, and those of the bonding angles 0.5° to 2°. Hopefully some of the earlier, but fundamental, work will be repeated with the accuracy presently possible. Despite the limitations mentioned above, X-ray diffraction methods still yield the most accurate information about the geometry of complex molecules.

II. σ-BONDED IRON CARBON COMPOUNDS

BONDING

Compounds containing iron-carbon bonds are sensitive to homolytic cleavage, producing a free organic radical and the metal in a lower oxidation state. This instability is explained by small energy differences between the filled *d* orbitals and the valence *s* and *p* orbitals of the metal used in bonding to the carbon. As a result, high energy *d* electrons can transfer to antibonding orbitals of the Fe-C bond. In compounds of higher oxidation state, electrons of the Fe-C bond may move into vacant metal *d* orbitals. Both formalisms result in a weakening of the metal-carbon bond.

From this scheme, Fe-C bonds may be stabilized in two ways. Often ligands with acceptor properties (such as carbon monoxide, η^5-cyclopentadienyl, phosphines, arsines, *etc.*) are attached to the metal, along with additional ligands to complete a stable electronic configuration. Secondly, the Fe-C bond may be strengthened by altering the effective electronegativity of the carbon by using different hybridization states of the carbon (sp^3 - sp^2 - sp). Attaching strongly electronegative substituents to the carbon (*e.g.* fluorine) gives similar results.

All effects mentioned are reflected in the observed Fe-C bond lengths. In addition, the geometry, the oxidation state of the iron, as well as the steric arrangement and the electronic properties of the other ligands, may influence the observed bond lengths. Average values stated below should only be accepted within the limits of these considerations.

STRUCTURAL DETAILS OF FE-C σ-BONDS

Few X-ray structures containing Fe-C(sp^3) bonds have been reported so far. The observed bond lengths range from 2.08 to 2.16 Å (see Table 1), with the average value being 2.11 Å.

Introducing fluorine substituents at the bonded carbon (*e.g.* Fe-CF_2-) shortens this value remarkably to 2.07 Å *(131)*. A similar value (2.06 Å) has been reported for an acetic acid dicarbonylcyclopentadieneiron complex, the acetic acid being the σ-bonded group *via* the α-carbon *(23,330)* (see Figure 1). In the latter compound, an interaction between Fe and the C atom of the carboxylic group (Fe-C : 2.85 Å) is suggested. This is supported by the unusually low pK value of the acid and correspondingly lengthened C-O bonds (1.32 Å) in the carboxylic group.

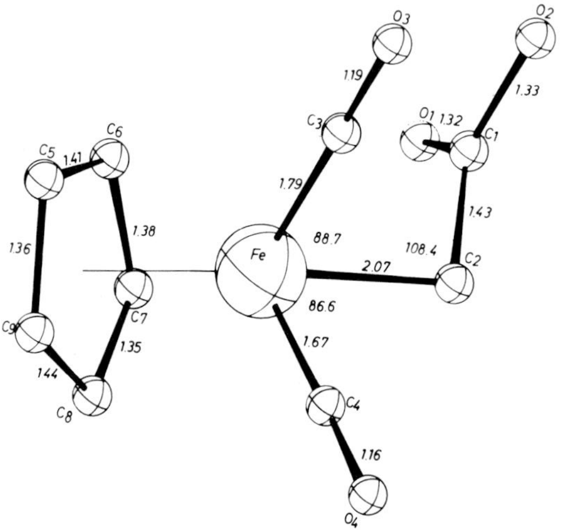

Fig. 1: The molecular structure of $(\eta^5\text{-}C_5H_5)Fe(CO)_2CH_2COOH$

Interestingly, perfluorobutadiene is found to be σ-bonded to a tetracarbonyliron moiety *(359)*. The average Fe-C distance in this planar ring system (see Table 1, Nr. 3) is 2.00 Å.

Among the earlier published structures containing Fe-C (sp^2) and Fe-C(sp) bonds are those of hexamethylisocyanidoferrous chloride *(537)* and β-tetramethyl ferrocyanide *(379)*. This class of compounds, which include iron-σ-olefin, iron-σ-arene, iron-carbene and ferracyclopentadiene moieties, is listed in Table 2. The observed Fe-C(sp^2) distances in these compounds range from 1.89 Å to 2.11 Å, clustering around 1.98 Å. A typical example of this class, the structure of *trans*-1,4-bis[dicarbonyl-(η^5-cyclopentadienyl)iron]buta-1,3-diene, determined independendly by two groups, is shown in Figure 2.

The olefinic bond in the butadiene system is not altered by the end-on bonding to the transition metal (C=C in 1,3-butadiene: 1.337(5) Å).

Examples of structures containing Fe-(C=C) groups coordinated to a *second* iron atom within one molecule are also known (see Table 2). The reported Fe-C distances range from 1.96 to 2.09 Å in these compounds. The participation of the iron atom

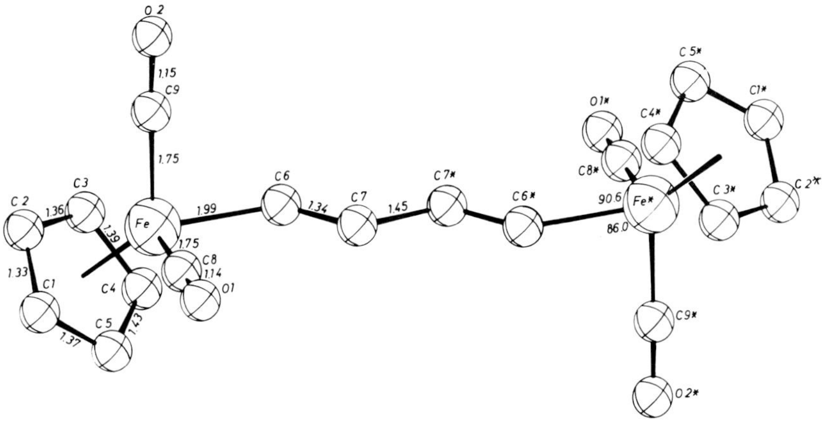

Fig. 2: The molecular structure of trans-1,4-bis[dicarbonyl-(η^5-cyclopentadienyl)iron]buta-1,3-diene (Refs. 141,238).

in an η^3-ferra-allylic system, implying a partial multiple bond order of the Fe-C bond (see Figure 3), has been suggested in a compound where this Fe-C distance has been reduced to 1.89 Å.

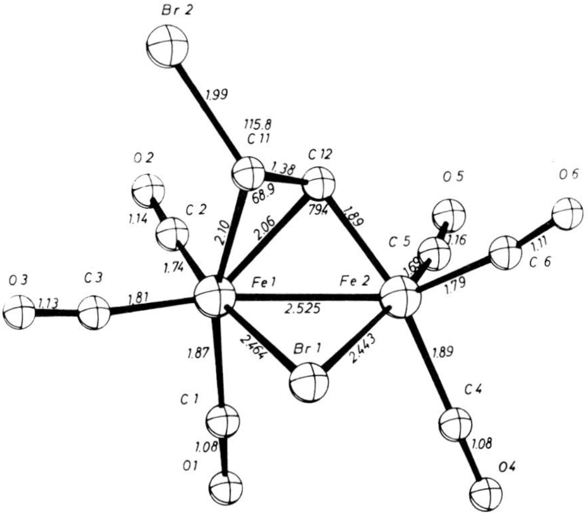

Fig. 3: The molecular structure of μ-[1-η:1,2-η(trans-2-bromovinyl)]-μ-bromo-bis(tricarbonyliron)-(Fe-Fe) (Ref. 430).

Multiple bond character within iron-σ-vinyl linkages has been claimed even when the vinyl group is not π-bonded to a second Fe atom *(130,154)*.

Ferracyclopentadienes, in which an iron atom replaces one carbon atom of a cyclopentadienyl system (see below), are stabilized by being coordinated to a second iron atom. The C-Fe-C angles within the five-membered ring range from 81° to 82.3°. Substituents at the carbon atoms of the *cis*-diene fragment of the ring system do not significantly deviate from the best plane of the diene, suggesting a conjugated character of this system. The ring iron atom is displaced by 0.10 to 0.28 Å from the basal plane defined by the terminal diene carbon atoms and two carbonyl carbon atoms. A typical ferracyclopentadiene structure is shown in Figure 4.

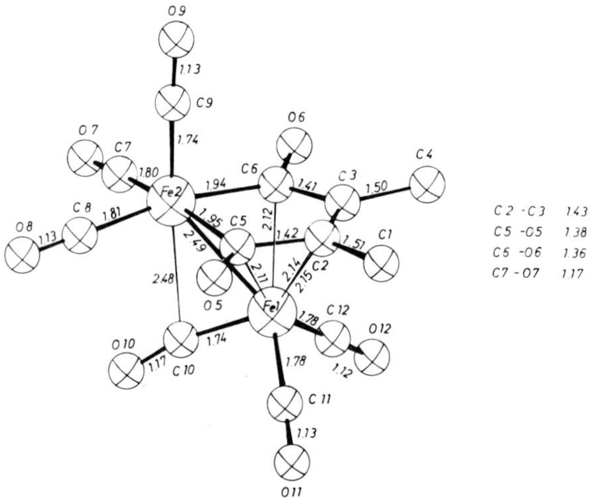

Fig. 4: The molecular structure of (2,5-dihydroxy-3,4-dimethyltricarbonylferracyclopenta-2,4-diene)tricarbonyliron (Refs. 364,363).

X-ray structural analyses of two binuclear iron carbene complexes have been completed *(293,449,488)*. The structure of μ-diphenylvinylidene-bis(tetracarbonyliron)-(Fe-Fe) (see Table 2, Nr.37) is shown with its molecular dimensions in Figure 5. The sp^2-hybridized atom C(9) bridges two iron atoms symmetrically. The system Fe(1)-Fe(2)-C(9)-C(10) is planar and twisted slightly with respect to the moiety C(9)-C(10)-C(11)-C(17). The structure of bis(μ-phenyloxycarbene)-bis(tricarbonyliron)-(Fe-Fe) (see Table 2, Nr.25) is shown in Figure 6. The atomic arrangement possesses a crystallographic mirror plane passing through Fe(1)-Fe(2). The sp^2-hybridized carbon atom C(5) is coplanar with its neighbours, and the Fe(2)-C(5)-O(5) angle is 114.6°.

Interesting structural frameworks were found in (μ-cyclo-

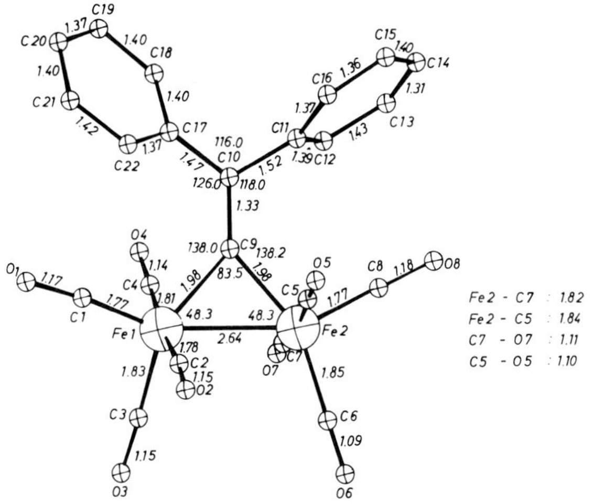

Fig. 5: The structure of μ-diphenylvinylidene-bis(tetracarbonyliron)-(Fe-Fe) (Ref. 488).

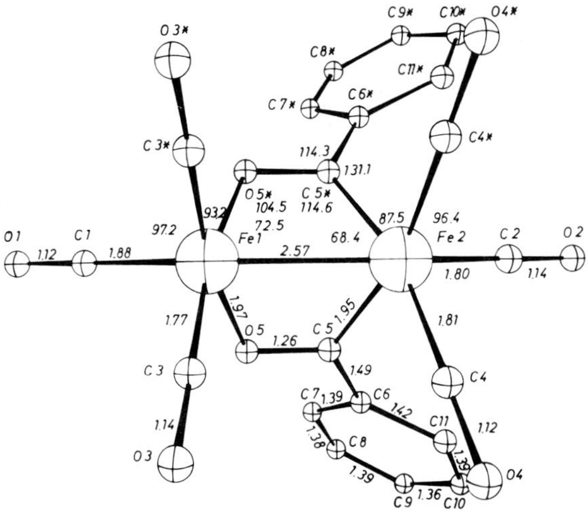

Fig. 6: The molecular arrangement of bis(μ-phenyloxycarbene)-bis(tricarbonyliron)-(Fe-Fe) (Refs. 293,449).

undeca-1,2-diene)heptacarbonyldiiron *(450)* as well as in (μ-allene)hexacarbonyl-triphenylphosphine-diiron *(237)*. Both products were obtained in the reaction of an allene with enneacarbonyldiiron. In the former, an eleven-membered ring is bonded to a heptacarbonyldiiron fragment through formation of a σ-bond between one iron and the central carbon atom of an allylic group η^3-bonded to the second iron. The structure is shown in Figure 13. The Fe-C bond is not found in the plane of the allylic group, but is bent towards the π-bonded Fe(2) by about 30°. This effect is common to η^3-bonded allylic systems, therefore no conclusions can be drawn about the exact hybridization of the central *(meso)* carbon atom (see discussion of η^3-allyl compounds below). In the latter compound mentioned above *(237)*, in which the allyl group is not incorporated into a cyclic framework, similar structural features are evident.

III. π-BONDED ORGANOIRON COMPLEXES

Although π-bonded organometallic compounds are among the longest known organometallic compounds, since the discovery of Zeise's salt $[PtCl_3(C_2H_4)]^-$ in 1827, their structural and bonding properties are still a matter of discussion and controversy. In recent years, diffraction methods have provided sufficient material on which theories can be developed regarding the metal to ligand interaction. Yet, clearly, diffraction methods alone are not sufficient to answer the arising questions.

The bonding theory of alkenes (and alkynes) to transition metals is based on three principal observations:

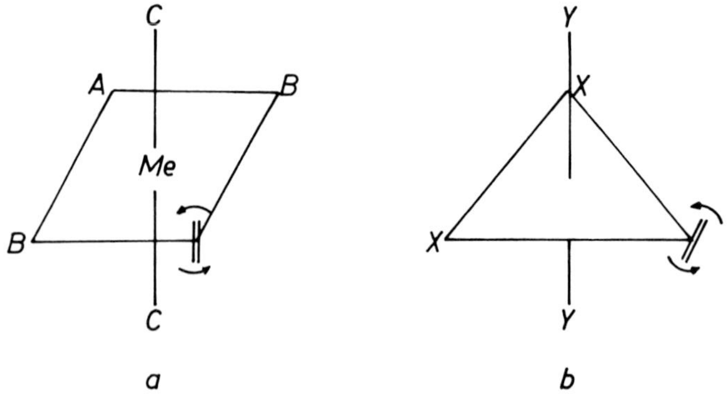

Fig. 7.

(1) The alkene is bonded with its π electron system vertical (a) or horizontal (b) to the coordination plane of the transition metal (see Figure 7).
The observed geometries may be correlated with the population (oxidation state) of the *d* orbitals on the metal and their energy levels relative to those of the π^* orbitals on the olefin (see below). Both arrangements represent ideal descriptions; small amounts of tilt *(362,615)* and, sometimes, even full rotation of the olefin seem to be observed *(218,371,511, 543)*. The hybridization of the metal and the charge distribution between metal and carbon have been studied by MO calculations *(615)*. The results of these calculations are in partial agreement with recent ESCA studies, which suggest a small partial negative charge at the olefin. For iron(O), as well as for other zerovalent transition metals, only form (b) and slight deviations thereof have been reported. In trigonal bipyramidal iron(O) complexes the equatorial attachment of the olefin (form (b)) affords better π-bonding than an apical attachment, because of the symmetry of the *d* orbitals in the equatorial plane.

(2) The free olefinic bond distance, which has the accepted value of 1.337 Å, increases by 0.03 to 0.07 Å upon coordination. If 1,3-dienes are complexed, the bond alternation of the free ligand is modified. The extend of these changes depends on the oxidation state of the transition metal and, importantly, on the nature of other ligands. The lengthening can be explained by a rehybridization of the carbon atoms of the complexed alkenes or alkynes. This rehybridization is towards sp^3 hybridization in complexed alkenes, and towards sp^2 hybridization in complexed alkynes, along with a mutual *cis*-bending of the substituents at the complexed bond.

(3) The Dewar-Chatt-Duncanson (DCD) description of olefin complexation by a σ-π-bonding model, which is based on the electroneutrality principle and symmetry arguments, is still the basis for more detailed bonding theories. This description was introduced by Dewar *(248)* for silver olefin complexes, and extended by Chatt and Duncanson *(125)* to platinum-olefin complexes. The main features are shown in Figure 8.

The filled π-orbital (ψ_1) of the olefin overlaps with an empty transition metal hybrid orbital to form a σ-bond, the empty π^*-orbital (ψ_2) has the appropriate symmetry to overlap with a filled metal orbital to form a π-bond. McWeeny, Mason and Towe *(478)* have proposed that the charge distribution of the complexed ligand is identical with that of the free ligand in its first excited (triplet) state. The Dewar-Chatt-Duncanson scheme of bonding is reminiscent of that developed for the bonding of carbonyl groups to transition metals. The olefinic double bond has only one antibonding orbital available for π-

bonding, in contrast to carbon monoxide, where the backbonding is through two orthogonal antibonding orbitals. This difference explains the greater amount of backbonding in carbonyl compounds, and is supported by shorter metal-carbon(carbonyl)

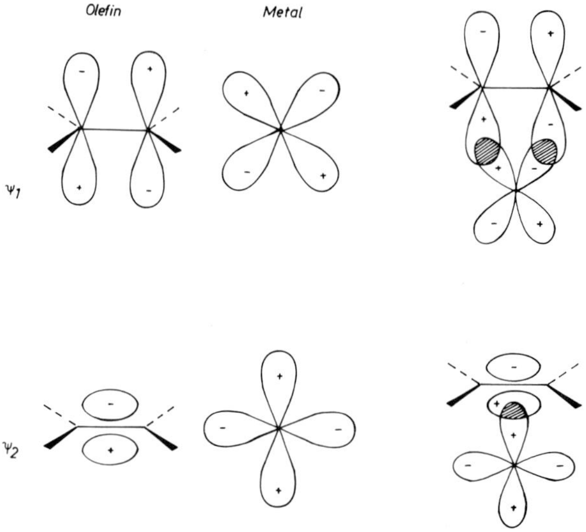

Fig. 8: Dewar-Chatt-Duncanson model for olefin complexation at transition metals.

distances.

In the DCD bonding scheme, metal and olefin act simultaneously as Lewis acids and bases. These donor and acceptor properties depend on the additional ligands, the oxidation state of the metal, and on the substituents of the olefin. In the special cases of octahedral and square planar complexes, ligands in the *trans*-position can have pronounced electronic effects. IR and Raman spectroscopic studies on Zeise's salts *(357)* emphasized the importance of the σ and π contribution to the bonding. By analogy between Pt-olefin bonds and ethylene oxide, a nonplanarity of the complexed olefin, intermediate between that of the C-C part of ethylene oxide and ethylene was further suggested.

Often the question arises over which hybrid orbitals of the transition metal are best incorporated in the DCD model. While the original DCD model incorporated dsp^2 platinum hybridization, results of recent molecular orbital calculations have suggested that the symmetry equivalent d^2p^2 orbital combination may be a better choice *(615)*. This suggestion may be extended to other transition metals. For trigonal bipyramidal iron complexes the d^2p^3 orbital combination could be chosen instead of the usual dsp^3 hybridization.

A. *COMPLEXES OF IRON WITH NON-CONJUGATED OLEFIN SYSTEMS*

As mentioned previously all present structural evidence for zerovalent iron olefin compounds indicates that the ethylene moiety lies in the trigonal plane of the ideal dsp^3 hybridization scheme. (Ethylene)tetracarbonyliron, the simplest compound of this type, has been investigated *(236)*.

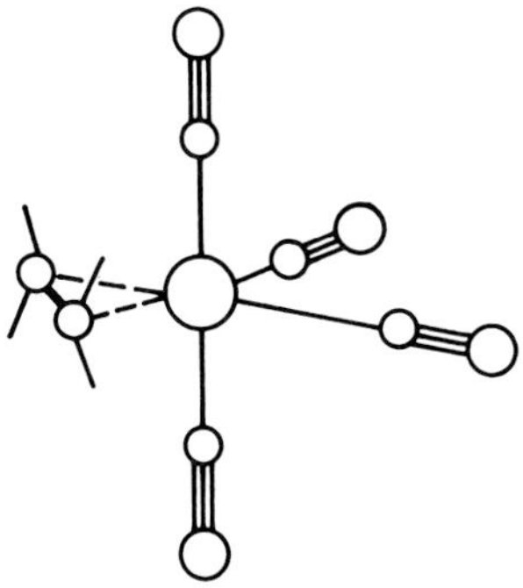

Fig. 9: Structure of (η^2-ethylene)tetracarbonyliron (Ref. 236).

In an electron diffraction study, which assumed trigonal bipyramidal coordination, the C=C and Fe-C(olefin) bond lengths were found to be 1.46(6) and 2.12(3) Å, respectively.

A similar structure, determined by a low-temperature X-ray study, is shown in Figure 10, which illustrates the molecular geometry of (η^2-acrylonitrile)tetracarbonyliron *(458, 459)*. The changes in the geometry of the acrylonitrile ligand, caused by the complexation, are evident by comparing the bond distances and angles of the free and the complexed species given in Figure 10. The ethylene moiety [C(5), C(6)] does not deviate significantly from the trigonal plane [C(1), C(3),C(5),C(6)]. The planar heavy-atom skeleton of the acrylonitrile group forms a dihedral angle of 76.4° with the trigonal plane. This 'bending back' is a common feature for substituents on complexed ethylenes.

The structure of two tetracarbonyliron complexes of fumaric acid have been shown by X-ray methods to have the same basic geometry (Table 3, Nr. 2). In the racemic structure, the ethylenic group is reported to be rotated by 11° about the coordination bond with respect to the trigonal plane *(523)*. In the optically active crystals, which contain three independent molecules per asymmetric unit, rotations of 17°, 17°

and 0° were observed *(524)*. This inconsistency was rationalized by hydrogen bonding effects. Often small differences in bonding energy or crystal packing forces may cause sizeable distortions in molecules.

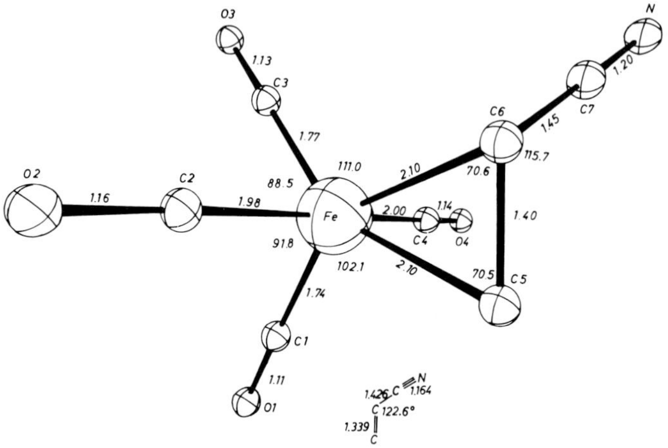

Fig. 10: Molecular structure of (η^2-acrylonitrile)tetracarbonyliron (Refs. 458,459).

The structure of (1,5-cyclooctadiene)-bis(tetracarbonyliron) *(427)* is shown in Figure 11. The 1.5-cyclooctadiene

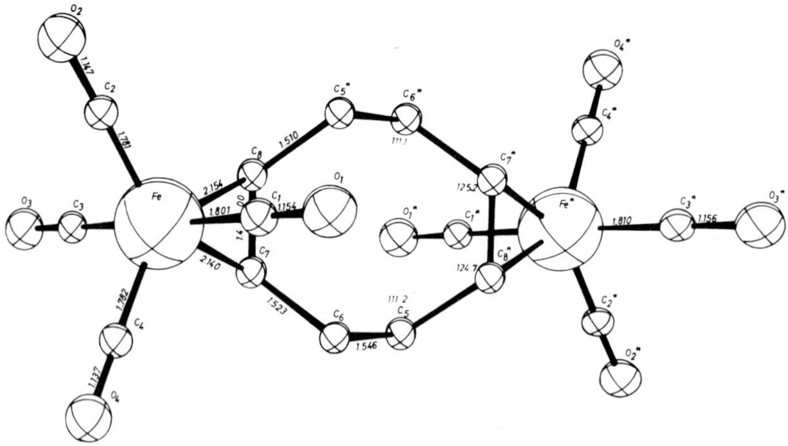

Fig. 11: The molecular structure of μ-[1,2-η:5,6-η(1,5-cyclooctadiene]-bis(tetracarbonyliron) (Ref. 427).

ring, which is in its chair conformation, is complexed to two

$Fe(CO)_4$ units. While the crystallographic symmetry is $\bar{1}$ (C_i), the molecular symmetry deviates only slightly from 2/m (C_{2h}) with the ethylene fragments lying in the respective trigonal planes.

The iron coordination in (bicyclo[3.2.1]octadienyl)tricarbonyliron tetrafluoroborate (see Table 3, Nr. 10) can be considered octahedral if the allyl moiety is assigned two coordination positions. The Fe-C(olefinic) distances (2.24 and 2.26 Å) are unusually long. This may be due to the cationic nature of the complex and the presence of a carbonyl group *trans* to the olefin.

An unusual geometry is observed in (tetrafluorobenzobicyclo[2.2.2]octatriene)tricarbonyliron (see Table 3, Nr. 12). The geometry was described as sqare pyramidal with the midpoints of the two ethylene moieties and two carbonyl carbon atoms defining the basal plane *(380)*. Apparently the ethylene groups are perpendicular to the basal plane. The Fe-C and C-C distances do not differ significantly from those often found in Fe(O) complexes with other geometries.

B. (η^3-ALLYL)IRON COMPLEXES

Structural and bonding characteristics of η^3-allyl transition metal complexes have been reviewed recently *(130,148)*. Coordinated η^3-allyl fragments may be described as three- or four-electron donors, assuming either an allyl radical or an allyl anion, respectively. Formally, two coordination sites are assigned to the allyl anion. A common bonding scheme (see Figure 12) for these complexes assumes: (1) a σ-bond between the filled bonding orbital of the allyl group and an empty hybrid orbital of the metal atom; (2) a bond between the filled nonbonding orbital of the allyl group and an empty hybrid orbital on the metal atom; (3) a 'back' bond formed by the empty antibonding orbital of the allyl group and a filled hybrid orbital of the metal atom. Theoretical calculations have cast doubt on the importance of the back-bonding contributions *(428)*.

A fully symmetric attachment of the η^3-allyl ligand to the transition metal is certainly an idealized concept. Intermediates between the symmetric form (A) and the σ-π-bonded form (B) are well known.

A B

The degree of asymmetry depends on the nature of the other ligands about the metal. This ligand dependence enables these compounds to be important in organometallic catalysis.

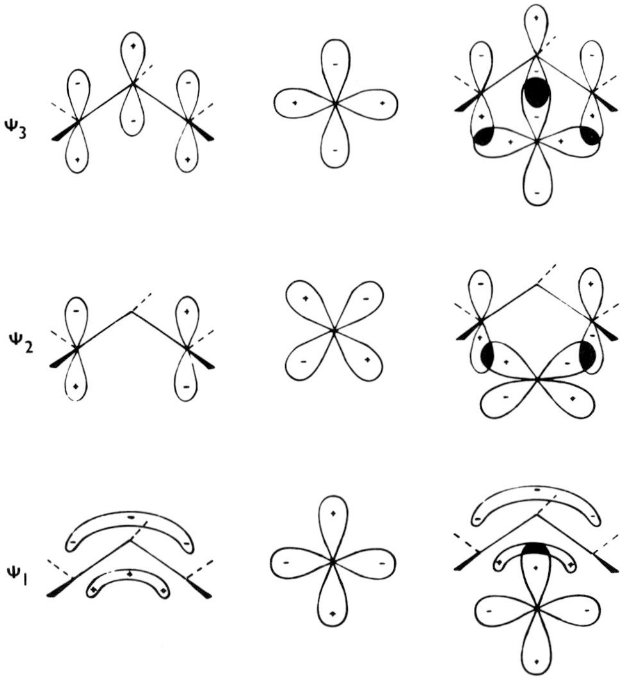

Fig. 12.

All (η^3-allyl)Fe structures presently available are given in Table 4. General structural features of complexes with η^3-allyl groups will now be discussed. When an η^3-allyl group and two other ligands (L) define a square-planar coordination sphere for a transition metal (M), the dihedral angle formed by the plane of the η^3-allyl carbon atom skeleton and the plane of ML_2 is approximately 110° *(413)*. Analogous situations may be seen in octahedral (η^3-allyl)iron complexes. Published C-C-C angles within the η^3-allyl groups range from 110° to 135°. While some workers have attributed this large variation to thermal motion effects *(193)*, no definitive investigations have been reported. Carbocyclic groups which contain η^3-allyl fragments have also been investigated (see Table 4 and Figures 13,14).

The stereochemically non-rigid molecules (1,3,5,7-tetramethylcyclooctatetraene)pentacarbonyldiiron *(189)* (see Figure 15) and (cyclooctatetraene)pentacarbonyldiiron *(295)* have been

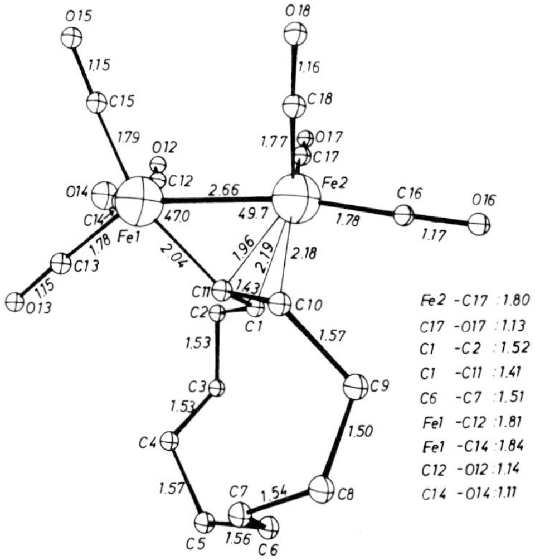

Fig. 13: The structure of μ-(cycloundeca-1,2-diene)heptacarbonyldiiron (Ref. 450).

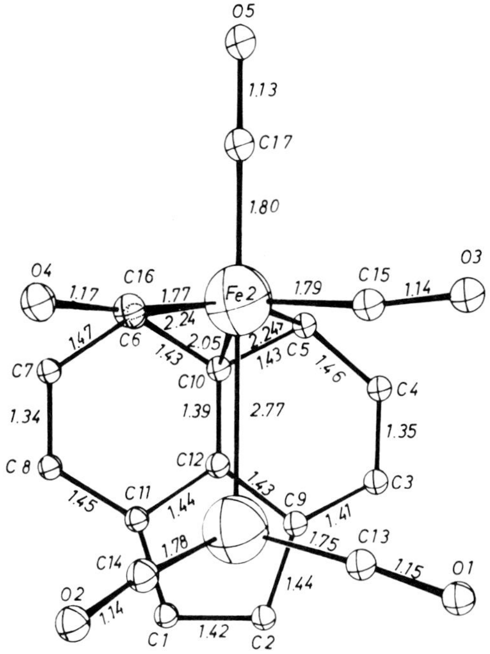

Fig. 14: The structure of (acenaphthylene)pentacarbonyldiiron (Refs. 145,137).

included in this class. Another example is an azulene complex, which contains an η^3-allyl fragment in the seven-membered ring and an η^5-cyclopentadienyl group bonded to a second iron atom (see Table 4 and Figure 16) *(129,132)*.

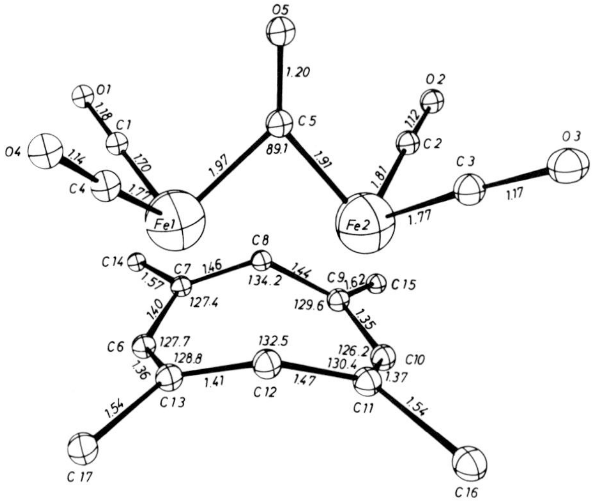

Fig. 15: The molecular structure of (1,3,5,7-tetramethylcyclooctatetraene)pentacarbonyldiiron (Ref. 189).

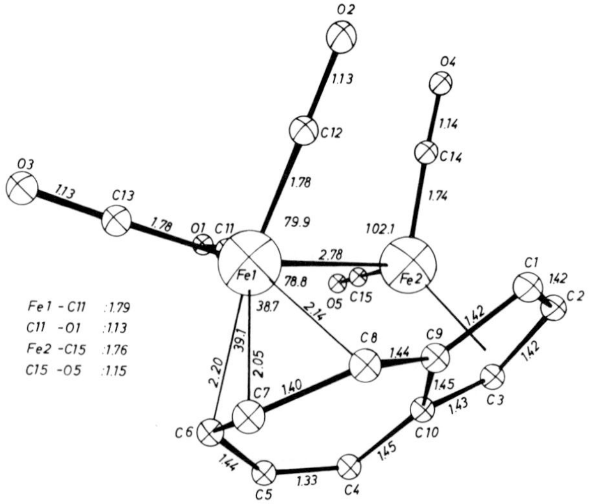

Fig. 16: The structure of (azulene)pentacarbonyldiiron (Refs. 129,132).

C. (CONJUGATED DIENE)IRON COMPLEXES

cis-1,3-diene fragments of open chain and cyclic ligands form π complexes with transition metals. Two bonding arrangements have been proposed for *cis*-1,3-diene complexes. Form *A* emphasizes a π-donation from the ligand to the metal. Form *B* emphasizes σ- and π-bonding contributions from the terminal and central carbon atoms, respectively, and supposedly indicates a stronger interaction than does Form *A*.

A B

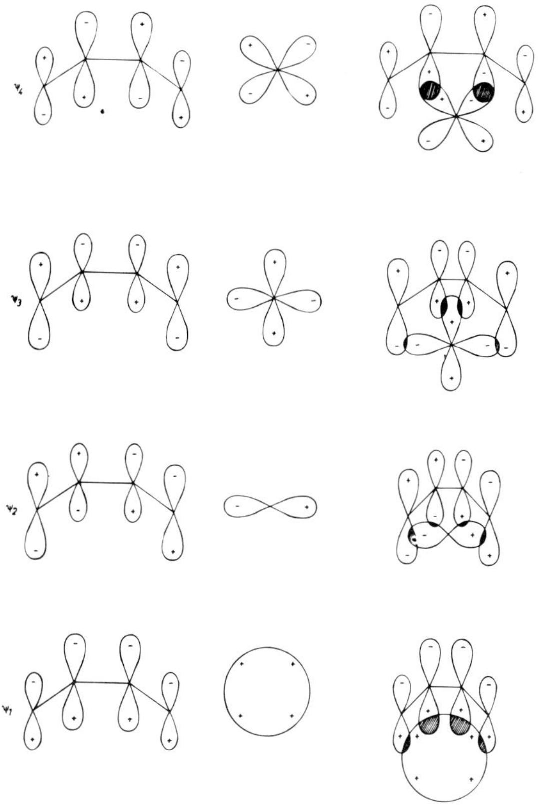

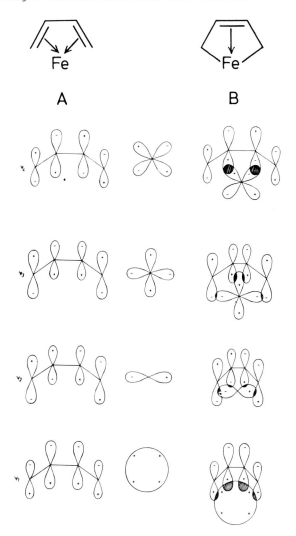

Fig. 17: Bonding in (4-electron donor)-metal systems.

A molecular orbital description of bonding has also been given, where two molecular orbitals could be formed by filled ligand π orbitals and empty metal hybrid orbitals (see Figure 17). The two π* orbitals of the ligand and filled *d* orbitals of the metal could form two molecular orbitals for back-donation. The molecular orbital description can be compared to a linear combination of forms *A* and *B*, and thus allows for a

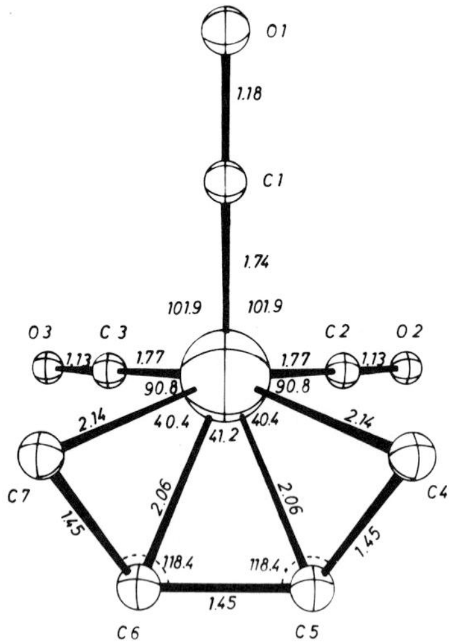

Fig. 18: The structure of (η^4-butadiene)tricarbonyliron (Refs. 485,486).

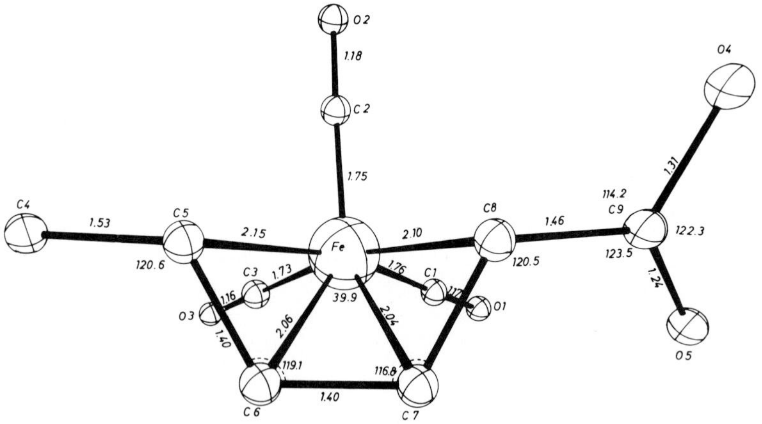

Fig. 19: The molecular structure of (sorbic acid)tricarbonyliron, average distances of two independent molecules are given (Ref. 280).

smooth transition between them *(35,130,413)*. The stability of η^4-1,3-diene complexes is greatly enhanced when additional ligands about the transition metal allow for more back-donation to the diene, therefore favouring form *B*. For example, poor acceptor ligands, such as η^5-C_5H_5 groups, favour form *B*, and good acceptor ligands, such as carbonyl groups, favour form *A*.

The structure of (η^4-butadiene)tricarbonyliron has been investigated by gas phase electron diffraction *(236)* and single-crystal X-ray *(485,486)* diffraction techniques. The coordination geometry (see Figure 18) is square-pyramidal with the two carbonyl carbon atoms and the mid-points of the diene 'double' bonds defining the basal plane. The plane of the diene moiety is approximately perpendicular to the basal plane.

The crystal structure of (sorbic acid)tricarbonyliron (see Figure 19), which contains two independent molecules per asymmetric unit, permits a comparison of the geometry of two η^4-diene complexes of comparable precision *(280)*. The substituents at the 1,3-diene fragment do not lie in the 1,3-diene plane. The terminal C-C bonds are twisted by 8.5° about the central C-C bond. This twist was thought to increase the overlap between the ligand π orbitals and the iron orbitals. Seemingly this twist can be related to the frequently observed bending-back distortion of η^2-olefin complexes (see above).

Evidence for bond alteration in the hydrocarbon-1,3-dienes is not consistent (see Table 5). In bis(η^4-butadiene)-monocarbonyliron, the C-C bonds are all 1.40 Å within experimental error. In bis(η^4-1,3-cyclohexadiene)monocarbonyliron *(429)* (see Figure 20), the two sets of bond lengths show an inequality that is on the borderline of significance; yet the average value for all C-C bond lengths is 1.402 Å.

In an interesting series of structures on π-bonded aromatic vinyl compounds, a significant loss of π-electron delocalization in the aromatic moieties is reported *(242)*. Two examples are schown in Figure 21. Conformational changes upon complexation are frequently observed. For example, vitamin-A-aldehyde is coordinated to an $Fe(CO)_3$ species through a *cis*-1,3-diene fragment, even though the polyene chain of the free ligand possesses an all-*trans* conformation. Hetero-1,3-diene-systems, such as azomethines or conjugated ketones, are also bonded as *cis*-1,3-diene entities. This has been shown in the structure of [η^4-(1-aza-1,3-butadiene)]tricarbonyliron *(165)* and others.

Cyclooctatetraene (COT) complexes of transition metals are of interest because COT ligands can act as four, six and eight electron donors and achieve a variety of conformations (see Figures 22 and 23). In bis(cyclooctatetraene)iron, one ligand is a six-electron donor and the other is a four-electron donor. This has been observed in solution (see

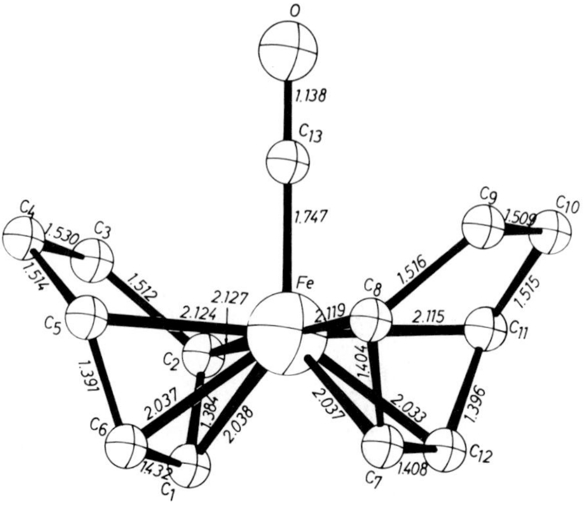

Fig. 20: The molecular structure of bis(η^4-1,3-cyclohexadiene)monocarbonyliron (Ref. 429).

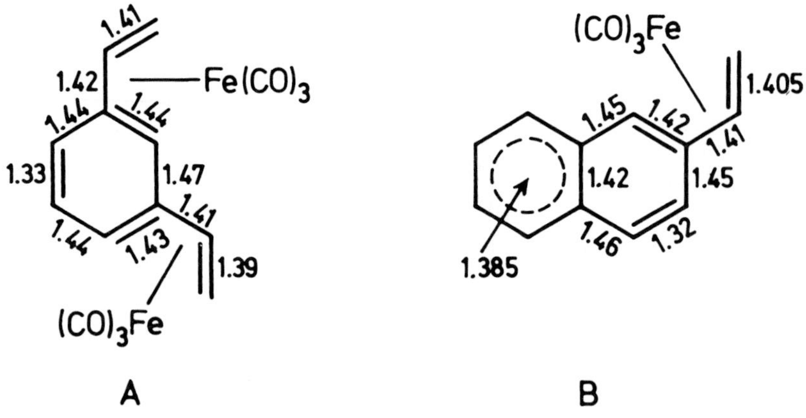

Fig. 21: Bond distances in vinylarene-Fe(CO)$_3$ complexes. In B average distances of two independent molecules are given.

volume 2) and in the solid *(8)*. The four-electron donor bonds to the iron *via* an η^4-1,3-diene fragment; the dihedral conformation for this ligand is similar to that found in (cyclooctatetraene)tricarbonyliron (see Figure 23). Several crystal structures of COT-dimers complexed to Fe(CO)$_3$ groups have also been reported to contain *cis*-1,3-diene entities (see

Table 5, refs. *544,552*). In the reaction of bullvalene with $Fe_2(CO)_9$, one such complex was isolated, which contains a

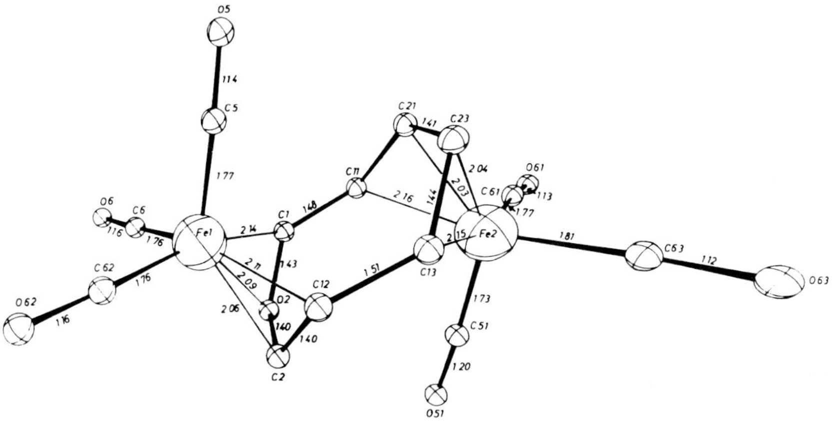

Fig. 22: (Cyclooctatetraene)-bis(tricarbonyliron) (Ref. 250).

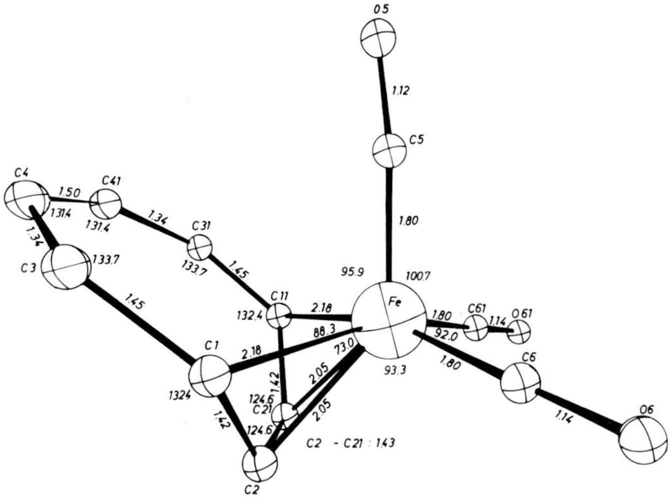

Fig. 23: (Cyclooctatetraene)tricarbonyliron (Ref. 251).

bicyclic ring system with a 1,3-diene moiety bonded to one $Fe(CO)_3$ group. The second $Fe(CO)_3$ group is attached to the 1,4-diene part of the ligand. In the latter, the Fe-C distances are somewhat longer than those in conjugated diene systems, indicating a weaker interaction (see Table 5, ref. *559*).

Ferracyclopentadiene systems are complexed to a second

$Fe(CO)_3$ group. One example of this type of compound is shown in Figure 4.

Various seven-membered ring systems, several of which contain heteroatoms, coordinate to tricarbonyliron groups *via* *cis*-(η^4-1,3-diene) fragments (see Table 5).

CYCLOBUTADIENE- AND TRIMETHYLENEMETHANE COMPLEXES

An electron diffraction study *(236)* has established the structure of (cyclobutadiene)tricarbonyliron. The C_4 ring, assumed to be square, has a C-C bond length of 1.456(15) Å and the Fe-C distance was found to be 2.063(10) Å. X-ray investigations of cyclobutadiene transition metal complexes also do not support large deviations from a square cyclobutadiene ring. In (tetraphenylcyclobutadiene)tricarbonyliron, obtained by the reaction of tolane with pentacarbonyliron, the cyclobutadiene ring is found to be planar with the C-C distances averaging 1.459 Å (see Figure 24). This value agrees with the

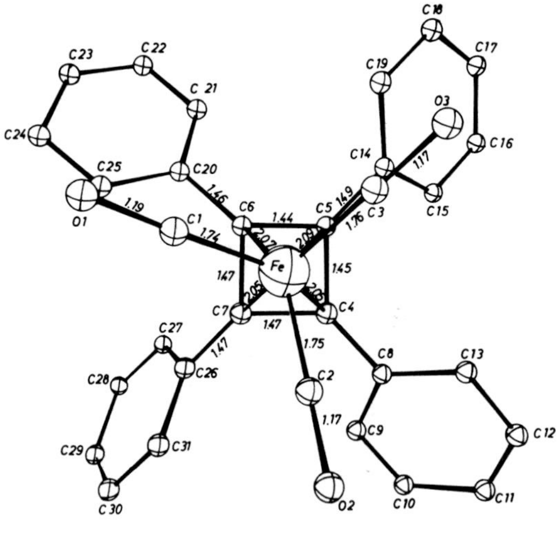

Fig. 24: The structure of (tetraphenylcyclobutadiene)tricarbonyliron (Ref. 256).

corresponding C-C distances found in other transition metal cyclobutadiene complexes. The attached phenyl groups are bent away from the iron by an average angle of 10.8°. Slight, but probably not significant, alterations in bond lengths for the C_4 skeleton of a benzocyclobutadiene derivative complexed to tricarbonyliron have been reported *(242)*.

The unisolated, free trimethylenemethane and its derivatives form stable compounds with the $Fe(CO)_3$ entity. The ligand acts as a four-electron donor to the iron using its lowest bonding orbital and two degenerate nonbonding orbitals for overlap with the iron orbitals of appropriate symmetry; in addition a fourth (antibonding) orbital is available for back-bonding *(139)*. This bonding situation is very similar to that of *cis*-1,3-dienes described above.

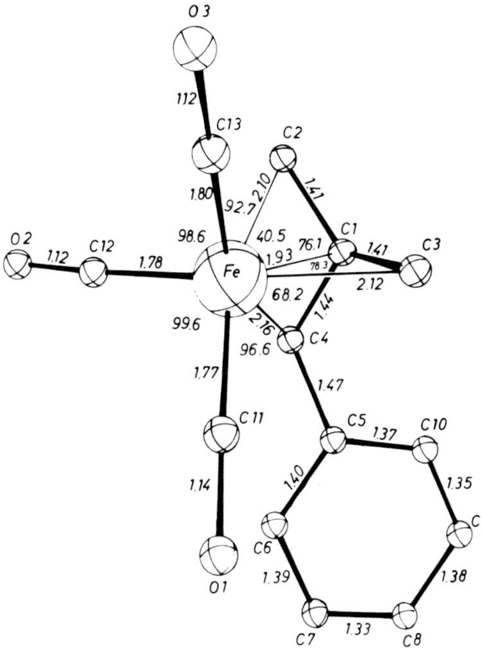

Fig. 25: The structure of (phenyltrimethylenemethane)tricarbonyliron (Refs. 134,139).

The ligand is nonplanar, the middle C atom, although having the shortest Fe-C distance, is displaced by 0.315 Å from the plane of the other three carbon atoms in a direction away from the iron. The pyramidal ligand and the $Fe(CO)_3$ moiety adopt a staggered conformation.

IV. FERROCENES AND RELATED COMPOUNDS

An excellent review on transition metal π-complexes with aromatic systems has been published recently *(614)*. For this reason, only a brief discussion of the subject and a literature survey (Table 6) is included.

At the beginning of ferrocene chemistry, the stereochemistry of these compounds was of fundamental interest because a new type of bonding was recognized. Subsequent structural studies have revealed conformational ambiguities in ferrocene-like complexes, as well as subtle differences from idealized bonding models. Ferrocene derivatives have been found in the eclipsed and staggered conformations and in orientations intermediate between these two conformations. Occasionally in structures with two cyclopentadienyl rings complexed to the same transition metal, a significant tilt between the two cyclopentadienyl planes has been observed. While tilts as large as 23.3° have been observed, calculations indicate only tilts as large as 45° may give rise to configurations with insufficient overlap for bonding *(40)*. Furthermore, non-equivalent C-C bond lengths in η^5-cyclopentadienyl ligands, illustrated in Figure 26, have been frequently found. Although these distortions have been recognized to be only on the borderline of significance, trends may soon be realized.

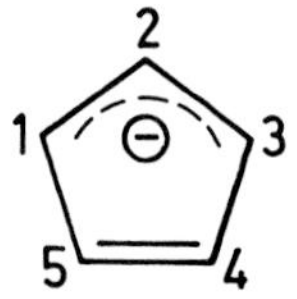

Fig. 26.

The mean iron-(ligand plane) distance in unperturbed cyclopentadienyl systems is 1.65 Å, the average Fe-C distance is 2.04 Å. The C-C distances average 1.419 Å and range from 1.356 Å to 1.458 Å.

An interesting structure containing a planar five-electron donating pentadienyl system within a six-membered ring is shown in Figure 27. The distances found in this compound *(391)* are very similar to those found in cyclopentadienyl complexes.

V. IRON COMPOUNDS WITH MISCELLANEOUS LIGANDS.

In Tables 7,8, and 9, structural information on compounds containing nitrogen, oxygen, and sulfur ligands is summarized. The Fe-O and Fe-N distances are given together with the corresponding literature references. One specific example of an octahedral Fe-O complex, the structure of ferric acetylacetonate, is shown in detail in Figure 28 *(386)*. As in similar compounds, the acetylacetonate moieties were found to be essentially planar.

In Table 10, compounds of boron, silicon, phosphorus,

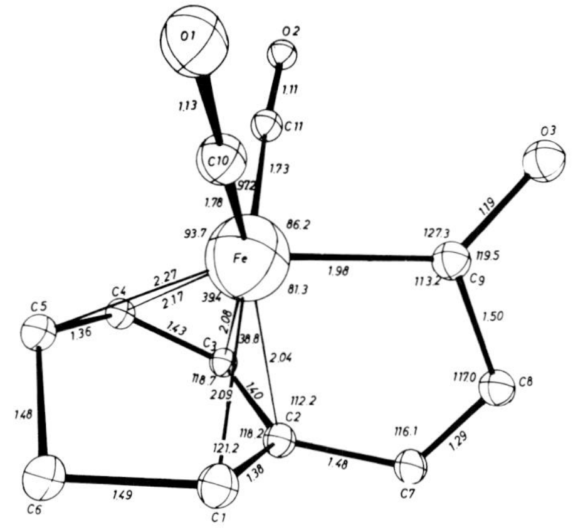

Fig. 27: The molecular structure of dicarbonyl-3-[η^5-(2-cyclohexadienyl)]-σ-propenoyliron (Ref. 391).

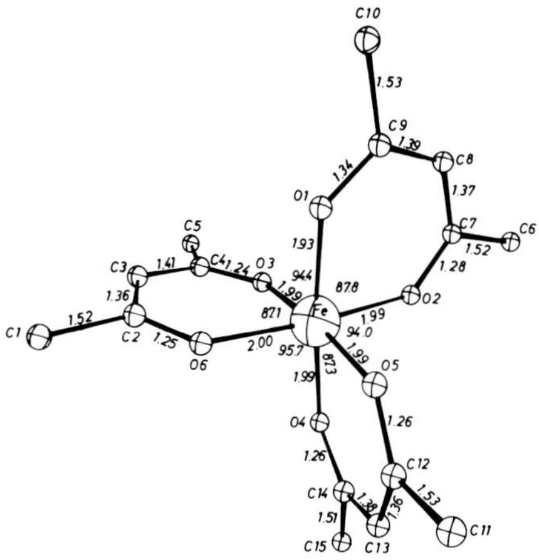

Fig. 28: The molecular structure of tris(2,4-pentanedionato)-iron (Ref. 386).

arsenic, antimony *etc.* containing ligands are summarized. Table 11 gives a survey of iron-metal compounds. For a detailed discussion of structure investigations on compounds containing Fe-Fe bonds and Fe-metal bonds see the chapter by L.F. Dahl *et al.*. Table 12 gives structural information about compounds of biological interest.

We are deeply indebted to Frau Manuele Biermann of our institute for compiling and editing the numerical part and tables of this report.

Table 1: Compounds Containing Fe-C(sp^3) σ-Bonds.

No.	Formula	Distance [Å] Fe-C	*Ref.*
1	$[Me_4N]_2[Fe_6(CO)_{16}C]$	1.82-1.97 (carbido)	*163,155*
2	1,6-(η-C_5H_5)$_2$-1,6,2,3-$Fe_2C_2B_6H_8$ 1,6-bis(η^5-cyclopentadienyl)-1,6-diferra-2,3-dicarba-closo-decaborane(8)	1.962(5) 1.993(5) 1.873(5)	*108*
3	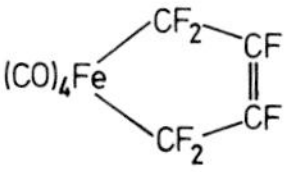	2.00(2)	*359*
4	(η^5-C_5H_5)Fe(CO)$_2$-CH_2-CO_2H	2.06(2)	*23,330*
5	$(HCF_2-CF_2)_2Fe(CO)_4$	2.068(14)	*131*
6	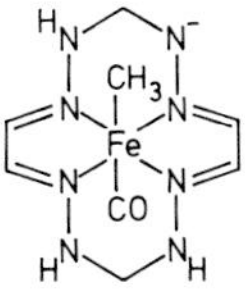	2.077(5)	*325*
7	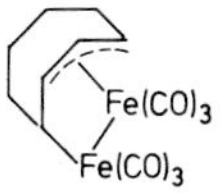	2.08(12)	*191*
8	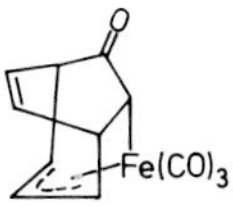	2.080(14)	*604*
9	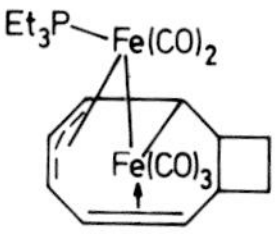	2.085	*211*
10	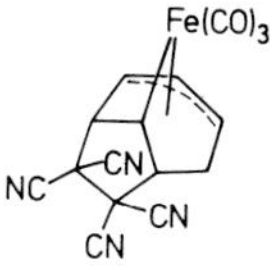	2.09(2) 2.09(3)	*606* *327*
11	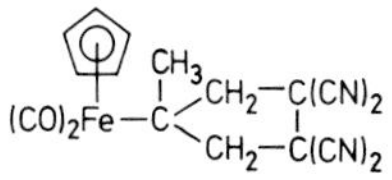	2.0961(21)	*162*
12	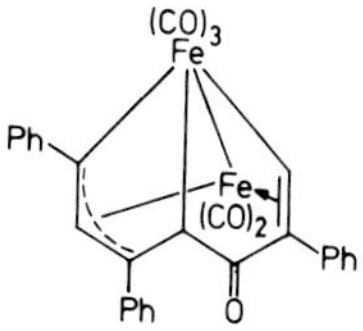	2.097(10)	*418*
13	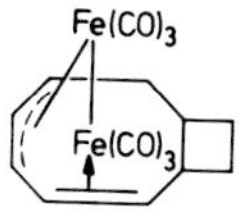	2.099(4)	*197*

Table 1 (continued)

No.	Formula	Distance [Å] Fe-C	*Ref.*
14	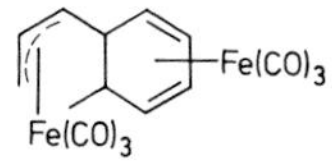	2.10(1)	*200*
15	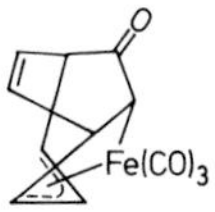	2.110(2)	*207*
16	$(\eta^5\text{-}C_5H_5)Fe(CO)_2(\eta^1\text{-}C_5H_5)$	2.11(2)	*55*
17	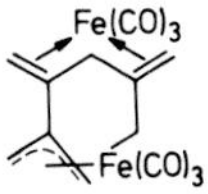	2.12(2)	*625*
18	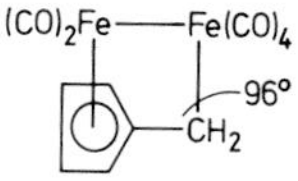	2.123(15)	*482*
19	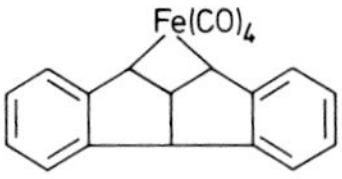	2.141 2.137	*493,300*
20	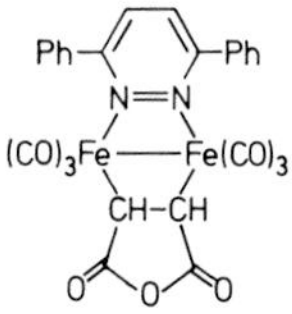	2.14(1) 2.15(1)	*517*
21	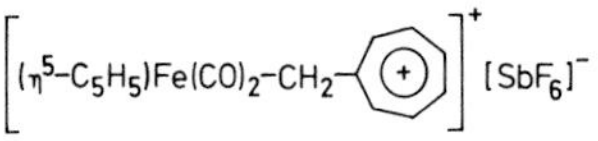	2.16(5)	*150*
22	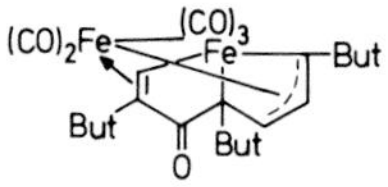	2.160(6)	*554*
23	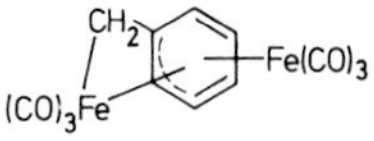		*598*
24	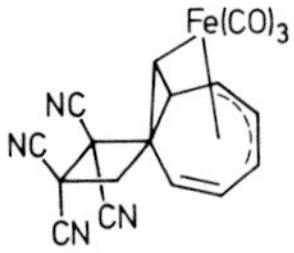		*519*

Table 2: Compounds Containing Fe-C(sp^2) and Fe-C(sp) σ-Bonds.

No.	Formula	Distance [Å] Fe-C	*Ref.*
1	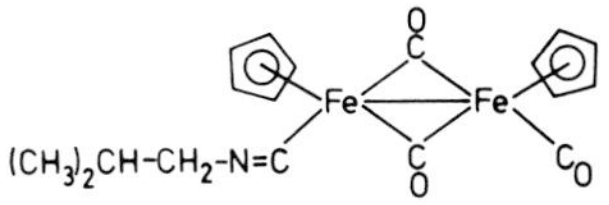	1.846(15)	*116*
2	$(\eta^5-C_5H_5)_2Fe_2(CO)_3[CN(CH_3)_3]$	1.94(1) 1.92(1) 1.86(1) 1.88(1) non-symmetrical bridge	*2*
3	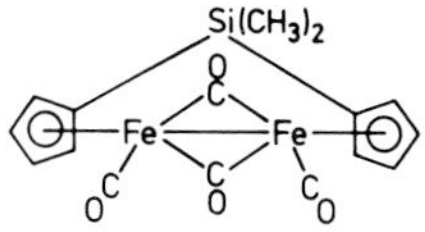	1.92(1) 1.89(1) 1.87(1) 1.88(1)	*607*
4	Hexamethylisonitrile-iron(II) tetrachloroferrate(III)	1.874(3) 1.878(3) 1.871(4)	*184*
5	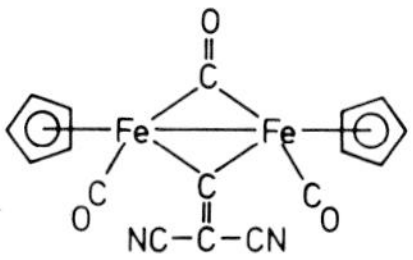	av. 1.94(3) av. 1.87(4)	*420*
6	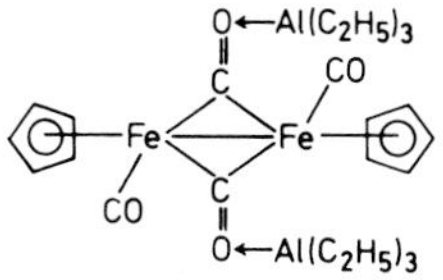	1.88(2)	*416*
7	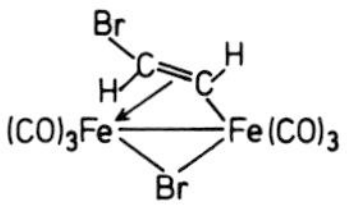	1.888(14)	*430*
8	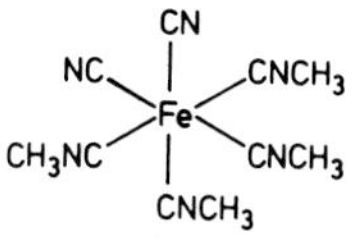	1.98(6) 2.05(6) 1.88(6)	*618*
9	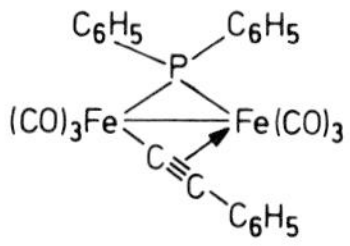	1.891(6)	*518*
10	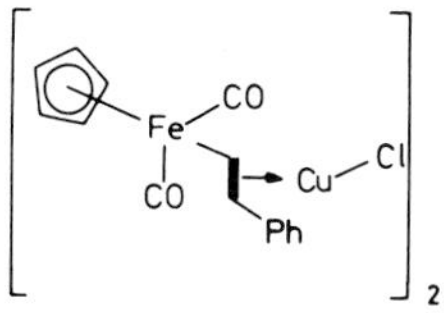	1.906(10) 1.89(2)	*171* *92*

Table 2 (continued)

No.	Formula	Distance [Å] Fe-C	*Ref.*
11	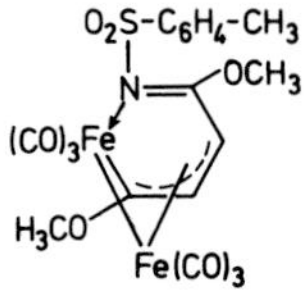	1.91(1)	*553*
12	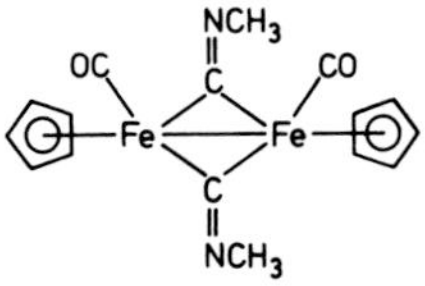	1.947(7) 1.917(7) 1.937(7) 1.948(7)	*201*
13	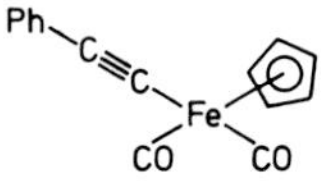	1.920(6)	*297*
14	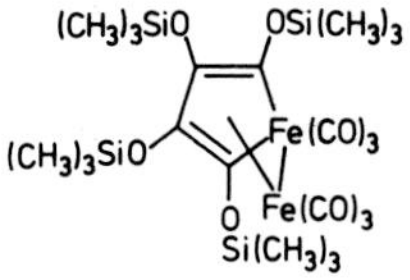	1.92(2)	*60*
15	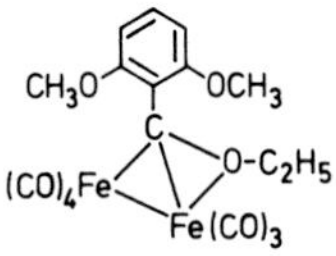	2.000(7) 1.920(8)	*382*
16	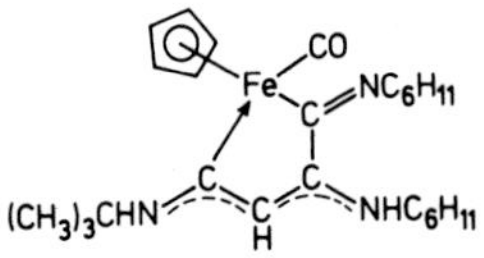	1.926 1.973	*624*
17	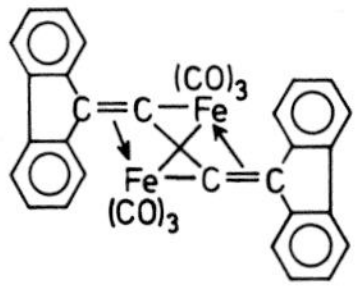	1.93(2)	*88*
18	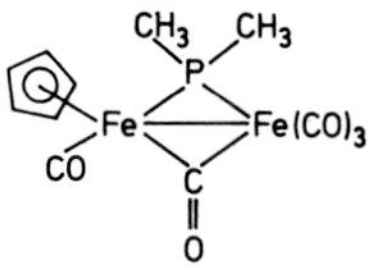	1.93	*599*
19	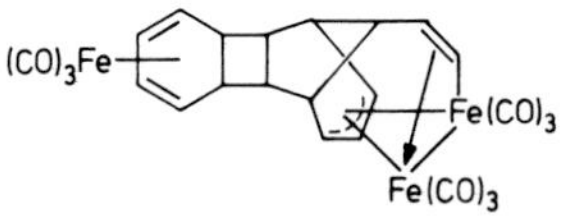	1.93(2)	*603*
20	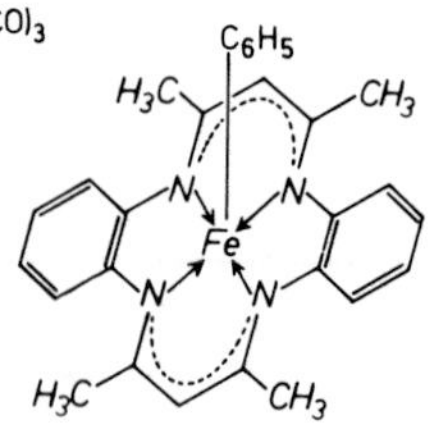	1.933(3)	*323*

Table 2 (continued)

No.	Formula	Distance [Å] Fe-C	*Ref.*
21	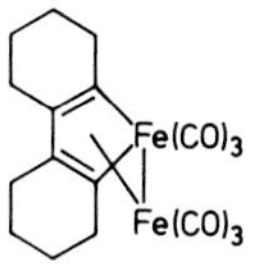	1.935(13) 1.987(14)	*126*
22	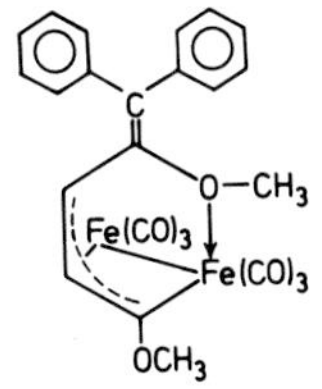	1.94	*395*
23	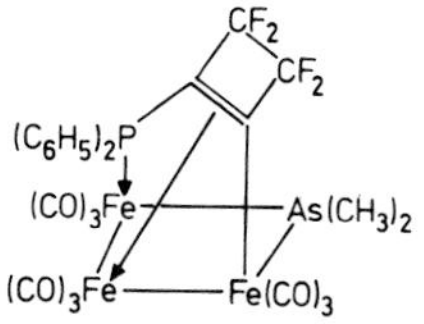	1.94	*276*
24	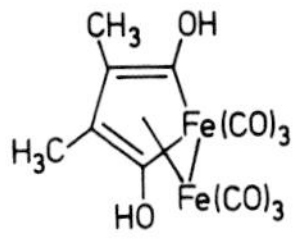	1.943(7)	*364,363*
25	(see Fig. 6) 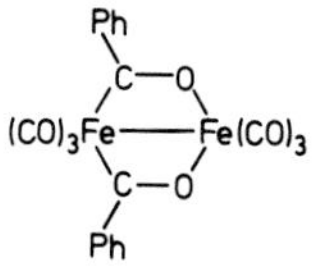	1.945(6)	*293,449*
26	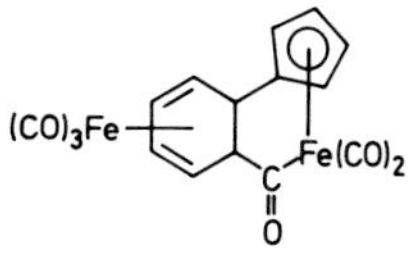	1.9596(30) 1.960(3)	*164* *494*
27	$(\eta^5\text{-}C_5H_5)Fe(CO)(PPh_3)(\sigma\text{-}CO\text{-}Ph)$	1.96	*124,563*
28	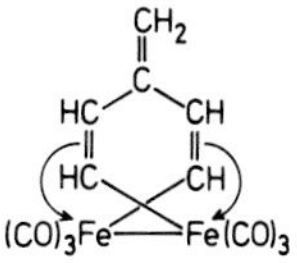	1.960(27) 2.024(31)	*531*
29	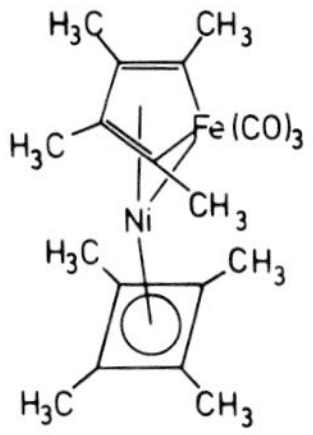	1.965(11)	*288*

Table 2 (continued)

No.	Formula	Distance [Å] Fe-C	*Ref.*
30	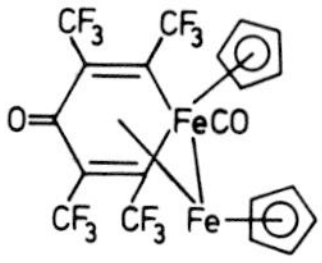	1.97(1) 1.97(1) 2.00 2.00 2.00 2.00	*226*
31	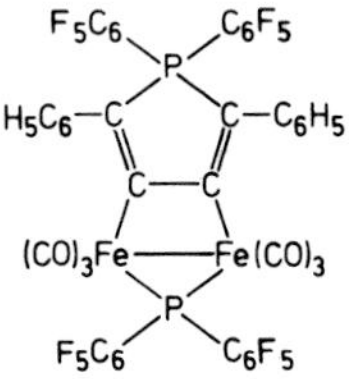	1.99(1) 1.97(1)	*586*
32	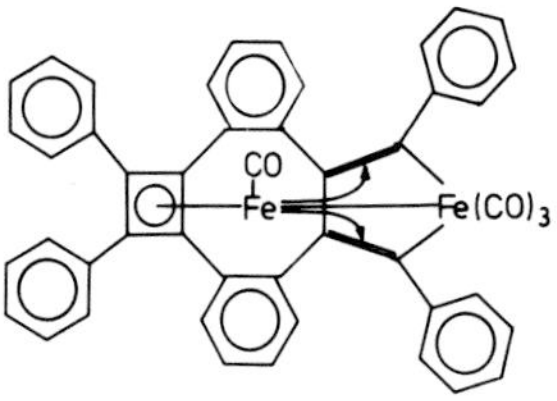	1.990(15) 1.978(16)	*287*
33	$\{Fe_3(CO)_7[Ph_2PC(CO_2Me)C(CF_3)C_2(CF_3)](PPh_2)\}\cdot 2C_6H_6$	2.061(9) 1.978(9)	*512*
34	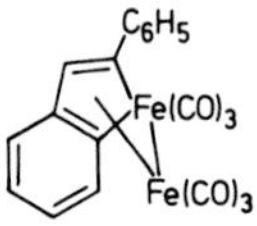	1.978(1) 1.970(1)	*246*
35	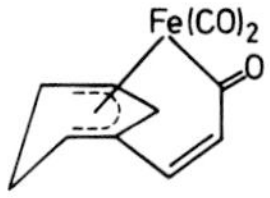	1.979(5)	*390,391*
36	$(\eta^5\text{-}C_5H_5)Fe(CO)(PPh_3)(\sigma\text{-}\alpha\text{-}C_4H_3S)$	1.98(3)	*17*
37	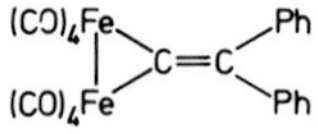	1.98(1)	*488*
38	$[Fe_4(\eta^5\text{-}C_5H_5)_4(CO)_4]^+[PF_6]^-$	av. 1.984	*594*
39	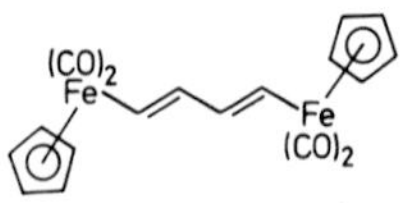	2.03 1.987(5)	*90,92,159, 421*
40	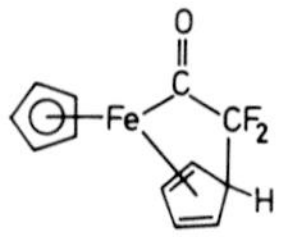	1.99(1)	*226*

Table 2 (continued)

No.	Formula	Distance [Å] Fe-C	*Ref.*
41	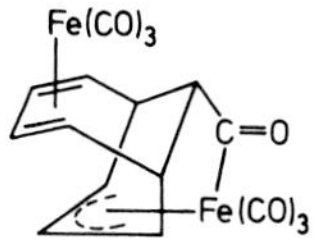	1.991(10)	*604*
42	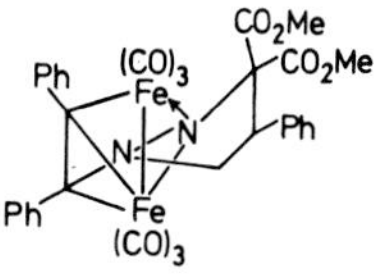	1.994	*433*
43	$(\eta^5\text{-}C_5H_5)Fe(CO)_2$-[-C=C($CH_3$)-S(O)-O$CH_2$]	1.996(8)	*154,147*
44	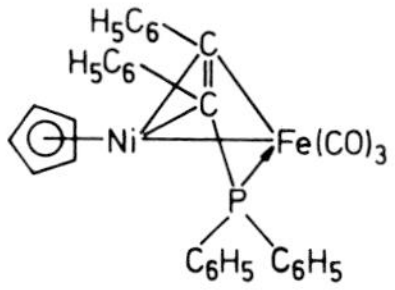	2.000(9)	*43*
45	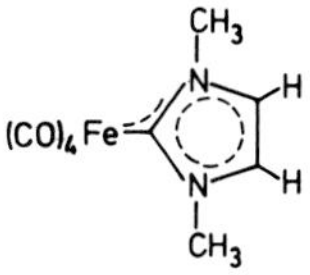	2.007(5)	*381*
46	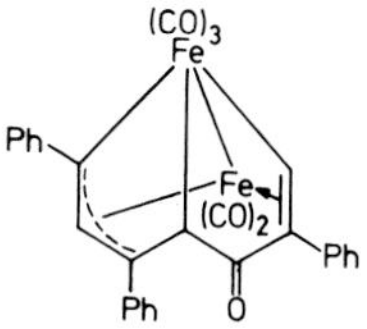	2.01 2.09	*418*
47	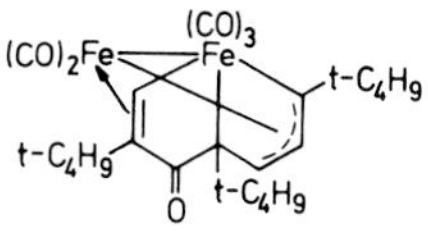	2.017(6) 2.125(7)	*554*
48	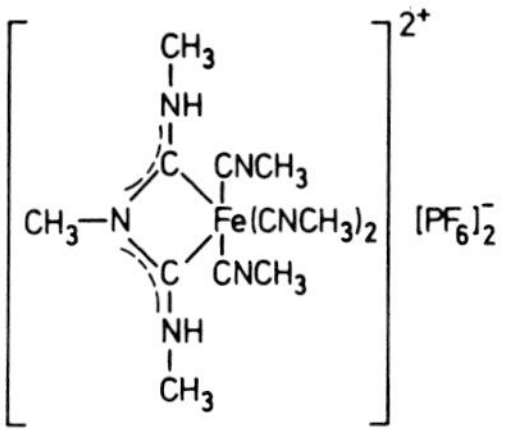	2.02(4) 2.03(4)	*483*
49	(see Fig. 13) 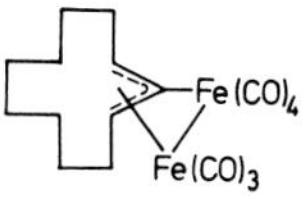	2.02(1)	*450*
50		2.021(3) 2.022(3)	*59*

Table 2 (continued)

No.	Formula	Distance [Å] Fe-C	*Ref.*
51	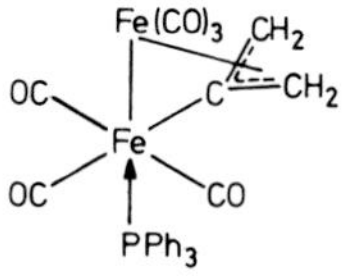	2.03(2)	*237*
52	$Fe_3(CO)_8(C_6H_5C_2C_6H_5)_2$	2.031(1) 2.063(1)	*257*
53	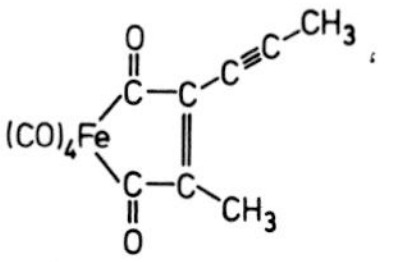	2.035 2.012	*526*
54	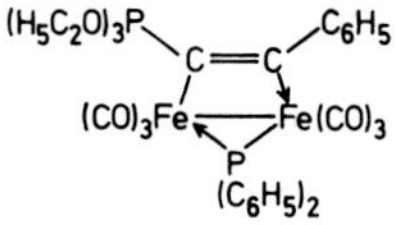	2.08(1) 2.04(1)	*621*
55	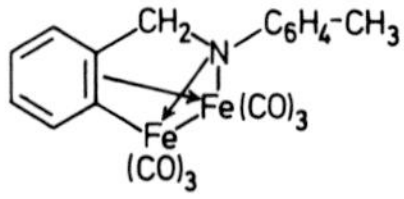	2.06	*32*
56	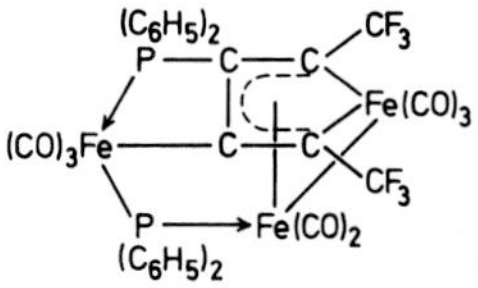	2.079	*509*
57	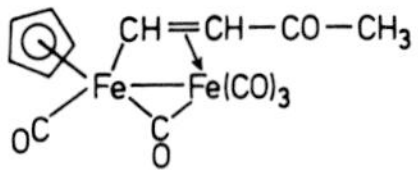	2.09(1)	*12,15*
58	$(\eta^5\text{-}C_5H_5)Fe(CO)(PPh_3)(\sigma\text{-}C_6H_5)$	2.11	*28*
59	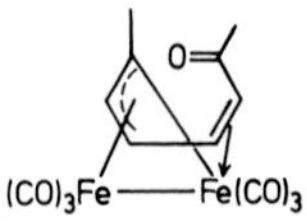	2.241(2) 2.189(3)	*26*
60	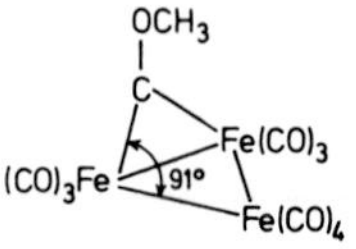	2.70(1)	*566*
61	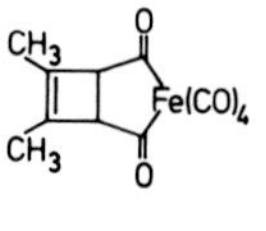		*247*

Table 2 (continued)

No.	Formula	Distance [Å] Fe-C	*Ref.*
62	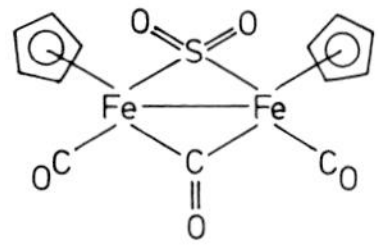	1.925(4) 1.939(4)	*160*
63	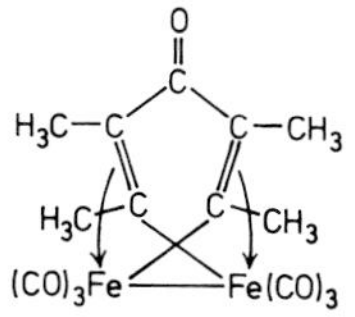	1.937(21) 1.932(21)	*533*
64	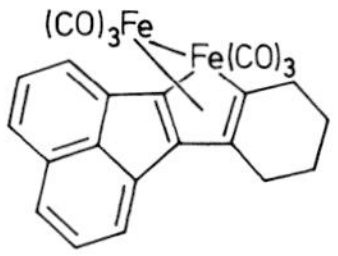	1.948 2.001	*389*
65	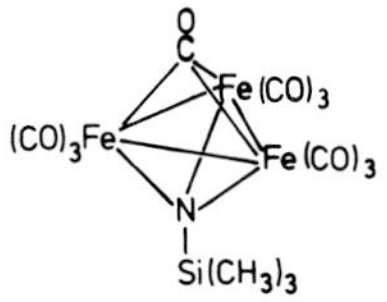	2.070 2.052	*42*

Table 3: Complexes Containing η^2-coordinated Olefinic and Acetylenic Compounds.

No.	Formula	Distance [Å] Fe-C	C-C	*Ref.*
1	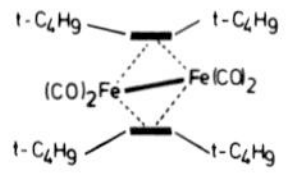	2.048 2.113 2.049 2.116	1.283	*503*
	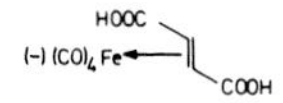 3 independent molecules in asym. unit	2.09(3) 2.06(3) 2.03(3)	1.30(4) 1.40(4) 1.40(4)	*187,524*
2	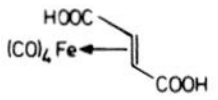	2.04(3)	1.42(4)	*523,186*
3	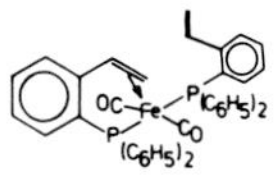	2.10(1) 2.10(1)	1.45(2)	*54*
4	$(C_6H_5C_2C_6H_5)Fe_3(CO)_9$	1.95(2) 1.95(2) 2.04(2) 2.10(2) 2.05(2)	1.41(2)	*77*
5	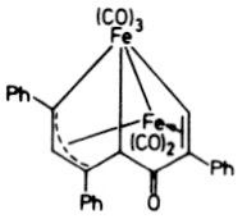	1.97(1) 2.23(1)	1.42(1.0)	*418*
6	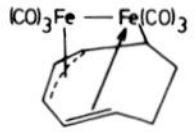	2.18(1.3) 2.22(1.2)	1.40(1.8)	*191*
7	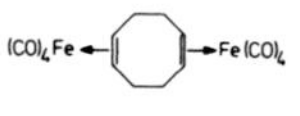	2.140(6) 2.154(5)	1.400(9)	*427*
8	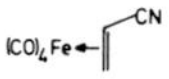	2.10(1) 2.09(1)	1.40(2)	*458,459*
9	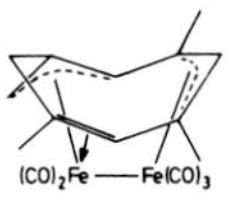	2.214(7)	1.386(11)	*190*
10	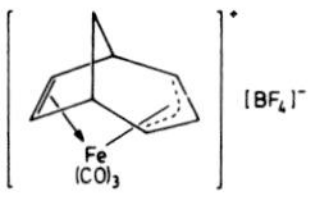	2.24(3) 2.26(3)	1.39(3)	*462*
11	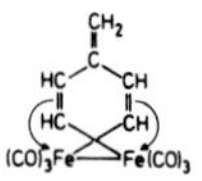	2.256(27) 2.104(27) 2.257(28) 2.071(31)	1.325(41) 1.411(38)	*531*
12	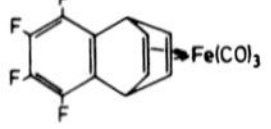	2.150(25)	1.380(29) 1.400(29)	*380*
13	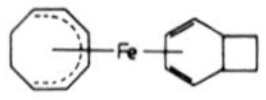	2.098(10) 2.103(9) 2.050(10) 2.045(10) 2.163(10) 2.149(9)	1.433(16) 1.436(15) 1.428(15) 1.397(16) 1.394(16)	*384*
14	$[As(CH_3)_2]C{=}C{-}CF_2{-}CF_2[As(CH_3)_2Fe_3(CO)_9]$	2.05(3) 2.16(3)	1.47(4)	*272*
15	$(CH_3)_2As{-}C{=}C[As(CH_3)_2]{-}CF_2{-}CF_2[Fe(CO)_3]_2$	2.09(3) 1.99(3)	1.51(4)	*271*

Table 3 (continued)

No.	Formula	Distance [Å]		*Ref.*
		Fe-C	C-C	
16	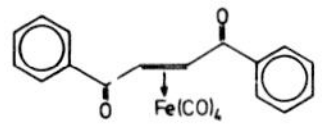	2.10	1.47	*19*
17	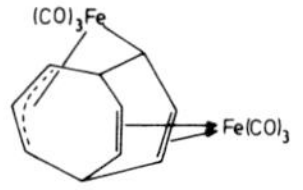	2.134(10) 2.186(10) 2.140(11) 2.145(10)	1.418(13) 1.377(14)	*383*
18	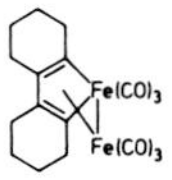	2.087(14) 2.172(12) 2.189(13) 2.066(12)		*126*
19	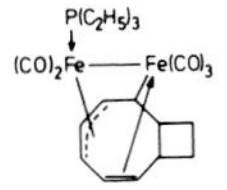	2.261(6) 2.167(6)	1.40	*211*
20	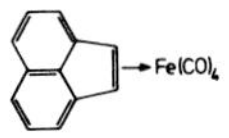	2.156(4) 2.146(3)	1.421(5)	*210*
21	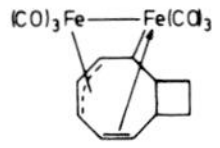	2.199(4) 2.214(4)		*197*
22	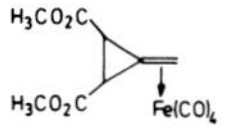	2.092(7) 2.024(5)	1.401(9)	*616*
23	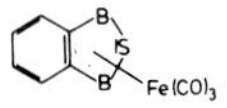	2.220 2.218		*567*
24	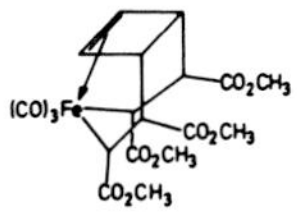			*548*
25	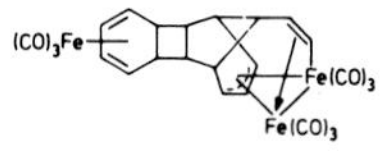	2.10(2) 2.16(2)	1.40(3)	*603*
26	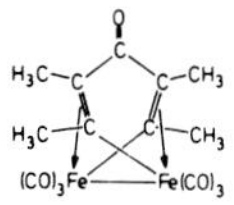	2.185(19) 2.057(21) 2.186(19) 2.083(20)	1.400(30) 1.435(28)	*533*
27	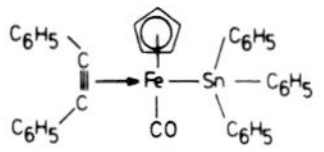	2.02 2.05	1.3	*565*
28	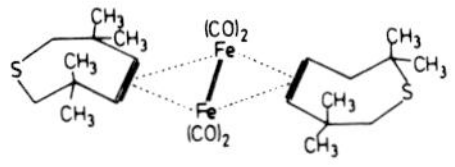	2.011/2.060 2.061/2.082 2.062/2.016 2.073/2.069	1.277 1.267	*557*
29	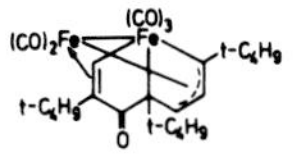	1.950(7) 2.254(7)	1.405(9)	*554*
30	$\{(C_6H_5)_2P\text{-}C{=}C[(CH_3)_2As]\text{-}CF_2\text{-}CF_2\}_2Fe_2(CO)_4$	2.015(6) 2.023(6)	1.476(8)	*277*
31	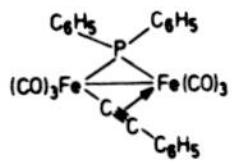	2.125(8) 2.304(7)	1.232(10)	*518*

Table 3 (continued)

No.	Formula	Distance [Å] Fe-C	C-C	Ref.
32	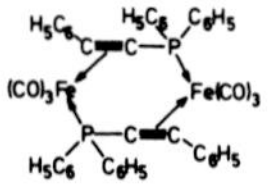	2.046(8) 2.076(8) 2.064(8) 2.068(8)	1.260(11) 1.273(11)	*513*
33	$[Fe_3(CO)_7\{Ph_2PC(CO_2Me)C(CF_3)C_2(CF_3)\}-(PPh_2)]\cdot 2C_6H_6$	2.039(9) 2.077(9)	1.433(11)	*512*
34	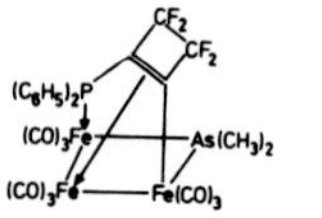	av. 1.96	1.40	*276*
35	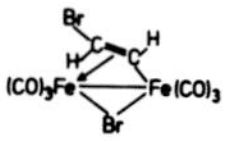	2.105(14) 2.058(14)	1.377(19)	*430*
36	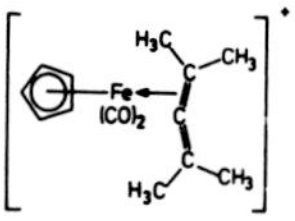	2.237 2.063	1.367	*302*
37	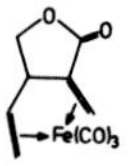	2.114(3) 2.105(4) 2.205(4) 2.201(4)	1.402(5) 1.369(6)	*328*
38	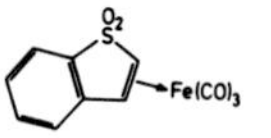	2.078(14) 2.112(12)	1.430(26)	*336*
39	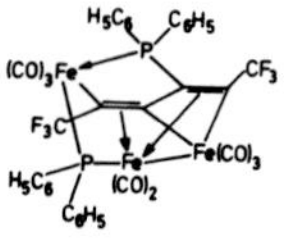	2.02(2) 2.03(2)	1.39(3) 1.39(2)	*122*
40	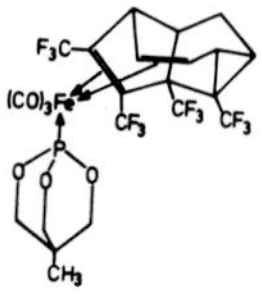	2.181(6) 2.246(6) 2.153(5) 2.047(5)	1.395(8)	*81*
41	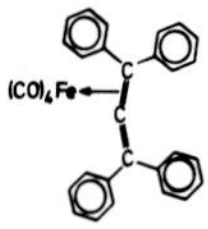	2.04(1) 2.06(1)	1.35(2) 1.35(2) 1.33(2)	*87*
42	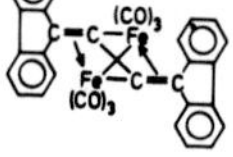	1.94(1) 2.04(1)	1.39(2)	*88*

Table 4: η^3-Allyl-Iron Complexes

No.	Formula	Distance [Å] Fe-C	C-C	Angle	*Ref.*
1	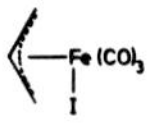	2.26(2) 2.09(2) 2.34(2)	1.35(3) 1.43(3)	121° (3)	*489,490*
2	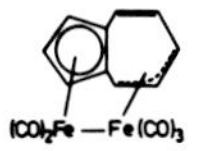	2.200(8) 2.050(8) 2.142(8)	1.391(12) 1.432(12)	122.8°	*132,129*
3	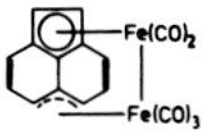	2.241(8) 2.051(7) 2.242(8)	1.431(11) 1.431(11)	123.8° (7)	*145,137*
4	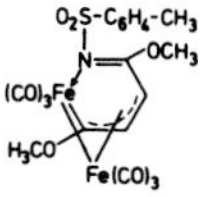	2.116(13) 2.077(10) 2.138(8)	1.447(16) 1.469(16)	116° (1)	*552*
5	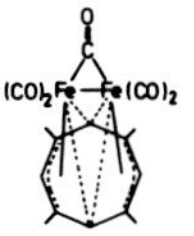 (see Fig. 15)	2.11(2) 2.12(2) 2.12(2) 2.07(2) 2.11(2) 2.13(2)	1.41(2) 1.46(2) 1.45(2) 1.47(2)	128° (2) 130° (2) 126° (2) 131° (2) 129° (2) 128° (2)	*189*
6	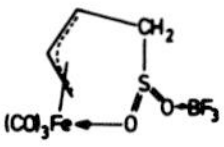	2.194(18) 2.091(16) 2.202(17)	1.420(24) 1.440(23)	124.3° (1.6)	*146,147*
7	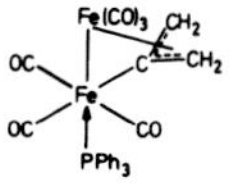	2.18 1.93 2.19	1.41(2) 1.45(2)	116° (1.3)	*237*
8	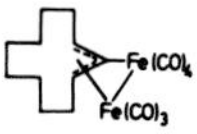	2.20(1) 1.96(1) 2.17(1)	1.41(1) 1.42(1)	114.6° (8)	*450*
9	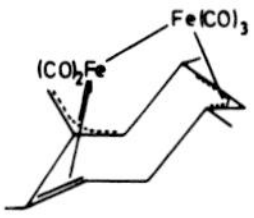	2.163(7) 2.094(8) 2.159(8) 2.117(7) 2.080(6) 2.173(7)	1.407(11) 1.424(11) 1.409(10) 1.420(10)	116.6° (7) 127.6° (6)	*190*
10	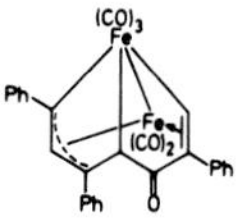	1.97(1) 2.11(1) 2.20(1)	1.41(1) 1.43(1)	116° (1)	*418*
11	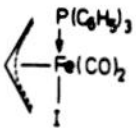	2.20(2) 2.22(2) 2.20(2)	1.40(3) 1.42(3)	118° (2)	*491*
12	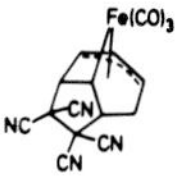	2.13(4) 2.11(2) 2.09(2)	1.40(4) 1.49(4)	115° (2)	*327*
		2.11(1) 2.08(2) 2.17(1)	1.44(3) 1.37(2)		*606*
13	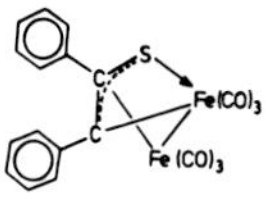	2.089(6) 2.061(6)	1.393(9)		*558*

Table 4 (continued)

No.	Formula	Distance [Å] Fe-C	C-C	Angle	*Ref.*
14	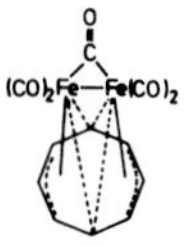	2.136(1) 2.113(1) 2.123(1)	1.398(2) 1.410(2)		*295*
15	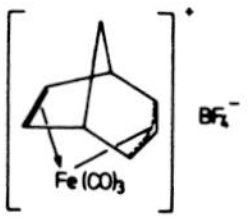	2.21(3) 2.17(3) 2.09(3)	1.39(3) 1.44(3)	119° (4)	*462*
16	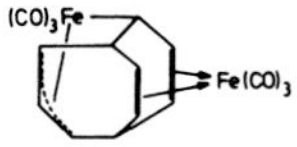	2.146(11) 2.054(9) 2.174(9)	1.427(15) 1.392(14)		*383*
17	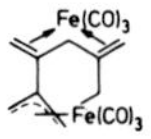	2.19(3) 2.13(3) 2.20(3)	1.48(4) 1.45(4)		*625*
18	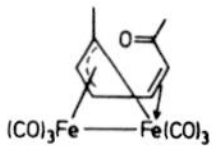	2.088(2) 2.078(2) 2.081(3)	1.404(3) 1.413(4)	115.4° (2)	*26*
19	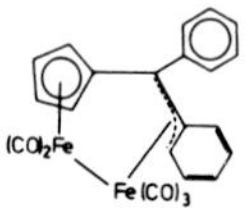	2.11 2.16 2.22	1.43 1.43		*53*
20	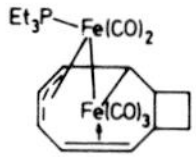	2.145(6) 2.040(6) 2.125(6)			*211*
21	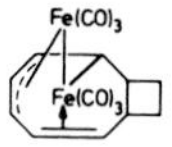	2.120(4) 2.042(4) 2.140(4)			*197*
22	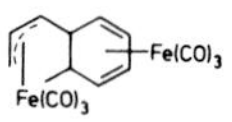	2.12(1) 2.14(2) 2.06(1)	1.37(2) 1.35(2)		*200*
23	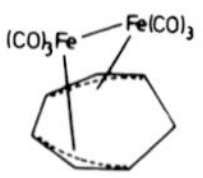	2.141(4) 2.043(3) 2.178(3) 2.168(3) 2.041(4) 2.116(4)	1.397(5) 1.412(5) 1.394(5) 1.402(5)		*192*
24	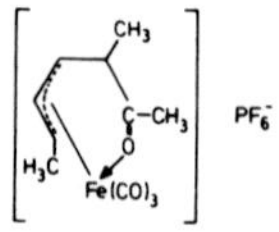	2.210(12) 2.087(13) 2.213(14)	1.384(15) 1.404(16)		*326*
25	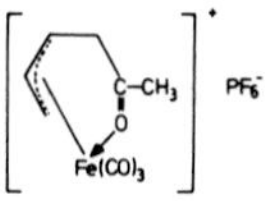	2.141(10) 2.063(9) 2.144(8)	1.380(15) 1.415(15)		*346*
26	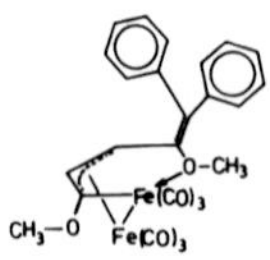	2.09 2.07 2.09			*395*

Table 4 (continued)

No.	Formula	Distance [Å] Fe-C	C-C	Angle	*Ref.*
27	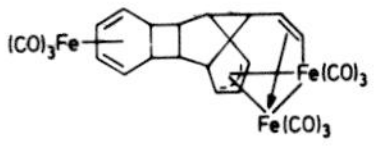	2.16(2) 2.48(2) 2.12(2) 2.40(2)			*603*
28	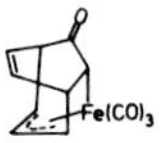	2.131(19) 2.037(16) 2.104(14)	1.38(3) 1.35(2)		*604*
29	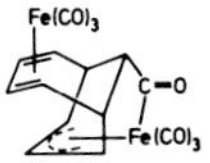	2.145(9) 2.038(10) 2.173(10)			*604*
30	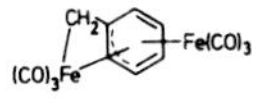				*598*
31	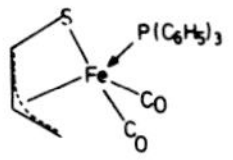	2.057 2.089 2.130	1.330 1.390		*583*
32	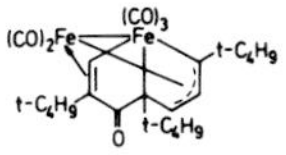	1.989(6) 2.049(7) 2.103(7)	1.371(10) 1.400(10)		*554*
33		2.137(3) 2.069(2) 2.195(3)	1.399(4) 1.382(4)		*207*
34	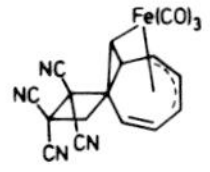				*519*

Table 5: Complexes Containing η^4-coordinated Ligands.

No.	Formula	Distance [Å] Fe-C	C-C	Ref.
1	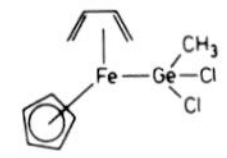	2.13 2.02	1.36 1.49	*18*
2	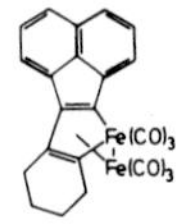		1.420 1.449 1.416	*389*
3	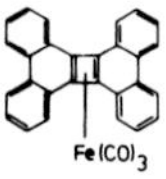	2.097 2.093 2.053 2.059		*389*
4	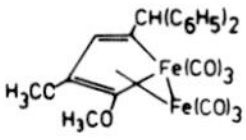	2.08 2.21		*31*
5		2.14(4) 2.06(3) 2.06(3) 2.14(4)	1.45(5) 1.46(5) 1.45(5)	*486,485*
6	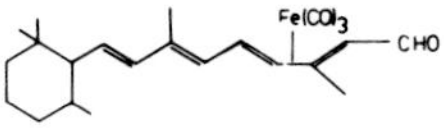	2.18(2) 2.09(2) 2.09(2) 2.13(2)	1.39(3) 1.45(3) 1.49(3)	*66,466*
7	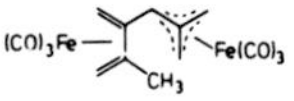	2.07(1) 2.09(1) 2.10(1) 2.13(1)	1.44(2) 1.44(1) 1.45(2)	*625*
8	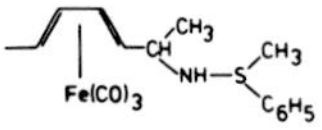	2.14 2.09 2.05 2.18	1.46 1.45 1.45	*460*
9	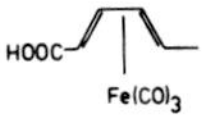 2 molecules per asym. unit	2.146(7) 2.061(7) 2.039(7) 2.104(7) 2.158(7) 2.063(6) 2.038(7) 2.117(7)	1.395(10) 1.399(10) 1.425(10) 1.403(9) 1.402(10) 1.436(9)	*280*
10	$(C_8H_8)_2Fe$	2.11(2) 1.97(2) 2.03(2) 2.27(2)	1.28(4) 1.36(4) 1.48(4)	*8*
11	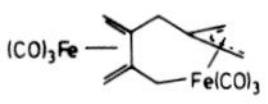	2.15(3) 2.16(3) 2.15(3) 2.03(3)	1.56(4) 1.35(5) 1.55(4)	*625*
12	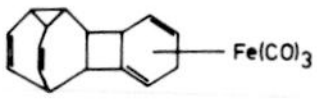	2.122(15) 2.027(16) 2.038(16) 2.128(13)	1.40(2) 1.38(2) 1.41(3)	*544*
13	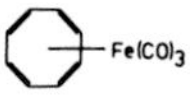	2.18(1) 2.05(1) 2.05(1) 2.18(1)	1.42(2) 1.42(1) 1.42(2)	*251*

Table 5 (continued)

No.	Formula	Distance [Å] Fe-C	C-C	Ref.
14	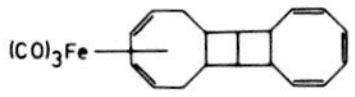	2.12(1) 2.07(1) 2.10(1) 2.14(1)	1.41(5) 1.42(5) 1.46(5)	*552*
15	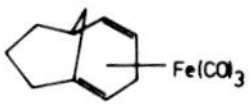	2.17 2.05	1.41(2) 1.43(2) 1.43(2)	*52*
		2.12 2.03 2.07	1.36 1.44	*165*
16	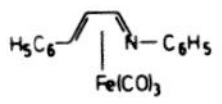	2.091 2.149 2.078	1.43 1.44	*166*
		2.134(6) 2.068(6) 2.086(6)	1.433(9) 1.413(7)	*167*
17	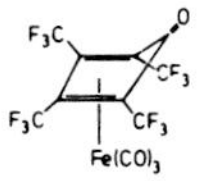	2.043(20) 1.986(22) 2.006(19) 2.121(22)	1.365(24) 1.395(42) 1.416(33)	*35*
18	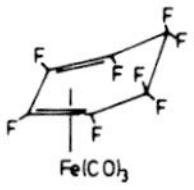	1.993(8) 2.060(8) 2.060(8) 1.993(8)	1.397(11) 1.376(12) 1.397(11)	*133,128*
19	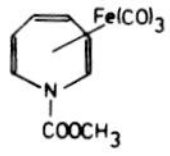	2.091(8) 2.145(10) 2.041(9) 2.059(10)	1.398(14) 1.409(13) 1.440(12)	*396,522*
20	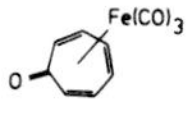	2.114(9) 2.067(10) 2.042(10) 2.149(10)	1.442(13) 1.396(13) 1.435(14)	*255*
21	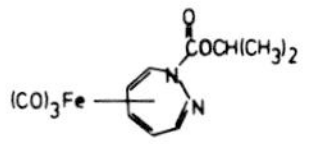	2.08(1) - - 2.12(1)	1.38(1) 1.41(1) 1.43(1)	*10*
22	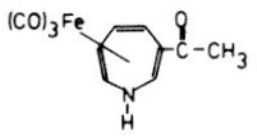	2.182(7) 2.063(9) 2.055(9) 2.168(9)	1.392(13) 1.457(10) 1.464(13)	*602*
23	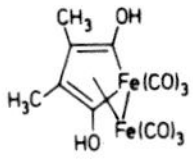	2.111(8) 2.150(8) 2.142(8) 2.121(8)	1.424(10) 1.430(10) 1.413(10)	*364,363*
24	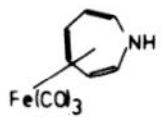	2.03 2.193 2.203 2.043	1.418 1.406 1.404	*372*
		2.203 2.193 2.043 2.030	1.406(10) 1.418(10) 1.404(10)	*314*

Table 5 (continued)

No.	Formula	Distance [Å] Fe-C	C-C	*Ref.*
25	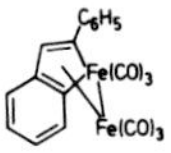	2.127(12) 2.209(12) 2.188(10) 2.132(10)	1.440(15) 1.442(15) 1.346(16)	*246*
26	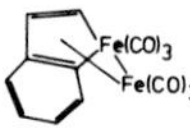			*41*
27	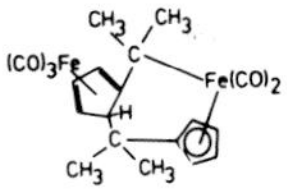	2.08 2.01 2.09 2.26	1.42 1.41 1.42	*53*
28	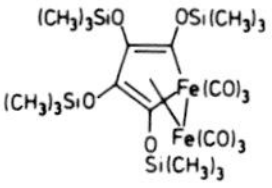	av. 2.13(2) av. 2.19(2)		*60*
29	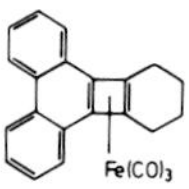	2.104 2.101 2.043 2.052		*389*
30	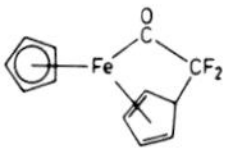	av. 2.18(2) av. 2.06(2)		*226*
31	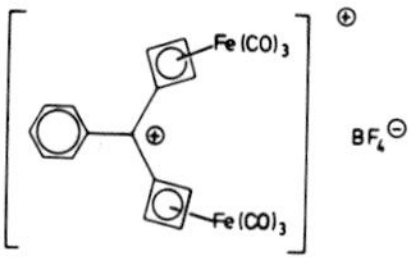	2.04 2.10 2.05 2.04 2.03 2.03 2.09 2.03	1.47 1.43 1.44 1.47 1.45 1.41 1.41 1.48	*244*
32	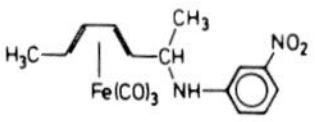	2.156(2) 2.063(2) 2.055(2) 2.117(2)	1.407(4) 1.407(3) 1.417(3)	*387*
33	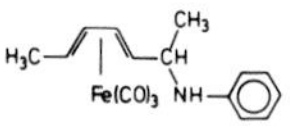	2.156(2) 2.068(2) 2.054(2) 2.138(2)	1.413(2) 1.401(3) 1.426(3)	*387*
34	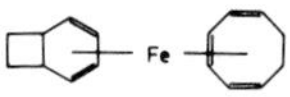	2.029(8) 2.090(9) 2.038(8) 2.110(9)	1.407(13) 1.423(14) 1.403(13)	*384*
35	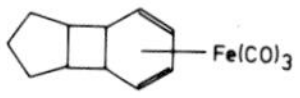	2.132(2) 2.053(2) 2.053(2) 2.139(2)	1.417(3) 1.411(4) 1.412(3)	*195*
36	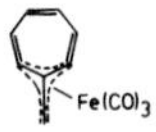	1.946(2) 2.175(3) 2.192(3) 2.120(3)		*158*
37	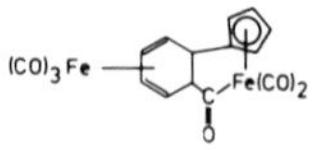	2.097(3) 2.048(3) 2.049(3) 2.098(3)		*164*
		2.098(3) 2.049(3) 2.048(3) 2.097(3)	1.428(4) 1.387(5) 1.429(4)	*494*

Table 5 (continued)

No.	Formula	Distance [Å] Fe-C	C-C	*Ref.*
38	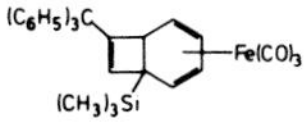	av. 2.085	av. 1.414	*185*
		2.153 2.072 2.176 2.082	1.407 1.436 1.419	*166*
39		2.185(9) 2.076(8) 2.079(8) 2.165(9)	1.429(10) 1.419(10) 1.429(10)	*169*
		2.15 2.13 2.13 2.15	1.42 1.42 1.42	*436*
40		2.152(4) 2.067(4) 2.031(5)		*166,168*
41	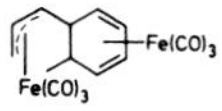	2.10(1) 2.04(1) 2.04(1) 2.12(1)	1.41(1) 1.37(2) 1.40(2)	*200*
42	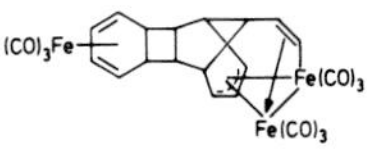			*603*
43	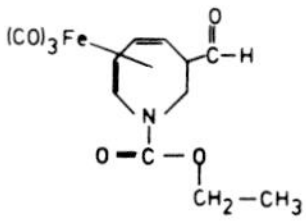	2.151(4) 2.053(4) 2.051(4) 2.096(4)	1.431(5) 1.406(6) 1.431(5)	*622*
44	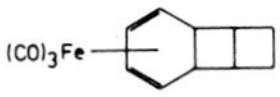	2.116(6) 2.039(6) 2.052(7) 2.114(5)	1.401(8) 1.386(7) 1.404(7)	*209*
45	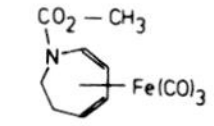	av. 2.110(4) 2.070(4) 2.048(4) 2.128(5)	av. 1.428(8) 1.425(7) 1.425(9)	*170*
46	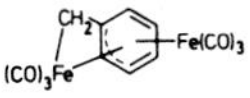			*598*
47	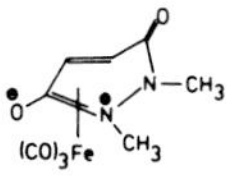	2.11 2.04 2.36 (Fe-N) 2.05		*500*
48	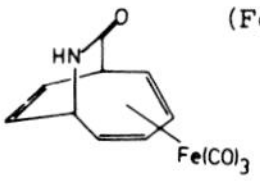			*519*
49	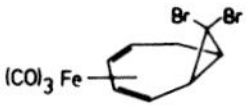	2.119 2.033 2.050 2.122	1.413 1.394 1.408	*569*
50	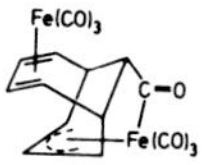	2.117(10) 2.018(16) 2.028(15) 2.138(14)		*604*

Table 5 (continued)

No.	Formula	Distance [Å] Fe-C	C-C	Ref.
51	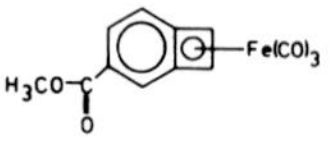			*510*
52	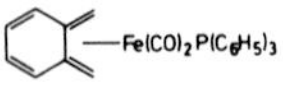			*510*
53	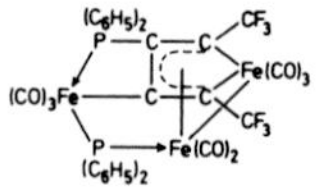			*509*
54	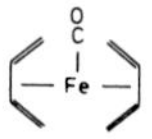	2.16(1) 2.09(1)	1.43(2) 1.46(1)	*241* *617*
55	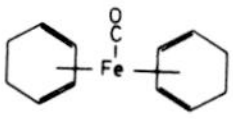	2.127(3) 2.038(4) 2.037(4) 2.124(3) 2.115(4) 2.033(4) 2.037(4) 2.119(3)	1.384(6) 1.432(6) 1.391(6) 1.396(7: 1.408(6) 1.404(6)	*429,428*
56	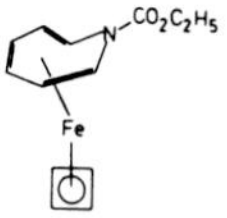			*241*
57	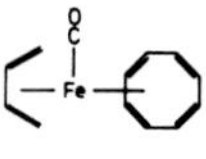	2.05 2.05 2.14 2.14 2.14 2.14 2.04 2.04	1.40 1.43 1.40 1.41 1.40 1.41	*48*
58	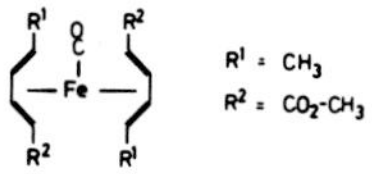	2.187 2.177 2.157 2.145 2.048 2.069 2.063 2.041	1.418 1.410 1.398 1.385 1.391 1.413	*431*
59		2.091(5) 2.025(5)	1.354(8) 1.361(6)	*431*
60				*421*
61	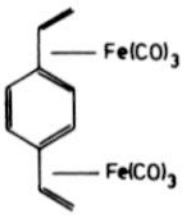 and related compounds		1.38 1.42 1.43	*242*
62	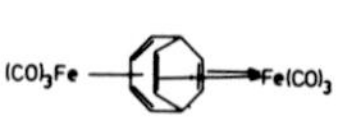	2.100(14) 2.038(16) 2.041(15) 2.097(14) 2.122(13) 2.131(13) 2.172(13) 2.175(13)	1.41(2) 1.36(2) 1.42(2) 1.40(2) 1.39(2)	*559*
63	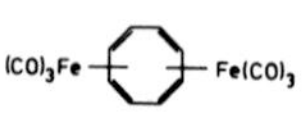	2.06 2.15	1.39(3) 1.49(3) 1.44(3)	*250*

Table 5 (continued)

No.	Formula	Distance [Å] Fe-C	C-C	*Ref.*
64	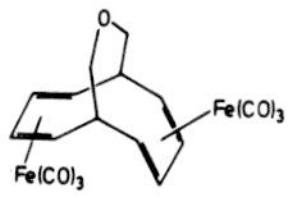	2.103(3) 2.032(3) 2.043(3) 2.114(2) 2.125(2) 2.039(2) 2.037(2) 2.115(2)		*67,64*
65	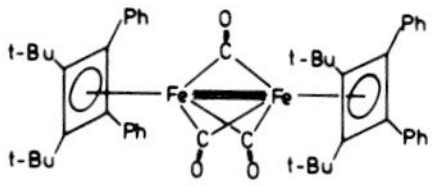	av. 2.062	av. 1.468	*495*
66	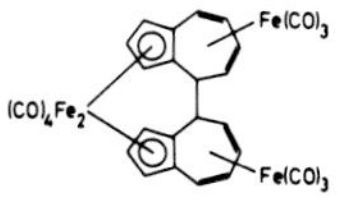	2.138(6) 2.045(6) 2.055(7) 2.134(6)	1.444(9) 1.371(9) 1.434(9)	*142,138*
67	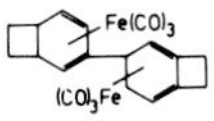	2.112(3) 2.076(3) 2.051(3) 2.128(3)	1.435(4) 1.415(4) 1.418(4)	*208*
68	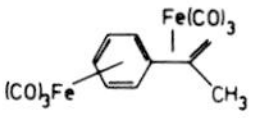	2.00 2.04 2.10 2.19 2.21 2.15 2.04 2.13	1.42 1.44 1.50 1.44 1.46 1.46	*356*
69	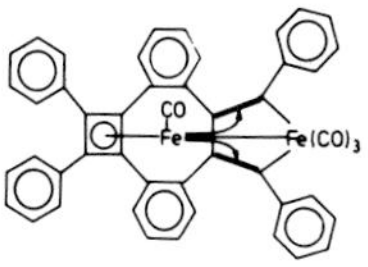	2.109(15) 2.098(16) 2.065(16) 2.074(16)	1.370(19) 1.421(19) 1.393(18)	*287*
70	$Fe_3(CO)_8(PhC_2Ph)_2$ (ferracyclopentadiene-ring)	2.191(15) 2.093(15) 2.169(15) 2.165(16)	1.435(21) 1.456(22) 1.459(21)	*257*
71	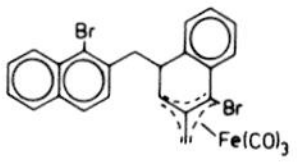	2.10 2.10 2.09 1.81		*437*
72	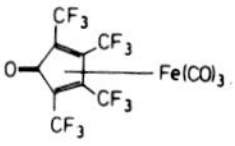			*35*

Table 6: η^5-Cyclopentadienyl and η^5-Heterocyclopentadienyl Iron Complexes.

No.		Formula	*Ref.*
1	$C_5H_7B_3FeO_3$	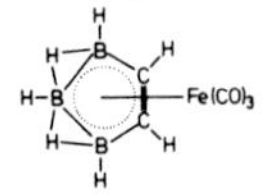	*84*
2	$C_6H_6Cl_6FeOSi_2$	$(\eta^5\text{-}C_5H_5)FeH(SiCl_3)_2(CO)$	*461*
3	$C_7H_5Br_3FeO_2Sn$	$(\eta^5\text{-}C_5H_5)Fe(CO)_2SnBr_3$	*98,481*
4	$C_7H_5Cl_3FeO_2Sn$	$(\eta^5\text{-}C_5H_5)Fe(CO)_2SnCl_3$	*98,332*
5	$C_7H_{16}B_9Fe$	$(\eta^5\text{-}C_5H_5)Fe(B_9C_2H_{11})$	*626*
6	$C_8H_5F_6FeO_3P$	$[(\eta^5\text{-}C_5H_5)Fe(CO)_3]^+[PF_6]^-$	*334*
7	$C_9H_5F_6FeO_2P$	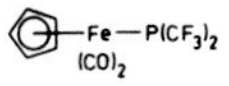	*46,47*
8	$C_9H_5F_6FeO_3P$	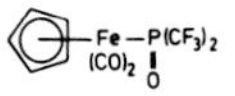	*46,47*
9	$C_9H_8FeO_4$	$(\eta^5\text{-}C_5H_5)Fe(CO)_2\text{-}CH_2\text{-}CO_2H$	*23,330*
10	$C_{10}H_8Cl_2FeO_4S_2$	$(\eta^5\text{-}C_5H_4SO_2Cl)_2Fe$	*576*
11	$C_{10}H_8S_3Fe$	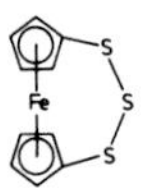	*234*
12	$C_{10}H_{10}Fe$	$(\eta^5\text{-}C_5H_5)_2Fe$	Neutron diffraction *619* *268,269,292*
13	$C_{10}H_{10}FeI_3$	$[(\eta^5\text{-}C_5H_5)_2Fe]I_3$	*62*
14	$C_{10}H_{10}Fe_2N_2O_2$	$[(\eta^5\text{-}C_5H_5)Fe(\mu\text{-}NO)]_2$	*107*
15	$C_{10}H_{14}Cl_2FeGe$	$(\eta^5\text{-}C_5H_5)(\eta^4\text{-}C_4H_6)Fe\text{-}GeCl_2(CH_3)$	*18,16*
16	$C_{11}H_5CoFeO_6$	$(\eta^5\text{-}C_5H_5)(CO)Fe(\mu\text{-}CO)_2Co(CO)_3$	*114*
17	$C_{11}H_5CoFeHgO_6$	$[(\eta^5\text{-}C_5H_5)(CO)_2Fe]HgCo(CO)_4$	*93*
18	$C_{11}H_{10}FeO_4S$	$(\eta^5\text{-}C_5H_5)Fe(CO)_2\text{-}(C_4H_5SO_2)$	*154,147*
19	$C_{12}H_5FeMnO_7$	$(\eta^5\text{-}C_5H_5)(CO)_2Fe\text{-}Mn(CO)_5$	*344*
20	$C_{12}H_6Fe_2O_6$	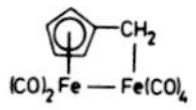	*482*
21	$C_{12}H_{10}FeO_2$	$(\eta^5\text{-}C_5H_5)Fe(CO)_2(\eta^1\text{-}C_5H_5)$	*55*
22	$C_{12}H_{10}FeO_4$	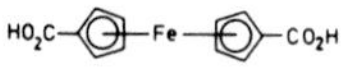	*515*
23	$C_{12}H_{10}F_2FeO$	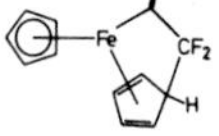	*226*
24	$C_{12}H_{11}Fe_2O_5P$	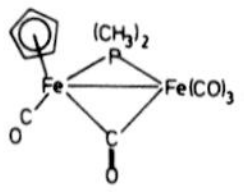	*599*
25	$C_{12}H_{13}FeNO$	$(\eta^5\text{-}C_5H_5)Fe(\eta^5\text{-}C_5H_4\text{-}CH_2\text{-}NH\text{-}CHO)$	*339*
26	$C_{12}H_{14}FeI_3$	$[(\eta^5\text{-}C_5H_4\text{-}CH_3)_2Fe]I_3$	*49*
27	$C_{13}H_{10}Cl_2FeO_2Sn$	$(\eta^5\text{-}C_5H_5)(CO)_2Fe\text{-}SnCl_2(C_6H_5)$	*333*
28	$C_{13}H_{10}Fe_2O_5S$	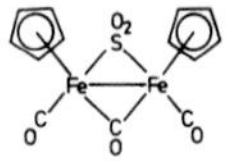	*160*
29	$C_{13}H_{11}FeN$	$(\eta^5\text{-}C_5H_5)Fe(\eta^5\text{-}C_5H_4\text{-}CH{=}CH\text{-}CN)$	*80*

Table 6 (continued)

No.		Formula	Ref.
30	$C_{13}H_{12}OFe$	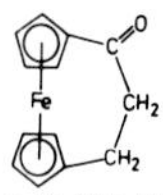	*398*
31	$C_{14}H_{10}BF_4Fe_2IO_4$	$\{[(\eta^5\text{-}C_5H_5)Fe(CO)_2]_2I\}BF_4$	*199*
32	$C_{14}H_{10}Cl_2Fe_2GeO_4$	$[(\eta^5\text{-}C_5H_5)Fe(CO)_2]_2GeCl_2$	*104,105*
33	$C_{14}H_{10}Cl_9Fe_2O_4Sb_3$	$\{[(\eta^5\text{-}C_5H_5)Fe(CO)_2]_2SbCl_2\}Sb_2Cl_7$	*278*
34	$C_{14}H_{10}Cl_2Fe_2O_4Sn$	$[(\eta^5\text{-}C_5H_5)Fe(CO)_2]_2SnCl_2$	*507*
35	$C_{14}H_{10}Cl_8FeO_2$	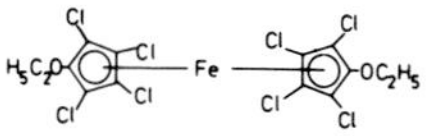	*91,304*
36	$C_{14}H_{10}FeN_2$	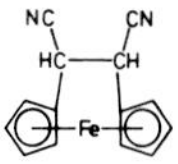	*623*
37	$C_{14}H_{10}FeN_2$	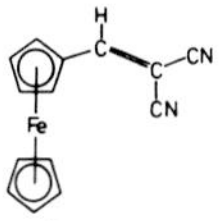	*434*
38	$C_{14}H_{10}Fe_2N_2O_8Sn$	$[(\eta^5\text{-}C_5H_5)Fe(CO)_2]_2Sn(ONO)_2$	*68,72,69*
39	$C_{14}H_{10}Fe_2O_4$	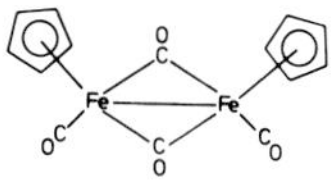	*188,484,99*
40	$C_{14}H_{10}Fe_2O_6$	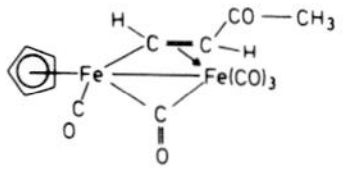	*12,15*
41	$C_{14}H_{10}Fe_2O_6S$	$[\eta^5\text{-}C_5H_5Fe(CO)_2]_2(\mu\text{-}SO_2)$	*159*
42	$C_{14}H_{14}FeO$	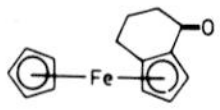	*296*
43	$C_{14}H_{14}FeO_2$	$(\eta^5\text{-}C_5H_4\text{-}CO\text{-}CH_3)_2Fe$	*516*
44	$C_{14}H_{16}BF_4Fe_2O_2S_2$	$\{[(\eta^5\text{-}C_5H_5)FeCO]_2(\mu\text{-}SCH_3)_2\}^+[BF_4]^-$	*180*
45	$C_{14}H_{17}FeO_2$	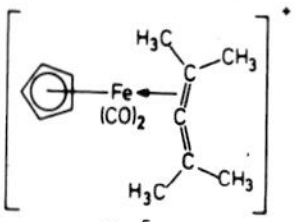	*302*
46	$C_{14}H_{18}FeNO$	$(\eta^5\text{-}C_5H_5)Fe[\eta^5\text{-}C_5H_4\text{-}N(O)\text{-}C(CH_3)_3]$	*301*
47	$C_{14}H_{20}Fe_2S_4$	$[(\eta^5\text{-}C_5H_5)Fe]_2(\mu\text{-}SC_2H_5)_2(\mu\text{-}S_2)$	*588*
48	$C_{15}H_8Fe_2O_5$	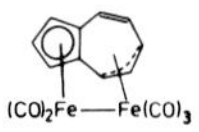	*132*
49	$C_{15}H_{10}FeO_2$	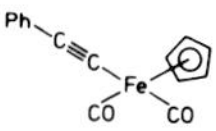	*297*
50	$C_{15}H_{13}F_6FeO_2Sb$	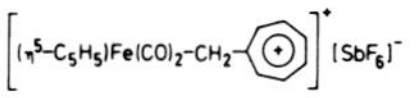	*150*
51	$C_{15}H_{16}FeO$	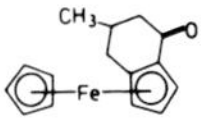	*442*
52	$C_{15}H_{20}As_2FeI_2NiO$	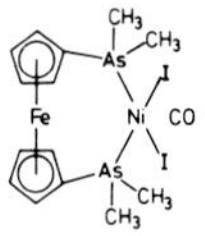	*529*

Table 6 (continued)

No.		Formula	*Ref.*
53	$C_{16}H_{10}FeN_4$	$[(\eta^5\text{-}C_5H_5)_2Fe][(CN)_2C{=}C(CN)_2]$	*3*
54	$C_{16}H_{12}Fe$	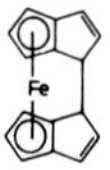	*161*
55	$C_{16}H_{12}FeN_3O_7$	$[(\eta^5\text{-}C_5H_5)_2Fe]^+[C_6H_2(NO_2)_3O]^-$	*525*
56	$C_{16}H_{13}Cl_9FeO_6$	$[(\eta^5\text{-}C_5H_5)_2Fe]^+[3CCl_3\text{-}CO_2H]^-$	*556*
57	$C_{16}H_{13}CoFeO_4$	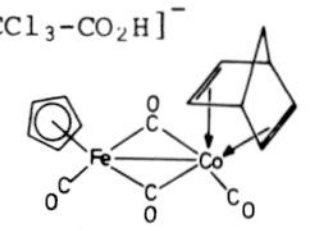	*115*
58	$C_{16}H_{14}Fe$	$(\eta^5\text{-}C_5H_5)Fe(\eta^5\text{-}C_5H_4\text{-}C_6H_5)$	*407*
59	$C_{16}H_{14}Fe_2SiO_4$	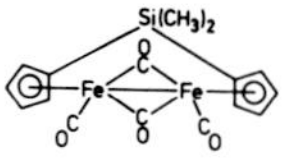	*607*
60	$C_{16}H_{16}Fe_2N_2O_2$	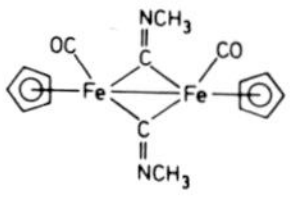	*201*
61	$C_{16}H_{16}Fe_2O_4Pb$	$[(\eta^5\text{-}C_5H_5)Fe(CO)_2]_2[\mu\text{-}Pb(CH_3)_2]$	*73*
62	$C_{16}H_{16}Fe_2O_4Sn$	$[(\eta^5\text{-}C_5H_5)Fe(CO)_2]_2[\ \text{-}Sn(CH_3)_2]$	*69,70*
63	$C_{16}H_{20}Fe$	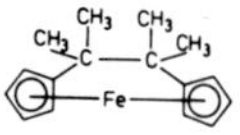	*438,521*
64	$C_{17}H_8Fe_2O_5$	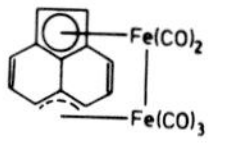	*145*
65	$C_{17}H_{10}Fe_2N_2O_3$	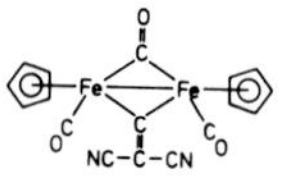	*420*
66	$C_{17}H_{10}Fe_2O_6$	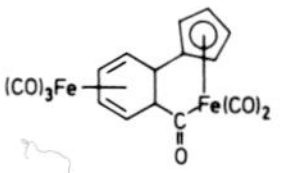	*164*
67	$C_{17}H_{12}FeN_4O_2$	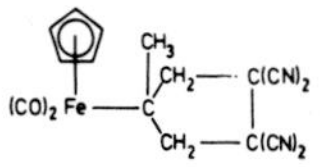	*162*
68	$C_{18}H_{14}FeO_4$	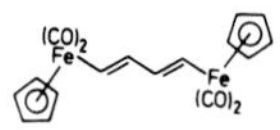	*238,136,141*
69	$C_{18}H_{16}FeO$	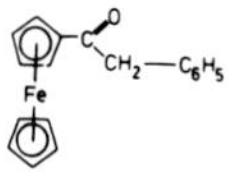	*337*
70	$C_{18}H_{19}Fe_2NO_3$	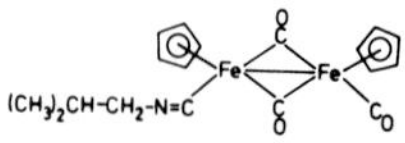	*116*
71	$C_{18}H_{19}Fe_2NO_3$	$(\eta^5\text{-}C_5H_5)_2Fe_2(CO)_3[CNC(CH_3)_3]$	*2*
72	$C_{18}H_{20}Fe$	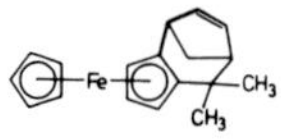	*113*
73	$C_{18}H_{22}Fe_2Ge_2O_5$	$[(\eta^5\text{-}C_5H_5)Fe(CO)_2]_2[\mu\text{-}Ge(CH_3)_2OGe(CH_3)_2]$	*1*
74	$C_{19}H_{16}FeO_2$	$(\eta^5\text{-}C_5H_4\text{-}CO\text{-}CH_3)Fe(\eta^5\text{-}C_5H_4\text{-}CO\text{-}C_6H_5)$	*110*
75	$C_{20}H_{10}F_{12}Fe_2O$	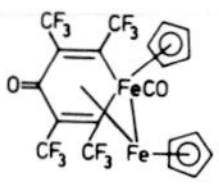	*226*

Table 6 (continued)

No.		Formula	Ref.
76	$C_{20}H_{16}FeO$	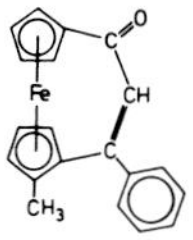	*441*
77	$C_{20}H_{16}Fe_2$	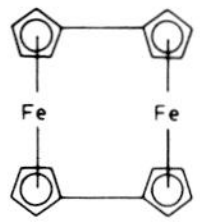	*143*
78	$C_{20}H_{16}Fe$	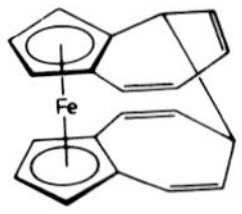	*140,135*
79	$C_{20}H_{16}CL_2Fe_2$	$[(\eta^5\text{-}C_5H_4Cl)Fe(\eta^5\text{-}C_5H_4)]_2$	*403,404,406*
80	$C_{20}H_{18}Fe_2$	$[(\eta^5\text{-}C_5H_5)Fe(\eta^5\text{-}C_5H_4)]_2$	*407,400,472,405*
81	$C_{20}H_{20}Fe_4S_4$	$[(\eta^5\text{-}C_5H_5)Fe]_4(\mu_3\text{-}S)_4$	*561,611*
82	$C_{20}H_{20}Fe_4S_6$	$[(\eta^5\text{-}C_5H_5)Fe]_4(\mu_3\text{-}S_2)_2(\mu_3\text{-}S\ )_2$	*601*
83	$C_{20}H_{22}Fe_2N_2O_4$	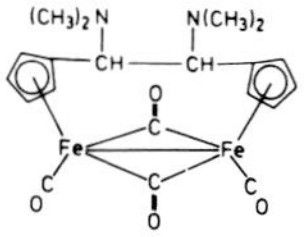	*578*
84	$C_{20}H_{38}FeSi_4$	$[(\eta^5\text{-}C_5H_4\text{-}Si(CH_3)_2\text{-}Si(CH_3)_3]_2Fe$	*358*
85	$C_{21}H_{18}Fe_2O$	$[(\eta^5\text{-}C_5H_5)Fe(\eta^5\text{-}C_5H_4)]_2CO$	*596*
86	$C_{21}H_{20}Fe_2O_5$	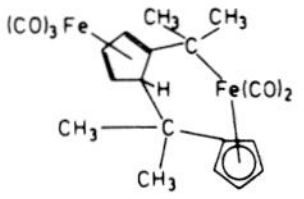	*53*
87	$C_{21}H_{24}FeO_4S$	$[\eta^5\text{-}C_5(CH_3)_5]Fe(CO)_2SO_2\text{-}CH_2\text{-}CH{=}CH\text{-}C_6H_5$	*152*
88	$C_{22}H_{15}ClFe_2MoO_7Sn$	$[(\eta^5\text{-}C_5H_5)Fe(CO)_2]_2(\mu_3\text{-}SnCL)[(\eta^5\text{-}C_5H_5)Mo(CO)_3]$	*508*
89	$C_{22}H_{18}Fe$	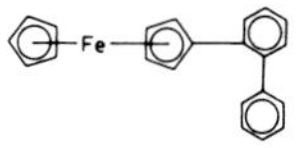	*7,597,620*
90	$C_{22}H_{22}F_6FeP$	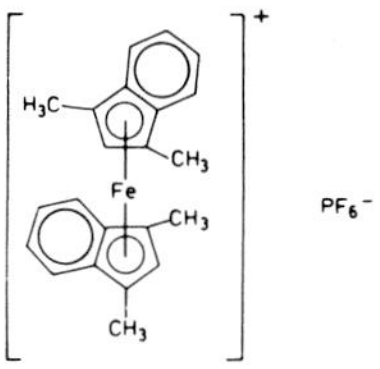	*592*
91	$C_{22}H_{24}FeO$	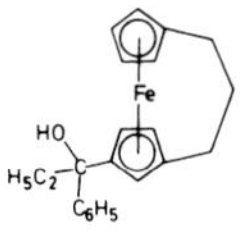	*440*
92	$C_{22}H_{27}FeNO_2$	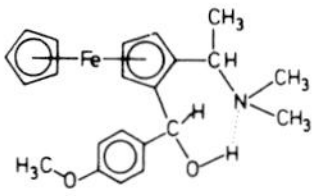	*50*
93	$C_{23}H_{14}Fe_2O_5$	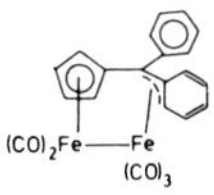	*53*
94	$C_{23}H_{18}CoFeO_5P$	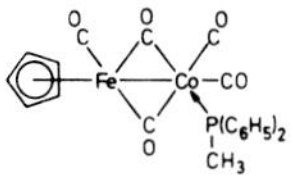	*225*

Table 6 (continued)

No.		Formula	Ref.
95	$C_{24}H_{16}Fe_2$	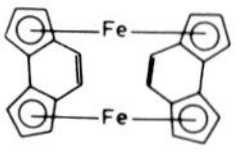	*319*
96	$C_{24}H_{18}FeO_2$	$(\eta^5\text{-}C_5H_4\text{-}CO\text{-}C_6H_5)_2Fe$	*581*
97	$C_{24}H_{20}Fe_2S_2O_2$	$[(\eta^5\text{-}C_5H_5)Fe(CO)]_2(\mu\text{-}S\text{-}C_6H_5)_2$	*291*
98	$C_{24}H_{20}Fe_4O_4$	$[(\eta^5\text{-}C_5H_5)Fe]_4(\mu_3\text{-}CO)_4$	*502*
99	$C_{24}H_{20}F_6Fe_4O_4P$	$\{[(\eta^5\text{-}C_5H_5)Fe]_4(\mu_3\text{-}CO)_4\}^+PF_6^-$	*594*
100	$C_{24}H_{20}Fe_2O_4Sn$	$[(\eta^5\text{-}C_5H_5)Fe(CO)_2]_2[\mu\text{-}Sn(\eta^1\text{-}C_5H_5)_2]$	*74,71,69*
101	$C_{24}H_{22}Fe_2O_4$	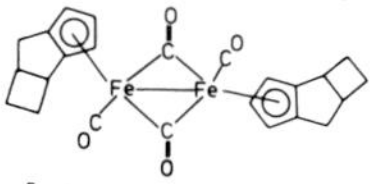	*198*
102	$C_{24}H_{22}Fe_2O_4$	$[(\eta^5\text{-}CH_3O\text{-}CO\text{-}C_5H_4)Fe(\eta^5\text{-}C_5H_4)]_2$	*409*
103	$C_{24}H_{26}Fe_2$	$[(\eta^5\text{-}C_2H_5\text{-}C_5H_4)Fe(\eta^5\text{-}C_5H_4)]_2$	*401*
104	$C_{24}H_{24}Fe_2$	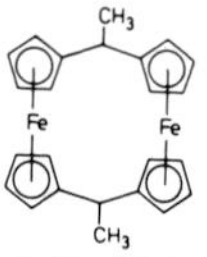	*477,476*
105	$C_{25}H_{20}FeO_2Sn$	$(\eta^5\text{-}C_5H_5)Fe(CO)_2Sn(C_6H_5)_3$	*96*
106	$C_{26}H_{20}Fe_2O_4Sn$	$[(\eta^5\text{-}C_5H_5)Fe(CO)_2]_2[\mu\text{-}Sn(C_6H_5)_2]$	*69*
107	$C_{26}H_{20}Fe_2O_8S_2Sn$	$[(\eta^5\text{-}C_5H_5)Fe(CO)_2]_2[\mu\text{-}Sn(SO_2\text{-}C_6H_5)_2]$	*97*
108	$C_{26}H_{28}Fe$	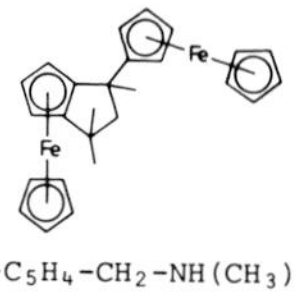	*370*
109	$C_{26}H_{36}Cl_4Fe_2N_2Zn\cdot H_2O$	$\{(\eta^5\text{-}C_5H_5)Fe[\eta^5\text{-}C_5H_4\text{-}CH_2\text{-}NH(CH_3)_2]\}_2^+ZnCl_4^{2-}$	*313*
110	$C_{26}H_{39}FeN_3O$	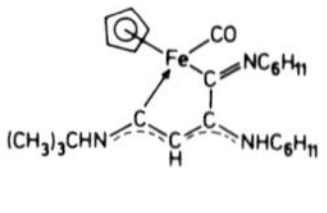	*624*
111	$C_{26}H_{40}Al_2Fe_2O_4$	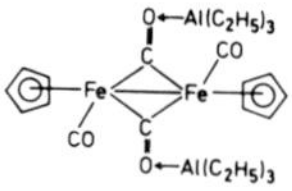	*416*
112	$C_{26}H_{42}Fe$	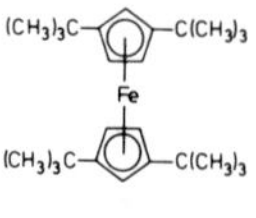	*410*
113	$C_{27}H_{20}Fe_3O_6Sn$	$[(\eta^5\text{-}C_5H_5)Fe(CO)_2]_3(\mu_3\text{-}Sn\text{-}C_6H_5)$	*69*
114	$C_{28}H_{20}Cl_{10}Fe_4O_8Sb_2$	$[(\eta^5\text{-}C_5H_5)(CO)_2Fe\text{-}Cl]_4(\mu_3\text{-}SbCl_3)_2$	*279*
115	$C_{28}H_{23}FeOPS$	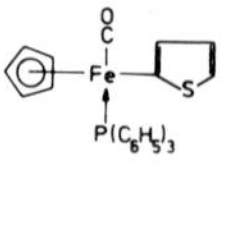	*17*

Table 6 (continued)

No.		Formula	Ref.
116	$C_{30}H_{20}Cl_2Cu_2Fe_2O_4$	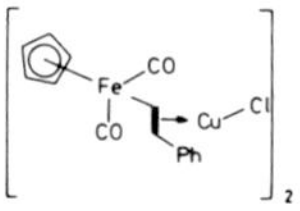	92,171
117	$C_{30}H_{25}FeOP$	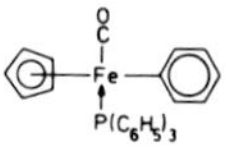	28,562
118	$C_{30}H_{26}Fe_3$	$(\eta^5\text{-}C_5H_5)Fe(\eta^5\text{-}C_5H_4)_2Fe(\eta^5\text{-}C_5H_4)_2Fe(\eta^5\text{-}C_5H_5)$	402,405
119	$C_{31}H_{25}FeO_2P$	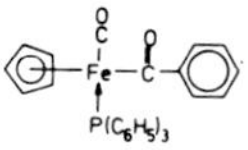	124,563
120	$C_{31}H_{25}Fe_2O_6P$	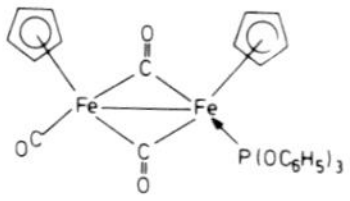	202
121	$C_{32}H_{20}Cl_2Fe_4O_{10}$	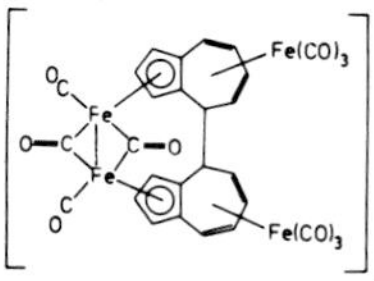 $C_2H_4Cl_2$	138
122	$C_{33}H_{24}FeO_9$	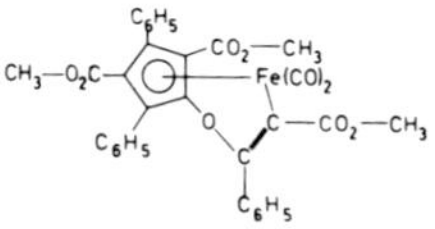	221
123	$C_{33}H_{30}Fe_3$	$[(\eta^5\text{-}C_5H_4)Fe(\eta^5\text{-}C_5H_4\text{-}CH_2\text{-})]_3$	457
124	$C_{33}H_{38}FeN_2O_4 \cdot H_2O$	$(\eta^5\text{-}CH_3\text{-}C_5H_4)Fe(\eta^5\text{-}CH_3\text{-}C_5H_3\text{-}CO_2H)$ absolute configuration of quinidine salt	119
125	$C_{35}H_{34}Fe_3O$	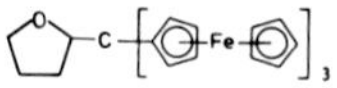	343
126	$C_{38}H_{30}FeOSn$	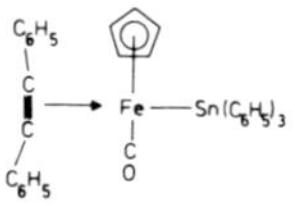	565
127	$C_{38}H_{32}Fe_2O_{10}S_2Sn_2$	$[(\eta^5\text{-}C_5H_5)Fe(CO)_2Sn(C_6H_5)(OSO\text{-}C_6H_5)(OH)]_2$	546
128	$C_{38}H_{34}F_6FeOP_2Sn$	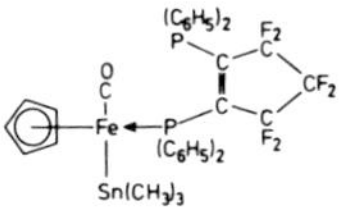	545,274
129	$C_{41}H_{35}FeIO_6P_2$	$(\eta^5\text{-}C_5H_5)Fe[P(OC_6H_5)_3]_2I$	14,13,11,123
130	$C_{43}H_{32}Cl_8Fe_7O_{12}Sb_2$	$\{[(\eta^5\text{-}C_5H_5)Fe(CO)_2]_3(\mu_3\text{-}SbCl)\}_2^+[FeCl_4]^{2-} \cdot CH_2Cl_2$	593
131	$C_{40}H_{34}F_6Fe_2O_4P_3Rh$	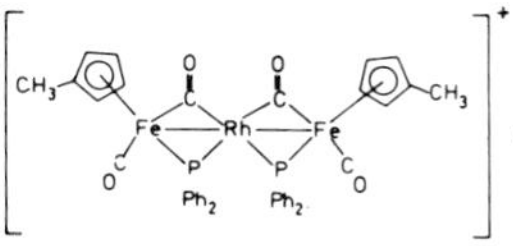 PF_6^-	340,468
132	$C_{46}H_{39}Au_2BF_4FeP_2$	$\{(\eta^5\text{-}C_5H_5)Fe[\eta^5\text{-}C_5H_4\text{-}Au(PPh_3)\text{-}Au(PPh_3)]\}^+BF_4^-$	20,21
133	$C_{52}H_{40}Fe_4O_6P_2$	$\{[(\eta^5\text{-}C_5H_5)Fe(CO)](\mu\text{-}CO)_2[(\eta^5\text{-}C_5H_5)Fe]\}_2\text{-}[\mu\text{-}(C_6H_5)_2P\text{-}C{\equiv}C\text{-}P(C_6H_5)_2]$	120

Table 7: Organoiron Complexes Containing Fe-N Bonds.

No.		Formula	Fe-N Distance [Å]	Ref.
1	$C_{16}H_{36}F_6Fe_2N_5OPS_4$	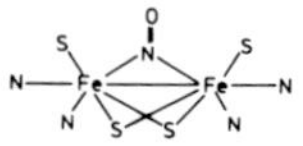 $[(FeL)_2NO]PF_6$ L = *N,N'*-dimethyl-*N,N'*-bis(β-mercapto ethyl)ethylenediamine	2.104(5) 2.086(5) bridge 1.193(8)	*411*
2	$C_8FeN_5OS_4$	$(NO)Fe[S_2C_2(CN)_2]_2$	1.56	*540*
3	$C_{29}H_{20}F_6FeN_2O_2P_2$	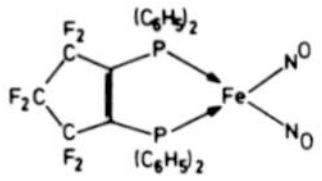	1.645(7) 1.661(7)	*348*
4	$C_{36}H_{30}FeN_2O_2P_2$	$Fe(NO)_2[P(C_6H_5)_3]_2$	1.650(7)	*6*
5	$C_4H_{10}Fe_2N_4O_4S_2$	$[Fe(NO)_2]_2(\mu\text{-}SC_2H_5)_2$	1.660(8)	*589*
6	$C_{10}H_{20}FeN_3OS_4$	$Fe(NO)[S_2C\text{-}N(C_2H_5)_2]_2$	1.690(4)	*176*
7	$C_{44}H_{35}BF_4FeN_2O_2P_2$	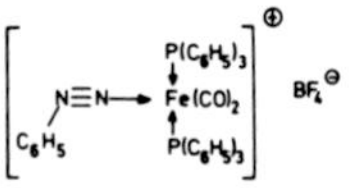	1.702(6)	*349*
8	$C_6H_{12}FeN_3OS_4$	$Fe(NO)[S_2C\text{-}N(CH_3)_2]_2$	(at -80°C) 1.720(5) 1.705(16)	*229* *227*
9	$C_{19}H_{15}FeN_2O_3P$	$Fe(NO)_2(CO)[P(C_6H_5)_3]$ (CO + NO ligands are disordered)	Fe-N/C av. 1.709	*6*
10	$C_{10}H_{10}Fe_2N_2O_2$	$[(\eta^5\text{-}C_5H_5)Fe]_2(\mu\text{-}NO)_2$	av. 1.768(9)	*107*
11	$C_5H_6FeN_4O_3$	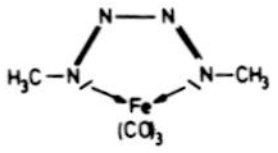	1.830(3)	*259*
12	$C_8H_6Fe_2N_2O_6$	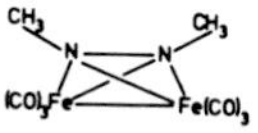	1.873(4) 1.882(4)	*262*
13	$C_{39}H_{27}Fe_2N_2O_5P$	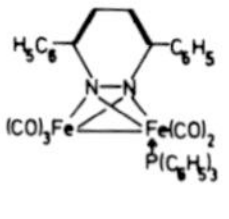	1.89 1.91 1.95 1.93	*435*
14	$C_{16}H_{30}FeN_8O_6$	bis(dimethylglyoximato)-diimidazoleiron(II)-dimethanol	1.893(6) 1.918(6) axial 1.985(5)	*82*
15	$C_{22}H_{24}Cl_2FeN_8O_8$	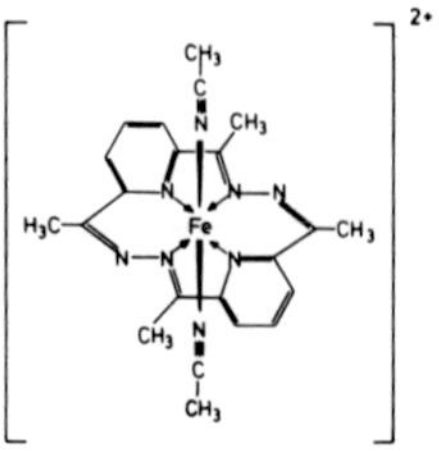 $[ClO_4]_2^-$	1.892 1.899 (CH_3CN) 1.938	*324*

Table 7 (continued)

No.		Formula	Fe-N Distance [Å]	*Ref.*
16	$C_{13}H_9Fe_3NO_{10}Si$	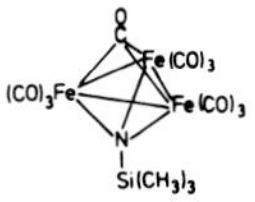	1.920 1.870 1.907	*42*
17	$C_{23}H_{24}Cl_2FeN_6O_8$	$[Fe(py_3TPN)][ClO_4]_2$ py_3TPN = 1,1,1-tris(pyridine-2-aldiminomethyl)ethane	1.89(2) 1.99(2) 1.94(2) 2.01(1) 1.90(1) 2.01(2)	*299*
18	$C_{12}H_{22}FeN_8O$	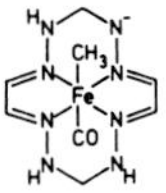	1.900(2)	*325*
19	$C_{28}H_{27}FeN_4$	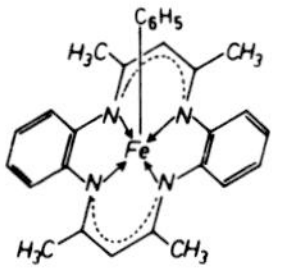	1.904(2) 1.906(2) 1.909(3) 1.915(3)	*323*
20	$C_{18}H_8Fe_2N_2O_6$	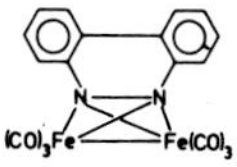	1.910(6) 1.918(6)	*263*
21	$C_{18}H_{30}FeN_8O_6$	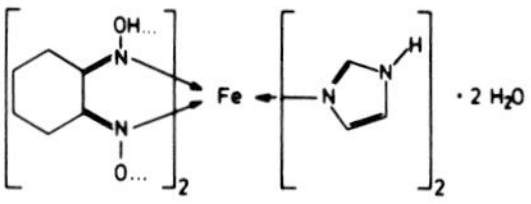	1.910(2) 1.970(1) 2.050(2)	*538*
22	$C_8H_{18}Fe_4N_6O_4S_2$	$[Fe(NO)]_4(\mu_3\text{-}S)_2[\mu_3\text{-}N\text{-}C(CH_3)_3]_2$	1.914(3) 1.908(3)	*306*
23	$C_{18}H_{54}FeN_3Si_6$	$\{[(CH_3)_3Si]_2N\}_3Fe$	1.917(4)	*631*
24	$C_{29}H_{18}Fe_2N_2O_6$	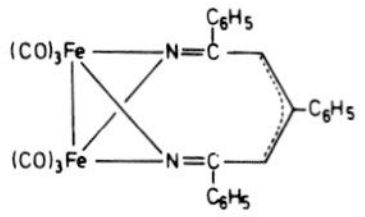	1.919 1.921 1.960 1.969	*121*
25	$C_{56}H_{44}FeN_{12}$	[(Phthalocyanine)(4-methylpyridine)$_2$Fe] •2(4-methylpyridine)	av. 1.92 (phthalocyanine) av. 2.00 (methylpy)	*422*
26	$C_{44}H_{28}FeN_5O$	(Nitrosyl)(α,β,γ,δ-tetraphenyl-porphinato)iron(II)	1.928(6) 2.001(3)	*555*
27	$C_{36}H_{28}Fe_2N_2O_6$	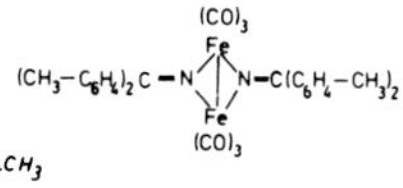	1.94	*86*
28	$C_{23}H_{26}FeN_6O$	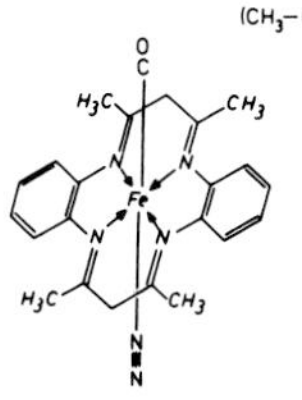	1.94 (axial) 2.12	*322*
29	$C_{35}H_{20}Fe_3N_4O_9$	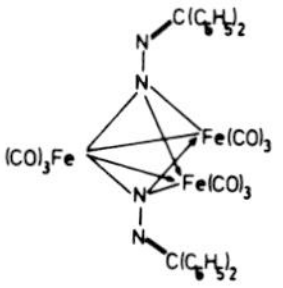	av. 1.95	*33*

Table 7 (continued)

No.	Formula		Fe-N Distance [Å]	Ref.
30	$C_{24}H_{27}B_2F_8FeN_7$	$\{[(NC_5H_4\text{-}CH{=}N\text{-}CH_2\text{-}CH_2\text{-})_3N]Fe\}^{2+}[BF_4]_2^-$	(imine) 1.95 av. (pyridine) 1.96 1.97 1.98	*480,454*
31	$C_{20}H_{13}Fe_2NO_6$	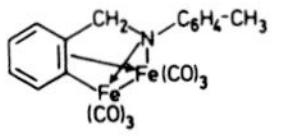	1.950 1.960	*32*
32	$C_{18}H_{10}Fe_2N_2O_6$	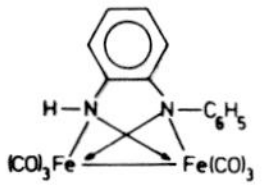	1.980 2.000 2.020	*32*
			1.979(11) 2.024(7) 1.999(9) 2.017(9)	*34*
33	$C_{30}H_{28}Cl_2FeN_4O_8S_2 \cdot CH_3OH$	$[Fe(N_4S_2)]^{2+}(ClO_4)_2^- \cdot CH_3OH$ N_2S_4 = 15,18-Dithia-1,5,8,12-tetraaza-3,4:9,10:13,14:19,20-tetrabenzo-cycloeicosa-1,11-diene	1.950(2) 1.960(2) 1.980(2) 2.000(2)	*585*
34	$C_{48}H_{40}As_2FeN_{15}$	$[(C_6H_5)_4As]_2Fe(N_3)_5$	1.960(4) 1.970(1) 2.040(2)	*267*
35	$C_{44}H_{28}FeN_6O_{12}Sb_2 \cdot 8H_2O$	[*l*-Tris(1,10-phenanthroline)-iron(II)]bis[antimony(III)-*d*-tartate]-octahydrate $[Fe(C_{12}H_8N_2)_3][Sb(C_4H_2O_6)_2] \cdot 8H_2O$	1.96(1) 1.98(1) 1.97(1)	*628,587*
36	$C_9H_6Fe_2N_2O_7$	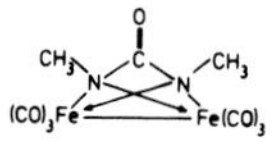	1.965(10)	*260*
37	$C_{33}H_{24}Fe_2N_2O_{10}$	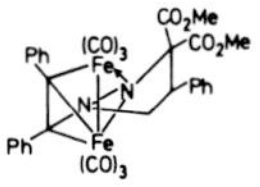	1.971 2.011	*433*
38	$C_{19}H_{14}B_2Cl_2F_5FeN_6O_3P$	$[\{FB(ONCHC_5H_3N)_3P\}Fe]^+[BF_4]^- \cdot CH_2Cl_2$ {[Fluoroboro-tris(2-aldoximo-6-pyridyl)phosphine]iron(II)}-tetrafluoroborate	1.972(7) - 1.984(8) (pyridyl) 1.921(7) - 1.943(8) (aldoximo)	*157*
39	$C_{36}H_{26}Cl_3FeN_6O_{13}$	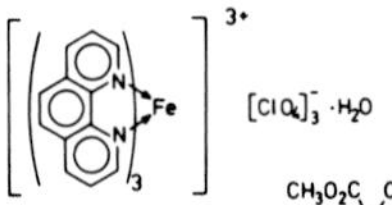	1.973	*38*
40	$C_{17}H_{14}FeN_2O_8$	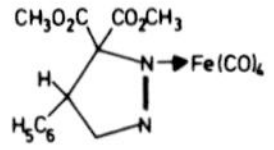	1.979(2)	*432*
41	$C_6H_4Fe_2N_2O_6$	$[H_2NFe(CO)_3]_2$	1.98(2) 2.01(2) 1.98(2) 1.96(2)	*222*
42	$C_{20}H_{12}Cl_2Fe_2N_2O_8$	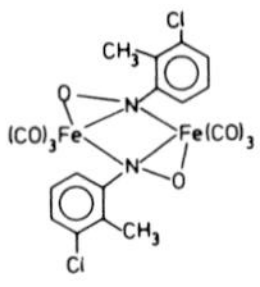	1.980(1) 2.020(1)	*44,45*

Table 7 (continued)

No.	Formula		Fe-N Distance [Å]	Ref.
43	$C_{50}H_{36}ClFeN_8 \cdot CH_3OH$	bis(imidazole)-α,β,γ,δ-tetra-phenyl-porphinatoiron(III) chloride	1.989 1.957(4) (imid) 1.991(5)	*217,179*
44	$C_{26}H_{16}FeN_6O_2S_2$ $\cdot CHCl_3 \cdot C_2H_5OH \cdot H_2O$	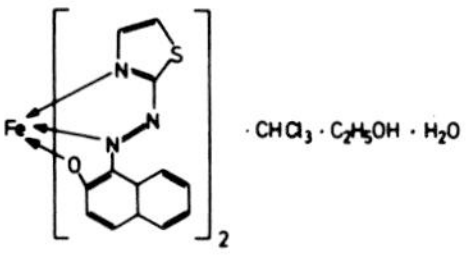	1.90 1.93	*426*
45	$C_{19}H_{10}Fe_2N_2O_7$	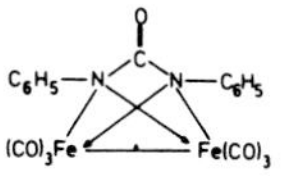	1.99 2.01	*392*
			1.993(9) 2.004(8) 2.000(8) 1.985(8)	*532*
46	$C_{36}H_{30}Fe_2N_4O_6$	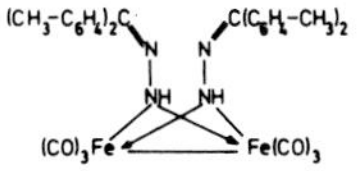	1.990 2.020	*30*
47	$C_{49}H_{35}Cl_3FeN_7O$	(Nitrosyl)(α,β,γ,δ-tetraphenyl-porphinato)(1-methylimidazole)iron	1.993- 2.180	*528*
48	$C_{26}H_{14}FeN_2O_9$	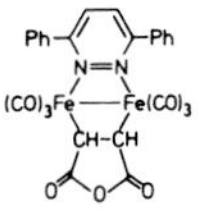	2.00(1) 2.01(1)	*517*
49	$C_{42}H_{52}N_8O_4FeCl \cdot 2CHCl_3$	bis(imidazole)octaethylporphinato-iron(III) perchlorate	2.01	*584*
50	$C_{34}H_{32}ClFeN_4O_4$	α-Chlorohemin	2.001 2.003 2.011 2.013	*424*
51	$C_{12}H_5Fe_2NO_6S$	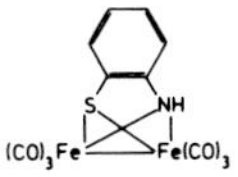	2.002(4)	*79*
52	$C_8H_4FeN_2O_4$	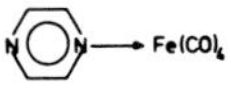	2.031(2)	*204*
53	$C_{19}H_{15}Fe_2NO_{10}S$	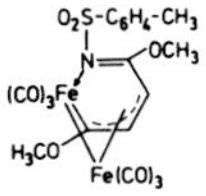	2.040	*553*
54	$C_{42}H_{40}FeN_5O_6S$	(*p*-Nitrobenzenethiolato)(proto-porphyrin IX dimethylester)iron(III)	2.04 2.07 2.09 2.09	*423*
55	$C_9H_5FeNO_4$	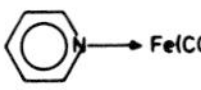	2.046(5)	*204*

Table 7 (continued)

No.		Formula	Fe-N Distance [Å]	Ref.
56	$C_9H_8FeN_2O_5$	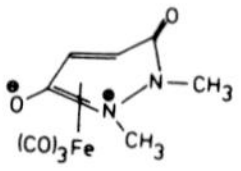	2.05	*500*
57	$C_{18}H_{13}FeNO_3$	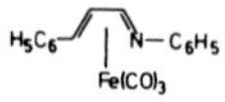	2.063(5) 2.057 2.01	*167* *166* *165*
58	$C_{30}H_{28}ClFeN_2O_2$	Fe(SANE)$_2$Cl SANE = *N*-(2-phenylethyl)salicylaldimine	2.14(1) 2.12(1)	*64*
59	$C_{88}H_{56}Fe_2N_2O$	(μ-oxo)bis[α,β,γ,δ-tetraphenyl-porphinatoiron(III)] O(Fe-TPP)$_2$	2.087(3)	*365*
60	$C_{22}H_{20}FeN_6S_2$	Fe(C_5H_5N)$_4$(NCS)$_2$	(NCS) 2.088(4) (pyridine) {2.241(4), 2.268(4)}	*574*
61	$C_{20}H_{24}ClFeN_2O_2$	(chloro)bis(*N*-*n*-propylsalicylald-iminato)iron(III)	2.09(2) 2.096	*235,231*
62	$C_{22}H_{16}FeN_6S_2$	Fe(bipy)$_2$(NCS)$_2$	(NCS) 2.09(3) (bipy) {2.17(2), 2.17(2)}	*425*
63	$C_{54}H_{50}FeN_6$	bis(piperidine)(α,β,γ,δ-tetra-phenylporphinato)iron(II)	(ax.) 2.127(3) (eq.) {2.008(3), 2.000(3)}	*541*
64	$C_{33}H_{30}Cl_2Fe_2N_4O_5$	μ-O[Fe-salen^a]$_2$•CH_2Cl_2	2.11 2.13	*25*
65	$C_{32}H_{28}Fe_2N_4O_5$	μ-O[Fe-salen^a]$_2$	2.12(1)	*233*
66	$C_{27}H_{30}Cl_2Fe_2N_2O_4$	[Fe(SALPA)Cl]$_2$•$C_6H_5CH_3$ SALPA = *N*-(3-hydroxypropyl)salicyald-imine	2.06(1)	*64*
67	$C_{16}H_{32}ClFeIN_4$	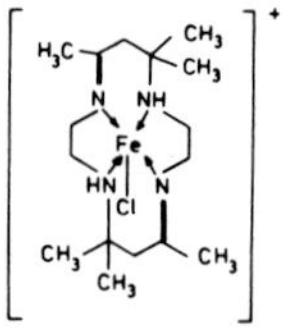 I^-	2.13(1) 2.17(1)	*321*
68	$C_{12}H_{18}Cl_8Fe_3N_6$	[(CH_3-C≡N)$_6$Fe]$^{2+}$[$FeCl_4$]$_2^-$	2.13(2) 2.24(1)	*182*
			2.111(9) 2.240(11)	*580*
69	$C_{10}H_{12}FeLiN_2O_8$•$3H_2O$	{Li^+[Fe(EDTA^b)]$^-$}•$3H_2O$	2.14(5) 2.18(5)	*506*
70	$C_{40}H_{48}Fe_2N_4O_5$	(μ-oxo)bis[(*N*-*n*-propylsalicylidene-iminato)iron(III)]	2.14(2) 2.14(2) 2.13(2) 2.14(2)	*232*
71	$C_{12}H_{30}Br_2FeN_4$	Fe{N[-CH_2-CH_2-N(CH_3)$_2$]$_3$}Br_2	2.150(1) 2.210(1)	*600*
72	$C_2H_{10}FeN_6O_4S_3$	[Fe(NH_2-NH-CS-NH_2)$_2SO_4$] (polymeric)	2.164(6) 2.195(6)	*496*

Table 7 (continued)

No.	Formula		Fe-N Distance [Å]	*Ref.*
73	$C_{26}H_{18}ClFeN_2O_2S_2$	(chloro)[bis(salicylideniminephenyl)-disulfido]iron(III)	2.166(12) 2.204(10)	*65*
74	$C_{27}H_{24}Cl_3FeN_4O_3 \cdot 3H_2O$	$\left[N \left(CH_2-CH_2-N-CH-\bigcirc-Cl \right)_3 Fe^{3+} \right]$	2.18	*36*
75	$C_{30}H_{50}Cl_4Fe_2N_{10}O_{19}$	$[(H_2O)LFe-O-FeL(H_2O)]^{4+}[ClO_4]_4^-$	2.2	
	$C_{17}H_{23}ClFeN_7O_4S_2$	$[FeL(NCS)_2]ClO_4$	2.23(5) (NCS) 2.01(2)	*297*
		L = 2,13-dimethyl-3,6,9,12,18-pentaazabicyclo[12.3.1]octadeca-1(18),2,12,14,16-pentaene		
76	$C_{22}H_{40}Fe_2N_6O_{15} \cdot 6H_2O$	$\{[NH_3-CH_2-CH_2-NH_3]^{2+}[LFe(III)-O-Fe(III)L]^{2-}\} \cdot 6H_2O$	2.200 2.270	*456*
		L = $\overset{HO-CH_2-CH_2}{\underset{O_2C-CH_2}{}}N-CH_2-CH_2-N\overset{CH_2-CO_2^-}{\underset{CH_2-CO_2^-}{}}$		
77	$C_{16}H_{36}Fe_2N_4S_4$	$(\mu-L)_2Fe_2$ L = N,N'-dimethyl-N,N'-bis(β-mer-captoethyl)ethylenediamine	2.20(1) 2.32(1)	*377*
78	$C_{10}H_{15}FeN_2O_9$	$[Fe(HEDTA^b)H_2O]$	2.22 2.325	*412* *360*
79	$C_{18}H_{40}Fe_2N_4S_4$	$(\mu-L)_2Fe_2$ L = N,N'-dimethyl-N,N'-bis(β-mer-captoethyl)-1,3-propanediamine	2.207(7) 2.337(7)	*377*
80	$C_{43}H_{30}Fe_5N_6O_{13}$	$[Fe(C_5H_5N)_6]^{2+}[Fe_4(CO)_{13}]^{2-}$	2.25(3) 2.28(3) 2.29(3) 2.26(3) 2.22(3) 2.25(3)	*258*
81	$C_{29}H_{40}CaFe_2N_4O_{18} \cdot 8H_2O$	$Ca^{2+}[FeL(H_2O)]_2^- \cdot 8H_2O$ L = trans-1,2-diaminocyclohexane-N,N'-tetraacetate	2.290(4)	*175*
82	$C_{32}H_{24}Cl_2FeN_8O_8$	$\left[Fe\left(\bigcirc_N^N \right)_4 \right]^{2+} [ClO_4]_2^-$	2.284(11) 2.188(10) 2.213(10) 2.184(11) 2.335(10) 2.378(11) 2.465(11) 2.756(11)	*568*
83	$C_{10}H_{14}FeLiN_2O_9 \cdot 2H_2O$	$\{Li^+[Fe(EDTA^b)(H_2O)]^-\} \cdot 2H_2O$	av. 2.325	*447*
84	$C_{10}H_{14}FeN_2O_9Rb \cdot 2H_2O$	$\{Rb^+[Fe(EDTA^b)(H_2O)]^-\} \cdot 2H_2O$	av. 2.317	*447*
85	$C_{10}H_{14}FeN_2NaO_9 \cdot 2H_2O$	$\{Na^+[Fe(EDTA^b)(H_2O)]^-\} \cdot 2H_2O$	av. 2.26	*505*
86	$C_{11}H_{18}FeN_5O_8 \cdot 3H_2O$	$\{(CN_3H_6)^+[Fe(EDTA^b)(H_2O)]^-\} \cdot 2H_2O$	av. 2.38	*501*
87	$C_{16}H_{46}Fe_2N_{10}OI_4$	(μ-oxo)bis[tetraethylene penta-amineiron(III)]iodide	2.32(5) 2.36(4) 2.30(3) 2.12(4) 2.13(4)	*177*

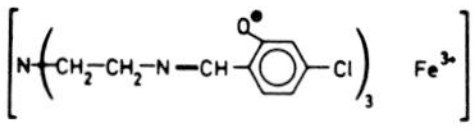

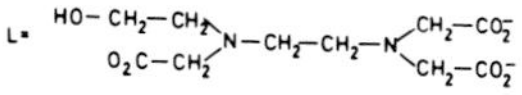

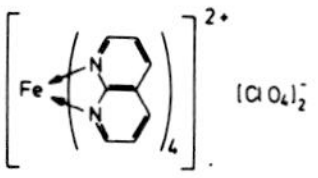

Table 7 (continued)

No.		Formula	Fe-N Distance [Å]	*Ref.*
88	$C_4H_{30}Fe_2N_2O_{22}S_2$	$\{[Fe(H_2O)_6]^{2+}$ $[Fe(H_2O)_4(H_3N^+-CH_2-CO_2^-)_2][SO_4]_2^{2-}\}$		*453*
89	$Fe_4N_4O_4S_4$	$[Fe(NO)]_4(\mu_3-S)_4$		*306*
90	$C_{20}H_{22}Cl_2Fe_2N_2O_4$	$[Fe(Sal{=}N-(CH_2)_3O)Cl]_2$, and other Sal=N-R complexes of iron(III) Sal=N-R = 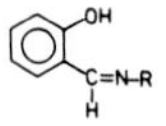		*284*
91	$C_{12}H_{18}B_2F_2FeN_6O_6 \cdot \frac{1}{2}C_6H_{12}$	$FeL \cdot \frac{1}{2}C_6H_{12}$ L = 1,8-bis(fluoroboro)-2,7,9,14,15,20-hexaoxa-3,6,10,13,16,19-hexaaza-4,5,11,12,17,18-hexamethylbicyclo[6.6.6]eicosa-3,5,10,12,16,18-hexaene		*455*
92	$C_{17}H_8Fe_2N_2O_7$	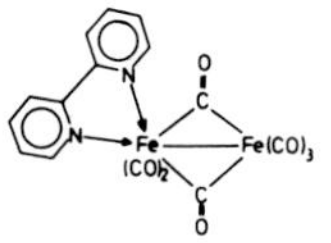		*203*
93	$C_{18}H_{30}F_{12}FeN_6P_2$	$[FeL(CH_3CN)_2]^{2+}[PF_6]_2^-$ L = 2,3,9,10-tetramethyl-1,4,8,11-tetraaza-cyclotetradeca-1,3,8,10-tetraene		*571*
94	$C_{11}H_{21}Cl_2FeN_7O_5$	$[Cl-FeL(H_2O)]^+Cl^- \cdot 2H_2O$ L = 2,6-diacetylpyridine-bis(semicarbazone)		*613*

a 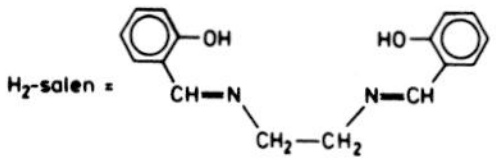

b 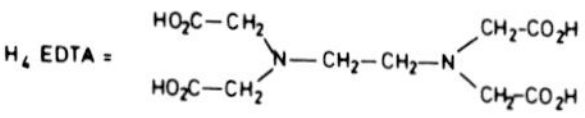

Table 8: Organoiron Complexes Containing Fe-O Bonds.

No.		Formula	Fe-O Distance [Å]	*Ref.*
1	$C_{88}H_{56}Fe_2N_8O$	(μ-Oxo)bis[α,β,γ,δ-tetraphenyl-porphinatoiron(III)] $O(Fe\text{-}TPP)_2$	1.763(1)	*365*
2	$C_{16}H_{46}Fe_2N_{10}OI_4$	(μ-Oxo)bis[tetraethylenepenta-amineiron(III)]iodide	1.77(1)	*177*
3	$C_{32}H_{28}Fe_2N_4O_5$	μ-O[Fe-salen[a]]$_2$ (bridge)	1.93(1) 1.78(1)	*233*
4	$C_{22}H_{40}Fe_2N_6O_{15}\cdot 6H_2O$	$\{[NH_3\text{-}CH_2\text{-}CH_2\text{-}NH_3]^{2+}[LFe(III)\text{-}O\text{-}Fe(III)L]^{2-}\}\cdot 6H_2O$ L= 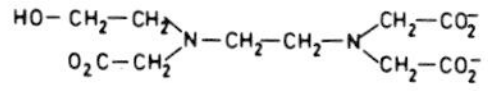	1.79(1) 1.80(1) 2.03(1) (average)	*456*
5	$C_{33}H_{30}Cl_2Fe_2N_4O_5$	μ-O[Fe-salen[a]]$_2\cdot CH_2Cl_2$	1.791(9) 1.797(9) 1.926(9) 1.924(9) 1.924(9) 1.916(9)	*25,174*
6	$C_{30}H_{28}ClFeN_2O_4$	$Fe(SANE)_2Cl$ SANE = *N*-(2-phenylethyl)sali-cylaldimine	1.85(1) 1.93(1) 1.98(1)	*64*
7	$C_{27}H_{30}Cl_2Fe_2N_2O_4$	$[Fe(SALPA)Cl]_2\cdot C_6H_5CH_3$ SALPA = *N*-(3-hydroxypropyl)sali-cylaldimine	1.86(1) 1.88(1)	*64*
8	$C_{26}H_{18}ClFeN_2O_2S_2$	(Chloro)[bis(salicylidenimine-phenyl)disulfido]iron(III)	1.871(8) 1.909(7)	*65*
9	$C_{16}H_{14}ClFeN_2O_2\cdot CH_3NO_2$	[Cl-Fe-salen[a]]	1.879(10) 1.885(11)	*308*
10	$C_{16}H_{14}ClFeN_2O_2$	[Cl-Fe-salen[a]]$_2$	1.898(7) 1.978(7)	*309*
11	$C_{20}H_{24}ClFeN_2O_2$	(Chloro)bis(*N*-*n*-propylsalicylal-diminato)iron(III)	1.887 1.89(2)	*230,231*
12	$C_{18}H_{48}Cl_7Fe_3N_6O_{44}$	(bridge-O) $\{[Fe(III)(alanine)_2H_2O]_3O\}(ClO_4)_7$ (H_2O)	1.92 2.07	*369*
13	$C_{40}H_{48}Fe_2N_4O_5$	(μ-Oxo)bis[bis(*N*-*n*-propylsali-cylideneiminato)iron(III)]	1.93(1) 1.92(1) 1.94(2) 1.92(1)	*232*
14	$C_{42}H_{38}Fe_2N_6O_5$	μ-O[Fe-salen[a]]$_2\cdot py_2$ (bridge-O) (salen-O)	1.79 1.93	*310*
15	$C_{26}H_{16}FeN_6O_2S_2\cdot CHCl_3\cdot C_2H_5OH\cdot H_2O$	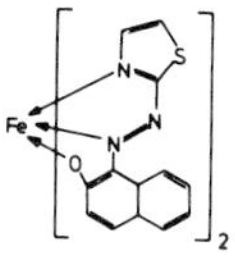 $\cdot CHCl_3\cdot C_2H_5OH\cdot H_2O$	1.93 2.01	*426*
16	$C_{10}H_{14}ClFeO_4$	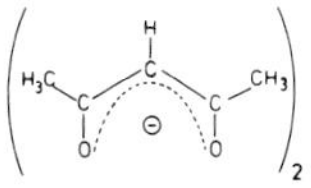 FeCl	1.95(1)	*451*
17	$C_{26}H_{15}F_{12}FeOPS_4$	$\{[(C_6H_5)_3PO]Fe[S_2C_2(CF_3)_2]_2\}^-$	1.957(15)	*289*

Table 8 (continued)

No.		Formula	Fe-O Distance [Å]	Ref.
18	$C_{20}H_{10}Fe_2O_8$	(see Fig. 6) 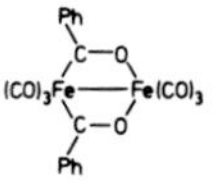	1.967(5)	*449*
19	$C_{20}H_{15}Fe_2O_7P$	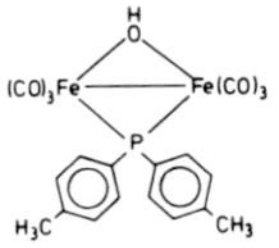	1.969(6) 1.974(6)	*591*
20	$C_{21}H_{24}FeN_3O_9$	$Fe(C_6H_5\text{-}CO\text{-}NH\text{-}O)_3 \cdot 3H_2O$	av. 1.98(1) av. 2.06(1)	*452*
21	$C_{10}H_{15}FeN_2O_9$	[Fe(HEDTA[b])H_2O]	1.98	*412*
22	$C_{11}H_{13}F_6FeO_4P$	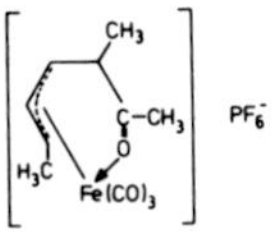	1.987(10)	*326*
23	$C_{15}H_{21}O_{10}AgClFe$	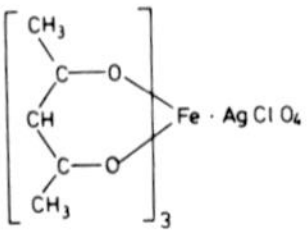	av. 1.99	*497*
24	$C_{18}H_{15}FeN_6O_6$	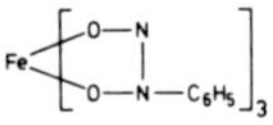	1.99(1) 2.01(1) 2.00(1) 2.01(1) 1.99(1) 2.00(1)	*354*
25	$C_{15}H_{21}FeO_6$	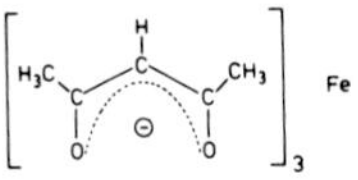	1.992(6)	*386*
26	$C_9H_9F_6FeO_4P$	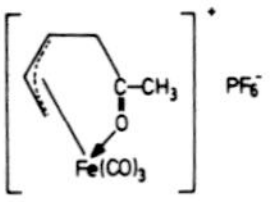	1.995(5)	*346*
27	$C_{11}H_{18}FeN_5O_8 \cdot 3H_2O$	$\{(CN_3H_6)^+$[Fe(EDTA[b])$(H_2O)]^-\} \cdot 2H_2O$	av. 2.0	*501*
28	$C_7H_6BF_3FeO_5S$	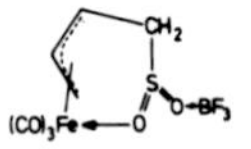	2.00(1)	*144,146*
29	$C_{114}H_{90}Ag_3FeO_6P_6S_6$	$\{Ag[P(C_6H_5)_3]_2\}_3Fe(O_2C_2S_2)_3$	2.003(6)	*216,366*
30	$C_8H_{24}Cl_6Fe_2O_4S_4$	*trans*-$\{Cl_2Fe[OS(CH_3)_2]_4\}^+[FeCl_4]^-$	2.006(6)	*57,56*
31	$C_{21}H_{15}FeO_6$	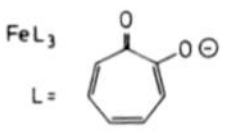	2.008(3)	*342*
32	$C_{12}H_8FeO_4$	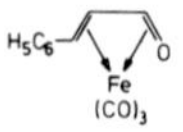	2.013(3)	*166,168*
33	$C_{28}H_{40}CaFe_2N_4O_{18} \cdot 8H_2O$	$Ca^{2+}[FeL(H_2O)]_2^- \cdot 8H_2O$ L = *trans*-1,2-diaminocyclohexane-*N,N'*-tetraacetate	2.017(17) 2.092(5)	*175*
34	$C_{10}H_{14}FeN_2NaO_9 \cdot 2H_2O$	$\{Na^+$[Fe(EDTA[b])$(H_2O)]^-\} \cdot 2H_2O$	av. 2.05 (H_2O) 2.11	*505*

Table 8 (continued)

No.		Formula	Fe-O Distance [Å]	*Ref.*
35	$C_{10}H_{12}FeLiN_2O_8 \cdot 3H_2O$	$\{Li^+[Fe(EDTA^b)]^-\} \cdot 3H_2O$	2.04 1.87 1.94 1.8	*506*
36	$C_{18}H_{14}Fe_2O_{10}$	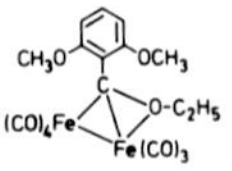 	2.040(5)	*382*
37	$C_{25}H_{18}Fe_2O_8$	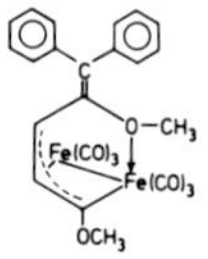 	2.07	*395*
38	$C_2H_6Cl_2FeN_2O_2$	*Catena*-di(μ-chloro)bis(formamido-*O*) iron(II)	2.13(1)	*181*
39	$C_{22}H_{16}ClFeN_8O_8 \cdot 6H_2O$	Bis(10-methylisoalloxazine)ferrous perchlorate hexahydrate	2.165(5)	*305*
40	$C_2H_{10}FeN_6O_4S_2$	$[Fe(NH_2\text{-}NH\text{-}CS\text{-}NH_2)_2SO_4]$ (polymeric)	2.187(5) 2.165(4)	*496*
41	$C_2H_2FeO_2S \cdot H_2O$	$Fe(S\text{-}CH_2\text{-}CO_2) \cdot H_2O$	2.207(7) 2.186(7) 2.126(9)	*394*
42	$C_2FeO_4 \cdot 2H_2O$	Iron(II)oxalate dihydrate	2.22 (H_2O) 2.11	*241*
43	$C_{20}H_{12}Cl_2Fe_2N_2O_8$	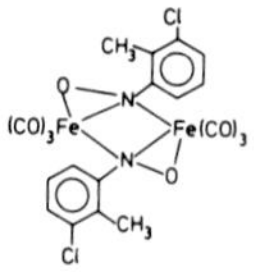 	1.91(1) 2.73(1)	*44,45*
44	$C_{12}H_{24}ClFe_3O_{20}$	$[Fe_3(CH_3\text{-}CO_2)_6O \cdot 3H_2O]^+[ClO_4]^-$		*22*
45	$C_{11}H_{21}Cl_2FeN_7O_5$	$[Cl\text{-}FeL(H_2O)]^+Cl^- \cdot 2H_2O$ L = 2,6-diacetylpyridine-bis(semi-carbazone)		*613*
46	$C_{20}H_{22}Cl_2Fe_2N_2O_4$	$[Fe(Sal{=}N\text{-}(CH_2)_3O)Cl]_2$, and other Sal=N-R complexes of iron(II) Sal=N-R = 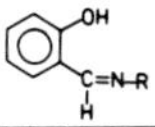 		*284*

a 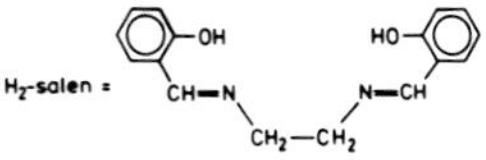

b 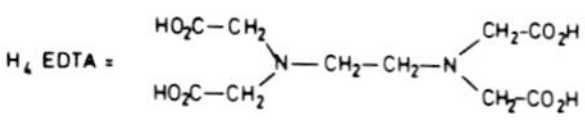

Table 9: Organoiron Complexes Containing Fe-S Bonds.

No.		Formula	Fe-S Distance [Å]	Ref.
1	$C_{16}H_{16}Fe_3O_8S_2$	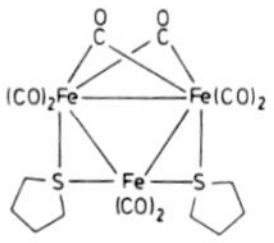	2.206(2) 2.142(2)	*206*
2	$C_8H_{44}B_{20}FeN_2S_2$	$[N(CH_3)_4]_2^+[Fe(B_{10}H_{10}S)_2]^{2-}$	2.155(3)	*235*
3	$C_{24}H_{36}F_{12}FeNS_4$	$[(n\text{-}C_4H_9)_4N]^+\{Fe[S_2C_2(CF_3)_2]_2\}^-$	av. 2.17	*290*
4	$C_{24}H_{21}FeS_7$	$[Fe(CH_3\text{-}C_6H_4\text{-}CS_3)(CH_3\text{-}C_6H_4\text{-}CS_2)_2]$	2.184(7) 2.238(7) 2.297(5) 2.336(6) 2.343(7) 2.296(6)	*212,213*
5	$C_{32}H_{56}Fe_2N_2S_6$	$[(C_2H_5)_4N]_2^+\{(\mu\text{-}S)_2Fe_2[(SCH_2)_2C_6H_4]_2\}^{2-}$	2.185(2) 2.306(1) 2.232(1) 2.303(1)	*471*
6	$C_{13}H_{10}Fe_2O_5S$	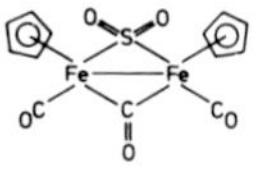	2.187(1) 2.181(1)	*160*
7	$C_{14}H_{20}F_6FeN_2S_6$	$Fe[S_2CN(C_2H_5)_2]_2[S_2C_2(CF_3)_2]$	2.195(3) 2.310(3)	*397*
8	$C_{26}H_{15}F_{12}FeOPS_4$	$\{[(C_6H_5)_3PO]Fe[S_2C_2(CF_3)_2]_2\}^-$	2.199(15) 2.225(15) 2.239(15) 2.231(15)	*289*
9	$C_{44}H_{68}Fe_2N_2S_6$	$[(C_2H_5)_4N]_2^+[(\mu\text{-}S)_2Fe_2(S\text{-}C_6H_4\text{-}p\text{-}CH_3)_4]^{2-}$	2.200 2.202 2.312 2.312	*471*
10	$C_{20}H_{20}Fe_4S_4$	$[(\eta^5\text{-}C_5H_5)Fe]_4(\mu_3\text{-}S)_4$	2.204(4) 2.250(8)	*611*
			av. 2.206(2) av. 2.256(3)	*561*
11	$Fe_4N_4O_4S_4$	$[Fe(NO)]_4(\mu_3\text{-}S)_4$	av. 2.217(2)	*306*
12	$C_{21}H_{24}FeO_4S$	$[\eta^5\text{-}C_5(CH_3)_5](CO)_2Fe\text{-}SO_2\text{-}CH_2\text{-}CH{=}CH\text{-}C_6H_5$	2.218(2)	*152*
13	$C_{48}H_{72}Fe_2N_{10}S_8$	$[(n\text{-}C_4H_9)_4N]_2^+\{Fe_2[S_2C_2(CN)_2]_4\}^{2-}$	2.20(3) 2.20(3) 2.25(3) 2.28(3) bridge 2.46	*341*
14	$C_{16}H_{30}Fe_2S_{14}$	$(\mu\text{-}SC_2H_5)_2[Fe(S_2C\text{-}S\text{-}C_2H_5)]_2(\mu\text{-}S_2C\text{-}S\text{-}C_2H_5)_2$	bridge S 2.22(1) bridge S_2 2.28(1) 2.34(1)	*214*
15	$C_8H_{18}Fe_4N_6O_4S_2$	$[Fe(NO)]_4(\mu_3\text{-}S)_2[\mu_3\text{-}N\text{-}C(CH_3)_3]_2$	2.224(2) 2.222(2)	*306*
16	$C_4H_{10}Fe_2N_4O_4S_2$	$(\mu\text{-}SC_2H_5)_2[Fe(NO)_2]_2$	2.26(2) 2.28(7)	*589*

Table 9 (continued)

No.		Formula	Fe-S Distance [Å]	Ref.
17	$C_8FeN_5OS_4$	$Fe(NO)[S_2C_2(CN)_2]_2$	2.26(6) 2.26(6) 2.28(6) 2.29(6)	540
18	$C_6Fe_2O_6S_2$	$(\mu\text{-}S_2)[Fe(CO)_3]_2$	2.228(2)	609,610
19	$C_{10}H_{20}FeN_3OS_4$	$Fe(NO)[S_2C\text{-}N(C_2H_5)_2]_2$	2.28(1) 2.30(1) 2.30(1) 2.26(1)	176
20	$C_9Fe_3O_9S_2$	$(\mu_3\text{-}S)_2[Fe(CO)_3]_3\cdot(\mu\text{-}S_2)[Fe(CO)_3]$	2.228(9) 2.225(9) 2.233(9) 2.220(7) 2.212(9)	610
21	$C_{10}H_{20}ClFeN_2S_4$	$Cl\text{-}Fe[S_2C\text{-}N(C_2H_5)_2]_2$	2.29(1) 2.31(1) 2.30(1) 2.30(1)	373,375
22	$C_{14}H_{16}BF_4Fe_2O_2S_2$	$\{(\mu\text{-}SCH_3)_2[(\eta^5\text{-}C_5H_5)FeCO]_2\}^+[BF_4]^-$	2.233(4) 2.235(4)	180
23	$C_{30}H_{28}Cl_2FeN_4O_8S_2\cdot CH_3OH$	$[Fe(N_4S_2)]^{2+}(ClO_4)_2^-\cdot CH_3O$ N_4S_2 = 15,18-Dithia-1,5,8,12-tetraaza-3,4:9,10:13,14:19,20-tetrabenzocyclo-eicosa-1,11-diene	2.236(9) 2.246(9)	585
24	$C_{44}H_{68}Fe_4N_2S_8$	$[(C_2H_5)_4N]_2^+[(\mu_3\text{-}S)_4(FeS\text{-}CH_2\text{-}C_6H_5)_4]^{2-}$	2.239(4)	27
25	$C_{14}H_6Fe_4O_{12}S_3$	$(\mu_4\text{-}S)\{(\mu\text{-}SCH_3)\ [Fe(CO)_3]_2\}_2$	2.248(8) 2.274(7)	178
26	$C_{10}H_{10}Fe_2O_6S_2$	$(\mu\text{-}SC_2H_5)_2[Fe(CO)_3]_2$	2.259(7)	219
27	$C_{20}H_{10}Fe_2O_6S_2$	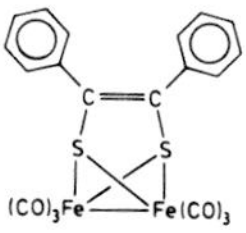	2.259(3)	93,94,608
28	$C_{24}H_{20}Fe_2S_2O_2$	$(\mu\text{-}SC_6H_5)_2[(\eta^5\text{-}C_5H_5)FeCO]_2$	2.262(6)	291
29	$C_{32}H_{44}FeN_2S_8$	$[(CH_3)_4N]_2^+[(\mu_3\text{-}S)_4Fe(SC_6H_5)_4]^{2-}$	av. 2.263(3)	539
30	$C_{60}H_{40}As_2FeN_6S_6$	$[(C_6H_5)_4As]_2^+\{Fe[S_2C_2(CN)_2]_3\}^{2-}$	2.264(1) 2.258(1) 2.271(1)	564
31	$C_{40}H_{88}Fe_2N_2S_8$	$[(n\text{-}C_4H_9)_4N]_2^+$ 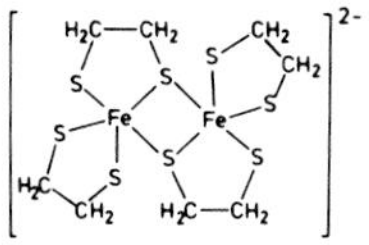	2.265(3) 2.236(3) 2.503(3) 2.220(3) 2.247(3)	573
32	$C_{14}H_{20}Fe_2S_4$	$(\mu\text{-}S_2)[(\eta^5\text{-}C_5H_5)Fe]_2(\mu\text{-}S\text{-}C_2H_5)_2$	2.273(2) 2.289(2) 2.129(2) 2.275(2) 2.285(2) 2.129(2)	588

Table 9 (continued)

No.	Formula		Fe-S Distance [Å]	Ref.
33	$C_{18}H_{10}Fe_2O_6S_2$	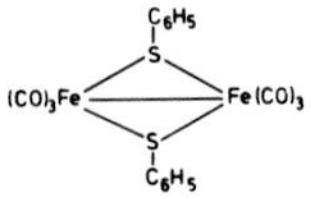	2.275(2) 2.261(2) 2.281(2) 2.264(2)	*355*
34	$C_{16}H_{36}F_6Fe_2N_5OPS_4$	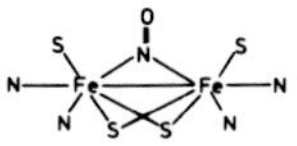 $[(FeL)_2NO]PF_6$	bridge 2.276(2) 2.245(2) 2.298(2)	*411*
	L = *N,N'*-dimethyl-*N,N'*-bis(β-mercaptoethyl)ethylenediamine			
35	$C_{14}H_{10}Fe_2O_6S$	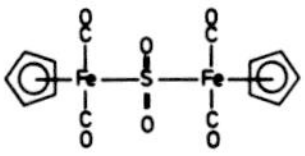	2.2790(6) 2.2814(7)	*159*
36	$C_{12}H_5Fe_2NO_6S$	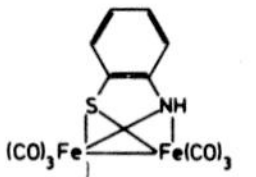	2.283(2)	*79*
37	$C_{74}H_{74}FeO_{12}P_2S_6$	$[(C_6H_5)_3(C_6H_5\text{-}CH_2)P]_2^+\{Fe[S_2CC(CO_2C_2H_5)_2]_3\}^{2-}$	2.289(2) 2.305(2) 2.301(2)	*367*
38	$C_{15}H_{27}FeS_9$	$Fe[S_2C\text{-}S\text{-}C(CH_3)_3]_3$	2.293(3) 2.296(3) 2.297(3) 2.310(3) 2.291(3) 2.294(2)	*445*
39	$C_{25}H_9F_{24}Fe_4S_{11}O_6$	$\{(\mu\text{-}SCH_3)_3[Fe(CO)_3]_2\}^+\{Fe_2[S_2C_2(CF_3)_2]_4\}^-$	cation 2.295(4) 2.305(3) 2.310(3) 2.310(4) 2.303(4) 2.307(4) anion av. 2.206(3) 2.311(5)	*560*
40	$C_6H_{12}FeN_3OS_4$	$ON\text{-}Fe[S_2C\text{-}N(CH_3)_2]_2$	2.295(2) 2.298(2) 2.308(2) 2.294(2)	*229,227*
41	$C_{15}H_{24}ClFeN_3O_4S_6$	$[Fe(S_2C\text{-}N$$)_3]^+[ClO_4]^-$	av. 2.300(2)	*463*
42	$C_9H_{15}FeO_3S_6$	$Fe(S_2C\text{-}OC_2H_5)_3$	2.308(3) 2.326(3)	*376*
43	$C_{27}H_{54}FeN_3S_6$	$Fe[S_2C\text{-}N(C_4H_9)_2]_3$	av. 2.418(6)	*374*
44	$C_{24}H_{24}FeN_3S_6$	$Fe[S_2C\text{-}N(CH_3)(C_6H_5)]_3$	2.308(9) 2.317(9) 2.307(8) 2.328(9) 2.280(9) 2.334(8)	*350*

Table 9 (continued)

No.	Formula		Fe-S Distance [Å]	*Ref.*
45	$C_{15}H_{24}FeN_3S_6$	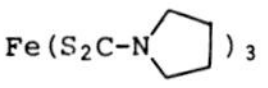	2.41(1) 2.40(1) 2.44(1) 2.41(1) 2.38(1) 2.40(1)	*350*
46	$C_{23}H_{19}FeO_2PS$	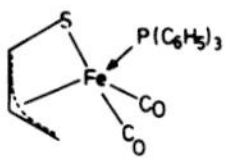	2.313	*583*
47	$C_{42}H_{40}FeN_5O_6S$	(*p*-Nitrobenzenethiolato)(protoporphyrin IX dimethylester)iron(II)	2.32	*423*
48	$C_{14}H_{20}FeN_2O_2S_4$	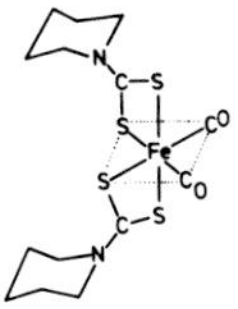	*trans* to CO { 2.337(6) 2.340(6) axial { 2.309(5) 2.308(5)	*547*
49	$C_{11}H_{10}B_2FeO_3S$	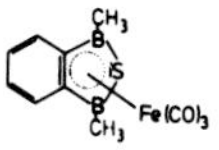	2.342	*567*
50	$C_8H_{24}FeN_2P_4S_4$	$Fe[S-P(CH_3)_2-N-P(CH_3)_2-S]_2$	2.356(3) 2.364(3) 2.339(3) 2.380(3)	*149,153*
51	$C_{15}H_{30}FeN_3S_6$	$Fe[S_2C-N(C_2H_5)_2]_3$	2.364(3) 2.362(3) 2.351(3) 2.356(3) 2.361(3) 2.352(3)	*444*
52	$C_{18}H_{28}Cl_2FeMoS_2$	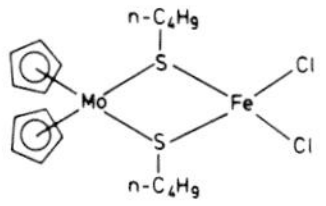	2.384(5) 2.387(5)	*111,112*
53	$C_2H_{10}FeN_6O_4S_2$	$[Fe(NH_2-NH-CS-NH_2)_2SO_4]$ (polymeric)	2.427(2) 2.425(2)	*496*
54	$C_2H_2FeO_2S\cdot H_2O$	$Fe(S-CH_2-CO_2)\cdot H_2O$	2.436(7) 2.432(5)	*394*
55	$C_{15}H_{24}FeN_3O_3S_6\cdot CH_2Cl_2$	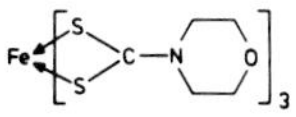	2.452(4) 2.431(4) 2.424(4) 2.423(4) 2.432(4) 2.420(4)	*352*
56	$C_{16}H_{36}Fe_2N_4S_4$ L = *N,N'*-dimethyl-*N,N'*-bis(β-mercaptoethyl)ethylenediamine	$(\mu-L)_2Fe_2$	2.471(5) 2.379(5) 2.304(5)	*377*

Table 9 (continued)

No.	Formula		Fe-S Distance [Å]	*Ref.*
57	$C_{18}H_{40}Fe_2N_4S_4$	$(\mu\text{-}L)_2Fe_2$ L = *N,N'*-dimethyl-*N,N'*-bis(β-mercaptoethyl)-1,3-propane-diamine	2.490(3) 2.411(2) 2.325(2)	*377*
58	$C_{26}H_{18}ClFeN_2O_2S_2$	(Chloro)-bis(salicylideniminephenyl)(disulfido)iron(III)	2.536(4)	*65*
59	$C_{10}H_{20}FeIN_2S_4$	$I\text{-}Fe[S_2C\text{-}N(C_2H_5)_2]_2$	2.82(2)	*351*
60	$C_{28}H_{52}Fe_4NNa_5O_8S_8 \cdot 5(C_5H_9NO)$	$Na_5^+[(C_4H_9)_4N]^+\{(\mu_3\text{-}S)_4[Fe(S\text{-}CH_2\text{-}CH_2\text{-}CO_2)]_4\}^{6-}$ •5(*N*-methyl pyrrolidone)		*118*
61	$C_{16}H_{20}Fe_2O_4S_2$	[2,3-η:μ-*S*(CH_3-CH=C(C_2H_5)-C(=S)H)]$_2$[Fe(CO)$_2$]$_2$		*527*
62	$C_{16}F_{24}Fe_3S_{10} \cdot \frac{1}{2}S_8$	$(\mu_3\text{-}S)(\mu\text{-}S)[\mu_3\text{-}S_2C_2(CF_3)_2][FeS_2C_2(CF_3)_2]_3 \cdot \frac{1}{2}S_8$		*312*

Table 10: Organoiron Compounds Containing Iron-(Main Group Element) Bonds.

No.	Formula		Distance [Å]	*Ref.*
	A. Boron		*Fe-B*	
1	$C_{12}H_{18}B_6Fe_2$	1,6-bis(η^5-cyclopentadienyl)--1,6-diferra-2,3-dicarba-*closo*-decaborane(8)	2.096(6) 2.196(6) 2.219(6) 1.930(6) 2.089(6) 2.083(6)	*108*
2	$C_8H_{44}B_{20}FeN_2S_2$	$[(CH_3)_4N]_2^+[Fe(B_{10}H_{10}S)_2]^{2-}$	2.15(1)	*235*
3	$C_{11}H_{10}B_2FeO_3S$	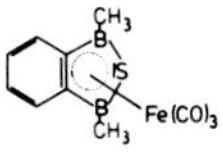	2.277 2.282	*567*
4	$C_{20}H_{48}B_7FeNO_4$	$[(n\text{-}C_4H_9)_4N]^+[(\eta^2\text{-}B_7H_{12})Fe(CO)_4]^-$	av. 2.20(2)	*368*
	B. Silicon		*Fe-Si*	
1	$C_6H_6Cl_6FeOSi_2$	$(\eta^5\text{-}C_5H_5)FeH(SiCl_3)_2CO$	2.252(3)	*461*
	C. Germanium		*Fe-Ge*	
1	$C_{10}H_{14}Cl_2FeGe$	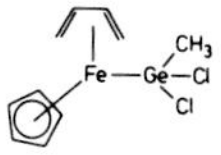	2.28	*18*
2	$C_{14}H_{10}Cl_2Fe_2GeO_4$	$[(\eta^5\text{-}C_5H_5)Fe(CO)_2]_2(\mu\text{-}GeCl_2)$	2.357(4)	*105*
3	$C_{18}H_{22}Fe_2Ge_2O_5$	$[(\eta^5\text{-}C_5H_5)Fe(CO)_2]_2[\mu\text{-}Ge(CH_3)_2\text{-}OGe(CH_3)_2]$	2.372	*1*
4	$C_{12}H_{18}Fe_2Ge_3O_6$	$[\mu\text{-}Ge(CH_3)_2]_3[Fe(CO)_3]_2$	2.398(5)	*89,281*
5	$C_{31}H_{20}Fe_2Ge_2O_7$	$(\mu\text{-}CO)[\mu\text{-}Ge(C_6H_5)_2]_2[Fe(CO)_3]_2$	2.416(3) 2.432(3) 2.402(3) 2.440(3)	*282*
6	$C_{16}H_{10}Cl_4Co_2FeGe_2O_6$	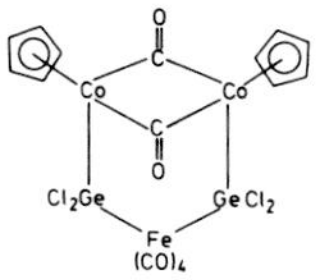	2.433(8) 2.438(4)	*58* *283*
7	$C_{16}H_{20}Fe_2Ge_2O_8$	$[\mu\text{-}Ge(C_2H_5)_2]_2[Fe(CO)_4]_2$	av. 2.492	*630*
	D. Tin		*Fe-Sn*	
1	$C_7H_5Br_3FeO_2Sn$	$(\eta^5\text{-}C_5H_5)Fe(CO)_2SnBr_3$	2.465(3) 2.462(2)	*98* *481*
2	$C_7H_5Cl_3FeO_2Sn$	$(\eta^5\text{-}C_5H_5)Fe(CO)_2SnCl_3$	2.466(2)	*98*
3	$C_{13}H_{10}Cl_2FeO_2Sn$	$(\eta^5\text{-}C_5H_5)(CO)_2Fe\text{-}SnCl_2(C_6H_5)$	2.467(2)	*333*
4	$C_{26}H_{20}Fe_2O_8S_2Sn$	$[(\eta^5\text{-}C_5H_5)Fe(CO)_2]_2[\mu\text{-}Sn(SO_2\text{-}C_6H_5)_2]$	2.490(10) 2.507(10)	*97*
5	$C_{14}H_{10}Cl_2Fe_2O_4Sn$	$[(\eta^5\text{-}C_5H_5)Fe(CO)_2]_2(\mu\text{-}SnCl_2)$	2.492(8)	*507*
6	$C_{38}H_{32}Fe_2O_{10}S_2Sn_2$	$[(\eta^5\text{-}C_5H_5)Fe(CO)_2Sn(C_6H_5)(SO_2\text{-}C_6H_5)(OH)]_2$	2.499(1)	*546*

Table 10 (continued)

No.		Formula	Distance [Å]	*Ref.*
7	$C_{38}H_{34}F_6FeOP_2Sn$	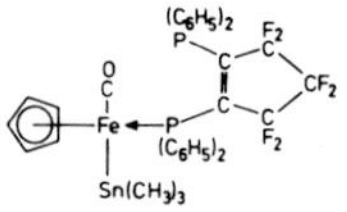	2.536(3) 2.560(5)	*274* *545*
8	$C_{25}H_{20}FeO_2Sn$	$(\eta^5\text{-}C_5H_5)Fe(CO)_2Sn(C_6H_5)_3$ (two independent molecules)	2.533(5) 2.540(5)	*96*
9	$C_{16}Fe_4O_{16}Sn$	$(\mu_4\text{-}Sn)[Fe(CO)_4]_4$	2.540	*448*
10	$C_{38}H_{30}FeOSn$	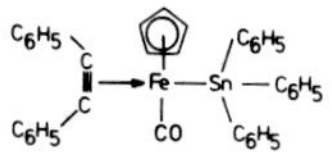	2.56	*565*
11	$C_{14}H_{10}Fe_2N_2O_8Sn$	$[(\eta^5\text{-}C_5H_5)Fe(CO)_2]_2[\mu\text{-}Sn(ONO)_2]$	2.563	*69,72,70*
12	$C_{16}H_{16}Fe_2O_4Sn$	$[(\eta^5\text{-}C_5H_5)Fe(CO)_2]_2[\mu\text{-}Sn(CH_3)_2]$	2.602 2.608	*69,72,70*
13	$C_{24}H_{20}Fe_2O_4Sn$	$[(\eta^5\text{-}C_5H_5)Fe(CO)_2]_2[\mu\text{-}Sn(\eta^1\text{-}C_5H_5)_2]$	2.568 2.573	*74,71*
14	$C_{22}H_{15}ClFe_2MoO_7Sn$	$[(\eta^5\text{-}C_5H_5)Fe(CO)_2]_2(\mu\text{-}SnCl)\text{-}$ $[(\eta^5\text{-}C_5H_5)Mo(CO)_3]$	2.583(7) 2.598(7)	*508*
15	$C_{20}H_{12}Fe_4O_{16}Sn_3$	$(\mu_4\text{-}Sn)\{[\mu\text{-}Sn(CH_3)_2][Fe(CO)_4]_2\}_2$	2.625(8) 2.747(8)	*582*
16	$C_{12}H_{12}Fe_2O_8Sn_2$	$[\mu\text{-}Sn(CH_3)_2]_2[Fe(CO)_4]_2$ (two distinct molecules)	2.631(11) 2.647(8)	*316*
	E. Lead		*Fe-Pb*	
1	$C_{16}H_{16}Fe_2O_4Pb$	$[(\eta^5\text{-}C_5H_5)Fe(CO)_2]_2[\mu\text{-}Pb(CH_3)_2]$	av. 2.708	*73*
	F. Phosphorus		*Fe-P*	
1	$C_{41}H_{35}FeIO_6P_2$	$(\eta^5\text{-}C_5H_5)Fe[P(OC_6H_5)_3]_2I$	2.15(1)	*11,123,13*
2	$C_{41}H_{33}FeIO_6P_2$	"isomer" of No. *F*. 1: 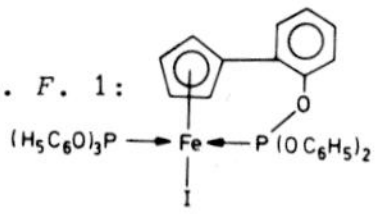	2.14	*11,14*
3	$C_{31}H_{25}FeO_2P$	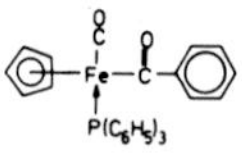	2.17 2.16	*563* *124*
4	$C_{28}H_{23}FeOPS$	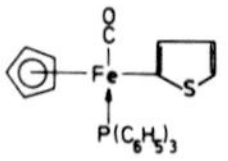	2.22	*17*
5	$C_{42}H_{34}FeO_2P_2$	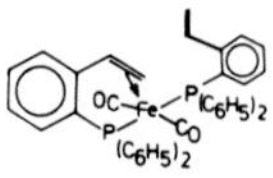	2.207(3) 2.234(3)	*54*
6	$C_{23}H_{20}FeO_2PI$	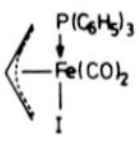	2.23(1)	*491*
7	$C_{30}H_{25}FeOP$	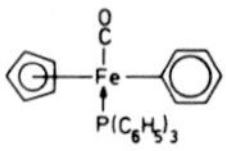	2.23 2.24	*562* *28*

Table 10 (continued)

No.	Formula		Distance [Å]	Ref.
8	$C_{33}H_{33}Fe_3O_9P_3$	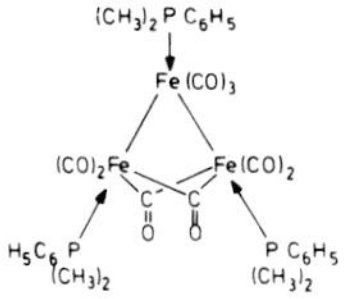	2.242(9) 2.232(9) 2.236(9)	*542*
9	$C_{29}H_{15}Fe_3PO_{11}$	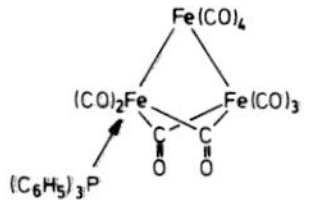 A 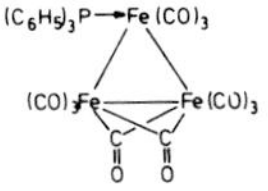 B	A 2.24(1) B 2.25(1)	*223,224*
10	$C_{29}H_{20}F_6FeN_2O_2P_2$	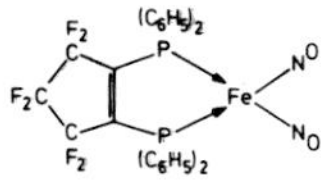	2.240(2) 2.248(2)	*348*
11	$C_{16}H_{11}FeO_4P$	$(CO)_4Fe \leftarrow PH(C_6H_5)_2$	2.237(2)	*415*
12	$C_{39}H_{28}Cl_2Fe_2O_8P_2Pd_2$	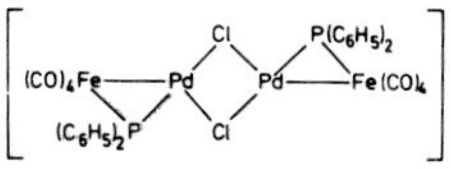 $C_6H_5-CH_3$	2.24(2)	*414*
13	$C_{27}H_{19}Fe_2O_6P$	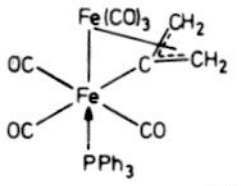	2.25	*237*
14	$C_{12}H_{12}Fe_2O_8P_2$	$[\mu\text{-}P(CH_3)_2\text{-}P(CH_3)_2][Fe(CO)_4]_2$	2.260(5)	*393*
15	$C_{10}H_{12}Fe_2I_2O_6P_2$	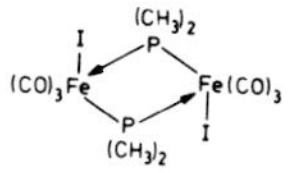	2.293(6) 2.312(6)	*228*
16	$C_{52}H_{40}Fe_4O_6P_2$	$\{[(\eta^5\text{-}C_5H_5)Fe(CO)](\mu\text{-}CO)_2[(\eta^5\text{-}C_5H_5)Fe]\}_2[\mu\text{-}(C_6H_5)_2P\text{-}C{\equiv}C\text{-}P(C_6H_5)_2]$		*120*
17	$C_4H_{26}B_{18}FeP_2$	$Fe(B_9H_9CHPCH_3)_2$		*590*
18	$C_{19}H_{15}FeN_2O_3P$	$Fe(NO)_2(CO)[P(C_6H_5)_3]$	2.260(3)	*6*
19	$C_{36}H_{30}FeN_2O_2P_2$	$Fe(NO)_2[P(C_6H_5)_3]_2$	2.267(2)	*6*
20	$C_9H_{12}FeO_9P_4$	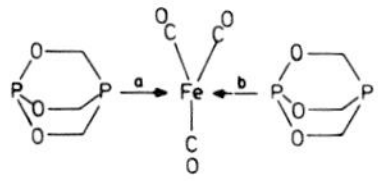	a 2.190(4) b 2.116(4)	*9*
21	$C_{32}H_{22}Fe_2O_7P_2$	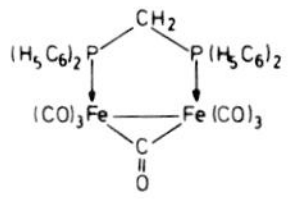	2.250(3) 2.257(3)	*205*
22	$C_{12}H_{11}Fe_2O_5P$	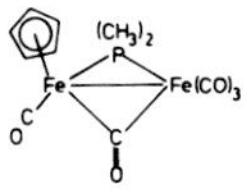	av. 2.194	*599*
23	$C_{20}H_{15}Fe_2O_7P$	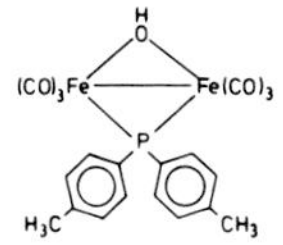	2.239(3) 2.236(3)	*591*

Table 10 (continued)

No.	Formula	Distance [Å]	*Ref.*
24	$C_{23}H_{19}FeO_2PS$ 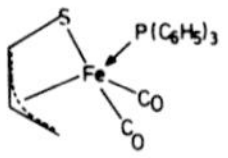	2.238	*583*
25	$C_{46}H_{10}F_{20}Fe_2O_6P_2$ 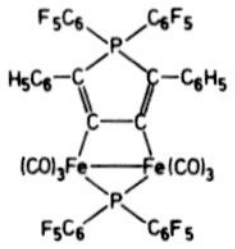	2.233(3) 2.212(4)	*586*
26	$C_{16}H_{30}Fe_2HgN_2O_6P_2$ $(\mu\text{-}Hg)\{Fe(CO)_2(NO)[P(C_2H_5)_3]\}_2$	2.223(3)	*577*
27	$C_{40}H_{32}As_2F_8Fe_2O_4P_2$ $\{(C_6H_5)_2P\text{-}C{=}C[As(CH_3)_2]\text{-}CF_2\text{-}CF_2\}_2\text{-}Fe_2(CO)_4$	2.270(2)	*277*
28	$C_8H_{12}F_3FeP$ 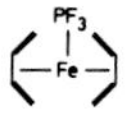	2.024	*431*
29	$C_9H_5F_6FeO_3P$ 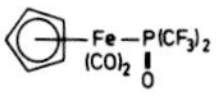	2.191(3)	*46,47*
30	$C_9H_5F_6FeO_2P$ 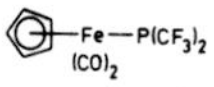	2.265(3)	*46,47*
31	$C_{34}H_{25}FeNiO_3P$ 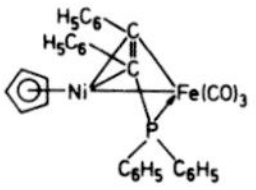	2.220(3)	*43*
32	$C_{38}H_{20}F_6Fe_3O_8P_2$ 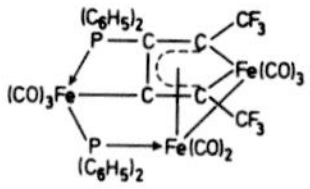	2.366 2.283	*509*
33	$C_{40}H_{34}F_6Fe_2O_4P_3Rh$ 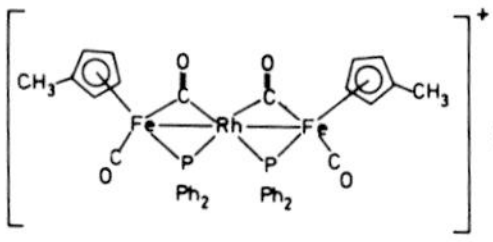 PF_6^-.	2.241(3) 2.222(3)	*468,340*
34	$C_{26}H_{15}Fe_2O_6P$ 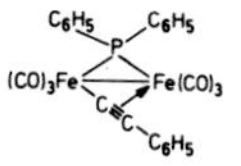	2.213(2) 2.224(2)	*518*
35	$C_{46}H_{30}Fe_2O_6P_2$ 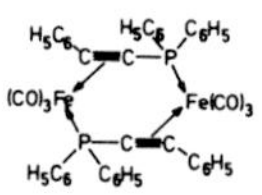	2.287(2) 2.298(2)	*513*
36	$C_{39}H_{27}Fe_2N_2O_5P$ 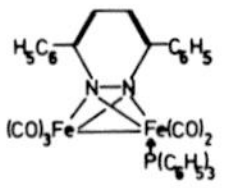	2.06	*435*
37	$C_{67}H_{64}BBrCl_2FeP_4$ 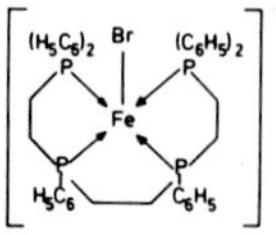 $[B(C_6H_5)_4]^- \cdot CH_2Cl_2$	2.299(5) 2.248(5) 2.181(5) 2.215(5)	*29*
38	$C_{22}H_{16}AsF_4FeO_4P$ 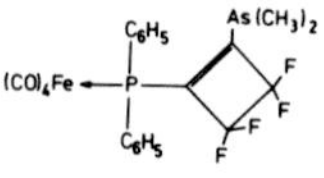	2.224(3)	*275*
39	$C_{27}H_{16}AsF_4Fe_3O_9P$ 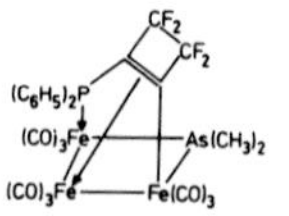	2.252 2.243	*276*

(two molecules in the asymmetric unit)

Table 10 (continued)

No.		Formula	Distance [Å]	Ref.
40	$C_{40}H_{62}FeO_8P_4$	*cis*-$H_2Fe[P(C_6H_5)(OC_2H_5)_2]_4$	2.134(2) 2.119(2) 2.153(2) 2.151(2)	*335*
41	$C_9H_{18}FeO_9P_2$	*trans*-$[P(OCH_3)_3]_2Fe(CO)_3$	2.155(1)	*317*
42	$C_{44}H_{35}BF_4FeN_2O_2P_2$	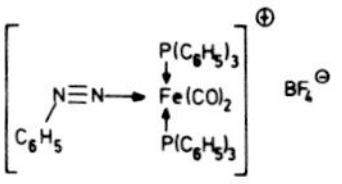	2.261(2) 2.266(2)	*349*
43	$C_{37}H_{20}F_6Fe_3O_7P_2$	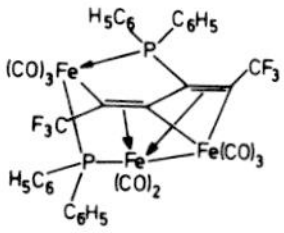	2.241(6) 2.173(9) 2.229(5)	*122*
44	$C_{21}H_{27}Fe_2O_5P$	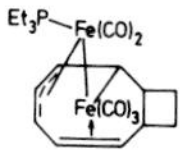	2.220(2)	*211*
45	$C_{31}H_{25}Fe_2O_6P$	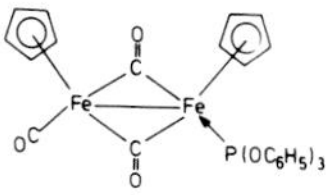	2.126(4) 2.125(4)	*202*
46	$C_{28}H_{22}FeO_3P_2$	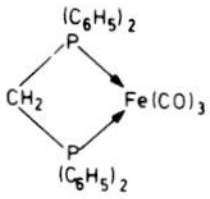	2.209(3) 2.225(3)	*196*
47	$C_{32}H_{30}Fe_2O_9P_2$	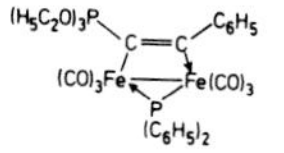	2.226(3) 2.206(4)	*621*
	G. Arsenic		*Fe-As*	
1	$C_{18}H_{12}As_2F_4Fe_3O_{10}$	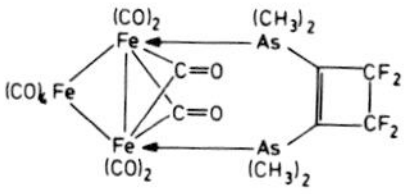	2.300(6) 2.301(7)	*549*
2	$C_{13}H_{16}As_2FeO_3$	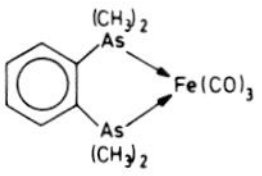	2.36 2.31	*90*
3	$C_{10}H_{12}As_4Fe_2O_6$	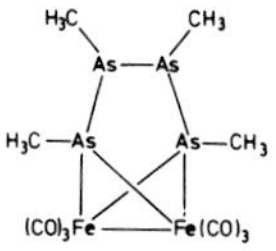	2.336 2.311	*307*
4	$C_{14}H_{12}As_2F_4Fe_2O_6$	$\{(CH_3)_2As-C{=}C[As(CH_3)_2]-CF_2-CF_2\}-$ $[Fe(CO)_3]_2$	2.350(5) 2.470(5)	*271,270*
5	$C_{17}H_{12}As_2F_4Fe_3O_9$	$[(CH_3)_2As-C{=}C-CF_2-CF_2][As(CH_3)_2-$ $Fe_3(CO)_9]$	2.371(4) 2.322(4) 2.360(4)	*272,273*
6	$C_{27}H_{16}AsF_4Fe_3O_9P$	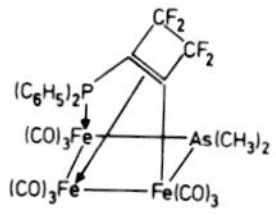 (two molecules in the asymmetric unit)	av. 2.387	*276*

Table 10 (continued)

No.		Formula	Distance [Å]	*Ref.*
7	$C_{40}H_{32}As_2F_8Fe_2O_4P_2$	$\{(C_6H_5)_2P\text{-}\overline{C{=}C[As(CH_3)_2]\text{-}CF_2\text{-}C}F_2\}_2\text{-}$ $Fe_2(CO)_4$	2.449(1) 2.363(1)	*277*
8	$C_7H_9AsFeO_4$	$(CO)_4Fe \leftarrow As(CH_3)_3$	2.30	*443*
9	$C_{16}As_2F_{10}FeO_4$	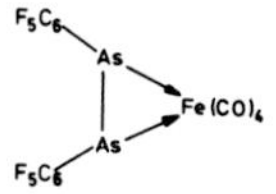 	2.510	*285*
	H. Antimony		*Fe-Sb*	
1	$C_7H_9FeO_4Sb$	$(CO)_4Fe \leftarrow Sb(CH_3)_3$	2.49	*443*
2	$C_{14}H_{10}Cl_9Fe_2O_4Sb_3$	$\{[(\eta^5\text{-}C_5H_5)Fe(CO)_2]_2SbCl_2\}Sb_2Cl_7$	av. 2.440	*278*
3	$C_{22}H_{15}FeO_4Sb$	$(CO)_4Fe \leftarrow Sb(C_6H_5)_3$	2.472(1)	*102*
4	$C_{43}H_{32}Cl_8Fe_7O_{12}Sb_2$	$\{[(\eta^5\text{-}C_5H_5)Fe(CO)_2]_3(\mu_3\text{-}SbCl)\}_2^+$ $[FeCl_4]^{2-}\cdot CH_2Cl_2$	av. 2.54	*593*

Table 11: Organoiron Compounds Containing Iron-Metal Bonds.

No.		Formula	Distance [Å]	*Ref.*
	A. Iron		*Fe-Fe*	
1	$C_{51}H_{56}Fe_2O_3$	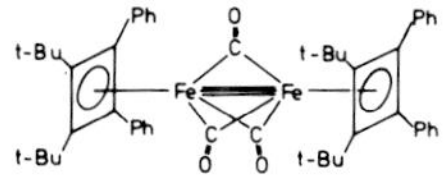	2.177(3)	*495*
2	$C_{24}H_{36}Fe_2O_4$	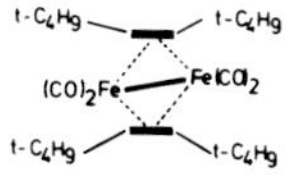	2.215	*503*
3	$C_{24}H_{32}Fe_2O_4S_2$	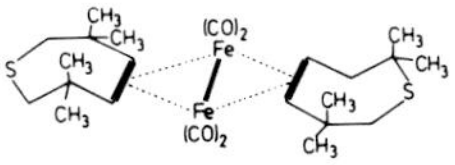	2.225	*557*
4	$C_{10}H_{10}Fe_2N_2O_2$	$[(\eta^5\text{-}C_5H_5)Fe]_2(\mu\text{-}NO)_2$	2.326(4)	*107*
5	$C_{18}H_{10}Fe_2N_2O_6$	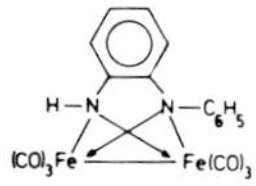	2.37 2.372(2)	*32* *34*
6	$C_9H_6Fe_2N_2O_7$	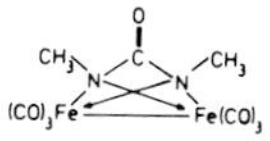	2.391(7)	*260*
7	$C_{29}H_{18}Fe_2N_2O_6$	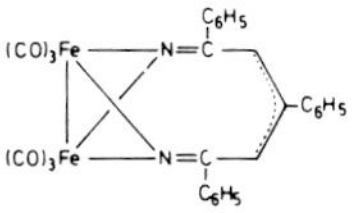	2.392	*121*
8	$C_{36}H_{30}Fe_2N_4O_6$	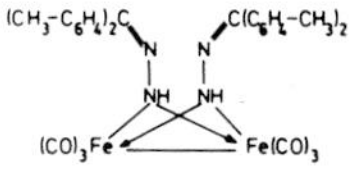	2.40	*30*
9	$C_{36}H_{28}Fe_2N_2O_6$	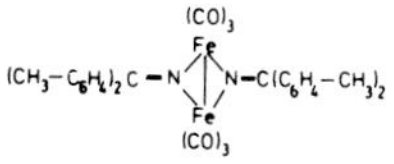	2.403	*86*
10	$C_{12}H_5Fe_2NO_6S$	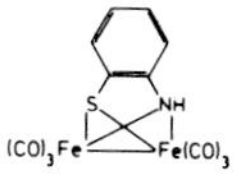	2.411(1)	*79*
11	$C_{19}H_{10}Fe_2N_2O_7$	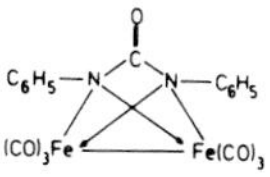	2.416(3)	*532*
12	$C_{20}H_{13}Fe_2NO_6$	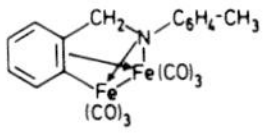	2.43	*32*
13	$C_{35}H_{20}Fe_3N_4O_9$	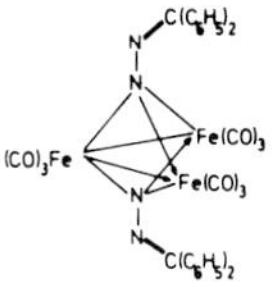	2.43 2.46 (3.06)	*33*
14	$C_{11}H_6Fe_3N_2O_9$	$(\mu_3\text{-}N\text{-}CH_3)_2[Fe(CO)_3]_3$	2.436(7) 2.488(7) (3.044)	*261*
15	$C_{33}H_{24}Fe_2N_2O_{10}$	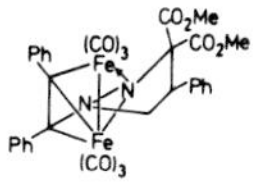	2.459	*433*
16	$C_{18}H_{16}Fe_2O_6$	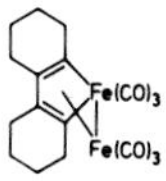	2.462(3)	*126*

Table 11 (continued)

No.	Formula		Distance [Å]	*Ref.*
17	$C_{16}H_{36}F_6Fe_2N_5OPS_4$	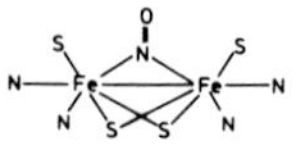 $[(FeL)_2NO]PF_6$ L = *N,N'*-dimethyl-*N,N'*-bis(β-mercaptoethyl)ethylenediamine	2.468(2)	*411*
18	$C_{20}H_{22}Cl_2Fe_2N_2O_4$	$[C_6H_4OCHN(CH_2)_3O]_2Fe_2Cl_2$	2.470	*63*
19	$C_{23}H_{10}Fe_3O_9$	$(C_6H_5-C{\equiv}C-C_6H_5)Fe_3(CO)_9$	2.480 2.500 2.579	*77*
20	$C_{26}H_{40}Al_2Fe_2O_4$	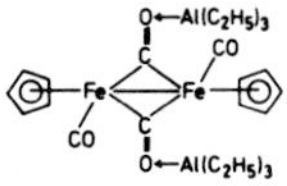	2.491(8)	*498,416*
21	$C_{14}H_{10}Fe_2O_4$	$(\eta^5-C_5H_5)_2Fe_2(CO)_4$	2.49(2)	*484*
22	$C_{12}H_8Fe_2O_8$	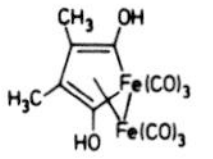	2.49	*364*
23	$C_{11}H_8Fe_2N_2O_6$	$(C_5H_8N_2)Fe_2(CO)_6$	2.490	*264*
24	$C_{48}H_{28}Fe_2O_4$	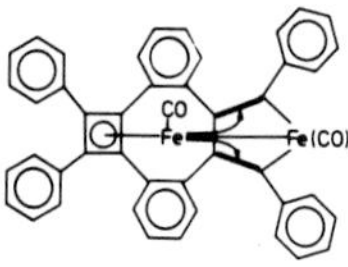	2.494(5)	*287*
25	$C_{24}H_{30}Fe_2O_6$	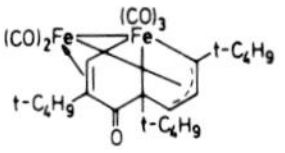	2.496(2)	*554*
26	$C_{20}H_{10}Fe_2O_6S_2$	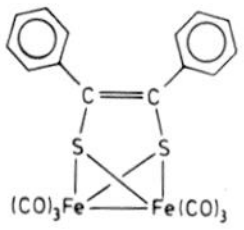	2.50	*93*
27	$C_{22}H_{36}Fe_2O_{10}Si_4$	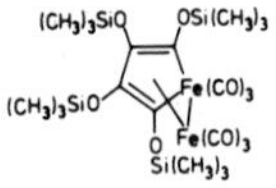	2.500(3)	*60*
28	$C_{43}H_{30}Fe_5N_6O_{13}$	$[Fe(C_5H_5N)_6]^{2+}[Fe_4(CO)_{13}]^{2-}$	2.570(6) 2.600(5) 2.580(5) 2.500(6) 2.500(6) 2.500(6)	*258*
29	$C_{30}H_{18}Fe_2O_6$	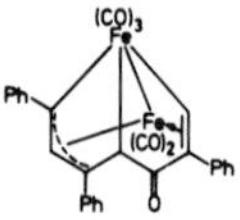	2.501(3)	*418*
30	$C_{24}H_{20}Fe_4O_4$	$[(\eta^5-C_5H_5)Fe]_4(\mu_3-CO)_4$	2.506 - 2.530	*502*
31	$C_{24}H_{20}F_6Fe_4O_4P$	$\{[(\eta^5-C_5H_5)Fe]_4(\mu_3-CO)_4\}^+[PF_6]^-$	2.506(2) 2.467(2) 2.478(2) 2.484(2)	*594*

Table 11 (continued)

No.	Formula		Distance [Å]	*Ref.*
32	$C_{15}H_{12}Fe_2O_7$	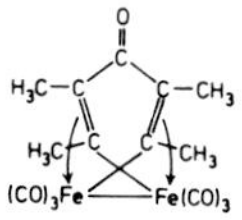	2.507(8)	*533*
33	$C_{18}H_8Fe_2N_2O_6$	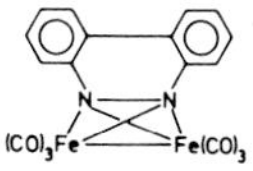	2.508(4)	*263*
34	$C_{20}H_{22}Fe_2O_4N_2$	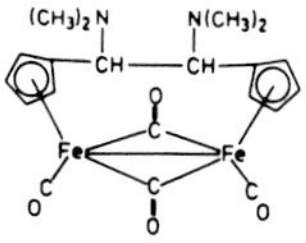	2.510	*473*
35	$C_{17}H_{10}Fe_2N_2O_3$	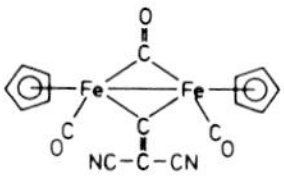	2.511(4)	*420*
36	$C_{20}H_{15}Fe_2O_7P$	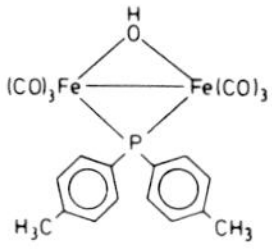	2.511(2)	*591*
37	$C_{16}H_{14}Fe_2SiO_4$	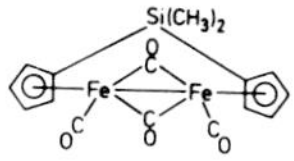	2.512(3)	*607*
38	$C_{18}H_{10}Fe_2O_6S_2$	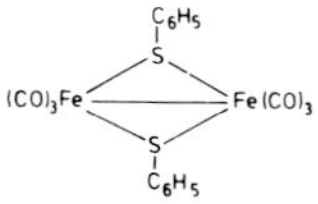	2.516(2)	*355*
39	$C_{30}H_{16}Fe_4O_{10}$	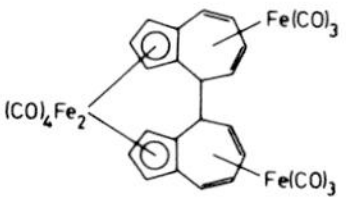	2.519(2)	*138*
40	$C_{18}H_{19}Fe_2NO_3$	$(\eta^5\text{-}C_5H_5)_2Fe_2(CO)_3[C{=}N\text{-}C(CH_3)_3]$	2.523(2)	*2*
41	$C_{18}H_{19}Fe_2NO_3$	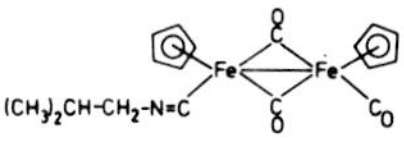	2.524(3)	*116*
42	$C_8H_2Br_2Fe_2O_6$	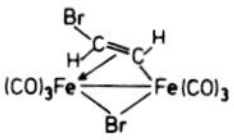	2.525(3)	*430*
43	$C_{24}H_{14}Fe_2O_6$	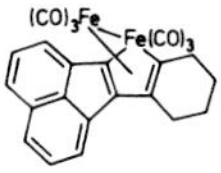	2.526	*389*
44	$C_{12}H_6Fe_2O_6$	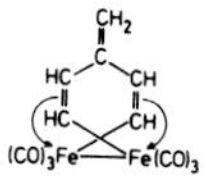	2.527(6)	*531*
45	$C_{18}H_{12}As_2F_4Fe_3O_{10}$	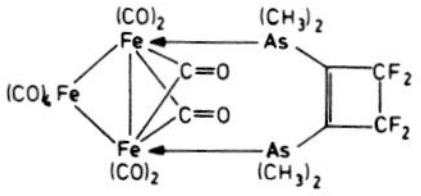	2.530 2.650 2.650	*549*
46	$C_{37}H_{20}F_6Fe_3O_7P_2$	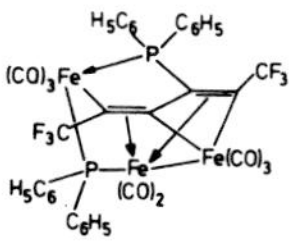	2.532(11) 2.665(8)	*122*

Table 11 (continued)

No.	Formula		Distance [Å]	*Ref.*
47	$C_{39}H_{27}Fe_2N_2O_5P$	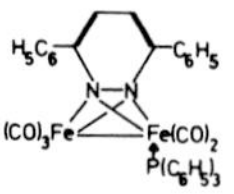	2.53	*435*
48	$C_{14}H_{10}Fe_2O_4$	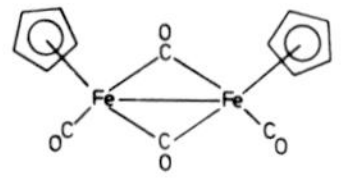	2.531(2) 2.533	*99,101*
49	$C_{20}H_{10}Fe_2O_6S$	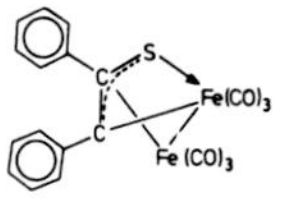	2.533(1)	*558*
50	$C_{14}H_{10}Fe_2O_4$	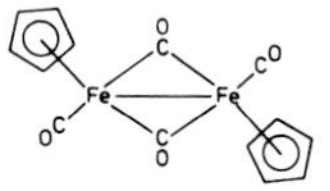	2.534(2)	*100*
51	$C_{18}H_{14}Fe_2O_{10}$	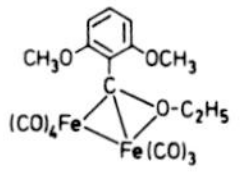	2.535(2)	*382*
52	$C_{13}H_9Fe_3NO_{10}Si$	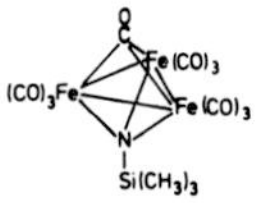	av. 2.535(2)	*42*
53	$C_{16}H_{16}Fe_2N_2O_2$	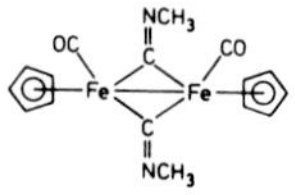	2.538(1)	*201*
54	$C_{18}H_{10}Fe_2O_8Rh_2$	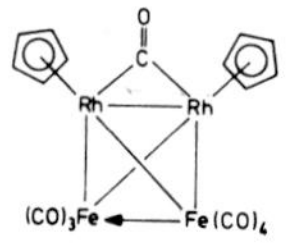	2.539(7)	*148,156*
55	$C_{52}H_{40}Fe_4O_6P_2$	$\{[(\eta^5\text{-}C_5H_5)Fe(CO)](\mu\text{-}CO)_2[(\eta^5\text{-}C_5H_5)Fe]\}_2\text{-}[\mu\text{-}(C_6H_5)_2P\text{-}C{\equiv}C\text{-}P(C_6H_5)_2]$	2.540	*120*
56	$C_{25}H_{18}Fe_2O_8$	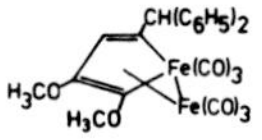	2.54	*31*
57	$C_{33}H_{33}Fe_3O_9P_3$	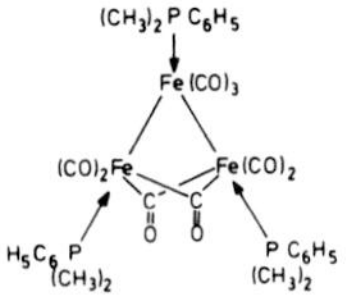	2.540(7) 2.688(7) 2.689(7)	*474,542*
58	$C_{39}H_{23}F_6Fe_3O_9P_2 \cdot 2C_6H_6$	$\{Fe_3(CO)_7[Ph_2PC(CO_2Me)C(CF_3)C_2\text{-}(CF_3)](PPh_2)\} \cdot 2C_6H_6$	2.543(2) 2.683(2) 2.679(2)	*512*
59	$C_{31}H_{25}Fe_2O_6P$	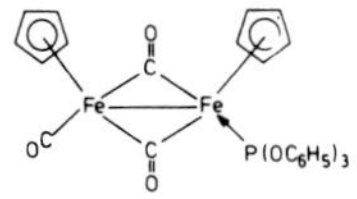 (two molecules in the asymmetric unit)	2.543(3) 2.548(3)	*202*
60	$C_{14}H_{10}Fe_2O_6$	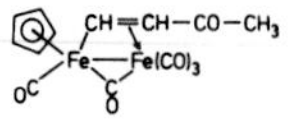	2.556	*15*
61	$C_6Fe_2O_6S_2$	$(\mu\text{-}S_2)[Fe(CO)_3]_2$	2.550	*609,610*
62	$C_{15}Fe_5O_{15}S_4$	$(\mu_3\text{-}S)_2[Fe(CO)_3]_3 \cdot (\mu\text{-}S_2)[Fe(CO)_3]_2$ A B	A { 3.371(10), 2.582(9), 2.609(10) B 2.545(11)	*610*

Table 11 (continued)

No.	Formula		Distance [Å]	Ref.
63	$C_{25}H_{24}Fe_6N_2O_{16}$	$[(CH_3)_4N]_2^+[Fe_6(CO)_{16}C]^{2-}$	(bridged) 2.533(10) - 2.632(10) (non bridged) 2.646(10) - 2.743(10)	*163, 155*
64	$C_{16}H_5Fe_3O_{11}Rh$	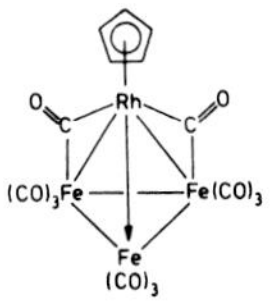	2.553(3) 2.586(3) 2.594(3)	*151*
65	$C_{24}H_{22}Fe_2O_4$	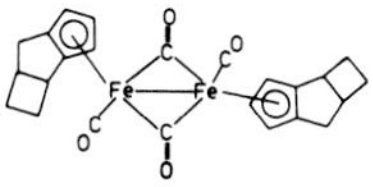	2.553(2)	*198*
66	$C_{38}H_{20}F_6Fe_3O_8P_2$	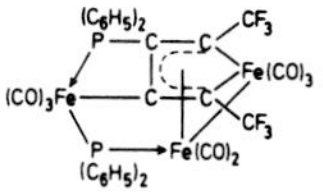	2.554	*509*
67	$C_{20}H_{10}Fe_2O_8$	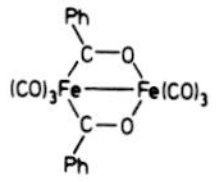	2.568	*449,293*
68	$C_{12}H_{18}B_6F_2$	1,6-bis(η^5-cyclopentadienyl)-1,6-diferra-2,3-dicarba-*closo*-decaborane(8)	2.571(1)	*108*
69	$C_{16}H_{16}B_2Fe_2O_4$	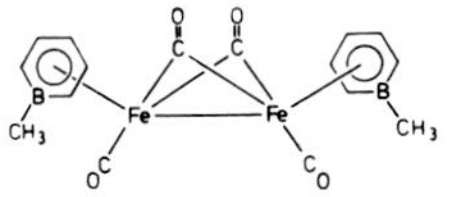	2.574(2)	*385*
70	$C_{17}H_{17}Fe_3NO_{11}$	$[(C_2H_5)_3NH]^+[HFe_3(CO)_{11}]^-$	2.577(3) 2.685(3) 2.696(3)	*220*
71	$C_{20}H_{10}F_{12}Fe_2O_2$	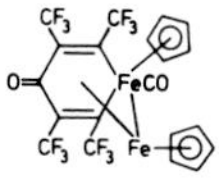	2.590(2)	*226*
72	$C_{34}H_{16}Fe_2O_6$	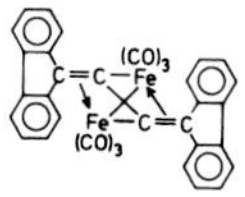	2.596(4)	*88*
73	$C_{13}H_{10}Fe_2O_5S$	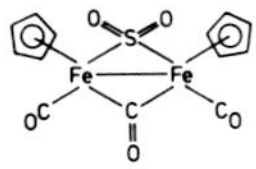 (two molecules in the asymmetric unit)	2.597(1) 2.584(1)	*160*
74	$C_{26}H_{14}Fe_2N_2O_9$	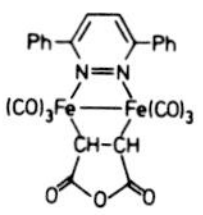	2.597(3)	*517*
75	$C_{26}H_{15}Fe_2O_6P$	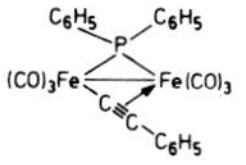	2.597(2)	*518*
76	$C_{17}H_8Fe_2N_2O_7$	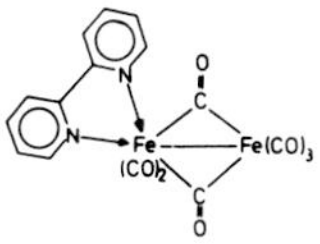	2.611	*203*

Table 11 (continued)

No.	Formula		Distance [Å]	Ref.
77	$C_{16}H_{16}Fe_3O_8S_2$	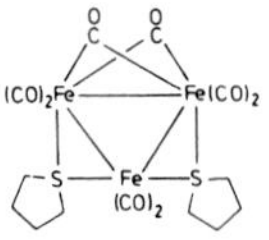	2.645(2) 2.611(2)	*206*
78	$C_{12}H_{11}Fe_2O_5P$	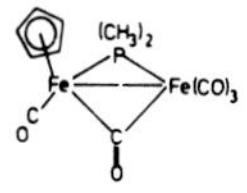 (two independent molecules)	2.615 2.638	*599*
79	$C_{16}H_{30}Fe_2S_{14}$	$(\mu\text{-}SC_2H_5)_2[Fe(S_2C\text{-}S\text{-}C_2H_5)]_2\text{-}(\mu\text{-}S_2C\text{-}S\text{-}C_2H_5)_2$	2.618(2)	*215*
80	$C_{25}H_{18}Fe_2O_8$	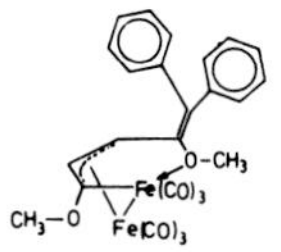	2.62	*395*
81	$Fe_4N_4O_4S_4$	$(\mu_3\text{-}S)_4[Fe(NO)]_4$	av. 2.634	*306*
82	$C_{22}H_{10}Fe_2O_8$	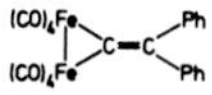	2.635(3)	*488*
83	$C_{19}H_{15}Fe_2NO_{10}S$	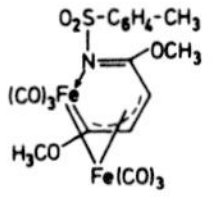	2.636(2)	*553*
84	$C_8H_{18}Fe_4N_6O_4S_2$	$(\mu_3\text{-}S)_2[\mu_3\text{-}NC(CH_3)_3]_2[Fe(NO)]_4$	2.642(1) 2.562(1) 2.496(1)	*306*
85	$C_{20}H_{20}Fe_4S_6$	$[(\eta^5\text{-}C_5H_5)Fe]_4(\mu_3\text{-}S)_2(\mu_3\text{-}S_2)_2$	2.64	*601*
86	$C_{14}H_{10}Fe_2O_7$	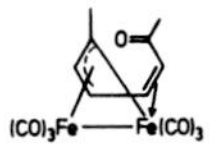	2.642(1)	*26*
87	$C_{18}H_{18}Fe_2O_7$	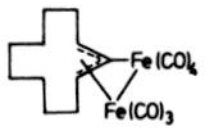	2.645(2)	*450*
88	$C_{20}H_{20}Fe_4S_4$	$(\mu_3\text{-}S)_4[(\eta^5\text{-}C_5H_5)Fe]_4$	(non bridged) 2.650(6) (bridged, av.) 3.365(6)	*611*
89	$C_{31}H_{20}Fe_2Ge_2O_7$	$(\mu\text{-}CO)[\mu\text{-}Ge(C_6H_5)_2]_2[Fe(CO)_3]_2$	2.666(3)	*282*
90	$C_{12}Fe_3O_{12}$	$Fe_3(CO)_{12}$	2.668(7) 2.560(6) 2.678(5)	*612*
91	$C_{32}H_{30}Fe_2O_9P_2$	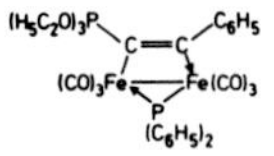	2.671(2)	*621*
92	$C_{27}H_{16}AsF_4Fe_3O_9P$	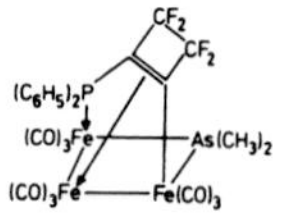	2.676 2.866	*276*
93	$C_{12}H_6Fe_2O_6$	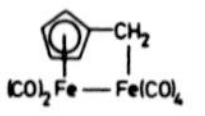	2.679	*482*

Table 11 (continued)

No.		Formula	Distance [Å]	*Ref.*
94	$C_{10}H_{12}As_4Fe_2O_6$	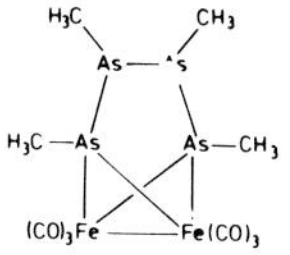	2.680	*307*
95	$C_{25}H_{16}Fe_3O_9$	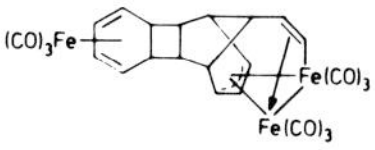	2.684(4)	*603*
96	$C_{32}H_{56}Fe_2N_2S_6$	$[(C_2H_5)_4N]_2^+\{(\mu\text{-}S)_2Fe_2[(SCH_2)_2C_6H_4]_2\}^{2-}$	2.691 - 2.776	*471*
97	$C_{46}H_{10}F_{20}Fe_2O_6P_2$	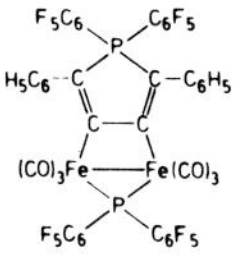	2.697(2)	*586*
98	$C_{32}H_{22}Fe_2O_7P_2$	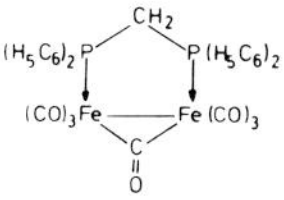	2.709(2)	*205*
99	$C_4H_{10}Fe_2N_4O_4S_2$	$[Fe(NO)_2]_2(\mu\text{-}SC_2H_5)_2$	2.72(0)	*589*
100	$C_{17}H_{16}Fe_2O_5$	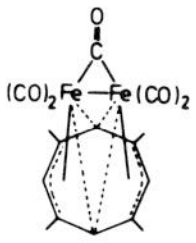	2.724(4)	*189*
101	$C_{12}H_{18}Fe_2Ge_3O_6$	$[\mu\text{-}Ge(CH_3)_2]_3[Fe(CO)_3]_2$	2.750(11)	*281,89*
102	$C_{25}H_9F_{24}Fe_4S_{11}O_6$	$\{(\mu\text{-}SCH_3)_3[Fe(CO)_3]_2\}^+$ $\{Fe_2[S_2C_2(CF_3)_2]_4\}^-$ two cation - anion pairs in the unit cell	3.062(4) 2.756(4) 2.777(3)	*560*
103	$C_{27}H_{15}Fe_2O_9PPt$	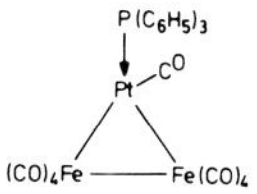	2.758(8)	*469*
104	$C_{23}H_{14}Fe_2O_5$	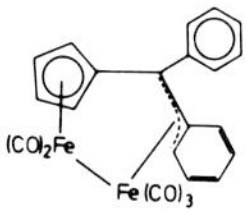	2.765	*53*
105	$C_{17}H_{16}Fe_2O_5$	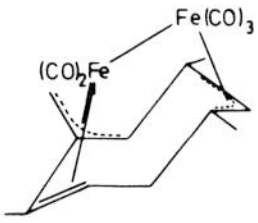	2.766(1)	*190*
106	$C_{12}H_8Fe_2(CO)_5$	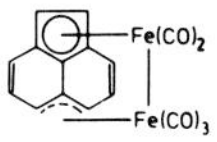	2.769(3)	*145,137*
107	$C_{16}H_{12}Fe_2O_6$	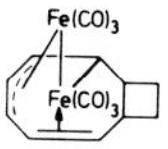	2.786(2)	*197*
108	$C_{80}H_{60}Fe_2N_2O_8P_4 \cdot 2CH_3CN$	$\{[(C_6H_5)_3P]_2N\}_2^+[Fe_2(CO)_8]^{2-}\cdot \cdot 2CH_3CN$	2.787(2)	*127*
109	$C_{14}F_4Fe_2O_8$		2.797(1)	*59*

Table 11 (continued)

No.		Formula	Distance [Å]	*Ref.*
110	$C_{21}H_{27}Fe_2O_5P$	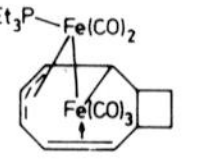	2.804(1)	*211*
111	$C_{13}H_8Fe_2O_6$	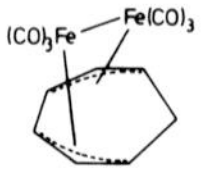	2.866(1)	*192*
112	$C_{40}H_{32}As_2F_8Fe_2O_4P_2$	$\{[(C_6H_5)_2P\text{-}C{=}C[As(CH_3)_2]\text{-}CF_2\text{-}CF_2]\}_2\text{-}Fe_2(CO)_4$	2.869(1)	*277*
113	$C_{16}Fe_4O_{16}Sn$	$(\mu_4\text{-}Sn)[Fe(CO)_4]_4$	2.870	*448*
114	$C_{14}H_{12}As_2F_4Fe_2O_6$	$\{(CH_3)_2As\text{-}C{=}C[As(CH_3)_2]\text{-}CF_2\text{-}CF_2\}\text{-}[Fe(CO)_3]_2$	2.89(1)	*271,270*
115	$C_{17}H_{13}As_2F_4Fe_3O_9$	$[(CH_3)_2As\text{-}C{=}C\text{-}CF_2\text{-}CF_2][As(CH_3)_2\text{-}Fe_3(CO)_9]$	2.917(5) 2.667(5)	*273* *272*
116	$C_{14}H_{16}BF_4Fe_2O_2S_2$	$\{[(\eta^5\text{-}C_5H_5)FeCO]_2(\mu\text{-}SCH_3)_2\}^+[BF_4]^-$	2.925(4)	*180*
117	$C_{16}H_{36}Fe_2N_4S_4$	$(\mu\text{-}L)_2Fe_2$	3.206(5)	*377*
	L = *N,N'*-dimethyl-*N,N'*-bis(β-mercaptoethyl)ethylenediamine			
118	$C_{18}H_{40}Fe_2N_4S_4$	$(\mu\text{-}L)_2Fe_2$	3.371(2)	*377*
	L = *N,N'*-dimethyl-*N,N'*-bis(β-mercaptoethyl)-1,3-propanediamine			
119	$C_{12}H_4Fe_3O_{11}$	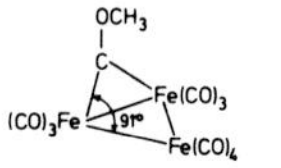		*566*
	B. Cobalt		*Fe-Co*	
1	$C_{18}H_{28}Co_3FeO_{18}P_3$	$\{H[FeCo_3(CO)_9][P(OCH_3)_3]_3\}$	2.560(2)	*378*
2	$C_{23}H_{18}CoFeO_5P$	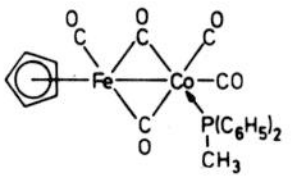	2.540(4)	*225*
3	$C_{16}H_{13}CoFeO_4$	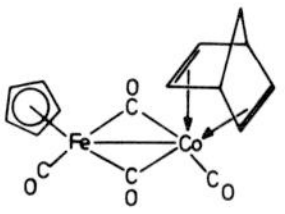	2.520(1)	*115*
4	$C_{11}H_5CoFeO_6$	$(\eta^5\text{-}C_5H_5)(CO)Fe(\mu\text{-}CO)_2Co(CO)_3$	2.545(1)	*114*
5	$C_{44}H_{30}CoFeNO_8P_2$	$\{[(C_6H_5)_3P]_2N\}^+[FeCo(CO)_8]^-$	2.835(3)	*127*
	C. Nickel		*Fe-Ni*	
1	$C_{34}H_{25}FeNiO_3P$	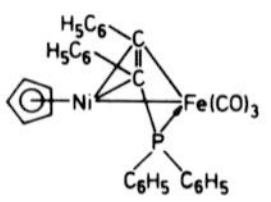	2.440(2)	*43*
2	$C_{19}H_{24}FeNiO_3$	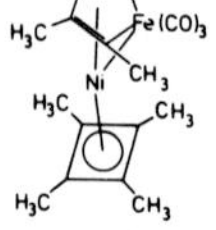	2.449(3)	*288*
	D. Ruthenium		*Fe-Ru*	
1	$C_{13}H_2FeO_{13}Ru_3$	$H_2FeRu_3(CO)_{13}$	2.620(1) 2.640(1) 2.670(1) 2.670(1) 2.270(1) 2.690(1)	*315*

Table 11 (continued)

No.	Formula		Distance [Å]	Ref.
	E. Rhodium		*Fe-Rh*	
1	$C_{16}H_5Fe_3O_{11}Rh$	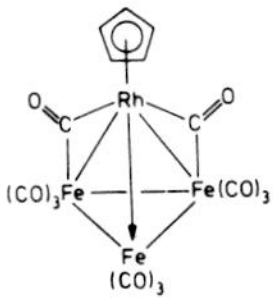	2.568(3) 2.615(3) 2.607(3)	*151*
2	$C_{18}H_{10}Fe_2O_8Rh_2$	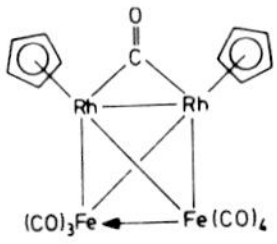	2.570(5) 2.598(5)	*148,156*
3	$C_{40}H_{34}F_6Fe_2O_4P_3Rh$	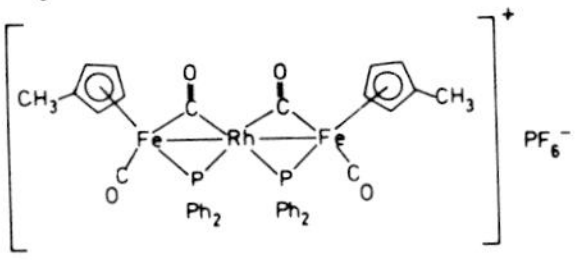	2.674(1) 2.659(2)	*468*
			2.671 2.660	*340*
	F. Palladium		*Fe-Pd*	
1	$C_{39}H_{28}Cl_2Fe_2O_8P_2Pd_2$	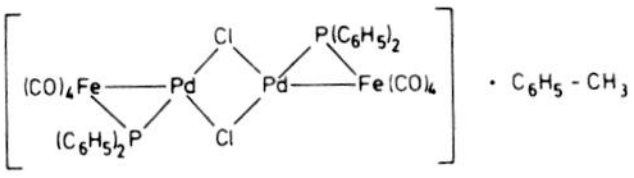	2.59(1)	*414*
	G. Molybdenum		*Fe-Mo*	
1	$C_{18}H_{28}Cl_2FeMoS_2$	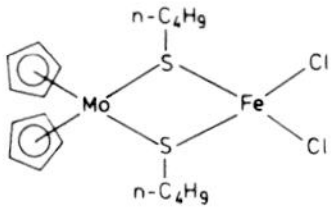	(3.660)	*111,113*
	H. Manganese		*Fe-Mn*	
1	$C_{12}H_5FeMnO_7$	$(\eta^5\text{-}C_5H_5)(CO)_2Fe\text{-}Mn(CO)_5$ (two crystallographically independent molecules)	2.840 2.845	*344*
	I. Mercury		*Fe-Hg*	
1	$C_4Br_2FeHg_2O_4$	$(BrHg)_2Fe(CO)_4$	2.440(7) 2.590(7)	*37*
2	$C_{11}H_5CoFeHgO_6$	$[(\eta^5\text{-}C_5H_5)(CO)_2Fe]HgCo(CO)_4$	2.49	*93*
3	$C_{16}H_{30}Fe_2HgN_2O_6P_2$	$(\mu\text{-}Hg)\{Fe(CO)_2(NO)[P(C_2H_5)_3]\}_2$	2.534(2)	*577*
4	$C_{14}H_{10}Cl_2FeHg_2N_2O_4$	$[(C_5H_5N)HgCl]_2[\mu\text{-}Fe(CO)_4]$	2.552(8)	*39*
	J. Platinum		*Fe-Pt*	
1	$C_{59}H_{45}FeO_{14}P_3Pt_2$	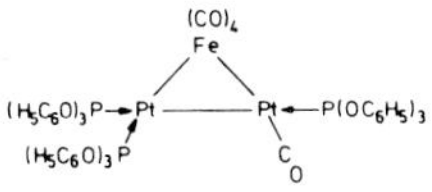	2.583(6) 2.550(5)	*4,5*
2	$C_{27}H_{15}Fe_2O_9PPt$	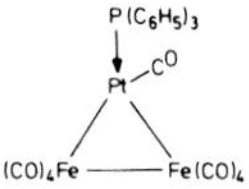	2.597(5) 2.530(5)	*469*

Table 12: Biorganic iron compounds and relevant model systems.

		Ref.
1	α-Chlorohemin	*424*
2	α,β,γ,δ-Tetraphenylporphine diacid, $[H_4TPP]^{2+}[Cl^-, FeCl_4^-]$	*579*
3	Ferric hydroxide tetraphenylporphine monohydrate	*294*
4	Methoxyiron(III) mesoporphyrin-IX dimethyl ester	*361*
5	Chloroiron(III) tetraphenylporphine	*362*
6	(μ-Oxo)[bis(α,β,γ,δ-tetraphenylporphinato-iron(III))]	*298,365*
7	Bis(imidazole)-α,β,γ,δ-tetraphenylporphinato-iron(III) chloride	*217*
8	Di[cyclohexane-1,2-dioximato(1-)]diimidazole-iron(II)	*538*
9	Bis(imidotetramethyldithiodiphosphino-SS)-iron(II), a model for Rubredoxin	*149,153*
10	Iron(II) sulphate pentahydrate glycine	*453*
11	Ferrichrome-A tetrahydrate (fungal metabolite including a hexapeptide ring)	*627*
12	Atrovenetin, orange trimethyl ether ferrichloride	*520*
13	(Vitamin-A aldehyde)tricarbonyliron	*66*
14	$[(\eta^5\text{-}C_5H_5)_2Mo](\mu\text{-}S\text{-}n\text{-}C_4H_9)_2[FeCl_2]$, a model for Nitrogenase System	*111,112*
15	Ferroverdin	*117*
16	Ferrioxamine E	*534*
17	(Nitrosyl)(α,β,γ,δ-tetraphenylporphinato)-(1-methylimidazole)iron	*528*
18	(*N*-methylimidazole)(dioxygen)(*meso*-tetra-α,β,γ,δ-*o*-pivalylamidephenyl-porphinato)iron(III)	*551*
19	Ferrichrysin	*504*
20	Bis(10-methylisoalloxazine)iron(II) perchlorate hexahydrate	*305*

REFERENCES

1. Adams, R.D., Cotton, F.A., and Frenz, B.A., *J. Organometal. Chem., 73,* 93 (1974).
2. Adams, R.D., Cotton, F.A., and Troup, J.M., *Inorg. Chem., 13,* 257 (1974).
3. Adman, E., Rosenblum, M., Sullivan, S., and Margulis, T.N., *J. Amer. Chem. Soc., 89,* 4540 (1967).
4. Albano, V.G., Ciani, G., Bruce, M.I., Shaw, G., and Stone, F.G.A., *J. Organometal. Chem., 42,* C 99 (1972).
5. Albano, V.G., and Ciani, G., *J. Organometal. Chem. 66,* 311 (1974).
6. Albano, V.G., Araneo, A., Bellon, P.L., Ciani, G., and Manassero, M., *J. Organometal. Chem., 67,* 413 (1974).
7. Allen, F.H., Trotter, J., and Williston, C.S., *J. Chem. Soc. A, 1970,* 907.
8. Allegra, G., Colombo, A., Immirzi, A., and Bassi, I.W., *J. Amer. Chem. Soc., 90,* 4455 (1968).
9. Allison, D.A., Clardy, J., and Verkade, J.G., *Inorg. Chem., 11,* 2804 (1972).
10. Allmann, R., *Angew. Chem., 82,* 982, (1970); *Angew. Chem. Int. Ed. Engl., 9,* 958 (1970).
11. Andrianov, V.G., Chapovskii, Yu.A., Semion, V.A., and Struchkov, Yu.T., *Chem. Commun., 1968,* 282.
12. Andrianov, V.G., and Struchkov, Yu.T., *Chem. Commun., 1968,* 1590.
13. Andrianov, V.G., and Struchkov, Yu.T., *Zh. Strukt. Khim., 9,* 240 (1968); *J. Struct. Chem., 9,* 182 (1968).
14. Andrianov, V.G., and Struchkov, Yu.T., *Zh. Strukt. Khim., 9,* 503 (1968); *J. Struct. Chem., 9,* 426 (1968).
15. Andrianov, V.G., and Struchkov, Yu.T., *Zh. Strukt. Khim., 9,* 845 (1968); *J. Struct. Chem., 9,* 737 (1968).
16. Andrianov, V.G., Martynov, V.P., Anisimov, K.N., Kolobova, N.E., and Skripkin, V.V., *J. Chem. Soc. D, Chem. Commun., 1970,* 1252.
17. Andrianov, V.G., Sergeeva, G.N., Struchkov, Yu.T., Anisimov, K.N., Kolobova, N.E., and Beschastnov, A.S., *Zh. Strukt. Khim., 11,* 168 (1970); *J. Struct. Chem., 11,* 163 (1970).
18. Andrianov, V.G., Martynov, V.P., and Struchkov, Yu.T., *Zh. Strukt. Khim., 12,* 866 (1971); *J. Struct. Chem., 12,* 793 (1971).
19. Andrianov, V.G., Struchkov, Yu.T., Rybinskaya, M.I., Rybin, L.V., and Gubenko, N.T., *Zh. Strukt. Khim., 13,* 86 (1972); *J. Struct. Chem., 13,* 74 (1972).
20. Andrianov, V.G., Struchkov, Yu.T., and Rossinskaya, E.R., *J. Chem. Soc. Chem. Commun., 1973,* 338.
21. Andrianov, V.G., Struchkov, Yu.T., and Rossinskaya,

E.R., *Zh. Strukt. Khim., 15,* 74 (1974); *J. Struct. Chem., 15,* 65 (1974).

22. Anzenhofer, K., and de Boer, J.J., *Rec. Trav. Chim. Pays-Bas, 88,* 286 (1969).
23. Ariyaratne, J.K.P., Bierrum, A.M., Green, M.L.H., Ishaq, M., Prout, C.K., and Swanwick, M.G., *J. Chem. Soc. A, 1969,* 1309.
24. Astakhova, I.S., and Struchkov, Yu.T., *Zh. Strukt. Khim., 11,* 472 (1970); *J. Struct. Chem., 11,* 432 (1970).
25. Atovmyan, L.O., D'yachenko, O.A., and Soboleva, S.V., *Zh. Strukt. Khim., 11,* 557 (1970); *J. Struct. Chem., 11,* 517 (1970).
26. Aumann, R., Averbeck, H., and Krüger, C., *Chem. Ber., 108,* 3336 (1975).
27. Averill, B.A., Herskovitz, T., Holm, R.H., and Ibers, J.A., *J. Amer. Chem. Soc., 95,* 3523 (1973).
28. Avoyan, R.L., Chapovskii, Yu.A., and Struchkov, Yu.T., *Zh. Strukt. Chem. 7,* 900 (1966); *J. Struct. Chem., 7,* 839 (1966).
29. Bacci, M., and Ghilardi, C.A., *Inorg. Chem., 13,* 2398 (1974).
30. Bagga, M.M., Baikie, P.E., Mills, O.S., and Pauson, P.L., *Chem. Commun., 1967,* 1106.
31. Bagga, M.M., Ferguson, G., Jeffreys, J.A.D., Mansell, C.M., Pauson, P.L., Robertson, I.C., and Sime, J.G., *J. Chem. Soc. D, Chem. Commun., 1970,* 672.
32. Baikie, P.E., and Mills, O.S., *Chem. Commun., 1966,* 707.
33. Baikie, P.E., and Mills, O.S., *Chem. Commun., 1967,* 1228.
34. Baikie, P.E., and Mills, O.S., *Inorg. Chim. Acta, 1,* 55 (1967).
35. Bailey, N.A., and Mason, R., *Acta Crystallogr., 21,* 652 (1966).
36. Bailey, N.A., Cook, D.F., Cummins, D., and McKenzie, E.D., *Inorg. Nucl. Chem. Lett., 11,* 51 (1975).
37. Baird, H.W., and Dahl, L.F., *J. Organometal. Chem., 7,* 503 (1967).
38. Baker, J., Engelhardt, L.M., Figgis, B.N., and White, A.H., *J. Chem. Soc. Dalton Trans., 1975,* 530.
39. Baker, R.W., and Pauling, P., *J. Chem. Soc. D, Chem. Commun., 1970,* 573.
40. Ballhausen, C.J., and Dahl, J.P., *Acta Chem. Scand., 15,* 1333 (1961).
41. Barnett, B.L., and Davis, R.E., *Amer. Cryst. Assoc., Winter Meeting 1970,* New Orleans, Abstracts, p.45.
42. Barnett, B.L., and Krüger, C., *Angew. Chem., 83,* 969

(1971); *Angew. Chem. Int. Ed. Engl., 10,* 910 (1971).
43. Barnett, B.L., and Krüger, C., *Cryst. Struct. Commun., 2,* 347 (1973).
44. Barrow, M.J., and Mills, O.S., *Angew. Chem., 81,* 898 (1969); *Angew. Chem. Int. Ed. Engl., 8,* 879 (1969).
45. Barrow, M.J., and Mills, O.S., *J. Chem. Soc. A, 1971,* 864.
46. Barrow, M.J., Sim, G.A., Dobbie, R.C., and Mason, P.R., *J. Organometal. Chem., 69,* C 4 (1974).
47. Barrow, M.J., and Sim, G.A., *J. Chem. Soc. Dalton Trans., 1975,* 291.
48. Bassi, I.W., and Scordamaglia, R., *J. Organometal. Chem., 37,* 353 (1972).
49. Bats, J.W., de Boer, J.J., and Bright, D., *Inorg. Chim. Acta, 5,* 605 (1971).
50. Battelle, L.F., Bau, R., Gokel, G.W., Oyakawa, R.T., and Ugi, I.K., *J. Amer. Chem. Soc., 95,* 482 (1973).
51. Beckett, R., Heath, G.A., Hoskins, B.F., Kelly, B.P., Martin, R.L., Roos, I.A.G., and Weickhardt, P.L., *Inorg. Nucl. Chem. Lett., 6,* 257 (1970).
52. Beddoes, R.L., Lindley, P.F., and Mills, O.S., *Angew. Chem., 82,* 293 (1970); *Angew. Chem. Int. Ed. Engl., 9,* 304 (1970).
53. Behrens, U., and Weiss, E., *J. Organometal. Chem., 73,* C 64 (1974).
54. Bennett, M.A., Robertson, G.B., Tomkins, I.B., and Whimp, P.O., *J. Chem. Soc. D, Chem. Commun. 1971,* 341.
55. Bennett, M.J., Jr., Cotton, F.A., Davison, A., Faller, J.W., Lippard, S.J., and Morehouse, S.M., *J. Amer. Chem. Soc., 88,* 4371 (1966).
56. Bennett, M.J., Cotton, F.A., and Weaver, D.L., *Nature, 212,* 286 (1966).
57. Bennett, M.J., Cotton, F.A., and Weaver, D.L., *Acta Crystallogr., 23,* 581 (1967).
58. Bennett, M.J., Brooks, W., Elder, M., Graham, W.A.G., Hall, D., and Kummer, R., *J. Amer. Chem. Soc., 92,* 208 (1970).
59. Bennett, M.J., Graham, W.A.G., Stewart, P.R., Jr., and Tuggle, L.M., *Inorg. Chem., 12,* 2944 (1973).
60. Bennett, M.J., Graham, W.A.G., Smith, R.A., and Stewart, R.P., Jr., *J. Amer. Chem. Soc., 95,* 1684 (1973).
61. Bernal, J., and Sequeira, A., *Amer. Cryst. Assoc., Summer Meeting 1967,* Abstracts, p.75.
62. Bernstein, T., and Herbstein, F.H., *Acta Crystallogr., B 24,* 1640 (1968).
63. Bertrand, J.A., Breece, J.L., Kalyanaraman, A.R., Long, G.J., and Baker, W.A., Jr., *J. Amer. Chem. Soc.,*

92, 5233 (1970).

64. Bertrand, J.A., Breece, J.L., and Eller, P.G., *Inorg. Chem.*, *13*, 125 (1974).
65. Bertrand, J.A., and Breece, J.L., *Inorg. Chim. Acta*, *8*, 267 (1974).
66. Birch, A.J., Fitton, H., Mason, R., Robertson, G.B., and Stangroom, J.E., *Chem. Commun.*, *1966*, 613.
67. Birnbaum, G.I., *J. Amer. Chem. Soc.*, *94*, 2455 (1972); *Amer. Cryst. Assoc.*, *Summer Meeting 1970*, Ottawa, Abstracts, p.81.
68. Bir'yukov, B.P., Struchkov, Yu.T., Anisimov, K.N., Kolobova, N.E., and Skripkin, V.V., *Chem. Commun.*, *1967*, 750.
69. Bir'yukov, B.P., Anisimov, K.N., Struchkov, Yu.T., Kolobova, N.E., and Skripkin, V.V., *Zh. Strukt. Khim.*, *8*, 556 (1967); *J. Struct. Chem.*, *8*, 498 (1967).
70. Bir'yukov, B.P., Struchkov, Yu.T., Anisimov, K.N., Kolobova, N.E., and Skripkin, V.V., *Chem. Commun.*, *1968*, 159.
71. Bir'yukov, B.P., Struchkov, Yu.T., Anisimov, K.N., Kolobova, N.E., and Skripkin, V.V., *Chem. Commun.*, *1968*, 1193.
72. Bir'yukov, B.P., and Struchkov, Yu.T., *Zh. Strukt. Khim.*, *9*, 488 (1968); *J. Struct. Chem.*, *9*, 412 (1968).
73. Bir'yukov, B.P., Struchkov, Yu.T., Anisimov, K.N., Kolobova, N.E., and Skripkin, V.V., *Zh. Strukt. Khim.*, *9*, 922 (1968); *J. Struct. Chem.*, *9*, 821 (1968).
74. Bir'yukov, B.P., and Struchkov, Yu.T., *Zh. Strukt. Khim.*, *10*, 95 (1969); *J. Struct. Chem.*, *10*, 86 (1969).
75. Bjåmer Birnbaum, K., Altman, J., Maymon, T., and Ginsburg, D., *Tetrahedron Lett.*, *1970*, 2051.
76. Blake, A.B., and Fraser, L.R., *J. Chem. Soc. Dalton Trans.*, *1975*, 193.
77. Blount, J.F., Dahl, L.F., Hoogzand, C., and Hübel, W., *J. Amer. Chem. Soc.*, *88*, 292 (1966).
78. Bok, L.D.C., Leipoldt, J.G., and Basson, S.S., *Z. anorg. allg. Chem.*, *389*, 307 (1972).
79. le Borgne, G., and Grandjean, D., *Acta Crystallogr.*, *B 29*, 1040 (1973).
80. Borovyak, T.E., Shklover, V.E., Gusev, A.I., Gubin, S.P., Koridze, A.A., and Struchkov, Yu.T., *Zh. Strukt. Khim.*, *11*, 1087 (1970); *J. Struct. Chem.*, *11*, 1012 (1970).
81. Bottrill, M., Goddard, R., Green, M., Hughes, R.P., Lloyd, M.K., Lewis, B., and Woodward, P., *J. Chem. Soc. Chem. Commun.*, *1975*, 253.
82. Bowman, K., Gaughan, A.P., and Dori, Z., *J. Amer. Chem. Soc.*, *94*, 727 (1972).

83. Bradley, D.C., Hursthouse, M.B., Rodesiler, P.F., *Chem. Commun., 1969,* 14.
84. Brennan, J.P., Grimes, R.N., Schaeffer, R., and Sneddon, L.G., *Inorg. Chem., 12,* 2266 (1973).
85. Bright, D., and Mills, O.S., *Chem. Commun., 1966,* 211.
86. Bright, D., and Mills, O.S., *Chem. Commun., 1967,* 245.
87. Bright, D., and Mills, O.S., *J. Chem. Soc. A, 1971,* 1979.
88. Bright, D., and Mills, O.S., *J. Chem. Soc. Dalton Trans., 1972,* 2465.
89. Brooks, E.H., Elder, M., Graham, W.A.G., and Hall, D., *J. Amer. Chem. Soc., 90,* 3587 (1968).
90. Brown, D.S., and Bushnell, G.W., *Acta Crystallogr., 22,* 296 (1967).
91. Brown, G.M., Hedberg, F.L., and Rosenberg, H., *Amer. Cryst. Assoc., Winter Meeting 1973,* Gainesville (Florida), Abstracts, p.90.
92. Bruce, M.I., Clark, R., Howard, J., and Woodward, P., *J. Organometal. Chem., 42,* C 107 (1972).
93. Bryan, R.F., and Weber, H.P., *Acta Crystallogr., 21,* A 138 (1966).
94. Bryan, R.F., and Weber, H.P., *Chem. Commun., 1966,* 329.
95. Bryan, R.F., *Amer. Cryst. Assoc., Winter Meeting 1967,* Abstracts, p.49.
96. Bryan, R.F., *J. Chem. Soc. A, 1967,* 192.
97. Bryan, R.F., and Manning, A.R., *Chem. Commun., 1968,* 1220.
98. Bryan, R.F., Greene, P.T., Melson, G.A., Stokely, P.F., and Manning, A.R., *J. Chem. Soc. D, Chem. Commun., 1969,* 722.
99. Bryan, R.F., Greene, P.T., Field, D.S., and Newlands, M.J., *J. Chem. Soc. D, Chem. Commun., 1969,* 1477.
100. Bryan, R.F., and Greene, P.T., *J. Chem. Soc. A, 1970,* 3064.
101. Bryan, R.F., Greene, P.T., Newlands, M.J., and Field, D.S., *J. Chem. Soc. A, 1970,* 3068.
102. Bryan, R.F., and Schmidt, W.C., Jr., *J. Chem. Soc. Dalton Trans., 1974,* 2337.
103. Burlitch, J.M., Petersen, R.B., Conder, H.L., and Robinson, W.R., *J. Amer. Chem. Soc., 92,* 1783 (1970).
104. Bush, M.A., and Woodward, P., *Chem. Commun., 1967,* 166.
105. Bush, M.A., and Woodward, P., *J. Chem. Soc. A, 1967,* 1833.
106. Butler, W.M., Enemark, J.H., *J. Organometal. Chem., 49,* 233 (1973).
107. Calderôn, J.L., Fontana, S., Frauendorfer, E., Day, V.W., and Iske, S.D.A., *J. Organometal. Chem., 64,*

C 16 (1974).

108. Callahan, K.P., Evans, W.J., Lo, F.Y., Strouse, C.E., and Hawthorne, M.F., *J. Amer. Chem. Soc., 97,* 296 (1975).

109. Calvarin, G., Bouvaist, J., and Weigel, D., *C.R. Acad. Sci., Ser. C, 268,* 2288 (1969).

110. Calvarin, G., and Weigel, D., *Acta Crystallogr., B 27,* 1253 (1971).

111. Cameron, T.S., and Prout, C.K., *J. Chem. Soc. Chem. Commun., 1971,* 161.

112. Cameron, T.S., and Prout, C.K., *Acta Crystallogr., B 28,* 453 (1972).

113. Cameron, T.S., Maguire, J.F., Turbitt, T.D., and Watts, W.E., *J. Organometal. Chem., 49,* C 79 (1973).

114. Campbell, I.L.C., and Stephens, F.S., *J. Chem. Soc. Dalton Trans., 1975,* 22.

115. Campbell, I.L.C., and Stephens, F.S., *J. Chem. Soc. Dalton Trans., 1975,* 226.

116. Campbell, I.L.C., and Stephens, F.S., *J. Chem. Soc. Dalton Trans., 1975,* 982.

117. Candeloro, S., Grdenić, D., Taylor, N., Thompson, B., Viswamitra, M., and Crowfoot Hodgkin, D., *Nature, 224,* 589 (1969).

118. Carrell, H.L., and Glusker, J.P., *Amer. Cryst. Assoc., 25 th Anniversary Meeting 1975,* Charlottesville (Virginia), Abstracts, p.12.

119. Carter, O.L., McPhail, A.T., and Sim, G.A., *J. Chem. Soc. A, 1967,* 365.

120. Carty, A.J., Ng, T.W., Carter, W., Palenik, G.J., and Birchall, T., *J. Chem. Soc. D, Chem. Commun., 1969,* 1101.

121. Carty, A.J., Madden, D.P., Mathew, M., Palenik, G.J., and Birchall, T., *J. Chem. Soc. D, Chem. Commun., 1970,* 1664.

122. Carty, A.J., Ferguson, G., Paik, H.N., and Restivo, R., *J. Organometal. Chem., 74,* C 14 (1974).

123. Chapovskii, Yu.A., Andrianov, B.G., Struchkov, Yu.T., and Semion, V.A., *Zh. Strukt. Khim., 8,* 559 (1967); *J. Struct. Chem., 8,* 501 (1967).

124. Chapovskii, Yu.A., Semion, V.A., Andrianov, V.G., and Struchkov, Yu.T., *Zh. Strukt. Khim., 9,* 1100 (1968); *J. Struct. Chem., 9,* 990 (1968).

125. Chatt, J., and Duncanson, L.A., *J. Chem., Soc., 1953,* 2939.

126. Chin, H.B., and Bau, R., *J. Amer. Chem. Soc., 95,* 5068 (1973).

127. Chin, H.B., Smith, M.B., Wilson, R.D., and Bau, R., *J. Amer. Chem. Soc., 96,* 5285 (1974).

128. Churchill, M.R., and Mason, R., *Proc. Chem. Soc., 1964,* 226.
129. Churchill, M.R., *Chem. Commun., 1966,* 450.
130. Churchill, M.R., and Mason, R., *Advan. Organometal. Chem., 5,* 93 (1967).
131. Churchill, M.R., *Inorg. Chem., 6,* 185 (1967).
132. Churchill, M.R., *Inorg. Chem., 6,* 190 (1967).
133. Churchill, M.R., and Mason, R., *Proc. Roy. Soc.(London) A, 301,* 433 (1967).
134. Churchill, M.R., and Gold, K., *Chem. Commun., 1968,* 693.
135. Churchill, M.R., and Wormald, J., *Chem. Commun., 1968,* 1033.
136. Churchill, M.R., Wormald, J., Giering, W.P., and Emerson, G.F., *Chem. Commun., 1968,* 1217.
137. Churchill, M.R., and Wormald, J., *Chem. Commun., 1968,* 1597.
138. Churchill, M.R., and Bird, P.H., *J. Amer. Chem. Soc., 90,* 3241 (1968).
139. Churchill, M.R., and Gold, K., *Inorg. Chem., 8,* 401 (1969).
140. Churchill, M.R., and Wormald, J., *Inorg. Chem., 8,* 716 (1969).
141. Churchill, M.R., and Wormald, J., *Inorg. Chem., 8,* 1936 (1969).
142. Churchill, M.R., and Bird, P.H., *Inorg. Chem., 8,* 1941 (1969).
143. Churchill, M.R., and Wormald, J., *Inorg. Chem., 8,* 1970 (1969).
144. Churchill, M.R., Wormald, J., Young, D.A.T., and Kaesz, H.D., *J. Amer. Chem. Soc., 91,* 7201 (1969).
145. Churchill, M.R., and Wormald, J., *Inorg. Chem., 9,* 2239 (1970).
146. Churchill, M.R., and Wormald, J., *Inorg. Chem., 9,* 2430 (1970).
147. Churchill, M.R., Wormald, J., Ross, D.A., Thomasson, J.E., and Wojcicki, A., *J. Amer. Chem. Soc., 92,* 1795 (1970).
148. Churchill, M.R., and Veidis, M.V., *J. Chem. Soc. D, Chem. Commun., 1970,* 529.
149. Churchill, M.R., and Wormald, J. *J. Chem. Soc. D, Chem. Commun., 1970,* 703.
150. Churchill, M.R., and Fennessey, J.P., *J. Chem. Soc. D, Chem. Commun., 1970,* 1056.
151. Churchill, M.R., and Veidis, M.V., *J. Chem. Soc. D, Chem. Commun., 1970,* 1470.
152. Churchill, M.R., and Wormald, J., *Inorg. Chem., 10,* 572 (1971).

153. Churchill, M.R., and Wormald, J., *Inorg. Chem.*, *10*, 1778 (1971).
154. Churchill, M.R., and Wormald, J., *J. Amer. Chem. Soc.*, *93*, 354 (1971).
155. Churchill, M.R., Wormald, J., Knight, J., and Mays, M.J., *J. Amer. Chem. Soc.*, *93*, 3073 (1971).
156. Churchill, M.R., and Veidis, M.V., *J. Chem. Soc. A*, *1971*, 2170.
157. Churchill, M.R., and Reis, A.H., Jr., *Inorg. Chem.*, *11*, 2299 (1972).
158. Churchill, M.R., and DeBoer, B.G., *Inorg. Chem.*, *12*, 525 (1973).
159. Churchill, M.R., DeBoer, B.G., and Kalra, K.L., *Inorg. Chem.*, *12*, 1646 (1973).
160. Churchill, M.R., and Kalra, K.L., *Inorg. Chem.*, *12*, 1650 (1973).
161. Churchill, M.R., and Lin, K.-K.G., *Inorg. Chem.*, *12*, 2274 (1973).
162. Churchill, M.R., and Chang, S.W.-Y.N., *J. Amer. Chem. Soc.*, *95*, 5931 (1973).
163. Churchill, M.R., and Wormald, J., *J. Chem. Soc. Dalton Trans.*, *1974*, 2410.
164. Churchill, M.R., and Chang, S.W.-Y., *Inorg. Chem.*, *14*, 1680 (1975).
165. de Cian, A., and Weiss, R., *Chem. Commun.*, *1968*, 348.
166. de Cian, A., and Weiss, R., *Nat. Proprietes Liason Coordin. Coll. Internation. Paris*, 261 (1970).
167. de Cian, A., and Weiss, R., *Acta Crystallogr.*, *B 28*, 3264 (1972).
168. de Cian, A., and Weiss, R., *Acta Crystallogr.*, *B 28*, 3273 (1972).
169. de Cian, A., L'Huillier, P.M., and Weiss, R., *Bull. Soc. Chim. Fr.*, *1973*, 451.
170. de Cian, A., L'Huillier, P.M., and Weiss, R., *Bull. Soc. Chim. Fr.*, *1973*, 457.
171. Clark, R., Howard, J., and Woodward, P., *J. Chem. Soc. Dalton Trans.*, *1974*, 2027.
172. Clearfield, A., Singh, P., and Bernal, I., *Amer. Cryst. Assoc., Winter Meeting 1970*, New Orleans, Abstracts, p.65.
173. Clearfield, A., Singh, P., and Bernal, I., *J. Chem. Soc. D, Chem. Commun.*, *1970*, 389.
174. Coggon, P., McPhail, A.T., Mabbs, F.E., and McLachlan, V.N., *J. Chem. Soc. A*, *1971*, 1014.
175. Cohen, G.H., and Hoard, J.L., *J. Amer. Chem. Soc.*, *88*, 3228 (1966).
176. Colapietro, M., Domenicano, A., Scaramuzza, L., Vaciago, A., and Zambonelli, L., *Chem. Commun.*, *1967*, 583.

177. Coda, A., Kamenar, B., Prout, K., Carruthers, J.R., and Rollett, J.S., *Acta Crystallogr.*, *B 31*, 1438 (1975).
178. Coleman, J.M., Wojcicki, A., Pollick, P.J., and Dahl, L.F., *Inorg. Chem.*, *6*, 1236 (1967).
179. Collins, D.M., Countryman, R., and Hoard, J.L., *J. Amer. Chem. Soc.*, *94*, 2066 (1972).
180. Connelly, N.G., and Dahl, L.F., *J. Amer. Chem. Soc.*, *92*, 7472 (1970).
181. Constant, G., Daran, J.-C., and Jeannin, Y., *J. Inorg. Nucl. Chem.*, *33*, 4209 (1971).
182. Constant, G., Daran, J.-C., and Jeannin, Y., *J. Organometal. Chem.*, *44*, 353 (1972).
183. Constant, G., Daran, J.-C., and Jeannin, Y., *1st Europ. Crystallogr. Meeting 1973*, Bordeaux, Abstracts, Group A_6.
184. Constant, G., Daran, J.-C., and Jeannin, Y., *J. Inorg. Nucl. Chem.*, *35*, 4083 (1973).
185. Cooke, M., Howard, J.A.K., Russ, C.R., Stone, F.G.A., and Woodward, P., *J. Organometal. Chem.*, *78*, C 43 (1974).
186. Corradini, P., Pedone, C., and Sirigu, A., *Chem. Commun.*, *1966*, 341.
187. Corradini, P., Pedone, C., and Sirigu, A., *Chem. Commun.*, *1968*, 275.
188. Cotton, F.A., and Yagupsky, G., *Inorg. Chem.*, *6*, 15 (1967).
189. Cotton, F.A., and LaPrade, M.D., *J. Amer. Chem. Soc.*, *90*, 2026 (1968).
190. Cotton, F.A., and Takats, J., *J. Amer. Chem. Soc.*, *90*, 2031 (1968).
191. Cotton, F.A., and Edwards, W.T., *J. Amer. Chem. Soc.*, *91*, 843 (1969).
192. Cotton, F.A., DeBoer, B.G., and Marks, T.J., *J. Amer. Chem. Soc.*, *93*, 5069 (1971).
193. Cotton, F.A., and Pipal, J.R., *J. Amer. Chem. Soc.*, *93*, 5441 (1971).
194. Cotton, F.A., and Troup, J.M., *J. Amer. Chem. Soc.*, *95*, 3798 (1973).
195. Cotton, F.A., Day, V.W., Frenz, B.A., Hardcastle, K.I., and Troup, J.M., *J. Amer. Chem. Soc.*, *95*, 4522 (1973).
196. Cotton, F.A., Hardcastle, K.I., and Rusholme, G.A., *J. Coord. Chem.*, *2*, 217 (1973).
197. Cotton, F.A., Frenz, B.A., Deganello, G., and Shaver, A., *J. Organometal. Chem.*, *50*, 227 (1973).
198. Cotton, F.A., Frenz, B.A., Troup, J.M., and Deganello, G., *J. Organometal. Chem.*, *59*, 317 (1973).

199. Cotton, F.A., Frenz, B.A., and White, A.J., *J. Organometal. Chem.*, *60*, 147 (1973).
200. Cotton, F.A., Frenz, B.A., and Troup, J.M., *J. Organometal. Chem.*, *61*, 337 (1973).
201. Cotton, F.A., and Frenz, B.A., *Inorg. Chem.*, *13*, 253 (1974).
202. Cotton, F.A., Frenz, B.A., and White, A.J., *Inorg. Chem.*, *13*, 1407 (1974).
203. Cotton, F.A., and Troup, J.M., *J. Amer. Chem. Soc.*, *96*, 1233 (1974).
204. Cotton, F.A., and Troup, J.M., *J. Amer. Chem. Soc.*, *96*, 3438 (1974).
205. Cotton, F.A., and Troup, J.M., *J. Amer. Chem. Soc.*, *96*, 4422 (1974).
206. Cotton, F.A., and Troup, J.M., *J. Amer. Chem. Soc.*, *96*, 5070 (1974).
207. Cotton, F.A., and Troup, J.M., *J. Organometal. Chem.*, *76*, 81 (1974).
208. Cotton, F.A., and Troup, J.M., *J. Organometal. Chem.*, *77*, 83 (1974).
209. Cotton, F.A., and Troup, J.M., *J. Organometal. Chem.*, *77*, 369 (1974).
210. Cotton, F.A., and Lahuerta, P., *Inorg. Chem.*, *14*, 116 (1975).
211. Cotton, F.A., and Hunter, D.L., *J. Amer. Chem. Soc.*, *97*, 5739 (1975).
212. Coucouvanis, D., and Lippard, S.J., *J. Amer. Chem. Soc.*, *90*, 3281 (1968).
213. Coucouvanis, D., and Lippard, S.J., *J. Amer. Chem. Soc.*, *91*, 307 (1969).
214. Coucouvanis, D., Lippard, S.J., and Zubieta, J.A., *J. Amer. Chem. Soc.*, *91*, 761 (1969).
215. Coucouvanis, D., Lippard, S.J., and Zubieta, J.A., *Inorg. Chem.*, *9*, 2775 (1970).
216. Coucouvanis, D., and Piltingsrud, D., *J. Amer. Chem. Soc.*, *95*, 5556 (1973).
217. Countryman, R., Collins, D.M., and Hoard, J.L., *J. Amer. Chem. Soc.*, *91*, 5166 (1969).
218. Cramer, R., *J. Amer. Chem. Soc.*, *86*, 217 (1964).
219. Dahl, L.F., and Wei, C.-H., *Inorg. Chem.*, *2*, 328 (1963).
220. Dahl, L.F., and Blount, J.F., *Inorg. Chem.*, *4*, 1373 (1965).
221. Dahl, L.F., Doedens, R.J., Hübel, W., and Nielsen, J., *J. Amer. Chem. Soc.*, *88*, 446 (1966).
222. Dahl, L.F., Costello, W.R., and King, R.B., *J. Amer. Chem. Soc.*, *90*, 5422 (1968).
223. Dahm, D.J., and Jacobson, R.A., *Chem. Commun.*, *1966*,

496.

224. Dahm, D.J., and Jacobson, R.A., *J. Amer. Chem. Soc.*, *90*, 5106 (1968).

225. Davey, G., and Stephens, F.S., *J. Chem. Soc. Dalton Trans.*, *1974*, 698.

226. Davidson, J.L., Green, M., Stone, F.G.A., and Welch, A.J., *J. Chem. Soc. Chem. Commun.*, *1975*, 286.

227. Davies, G.R., Mais, R.H.B., and Owston, P.G., *Chem. Commun.*, *1968*, 81.

228. Davies, G.R., Mais, R.H.B., Owston, P.G., and Thompson, D.T., *J. Chem. Soc. A*, *1968*, 1251.

229. Davies, G.R., Jarvis, J.A.J., Kilbourn, B.T., Mais, R.H.B., and Owston, P.G., *J. Chem. Soc. A*, *1970*, 1275.

230. Davies, J.E., and Gatehouse, B.M., *J. Chem. Soc. D, Chem. Commun.*, *1970*, 1166.

231. Davies, J.E., and Gatehouse, B.M., *Acta Crystallogr.*, *B 28*, 3641 (1972).

232. Davies, J.E., and Gatehouse, B.M., *Cryst. Struct. Commun.*, *1*, 115 (1972).

233. Davies, J.E., and Gatehouse, B.M., *Acta Crystallogr.*, *B 29*, 1934 (1973).

234. Davis, B.R., and Bernal, I., *J. Cryst. Mol. Struct.*, *2*, 107 (1972).

235. Davis, B.R., and Bernal, I., *J. Cryst. Mol. Struct.*, *2*, 261 (1972).

236. Davis, M.I., and Speed, C.S., *J. Organometal. Chem.*, *21*, 401 (1970).

237. Davis, R.E., *Chem. Commun.*, *1968*, 248.

238. Davis, R.E., *Chem. Commun.*, *1968*, 1218.

239. Davis, R.E., *Amer. Cryst. Assoc.*, *Winter Meeting 1969*, Abstracts, p.78.

240. Davis, R.E., *Amer. Cryst. Assoc.*, *Winter Meeting 1970*, New Orleans, Abstracts, p.45.

241. Davis, R.E., Cupper, G.L., and Simpson, H.D., *Amer. Cryst. Assoc.*, *Summer Meeting 1970*, Ottawa, Abstracts, p.80.

242. Davis, R.E., and Pettit, R., *J. Amer. Chem. Soc.*, *92*, 716 (1970).

243. Davis, R.E., and Simpson, H.D., *Amer. Cryst. Assoc.*, *Winter Meeting 1971*, Columbia (South Carolina), Abstracts, p.66.

244. Davis, R.E., Simpson, H.D., Grice, N., and Pettit, R., *J. Amer. Chem. Soc.*, *93*, 6688 (1971).

245. Davis, R.E., Barnett, B.L., Amiet, R.G., Merk, W., McKennis, J.S., and Pettit, R., *J. Amer. Chem. Soc.*, *96*, 7108 (1974).

246. Degrève, Y., Meunier-Piret, J., van Meerssche, M., and Piret, P., *Acta Crystallogr.*, *23*, 119 (1967).

247. Deppisch, B., *Acta Crystallogr., A 31,* S 137 (1975).
248. Dewar, M.J.S., *Bull. Soc. Chim. Fr., 1951,* C 71.
249. Deyrieux, R., and Peneloux, A., *Bull. Soc. Chim. Fr., 1969,* 2675.
250. Dickens, B., and Lipscomb, W.N., *J. Amer. Chem. Soc., 83,* 489 (1961).
251. Dickens, B., and Lipscomb, W.N., *J. Amer. Chem. Soc., 83,* 4862 (1961).
252. Dickens, B., and Lipscomb, W.N., *J. Chem. Phys., 37,* 2084 (1962).
253. Dighe, S.V., and Orchin, M., *J. Amer. Chem. Soc., 86,* 3895 (1964).
254. Dodge, R.P., and Schomaker, V., *Nature, 186,* 798 (1960).
255. Dodge, R.P., *J. Amer. Chem. Soc., 86,* 5429 (1964).
256. Dodge, R.P., and Schomaker, V., *Acta Crystallogr., 18,* 614 (1965).
257. Dodge, R.P., and Schomaker, V., *J. Organometal. Chem., 3,* 274 (1965).
258. Doedens, R.J., and Dahl, L.F., *J. Amer. Chem. Soc., 88,* 4847 (1966).
259. Doedens, R.J., *Chem. Commun., 1968,* 1271.
260. Doedens, R.J., *Inorg. Chem., 7,* 2323 (1968).
261. Doedens, R.J., *Inorg. Chem., 8,* 570 (1969).
262. Doedens, R.J., and Ibers, J.A., *Inorg. Chem., 8,* 2709 (1969).
263. Doedens, R.J., *Inorg. Chem., 9,* 429 (1970).
264. Doedens, R.J., and Little, R.G., *Amer. Cryst. Assoc., Summer Meeting 1971,* Ames (Iowa), Abstracts, p.75.
265. Donohue, J., and Caron, A., *Acta Crystallogr., 17,* 663 (1964).
266. Donohue, J., and Caron, A., *J. Phys. Chem., 70,* 603 (1966).
267. Drummond, J., and Wood, J.S., *J. Chem. Soc. D, Chem. Commun., 1969,* 1373.
268. Dunitz, J.D., Orgel, L.E., and Rich, A., *Acta Crystallogr., 9,* 373 (1956).
269. Eiland, P.F., and Pepinsky, R., *J. Amer. Chem. Soc., 74,* 4971 (1952).
270. Einstein, F.W.B., Cullen, W.R., and Trotter, J., *J. Amer. Chem. Soc., 88,* 5670 (1966).
271. Einstein, F.W.B., and Trotter, J., *J. Chem. Soc. A, 1967,* 824.
272. Einstein, F.W.B., and Svensson, A.-M., *J. Amer. Chem. Soc., 91,* 3663 (1969).
273. Einstein, F.W.B., Pilotti, A.-M., and Restivo, R., *Inorg. Chem., 10,* 1947 (1971).
274. Einstein, F.W.B., and Restivo, R., *Inorg. Chim. Acta,*

5, 501 (1971).
275. Einstein, F.W.B., and Jones, R.D.G., *J. Chem. Soc. Dalton Trans., 1972,* 442.
276. Einstein, F.W.B., and Jones, R.D.G., *J. Chem. Soc. Dalton Trans., 1972,* 2563.
277. Einstein, F.W.B., and Jones, R.D.G., *Inorg. Chem., 12,* 255 (1973).
278. Einstein, F.W.B., and Jones, R.D.G., *Inorg. Chem., 12,* 1690 (1973).
279. Einstein, F.W.B., and MacGregor, A.C., *J. Chem. Soc. Dalton Trans., 1974,* 778.
280. Eiss, R., *Inorg. Chem., 9,* 1650 (1970).
281. Elder, M., and Hall, D., *Inorg. Chem., 8,* 1424 (1969).
282. Elder, M., *Inorg. Chem., 8,* 2703 (1969).
283. Elder, M., and Hutcheon, W.L., *J. Chem. Soc. Dalton Trans., 1972,* 175.
284. Eller, P.G., Breece, J.L., and Bertrand, J.A., *Amer. Cryst. Assoc., Winter Meeting 1973,* Gainesville (Florida), Abstracts, p.70.
285. Elmes, P.S., Leverett, P., and West, B.O., *J. Chem. Soc. D, Chem. Commun., 1971,* 747.
286. Epstein, E.F., and Bernal, I., *Amer. Cryst. Assoc., Winter Meeting 1969,* Abstracts, p.79.
287. Epstein, E.F., and Dahl, L.F., *J. Amer. Chem. Soc., 92,* 493 (1970).
288. Epstein, E.F., and Dahl, L.F., *J. Amer. Chem. Soc., 92,* 502 (1970).
289. Epstein, E.F., Bernal, I., and Balch, A.L., *J. Chem. Soc. D, Chem. Commun., 1970,* 136.
290. Epstein, E.F., and Bernal, I., *Amer. Cryst. Assoc., Winter Meeting 1971,* Columbia (South Carolina), Abstracts, p.61.
291. Ferguson, G., Hannaway, C., and Islam, K.M.S., *Chem. Commun., 1968,* 1165.
292. Fischer, D.W., *Acta Crystallogr., 17,* 619 (1964).
293. Fischer, E.O., Kiener, V., Bunbury, D.St.P., Frank, E., Lindley, P.F., and Mills, O.S., *Chem. Commun., 1968,* 1378.
294. Fleischer, E.B., Miller, C.K., and Webb, L.E., *J. Amer. Chem. Soc., 86,* 2342 (1964).
295. Fleischer, E.B., Stone, A.L., Dewar, R.B.K., Wright, J.D., Keller, C.E., and Pettit, R., *J. Amer. Chem. Soc., 88,* 3158 (1966).
296. Fleischer, E.B., and Hawkinson, S.W., *Acta Crystallogr., 22,* 376 (1967).
297. Fleischer, E., and Hawkinson, S., *J. Amer. Chem. Soc., 89,* 720 (1967).
298. Fleischer, E.B., and Srivastava, T.S., *J. Amer. Chem.*

Soc., *91*, 2403 (1969).

299. Fleischer, E.B., Gebala, A.E., Swift, D.R., and Tasker, P.A., *Inorg. Chem.*, *11*, 2775 (1972).

300. Flippen, J.L., *Inorg. Chem.*, *13*, 1054 (1974).

301. Forrester, A.R., Hepburn, S.P., Dunlop, R.S., and Mills, H.H., *J. Chem. Soc. D, Chem. Commun.*, *1969*, 698.

302. Foxman, B.M., *J. Chem. Soc. Chem. Commun*, *1975*, 221.

303. Freeman, H.C., Milburn, G.H.W., Nockolds, C.E., Hemmerich, P., and Knauer, K.H., *J. Chem. Soc. D, Chem. Commun.*, *1969*, 55.

304. Frenz, B.A., Troup, J.M., and Cotton, F.A., *Amer. Cryst. Assoc.*, *Winter Meeting 1973*, Gainesville (Florida), Abstracts, p.90.

305. Fritchie, C.J., Jr., and Wade, T.D., *Amer. Cryst. Assoc.*, *Summer Meeting 1974*, Pennsylvania State University, Abstracts, p.239.

306. Gall, R.S., Chu, C.T.-W., and Dahl, L.F., *J. Amer. Chem. Soc.*, *96*, 4019 (1974).

307. Gatehouse, B.M., *J. Chem. Soc. D, Chem. Commun.*, *1969*, 948.

308. Gerloch, M., and Mabbs, F.E., *J. Chem. Soc. A*, *1967*, 1598.

309. Gerloch, M., and Mabbs, F.E., *J. Chem. Soc. A*, *1967*, 1900.

310. Gerloch, M., McKenzie, E.D., and Towl, A.D.C., *Nature*, *220*, 906 (1968).

311. Gerloch, M., McKenzie, E.D., and Towl, A.D.C., *J. Chem. Soc. A*, *1969*, 2850.

312. Gerst, K., and Nordman, C.E., *Amer. Cryst. Assoc.*, *Summer Meeting 1974*, Pennsylvania State University, Abstracts, p.225.

313. Gibbons, C.S., and Trotter, J., *J. Chem. Soc. A*, *1971*, 2659.

314. Gieren, A., and Hoppe, W., *Acta Crystallogr.*, *B 28*, 2766 (1972).

315. Gilmore, C.J., and Woodward, P., *J. Chem. Soc. D, Chem. Commun*, *1970*, 1463.

316. Gilmore, C.J., and Woodward, P., *J. Chem. Soc. Dalton Trans.*, *1972*, 1387.

317. Ginderow, D., *Acta Crystallogr.*, *B 30*, 2798 (1974).

318. Ginsberg, A.P., and Robin, M.B., *Inorg. Chem.*, *2*, 817 (1963).

319. Gitany, R., Paul, I.C., Acton, N., and Katz, T.J., *Tetrahedron Lett.*, *1970*, 2723.

320. Goddard, R., Howard, J., and Woodward, P., *J. Chem. Soc. Dalton Trans.*, *1974*, 2025.

321. Goedken, V.L., Molin-Case, J., and Christoph, G.G.,

Inorg. Chem., 12, 2894 (1973).

322. Goedken, V.L., Molin-Case, J., and Whang, Y.-a., *J. Chem. Soc. Chem. Commun., 1973,* 337.

323. Goedken, V.L., Peng, S.-M., and Park, Y.-a., *J. Amer. Chem. Soc., 96,* 284 (1974).

324. Goedken, V.L., Park, Y.-a., Peng, S.-M., and Molin Norris, J., *J. Amer. Chem. Soc., 96,* 7693 (1974).

325. Goedken, V.L., and Peng, S.-M., *J. Amer. Chem. Soc., 96,* 7826 (1974).

326. Greaves, E.O., Knox, G.R., Pauson, P.L., Toma, S., Sim, G.A., and Woodhouse, D.I., *J. Chem. Soc. Chem. Commun., 1974,* 257.

327. Green, M., Tolson, S., Weaver, J., Wood, D.C., and Woodward, P., *J. Chem. Soc. D, Chem. Commun., 1971,* 222.

328. Green, M., Hughes, R.P., and Welch, A.J., *J. Chem. Soc., Chem. Commun., 1975,* 487.

329. Green, M.L.H., and Nagy, P.L.I., *Advan. Organometal. Chem., 2,* 325 (1964).

330. Green, M.L.H., Ariyaratne, J.K.P., Bjerrum, A.M., Ishaq, M., and Prout, C.K., *Chem. Commun., 1967,* 430.

331. Greene, P.T., and Bryan, R.F., *Inorg. Chem., 9,* 1464 (1970).

332. Greene, P.T., and Bryan, R.F., *J. Chem. Soc. A, 1970,* 1696.

333. Greene, P.T., and Bryan, R.F., *J. Chem. Soc. A, 1970,* 2261.

334. Gress, M.E., and Jacobson, R.A., *Inorg. Chem., 12,* 1746 (1973).

335. Guggenberger, L.J., Titus, D.D., Flood, M.T., Marsh, R.E., Orio, A.A., and Gray, H.B., *J. Amer. Chem. Soc., 94,* 1135 (1972).

336. Guilard, R., and Dusausoy, Y., *J. Organometal. Chem., 77,* 393 (1974).

337. Gyepes, E., and Hanic, F., *Cryst. Struct. Commun., 4,* 229 (1975).

338. Hackert, M.L., and Jacobson, R.A., *Acta Crystallogr., B 27,* 1658 (1971).

339. Hall, L.H., and Brown, G.M., *Acta Crystallogr., B 27,* 81 (1971).

340. Haines, R.J., Mason, R., Zubieta, J.A., and Nolte, C.R., *J. Chem. Soc. Chem. Commun., 1972,* 990.

341. Hamilton, W.C., and Bernal, I., *Inorg. Chem., 6,* 2003 (1967).

342. Hamor, T.A., and Watkin, D.J., *J. Chem. Soc. D, Chem. Commun., 1969,* 440.

343. Hanic, F., Ŝevĉik, J., and McGandy, E.L., *Chem. Zvesti, 24,* 81 (1970).

344. Hansen, P.J., and Jacobson, R.A., *J. Organometal. Chem.*, *6*, 389 (1966).
345. Hanson, A.W., *Acta Crystallogr.*, *15*, 930 (1962).
346. Hardy, A.D.U., and Sim, G.A., *J. Chem. Soc. Dalton Trans.*, *1972*, 2305.
347. Harrison, W., and Trotter, J., *Amer. Cryst. Assoc., Summer Meeting 1970*, Ottawa, Abstracts, p.85.
348. Harrison, W., and Trotter, J., *J. Chem. Soc. A*, *1971*, 1542.
349. Haymore, B.L., and Ibers, J.A., *Inorg. Chem.*, *14*, 1369 (1975).
350. Healy, P.C., and White, A.H., *J. Chem. Soc. Dalton Trans.*, *1972*, 1163.
351. Healy, P.C., White, A.H., and Hoskins, B.F., *J. Chem. Soc. Dalton Trans.*, *1972*, 1369.
352. Healy, P.C., and Sinn, E., *Inorg. Chem.*, *14*, 109 (1975).
353. Heimbach, P., and Traunmüller, R., *Justus Liebigs Ann. Chem.*, *727*, 208 (1969).
354. van der Helm, D., Merritt, L.L., Jr., Degeilh, R., and MacGillavry, C.H., *Acta Crystallogr.*, *18*, 355 (1965).
355. Henslee, W., and Davis, R.E., *Cryst. Struct. Commun.*, *1*, 403 (1972).
356. Herbstein, F.H., and Reisner, M.G., *J. Chem. Soc. Chem. Commun.*, *1972*, 1077.
357. Hiraishi, J., Finseth, D., and Miller, F.A., *Spectrochim. Acta*, *25 A*, 1657 (1969).
358. Hirotsu, K., Higuchi, T., and Shimada, A., *Bull. Chem. Soc. Jap.*, *41*, 1557 (1968).
359. Hitchcock, P.B., and Mason, R., *Chem. Commun.*, *1967*, 242.
360. Hoard, J.L., Kennard, C.H.L., and Smith, G.S., *Inorg. Chem.*, *2*, 1316 (1963).
361. Hoard, J.L., Hamor, M.J., Hamor, T.A., and Caughey, W.S., *J. Amer. Chem. Soc.*, *87*, 2312 (1965).
362. Hoard, J.L., Cohen, G.H., and Glick, M.D., *J. Amer. Chem. Soc.*, *89*, 1992 (1967).
363. Hock, A.A., and Mills, O.S., *Proc. Chem. Soc.*, *1958*, 233.
364. Hock, A.A., and Mills, O.S., *Acta Crystallogr.*, *14*, 139 (1961).
365. Hoffman, A.B., Collins, D.M., Day, V.W., Fleischer, E.B., Srivastava, T.S., and Hoard, J.L., *J. Amer. Chem. Soc.*, *94*, 3620 (1972).
366. Hollander, F.J., and Coucouvanis, D., *Inorg. Chem.*, *13*, 2381 (1974).
367. Hollander, F.J., Pedelty, R., and Coucouvanis, D.,

J. Amer. Chem. Soc., 96, 4032 (1974).

368. Hollander, O., Clayton, W.R., and Shore, S.G., *J. Chem. Soc. Chem. Commun., 1974,* 604.

369. Holt, E.M., Holt, S.L., Tucker, W.F., Asplund, R.O., and Watson, K.J., *J. Amer. Chem. Soc., 96,* 2621 (1974).

370. Horspool, W.M., Iball, J., Rafferty, M., and Scrimgeour, S.N., *J. Chem. Soc. Dalton Trans., 1974,* 401.

371. Holloway, C.E., Hulley, G., Johnson, B.F.G., and Lewis, J., *J. Chem. Soc. A, 1969,* 53.

372. Hoppe, W., Brodherr, N., Englmeier, Hp., Gassmann, J., Gieren, A., Hechtfischer, S., Preuss, L., Roehrl, M., Schaeffer, J., Schmidt, E., Steigemann, W., and Zechmeister, K., *Pure Appl. Chem., 18,* 465 (1969).

373. Hoskins, B.F., Martin, R.L., and White, A.H., *Nature, 211,* 627 (1966).

374. Hoskins, B.F., and Kelly, B.P., *Chem. Commun., 1968,* 1517.

375. Hoskins, B.F., and White, A.H., *J. Chem. Soc. A, 1970,* 1668.

376. Hoskins, B.F., and Kelly, B.P., *J. Chem. Soc. D, Chem. Commun., 1970,* 45.

377. Hu, W.-j., and Lippard, S.J., *J. Amer. Chem. Soc., 96,* 2366 (1974).

378. Huie, B.T., Knobler, C.B., and Kaesz, H.D., *J. Chem. Soc. Chem. Commun., 1975,* 684.

379. Hulme, R., and Powell, H.M., *J. Chem. Soc., 1957,* 719.

380. Hursthouse, M.B., Massey, A.G., Tomlinson, A.J., and Urch, D.S., *J. Organometal. Chem., 21,* P 51 (1970).

381. Huttner, G., and Gartzke, W., *Chem. Ber., 105,* 2714 (1972).

382. Huttner, G., and Regler, D., *Chem. Ber., 105,* 2726 (1972).

383. Huttner, G., and Regler, D., *Chem. Ber., 105,* 3936 (1972).

384. Huttner, G., and Bejenke, V., *Chem. Ber., 107,* 156 (1974).

385. Huttner, G., and Gartzke, W., *Chem. Ber., 107,* 3786 (1974).

386. Iball, J., and Morgan, C.H., *Acta Crystallogr., 23,* 239 (1967).

387. Immirzi, A., *J. Organometal. Chem., 76,* 65 (1974).

388. van Ingen Schenau, A.D., Verschoor, G.C., and Romers, C., *2nd Europ. Crystallogr. Meeting 1974,* Keszthely (Hungary), Abstracts, p.370.

389. Irngartinger, H., *2nd Europ. Crystallogr. Meeting 1974,* Keszthely (Hungary), Abstracts, p.372.

390. Janse-van Vuuren, P., Fletterick, R.J., Meinwald, J., and Hughes, R.E., *J. Chem. Soc. D, Chem. Commun.*, *1970*, 883.
391. Janse Van Vuuren, P., Fletterick, R.J., Meinwald, J., and Hughes, R.E., *J. Amer. Chem. Soc.*, *93*, 4394 (1971).
392. Jarvis, J.A.J., Job, B.E., Kilbourn, B.T., Mais, R.H.B., Owston, P.G., and Todd, P.F., *Chem. Commun.*, *1967*, 1149.
393. Jarvis, J.A.J., Mais, R.H.B., Owston, P.G., and Thompson, D.T., *J. Chem. Soc. A*, *1968*, 622.
394. Jeannin, S., Jeannin, Y., and Lavigne, G., *J. Organometal. Chem.*, *40*, 187 (1972).
395. Jeffreys, J.A.D., Willis, C.M., Robertson, I.C., Ferguson, G., and Sime, J.G., *J. Chem. Soc. Dalton Trans.*, *1973*, 749.
396. Johnson, S.M., and Paul, I.C., *J. Chem. Soc. B*, *1970*, 1783.
397. Johnston, D.L., Rohrbaugh, W.L., and DeW. Horrocks, W., Jr., *Inorg. Chem.*, *10*, 1474 (1971).
398. Jones, N.D., Marsh, R.E., and Richards, J.H., *Acta Crystallogr.*, *19*, 330 (1965).
399. Kaluski, Z.L., Avoyan, R.L., and Struchkov, Yu.T., *Zh. Strukt. Khim.*, *3*, 599 (1962); *J. Struct. Chem.*, *3*, 573 (1962).
400. Kaluski, Z.L., Struchkov, Yu.T., and Avoyan, R.L., *Zh. Strukt. Khim.*, *5*, 743 (1964); *J. Struct. Chem.*, *5*, 683 (1964).
401. Kaluski, Z.L., and Struchkov, Yu.T., *Zh. Strukt. Khim.*, *6*, 104 (1965); *J. Struct. Chem.*, *6*, 90 (1965).
402. Kaluski, Z.L., and Struchkov, Yu.T., *Zh. Strukt. Khim.*, *6*, 316 (1965); *J. Struct. Chem.*, *6*, 296 (1965).
403. Kaluski, Z.L., and Struchkov, Yu.T., *Zh. Strukt. Khim.*, *6*, 475 (1965); *J. Struct. Chem.*, *6*, 456 (1965).
404. Kaluski, Z.L., and Struchkov, Yu.T., *Zh. Strukt. Khim.*, *6*, 745 (1965); *J. Struct. Chem.*, *6*, 705 (1965).
405. Kaluski, Z.L., *Acta Crystallogr.*, *21*, A 119 (1966).
406. Kaluski, Z., and Struchkov, Yu.T., *Bull. Acad. Pol. Sci. Ser. Sci. Chim.*, *14*, 719 (1966).
407. Kaluski, Z.L., Avoyan, R.L., and Struchkov, Yu.T., *Zh. Strukt. Khim.*, *7*, 131 (1966); *J. Struct. Chem.*, *7*, 130 (1966).
408. Kaluski, Z.L., and Struchkov, Yu.T., *Zh. Strukt. Khim.*, *7*, 283 (1966); *J. Struct. Chem.*, *7*, 278 (1966).
409. Kaluski, Z., and Struchkov, Yu.T., *Bull. Acad. Pol. Sci. Ser. Sci. Chim.*, *16*, 557 (1968).
410. Kaluski, Z.L., Gusev, A.I., Kalinin, A.E., and Struchkov,, Yu.T., *Zh. Strukt. Khim.*, *13*, 950 (1972);

J. Struct. Chem., *13*, 888 (1972).
411. Karlin, K.D., Lewis, D.L., Rabinowitz, H.N., and Lippard, S.J., *J. Amer. Chem. Soc.*, *96*, 6519 (1974).
412. Kennard, C.H.L., *Inorg. Chim. Acta.*, *1*, 347 (1967).
413. Kettle, S.F.A., and Mason, R., *J. Organometal. Chem.*, *5*, 573 (1966).
414. Kilbourn, B.T., and Mais, R.H.B., *Chem. Commun.*, *1968*, 1507.
415. Kilbourn, B.T., Raeburn, U.A., and Thompson, D.T., *J. Chem. Soc. A*, *1969*, 1906.
416. Kim, N.E., Nelson, N.J., and Shriver, D.F., *Inorg. Chim. Acta*, *7*, 393 (1973).
417. Kimball, G.E., *J. Chem. Phys.*, *8*, 188 (1940).
418. King, G.S.D., *Acta Crystallogr.*, *15*, 243 (1962).
419. King, T.J., Logan, N., Morris, A., and Wallwork, S.C., *J. Chem. Soc. D, Chem. Commun.*, *1971*, 554.
420. Kirchner, R.M., and Ibers, J.A., *J. Organometal. Chem.*, *82*, 243 (1974).
421. Klanderman, K.A., *Dissert. Abstr.*, *25*, 6253 (1965).
422. Kobayashi, T., Kurokawa, F., Ashida, T., Uyeda, N., and Suito, E., *J. Chem. Soc. D, Chem. Commun.*, *1971*, 1631.
423. Koch, S., Tang, S.C., Holm, R.H., Frankel, R.B., and Ibers, J.A., *J. Amer. Chem. Soc.*, *97*, 916 (1975).
424. Koenig, D.F., *Acta Crystallogr.*, *18*, 663 (1965).
425. König, E., and Watson, K.J., *Chem. Phys. Lett.*, *6*, 457 (1970).
426. Kurahashi, M., Kawase, A., Hirotsu, K., Fukuyo, M., and Shimada, A., *Bull. Chem. Soc. Jap.*, *45*, 1940 (1972).
427. Krüger, C., *J. Organometal. Chem.*, *22*, 697 (1970).
428. Krüger, C., and Tsay, Y.-H., *Angew. Chem.*, *83*, 250 (1971); *Angew. Chem. Int. Ed. Engl.*, *10*, 261 (1971).
429. Krüger, C., and Tsay, Y.-H., *J. Organometal. Chem.*, *33*, 59 (1971).
430. Krüger, C., Tsay, Y.-H., Grevels, F.W., and Koerner von Gustorf, E., *Israel J. Chem.*, *10*, 201 (1972).
431. Krüger, C., and Tsay, Y.-H., *1st Europ. Crystallogr. Meeting 1973*, Bordeaux, Abstracts, Group A_6.
432. Krüger, C., *Chem. Ber.*, *106*, 3230 (1973).
433. Krüger, C., and Kisch, H., *J. Chem. Soc. Chem. Commun.*, *1975*, 65.
434. Krukonis, A.P., Silverman, J., and Yannoni, N.F., *Acta Crystallogr.*, *B 28*, 987 (1972).
435. Kuz'mina, L.G., Bokii, N.G., Struchkov, Yu.T., Arutyunyan, A.V., Rybin, L.V., and Rybinskaya, M.I., *Zh. Strukt. Khim.*, *12*, 875 (1971); *J. Struct. Chem.*, *12*, 801 (1971).
436. Kuz'mina, L.G., Struchkov, Yu.T., and Nekhaev, A.I.,

Zh. Strukt. Khim., *13*, 1115 (1972); *J. Struct. Chem.*, *13*, 1033 (1972).

437. Kuz'min, V.S., Zol'nikova, G.P., Struchkov, Yu.T., and Kritskaya, I.I., *Zh. Strukt. Khim.*, *15*, 162 (1974); *J. Struct. Chem.*, *15*, 153 (1974).

438. Laing, M.B., and Trueblood, K.N., *Acta Crystallogr.*, *19*, 373 (1965).

439. Lauher, J.W., and Ibers, J.A., *Inorg. Chem.*, *14*, 348 (1975).

440. Lecomte, C., Dusausoy, Y., Protas, J., Moise, C., and Tirouflet, J., *Acta Crystallogr.*, *B 29*, 488 (1973).

441. Lecomte, C., Dusausoy, Y., Protas, J., and Moise, C., *Acta Crystallogr.*, *B 29*, 1127 (1973).

442. Lecomte, C., Dusausoy, Y., Protas, J., Gautheron, B., and Broussier, R., *Acta Crystallogr.*, *B 29*, 1504 (1973).

443. Legendre, J.-J., Girard, C., and Huber, M., *Bull. Soc. Chim. Fr.*, *1971*, 1998.

444. Leipoldt, J.G., and Coppens, P., *Inorg. Chem.*, *12*, 2269 (1973).

445. Lewis, D.F., Lippard, S.J., and Zubieta, J.A., *Inorg. Chem.*, *11*, 823 (1972).

446. Lewis, J., and Nyholm, R.S., *23rd Int. Congr. Pure Appl. Chem. 1971*, Butterworth, London, 1971, Vol. 6, p.61.

447. Lind, M.D., Hoard, J.L., Hamor, M.J., and Hamor, T.A., *Inorg. Chem.*, *3*, 34 (1964).

448. Lindley, P.F., and Woodward, P., *J. Chem. Soc. A*, *1967*, 382.

449. Lindley, P.F., and Mills, O.S., *J. Chem. Soc. A*, *1969*, 1279.

450. Lindley, P.F., and Mills, O.S., *J. Chem. Soc. A*, *1970*, 38.

451. Lindley, P.F., and Smith, A.W., *J. Chem. Soc. D, Chem. Commun.*, *1970*, 1355.

452. Lindner, H.J., and Göttlicher, S., *Acta Crystallogr.*, *B 25*, 832 (1969).

453. Lindqvist, I., and Rosenstein, R., *Acta Chem. Scand.*, *14*, 1228 (1960).

454. Lingafelter, E.C., Bailey, M.F., Howe, N.L., Kirchner, R.M., Mealli, C., and Torre, L.P., *Amer. Cryst. Assoc., Summer Meeting 1970*, Ottawa, Abstracts, p.81.

455. Lingafelter, E.C., and Dunaj-Jurĉo, M., *Acta Crystallogr.*, *A 28*, S 79 (1972).

456. Lippard, S.J., Schugar, H., and Walling, C., *Inorg. Chem.*, *6*, 1825 (1967).

457. Lippard, S.J., and Martin, G., *J. Amer. Chem. Soc.*, *92*, 7291 (1970).

458. Luxmoore, A.R., and Truter, M.R., *Proc. Chem. Soc., 1961,* 466.
459. Luxmoore, A.R., and Truter, M.R., *Acta Crystallogr., 15,* 1117 (1962).
460. Maglio, G., Musco, A., Palumbo, R., and Sirigu, A., *J. Chem. Soc. D, Chem. Commun., 1971,* 100.
461. Manojlović-Muir, L., Muir, K.W., and Ibers, J.A., *Inorg. Chem., 9,* 447 (1970).
462. Margulis, T.N., Schiff, L., and Rosenblum, M., *J. Amer. Chem. Soc., 87,* 3269 (1965).
463. Martin, R.L., Rohde, N.M., Robertson, G.B., and Taylor, D., *J. Amer. Chem. Soc., 96,* 3647 (1974).
464. Mason, R., McKenzie, E.D., Robertson, G.B., and Rusholme, G.A., *Chem. Commun., 1968,* 1673.
465. Mason, R., *23rd. Int. Congr. Pure Appl. Chem. 1971,* Butterworth, London 1971, Vol. 6., p.31.
466. Mason, R., and Robertson, G.B., *J. Chem. Soc. A, 1970,* 1229.
467. Mason, R., Zubieta, J., Hsieh, A.T.T., Knight, J., and Mays, M.J., *J. Chem. Soc. Chem. Commun., 1972,* 250.
468. Mason, R., and Zubieta, J.A., *J. Organometal. Chem., 66,* 279 (1974).
469. Mason, R., and Zubieta, J.A., *J. Organometal. Chem., 66,* 289 (1974).
470. Mathew, M., and Palenik, G.J., *Inorg. Chem., 11,* 2809 (1972).
471. Mayerle, J.J., Denmark, S.E., DePamphilis, B.V., Ibers, J.A., and Holm, R.H., *J. Amer. Chem. Soc., 97,* 1032 (1975).
472. Macdonald, A.C., and Trotter, J., *Acta Crystallogr., 17,* 872 (1964).
473. McArdle, P., Manning, A.R., and Stevens, F.S., *J. Chem. Soc. D, Chem. Commun., 1969,* 1310.
474. McDonald, W.S., Moss, R.J., Raper, G., Shaw, B.L. Greatrex, R., and Greenwood, N.N., *J. Chem. Soc. D, Chem. Commun., 1969,* 1295.
475. McKechnie, J.S., and Paul, I.C., *J. Amer. Chem. Soc., 88,* 5927 (1966).
476. McKechnie, J.S., Bersted, B., Paul, I.C., and Watts, W.E., *J. Organometal. Chem., 8,* P 29 (1967).
477. McKechnie, J.S., Maier, C.A., Bersted, B., and Paul, I.C., *J. Chem. Soc. Perkin Trans. II, 1973,* 138.
478. McWeeny, R., Mason, R., and Towl, A.D.C., *Discuss. Faraday Soc., 47,* 20 (1969).
479. Meakin, P., Guggenberger, L.J., Jesson, J.P., Gerlach, D.H., Tebbe, F.N., Peet, W.G., and Muetterties, E.L., *J. Amer. Chem. Soc., 92,* 3482 (1970).
480. Mealli, C., and Lingafelter, E.C., *J. Chem. Soc. D,*

Chem. Commun., 1970, 885.

481. Melson, G.A., Stokely, P.F., and Bryan, R.F., *J. Chem. Soc. A, 1970,* 2247.

482. Meunier-Piret, J., Piret, P., and van Meerssche, M., *Acta Crystallogr., 19,* 85 (1965).

483. Miller, J., Balch, A.L., and Enemark, J.H., *J. Amer. Chem. Soc., 93,* 4613 (1971).

484. Mills, O.S., *Acta Crystallogr., 11,* 620 (1958).

485. Mills, O.S., and Robinson, G., *Proc. Chem. Soc., 1960,* 421.

486. Mills, O.S., and Robinson, G., *Acta Crystallogr., 16,* 758 (1963).

487. Mills, O.S., and Redhouse, A.D., *Chem. Commun., 1966,* 444.

488. Mills, O.S., and Redhouse, A.D., *J. Chem. Soc. A, 1968,* 1282.

489. Minasyan, M.Kh., Struchkov, Yu.T., Kritskaya, I.I., and Avoyan, R.L., *Zh. Strukt. Khim., 7,* 903 (1966); *J. Struct. Chem., 7,* 840 (1966).

490. Minasyants, M.Kh., and Struchkov, Yu.T., *Zh. Strukt. Khim., 9,* 665 (1968); *J. Struct. Chem., 9,* 577 (1968).

491. Minasyants, M.Kh., Andrianov, V.G., and Struchkov, Yu.T., *Zh. Strukt. Khim., 9,* 1055 (1968); *J. Struct. Chem., 9,* 939 (1968).

492. Molin Case, J.A., *Dissert. Abstr., B 28,* 2786 (1968).

493. Moriarty, R.M., Chen, K.-N., Yeh, C.-L., Flippen, J.L., and Karle, J., *J. Amer. Chem. Soc., 94,* 8944 (1972).

494. Moriarty, R.M., Chen, K.-N., Churchill, M.R., and Chang, S.W.-Y., *J. Amer. Chem. Soc., 96,* 3661 (1974).

495. Murahashi, S.-I., Mizoguchi, T., Hosokawa, T., Moritani, I., Kai, Y., Kohara, M., Yasuoka, N., and Kasai, N., *J. Chem. Soc. Chem. Commun., 1974,* 563.

496. Naik, D.V., and Palenik, G.J., *Chem. Phys. Lett., 24,* 260 (1974).

497. Nassimbeni, L.R., and Thackeray, M.M., *Inorg. Nucl. Chem. Lett., 9,* 539 (1973).

498. Nelson, N.J., Kime, N.E., and Shriver, D.F., *J. Amer. Chem. Soc., 91,* 5173 (1969).

499. Nesmeyanov, A.N., Astakhova, I.S., Zol'nikova, G.P., Kritskaya, I.I., and Struchkov, Yu.T., *J. Chem. Soc. D, Chem. Commun., 1970,* 85.

500. Nesmeyanov, A.N., Rybinskaya, M.I., Rybin, L.V., Arutyunyan, A.V., Kuz'mina, L.G., and Struchkov, Yu.T., *J. Organometal. Chem., 73,* 365 (1974).

501. Nesterova, Ya.M., Polynova, T.N., Martynenko, L.I., and Pechurova, N.I., *Zh. Strukt. Khim., 12,* 1110 (1971); *J. Struct. Chem., 12,* 1028 (1971).

502. Neuman, M.A., Trinh-Toan, and Dahl, L.F., *J. Amer.*

Chem. Soc., *94*, 3383 (1972).
503. Nicholas, K., Bray, L.S., Davis, R.E., and Pettit, R., *J. Chem. Soc. D, Chem. Commun.*, *1971*, 608.
504. Norrestam, R., and Stensland, B., *Acta Crystallogr.*, *A 28*, S 39 (1972).
505. Novozhilova, N.V., Polynova, T.N., Porai-Koshits, M.A., and Martynenko, L.I., *Zh. Strukt. Khim.*, *15*, 717 (1974); *J. Struct. Chem.*, *15*, 621 (1974).
506. Novozhilova, N.V., Polynova, T.N., Porai-Koshits, M.A., Pechurova, N.I., Martynenko, L.I., and Ali-Khadi, *Zh. Strukt. Khim.*, *14*, 745 (1973); *J. Struct. Chem.*, *14*, 694 (1973).
507. O'Connor, J.E., and Corey, E.R., *Inorg. Chem.*, *6*, 968 (1967).
508. O'Connor, J.E., and Corey, E.R., *J. Amer. Chem. Soc.*, *89*, 3930 (1967).
509. O'Connor, T., Carty, A.J., Mathew, M., and Palenik, G.J., *J. Organometal. Chem.*, *38*, C 15 (1972).
510. Oliver, J.D., and Davis, R.E., *Amer. Cryst. Assoc.*, *Winter Meeting 1971*, Columbia (South Carolina), Abstracts, p.61.
511. Orchin, M., and Schmidt, P.J., *Inorg. Chim. Acta Rev.*, *2*, 123 (1968).
512. Paik, H.N., Carty, A.J., Mathew, M., and Palenik, G.J., *J. Chem. Soc. Chem. Commun.*, *1974*, 946.
513. Paik, H.N., Carty, A.J., Dymock, K., and Palenik, G.J., *J. Organometal. Chem.*, *70*, C 17 (1974).
514. Palenik, G.J., *Amer. Cryst. Assoc.*, *Winter Meeting 1967*, Abstracts, p.62.
515. Palenik, G.J., *Inorg. Chem.*, *8*, 2744 (1969).
516. Palenik, G.J., *Inorg. Chem.*, *9*, 2424 (1970).
517. Patel, H.A., Carty, A.J., Mathew, M., and Palenik, G.J., *J. Chem. Soc. Chem. Commun.*, *1972*, 810.
518. Patel, H.A., Fischer, R.G., Carty, A.J., Naik, D.V., and Palenik, G.J., *J. Organometal. Chem.*, *60*, C 49 (1973).
519. Paquette, L.A., Ley, S.V., Broadhurst, M.J., Truesdell, D., Fayos, J., and Clardy, J., *Tetrahedron Lett.*, *1973*, 2943.
520. Paul, I.C., and Sim, G.A., *J. Chem. Soc.*, *1965*, 1097.
521. Paul, I.C., *Chem. Commun.*, *1966*, 377.
522. Paul, I.C., Johnson, S.M., Paquette, L.A., Barrett, J.H., and Haluska, R.J., *J. Amer. Chem. Soc.*, *90*, 5023 (1968).
523. Pedone, C., and Sirigu, A., *Acta Crystallogr.*, *23*, 759 (1967).
524. Pedone, C., and Sirigu, A., *Inorg. Chem.*, *7*, 2614 (1968).

525. Pettersen, R.C., *Dissert. Abstr., B 27,* 3894 (1967).
526. Pettersen, R.C., Cihonski, J.L., Young, F.R., and Levenson, R.A., *J. Chem. Soc. Chem. Commun., 1975,* 370.
527. Pfluger, C.E., *Acta Crystallogr., A 31,* S 136 (1975).
528. Piciulo, P.L., Rupprecht, G., and Scheidt, W.R., *J. Amer. Chem. Soc., 96,* 5293 (1974).
529. Pierpont, C.G., and Eisenberg, R., *Inorg. Chem., 11,* 828 (1972).
530. Pierrot, M., Kern, R., and Weiss, R., *Acta Crystallogr., 20,* 425 (1966).
531. Piret, P., Meunier-Piret, J., van Meerssche, M., and King, G.S.D., *Acta Crystallogr., 19,* 78 (1965).
532. Piron, J., Piret, P., and van Meerssche, M., *Bull. Soc. Chim. Belg., 76,* 505 (1967).
533. Piron, J., Piret, P., Meunier-Piret, J., and van Meerssche, M., *Bull. Soc. Chim. Belg., 78,* 121 (1969).
534. Poling, M., and van der Helm, D., *Amer. Cryst. Assoc., Spring Meeting 1974,* Berkeley (Calif.), Abstracts, p.111.
535. Polynova, T.N., Bokii, N.G., and Porai-Koshits, M.A., *Zh. Strukt. Khim., 6,* 878 (1965); *J. Struct. Chem., 6,* 841 (1965).
536. Powell, H.M., and Ewens, R.V.G., *J. Chem. Soc., 1939,* 286.
537. Powell, H.M., and Bartindale, G.W.R., *J. Chem. Soc., 1945,* 799.
538. Prout, C.K., and Wiseman, T.J., *J. Chem. Soc., 1964,* 497.
539. Que, L., Jr., Bobrik, M.A., Ibers, J.A., and Holm, R.H., *J. Amer. Chem. Soc., 96,* 4168 (1974).
540. Rae, A.I.M., *Chem. Commun., 1967,* 1245.
541. Radonovich, L.J., Bloom, A., and Hoard, J.L., *J. Amer. Chem. Soc., 94,* 2073 (1972).
542. Raper, G., and McDonald, W.S., *J. Chem. Soc. A, 1971,* 3430.
543. Reeves, L.W., *Can. J. Chem., 38,* 736 (1960).
544. Reid, K.I.G., and Paul, I.C., *J. Chem. Soc. D, Chem. Commun., 1970,* 1106.
545. Restivo, R., and Einstein, F.W.B., *Amer. Cryst. Assoc., Summer Meeting 1970,* Ottawa, Abstracts, p.79.
546. Restivo, R., and Bryan, R.F., *J. Chem. Soc. A, 1971,* 3364.
547. Ricci, J.S., Eggers, C.A., and Bernal, I., *Inorg. Chim. Acta, 6,* 97 (1972).
548. Riley, P.E., and Davis, R.E., *Amer. Cryst. Assoc., 25th Anniversary Meeting 1975,* Charlottesville (Virginia) Abstracts, p.11.

549. Roberts, P.J., Penfold, B.R., and Trotter, J., *Inorg. Chem., 9,* 2137 (1970).
550. Robinson, W.T., *Amer. Cryst. Assoc., Winter Meeting 1973,* Gainesville (Florida), Abstracts, p.70.
551. Robinson, W.T., Rodley, G.A., and Jameson, G.B., *Acta Crystallogr., A 31,* S 49 (1975).
552. Robson, A., and Truter, M.R., *J. Chem. Soc. A, 1968,* 794.
553. Rodrique, L., van Meerssche, M., and Piret, P., *Acta Crystallogr., B 25,* 519 (1969).
554. Sappa, E., Milone, L., and Andreetti, G.D., *Inorg. Chim. Acta, 13,* 67 (1975).
555. Scheidt, W.R., and Frisse, M.E., *J. Amer. Chem. Soc., 97,* 17 (1975).
556. Schlueter, A.W., and Gray, H.B., *Amer. Cryst. Assoc., Summer Meeting 1971,* Ames (Iowa), Abstracts, p.41.
557. Schmitt, H.-J., and Ziegler, M.L., *Z. Naturforsch., B 28,* 508 (1973).
558. Schrauzer, G.N., Rabinowitz, H.N., Frank, J.A.K., and Paul, I.C., *J. Amer. Chem. Soc., 92,* 212 (1970).
559. Schrauzer, G.N., Glockner, P., Reid, K.I.G., and Paul, I.C., *J. Amer. Chem. Soc., 92,* 4479 (1970).
560. Schultz, A.J., and Eisenberg, R., *Inorg. Chem., 12,* 518 (1973).
561. Schunn, R.A., Fritchie, C.J., Jr., and Prewitt, C.T., *Inorg. Chem., 5,* 892 (1966).
562. Semion, V.A., and Struchkov, Yu.T., *Zh. Strukt. Khim., 10,* 88 (1969); *J. Struct. Chem., 10,* 80 (1969).
563. Semion, V.A., and Struchkov, Yu.T., *Zh. Strukt. Khim., 10,* 664 (1969); *J. Struct. Chem., 10,* 563 (1969).
564. Sequeira, A., and Bernal, I., *J. Cryst. Mol. Struct., 3,* 157 (1973).
565. Shklober, V.E., Skripkin, V.V., Gusev, A.I., and Struchkov, Yu.T., *Zh. Strukt. Khim., 13,* 744 (1972); *J. Struct. Chem., 13,* 698 (1972).
566. Shriver, D.F., Lehman, D., and Strope, D., *J. Amer. Chem. Soc., 97,* 1594 (1975).
567. Siebert, W., Augustin, G., Full, R., Krüger, C., and Tsay, Y.-H., *Angew. Chem., 87,* 286 (1975); *Angew. Chem. Int. Ed. Engl., 14,* 262 (1975).
568. Singh, P., Clearfield, A., and Bernal, I., *J. Coord. Chem., 1,* 29 (1971).
569. Skarstad, P., Janse-van Vuuren, P., Meinwald, J., and Hughes, R.E., *J. Chem. Soc. Perkins Trans. II, 1975,* 88.
570. Smith, D.L., and Dahl, L.F., *J. Amer. Chem. Soc., 84,* 1743 (1962).
571. Smith, H.W., Svetich, G.W., and Lingafelter, E.C.,

Amer. Cryst. Assoc., Summer Meeting 1973, Storrs (Connecticut), Abstracts, p.174.

572. Smith, M.B., and Bau, R., *J. Amer. Chem. Soc., 95*, 2388 (1973).
573. Snow, M.R., and Ibers, J.A., *Inorg. Chem., 12*, 249 (1973).
574. Søtofte, I., and Rasmussen, S.E., *Acta Chem. Scand., 21*, 2028 (1967).
575. Sproul, G.D., and Stucky, G.D., *Inorg. Chem., 11*, 1647 (1972).
576. Starovskii, O.V., and Struchkov, Yu.T., *Zh. Strukt. Khim., 5*, 257 (1964); *J. Struct. Chem., 5*, 231 (1964).
577. Stephens, F.S., *J. Chem. Soc. Dalton Trans., 1972*, 2257.
578. Stephens, F.S., *J. Chem. Soc. A, 1970*, 1722.
579. Stone, A., and Fleischer, E.B., *J. Amer. Chem. Soc., 90*, 2735 (1968).
580. Stork-Blaisse, B.A., Verschoor, G.C., and Romers, C., *Acta Crystallogr., B 28*, 2445 (1972).
581. Struchkov, Yu.T., *Dokl. Akad. Nauk. SSSR, Fiz. Khim., 110*, 67 (1956).
582. Sweet, R.M., Fritchie, C.J., Jr., and Schunn, R.A., *Inorg. Chem., 6*, 749 (1967).
583. Takahashi, K., Iwanami, M., Tsai, A., Chang, P.L., Harlow, R.L., Harris, L.E., McCaskie, J.E., Pluger, C.E., and Dittmer, D.C., *J. Amer. Chem. Soc., 95*, 6113 (1973).
584. Takenaka, A., Sasada, Y., Watanabe, E.-i., Ogoshi, H., and Yoshida, Z.-i., *Chem. Letters, 1972*, 1235.
585. Tasker, P.A., and Fleischer, E.B., *J. Amer. Chem. Soc., 92*, 7072 (1970).
586. Taylor, N.J., Paik, H.N., Chieh, P.C., and Carty, A.J., *J. Organometal. Chem., 87*, C 31 (1975).
587. Templeton, D.H., Zalkin, A., and Ueki, T., *Acta Crystallogr., 21*, A 154 (1966).
588. Terzis, A., and Rivest, R., *Inorg. Chem., 12*, 2132 (1973).
589. Thomas, J.T., Robertson, J.H., and Cox, E.G., *Acta Crystallogr., 11*, 599 (1958).
590. Todd, L.J., Paul, I.C., Little, J.L., Welcker, P.S., and Peterson, C.R., *J. Amer. Chem. Soc., 90*, 4489 (1968).
591. Treichel, P.M., Dean, W.K., and Calabrese, J.C., *Inorg. Chem., 12*, 2908 (1973).
592. Treichel, P.M., Johnson, J.W., and Calabrese, J.C., *J. Organometal. Chem., 88*, 215 (1975).
593. Trinh-Toan, and Dahl, L.F., *J. Amer. Chem. Soc., 93*, 2654 (1971).

594. Trinh-Toan, Fehlhammer, W.P., and Dahl, L.F., *J. Amer. Chem. Soc.*, *94*, 3389 (1973).
595. Trotter, J., *Acta Crystallogr.*, *11*, 355 (1958).
596. Trotter, J., and MacDonald, A.C., *Acta Crystallogr.*, *21*, 359 (1966).
597. Trotter, J., and Williston, C.S., *J. Chem. Soc. A*, *1967*, 1379.
598. Troup, J.M., Frenz, B.A., and Cotton, F.A., *Amer. Cryst. Assoc.*, *Winter Meeting 1973*, Gainesville (Florida), Abstracts, p.41.
599. Vahrenkamp, H., *J. Organometal. Chem.*, *63*, 399 (1973).
600. Di Vaira, M., and Orioli, P.L., *Acta Crystallogr.*, *B 24*, 1269 (1968).
601. Vergamini, P.J., Ryan, R.R., and Kubas, G.J., *Amer. Cryst. Assoc.*, *Winter Meeting 1973*, Gainesville (Florida), Abstracts, p.45.
602. Waite, M.G., and Sim, G.A., *J. Chem. Soc. A*, *1971*, 1009.
603. Wang, A.H.-J., Paul, I.C., and Schrauzer, G.N., *J. Chem. Soc. Chem. Commun.*, *1972*, 736.
604. Wang, A.H.-J., Paul, I.C., and Aumann, R., *J. Organometal. Chem.*, *69*, 301 (1974).
605. Watanabe, Y., and Yamahata, K., *Sci. Pap. Inst. Phys. Chem. Res.*, *Tokyo*, *64*, 71 (1970).
606. Weaver, J., and Woodward, P., *J. Chem. Soc. A*, *1971*, 3521.
607. Weaver, J., and Woodward, P., *J. Chem. Soc. Dalton Trans.*, *1973*, 1439.
608. Weber, H.P., and Bryan, R.F., *J. Chem. Soc. A*, *1967*, 182.
609. Wei, C.H., and Dahl, L.F., *Inorg. Chem.*, *4*, 1 (1965).
610. Wei, C.H., and Dahl, L.F., *Inorg. Chem.*, *4*, 493 (1965).
611. Wei, C.H., Wilkes, G.R., Treichel, P.M., and Dahl, L.F., *Inorg. Chem.*, *5*, 900 (1966).
612. Wei, C.H., and Dahl, L.F., *J. Amer. Chem. Soc.*, *91*, 1351 (1969).
613. Wester, D., and Palenik, G.J., *J. Amer. Chem. Soc.*, *95*, 6505 (1973).
614. Wheatley, P.J., in Dunitz, J.D., and Ibers, J.A.(Eds.), *Perspectives in Structural Chemistry*, *Vol.I*, Wiley, New York 1967, p.1.
615. Wheelock, K.S., Nelson, J.H., Cusachs, L.C., and Jonassen, H.B., *J. Amer. Chem. Soc.*, *92*, 5110 (1970).
616. Whitesides, T.H., Slaven, R.W., and Calabrese, J.C., *Inorg. Chem.*, *13*, 1895 (1974).
617. Whiting, D.A., *Cryst. Struct. Comm.*, *1*, 379 (1972).
618. Wilford, J.B., Smith, N.O., and Powell, H.M., *J. Chem. Soc. A*, *1968*, 1544.

619. Willis, B.T.M., *Acta Crystallogr.*, *13*, 1088 (1960).
620. Williston, C.S., *Dissert. Abstr.*, *B 28*, 2801 (1968).
621. Wong, Y.S., Paik, H.N., Chieh, P.C., and Carty, A.J., *J. Chem. Soc. Chem. Commun.*, *1975*, 309.
622. Woodhouse, D.I., Sim, G.A., and Sime, J.G., *J. Chem. Soc. Dalton Trans.*, *1974*, 1331.
623. Yannoni, N.F., Krukonis, A.P., and Silverman, J., *Amer. Cryst. Assoc.*, *Summer Meeting 1970*, Ottawa, Abstracts, p.80.
624. Yamamoto, Y., Aoki, K., and Yamazaki, H., *J. Amer. Chem. Soc.*, *96*, 2647 (1974).
625. Yasuda, N., Kai, Y., Yasuoka, N., Kasai, N., and Kakudo, M., *J. Chem. Soc. Chem. Commun.*, *1972*, 157.
626. Zalkin, A., Templeton, D.H., and Hopkins, T.E., *J. Amer. Chem. Soc.*, *87*, 3988 (1965).
627. Zalkin, A., Forrester, J.D., and Templeton, D.H., *J. Amer. Chem. Soc.*, *88*, 1810 (1966).
628. Zalkin, A., Templeton, D.H., and Ueki, T., *Inorg. Chem.*, *12*, 1641 (1973).
629. Ziegler, M.L., *Angew. Chem.*, *80*, 239 (1968); *Angew. Chem. Int. Ed. Engl.*, *7*, 222 (1968).
630. Zimmer, J.-C., and Huber, M., *C.R. Acad. Sci. Ser. C*, *267*, 1685 (1968).
631. Hursthouse, M.B., and Rodesiler, P.F., *J. Chem. Soc. Dalton Trans.*, *1972*, 2100.

NMR SPECTROSCOPY OF ORGANOIRON COMPOUNDS

By TOBIN J. MARKS

Department of Chemistry
Northwestern University, Evanston, Illinois 60201

TABLE OF CONTENTS

ISBN 0-12-417101-X (v.1)

I. INTRODUCTION

Nuclear magnetic resonance spectroscopy is one of the physicochemical techniques responsible for the great flowering of transition metal organometallic chemistry during the last twenty years. Today, NMR is indispensable to the practicing organometallic chemist, but every year sees new developments in theory, instrumentation, and techniques which promise to reveal ever more of the intimate details of molecular structure and dynamics. This article endeavours to summarize the NMR spectroscopy of organic iron molecules, with a view not toward compiling massive amounts of data, but toward highlighting important trends which have developed in this rapidly advancing field and at the same time providing a thorough set of authoritative references so that the reader may easily pursue any subject further.

Space does not allow discussion of the basic theory of nuclear magnetic resonance; the reader is referred to a host of excellent texts *(13,19,40,93,212)*. Likewise, several useful review articles have been written which introduce NMR with an inorganic *(41,88,140,188)* or organometallic *(108,142,165)* slant. A number of publications are also available which compend spectroscopic data for large numbers of organometallic molecules *(84,99,114,165,120)*, and these are of great utility for locating the spectra of specific compounds.

II. PROTON MAGNETIC RESONANCE SPECTROSCOPY

A. DIAMAGNETIC MOLECULES IN SOLUTION

By far, the majority of organic iron molecules studied by NMR are diamagnetic, and most chemists will examine them in isotropic solutions. Valuable information about the compound under study can be obtained from the observation of chemical shifts and coupling constants in the proton magnetic resonance spectrum.

1. Chemical Shifts

The resonance position (chemical shift) of a proton in an organoiron molecule will depend on a number of factors such as screening by valence electrons and shielding by neighbouring magnetically anisotropic groups. Since none of these terms can be predicted with great accuracy for complex organometallic molecules, it is unwise to draw far-reaching theoretical conclusions about electronic structure and bonding based upon comparison of chemical shifts (except possibly in a closely related series of molecules). However, a great deal of empirical data is at hand which allows various types of iron-organic ligand systems to be identified with gratifying reliability.

The first generalization which can be made for iron organometallics is that protons on complexed ligands almost invariably resonate at higher field than in the free ligand. This is probably due to the diamagnetic anisotropy of the iron atom and attendant groups rather than to any major net flow of electron density onto the ligand. Also, the amount protons shift upon complexation is characteristic of the different types of metal-ligand bonding systems. For example (Table 1), in simple mono-olefin tetracarbonyliron complexes (1), this shift is usually ca. 2.3-5.5 ppm *(95,165)*, whereas in cationic complexes (2), the shift, presumably due to decreased electron density at the iron among other factors, is usually less, 1-2.3 ppm *(165)*. Magnetically anisotropic groups such as η^5-C_5H_5 can shift resonance positions by several ppm *(95)*. Tricarbonyliron 1,3-diene complexes (3) show a characteristic pattern in which protons on carbons 1 and 4 ("outer protons") are shifted 3.5-4.5 ppm upfield while protons on carbons 2 and 3 ("inner protons") are shifted only ca. 0.7-1.0 ppm *(99,165)*. Structural studies indicate this trend may reflect proximity to the iron atom. In π-allyl molecules (4) the same characteristic pattern prevails. The unique proton on the center carbon of the η^3-allyl fragment typically resonates at lower field, ca. τ 5.0-6.0, and the accompanying *syn* and *anti* protons invariably at higher field, τ 6.0-7.3 and τ 6.3-9.3, respectively *(69,99,165,199)*. Again greater proximity to the iron and other magnetically anisotropic ligands may explain the shift difference between *syn* and *anti*. These results are summarized in Table 1. As might be expected, shielding and deshielding substituents have the same general effect on neighbouring protons in complexes as in the free ligand, though the exact magnitude of induced shifts is unsystematic.

η^5-cyclopentadienyl protons (5) usually exhibit a resonance at τ 5.0-6.0 in neutral organoiron molecules *(165)* and at ca. 1-2 ppm to lower field in cationic molecules (6) *(165)*. The author has observed large (1-1.5 ppm) upfield shifts for a number of $(\eta^5$-$C_5H_5)Fe$ molecules in aromatic solvents *(175)*, and though this phenomenon sometimes vitiates spectral comparisons, it is useful for spectral simplification, and presumably arises due to short-lived "collision complexes" in which the cyclopentadienyl protons are close to the shielding region of the aromatic π-system *(160,223)*. In the series of molecules $(C_5H_5)Fe(CO)_2C_6H_4X$, a correlation was found between $\tau(C_5H_5)$, $\tilde{\nu}(CO)$ and Hammett σ constants for X *(17)*.

Arene iron complexes (7) resonate in the region of τ 5.0

Table 1: Typical Proton NMR Data for Organoiron Molecules *(Refs.)*

	Type of Complex		Chemical Shifts [τ] *(234)*	Coupling Constants [Hz]	
1	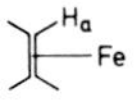		6.2-9.4 *(165,99,95)*		
2	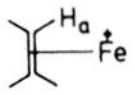		5.0-6.3 *(165,99)*		
3	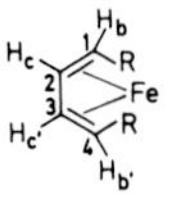	a	9.7 ($R = H_a$) *(165, 99,72)*	$J_{ab} \sim 2.5 (R=H_a)$	$J(^{13}C_1-H_a) \sim 160 (R=H_a)$ *(214)*
		b	7.2-8.3	$J_{ac} \sim 8.2$	
		c	4.2-5.2	$J_{bc} \sim 6.0-8.0$	$J(^{13}C_1-H_b) \sim 160$
				$J_{bc'} \sim 1.0$	
				$J_{cc'} \sim 4.0-5.0$	$J(^{13}C_2-H_c) \sim 170$
4	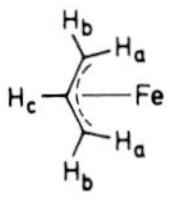	a	6.3-9.3 *(165,99)*	$J_{ab} \sim 0$	
		b	6.0-7.3	$J_{ac} \sim 4.0-10.0$	
		c	5.0-6.0	$J_{bc} \sim 8.0-14.0$	
5	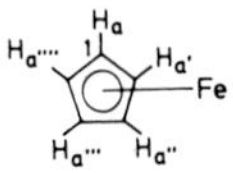		5.0-6.0 *(165)*	$J_{aa'} \sim 2.4$ *(71)*	$J(^{13}C_1-H_a) \sim 175$ *(71)*
				$J_{aa''} \sim 1.2$	$J(^{13}C_1-H_{a'}) \sim 6.3$
					$J(^{13}C_1-H_{a''}) \sim 7.2$
6	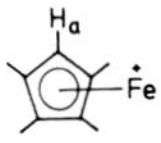		4.0-5.0 *(165)*		
7	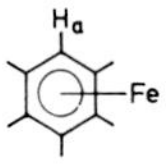		ca. 5.0 *(165)*		
8	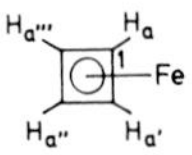		ca. 6.0 *(27-30,214)*	$J_{aa'} \sim 0.0$	$J(^{13}C_1-H_a) \sim 190$ *(27-30,214)*
				$J_{aa''} \sim 9.0$	
9	Fe-H		12.0-36.0 *(74,75,77, 90,121,165,183,198,254)*		
10	$Fe-CHR_2$		7.4-10.1 *(113,120)*		

(165), cyclobutadiene complexes (8) in the region of τ 6.0 *(27-30)*. As is usual for transition metal systems *(109,137)*, organoiron hydride protons (9) invariably resonate at fields greater than τ 10.0 (e.g. $(C_5H_5)_2FeH^+$ *(74)*, $[(C_5H_5)Fe(CO)_2]_2H^+$ *(77)* and related bimetallic species *(90,121)*, protonated olefin complexes *(254)*, H_2FeL_4 *(183)*). It has been observed *(75,198)* that the difference between $\tau(CH_3)$ and $\tau(CH_2)$ in substituted ethyl compounds, is a rough measure of the electro-

negativity of the substituent. This reasoning implies that the $(C_5H_5)Fe(CO)_2$ group *(113)* is approximately as electron-withdrawing as zinc *(199)*; of course the magnetic anisotropy of the other ligands on iron has been ignored in this simplified treatment. Data for iron alkyls (<u>10</u>) are given in Table 1.

It has also been reported that rare earth NMR "shift reagents" are of utility in studying organoiron molecules which have basic sites such as Cl, N_3, CN, F, $Fe{-}\overset{O}{\overset{\|}{C}}{-}R$ and $Fe{-}\overset{O}{\overset{\|}{C}}{-}Fe$ *(177,178,213)*. Besides permitting chemical, structural, and stereochemical *(218)* probing to be carried out via NMR spectroscopy in solution, this technique is also useful for spectral simplification. As might be anticipated, the shift reagents have been found to coordinate to standard organic basic sites on olefinic ligands *(101,227)*.

Proton chemical shift data for diamagnetic organoiron compounds are summarized in Table 1.

2. Coupling Constants

When interpreted with care, spin-spin coupling constants are probably more reliable than chemical shifts for deducing structure and bonding in organic iron systems *(119,219)*. Geminal and vicinal H-H couplings *(12,18,95,195)* and ^{13}C-H couplings *(12,93,136,195)* are quite sensitive to modest changes in dihedral angles and hybridization. Careful studies *(119,171,219,72)*, for example, have shown that <u>11</u> rather than <u>12</u> is the best description of diene-iron bonding. These results appear to be in good accord with structural data. Low

<u>11</u>

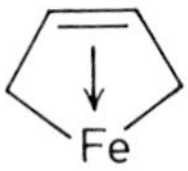

<u>12</u>

geminal methyl proton couplings in molecules such as $(C_5H_5)Fe(CO)_2CH_2D$ *(86)* have been interpreted in terms of iron to alkyl group back-donation. Spectral analyses have also been performed on metallocene *(71,217)* and cyclobutadiene *(27-30)* compounds; data have been analysed for an isoindene tricarbonyliron complex *(224)*. Table 1 presents typical values of proton-proton, and ^{13}C-proton coupling constants for the major structural types.

B. PARAMAGNETIC MOLECULES

Protons in paramagnetic molecules frequently exhibit displacements from diamagnetic resonance positions which are due to contact (hyperfine) and pseudocontact (dipolar) interactions with unpaired electron spin density *(41,78,89,140,141, 241,157,83)*. Similar processes also account for broadening of proton resonances *(41,78,89,140,141,241)*. If the spacially-dependent pseudocontact shift which arises from the magnetic anisotropy of the molecule can be accounted for or eliminated, then the remaining contact shift provides valuable information on electron delocalization and metal-ligand bonding. Ferricenium ions and related paramagnetic metallocenes have received considerable attention from NMR spectroscopists *(103-105,146,220,221)*. It appears that the pseudocontact contribution is small in ferricenium compounds *(103)* (but by no means negligible *(146)*), and that the direction of the ring proton contact shifts cannot be attributed to delocalization through metal-ring π-bonding orbitals. Rather it appears that unpaired spin density is delocalized principally through the σ-bonding system or *via* direct overlap of metal orbitals with ring protons *(105,141,220)*. Typical shift values for ferricenium ring protons are ca. 30 ppm downfield from the resonance position in ferrocene *(104,141)*. Substituent protons (e.g. on alkyl groups) are usually shifted upfield, and the shift rapidly attenuates with the number of intervening bonds *(104,141)*.

A number of ^{1}H-NMR studies have been carried out on paramagnetic coordination compounds of iron. Dynamic processes such as ligand exchange kinetics (L_2FeX_2, L = R_3P *(207)*, hexamethylphosphoramide *(256)* and 2-picoline *(255)*) and electron transfer (ferrocene-ferricenium ion *(81)*, phenanthroline complexes *(81,159)* and dithiocarbamates *(204)*) are examples. Several complexes of octahedral Fe(II) (dithiocarbamates *(208)*, pyrazolylborates *(132,133)*) and Fe(III) (dithiocarbamates *(111,112)*) exhibit temperature-dependent spin equilibria. Electron delocalization studies are also available for pyrazolylborate *(132,133)*, imidazole *(248)*, aminotroponiminate *(87)*, phenanthroline *(154)*, benzamide *(247)* and $(R_3P)_2FeX_2$ *(207)* complexes. Depending on the ligand system, delocalization appears to occur *via* σ or π mechanisms or *via* both.

The use of NMR to study structure and bonding in heme complexes of biological interest has been summarized *(252)*. Extensive investigations of electron delocalization *(155)*, magnetic anisotropy *(155)*, axial ligation dynamics *(155,238)* and electron spin relaxation *(158)* have been published for synthetic iron porphyrins. ^{1}H-NMR studies indicate that isotropic shifts in the cluster compounds $[Fe_4S_4(SR)_4]^{2-}$, which are models for high potential ferredoxins are predominantly contact in origin *(127)*.

Proton NMR has also found great use in the study of iron containing proteins *(206,252)*. Investigations have dealt with magnetic anisotropy in cytochrome-c *(129)*, the rate of electron transfer between ferri- and ferrocytochrome-c *(118)* and the conformational behaviour of normal and abnormal hemoglobins *(163)*.

C. *SPECTROSCOPY IN THE SOLID STATE AND IN ANISOTROPIC SOLVENT SYSTEMS*

Unlike solution spectra in which molecules are randomly tumbling, solid state NMR spectra are most strongly influenced by spacially dependent dipolar interactions between nuclear spins. Analysis of broad line spectra has afforded rough estimates of internuclear distances in simple molecules such as $(OC)_4FeH_2$ *(16,228)*. Variable temperature studies have revealed rapid cyclopentadienyl ring reorientation in ferrocene and substituted ferrocenes *(194)*. The barrier to ring "spinning" in solid ferrocene is estimated to be 2.3 kcal/mole *(194)*. Recently, fluxional processes *(vide infra)* have been detected in the solid state *(36,37,42)*. The broad line ^1H-NMR spectrum of the ferricenium ion infers that the molecule is magnetically isotropic *(103)*.

Anisotropic liquid crystal solvents offer unique media for structural studies by NMR *(80,186)*. Inter- but not intramolecular dipolar couplings are averaged to zero by solute motion in nematic mesophases. Thus, $(\eta^4\text{-}C_4H_4)Fe(CO)_3$ has been found to have a rigorously square (C_{4v}) cyclobutadiene ring *(253)*.

III. MAGNETIC RESONANCE OF NUCLEI OTHER THAN PROTONS

A. *BORON*

Excellent reviews are available on general aspects of ^{11}B-NMR *(91,125,226)*. Studies of bis(1,2-dicarbollide) and cyclopentadienyl-1,2-dicarbollide "sandwich" compounds of Fe(II) and Fe(III) have been reported *(123,250)*. The paramagnetic Fe(III) complexes provide the first known example of well-resolved, contact-shifted ^{11}B spectra *(123,250)*. The mode of electron spin delocalization is predominantly ligand-to-metal charge transfer. The spectrum of the carbadecaborane complex $[(B_{10}H_{10}CH)_2Fe(III)]^{3-}$ extends over 300 ppm and has not been described in detail *(130)*. Boron NMR spectra of R_2B-$Fe(CO)_2(C_5H_5)$ molecules *(201)*, $(C_5H_5)Fe(CO)_2$ derivatives of

$B_{10}H_{13}^-$ and 7,8-$B_9C_2H_{12}^-$ *(225)*, and iron pyrazolylborate complexes *(133)* have also been discussed. Generally coordination of boron to a diamagnetic transition metal results in a 10-20 ppm deshielding.

B. CARBON

The advent of Fourier transform techniques *(24,96,109)* heralds the day when ^{13}C spectra of organometallics will be recorded on a routine basis. There is already enough data

Table 2: Typical ^{13}C Chemical Shifts for Iron Organometallics

Compound	δ^a *(Ref.)*	
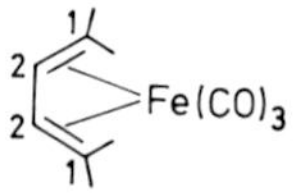	C_1 42 C_2 86	OC 209 *(219,214)*
$(\eta^4\text{-}C_4H_4)Fe(CO)_3$	HC 61	OC 209 *(214)*
$(\eta^5\text{-}C_5H_5)_2Fe$	HC 69 *(161)*	
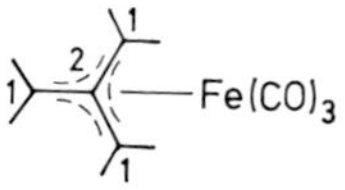	C_1 35 C_2 105	OC 212 *(92)*
$(C_3F_7)Fe(CO)_4I$		OC 199 *(161)*
$CH_3C(=O)Fe(C_5H_5)(CO)_2$	HC 87 *(161)*	
$(C_5H_5)Fe(CO)_2I$		OC 214 *(161)*
$Fe(CO)_5$		OC 210 *(161,20)*
$Fe(NO)_2(CO)_2$		OC 207 *(20)*
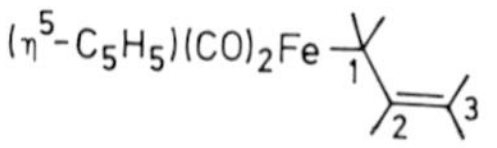	C_1 33, C_3 135 *(169)* C_2 172	

[a] Expressed in ppm relative to TMS. Positive shifts indicate resonance to low field of TMS. $\delta_{TMS} = \delta_{CS_2} + 193$ ppm.

available to afford a number of extensive ^{13}C reviews *(162, 187,230)* for organic molecules, and three for organometallics *(43,169,235)*. For organoiron molecules, it appears that, as in proton NMR, coordination to iron generally shifts free ligand carbon resonances to higher field. Data are available for (butadiene)tricarbonyliron *(214,219)*, numerous cyclopentadienyls *(106,143,148,161)*, indenyls *(147)*, (cyclobutadiene)-tricarbonyliron *(216)*, ferrocenyl carbenium ions *(23,203)*, (trimethylenemethane)tricarbonyliron *(92)* and of many other carbonyl compounds *(20,43,52,161,169,235)*. Resonances of olefin π-complexes invariably occur 33-93 ppm to low field of TMS. In (diene)tricarbonyliron complexes, "outer" carbons resonate at higher field than "inner" ones, a trend which is reminiscent of the ^{1}H-NMR spectra. Carbon resonances in terminal carbonyl groups appear at 193-220 ppm below TMS. Bridging carbonyls resonate at ca. 275 ppm *(107,235)*. Complexed carbenoid carbons resonate near 300 ppm *(43)*. In $(C_5H_5)Fe(CO)_2X$ molecules, carbonyl shifts depend almost linearly on the electronegativity of X *(106,143)*. Carbon chemical shifts are highly sensitive to diamagnetic and paramagnetic shielding effects. The former term arises from electronic screening in the ground state while the latter arises from magnetic field induced mixing of ground and excited electronic states. For carbon atoms bound directly to transition metal ions, the relative contributions of these terms vary considerably and are difficult to predict with accuracy. For these reasons, attempts to relate ^{13}C chemical shifts to bonding parameters are at best applicable to molecules in a closely related series *(94)*. Carbon chemical shifts have also been reported for several iron cyanide complexes *(76,126)*. ^{57}Fe ($I = 1/2$, 2.19% abundant) to ^{13}C coupling constants have been recently measured for several carbonyl compounds *(168)* and ferrocene derivatives *(149)*. Carbon magnetic resonance data are summarized in Table 2.

C. *NITROGEN AND OXYGEN*

To date, little material has been published on the NMR of these nuclei when incorporated in iron complexes (or any transition metal complexes). Two reviews of ^{14}N-NMR have appeared *(215,251)*. For ^{14}N nuclei directly bonded to iron *(21,126)*, high field shifts from the resonance position of the free ligand have been observed, while paramagnetic deshielding appears to play a larger role in cyano complexes *(126)*. Far larger shifts are observed in complexes with unpaired electrons *(126)*. Linewidths of ^{14}N spectra have been employed to study electron exchange between ferri- and ferrocyanide *(229)*. ^{17}O chemical shifts have only been reported for $Fe(CO)_5$ and

$Fe(CO)_2(NO)_2$ *(20)*.

D. FLUORINE

Excellent reviews have been written on the general aspects of ^{19}F-NMR *(98,135)*. The great "explosion" which has taken place in fluorocarbon organometallic chemistry has produced considerable ^{19}F-NMR data for iron-containing molecules. Several interesting trends have been revealed. It has been found for fluoroalkyl organometallics that resonances of fluorines on carbons directly attached to transition metals such as iron are anomalously shifted 60-70 ppm downfield from the "normal" resonance positions found in fluoroalkanes and for the same group more distant from a transition metal *(70)*. This effect has been attributed to paramagnetic deshielding, arising from the admixture of low-lying excited states involving fluorine *p*-electrons and empty metal *d*-orbitals *(210)*. An exception to this trend occurs for those fluorines attached to α-carbons also bearing two CF_3 groups - here the anomalous deshielding is not observed *(144)* (or it is cancelled by other effects).

It has been proposed that the relative ^{19}F shifts of *m*-FC_6H_4X and *p*-FC_6H_4X molecules provide a rough measure of the σ- and π-bonding ability of X *(205,232)*. Application to systems where X is $(\eta^5\text{-}C_5H_5)Fe(CO)_2$ or $(\eta^5\text{-}C_5H_5)Fe(CO)[(C_6H_5)_3P]$ *(17)* demonstrates that the iron group is electron-releasing by both σ and π mechanisms *(17)*, and that the phosphine substitution increases this electron-releasing power *(17)*. Chemical shift and coupling constant data have also been published for a large number of fluoroaryliron compounds *(26)*. That J(F-F-gem) in $(OC)_4Fe(F_2C{=}CFCl)$ has increased over the value in the free olefin (131 *vs* 78Hz) has been interpreted as indicating increased sp^3 character in hybridization of the olefinic carbon atoms upon complexation *(97)*.

Extensive ^{19}F data are available for PF_3 complexes *(135, 200)*.

E. PHOSPHORUS

For comprehensive reviews of ^{31}P-NMR, see references *73, 180* and *200*. In organoiron systems, all ^{31}P-NMR studies have been performed on complexes of formally trivalent phosphorus (phosphines) *(200)*. It is generally observed that the coordinated phosphine resonates to low field of the free phosphine. The chemical shift difference in ppm has been designated as the coordination shift. Since the magnitudes of these shifts involve both ground and excited state properties, they are difficult to interpret, however, some trends are at least partially explicable in terms of the σ-donor and π-acceptor ability of other ligands in the complex *(200)*. Coupling constant data such as J(P-P'),J(P-H;PH_3 complexes), J(P-F;PF_3 complexes) have been reported for a number of organoiron compounds *(200)*. ^{57}Fe to ^{31}P coupling constants have been measured for several phosphine substituted iron carbonyl molecules *(168)*.

F. IRON

It has been possible to measure ^{57}Fe chemical shifts by $^{13}C\{^{57}Fe\}$ double irraditation studies in iron organometallics enriched to 82% in ^{57}Fe *(149)*. Resonance positions are extremely sensitive to environment. For example, the iron resonance frequency in ferrocene shifts 1098 ppm upfield on protonation of the ferrocene and 216 ppm downfield on acetylation.

IV. STEREOCHEMICAL NONRIGIDITY

Chemists have been aware for some years that molecules are dynamic entities. That is, when circumstances permit, they are rapidly tumbling, flexing, twisting, bending, and vibrating. However, it has only been in the last few years that chemists have begun to discover and to appreciate the extent to which certain molecules are dynamic. In particular, there exist certain molecules which have more than one thermally accessible structure and which can pass rapidly among these structures. When these structures are chemically identical the rearrangement is degenerate and the molecules are called fluxional *(49,50,82,131,190)*. Other terms which have been used to encompass the larger class of molecules undergoing either degenerate or nondegenerate rapid rearrangements include "valence tautomers" *(50)*, "stereochemically nonrigid" *(50,190)*, and "stereochemically dynamic" *(164)* molecules.

The use of the adjective "rapid" to characterize these rearrangements is unfortunately not a rigorous means of classifying them. For nondegenerate tautomers one has some intuitive notion that the process is rapid if the rate of interconversion is fast enough to prevent separation by classical means at ambient temperatures. For fluxional molecules, even this working definition is meaningless since the molecule is passing among chemically identical configurations. However, since parts of the fluxional molecule are being permuted by the rearrangement process, the possibility exists of detecting the fluxionality by spectroscopic means. If the site interchange process is fast (*i.e.* the lifetime in each configuration is sufficiently short) relative to the timescale *(189)* of the spectroscopic technique, then the permuted sites will become indistinguishable to this spectroscopic technique. Since most spectroscopic studies to date have been accomplished with nuclear magnetic resonance, it is generally accepted that the term "rapid" implies a rate (ca. 10^{-1} - 10^{4} sec^{-1}) which is rapid on the NMR timescale. This does not mean that other methods such as vibrational spectroscopy, which has a much faster timescale, are not potentially useful. Indeed, the various physical techniques complement each other and frequently a combination of them is responsible for the discovery of a new nonrigid molecule, since the molecule may appear static to one form of spectroscopy (*e.g.* infrared) but dynamic to another (*e.g.* NMR).

Stereochemically nonrigid molecules constitute one area out of many in which chemists are becoming increasingly aware of the necessity of viewing the geometry and dynamic behaviour of molecules in terms of potential energy surfaces. The classical structures we know for molecules are no more than wells in these surfaces, and numerous properties of the molecule (*e.g.* how easily it is deformed) are reflected by the depth of the wells and the steepness of the sides. To rigorously describe a molecule, the entire potential energy surface containing all conceivable configurations of the molecule should be given; the more that is known about the shape of this surface, the more that is known about the molecule. Nonrigid molecules possess the characteristic that there are low energy pathways connecting some of the wells (the wells are identical if the molecule is fluxional). The activation energy for the rearrangement describes how high this pathway is, relative to the ground state of the molecule. The mechanism of the rearrangement describes the exact pathway taken by the molecule. It can be seen from potential energy surfaces that there is a very subtle interplay of structural and dynamic properties. This is perhaps the most fascinating aspect of nonrigid molecules.

In the past few years, there has been a tremendous flurry of activity in the area of nonrigid organometallic molecules *(50,51,237)*. This has been due both to the availability of the sophisticated instrumentation necessary to study these systems and also to the interest of many chemists in the structure, bonding, and reactions of organometallic molecules. The theoretical and computational means are presently available to calculate accurately NMR lineshapes as a function of exchange rate for complex systems *(131,134)*, and the possibility of obtaining reliable activation parameters for these processes makes studies even more attractive. The ultimate aim of these investigations is to learn something about the fundamental nature of structure and bonding in organometallic molecules and something about the fundamental nature of organometallic reactions. It follows that, especially for fluxional systems, the full understanding of the mechanism and energetics of the simplest possible types of organometallic reactions, *i.e.* those which are reversible, degenerate, and intramolecular, is potentially of great value.

Even the seemingly compact area designated as organoiron chemistry has seen such an exponential growth in examples of stereochemical nonrigidity during the last nine years, that it is impossible here to discuss the entire field. Several representative classes of compounds will be treated as models, and related molecules in the class as well as those outside the classes, will be referenced. Molecules have been classified by the type of dynamic process occurring.

A. SIGMATROPIC PROCESSES

These are tautomeric reorganizations involving repositioning of an iron moiety, which is bound in a *monohapto* manner. As early as 1956 *(209)*, it was proposed that the compound $(\eta^5\text{-}C_5H_5)Fe(CO)_2(\eta^1\text{-}C_5H_5)$, which exhibited two singlets in the ambient temperature ^{1}H-NMR spectrum, might be rearranging in such a manner as to render the magnetically nonequivalent protons on the σ-cyclopentadienyl ring equivalent on the NMR timescale. In 1966, Bennett *et al. (14)* reported X-ray and low temperature NMR studies on this compound. The structure in the solid state contained one $\eta^5\text{-}C_5H_5$ ring and one $\eta^1\text{-}C_5H_5$ ring. The NMR spectrum was temperature dependent, and at ca.-80°C the $\eta^1\text{-}C_5H_5$ ring displayed the AA'BB'X pattern to be expected for a σ-cyclopentadienyl metal compound. Upon raising the temperature (and thus increasing the rate of rearrangement) the low field portion of the olefinic multiplet collapsed more rapidly than the high field olefinic portion. It could be shown that this was not consistent with a random exchange of sites in the $\eta^1\text{-}C_5H_5$ ring (which presumably would occur if a *pentahapto* structure were an intermediate) but was consistent with either 1,2 or 1,3 shift processes which do not permute all sites at the same rate *(14,242)* as can be seen in scheme [1]. In both these cases, the olefinic protons are not permuted at the same rate, and this would be expected to lead

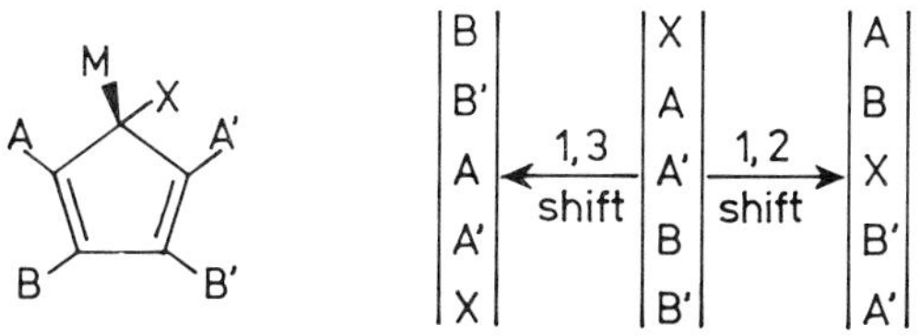

Scheme [1]

to an asymmetric collapse of the olefinic multiplet as the rearrangement rate increased. The problem, of course, is that unless the AA' and BB' multiplets can be assigned correctly, whether a 1,2 or 1,3 shift is occurring cannot be determined. Arguments based upon reasonable values for coupling constants *(14,38,63)* together with data for the η^1-indenyl analog *(66)* yield chemical shifts in accord with the 1,2-shift mechanism. That the indenyl molecule is rigid up to 120°C, mitigates against a 1,3-shift mechanism *(66)*. Orbital symmetry considerations support this mechanistic conclusion (formally a 1,5 shift), [2] *(231)*, as do ^{13}C-NMR studies *(45)*.

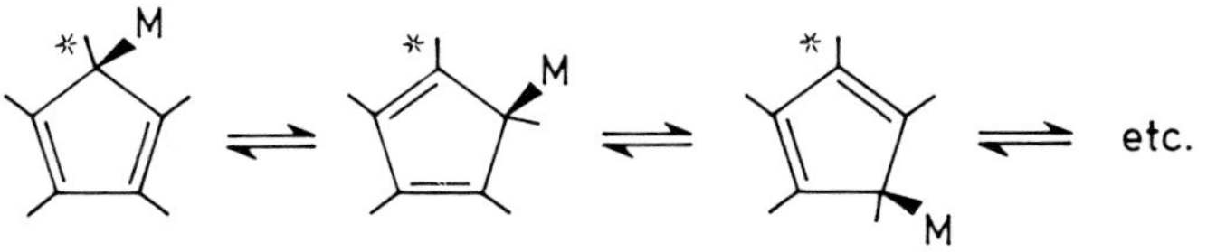

[2]

The two types of cyclopentadienyl rings do not rapidly interconvert at the highest accessible temperatures *(174)*. The ruthenium analog, $(\eta^5\text{-}C_5H_5)Ru(CO)_2(\eta^1\text{-}C_5H_5)$, exhibits analogous sigmatropic behaviour, with a slightly greater activation energy *(38,63)*. Computer simulation of the experimental lineshape changes further supports the predominance of the 1,2 shift pathway *(174)*.

B. REARRANGEMENTS INVOLVING π-BONDED SYSTEMS

From the time of initial synthesis, the structure of (cyclooctatetraene)tricarbonyliron has been cloaked in controversy *(172,197,216)*. The room temperature solution ^{1}H-NMR spectrum exhibited one type of proton, a result in apparent contradiction to the 1,2,3,4-*tetrahapto* structure found in the solid state *(79)*. Variable temperature proton NMR studies *(53,139,150)* revealed that an extremely rapid rearrangement process was taking place at room temperature. The exact nature of the lineshape changes occurring down to the lowest accessible temperatures (ca. -155°C) sparked considerable controversy. It was apparent that two questions had to be answered: What was the structure of $C_8H_8Fe(CO)_3$ in solution? What was the nature of the dynamic process? Arguments based on chemical shifts *(6)*, and results for both nonrigid *(5)* and rigid *(115)* substituted cyclooctatetraene complexes, provided strong support for 1,2,3,4-*tetrahapto* solution structure. Though certain types of rearrangement mechanisms were shown to be unlikely, it became evident that the slow exchange limit had not been reached even at -155°C and 100 MHz *(115)*. The problem was finally resolved with NMR studies on the isostructural *(57)* analog, $C_8H_8Ru(CO)_3$, which, due to a slightly higher barrier to rearrangement, had an instrumentally accessible slow exchange limit *(22,54)*. Computer-aided spectral simulation studies employing arguments similar to those already discussed for $(\eta^5\text{-}C_5H_5)Fe(CO)_2(\eta^1\text{-}C_5H_5)$, demonstrated that the rearrangement proceeded predominantly if not exclusively via 1,2 shifts of the metal atom *(22,54)* [3]. For $C_8H_8Fe(CO)_3$, it has also been shown that independent positional exchange of CO groups occurs *(222)*. Subsequent ^{1}H-NMR studies on $C_8H_8Os(CO)_3$

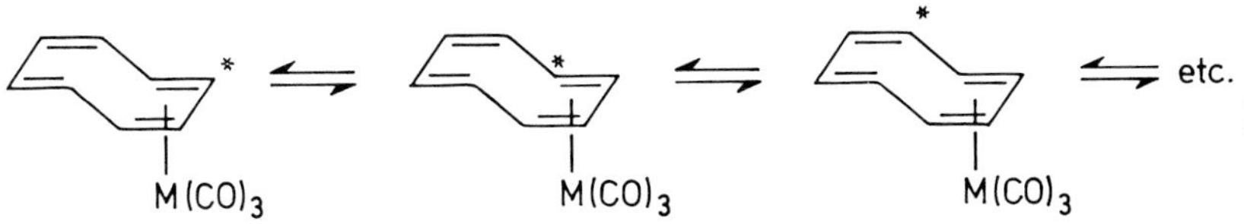

[3]

which rearranges even more slowly, support the conclusion of 1,2 shifts *(47)* as do ^{13}C-NMR studies on $C_8H_8Fe(CO)_3$ and C_8H_8-$Ru(CO)_3$ *(58)*. An interesting broadline NMR study has revealed that $C_8H_8Fe(CO)_3$ is also dynamic in the solid state, though the barrier to rearrangement is higher *(36,37)*. The tricarbonyliron complexes of benzo- and naphthocyclooctatetraene are also fluxional *(246)*, however the free energies of activation are 10-20 kcal/mole higher than in $C_8H_8Fe(CO)_3$.

The related cyclooctatetraene π-complex, *trans*-μ-(1-4-η: 5-8-η-C_8H_8)$[Fe(CO)_3]_2$ (13) is rigid at all accessible temperatures *(175)*. It has been found that (1-6-η-C_8H_8)(1-4-η-C_8H_8)Fe (14) *(4,39)* undergoes dynamic rearrangement which involves both exchange of proton environments on each cyclooctatetraene ring and exchange of environments between the two kinds of rings *(4,39)*. A single line is observed in the fast exchange limit ^{1}H-NMR spectrum.

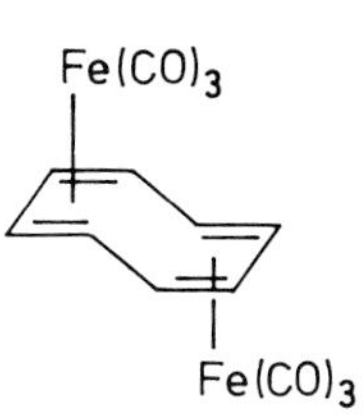

13

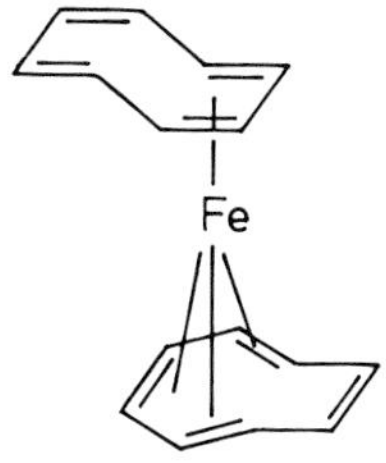

14

Two cycloheptatrienyltricarbonyliron complexes, $[C_7H_7Fe(CO)_3]^+$ *(166,243)* and $[C_7H_7Fe(CO)_3]^-$ *(167)*, appear to be fluxional, though detailed NMR studies have not been published. The extraordinary molecule, *trans*-μ-(1-3-η : 4-7-η-C_7H_7)$[(\eta^5\text{-}C_5H_5)Mo(CO)_2][Fe(CO)_3]$ (15) *(67)* is fluxional, and detailed analysis of the rearrangement process reveals that a synchronous movement (1,2 shifts) of *both* metal atoms around the C_7H_7 ring occurs *(67)*. The iron moiety in the cycloheptatrienyl complex $(\eta^5\text{-}C_5H_5)Fe(CO)(\eta^3\text{-}C_7H_7)$ also makes 1,2 hops about the seven-membered ring *(44)*. In the *tetrahapto* tricarbonyliron complex of N-carbethoxyazepine (16), the metal executes a rapid, degenerate rearrangement which interconverts the two enantiomorphous forms of the complex *(117)*. In linear polyene tricarbonyliron systems, Whitlock has observed far slower rates of iron migration *(173,245)*.

Though the processes are not degenerate, a variety of π-allyl iron complexes exhibit conformational equilibria, which are rapid on the NMR timescale at room temperature *(48,69, 199)*. These processes appear to involve rotation of the

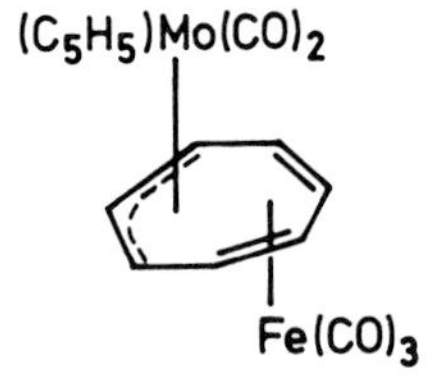

15

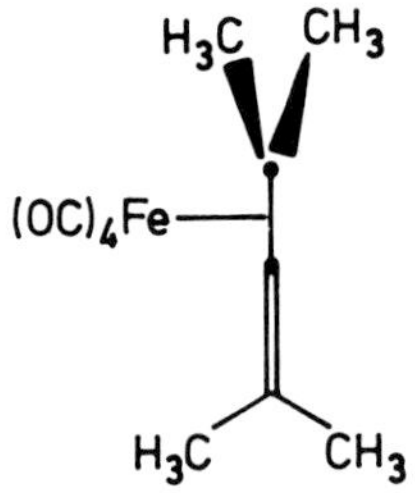

17

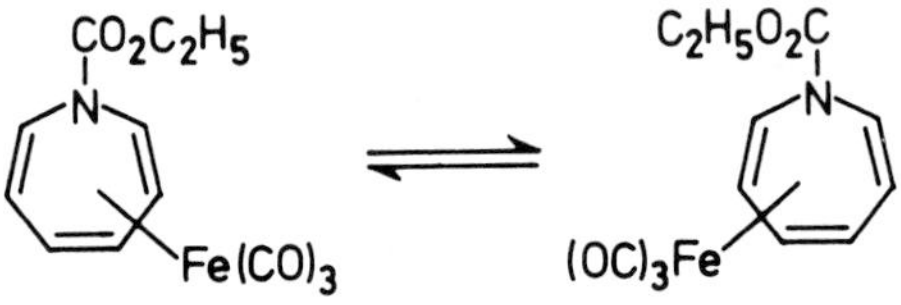

16 16

allylic moiety about an axis passing through the iron atom and approximately perpendicular to the plane defined by the three allylic carbon atoms *(48,69,199)*.

The tetracarbonyliron complex of tetramethylallene (17) is fluxional *(15)*. At ambient temperature, the tetracarbonyliron group has access to the four equivalent bonding sites on the allene skeleton, and rapidly passes among them. There is some controversy as to whether or not the related compound, $C_6H_8Fe_2(CO)_6$ (prepared from allene and $Fe_2(CO)_9$ or $Fe_3(CO)_{12}$) is fluxional *(15,196)*.

C. REARRANGEMENTS INVOLVING σ- AND π-BONDED SYSTEMS

Compounds have been relegated to this category somewhat arbitrarily, depending upon the bonding formalism preferred by the authors of the original literature. Two cyclooctatetraene complexes exhibit interesting behaviour. The sparingly soluble complex (η^8-C_8H_8)$Fe_2(CO)_5$ (18) *(100)* exhibits a single line in the ^{1}H-NMR spectrum at all accessible temperatures *(138)*. The 1,3,5,7-tetramethyl *(61,65)* and ruthenium *(54)*

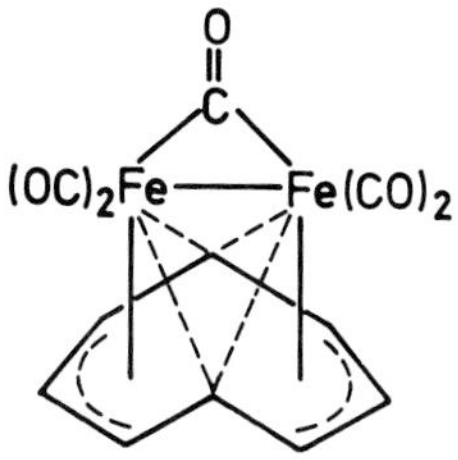

18

analogs are similarly "unstoppable". Indeed, 18 also rearranges rapidly in the solid state *(36,37)*. The precursor to this molecule, $C_8H_8Fe_2(CO)_6$ *(138)*, and its ruthenium *(54)* analog 19 undergo a rapid, degenerate rearrangement [4] which interconverts enantiomers. The cyclooctatriene analog has a

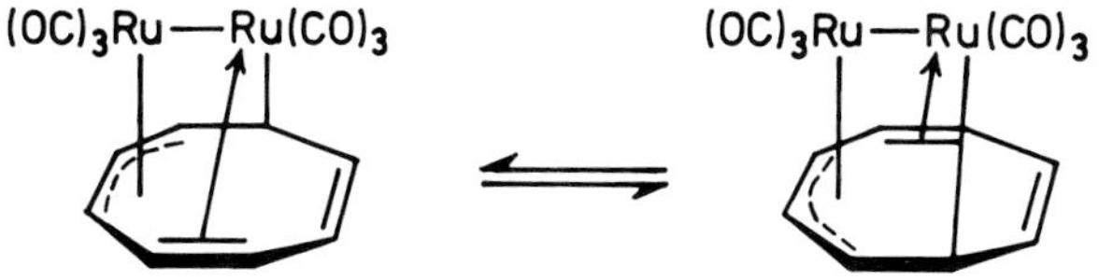

19 19 [4]

similar structure *(56)* and exhibits similar fluxional behaviour *(10,59,62)* as well as two processes involving motion of the carbonyl ligands *(59)*. The hexacarbonyldiiron derivative of bicyclo[6,1,0]nona-1,3,5-triene has an analogous molecular structure and appears to execute the same type of degenerate rearrangement even more rapidly *(233)*. "Pseudoferrocene" systems 20 also were proposed to undergo rapid reorganization of σ and π bonds [5] *(124)*. Considering the molecular structure *(179)* it is more likely that the rings are rapidly changing orientation with respect to one another about the η^5-C_5 centroid-Fe axis.

Aumann *(7,8,9)* has reported two exotic $C_{10}H_{10}Fe_2(CO)_6$ molecules derived from bullvalene which are fluxional by virtue of mobile iron to carbon σ and π bond networks. These are

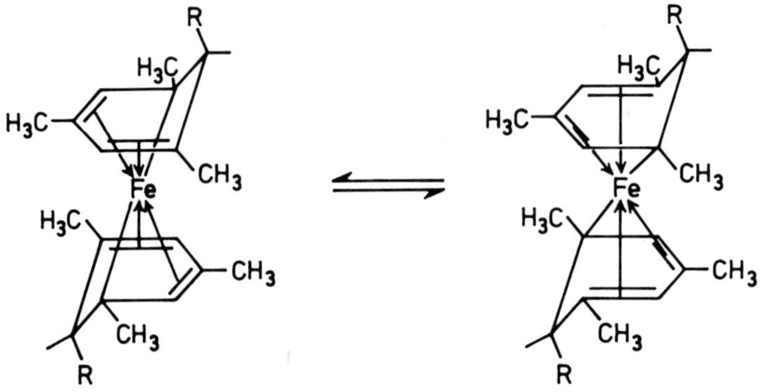

20 20 [5]

"diferratetracyclododecadiene" (21) *(7)* and "diferracyclododecatriene" (22) *(8)*.

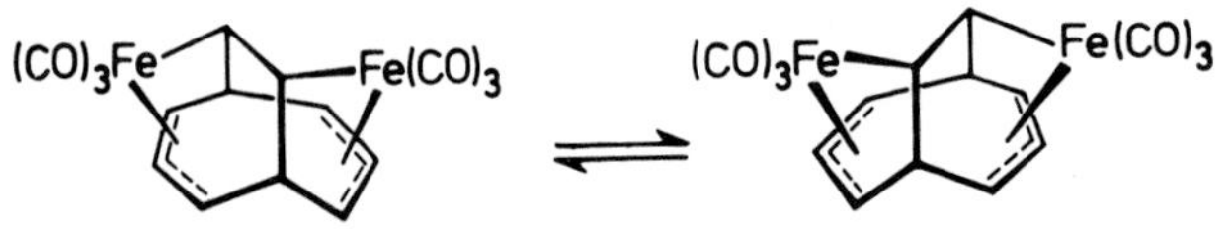

21 21 [6]

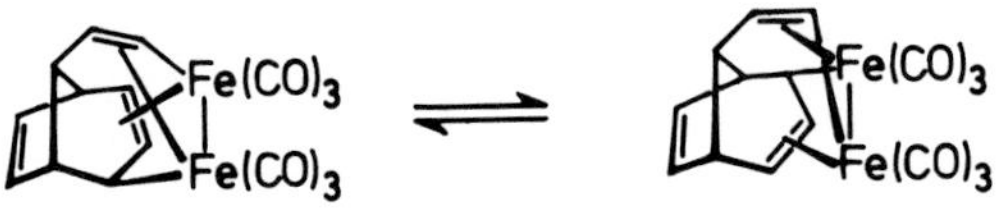

22 22 [7]

The description of the rapid σ-π rearrangement observed in protonated diene tricarbonyliron complexes has now been refined *(25,244,254)*. It appears to involve both rapid intramolecular interchange of the metal-bound hydride with the protons on the diene, as well as slower interchange of the various diene protons (eq. [8]).

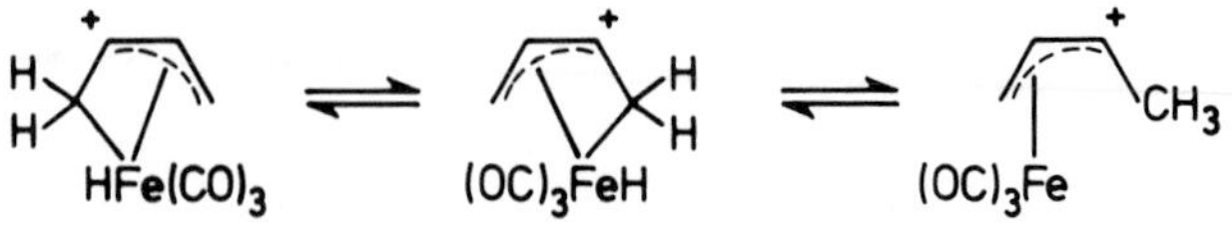

[8]

D. NONRIGID IRON COORDINATION SPHERES

A great many formally five-coordinate organic iron complexes are nonrigid. Early ^{13}C studies revealed that the carbonyls in pentacarbonyliron were magnetically equivalent at all accessible temperatures *(52)*. Recently, nonrigidity has been unambiguously demonstrated for an extensive series of phosphorus trifluoride substituted η^4-diene tricarbonyliron complexes *(35,239,240)*, e.g. 23 (equation [9]), and trimethylenemethane analogs *(46)* as well as simple diene and olefin carbonyliron complexes *(151,152,249)*. In the latter, olefin rotation appears to be coupled with rearrangement of the iron coordination sphere *(152,248)*. Similar kinds of dynamic polytopal *(190)* processes have been identified in the series $[HM(PF_3)_4]^-$, (M = Fe, Ru, Os) *(185,193)* and have been interpreted in terms of hydrogen hopping between the faces of the

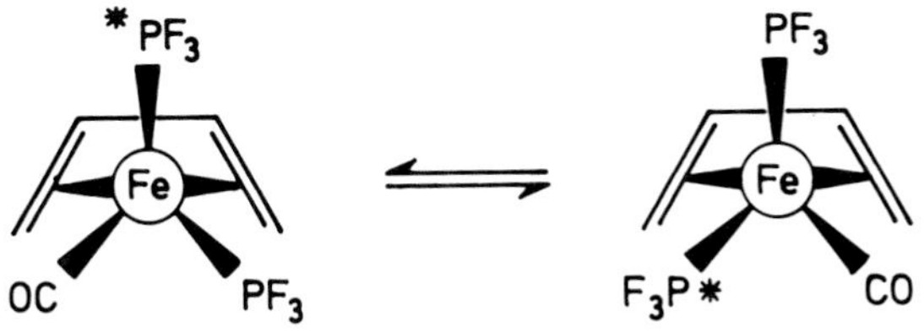

23 23 [9]

approximate tetrahedron defined by the four phosphorus atoms. Trigonal biyramidal phosphite complexes, $Fe[P(OR)_3]_5$, exhibit simultaneous intramolecular interchange of the two axial phosphites with two equatorial phosphites. This has been explained in terms of Berry pseudorotation *(182)*.

Though stereochemical nonrigidity may not be surprising for five coordination *(191)*, it would, *a priori*, seem less likely for 6-coordinate iron complexes *(128,191,192)*. Still, a number of examples have recently been reported. Complexes $Fe(R_1R_2dtc)_2(tfd)$ *(208)* and $[Fe(R_1R_2dtc)_3]^+BF_4^-$ *(85)* (dtc = N,N-dithiocarbamate, tfd = bis(perfluoromethyl)-1,2-dithietene), as well as H_2FeL_4 *(183,184)* where L = various phosphines and phosphites, are dynamic, and significant mechanistic information has been derived from NMR lineshape analyses. In the former systems *(85,208)* the so-called trigonal or Bailar "twist" mechanism *(11)* [10] is operative. The lat-

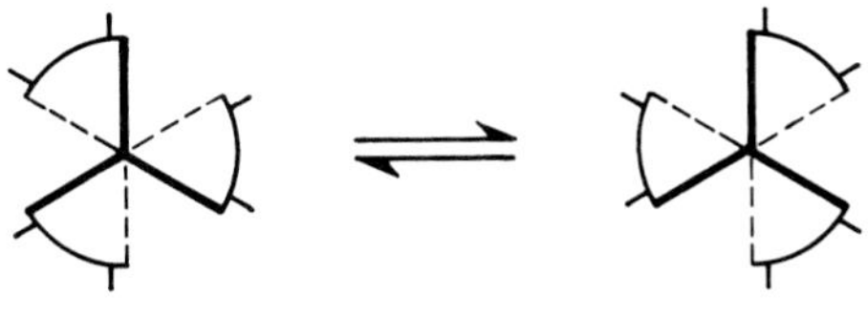

[10]

ter system appears on the basis of detailed lineshape analysis to involve H-hopping about the faces of the approximate tetrahedron represented by the four ligand phosphorus atoms *(183, 184)*. A similar process has been reported in compounds of the formula $Fe(CO)_4[E(CH_3)_3]_2$, E=Si, Ge, Sn, and $Fe(CO)_4[Si(CH_3)_{3-n}Cl_n]_2$ *(236)*. In the *cis*-isomers, scrambling of axial and equatorial carbonyl groups occurs *via* the *trans*-isomers. The cyclic compound $[(n\text{-}C_4H_9)_2SnFe(CO)_4]_2$ (24) exhibits a similar carbonyl interchange process *(116,176)* as do several other $X_2Fe(CO)_4$ systems *(102)*.

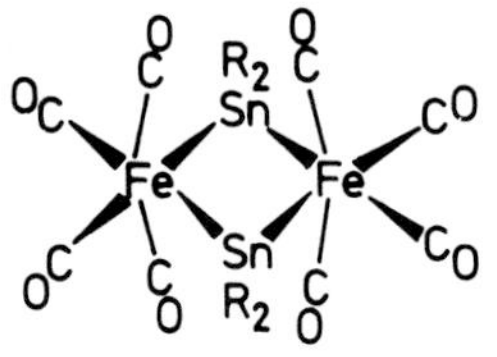

24

Rather than rearrangement of the iron coordination sphere, the ^{1}H-NMR temperature dependence observed for $(\eta^5\text{-}C_5H_5)Fe(CO)[S_2CN(i\text{-}C_3H_7)H]$ is due to restricted rotation about the C-N bond *(31)*.

E. INTERMETAL LIGAND TRANSFER

After years of controversy, the solution structure of $[(\eta^5\text{-}C_5H_5)Fe(CO)_2]_2$ (25) has been elucidated by a combination of variable temperature and solvent ^{1}H-NMR *(32-34)*, ^{13}C-NMR *(107,122)*, and infrared analyses *(34,170,181)*. Three isomers (25a, 25b, 25c) are in dynamic equilibrium [11] *(33,34)*. Similar bridge-terminal ligand interchange processes have been identified for isocyanides *(1,2)* and nitrosyls *(145)*, and are no doubt rather widespread. Where possible, two bridging ligands interchange synchronously with two terminal ligands *(2)*. Apparently, some torsional motion about the metal-metal bond also accompanies carbonyl transfer in molecules such as 25.

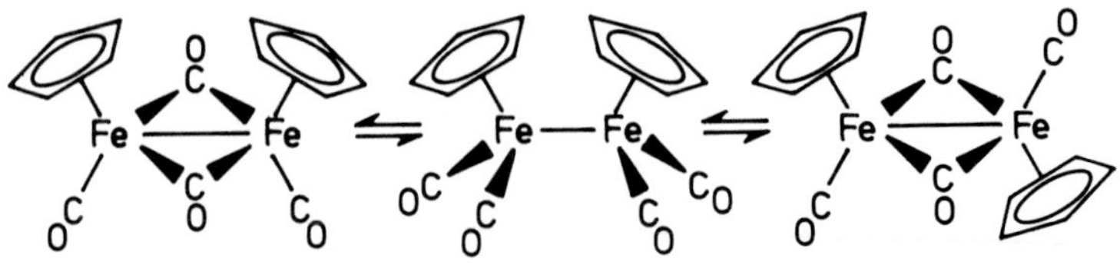 [11]

25a 25b 25c

However, studies with $[(C_5H_5)Fe(CO)_2]_2$-type compounds where the η^5-cyclopentadienyl rings are linked together demonstrate that rapid bridge-terminal carbonyl interchange in "*cis*" (25a) isomers can still occur, without isomerization to the "*trans*" (25c) species *(60)*.

Molecules of the type $[R_2Sn]_2Fe_2(CO)_7$ (26) undergo two types of fluxional processes *(116,176)*. The first involves deformation *(153)* of the FeSnFe bridges to effect interchange

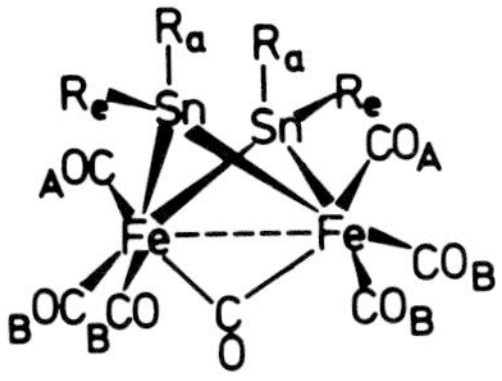

26

of R_e and R_a functionalities. This occurs in concert with a stereospecific rearrangement of the carbonyl ligands so as to replace the bridging carbonyl with one CO(A), and to simultaneously scramble CO(A) and CO(B) units (presumably *via* pseudorotation) on the opposite iron atom. At higher temperatures, breaking of the Fe-Sn bonds occurs to produce short-lived "stannylene" $R_2Sn \rightarrow Fe$ moieties *(116,176)*. Analogous Fe-Ge bond rupture has been reported for $(\eta^5\text{-}C_5H_5)_2Fe_2(CO)_3[\mu\text{-}Ge(CH_3)_2]$ *(3)*.

The solution structure of $Fe_3(CO)_{12}$ remains a mystery, and there is good evidence that phosphine and phosphite-substituted molecules are nonrigid *(211)*. The dynamic process most likely involves extremely rapid bridge$\rightleftharpoons$terminal equilibria of carbonyl groups *(68)*, since a single line is observed in the ^{13}C-NMR spectrum of $Fe_3(CO)_{12}$ down to -160°C.

Acknowledgements

I thank the National Science Foundation and the Alfred P. Sloan Foundation for support during the time this article was written.

REFERENCES

1. Adams, R.D., and Cotton, F.A., *J. Amer. Chem. Soc.*, *95*, 6589 (1973).
2. Adams, R.D., and Cotton, F.A., ref. *109*, Chapter 12.
3. Adams, R.D., Brice, M.D., and Cotton, F.A., *Inorg. Chem.*, *13*, 1080 (1974).
4. Allegra, G., Colomba, A., Immirzi, A., and Bassi, I. W., *J. Amer. Chem. Soc.*, *90*, 4455 (1968).
5. Anet, F.A.L., *J. Amer. Chem. Soc.*, *89*, 2491 (1967).
6. Anet, F.A.L., Kaesz, H.D., Maasbol, A., and Winstein, S., *J. Amer. Chem. Soc.*, *89*, 2489 (1967).
7. Aumann, R., *Angew. Chem.*, *83*, 176 (1971); *Angew. Chem. Int. Ed. Engl.*, *10*, 189 (1971).
8. Aumann, R., *Angew. Chem.*, *83*, 583 (1971); *Angew. Chem. Int. Ed. Engl.*, *10*, 560 (1971).
9. Aumann, R., *Chem. Ber.*, *108*, 1974 (1975).
10. Aumann, R., and Winstein, S., *Angew. Chem.*, *82*, 667 (1970); *Angew. Chem. Int. Ed. Engl.*, *10*, 638 (1970).
11. Bailar, J.C., Jr., *J. Inorg. Nucl. Chem.*, *8*, 165 (1958).
12. Barfield, M., and Grant, D.M., *Advan. Magn. Resonance*, *1*, 149 (1965).
13. Becker, E.D., *High Resolution NMR:Theory and Chemical Applications*, Academic Press, New York, 1969.
14. Bennett, M.J., Cotton, F.A., Davison, A., Faller, J. W., Lippard, S.J., and Morehouse, S.M., *J. Amer. Chem. Soc.*, *88*, 4371 (1966).
15. Ben-Shoshan, R., and Pettit, R., *J. Amer. Chem. Soc.*, *89*, 2231 (1967).
16. Bishop, E.O., Down, J.L., Emtage, P.R., Richards, R. E., and Wilkinson, G., *J. Chem. Soc.*, *1959*, 2484.
17. Bolton, E.S., Knox, G.R., and Robertson, C.G., *Chem. Commun.*, *1969*, 664.
18. Bothner-By, A.A., *Advan. Magn. Resonance*, *1*, 195 (1965).
19. Bovey, F.A., *Nuclear Magnetic Resonance Spectroscopy*, Academic Press, New York, 1969.
20. Bramley, R., Figgis, B.N., and Nyholm, R.S., *Trans. Faraday Soc.*, *58*, 1893 (1962).
21. Bramley, R., Figgis, B.N., and Nyholm, R.S., *J. Chem. Soc.A*, *1967*, 861.
22. Bratton, W.K., Cotton, F.A., Davison, A., Musco, A., and Faller, J.W., *Proc. Nat. Acad. Sci. U.S.*, *58*, 1324 (1967).
23. Braun, S., and Watts, W.E., *J. Organometal. Chem.* *84*, C33 (1975).
24. Breitmaier, E., Jung, G., and Voelter, W., *Angew.*

Chem., *83*, 659 (1971); *Angew. Chem. Int. Ed. Engl.*, *10*, 673 (1971).

25. Brookhart, M., and Harris, D.L., *Inorg. Chem.*, *13*, 1540 (1974).
26. Bruce, M.I., *J. Chem. Soc. A*, *1968*, 1459.
27. Brune, H.A., Hanebeck, H., and Hüther, H., *Tetrahedron*, *26*, 3099 (1970).
28. Brune, H.A., Hüther, H., Horlbeck, G., and Hanebeck, H., *Org. Magn. Resonance*, *3*, 737 (1971).
29. Brune, H.A., Hüther, H., Wolf, R., and Körber, I., *Org. Magn. Resonance*, *1*, 351 (1969).
30. Brune, H.A., and Wolff, H.P., *Z. Naturforsch.*, *23b*, 1184 (1968).
31. Brunner, H., Burgemeister, T., and Wachter, J., *Chem. Ber.*, *108*, 3349 (1975).
32. Bryan, R.F., Greene, P.T., Newlands, M.J. and Field, D.S., *J. Chem. Soc. A*, *1970*, 3068.
33. Bullitt, J.G., Cotton, F.A., and Marks, T.J., *J. Amer. Chem. Soc.*, *92*, 2155 (1970).
34. Bullitt, J.G., Cotton, F.A., and Marks, T.J., *Inorg. Chem.*, *11*, 671 (1972).
35. Busch, M.A., and Clark, R.J., *Inorg. Chem.*, *14*, 226 (1975).
36. Campbell, A.J., Fyfe, C.A., and Maslowsky, E., Jr., *J. Amer. Chem. Soc.*, *94*, 2690 (1972).
37. Campbell, A.J., Fyfe, C.A., Goel, R.G., Maslowsky, E., Jr., and Senoff, C.V., *J. Amer. Chem. Soc.*, *94*, 8387 (1972).
38. Campbell, C.H., and Green, M.L.H., *J. Chem. Soc. A*, *1970*, 1318.
39. Carbonaro, A., Segre, A.L., Greco, A., Tosi, C., and Dall'Asta, G., *J. Amer. Chem. Soc.*, *90*, 4453 (1968).
40. Carrington, A., and McLachlan, A.D., *Introduction to Magnetic Resonance*, Harper and Row, New York, 1967.
41. Chakravorty, A., in C.N.R. Rao and J.R. Ferraro (Eds.), *Spectroscopy in Inorganic Chemistry*, *Vol.I*, Academic Press, New York, 1970, p. 248.
42. Chierico, A., and Mognaschi, E.R., *J. Chem. Soc.*, *Faraday Trans. II*, *69*, 433 (1973).
43. Chisholm, M.H., and Godleski, S., *Prog. Inorg. Chem.* *20*, 299 (1976).
44. Ciappenelli, D., and Rosenblum, M., *J. Amer. Chem. Soc.*, *91*, 6876 (1969).
45. Ciappenelli, D.J., Cotton, F.A., and Kruczynski, L., *J. Organometal. Chem.*, *42*, 159 (1972).
46. Clark, R.J., Abraham, M.R., and Busch, M.A., *J. Organometal. Chem.*, *35*, C33 (1972).
47. Cooke, M., Goodfellow, R.J., Green, M., Maher, J.P.,

and Yandle, J.R., *Chem. Commun., 1970,* 565.
48. Cooke, M., Goodfellow, R.J., and Green, M., *J. Chem. Soc., A., 1971,* 16.
49. Cotton, F.A., *Chem. Brit., 4,* 345 (1968).
50. Cotton, F.A., *Accounts Chem. Res., 1,* 257 (1968).
51. Cotton, F.A., ref. *109,* Chapter 10.
52. Cotton, F.A., Danti, A., Waugh, J.S., and Fessenden, R.W., *J. Chem. Phys., 29,* 1427 (1958).
53. Cotton, F.A., Davison, A., and Faller, J.W., *J. Amer. Chem. Soc., 88,* 4507 (1966).
54. Cotton, F.A., Davison, A., Marks, T.J., and Musco, A., *J. Amer. Chem. Soc., 91,* 6598 (1969).
55. Cotton, F.A., and Edwards, W.T., *J. Amer. Chem. Soc., 90,* 5412 (1968).
56. Cotton, F.A., and Edwards, W.T., *J. Amer. Chem. Soc., 91,* 843 (1969).
57. Cotton, F.A., and Eiss, R., *J. Amer. Chem. Soc., 91,* 6593 (1969).
58. Cotton, F.A., and Hunter, D.L., *J. Amer. Chem. Soc., 98,* 1413 (1976).
59. Cotton, F.A., Hunter, D.L., and Lahuerta, P., *J. Amer. Chem. Soc., 97,* 1046 (1975).
60. Cotton, F.A., Hunter, D.L., Lahuerta, P., and White, A.J., *Inorg. Chem., 15,* 557 (1976).
61. Cotton, F.A., and LaPrade, M.D., *J. Amer. Chem. Soc., 90,* 2026 (1968).
62. Cotton, F.A., and Marks, T.J., *J. Organometal. Chem., 19,* 237 (1969).
63. Cotton, F.A. and Marks, T.J., *J. Amer. Chem. Soc., 91,* 7523 (1969).
64. Cotton, F.A., and Marks, T.J., *J. Organometal. Chem., 19,* 237 (1969).
65. Cotton, F.A., and Musco, A., *J. Amer. Chem. Soc., 90,* 1444 (1968).
66. Cotton, F.A., Musco, A., and Yagupsky, G., *J. Amer. Chem. Soc., 89,* 6136 (1967).
67. Cotton, F.A., and Reich, C.R., *J. Amer. Chem. Soc., 91,* 847 (1969).
68. Cotton, F.A., and Troup, J.M., *J. Amer. Chem. Soc., 96,* 4155 (1974).
69. Cotton, J.D., Doddrell, D., Heazlewood, R.L., and Kitching, W., *Austral. J. Chem., 22,* 1785 (1969).
70. Coyle, T.D., King, R.B., Pitcher, E., Stafford, S.L., Treichel, P.M., and Stone, F.G.A., *J. Inorg. Nucl. Chem., 20,* 172 (1961).
71. Crecely, R.W., Crecely, K.M., and Goldstein, J.H., *Inorg. Chem., 8,* 252 (1969).
72. Crews, P., *J. Amer. Chem. Soc., 95,* 636 (1973).

73. Crutchfield, M.M., Dungan, C.H., Letcher, J.H., Mark, V., and Van Wazer, J.R., *Topics Phosphorus Chem.*, *5*, 19 (1967).
74. Curphey, T.J., Santer, J.O., Rosenblum, M., and Richards, J.H., *J. Amer. Chem. Soc.*, *82*, 5249 (1960).
75. Dailey, B.P., and Shoolery, J.N., *J. Amer. Chem. Soc.*, *77*, 3977 (1955).
76. Davis, D.G., and Kurland, R.J., *J. Chem. Phys.*, *46*, 388 (1967).
77. Davison, A., McFarlane, W., Pratt, L., and Wilkinson, G., *J. Chem. Soc.*, *1962*, 3653.
78. deBoer, E., and Van Willigen, H., *Prog. Nucl. Magnetic Resonance Spectroscopy*, *2*, 111 (1967).
79. Dickens, B., and Lipscomb, W.N., *J. Chem. Phys.*, *37*, 2084 (1962).
80. Diehl, P., and Khetrapal, C.L., *NMR, Basic Principles and Progress*, *1*, 1 (1969).
81. Dietrich, M.W., and Wahl, A.C., *J. Chem. Phys.*, *38*, 1591 (1963).
82. Doering, W. von E., and Roth, W.R., *Angew. Chem.*, *75*, 27 (1963); *Angew. Chem. Int. Ed. Engl.*, *2*, 115 (1963).
83. Drago, R.S., Zink, J.I., and Perry, W.D., *J. Chem. Educ.*, *51*, 371, 464 (1974).
84. Dub, M., *Organometallic Compounds, Vol. I*, Springer-Verlag, Berlin, 1966.
85. Duffy, D.J., and Pignolet, L.H., *Inorg. Chem.*, *11*, 2843 (1972).
86. Duncan, J.D., Green, J.C., Green, M.L.H., and McLauchlan, K.A., *Chem. Commun.*, *1968*, 721.
87. Eaton, D.R., McClellan, W.R., and Weiher, J.F., *Inorg. Chem.*, *7*, 2040 (1968).
88. Eaton, D.R., in H.A.O. Hill and P. Day (Eds.), *Physical Methods in Advanced Inorganic Chemistry*, Interscience, London, 1968.
89. Eaton, D.R., and Phillips, W.D., *Advan. Magn. Resonance*, *1*, 103 (1965).
90. Fauvel, K., Mathieu, R., and Poilblanc, R., *Inorg. Chem.*, *15*, 976 (1976).
91. Eaton, G.R., and Lipscomb, W.N., *NMR Studies of Boron Hydrides and Related Compounds*, W.A. Benjamin, New York, 1969.
92. Emerson, G.F., Ehrlich, K., Giering, W.P., and Lauterbur, P.C., *J. Amer. Chem. Soc.*, *88*, 3172 (1966).
93. Emsley, J.W., Feeney, J., and Sutcliffe, L.H., *High Resolution Nuclear Magnetic Resonance Spectroscopy*, Pergamon Press, Oxford, 1966.
94. Evans, J., and Norton, J.R., *Inorg. Chem.*, *13*, 3024 (1974).

95. Faller, J.W., Johnson, B.V., and Schaeffer, C.D., Jr., *J. Amer. Chem. Soc., 98,* 1395 (1976).
96. Farrar, T.C., and Becker, E.D., *Pulse and Fourier Transform NMR,* Academic Press, New York, 1971.
97. Fields, R., Germain, M.M., Haszeldine, R.N., and Wiggans, P.W., *Chem. Commun., 1967,* 243.
98. Fields, R., *Annu. Rep. NMR Spectrosc., 5A,* 99 (1972).
99. Fischer, E.O., and Werner, H., *Metal π-Complexes,* Elsevier, Amsterdam, 1966.
100. Fleischer, E.B., Stone, A.L., Dewar, R.B.K., Wright, J.D., Keller, C.E., and Pettit, R., *J. Amer. Chem. Soc., 88,* 3158 (1966).
101. Foreman, M.J., and Leppard, D.G., *J. Organometal. Chem., 31,* C31 (1971).
102. Forster, A., Johnson, B.F.G., Lewis, J., Matheson, T.W., Robinson, B.H., and Jackson, W.G., *J. Chem. Soc. Chem. Commun., 1974,* 1042.
103. Fritz, H.P., Keller, H.J., and Schwarzhans, K.E., *J. Organometal. Chem., 7,* 105 (1967).
104. Fritz, H.P., Keller, H.J., and Schwarzhans, K.E., *J. Organometal. Chem., 6,* 652 (1966).
105. Fritz, H.P., Keller, H.J., and Schwarzhans, K.E., *Z. Naturforsch., 23b,* 298 (1968).
106. Gansow, O.A., Schexnayder, D.A., and Kimura, B.Y., *J. Amer. Chem. Soc., 94,* 3406 (1972).
107. Gansow, O.A., Burke, A.R., and Vernon, W.D., *J. Amer. Chem. Soc., 94,* 2550 (1972).
108. George, W.O., *Spectroscopic Methods in Organometallic Chemistry,* Chemical Rubber Co. Press, Cleveland, 1970.
109. Gillies, D.G., and Shaw, D., *Annu. Rep. NMR Spectrosc., 5A,* 560 (1973).
110. Ginsberg, A.P., *Transition Metal Chem., 1,* 111 (1965).
111. Golding, R.M., Tennant, W.C., Kanekar, C.R., Martin, R.L., and White, A.H., *J. Chem. Phys., 45,* 2688 (1966).
112. Golding, R.M., Tennant, W.C., Bailey, J.P.M., and Hudson, A., *J. Chem. Phys., 48,* 764 (1968).
113. Green, M.L.H., Nagy, P.L.I., *J. Organometal. Chem., 1,* 58 (1963).
114. Greenwood, N.N., *Spectroscopic Properties of Inorganic and Organometallic Compounds, Vol. 1-8,* Specialist Periodical Reports, The Chemical Society, London, 1967-1975.
115. Grubbs, R., Breslow, R., Herber, R., and Lippard, S.J., *J. Amer. Chem. Soc., 89,* 6864 (1967).
116. Grynkewich, G.W., and Marks, T.J., *Inorg. Chem., in press.*
117. Günther, H., and Wenzl, R., *Tetrahedron Lett., 1967,*

4155.
118. Gupta, R.K., Koenig, S.H., and Redfield, A.G., *J. Magnetic Resonance*, *7*, 66 (1972).
119. Gutowsky, H.S., and Jonas, J., *Inorg. Chem.*, *4*, 430 (1965).
120. Hagihara, N., Kumada, M., and Okawara, R., *Handbook of Organometallic Compounds*, Benjamin, New York, 1968.
121. Harris, D.C., and Gray, H.B., *Inorg. Chem.*, *14*, 1215 (1975).
122. Harris, D.C., Rosenberg, E., and Roberts, J., *J. Chem. Soc. Dalton Trans.*, *1974*, 2398.
123. Hawthorne, M.F., Young, D.C., Andrews, T.D., Howe, D.V., Pilling, R.L., Pitts, A.D., Reintjes, M., Warren, L.F., and Wagner, P.A., *J. Amer. Chem. Soc.*, *90*, 879 (1968).
124. Helling, J.F., and Braitsch, D.M., *J. Amer. Chem. Soc.*, *92*, 7209 (1970).
125. Henderson, W.G., and Mooney, E.F., *Annu. Rev. NMR Spectrosc.*, *2*, 219 (1969).
126. Herbison-Evans, D., and Richards, R.E., *Mol. Phys.*, *8*, 19 (1964).
127. Holm, R.H., Phillips, W.D., Averill, B.A., Mayerle, J.J., and Herskovitz, T., *J. Amer. Chem. Soc.*, *96*, 2109 (1974).
128. Holm, R.H., ref. *109*, Chapter 9.
129. Horrocks, W.DeW., Jr., and Greenberg, E.S., *Biochim. Biophys. Acta*, *322*, 38 (1973).
130. Hyatt, D.E., Little, J.L., Moran, J.T., Scholer, F.R., and Todd, L.J., *J. Amer. Chem. Soc.*, *89*, 3342 (1967).
131. Jackman, L.M., and Cotton, F.A., *Dynamic Nuclear Magnetic Resonance Spectroscopy*, Academic Press, New York, 1975.
132. Jesson, J.P., Trofimenko, S., and Eaton, D.R., *J. Amer. Chem. Soc.*, *89*, 3158 (1967).
133. Jesson, J.P., Trofimenko, S., and Eaton, D.R., *J. Amer. Chem. Soc.*, *89*, 3148 (1967).
134. Johnson, C.S.,Jr., *Advan. Magn. Resonance*, *1*, 33 (1965).
135. Jones, K., and Mooney, E.F., *Annu. Rep. NMR Spectrosc.*, *3*, 261 (1970).
136. Juan, C., and Gutowsky, H.S., *J. Chem. Phys.*, *37*, 2198 (1962).
137. Kaesz, H.D., and Saillant, R.B., *Chem. Rev.*, *72*, 231 (1972).
138. Keller, C.E., Emerson, G.F., and Pettit, R., *J. Amer. Chem. Soc.*, *87*, 1388 (1965).
139. Keller, C.E., Shoulders, B.A., and Pettit, R., *J. Amer. Chem. Soc.*, *88*, 4760 (1966).

140. Keller, H.J., *NMR, Basic Principles and Progress, 2,* 1 (1970).
141. Keller, H.J., and Schwarzhans, K.E., *Angew. Chem., 82,* 227 (1970); *Angew. Chem. Int. Ed. Engl., 9,* 196 (1970).
142. Kidd, R.G. in M. Tsutsui (Ed.), *Characterization of Organometallic Compounds, Part II,* Interscience, New York, 1971, p. 373.
143. Kimura, B.Y., and Gansow, O.A., *Abstr. 160th Natl. Meeting Amer. Chem. Soc., Chicago,* 1970, INOR 154.
144. King, R.B., Kapoor, R.N., and Houk, L.W., *J. Inorg. Nucl. Chem., 31,* 2179 (1969).
145. Kirchner, R.M., Marks, T.J., Kristoff, J.S., Ibers, J.A., *J. Amer. Chem. Soc., 95,* 6602 (1973).
146. Köhler, F.H., *J. Organometal. Chem., 69,* 145 (1974).
147. Köhler, F.H., *Chem. Ber., 107,* 570 (1974).
148. Köhler, F.H., and Matsubayashi, G.-E., *J. Organometal. Chem., 96,* 391 (1975).
149. Koridze, A.A., Petrovskii, P.V., Gubin, S.P., and Fedin, E.I., *J. Organometal. Chem., 93,* C26 (1975).
150. Kreiter, C.G., Maasbol, A., Anet, F.A.L., Kaesz, H.D., and Winstein, S., *J. Amer. Chem. Soc., 88,* 3444 (1966).
151. Kruczynski, L., and Takats, J., *J. Amer. Chem. Soc., 96,* 932 (1974).
152. Kruczynski, L., Lishingman, L.K.K., and Takats, J., *J. Amer. Chem. Soc., 96,* 4006 (1974).
153. Kummer, D., and Furrer, J., *Z. Naturforsch., B 26,* 162 (1971).
154. La Mar, G.N., and Van Hecke, G.R., *J. Amer. Chem. Soc., 91,* 3442 (1969).
155. La Mar, G.N., and Walker, F.A., *J. Amer. Chem. Soc., 95,* 1782 (1973).
156. La Mar, G.N., *J. Amer. Chem. Soc., 95,* 1662 (1973).
157. La Mar, G.N., Horrocks, W.DeW., Jr., Holm, R.H., *NMR of Paramagnetic Molecules, Principles and Applications,* Academic Press, New York, 1973.
158. La Mar, G.N., and Walker, F.A., *J. Amer. Chem. Soc., 95,* 6950 (1973).
159. Larsen, D.W., and Wahl, A.C., *J. Chem. Phys., 43,* 3765 (1965).
160. Laszlo, P., *Prog. Nucl. Magn. Resonance Spectrosc., 3,* 231 (1967).
161. Lauterbur, P.C., and King, R.B., *J. Amer. Chem. Soc., 87,* 3266 (1965).
162. Levy, G.C., and Nelson, G.L., *Carbon-13 Nuclear Magnetic Resonance for Organic Chemists,* Wiley-Interscience, New York, 1972.
163. Lindstrom, T.R., Ho, C., and Pisciotta, A.V., *Nature*

(London), New Biol., 237, 263 (1972).

164. Lippard, S.J., *Trans. New York Acad. Sci., 29,* 917 (1967).

165. Maddox, M.L., Stafford, S.L., and Kaesz, H.D., *Advan. Organometal. Chem., 3,* 1 (1965).

166. Mahler, J.E., Jones, D.A.K., and Pettit, R., *J. Amer. Chem. Soc., 86,* 3589 (1964).

167. Maltz, H., and Kelly, B.A., *Chem. Commun., 1971,* 1390.

168. Mann, B.E., *Chem. Commun., 1971,* 1173.

169. Mann, B.E., *Advan. Organometal. Chem., 12,* 135 (1974).

170. Manning, A.R., *J. Chem. Soc. A, 1968,* 1319.

171. Manuel, T.A., *Inorg. Chem., 3,* 510 (1964).

172. Manuel, T.A., and Stone, F.G.A., *J. Amer. Chem. Soc., 82,* 366 (1960).

173. Markezich, R.L., and Whitlock, H.W., Jr., *J. Amer. Chem. Soc., 93,* 5291 (1971).

174. Marks, T.J., Ph.D. Thesis, M.I.T., Cambridge, Mass., 1970.

175. Marks, T.J., unpublished observations.

176. Marks, T.J., and Grynkewich, G.W., *J. Organometal. Chem., 91,* C9 (1975).

177. Marks, T.J., Kristoff, J.S., Alich, A., and Shriver, D.F., *J. Organometal. Chem., 33,* C35 (1971).

178. Marks, T.J., Kristoff, J.S., Porter, R., and Shriver, D.F., in R. Stevers (Ed.), *NMR Shift Reagents,* Academic Press, New York, 1973.

179. Mathew, M., and Palenik, G.J., *Inorg. Chem., 11,* 2809 (1972).

180. Mavel, G., *Annu. Rep. NMR Spectrosc., 5B,* (1973).

181. Mc Ardle, P.A., and Manning, A.R., *J. Chem. Soc, A, 1969,* 1498.

182. Meaking, P., English, A.D., Ittel, S.D., and Jesson, J.P., *J. Amer. Chem. Soc., 97,* 1254 (1975).

183. Meakin, P., Muetterties, E.L., Tebbe, F.N., and Jesson, J.P., *J. Amer. Chem. Soc., 93,* 4701 (1971).

184. Meaking, P., Muetterties, E.L., and Jesson, J.R., *J. Amer. Chem. Soc., 95,* 75 (1973).

185. Meakin, P., Muetterties, E.L., and Jesson, J.P., *J. Amer. Chem. Soc., 94,* 5271 (1972).

186. Meiboom, S., and Snyder, L.C., *Accounts Chem. Res., 4,* 81 (1971).

187. Mooney, E.F., and Winson, P.H., *Annu. Rev. NMR Spectrosc., 1,* 244 (1968).

188. Muetterties, E.L., and Phillips, W.D., *Advan. Inorg. Chem. Radiochem., 4,* 231 (1962).

189. Muetterties, E.L., *Inorg. Chem., 4,* 769 (1965).

190. Muetterties, E.L., in M.L. Tobe (Ed.), *MTP International Review of Science, Inorganic Chemistry,*

Ser. 1., Vol. 9, University Park Press, Baltimore, 1972, p. 37.

191. Muetterties, E.L., *J. Amer. Chem. Soc., 91,* 1636 (1969).

192. Muetterties, E.L., *J. Amer. Chem. Soc., 91,* 4115 (1969).

193. Muetterties, E.L., *Abstr. 162nd Natl. Meeting. Amer. Chem. Soc., Washington, D.C.,* 1971, INOR 116.

194. Mulay, L.N., and Attalla, A., *J. Amer. Chem. Soc., 85,* 702 (1963).

195. Murrell, J.N., *Prog. Nucl. Magn. Resonance Spectrosc., 6,* 1 (1971).

196. Nakamura, A., *Bull. Chem. Soc. Japan, 39,* 543 (1966).

197. Nakamura, A., and Hagihara, N., *Bull. Chem. Soc. Japan, 32,* 880 (1959).

198. Narasimhan, P.T., and Rogers, M.T., *J. Amer. Chem. Soc., 82,* 5983 (1960).

199. Nesmeyanov, A.N., Ustynyuk, Yu. A., Kritskaya, I.I., and Shchembelov, G.A., *J. Organometal. Chem., 14,* 395 (1968).

200. Nixon, J.F., and Pidcock, A., *Annu. Rep. NMR Spectrosc., 2,* 346 (1969).

201. Nöth, H., and Schmid, G., *Allg. prakt. Chem., 17,* 610 (1966).

202. Ogilvie, F., Clark, R.J., and Verkade, J.G., *Inorg. Chem., 8,* 1904 (1969).

203. Olah, G.A., and Liang, G., *J. Org. Chem., 40,* 1849 (1975).

204. Palazzotto, M.C., and Pignolet, L.H., *Inorg. Chem., 13,* 1781 (1974).

205. Parshall, G.W., *J. Amer. Chem. Soc., 88,* 704 (1966).

206. Phillips, W.D., ref. *131,* Chapt. 11.

207. Pignolet, L.H., Forster, D., and Horrocks, W.DeW.,Jr., *Inorg. Chem., 7,* 828 (1968).

208. Pignolet, L.H., Lewis, R.A., and Holm, R.H., *J. Amer. Chem. Soc., 93,* 360 (1971).

209. Piper, T.S., and Wilkinson, G., *J. Inorg. Nucl. Chem., 3,* 104 (1956).

210. Pitcher, E., Buckingham, A.D., and Stone, F.G.A., *J. Chem. Phys., 36,* 124 (1962).

211. Pollick, P.J., and Wojcicki, A., *J. Organometal. Chem., 14,* 469 (1968).

212. Pople, J.A., Schneider, W.G., and Bernstein, H.J., *High-Resolution Nuclear Magnetic Resonance,* McGraw-Hill, New York, 1959.

213. Porter, R., Marks, T.J., and Shriver, D.F., *J. Amer. Chem. Soc., 95,* 3548 (1973).

214. Preston, H.G., Jr., and Davis, J.C., Jr., *J. Amer.*

Chem. Soc., *88*, 1585 (1966).
215. Randall, E.W., and Gillies, D.G., *Prog. Nucl. Magn. Resonance Spectrosc.*, *6*, 119 (1971).
216. Rausch, M.D., and Schrauzer, G.N., *Chem. Ind.* (London), *1959*, 957.
217. Rausch, M.D., and Siegel, A., *J. Organometal. Chem.*, *17*, 117 (1969).
218. Reger, D.L., *Inorg. Chem.*, *14*, 660 (1975).
219. Retcofsky, H.L., Frankel, E.N., and Gutowsky, H.S., *J. Amer. Chem. Soc.*, *88*, 2710 (1966).
220. Rettig, M.F., and Drago, R.S., *J. Amer. Chem. Soc.*, *91*, 1361 (1969).
221. Rettig, M.F., ref. *131*, Chapt. 6.
222. Rigatti, G., Boccalon, G., Ceccon, A., and Giacometti, G., *J. Chem. Soc., Chem. Commun.*, *1972*, 1165.
223. Ronayne, J., and Williams, D.H., *Annu. Rev. NMR Spectrosc.*, *2*, 83 (1969).
224. Roth, W.R., and Meier, J.D., *Tetrahedron Lett.*, *1967*, 2053.
225. Sato, F., Yamamoto, T., Wilkinson, J.R., and Todd, L.J., *J. Organometal. Chem.*, *86*, 243 (1975).
226. Schaeffer, R., *Prog. Boron Chem.*, *1*, 417 (1964).
227. Schurig, V., *Tetrahedron Lett.*, *1976*, 1269.
228. Sheldrick, G.M., *Chem. Commun.* *1967*, 751.
229. Shporer, M., Ron, G., Loewenstein, A., and Navon, G., *Inorg. Chem.*, *4*, 361 (1965).
230. Stothers, J.B., *Carbon-13 NMR Spectroscopy*, Academic Press, New York, 1972.
231. Su, C.-C., *J. Amer. Chem. Soc.*, *93*, 5653 (1971).
232. Taft, R.W., Price, E., Fox, I.R., Lewis, I.C., Andersen, K.K., and Davis, G.T., *J. Amer. Chem. Soc.* *85*, 709 3146 (1963).
233. Takats, J., *J. Organometal. Chem.*, *90*, 211 (1975).
234. Tiers, G.V.D., *J. Phys. Chem.*, *62*, 1151 (1958).
235. Todd, L.J., and Wilkinson, J.R., *J. Organometal. Chem.*, *77*, (1974).
236. Vancea, L., Pomeroy, R.K., and Graham, W.A.G., *J. Amer. Chem. Soc.*, *98*, 1407 (1976).
237. Vrieze, K., Vanleeuwen, P.W.N.M., *Prog. Inorg. Chem.*, *14*, 2 (1971).
238. Wang, J.T., Yeh, H.J.C., and Johnson, D.F., *J. Amer. Chem. Soc.*, *97*, 1968 (1975).
239. Warren, J.D., and Clark, R.J., *Inorg. Chem.*, *9*, 373 (1970).
240. Warren, J.D., Busch, M.A., and Clark, R.J., *Inorg. Chem.*, *11*, 452 (1972).
241. Webb, G.A., *Annu. Rep. NMR Spectrosc.*, *3*, 211 (1970).

242. Whitesides, G.M., and Fleming, J.S., *J. Amer. Chem. Soc., 89,* 2855 (1967).
243. Whitesides, T.H., and Budnik, R.A., *Chem. Commun., 1971,* 1514.
244. Whitesides, T.H., and Arhart, R.W., *Inorg. Chem., 14,* 209 (1975).
245. Whitlock, H.W., Jr., and Markezich, R.L., *J. Amer. Chem. Soc., 93,* 5290 (1971).
246. Whitlock, H.W., Jr., and Stucki, H., *J. Amer. Chem. Soc., 94,* 8594 (1972).
247. Wicholas, M., and Drago, R.S., *J. Amer. Chem. Soc., 91,* 5963 (1969).
248. Wicholas, M., Mustacich, R., Johnson, B., Smedley, T., and May, J., *J. Amer. Chem. Soc., 97,* 2113 (1975).
249. Wilson, S.T., Coville, N.J., Shapely, J.R., and Osborn, J.A., *J. Amer. Chem. Soc., 96,* 4038 (1974).
250. Wiersema, R.J., and Hawthorne, M.F., *J. Amer. Chem. Soc., 96,* 761 (1974).
251. Witanowski, M., and Webb, G.A., *Annu. Rep. NMR Spectrosc., 5A,* 395 (1972).
252. Wüthrich, K., *Struct. Bonding (Berlin), 8,* 53 (1970).
253. Yannoni, C.S., Ceasar, G.P., and Dailey, B.P., *J. Amer. Chem. Soc., 89,* 2833 (1967).
254. Young, D.A.T., Holmes, J.R., and Kaesz, H.D., *J. Amer. Chem. Soc., 91,* 6968 (1969).
255. Zumdahl, S.S., and Drago, R.S., *J. Amer. Chem. Soc., 89,* 4319 (1967).
256. Zumdahl, S.S., and Drago, R.S., *Inorg. Chem., 7,* 2162 (1968).

Mass Spectra

By JÖRN MÜLLER

Institut für Anorganische und Analytische Chemie der Technischen Universität, Berlin, Germany

TABLE OF CONTENTS

ISBN 0-12-417101-X (v.1)

I. INTRODUCTION

During the last ten years mass spectrometry has become a very efficient tool for solving analytical and structural problems in organometallic chemistry *(9,19,23,60,68,76)*. The general principles of this method have been well described in many books and reviews and will not be mentioned further here.

The application of mass spectrometry to organoiron compounds is restricted to complexes which on heating in a high vacuum can reach a vapour pressure of about 10^{-6} torr without complete decomposition. Many organoiron complexes are sufficiently volatile and stable, even several substances containing quite complicated ligands like vitamin-A tricarbonyliron *(5,21)*. However, thermal decomposition processes which take place in the inlet system or in the ion source of the spectrometer and which may be catalyzed by metal surfaces or by very thin metal deposits can yield species not originally present in the sample *(93,103)*. In the mass spectra of cyclopentadienyl carbonyliron derivatives ferrocene is found, and many tetracarbonyliron complexes are partially decomposed yielding pentacarbonyliron *(49,52)*. Under normal operating conditions tricarbonyl(norbornadiene-7-one)iron exhibits the expected mass spectrum; however, passing the sample through a suboven of 200° attached to the mass spectrometer results only in the appearance of the molecular ion of benzene, illustrating that mass spectra of organoiron complexes should be run at temperatures as low as possible *(65)*.

The difficulties arising from thermal decomposition during evaporation of organometallic species can be overcome by use of a field desorption ion source. This quickly developing technique even allows cationic complexes to be investigated by mass spectroscopy.

The peak at highest m/e value of a mass spectrum is assumed to be the molecular or parent ion (P^+). However, it must be mentioned that the molecular ions of several iron complexes are either insignificant or are totally absent in electron impact mass spectra. In the spectra of $LFe(CO)_4$ complexes with L = maleic anhydride or dimethylmaleate *(63)*, of $CpFe(CO)_2COPh$ *(52)* as well as of tricarbonyl-{5-[2-(5,5-dimethylcyclohexane-1,3-dionato)]cyclohexa-1,3-diene}iron *(4)* the peaks at heaviest masses correspond to the $(P - CO)^+$ ions; the π-allyl compounds $C_3H_5Fe(CO)_3X$ with X = NO_3, Cl, Br only show the $(P - CO)^+$ and $(P - X)^+$ ions instead of the molecular peaks *(92)*. The application of field desorption mass spectrometry, however, promises to make parent ions detectable even in such cases.

The isotopic pattern of a peak group in a mass spectrum will show whether the corresponding parent or fragment ion contains one or perhaps more iron atoms. Figure 1 shows the isotopic patterns of the species Fe to Fe_4. The exact formula of an ion can be determined by precise mass measurement using a double-focusing mass spectrometer.

The fragmentation pattern of a complex as derived from the observed fragment ions and the corresponding "metastable peaks" of the mass spectrum principally allow the structure of the compound to be evaluated. Unfortunately, for more complicated molecules this problem can only be solved empirically. In this chapter typical fragmentation processes of organometallic ions in the mass spectrometer are discussed and elucidated by examples of organoiron chemistry.

II. THE FRAGMENTATION OF ORGANOIRON IONS

As demonstrated by the values in Table 1 the ionization potentials of organoiron compounds are lower than those of the free ligands but differ only slightly from the ionization potentials of the free metal atom. Moreover, photoelectron spectra of complexes like ferrocene or pentacarbonyliron suggest that ionization primarily involves the removal of an electron associated with the metal atom - possibly from a molecular orbital involving considerable contribution from the metal atomic *d*-orbitals *(69,104)*. The positive charge of an ion is therefore assumed to be predominantly localized on the metal atom. These consider-

Table 1: Ionization potentials of some organoiron complexes and of the corresponding free ligands.

Compound	IP [eV]
$Fe(CO)_5$	8.14 *(35)*
$Fe(CO)_2(NO)_2$	8.45 *(36)*
$Fe(CO)(NO)_2P(OC_2H_5)_3$	7.50 *(36)*
1,3-cyclohexadiene-$Fe(CO)_3$	8.0 *(106)*
$FeCp_2$	7.15 *(77)*
CO	14.11
NO	9.25
$P(OC_2H_5)_3$	8.40
1,3-cyclohexadiene	8.40
$C_5H_5\bullet$	8.69
Fe	7.87

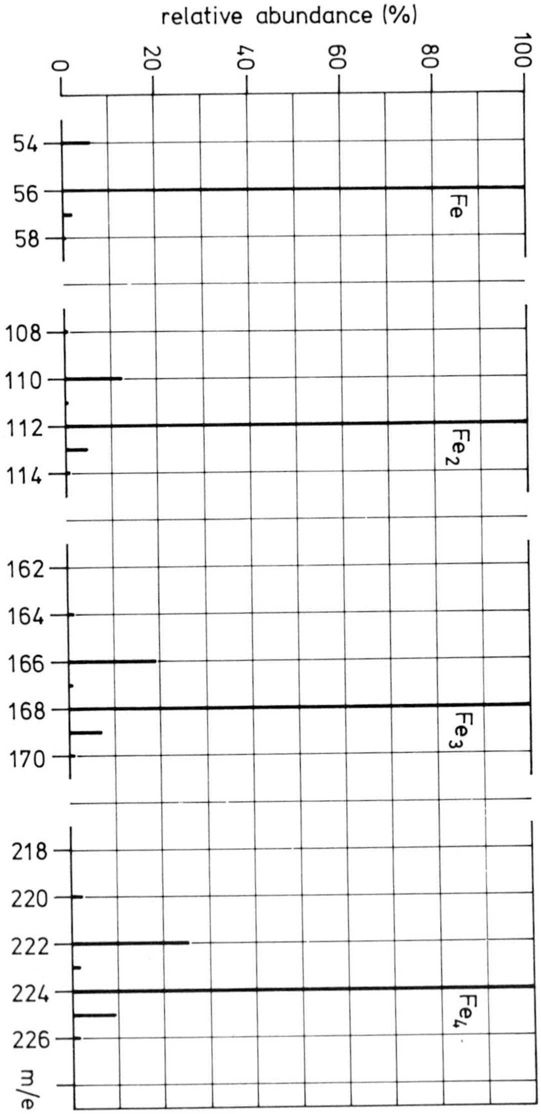

Fig. 1: The isotopic patterns of the species Fe, Fe_2, Fe_3, and Fe_4.

ations explain two important observations: 1. The fragmentation processes of organoiron ions are to a considerable extent controlled by the central iron atom; in many cases the behaviour of an organic molecule under electron impact is remarkably altered if this molecule is complexed by a transition metal. 2. A decomposition step usually leaves the positive charge on the metal-containing fragment.

Among the fragmentation processes of organoiron compounds a rough distinction between three types of reactions is possible: 1. Cleavage of metal-ligand and of metal-metal bonds; 2. Simple bond rupture within complexed ligands; 3. Fragmentation accompanied by rearrangements. Since ion structure determination is quite difficult it may occur that processes which are considered to be of type 2 in fact involve rearrangements. Although in many cases the decomposition of an ion occurs as a competition of all three fragmentation types these will be discussed separately.

A. CLEAVAGE OF METAL-LIGAND AND METAL-METAL BONDS

This simple decomposition mode of organoiron complexes is the most common one. Cleavage of a metal-ligand bond is to be expected if the ligand molecules or radicals are stable enough to be eliminated as neutral particles, if they do not contain functional groups which can be easily attacked, and if the metal-ligand bonds are weak relative to the bonds between the ligand atoms. Thus, the mass spectra of iron complexes with simple ligands like CO, NO, ethylene, acetylene, cyclopentadienyl etc. provide that the structural formulas of the compounds will be easily recognized from the masses of the consecutively lost neutral particles. Usually an iron-ligand bond rupture is indicated by the corresponding metastable peaks. In most cases the fragmentation ends up with the formation of the bare iron ion.

The spectra of the binary iron carbonyls are rather simple; pentacarbonyliron exhibits the series of the ions $Fe(CO)_n^+$ (n = 0 - 5), and metastable peaks prove that the parent ion decomposes by stepwise loss of all CO groups *(3, 35,47,105,107)*. The fragmentation of $Fe_2(CO)_9$ and $Fe_3(CO)_{12}$ is characterized by a competition between metal-ligand and metal-metal bond ruptures *(24,49,66,67)*. The spectrum of $Fe_3(CO)_{12}$ for example shows the whole series of the ions $Fe_3(CO)_{12}^+$ to Fe_3^+ as well as bi- and mononuclear fragments. In accord with the fact that the stability of metal cluster compounds usually increases with increasing number of metal atoms the proportion of bi- and trinuclear ions in the spectrum of $Fe_3(CO)_{12}$ is much higher than that of the binuclear species in the mass spectrum of $Fe_2(CO)_9$.

The molecular ion of ferrocene forms the base peak in the mass spectrum reflecting the high stability of the ferricenium cation *(26,38,72,77,100)*. The main fragments again produced by metal-ligand bond cleavage are $CpFe^{+}$ and Fe^{+}. Aromatic ligands with extended π-electron systems can stabilize a positive charge better than the cyclopentadienyl ring. This is demonstrated by the fragmentation pattern of bisindenyliron, $Fe(C_9H_7)_2$, which after ionization first loses one of its indenyl ligands giving the ion $C_9H_7Fe^{+}$ *(56)*. The metal-ligand bond cleavage of this fragment proceeds almost exclusively by expulsion of an iron atom whereby the $C_9H_7^{+}$ ion is produced; the ion Fe^{+} is observed only in low abundance. In addition the ion $C_{18}H_{14}^{+}$ arising from the coupling of two $C_9H_7\cdot$ radicals appears.

The number of binary organoiron complexes is restricted to a few examples and normally a complex contains more than one type of ligand. The question arises by what ligand properties the sequence of metal-ligand bond cleavages is influenced. Some general rules will be given although it is impossible to present a precisely true ligand series.

As a consequence of the positive charge mainly located on the iron atom the metal-to-ligand back donation is weakened, while a donor ligand is able to stabilize the positive charge on an ion. This means, in other words, that the probability of metal-ligand bond cleavage in an organoiron ion increases with increasing acceptor strength and with decreasing donor ability of a ligand and *vice versa*. One of the best known examples of this rule is the rupture of the metal-CO bond which in many carbonyliron complexes with additional ligands occurs in such a way that the molecular ion loses all of its carbonyl groups before any types of further fragmentation processes take place. It is beyond the scope of this chapter to discuss all known cases. The following equations only present some characteristic examples.

$$LFe(CO)_4^{+} \xrightarrow[\text{stepwise}]{-4CO} LFe^{+} \xrightarrow{-L} Fe^{+}$$

L = monoolefin *(63)*; carbene *(34,90)*; phosphine and phosphite *(6,36,51)*.

$$LFe(CO)_3^{+} \xrightarrow[\text{stepwise}]{-3CO} LFe^{+} \xrightarrow{-L} Fe^{+}$$

L = diene *(2,5,11,21,25,33,43,61,71,106)*

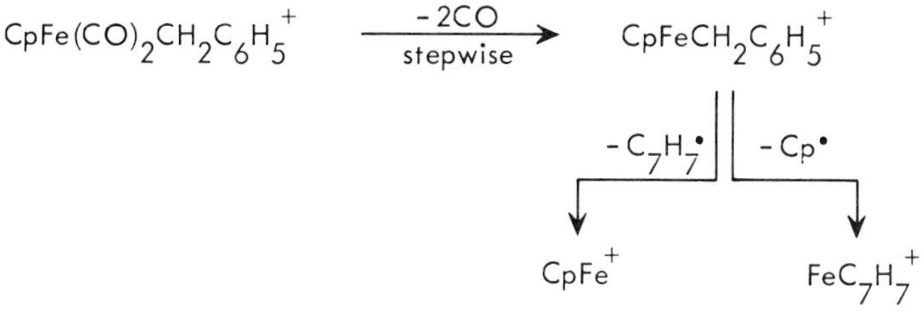

$$CpFe(CO)_2CH_2C_6H_5^+ \xrightarrow[\text{stepwise}]{-2CO} CpFeCH_2C_6H_5^+$$

$$\xrightarrow[-C_7H_7^\bullet]{} \Big\| \xleftarrow{-Cp^\bullet} \qquad (7)$$

$$CpFe^+ \qquad\qquad FeC_7H_7^+$$

$$H_2C{=\!=\!=}C{\Big\langle}^{CH_2}_{CH_2}{\Big\rangle}{+} {-}Fe(CO)_3 \xrightarrow[\text{stepwise}]{-3CO} H_2C{=\!=\!=}C{\Big\langle}^{CH_2}_{CH_2}{\Big\rangle}{-}Fe^+ \qquad (32)$$

$$\left[\begin{array}{c} \\ Fe(CO)_3 \\ Fe(CO)_3 \end{array} \right]^+ \xrightarrow[\text{stepwise}]{-6CO} C_4H_4Fe_2^+ \xrightarrow{-C_4H_4} Fe_2^+ \quad \begin{array}{c}(15,59,\\91)\end{array}$$

The facility of metal-ligand bond rupture decreases with increasing number of bonds between the metal and the ligand. Hence the Cp-Fe bond, for example, is not attacked before other two-, three-, or four-electron ligands have been lost. $(\eta^6\text{-Benzene})(\eta^4\text{-}1,3\text{-cyclohexadiene})$iron first loses the cyclic diene ligand as shown in eq. [1] (88). In mixed carbonyl nitrosyl complexes carbon monoxide as a two-electron ligand is lost prior to the three-electron ligand NO. As another consequence of this rule, a ligand in a terminal position is more easily lost than one in a bridging position.

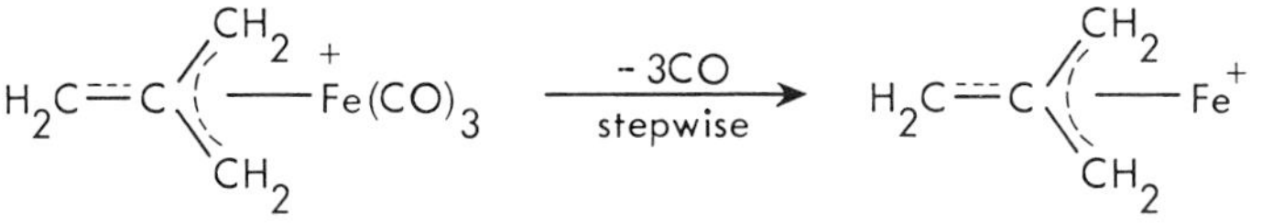

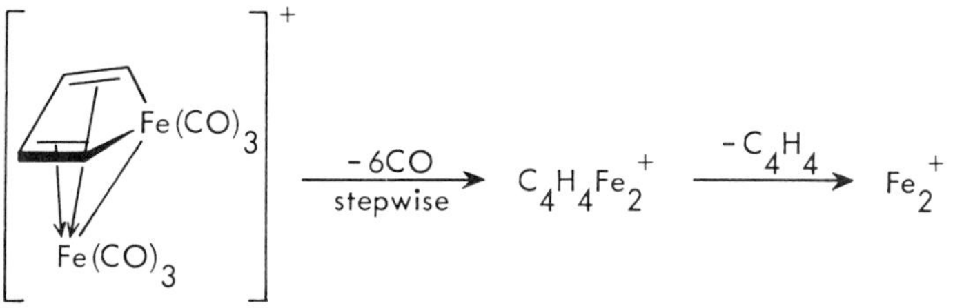

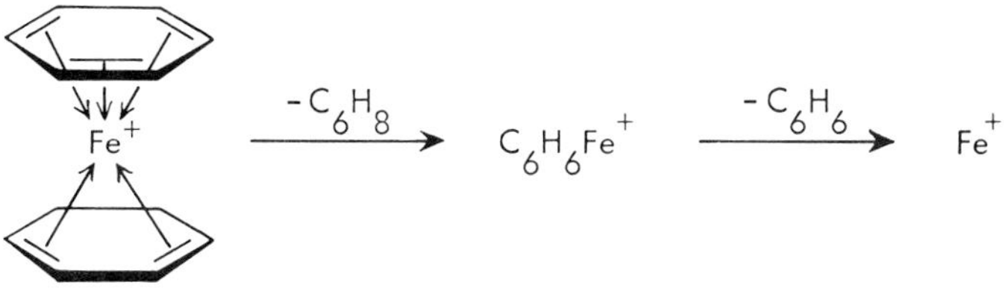

$$Fe^+ \xrightarrow{-C_6H_8} C_6H_6Fe^+ \xrightarrow{-C_6H_6} Fe^+ \qquad [1]$$

The role of halogen ligands in the fragmentation of organoiron ions is somewhat difficult to define because the rupture of the iron-halogen bond parallels the elimination of CO as well as the loss of the Cp ligand, as is shown in eq. [2] *(10,101)*.

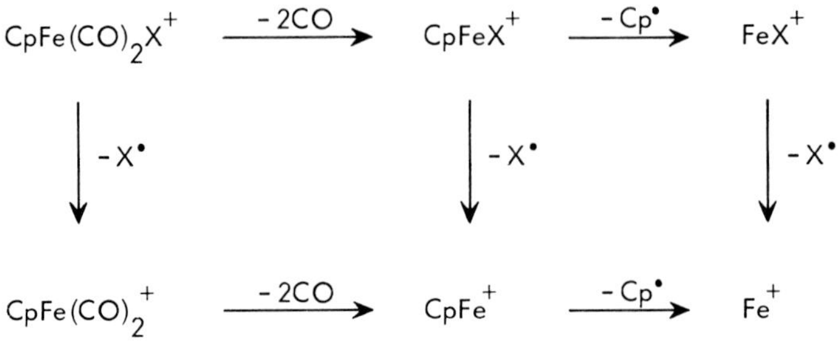

$$CpFe(CO)_2X^+ \xrightarrow{-2CO} CpFeX^+ \xrightarrow{-Cp^\bullet} FeX^+$$

$$\downarrow -X^\bullet \qquad\qquad \downarrow -X^\bullet \qquad\qquad \downarrow -X^\bullet \qquad\qquad [2]$$

$$CpFe(CO)_2^+ \xrightarrow{-2CO} CpFe^+ \xrightarrow{-Cp^\bullet} Fe^+$$

X = Cl, Br, J

B. *SIMPLE BOND RUPTURE WITHIN COMPLEXED LIGANDS*

Among the variety of fragmentations according to this
type only a few examples will be discussed.

π-Bonded cyclic olefins can undergo hydrogen abstraction. In the mass spectrum of tricarbonyl(1,3-cyclohexadiene)-
iron there are two series of ions, $C_6H_8Fe(CO)_n^+$ (n = 3,2,1)
and $C_6H_6Fe(CO)_n^+$ (n = 1,0) *(25,43,106)*. The loss of H_2 can
compete with the decarbonylation and hence must be a process
with very low energy requirement (eq. [3]). The formation
of an additional double bond and its coordination to the
electron-deficient iron atom is considered to be the driving
force of the observed dehydrogenation. This explanation can
be generalized in terms of the following rule: Many decomposition patterns of transition metal complexes are
determined by the tendency of the central metal atom to

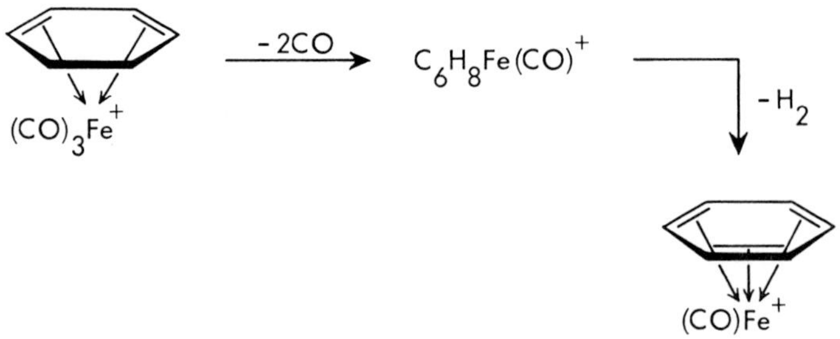

$$(CO)_3Fe^+ \xrightarrow{-2CO} C_6H_8Fe(CO)^+ \xrightarrow{-H_2} (CO)Fe^+ \qquad [3]$$

maintain its electron deficiency as low as possible (rule of
minimum electron deficiency). The loss of H_2 is in direct
contrast to the behaviour of free 1,3-cyclohexadiene in
which the base peak corresponds to the $C_6H_7^+$ ion. It can be
seen from eq. [4] that also π-bonded cyclohexene ligands
undergo dehydrogenation leading to π-bonded aromatic systems

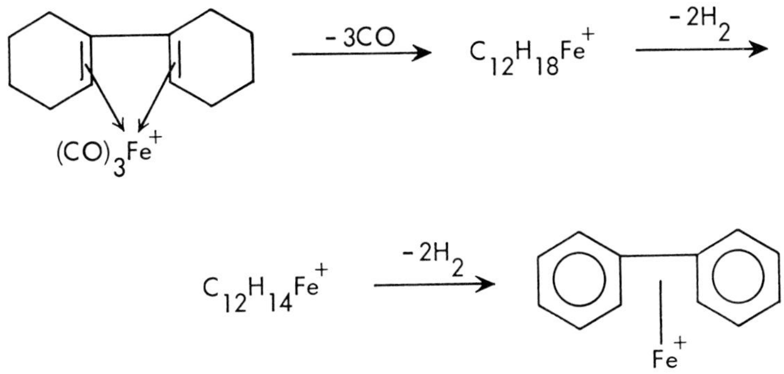

$$\text{(structure)} \xrightarrow{-3CO} C_{12}H_{18}Fe^{+} \xrightarrow{-2H_2}$$

[4]

$$C_{12}H_{14}Fe^{+} \xrightarrow{-2H_2} \text{(biphenyl-Fe}^{+}\text{ structure)}$$

(25). On the other hand, ions like $C_6F_6Fe^{+}$ are absent in
the spectrum of $C_6F_8Fe(CO)_3$ because their formation would
involve the rather unusual loss of two fluorine atoms or an
F_2 molecule *(44)*.

   In the spectra of metal carbonyl compounds ions formed
by carbon-oxygen fission are found. Usually this process
does not take place before several CO groups have been lost,
and in $Fe(CO)_5$ ions like $Fe(CO)C^{+}$ and FeC^{+} have rather low
abundances. However, in polynuclear complexes as for
example in $[CpFeCO]_4$ ions produced by carbon-oxygen bond
fission like $Cp_4Fe_4(CO)_nC^{+}$ (n = 2,0) have relative intensities similar to those of the "normal" fragments $Cp_4Fe_4(CO)_n^{+}$
(n = 3,2,1) *(48,57)*. This result is consistent with the
threeway bridging of the carbonyl groups in this complex
which strengthens the iron-carbon bonds and weakens the car-
bonoxygen bonds.

   Although organoiron molecular ions have an odd-electron
configuration the partial fragmentation of the complexed
ligands usually proceeds by elimination of neutral molecules.
Nevertheless, under certain circumstances radicals also are
lost. Like other simple bond fissions radical eliminations
are fast processes with high frequency factors and therefore
are not always accompanied by metastable peaks.

   If a complex contains groups X (X = halogen, alkyl,
aryl, alkoxyl, phenoxyl, dialkylamino) which are linked to
the iron atom by hetero atoms like N, P, As, O, or S, then
these groups can be lost as radicals. The radical elimination can even compete with the metal-CO fission as is
shown in eq. [5]. This fragmentation mode may be induced by
the ability of the four-coordinated phosphorus atom to take
over positive charge density from the central metal which
results in weakening of the P-X- and strengthening of the
M-CO-bonds.

Other examples are found in sulfur- and phosphorus-
bridged iron complexes *(28,45,46,53,94)*. The μ-mercapto
compounds $Fe_2(CO)_6(SR)_2$ (R = Me, Et, Bu, Ph) lose CO

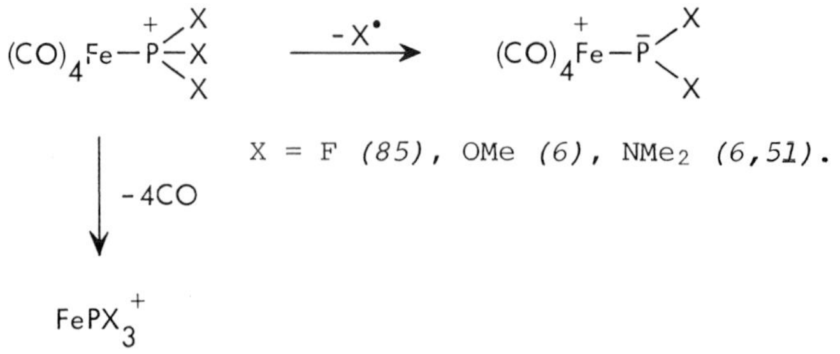

X = F *(85)*, OMe *(6)*, NMe_2 *(6,51)*.

[5]

preferentially with the preservation of the $Fe_2(SR)_2$ nucleus which shows further fragmentation by successive loss of $R^{\cdot}$ radicals (eq. [6]) *(28,53)*.

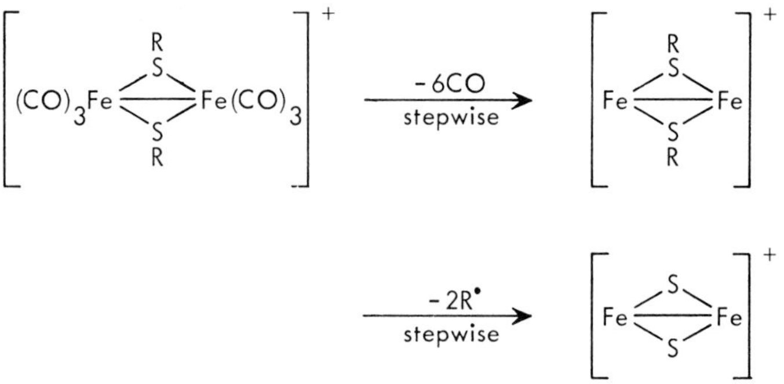

[6]

In a similar manner 1 is decarbonylated to give the ion $Fe_2(PMe_2)_2^+$, and several P-Me fissions then lead to the $Fe_2P_2^+$ nucleus *(46)*.

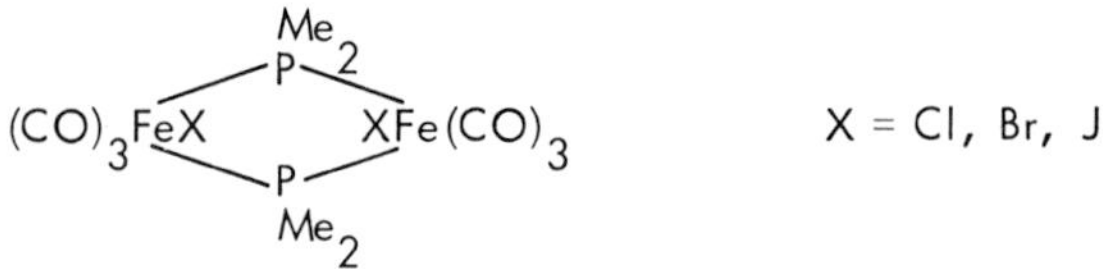

1

The elimination of an alkyl radical from the decarbonylated tetracarbonyl(pentafluorophenyl-alkoxycarbene)iron complex produces an acyliron ion (eq. [7]); with the corresponding pentafluorophenyl-dimethylaminocarbene complex an analogous fragmentation pattern is observed *(34)*. Another type of radical abstraction has been reported for several substituted ferrocenes (eq. [8]) *(70)*.

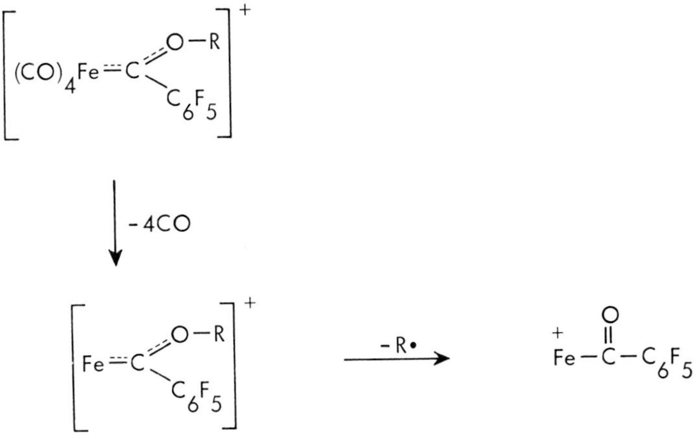

$$\left[(CO)_4 Fe = C \begin{smallmatrix} O-R \\ \\ C_6F_5 \end{smallmatrix} \right]^+$$

[7]

$\downarrow$ -4CO

$$\left[Fe = C \begin{smallmatrix} O-R \\ \\ C_6F_5 \end{smallmatrix} \right]^+ \xrightarrow{-R\cdot} \overset{+}{Fe} - \overset{O}{\overset{\|}{C}} - C_6F_5$$

$$\left[CpFe - \bigcirc - C \begin{smallmatrix} O \\ \\ R \end{smallmatrix} \right]^+ \xrightarrow{-R\cdot} \left[CpFe - \bigcirc - C \equiv O \right]^+$$

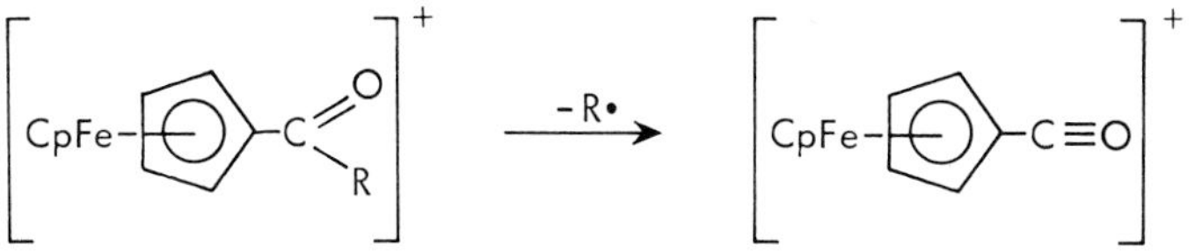

[8]

R = Me, Ph, p-MeOC$_6$H$_4$, OMe, NHMe

In the fragmentation of π-bonded cycloheptatriene, cyclohexadienyl, and cyclopentadiene complexes there exists the possibility of aromatization by loss of a radical from the methylene group of the cyclic ligands. It has been shown that a functional substituent bonded to the sp^3 center is preferentially lost if it takes the *exo*-position to the metal *(78)*. Thus, in such cases mass spectrometry serves to distinguish between *exo*- and *endo*- isomers. Eq. [9] presents an example from organoiron chemistry. The functional group in tricarbonyl[5-(hydroxymethyl)-5-methylcyclopentadiene]iron is eliminated from the molecular ion only in the case of the *exo*-hydroxymethyl isomer whereby the even-electron tricarbonyl(methylcyclopentadienyl)iron ion is produced *(74,89)*. The tendency towards the loss of the *exo*-methyl group is subordinate due to the minor stability of the methyl radical as compared to the hydroxymethyl radical.

C.   REARRANGEMENT PROCESSES

Rearrangement reactions play an important role in the mass spectroscopic fragmentation of organometallic molecules. They require sterically pretentious transition states and

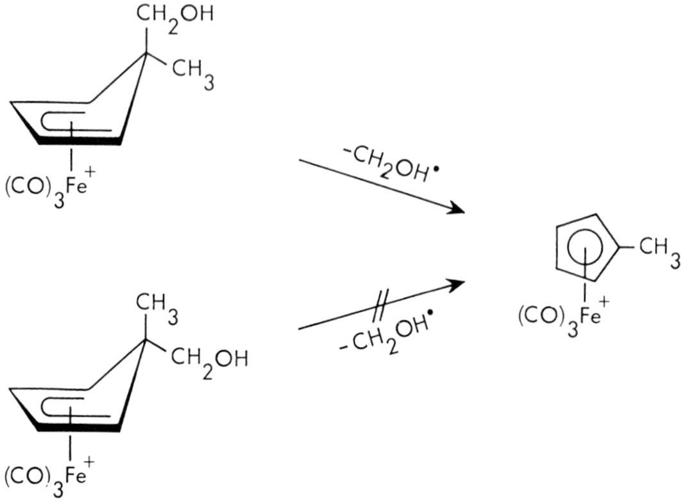

[9]

hence have low frequency factors; therefore, rearrangement processes are usually accompanied by metastable peaks. In this subchapter ring cleavage reactions will also be discussed although many of these do not necessarily involve rearrangements.

π-Bonded aromatic ligands and cyclic olefins undergo partial ring fragmentation mainly by loss of C_2 units. The most important ring cleavage process of cyclopentadienyliron ions gives fragments in which a cyclopropenium structure is probably attained (eq. [10]). The loss of C_3 units is also observed; however, the corresponding fragments are less abundant.

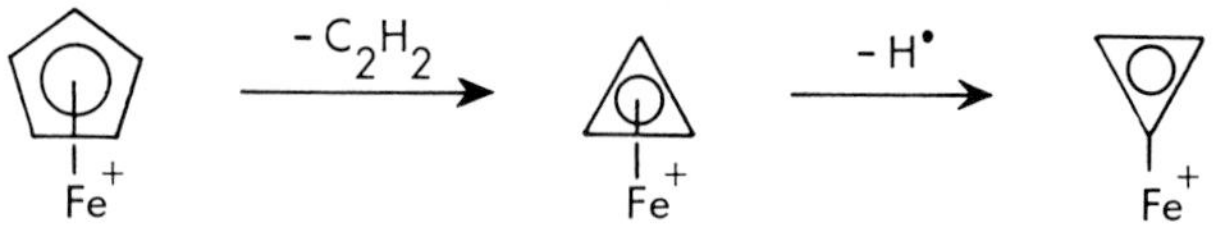

[10]

The partial ring fragmentation competes with the metal-ring fission, and the process which takes preponderance depends on the relative stability of the metal-ring bond. The correlation between partial ring fragmentation and stability is shown by a comparison between the decay of the cyclobutadiene complex 2 and that of its tetraphenyl derivative 3 *(33,61)*. No acetylene elimination is observed in the case of the less stable unsubstituted compound.

The partial ring cleavage of the π-pyrrolyl derivative $CpFeC_4H_4N$ occurs by loss of acetylene, HCN, and $C_2H_2N^{\cdot}$, and thus appears to be related to the decay of ferrocene *(22,55)*.

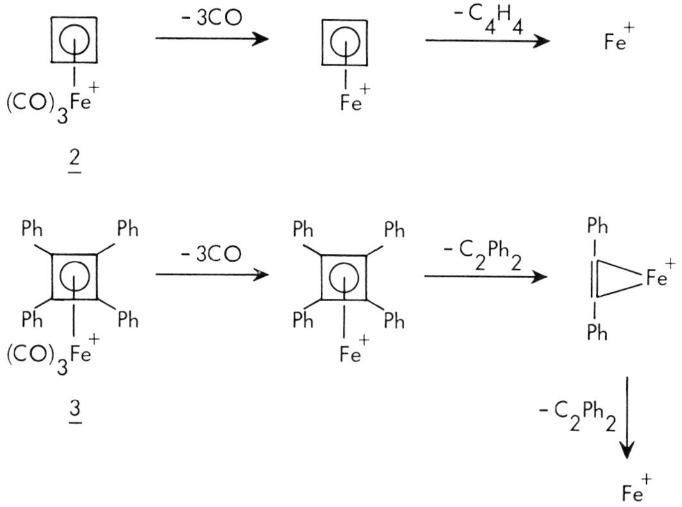

A similar type of fragmentation has been found for heterocyclic carbene ligands bonded to iron (eq. [11]) *(90)*.

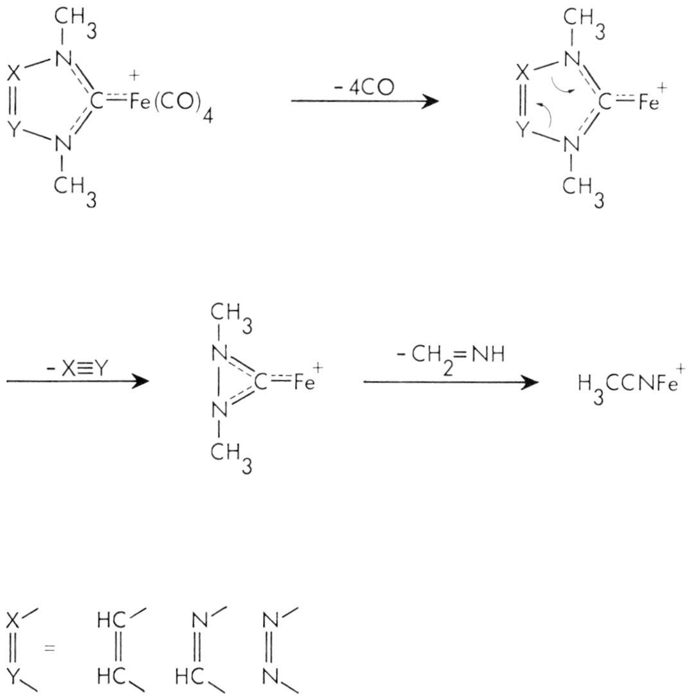

[11]

Complexes with π-bonded cyclic ligands containing a carbonyl group in the ring usually undergo an easy CO elimination, as is shown for cyclopentadienoneiron ions in eq. [12] and for the α-pyroneiron ion in eq. [13] *(11,18,98)*.

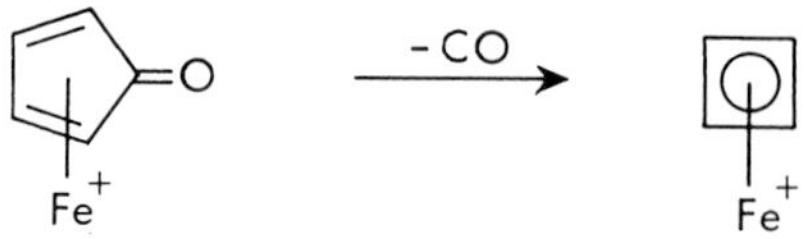

[12]

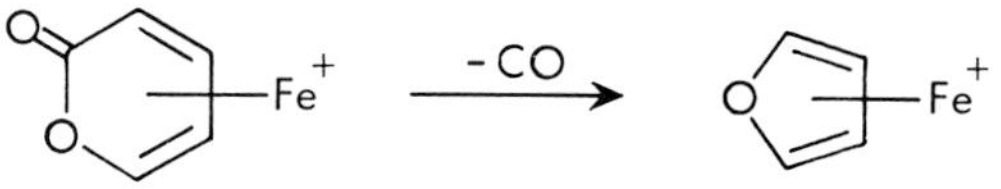

[13]

Only a little work has been done on possible structures of organoiron ions with cyclic ligands *(18,97)*. Mass spectrometric studies on deuterated divinylferrocenes have provided strong evidence for a rearrangement of the vinylcyclopentadienyl ligands to give π-bonded tropylium ions (eq. [14]) *(97)*. The ring expansion process transfers the

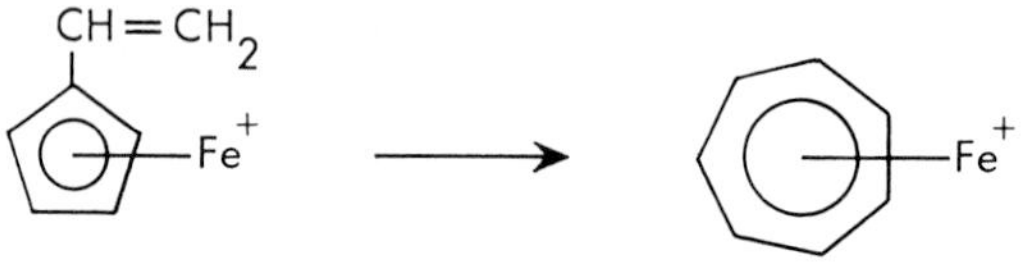

[14]

four electron ligand into a six electron donor and thus lowers the electron deficiency of the iron atom.

The fragmentation of several π-olefin complexes involves hydrogen transfer between two cyclic ligands. For example, the molecular ion of bis(cycloheptadienyl)iron, $C_7H_9FeC_7H_9$, in its first decomposition step exclusively loses a cycloheptadiene molecule with formation of a $C_7H_8Fe^+$ ion (rule of minimum electron deficiency!) *(87)*. Owing to this possible complication, one must be cautious in using the mass spectrometer for structural identification.

A majority of rearrangement processes involve hydrogen migration, and again only a small number of selected examples from organoiron mass spectrometry will be discussed. In most cases hydrogen rearrangements occur in connection with the elimination of rather stable neutral molecules like alkanes, olefins, H_2O, H_2S, alcohols, aldehydes, or ketones. Often such reactions of complexed ligands differ remarkably from the fragmentation of the free organic molecules, and it is therefore suggested that they are influenced by the presence of the central metal atom.

It has been found that complexes such as 4 which have two substituents at a common ring site and thus cannot aromatize by hydrogen loss such as shown in eq. [3], instead lose methane with great facility to give the ion 5 as the base species in the spectra. In these cases the loss of methane becomes a prominent process only after the loss of all three CO ligands has occurred *(25)*.

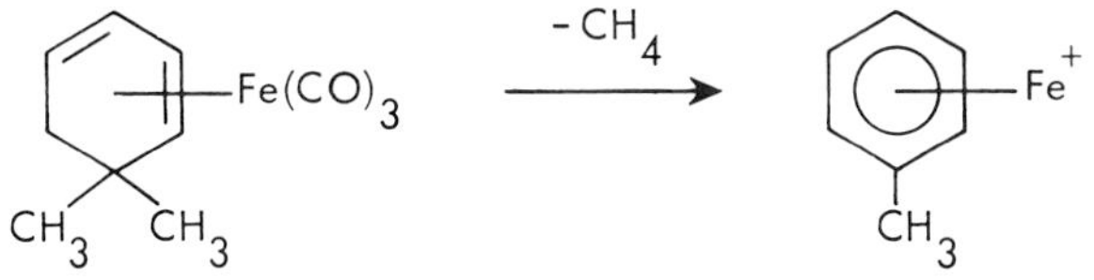

4 5

In the mass spectra of tetracarbonyliron complexes with π-bonded ethylenes the acetyleneiron ion is observed which is produced according to eq. [15] *(63)*. The loss of HX also occurs in π-allyl complexes $RC_3H_4Fe(CO)_3X$ (X = Cl, Br, I, and NO_3) from the decarbonylated ion $RC_3H_4FeX^+$ *(92)*.

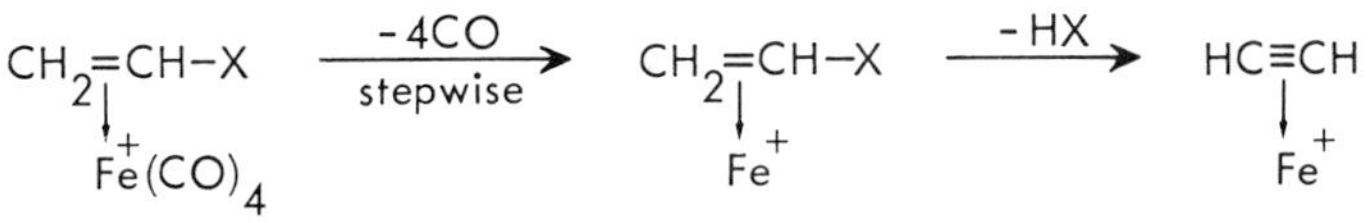 [15]

X = Cl, Br, CN, EtO, Ph

If alkyl chains with more than one carbon atom are linked to the central metal by a hetero atom the loss of an olefin molecule is observed in competition with the loss of the alkyl radical. An example is shown in eq. [16] *(45)*.

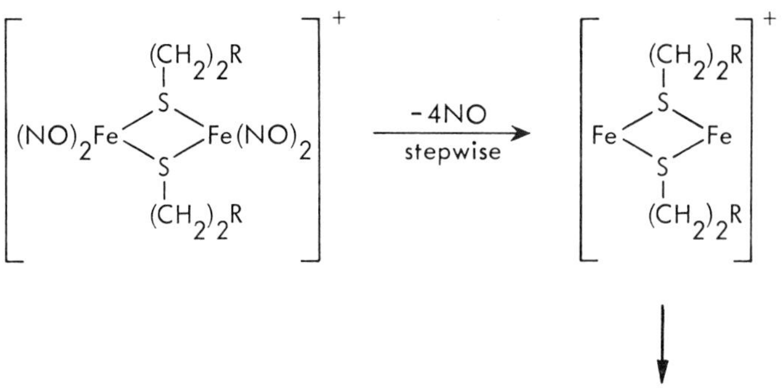 [16]

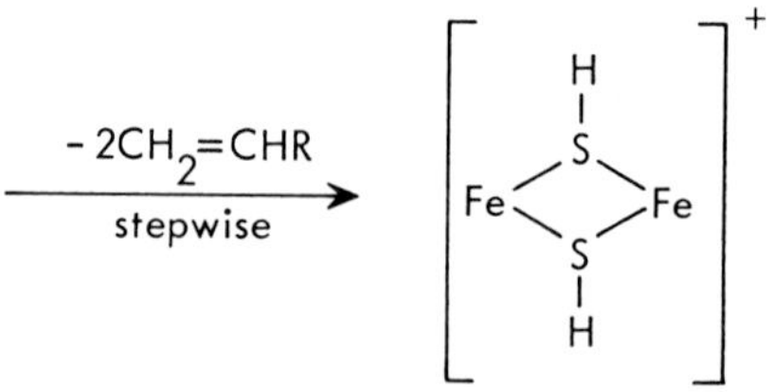

Hydrogen transfer from a cyclopentadienyl ring to a μ-mercapto ligand has been reported for 6 *(94)*. It is suggested that the product ion has the structure 7.

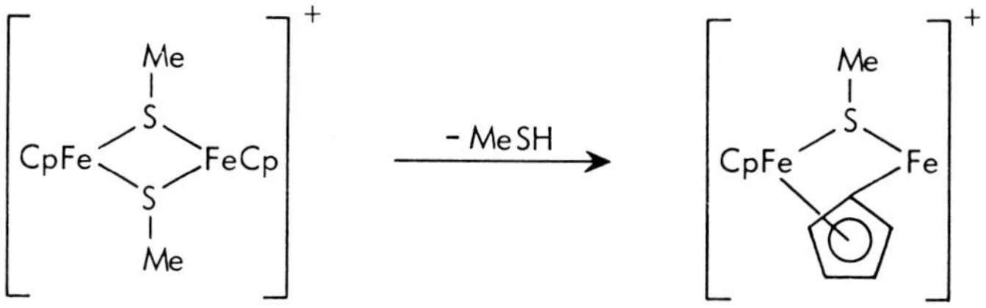

6 7

A characteristic fragmentation mode of organic alcohols is the loss of water. This has also been proved to be a prominent reaction in many ferrocene carbinols *(29-31)*. Deuteration studies on 8 indicate that upon loss of water from the molecular ion the hydrogen is transferred from one of the Cp rings, and this situation seems to be similar to the fragmentation of 6 *(29)*. The participation of a ring hydrogen atom was also proved for the water elimination of the decarbonylated fragment ion produced from tricarbonyl-(hydroxymethylcyclooctatetraene)iron *(1)*. In some cases of ferrocene alcohols good correlations between mass spectra and stereochemistry have been found. For example the *exo/endo*-isomer 9 loses two molecules of water in a two step process, while in the *endo/endo*-isomer 10 two water molecules are lost simultaneously, the $(P - H_2O)^+$ peak being very weak *(29)*.

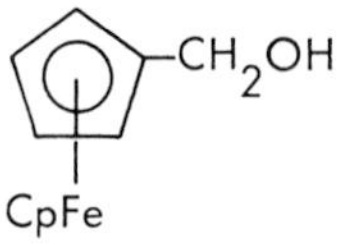

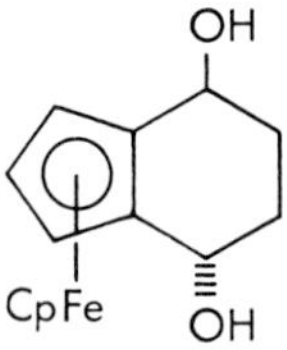

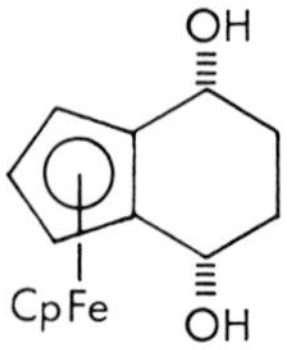

8 9 10

In the mass spectrum of tricarbonyl(methoxymethylcyclooctatetraene)iron a prominent ion appears which is formed by loss of formaldehyde from the carbonyl-free fragment *(1)*. There exists strong evidence that this decomposition [17] proceeds *via* a six-membered transition state like in the McLafferty rearrangement reported for benzyl ethers.

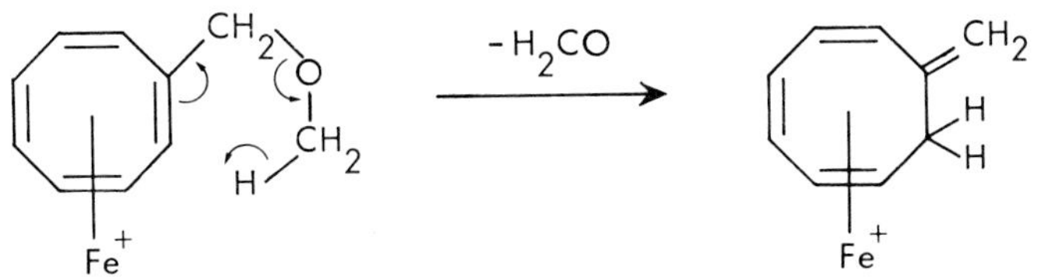

[17]

Several iron complexes with ester groups show an unusual aldehyde elimination which is not found in free organic esters *(39,95,96)*. One example is shown in eq. [18]. The aldehyde elimination is not restricted to iron complexes but seems to be of general importance in the decay of transition metal complexes with ester functions *(20,78,85,86)*. The loss of formaldehyde is also observed in trimethylphosphite complexes such as $(CO)_4FeP(OMe)_3$ *(6)*.

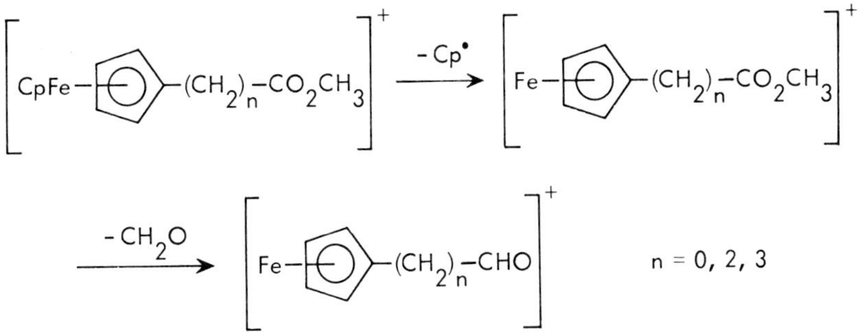

[18]

Few examples are known of the loss of radicals involving hydrogen migration. The mass spectrum of ferrocene shows a $(P - CH_3)^+$ peak which is also found in the spectra of other metallocenes *(77)*. Loss of a methyl radical has also been reported for the π-allyl derivative $C_3H_5Fe(CO)_3I$ and for some related compounds *(58,92)*; the spectrum of $C_3H_5Fe(CO)_3I$ exhibits a series of ions $C_2H_2Fe(CO)_n^+$ (n = 3,2,1, and 0), and the tendency for a methyl group to arise from a π-allyl group is further indicated by the presence of a fairly abundant CH_3Fe^+ ion.

The last class of rearrangement to be discussed involves migration of electronegative or nucleophilic groups from the ligands to the metal atom. Again the high positive charge density on the iron atom appears to be the triggering force

for these reactions. Groups which can be transferred in
this way are the halogens, oxygen, hydroxyl, alkoxyl,
cyanide, or amide; also alkyl and aryl groups are transferred if they are bonded to a carbonyl function.

Halogen atoms undergo this rearrangement very easily.
In the spectra of tetracarbonyl(vinylhalide)iron complexes
FeX^+ (X = Cl, Br) fragments appear, and the mass spectra of
tetracarbonyl(dihalogenoethylene)iron complexes show FeX_2^+
ions as the most intense peaks *(63)*. In many other cases
FeX_2 units (X = F, Cl, Br) are eliminated as neutral particles, and the positive charge remains on the ligand fragments
because the FeX_2 molecule has a higher ionization potential
than the bare Fe atom or a FeX particle. One of the primary
fragmentation processes of polychlorinated ferrocenes is the
loss of $FeCl_2$, and significant $FeCl^+$ peaks occur also in the
corresponding mass spectra *(102)*. The elimination of FeF_2
has been found in the decay of all iron complexes with
fluorocarbon ligands which have been investigated *(8,12-14,
16,17,50,54,73)*. Some of these and related processes are
found in the partial fragmentation patterns in eqs. [19] to
[21].

$$CpFe(CO)_2CH=CH-CF_3^+ \xrightarrow{\;-2CO\;} CpFe-CH=CH-CF_3^+$$

$$-FeF_2 \diagup \qquad \diagdown -CpFeF \qquad\qquad [19]$$

$$C_5H_5C_3FH_2^+ \qquad C_3F_2H_2^+$$

$$\left[(CO)_3Fe \overset{S}{\underset{S}{\diamond}} Fe(CO)_3 \;\Big|\; CF_2-CF_2 \right]^+ \xrightarrow{-3CO;\,-FeF_2} \left[\overset{F-C=S}{\underset{F-C=S}{|}} Fe(CO)_3 \right]^+ \qquad [20]$$

$$\left[(C_6F_5)_2P \overset{(CO)_3}{\underset{(CO)_3}{\diamond}} \overset{Fe}{\underset{Fe}{}} P(C_6F_5)_2 \right]^+ \xrightarrow{-6CO} \left[(C_6F_5)_2P \overset{Fe}{\underset{Fe}{}} P(C_6F_5)_2 \right]^+ \longrightarrow \qquad [21]$$

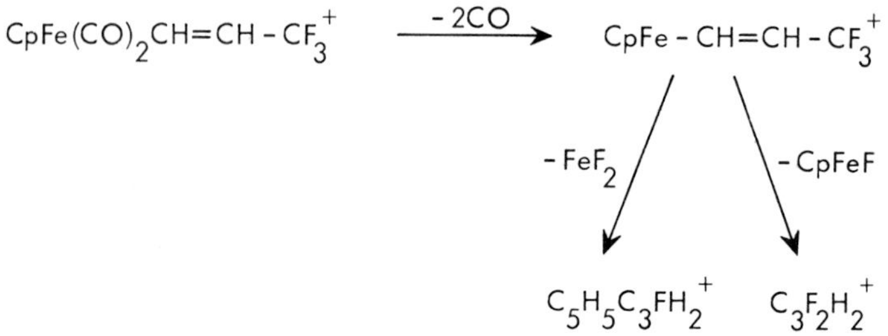

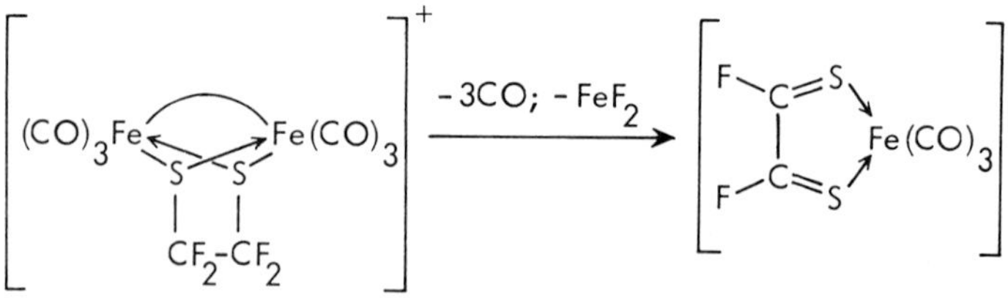

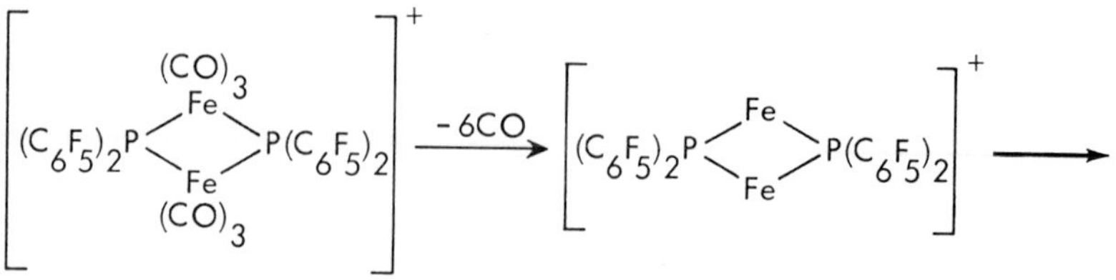

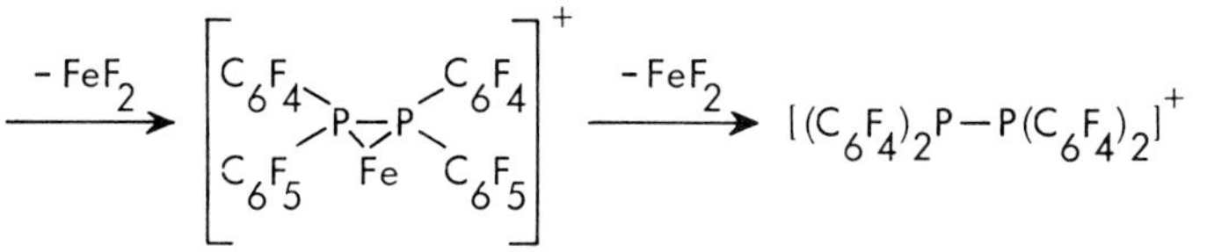

$$\xrightarrow{-FeF_2} \left[\begin{array}{c} C_6F_4 \\ C_6F_5 \end{array} \begin{array}{c} C_6F_4 \\ P \!=\! P \\ Fe \end{array} \begin{array}{c} C_6F_4 \\ C_6F_5 \end{array} \right]^{+} \xrightarrow{-FeF_2} [(C_6F_4)_2P-P(C_6F_4)_2]^{+} \quad [21]$$

Amide transfer to the iron atom occurs in the decay of tris(dimethylamino)phosphine and -arsine carbonyliron complexes (6,51,62); one of the rearrangement products of these compounds is the ion $FeNMe_2^+$. Another rearrangement ion is produced by the sequence of interesting fragmentation processes shown in eq. [22].

$$(CO)_4FeE(NMe_2)_3^{+}$$

$$\downarrow -4CO \qquad\qquad\qquad\qquad [22]$$

$$FeE(NMe_2)_3^{+} \xrightarrow{-H_3CN=CH_2} FeEH(NMe_2)_2^{+}$$

$$\xrightarrow{-EH_3} Fe(MeN=CH_2)_2^{+} \qquad E = P,\ As$$

Several cyclopentadienyl nitrosyl complexes undergo a rather unusual rearrangement by transfer of the oxygen atom of a nitrosyl ligand to the metal (eq. [23]). This kind of migration has been observed only when it starts from even-electron ions (75). The eliminated C_5H_5N unit in eq. [23] may possibly have the pyridine structure.

$$Cp_2Fe_2(NO)_2^{+} \xrightarrow{-NO} Cp_2Fe_2NO^{+} \xrightarrow{-C_5H_5N} CpFe_2O^{+} \quad [23]$$

Hydroxyl transfer to the metal is a characteristic reaction of complexes with alcohol ligands. For the series of primary ferrocene alcohols 11 the migration of the hydroxyl group to the iron atom decreases relative to cleavage giving $CpFe^+$, as n increases from 1 to 4 (29). Mass spectroscopic investigations of pairs of exo- and endo-isomers of ferrocene carbinols have shown that hydroxyl transfer depends on the stereochemistry of the complexes. Thus, for the endo-isomer 12 the most abundant peak after the parent ion is the $[P - CpFeOH]^+$ fragment, but this peak is very weak in the spectrum of the exo-isomer 13 (29).

Many examples are known of a general rearrangement mode

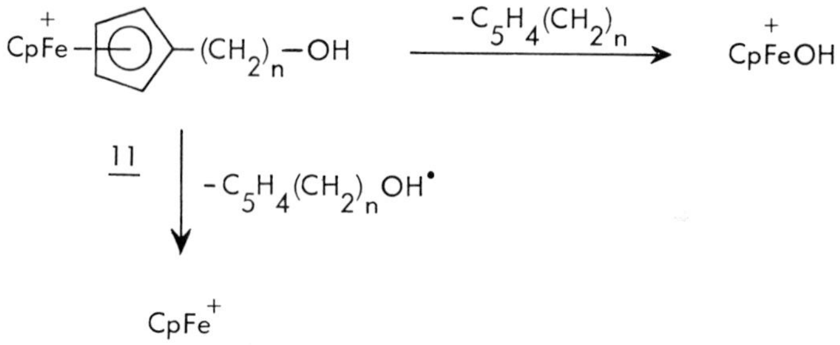

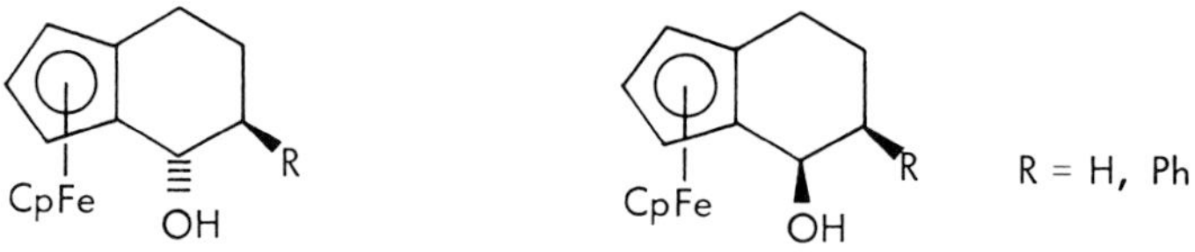

$$\overset{+}{CpFe}\!-\!\underset{11}{\boxed{\bigcirc}}\!-\!(CH_2)_n\!-\!OH \xrightarrow{-C_5H_4(CH_2)_n} \overset{+}{CpFeOH}$$

$$\Big\downarrow -C_5H_4(CH_2)_n OH^\bullet$$

$$\overset{+}{CpFe}$$

12   13   R = H, Ph

in which a substituent R is transferred from a CO-R group to
the metal. This reaction occurs in complexes in which the
CO-R group is directly linked to the metal atom (eq. [24])
as well as in compounds with the CO-R group bonded to a complexed ligand (eqs. [25] and [26]) *(52,70,71,96)*.

$$\overset{+}{CpFe}(CO)_2COR \xrightarrow{-2CO} \overset{+}{CpFe}\!-\!\overset{O}{\overset{\|}{C}}\!-\!R \xrightarrow{-CO} \overset{+}{CpFe}\!-\!R \qquad [24]$$

R = Me, Ph

$$\overset{+}{CpFe}\!-\!\boxed{\bigcirc}\!-\!\overset{O}{\overset{\|}{C}}\!-\!R \xrightarrow{-C_5H_4CO} \overset{+}{CpFe}\!-\!R \qquad [25]$$

R = Ph, p-MeOC$_6$H$_4$, OH, OMe, NHMe

$$\overset{+}{CpFe}(CO)_2CH_2CO_2CH_3 \xrightarrow{-2CO} \overset{+}{CpFe}\!\!\begin{matrix}CH_2\\O\text{-}C\!=\!O\\CH_3\end{matrix} \xrightarrow{-H_2C=C=O} \qquad [26]$$

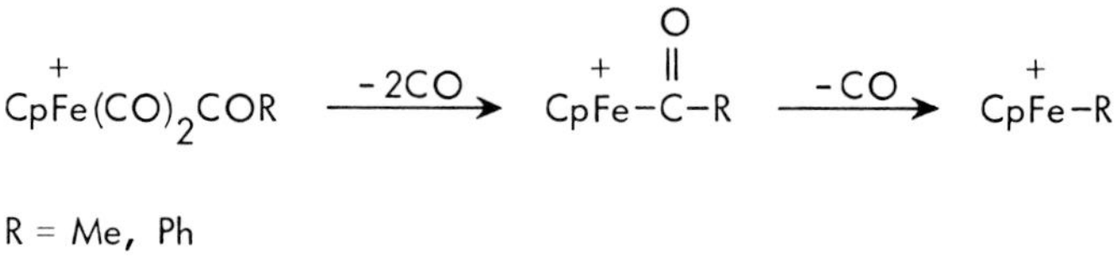

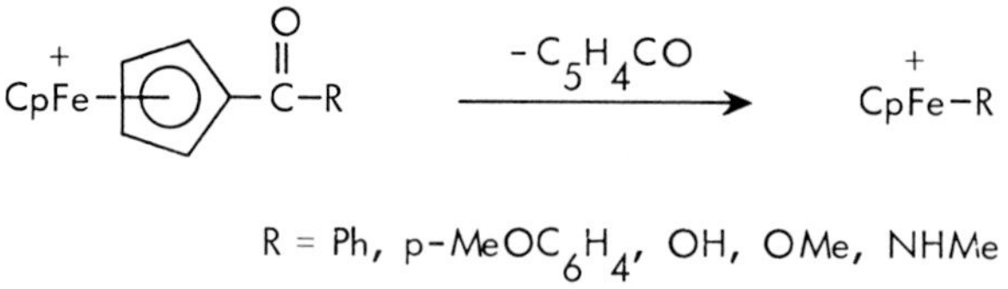

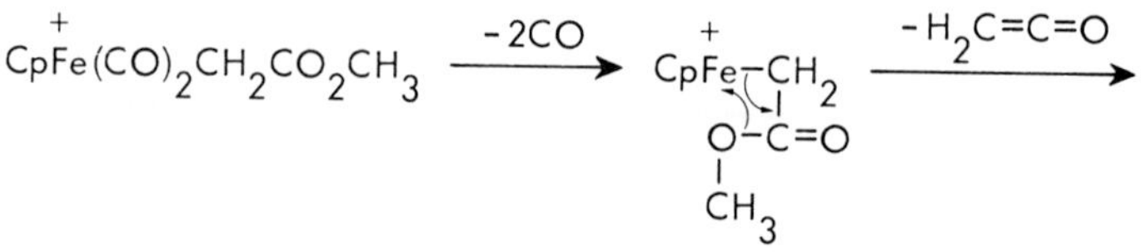

$$\longrightarrow \overset{+}{\mathrm{CpFe{-}OCH_3}} \xrightarrow{-H_2} \overset{+}{\mathrm{CpFe}}\mathrm{-}\overset{\mathrm{O}}{\overset{\|}{\mathrm{C}}}\mathrm{-H} \xrightarrow{-CO} \overset{+}{\mathrm{CpFe{-}H}}$$

Di- or polynuclear cyclopentadienyliron complexes show a kind of rearrangement which can also be described by the principle of the migration of nucelophilic groups. Thus, under electron impact of $Cp_2Fe_2(CO)_4$ a $[CpFe{-}FeCp]^+$ fragment is produced which shows further decay by loss of an iron atom with the formation of Cp_2Fe^+ *(101)*. Similarly the ion $Cp_3Fe_3^+$ arising from $[CpFe(CO)]_4$ loses an iron atom giving rise to the appearance of the $Cp_3Fe_2^+$ fragment to which the structure of a "triple-decker" sandwich has been attributed *(48,57)*.

Closing this chapter it should be mentioned that mass spectra of organoiron complexes can become quite complicated, because usually there exists a variety of competing fragmentation pathways from a molecular ion. It has been one of the purposes of this chapter to present some general rules which may be utilized by the organometallic chemist for the interpretation of mass spectra of known as well as of unknown substances.

III. ION-MOLECULE REACTIONS

Ion-molecule reactions (IMR) are chemical reactions induced by collisions between ions and neutral molecules in the gas phase. They are much more rapid than chemical reactions between molecules because of the strong polarization forces exerted by the ion on a molecule, by which the collision probability is considerably increased.

IMR can be studied in a conventional mass spectrometer with suitable equipment. In order to detect secondary ions as products of IMR it is necessary to increase the sample pressure in the ionization chamber of the spectrometer to some extent so that collisions between ions and molecules can occur. On working with low repeller (or drawing-out plate) voltages the IMR cross-sections can be remarkably increased. Recent development of the ion cyclotron resonance (ICR) technique has stimulated IMR investigations to a considerable extent.

Although much work has been done on IMR of small molecules (H_2, He, CO, N_2, C_2H_2, etc.) and of more complicated organic compounds, yet a limited number of publications concerning organometallic IMR has appeared, but there seems to be growing interest in this field *(27,37,40-42,64,79-84, 99-101)*. In this section some results from organoiron IMR involving the formation of polynuclear species and ligand displacement reactions will be discussed. For theoretical considerations the reader is referred to the corresponding literature.

Usually IMR of monomeric organometallic complexes yield bimetallic species. As mentioned in section II the mass spectrum of ferrocene exhibits the ions Cp_2Fe^+, $CpFe^+$, and Fe^+. Under special operating conditions the secondary ions $Cp_2Fe_2^+$ and $Cp_3Fe_2^+$ have been additionally observed *(99,100)*. It was shown by appearance potential measurements and by considerations of the shapes of ion efficiency curves that the ion $CpFe^+$ must act as the precursor of the IMR product $Cp_3Fe_2^+$ (eq. [27]); on the other hand, it was not possible to decide whether the secondary ion $Cp_2Fe_2^+$ is produced according to eq. [28] or to eq. [29].

$$CpFe^+ + Cp_2Fe \longrightarrow Cp_3Fe_2^+ \quad [27]$$

$$CpFe^+ + Cp_2Fe \longrightarrow Cp_2Fe_2^+ + Cp^\bullet \quad [28]$$

$$Cp_3Fe_2^+ \longrightarrow Cp_2Fe_2^+ + Cp^\bullet \quad [29]$$

When a mixture of ferrocene and nickelocene was introduced into the ion source of the instrument the secondary ion Cp_3FeNi^+ occurred besides the $Cp_3Fe_2^+$ and the $Cp_3Ni_2^+$ species *(100)*.

Pentacarbonyliron has been investigated using an ICR machine, and again the formation of binuclear species has been observed *(27,37)*. Within the limited mass range of the instrument the IMR products $Fe_2(CO)_4^+$ and $Fe_2(CO)_5^+$ occurred. Double-resonance experiments indicated the reactions in eq.

$$Fe^+ + Fe(CO)_5 \longrightarrow Fe_2(CO)_4^+ + CO \quad [30]$$

$$FeCO^+ + Fe(CO)_5 \longrightarrow Fe_2(CO)_4^+ + 2CO \quad [31]$$

$$FeCO^+ + Fe(CO)_5 \longrightarrow Fe_2(CO)_5^+ + CO \quad [32]$$

$$Fe(CO)_2^+ + Fe(CO)_5 \longrightarrow Fe_2(CO)_5^+ + 2CO \quad [33]$$

[30] to [33] to be responsible for the formation of these ions.

A large number of IMR products was formed from $[CpFe(CO)_2]_2$ which has been investigated by use of a conventional mass spectrometer *(101)*. The following secondary ions appeared: $Cp_4Fe_4(CO)_4^+$ (very weak), $Cp_3Fe_3(CO)_4^+$, $Cp_3Fe_2(CO)_4^+$, $Cp_2Fe_2(CO)_nCH_2^+$ (n = 4,3,2,1 and 0), and $Cp_2FeCH_2^+$. The formation reactions are shown in eqs. [34]

to [37], eq. [35] indicating the most important process.

$$Cp_2Fe_2(CO)_4 + Cp_2Fe_2^+ \longrightarrow Cp_4Fe_4(CO)_4^+ \qquad [34]$$

$$Cp_2Fe_2(CO)_4 + Cp_2Fe^+ \longrightarrow Cp_3Fe_2(CO)_4^+ + CpFe \qquad [35]$$

$$Cp_2Fe_2(CO)_4 + CpFe^+ \longrightarrow Cp_3Fe_3(CO)_4^+ \qquad [36]$$

$$Cp_3Fe_2(CO)_4^+ \longrightarrow Cp_2Fe_2CH_2(CO)_4^+ + C_4H_3^{\bullet} \qquad [37]$$

Binary mixtures of $Fe(CO)_5$ with CH_3F, H_2O, NH_3, and HCl were examined in an ICR instrument principally to delineate the occurrence of ligand displacement IMR processes. With CH_3F substitutions according to eq. [38] occurred, and additional reaction products were observed primarily from the $(CH_3)_2F^+$ ion in reactions [39] and [40] *(37)*.

$$Fe(CO)_n^+ + CH_3F \longrightarrow Fe(CH_3F)(CO)_{n-1}^+ + CO \qquad [38]$$

$$n = 1 - 4$$

$$CH_3Fe(CO)_5^+ + CH_3F \qquad [39]$$

$$(CH_3)_2F^+ + Fe(CO)_5$$

$$CH_3Fe(CO)_4^+ + CO + CH_3F \qquad [40]$$

With H_2O and NH_3 extensive ligand substitution takes place; in the case of H_2O the species $HFe(CO)_5^+$ and $HFe(CO)_4^+$ are also observed, derived from H_3O^+ by proton transfer.

Among the several products of ligand displacement reactions observed in a mixture of benzene and $Fe(CO)_5$, $Fe(C_6H_6)(CO)_2^+$ formed in reaction [41] by multiple displacement predominates at higher pressures *(37)*.

$$Fe(CO)_4^+ + C_6H_6 \longrightarrow \left[\begin{array}{c} OC \diagdown \diagup CO \\ Fe \\ \end{array} \right]^+ + 2CO \qquad [41]$$

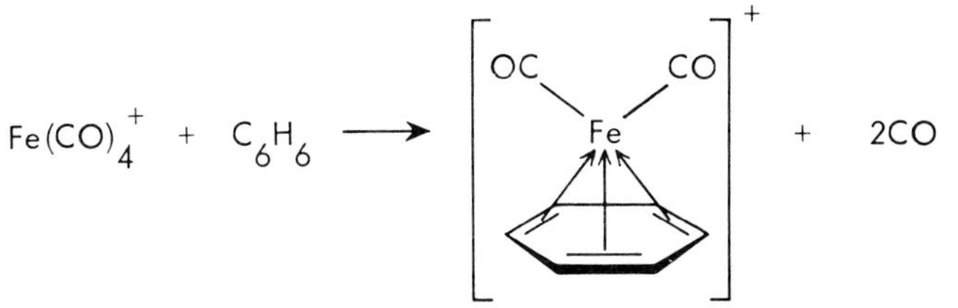

In summary, these results offer two important aspects:

1. IMR provide the possibility of investigating very elementary processes such as simple collisions between ions and molecules without complicating solvation phenomena.
2. From IMR studies we get valuable information relating to the formation and stability of metal-metal and metal-ligand bonds.

REFERENCES

1. Alsop, J.E., and Davis, R., *J. Chem. Soc. Dalton Trans., 1973,* 1686.
2. Amiet, R.G., Reeves, P.C., and Pettit, R., *Chem. Commun., 1967,* 1208.
3. Bidinosti, D.R., and McIntyre, N.S., *Can. J. Chem., 45,* 641 (1967).
4. Birch, A.J., Cross, P.E., Lewis, J., White, D.A., and Wild, S.B., *J. Chem. Soc. A, 1968,* 332.
5. Birch, A.J., and Fitton, H., *J. Chem. Soc. C, 1966,* 2060.
6. Braterman, P.S., *J. Organometal. Chem., 11,* 198 (1968).
7. Bruce, M.I., *Inorg. Nucl. Chem. Lett., 3,* 157 (1967).
8. Bruce, M.I., *J. Organometal. Chem., 10,* 495 (1967).
9. Bruce, M.I., *Advan. Organometal. Chem., 6,* 273 (1968).
10. Bruce, M.I., *Int. J. Mass Spectrom. Ion Phys., 1,* 141 (1968).
11. Bruce, M.I., *Int. J. Mass Spectrom. Ion Phys., 1,* 335 (1968).
12. Bruce, M.I., *Org. Mass Spectrom., 1,* 503 (1968).
13. Bruce, M.I., *Org. Mass Spectrom., 1,* 687 (1968).
14. Bruce, M.I., *Org. Mass Spectrom., 1,* 835 (1968).
15. Bruce, M.I., *Int. J. Mass Spectrom. Ion Phys., 2,* 349 (1969).
16. Bruce, M.I., *Org. Mass Spectrom., 2,* 63 (1969).
17. Bruce, M.I., *Org. Mass Spectrom., 2,* 997 (1969).
18. Bursey, M.M., Tibbets, F.E., and Little, W.F., *J. Amer. Chem. Soc., 92,* 1087 (1970).
19. Cais, M., and Lupin, M.S., *Advan. Organometal. Chem., 8,* 211 (1970).
20. Cais, M., Lupin, M.S., Maoz, N., and Sharvit, J., *J. Chem. Soc. A, 1968,* 3086.
21. Cais, M., and Maoz, N., *J. Organometal. Chem., 5,* 370 (1966).
22. Cataliotti, R., Foffani, A., and Pignataro, S., *Inorg. Chem., 9,* 2594 (1970).
23. Chambers, D.B., Glockling, F., and Light, J.R.C., *Quart. Rev. (London), 22,* 317 (1968).
24. Chisholm, M.H., Massey, A.G., and Thomson, N.R., *Nature, 211,* 67 (1966).
25. Dauben, W.G., and Lorber, M.E., *Org. Mass Spectrom., 3,* 211 (1970).
26. Denning, R.G., and Wentworth, R.A.D., *J. Amer. Chem. Soc., 88,* 4619 (1966).
27. Dunbar, R.C., Ennever, J.F., and Fackler Jr., J.P., *Inorg. Chem., 12,* 2734 (1973).

28. Edgar, K., Johnson, B.F.G., Lewis, J., Williams, I.G., and Wilson, J.M., *J. Chem. Soc. A, 1967*, 379.
29. Egger, H., *Monatsh. Chem., 97*, 602 (1966).
30. Egger, H., and Falk, H., *Monatsh. Chem., 97*, 1590 (1966).
31. Egger, H., and Falk, H., *Tetrahedron Lett., 1966*, 437.
32. Emerson, G.F., Ehrlich, K., Giering, W.P., and Lauterbur, P.C., *J. Amer. Chem. Soc., 88*, 3172 (1966).
33. Emerson, G.F., Watts, L., and Pettit, R., *J. Amer. Chem. Soc., 87*, 131 (1965).
34. Fischer, E.O., Beck, H.-J., Kreiter, C.G., Lynch, J., Müller, J., and Winkler, E., *Chem. Ber., 105*, 162 (1972).
35. Foffani, A., Pignataro, S., Cantone, B., and Grasso, F., *Z. Phys. Chem. (Frankfurt am Main), 45*, 79 (1965).
36. Foffani, A., Pignataro, S., Distefano, G., and Innorta, G., *J. Organometal. Chem., 7*, 473 (1967).
37. Foster, M.S., and Beauchamp, J.L., *J. Amer. Chem. Soc., 93*, 4924 (1971).
38. Friedman, L., Irsa, A.P., and Wilkinson, G., *J. Amer. Chem. Soc., 77*, 3689 (1955).
39. Gambino, O., Vaglio, G.A., and Cetini, G., *Org. Mass Spectrom., 6*, 1297 (1972).
40. Gilbert, J.R., Leach, W.P., and Miller, J.R., *J. Organometal. Chem., 30*, C41 (1971).
41. Gilbert, J.R., Leach, W.P., and Miller, J.R., *J. Organometal. Chem., 42*, C51 (1972).
42. Gilbert, J.R., Leach, W.P., and Miller, J.R., *J. Organometal. Chem., 56*, 295 (1973).
43. Haas, M.A., and Wilson, J.M., *J. Chem. Soc. B, 1968*, 104.
44. Hoehn, H.H., Pratt, L., Watterson, K.F., and Wilkinson, G., *J. Chem. Soc., 1961*, 2738.
45. Johnson, B.F.G., Lewis, J., Williams, I.G., and Wilson, J.M., *J. Chem. Soc. A, 1967*, 338.
46. Johnson, B.F.G., Lewis, J., Wilson, J.M., and Thompson, D.T., *J. Chem. Soc. A, 1967*, 1445.
47. Junk, G.A., and Svec, H.J., *Z. Naturforsch., 23b*, 1 (1968).
48. King, R.B., *Inorg. Chem., 5*, 2227 (1966).
49. King, R.B., *J. Amer. Chem. Soc., 88*, 2075 (1966).
50. King, R.B., *J. Amer. Chem. Soc., 89*, 6368 (1967).
51. King, R.B., *J. Amer. Chem. Soc., 90*, 1412 (1968).
52. King, R.B., *J. Amer. Chem. Soc., 90*, 1417 (1968).
53. King, R.B., *J. Amer. Chem. Soc., 90*, 1429 (1968).
54. King, R.B., *Appl. Spectrosc., 23*, 137 (1969).

55. King, R.B., *Appl. Spectrosc.*, *23*, 148 (1969).
56. King, R.B., *Can. J. Chem.*, *47*, 559 (1969).
57. King, R.B., *Chem. Commun.*, *1969*, 436.
58. King, R.B., *Org. Mass Spectrom.*, *2*, 401 (1969).
59. King, R.B., *Org. Mass Spectrom.*, *2*, 657 (1969).
60. King, R.B., *Fortschr. Chem. Forsch.*, *14*, 92 (1970).
61. King, R.B., and Efraty, A., *Org. Mass Spectrom.*, *3*, 1233 (1970).
62. King, R.B., and Korenowski, T.F., *Org. Mass Spectrom.*, *5*, 939 (1971).
63. Koerner von Gustorf, E., Henry, M.C., and McAdoo, D. J., *Liebigs Ann. Chem.*, *707*, 190 (1967).
64. Kraihanzel, C.S., Conville, J.J., and Sturm, J.E., *Chem. Commun.*, *1971*, 159.
65. Landesberg, J.M., and Sieczkowski, J., *J. Amer. Chem. Soc.*, *90*, 1655 (1968).
66. Lewis, J., and Johnson, B.F.G., *Accounts Chem. Res.*, *1*, 245 (1968).
67. Lewis, J., Manning, A.R., Miller, J.R., and Wilson, J.M., *J. Chem. Soc. A*, *1966*, 1663.
68. Litzow, M.R., and Spalding, T.R., *Mass Spectrometry of Inorganic and Organometallic Compounds*, Elsevier, Amsterdam, 1973.
69. Lloyd, D.R., and Schlag, E.W., *Inorg. Chem.*, *8*, 2544 (1969).
70. Mandelbaum, A., and Cais, M., *Tetrahedron Lett.*, *1964*, 3847.
71. Maoz, N., Mandelbaum, A., and Cais, M., *Tetrahedron Lett.*, *1965*, 2087.
72. McLafferty, F.W., *Anal. Chem.*, *28*, 306 (1956).
73. Miller, J.M., *J. Chem. Soc. A*, *1967*, 828.
74. Müller, H., and Herberich, G.E., *Chem. Ber.*, *104*, 2772 (1971).
75. Müller, J., *J. Organometal. Chem.*, *23*, C38 (1970).
76. Müller, J., *Angew. Chem.*, *84*, 725 (1972); *Angew. Chem. Internat. Edit. Engl.*, *11*, 653 (1972).
77. Müller, J., and D'Or, L., *J. Organometal. Chem.*, *10*, 313 (1967).
78. Müller, J., and Fenderl, K., *Chem. Ber.*, *103*, 3128 (1970).
79. Müller, J., and Fenderl, K., *Chem. Ber.*, *103*, 3141 (1970).
80. Müller, J., and Fenderl, K., *Chem. Ber.*, *104*, 2199 (1971).
81. Müller, J., and Fenderl, K., *Chem. Ber.*, *104*, 2207 (1971).
82. Müller, J., and Goll, W., *Chem. Ber.*, *106*, 1129 (1973).

83. Müller, J., and Goll, W., *Chem. Ber.*, *107*, 2084 (1974).
84. Müller, J., and Goll, W., *J. Organometal. Chem.*, *69*, C23 (1974).
85. Müller, J., and Göser, P., *Chem. Ber.*, *102*, 3314 (1969).
86. Müller, J., and Mertschenk, B., *J. Organometal. Chem.*, *34*, 165 (1972).
87. Müller, J., and Mertschenk, B., *Chem. Ber.*, *105*, 3346 (1972).
88. Müller, J., and Mertschenk, B., unpublished results.
89. Müller, J., Herberich, G.E., and Müller, H., *J. Organometal. Chem.*, *55*, 165 (1973).
90. Müller, J., Öfele, K., and Krebs, G., *J. Organometal. Chem.*, 383 (1974).
91. Nakamura, A., Kim, P.J., and Hagihara, N., *J. Organometal. Chem.*, *6*, 420 (1966).
92. Nesmeyanov, A.N., Nekrasov, Yu.S., Avakyan, N.P., and Kritskaya, I.I., *J. Organometal. Chem.*, *33*, 375 (1971).
93. Pignataro, S., and Lossing, F.P., *J. Organometal. Chem.*, *11*, 571 (1968).
94. Preston, F.J., and Reed, R.I., *Chem. Commun.*, *1966*, 51.
95. Roberts, D.T., Jr., Little, W.F., and Bursey, M.M., *J. Amer. Chem. Soc.*, *89*, 4917 (1967).
96. Roberts, D.T., Jr., Little, W.F., and Bursey, M.M., *J. Amer. Chem. Soc.*, *89*, 6156 (1967).
97. Roberts, D.T., Jr., Little, W.F., and Bursey, M.M., *J. Amer. Chem. Soc.*, *90*, 973 (1968).
98. Rosenblum, M., and Gatsonis, C., *J. Amer. Chem. Soc.*, *89*, 5074 (1967).
99. Schildcrout, S.M., *J. Amer. Chem. Soc.*, *95*, 3846 (1973).
100. Schumacher, E., and Taubenest, R., *Helv. Chim. Acta*, *47*, 1525 (1964).
101. Schumacher, E., and Taubenest, R., *Helv. Chim. Acta*, *49*, 1447 (1966).
102. Smithson, L.D., Bhattacharya, A.K., and Hedberg, F.L., *Org. Mass Spectrom.*, *4*, 383 (1970).
103. Svec, H.J., and Junk, G.A., *Inorg. Chem.*, *7*, 1688 (1968).

104. Turner, D.W., *Molecular Photoelectron Spectroscopy*, Wiley-Interscience, London, 1970, p. 361.

105. Winters, R.E., and Kiser, R.W., *Inorg. Chem.*, *3*, 699 (1964).

106. Winters, R.E., and Kiser, R.W., *J. Phys. Chem.*, *69*, 3198 (1965).

107. Winters, R.E., and Collins, J.H., *J. Phys. Chem.*, *70*, 2057 (1966).

Mössbauer Spectroscopy

By R.V. PARISH

The University of Manchester Institute of Science and Technology, Manchester, M60 1QD, England.

TABLE OF CONTENTS

ISBN 0-12-417101-X (v.1)

I. INTRODUCTION

Mössbauer spectroscopy provides information about the interaction of a nucleus with its environment. This information includes the electron density, the distribution of electronic charge, the magnitude of the local magnetic field, and the tightness of binding of the nucleus to its lattice site. All these factors are of value in examining the structure and bonding of compounds, and Mössbauer spectroscopy is a useful addition to the chemist's armoury. It is a "sporting" method in that the data have to be interpreted with care, and may sometimes be misleading; nevertheless, in conjunction with other techniques, especially IR and NMR spectroscopy, it can be most valuable. This section aims to demonstrate the utility of the Mössbauer method in organoiron chemistry. The approach will be illustrative rather than comprehensive, but all types of organoiron compounds are surveyed, and attempts have been made to cover important literature up to and including mid-1974. More comprehensive reviews of the subject are available *(64,67,73,142)*, and the technique and theory of Mössbauer spectroscopy have been described at various levels of sophistication *(11,64,67,69,73,75,105,111,113,116,135)*.

II. MÖSSBAUER SPECTROSCOPY

The Mössbauer technique is restricted to particular nuclei. Fortunately, ^{57}Fe is one of the most favourable, and data can usually be obtained very readily using the natural abundance (2.2%) of this isotope. The technique is basically absorption spectroscopy using gamma-rays, which are resonantly absorbed by the ^{57}Fe-nuclei of the sample. In order to provide gamma-rays of appropriate energy, an active isotope which decays to the excited state of ^{57}Fe is used as a source, the most convenient being ^{57}Co (half-life 270 days). The 14.4 keV gamma-rays emitted by such a source can, in principle, be absorbed by ground-state ^{57}Fe-nuclei in the sample. In practice, the hyperfine interactions of the nuclei with their environment can alter the transition energy by amounts considerably greater than the half-width of the radiation ($4.67 \cdot 10^{-9}$ eV), destroying the resonance. Resonance is restored, and the spectrum scanned, by modulating the energy of the gamma-ray by the Doppler effect: the source is mounted on a vibrator whose velocity can be controlled, and the energy, E, of the radiation incident on the sample is given by $E = E_0 \cdot (1 + v/c)$, where E_0 is the source transition energy, v and c are the velocities of the source and of light, respectively (v is positive for motion towards the absorber). The energy

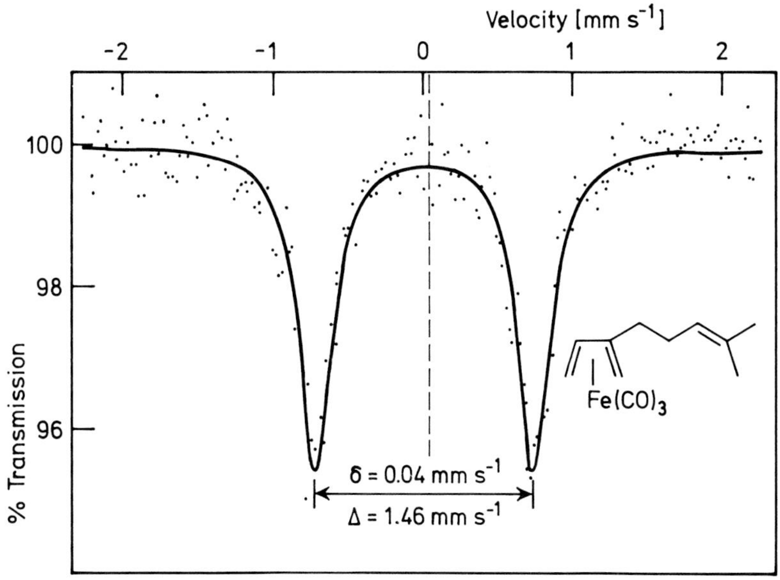

Fig. 1: Mössbauer spectrum of (myrcene)Fe(CO)$_3$ (117). The solid line represents the computed best fit for two independent Lorentzian peaks. The half-widths are 0,27 mm s^{-1}.

scale of a Mössbauer spectrum is thus a velocity scale, and data are normally presented in units of mm s^{-1}; for ^{57}Fe, 1 mm s^{-1} = 48•10^{-9} eV. The gamma-radiation transmitted by the sample is normally stored in a multi-channel analyser operating in phase with the vibrator, so that each channel represents a fraction of the velocity scale. A typical spectrum is shown in Figure 1. The spectra are characterised by two major parameters, the isomer shift, δ, and the quadrupole splitting, Δ, which are normally derived by least-squares fitting of Lorentzian absorption peaks to the experimental data.

A. THE ISOMER SHIFT

The isomer shift represents the velocity of maximum absorption or, in the case of a two-line spectrum such as that in Figure 1, the mid-point between the maxima. Measurements are made relative to a standard which is usually the centre of the doublet for sodium nitroprusside, $Na_2Fe(CN)_5NO•3H_2O$, or the centre of the six-line pattern of elemental iron (this pattern is due to the internal magnetic field in the metal). The latter is particularly convenient as the iron spectrum is frequently used to calibrate the velocity scale. All isomer shifts quoted here will refer to this standard, appropriate corrections being made to reported data where necessary. The isomer shift is somewhat temperature-dependent, becoming more positive as the temperature is lowered (the second-order Doppler shift). Since the magnitude of the temperature-dependence varies from compound to compound, small differences in isomer shift must be interpreted with care. Data will therefore be given only to 0.01 mm s^{-1}; quoted uncertainties are usually ± 0,02 mm s^{-1} or better, and are shown in the tables as the uncertainty in the last significant figure. Unless otherwise specified, all data refer to measurements at 80 K.

The isomer shift arises from the electrostatic interaction of the nucleus with the electron density which penetrates it. The correlation is negative in the case of ^{57}Fe, so that increasing isomer shift represents decreasing electron density at the nucleus. The only electrons which actually penetrate the nucleus are *s*-electrons (the contribution from $p_{\frac{1}{2}}$-electrons is usually neglected), so that an increase in isomer shift means a decrease in *s*-electron density. Such a change might occur, for instance, by decreasing the donor power of the ligands, e.g. the isomer shifts of the cations $Fe(NH_3)_6^{2+}$ and $Fe(H_2O)_6^{2+}$ are 1.02 and 1.25 mm s^{-1}, respectively. Electrons other than *s*-electrons act only indirectly by screening the nucleus from the *s*-electrons, so that this effect is much smaller than, and opposite to, that of the *s*-electrons. The contribution of, for example, the *d*-electrons is by no means negligible, however; it accounts for the difference between the characteristic isomer-shift ranges of high-spin ferrous (d^6) and ferric (d^5) complexes, which are 0.8 - 1.5 and 0.2 - 0.6 mm s^{-1}, respectively. Unfortunately, in low-spin systems, including organometallic derivatives, the isomer shift is relatively insensitive to change in oxidation state or ligands, but small, systematic changes can often be observed. For instance, the replacement of a carbonyl group by another ligand usually leads to an increase in isomer shift (of ca. 0.05 mm s^{-1}) which is attributable to a decrease in backdonation and an increased *d*-electron density on the iron atom. Thus, both forward (σ) and back (π) donation affect the isomer shift in the same sense.

B. QUADRUPOLE SPLITTING

The quadrupole splitting is manifested, for ^{57}Fe, as a doublet structure in the spectrum. The splitting arises from a quadrupole interaction of the nucleus with an electric field gradient, and is thus similar to the effect observed in nuclear quadrupole resonance spectroscopy. In the ^{57}Fe case, however, it is only the excited-state nucleus which possesses a

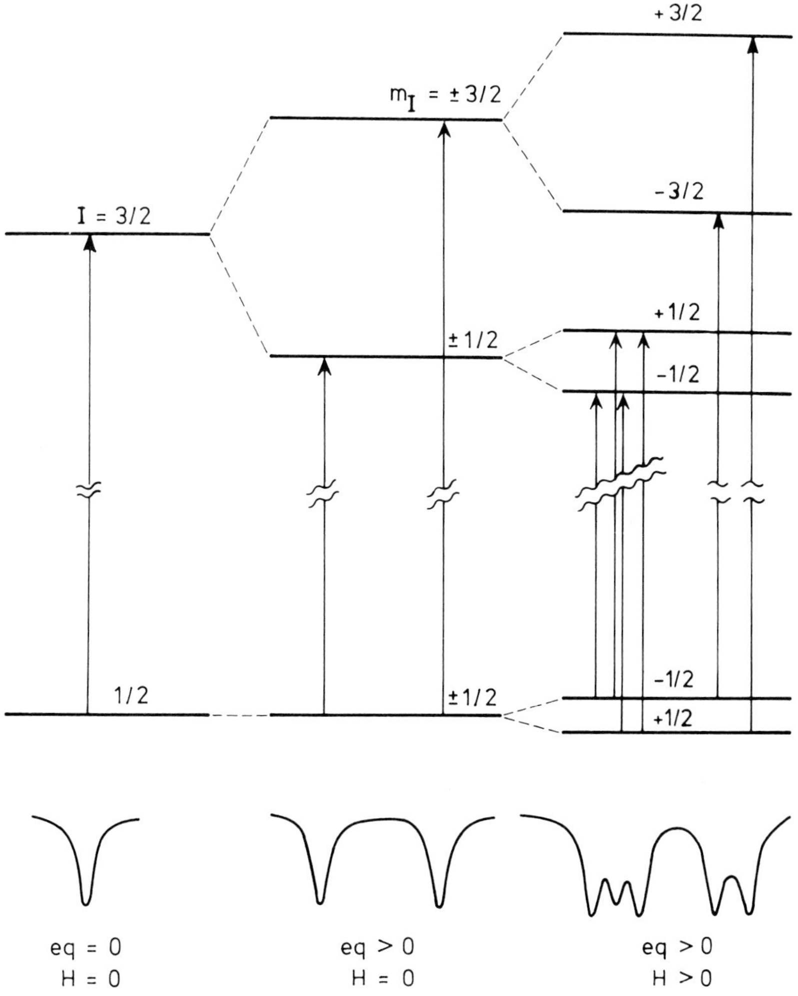

Fig. 2: Energy levels and transition for the ^{57}Fe nucleus. The excited-state quadrupole moment and the groundstate magnetic moment are positive, and the excited-state magnetic moment is negative.

quadrupole moment, so that the NQR technique cannot be used. The splitting, Δ, of the spectrum is half the quadrupole coupling constant, $\Delta = \frac{1}{2}e^2qQ(1 + 1/3\eta^2)^{\frac{1}{2}}$, where eQ is the nuclear quadrupole moment, eq is the principal component of the electric field gradient, and η measures the departure of the field gradient from axial symmetry. An electric field gradient arises when the electron density in the valence-shell of the iron atom has less than cubic symmetry. Charges beyond the iron atom also have an effect, but the field gradient varies with the inverse cube of the distance so that these effects can be ignored in first approximation. This also means that the contribution from 4*p*-electrons is considerably smaller than that from 3*d*-electrons, and only the latter need be considered. A non-cubic distribution of non-bonding electrons gives rise to a large quadrupole splitting (1-3 mm s^{-1}; 1 mm s^{-1} = 11.6 MHz), as in high-spin ferrous (d^6), low-spin ferric (d^5) and iron(O) (d^8) complexes. A ligand environment of less than cubic symmetry will also give an electric field gradient, so that an octahedral complex containing more than one type of ligand will show a quadrupole splitting. The majority of organometallic molecules have low symmetry, and large quadrupole splittings are often observed.

The sign of the electric field gradient cannot be determined from the simple measurement, but can be found by use of an oriented crystal sample, or by application of a large external magnetic field ($\geq$3 T). The latter technique is the more generally useful: in the presence of a large magnetic field the two lines of the quadrupole doublet split into a doublet and a triplet. For a positive electric field gradient, the doublet lies to higher energy (Figure 2). A positive value arises when there is a deficiency of(negative) charge along the principal (z) axis compared to the density along the other two axes. Electrons in d_{z^2}, d_{xz}, and d_{yz} orbitals thus give negative contributions, and those in d_{xy} and $d_{x^2-y^2}$ orbitals give positive contributions. For any given ligand, σ-donation gives a negative contribution to the field gradient along the metal-ligand bond axis, while π-acceptance gives a positive contribution. The quadrupole splitting thus represents the difference between σ- and π-bonding (contrast the isomer shift). In the tabulations which follow, the sign of the field gradient is shown explicitly where known; if no sign is given, its value has not been determined.

C. *MAGNETIC HYPERFINE SPLITTING*

In paramagnetic compounds, the unpaired electrons produce a magnetic field which can give additional splitting of the spectrum. Owing to relaxation effects, this magnetic hyperfine splitting is often not seen unless the sample is cooled to low temperatures or is magnetically dilute, but it can usually be revealed by the application of a small external magnetic field. The magnitude of the splitting gives information on the spin state and orbital degeneracy of the iron atom, and has been particularly useful in the examination of biological materials *(80,91)*.

D. *RECOIL-FREE FRACTION*

A nucleus emitting or absorbing a gamma-photon is likely to recoil with energy which must be derived from that of the photon. If this occurs, resonance will be lost. However, if the recoil energy is less than the phonon energy required to excite the crystal lattice, a certain fraction of the nuclei will not suffer recoil, and it is these which give rise to the observed spectrum. It follows that Mössbauer measurements must be made on solid samples, often at low temperatures (liquid nitrogen); liquids, solutions and gases must be frozen. Anisotropy in the binding of the Mössbauer atom in its lattice or molecule can in principle be detected by an asymmetry in the quadrupole doublet (Gol'danskii-Karyagin effect), although this effect has not yet been conclusively demonstrated for iron compounds *(67)*.

III. APPLICATIONS

As noted above, in organoiron compounds the isomer shift is relatively insensitive to the nature and number of the ligands. Also, the symmetry of the molecules is such that appreciable quadrupole splitting is nearly always found, and only rough generalisations can be made about its magnitude. It might seem, therefore, that measurements made on a single compound would be of little value in deducing the structure or bonding. However, compounds are rarely unique, and considerable headway can often be made by comparison with data for related compounds. Such comparisons, in conjunction with other physical data, usually lead to unequivocal assignments of structure.

Some generalizations can be made. In carbonyl compounds, for instance, five-coordinate structures invariably give large quadrupole splittings (> 1.5 mm s^{-1}), reflecting the formal d^8-configuration of the iron atom. Four- and six-coordinate species give much lower values, even when the "ligands" include other metal atoms. In these cases the formal non-bonding configurations are d^{10} or d^6, both of which have high symmetry, and the splitting represents the secondary effect of inequivalence in the metal-ligand bonding orbitals. The isomer shift appears to be more sensitive to π-bonding effects than σ-bonding, but anionic species often have lower isomer shifts than related neutral species, suggesting that the extra charge resides, in part at least, on the metal atom. The isomer shift also increases with increasing coordination number.

A. CARBONYLS AND SUBSTITUTED CARBONYLS

Carbonyl compounds are conveniently classified in terms of their nuclearity, and will be examined in order of increasing complexity.

Table 1: Derivatives of $Fe(CO)_4^{2-}$

	δ_{Fe} [mm s^{-1}][a]	Δ[mm s^{-1}]	*(Ref.)*
$Na_2Fe(CO)_4$	-0.18(1)	0.00(1)	*(58)*
$NaFe(CO)_3NO$	-0.10(1.3)	0.38(0.5)	*(106)*
$KFe(CO)_3NO$	-0.08(1.3)	+0.36(0.5)	*(106)*
$Fe(CO)_2(NO)_2$	+0.06(1.3)	-0.33(0.5)	*(106)*
$Fe(Ph_3P)_2(NO)_2$	+0.09(1.3)	-0.69(0.5)	*(106)*
$Fe(Ph_3As)_2(NO)_2$	+0.16(1)	0.59(1)	*(42)*
$Et_4N[Fe(CO)_4H]$	-0.17(1)	1.36(1)	*(58)*
$Fe(CO)_4H_2$	-0.18(1)	0.55(1)	*(8)*
$Fe(depb)_2H_2$	-0.04(1)	1.84(1)	*(8)*
$(NH_3)_3ZnFe(CO)_4$	-0.18(2)	0.80(2)	*(79)*
$enCdFe(CO)_4$	-0.09(2)	0.68(2)	*(79)*
$py_2CdFe(CO)_4$	-0.09(2)	0.81(2)	*(79)*
$bipyCdFe(CO)_4$	-0.08(2)	0.46(2)	*(79)*
$CdFe(CO)_4$	-0.05(2)	<0.15(2)	*(79)*
cis-$Fe(CO)_4Cl_2$	0.05(1)	0.26(1)	*(133)*
cis-$Fe(CO)_4Br_2$	0.06(1)	0.31(1)	*(133)*
cis-$Fe(CO)_4I_2$	0.06(1)	0.32(1)	*(133)*
cis-$Fe(CO)_4(Cl)SnCl_3$	0.03(2)	0.45(2)	*(52)*
cis-$Fe(CO)_4(Br)SnBr_3$	0.01(2)	0.46(2)	*(52)*
cis-$Fe(CO)_4(I)SnI_3$	0.03(2)	0.38(2)	*(52)*
cis-$Fe(CO)_4(SnCl_3)_2$	0.02(2)	0.20(2)	*(52)*
trans-$Fe(CO)_4(SnCl_3)_2$	0.02(2)	0.46(2)	*(52)*
$Me_2Sn[Fe(CO)_4]_2SnMe_2$	-0.09(2)	ca. 0.15	*(83)*
$Bu_2[Fe(CO)_4]_2SnBu_2$	-0.02	0.20	*(83)*
$Me_2Sn[Fe(CO)_4]_2Sn[Fe(CO)_4]_2SnMe_2$	-0.10(2)	0.30(2)	*(83)*

[a] In this and all subsequent Tables, numbers in parentheses are the reported uncertainties in the last significant figure.

1. Mononuclear Systems

Iron(-II) provides a relatively simple case, since it has a closed-shell $3d^{10}$ configuration and is usually tetrahedrally coordinated. The only contribution to the quadrupole splitting will come from non-equivalence in the metal-ligand bonds. Thus, $Fe(CO)_4^{2-}$ has no splitting and $Fe(CO)_3NO^-$ a small splitting (see Table 1). The positive sign for the electric field gradient in the latter case is consistent with CO being a better σ-donor and/or worse π-acceptor than NO^+, as would be expected. The data for $FeL_2(NO)_2$ (L = CO, Ph_3P) show that Ph_3P is still more σ-basic and/or less π-acidic than CO, and that the ON-Fe-NO bond angle is less than the tetrahedral value, which also indicates high π-acidity for NO^+.

With other $Fe(CO)_4$-derivatives, it is not possible to assign an unambiguous oxidation state to the iron atom. Representative data are included in Table 1. The constancy of isomer shift for the hydrides is quite striking, and these values are considerably lower than normally found for low-spin iron(II) complexes. The increase on replacing carbonyl by phosphine ligands is quite common, however, (representing reduced back-donation and hence greater *d*-electron density on the metal, *cf.* pp.178,190,194) and the association of the hydrogen atoms with the metal is clearly demonstrated by the quadrupole splittings. On the same basis, weak association between the Group II metal ammines and the $Fe(CO)_4^{2-}$ anion seems to occur in the zinc and cadmium derivatives, but the small quadrupole splitting of $CdFe(CO)_4$ and the organotin derivatives demonstrates strong covalent iron-metal bonds completing an octahedral coordination.

The halides $Fe(CO)_4X_2$ appear to be iron(II) complexes *(52)* rather than substitution-derivatives of pentacarbonyliron(O), $Fe(CO)_4(X_2)$, as had been suggested earlier *(33)*. These halides, the corresponding X_3Sn-derivatives, and several mono- and di-substituted compounds, $LFe(CO)_3X_2$ and $L_2Fe(CO)_2X_2$, give spectra characteristic of low-spin iron(II) and the data have been analysed in terms of the bonding characteristics of the individual ligands *(9)*.

All iron(O) complexes show a large quadrupole splitting, arising from the formal d^8 configuration which can never have cubic symmetry. In a trigonal bipyramidal molecule, such as $Fe(CO)_5$, these electrons would be accommodated in the d_{xz}, d_{yz}, d_{xy} and $d_{x^2-y^2}$ orbitals. Even after allowing for the effects of covalency, there will be a large relative deficiency of electron density in the d_{z^2} orbital, and the electric field gradient has been found to be positive as this would suggest *(30,85)*. The effects of the non-cubic ligand arrangement are superposed on this field gradient, giving a range of quadru-

pole splittings for the various derivatives. With the data
presently available for monosubstituted iron carbonyls, two
trends seem to occur. For a variety of tertiary phosphines
and arsines, both the isomer shift and the quadrupole splitting vary about the values for Fe(CO)$_5$, and correlate roughly
linearly (Table 2 and Figure 3) *(25)*. For the phosphines,
both values increase as the σ-donor power decreases and the π-
acceptor power increases. Decreased donation into the d_{z^2} orbital would increase the positive field gradient, as would increased back donation from the d_{xz} and d_{yz} orbitals. The isomer shift would also increase with decreased donation from the
ligand (into the 4s-orbital), although it is curious that the
observed values should lie on both sides of that for Fe(CO)$_5$.
    If the substituent is an olefin, the quadrupole splitting
is dramatically reduced and the isomer shift somewhat increased, relative to pentacarbonyliron. In these cases the
ligand lies in the equatorial plane of the bipyramid *(50,119)*,
and the decreased quadrupole splitting has been attributed
(50) to the inability of the olefin to accept charge from the
d_{xz} orbital (assuming that the z-axis and the electric field
gradient axis still lie along the "trigonal" axis of the
bipyramid). A range of quadrupole splitting values is found,
but the isomer shift is almost constant. There does seem to
be a trend towards an inverse correlation, the increasing

Table 2: Monosubstituted derivatives of Fe(CO)$_5$

		δ_{Fe}[mm s^{-1}]	Δ[mm s^{-1}]	*(Ref.)*
1	Fe(CO)$_5$	−0.09(1)	+2.57(1)	*(58)*
2	(Me$_2$N)$_3$PFe(CO)$_4$	−0.11	2.22	*(77)*
3	(EtO)$_3$PFe(CO)$_4$	−0.12(0.8)	2.31(0.8)	*(33)*
4	(OC)$_4$FePh$_2$PCH=CHPPh$_2$Fe(CO)$_4$	−0.11(1)	2.46(1)	*(45)*
5	Ph$_3$PFe(CO)$_4$	−0.07(0.8)	2.54(0.8)	*(33)*
6	ffosFe(CO)$_4$	−0.07(1)	2.61(1)	*(45)*
7	ffarsFe(CO)$_4$	−0.05(1)	2.79(1)	*(45)*
8	(OC)$_4$FeffarsFe(CO)$_4$	−0.05(1)	2.82(1)	*(45)*
9	(C̅O-CH=CH-CO-O̅)Fe(CO)$_4$	+0.01(0.8)	1.41(0.8)	*(33)*
10	(C̅O-CH=CH-CO-N̅Ph)Fe(CO)$_4$	−0.01(0.5)	1.54(0.5)	*(50)*
11	(*trans*-MeO$_2$C-CH=CH-CO$_2$Me)Fe(CO)$_4$	+0.01(0.5)	1.56(0.5)	*(50)*
12	(H$_2$C=CH-CO-NMe$_2$)Fe(CO)$_4$	0.00(0.5)	1.65(0.5)	*(50)*
13	(PhCH=CH-CHO)Fe(CO)$_4$	−0.01(0.8)	1.75(0.8)	*(33)*
14	(acenaphthalene)Fe(CO)$_4$	−0.01(0.8)	1.78(0.8)	*(33)*

See also Ref. 25.

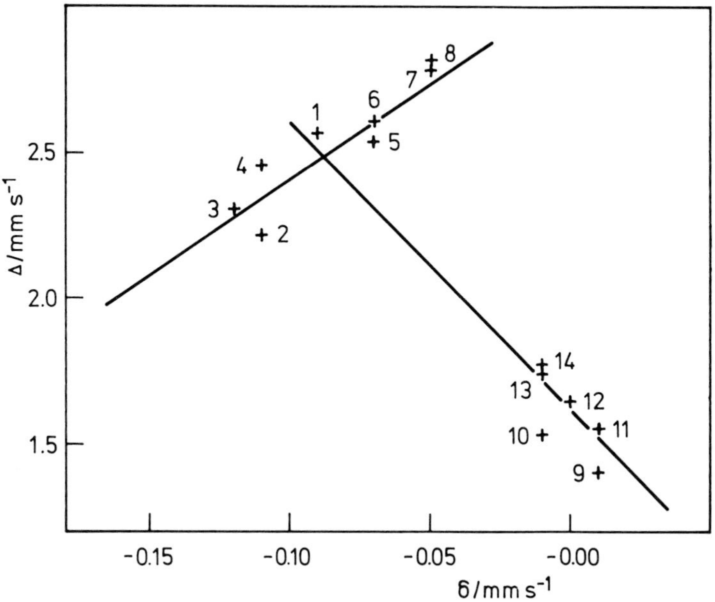

Fig. 3: Isomer shift vs. quadrupole splitting for compounds of the type LFe(CO)₄. The points are numbered to correspond with Table 2, and the lines are least-squares fits to points 1-8 and 9-14.

quadrupole splitting being associated with a decrease in isomer shift, which extrapolates to the pentacarbonyliron-values. In these cases, increasing donation from the olefin would augment the 4*s* and $3d_{x^2-y^2}$ populations, which would decrease the isomer shift and increase the quadrupole splitting. Increased backdonation from the d_{xy} orbital would decrease the quadrupole splitting. The observed trends seem to correlate with the nature of the substituents at the double bond of the olefin.

Trends similar to those described above are found for disubstituted complexes with both types of ligand *(19,25,32, 45,68,117)*.

2. Binuclear Systems

In the parent carbonyl, $Fe_2(CO)_9$ (1) each iron atom is octahedrally coordinated by carbonyl groups, three of which are bridging. If the bond angles were exactly octahedral, no electric field gradient would result from the inequivalence of the iron-carbon bonds. The small field gradient observed (see

Table 3) probably arises from the metal-metal bond, since it is found to be positive and to lie along the trigonal axis of

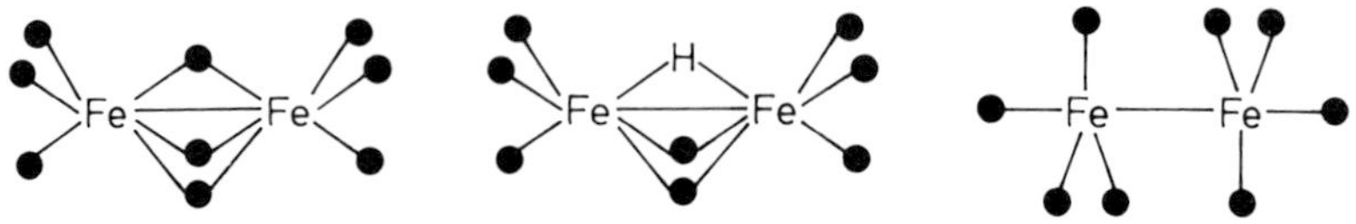

1 2 3

(In these and subsequent formulae, ● represent a carbonyl group)

the molecule *(62)*. The spectrum of the anion $Fe_2(CO)_8H^-$ is very similar to that of $Fe_2(CO)_9$, and a structure (2) in which one bridging carbonyl group is replaced by the hydride has been suggested *(58)*. In sharp contrast, the anion $Fe_2(CO)_8^{2-}$ shows a large quadrupole splitting, similar to that for $Fe(CO)_5$. The structures are closely related, one axial carbonyl group of the pentacarbonyl being replaced by an $Fe(CO)_4$-group in the anion (3) *(108)*.

A large number of compounds of the type $(OC)_3FeQ_2Fe(CO)_3$ has been examined, together with many in which some of the

Table 3: Binuclear carbonyliron derivatives

	δ_{Fe}[mm s^{-1}]	Δ[mm·s^{-1}]	*(Ref.)*
$Fe_2(CO)_9$	+0.17(1)	+0.42(1)	*(58)*
$Et_4N[Fe_2(CO)_8H]$	+0.07(1)	0.50(1)	*(58)*
$(Et_4N)_2[Fe_2(CO)_8]$	-0.08(1)	2.22(1)	*(58)*
$Fe_2(CO)_6(NH_2)_2$	+0.02	0.85	*(88)*
$Fe_2(CO)_6(PMe_2)_2$	-0.01(0.5)	0.65(0.5)	*(50)*
$[Fe_2(CO)_6(PMe_2)_2]^-$	-0.07(0.5)	1.29(0.5)	*(50)*
$[Fe_2(CO)_6(PMe_2)_2]^{2-}$	-0.16(0.5)	1.53(0.5)	*(50)*
$Fe_2(CO)_6(AsMe_2)_2$	+0.02(0.5)	0.81(0.5)	*(50)*
$[Fe_2(CO)_6(AsMe_2)_2]^-$	-0.08(0.5)	1.86(0.5)	*(50)*
syn-$Fe_2(CO)_6(SMe)_2$	+0.03(1)	1.00(1)	*(41)*
anti-$Fe_2(CO)_6(SMe)_2$	+0.03(1)	0.88(1)	*(41)*
$[Fe_2(CO)_6(SMe)_2]^-$	+0.02(0.5)	1.62(0.5)	*(50)*
$Fe_2(CO)_6(SPh)_2$	+0.06(0.5)	1.07(0.8)	*(90)*
$Fe_2(CO)_6[S(C_6F_5)]_2$	+0.06(0.8)	1.35(0.8)	*(90)*
$Fe_2(CO)_6(SePh)_2$	+0.04(0.8)	1.04(0.8)	*(90)*
$Fe_2(CO)_6[Se(C_6F_5)]_2$	+0.08(0.8)	1.16(0.5)	*(90)*
$Fe_2(CO)_6(PMe_2)_2I_2$	0.00(1)	0.99(1)	*(63)*

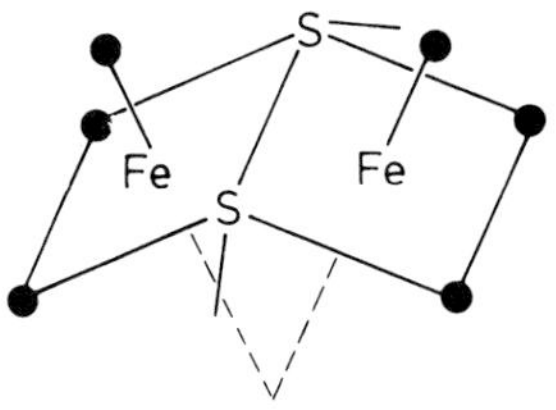

4

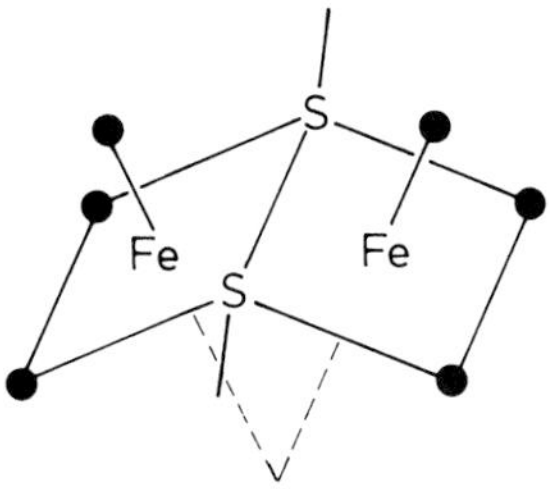

5

carbonyl groups have been replaced by other ligands (Table 3). All have modest quadrupole splittings in accordance with the basic octahedral coordination, the bent metal-metal bond occupying the sixth position. The order of isomer shifts is Q = R_2P < R_2As < RS ≲ RSe, which is the expected order of decreasing donor/acceptor power. For Q = MeS, slight differences are noted between the *syn* (4) and *anti* (5) isomers, the latter giving the smaller quadrupole splitting. This suggests that change in the disposition of the methyl group slightly alters the geometry of the FeS_2Fe ring, since in other pairs of geometrical isomers (see below) no difference in Mössbauer parameters is detectable.

Four types of substitution-derivatives have been obtained

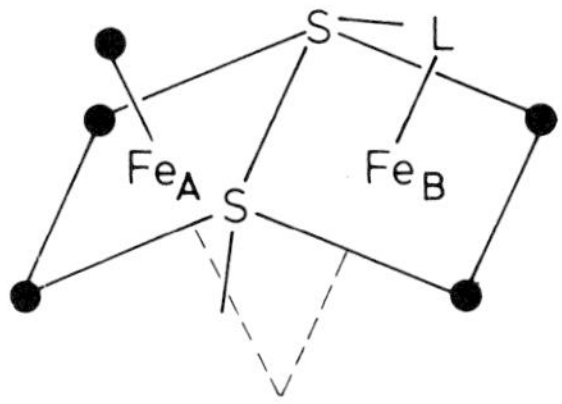

6

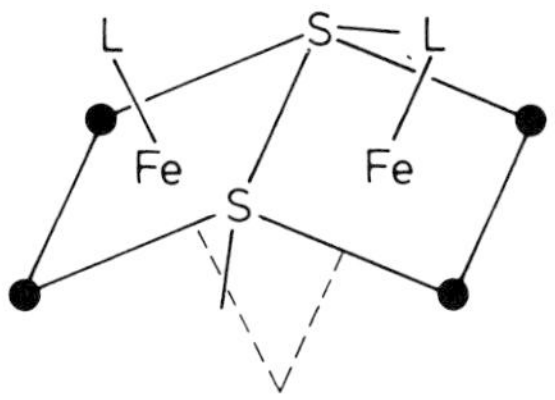

7

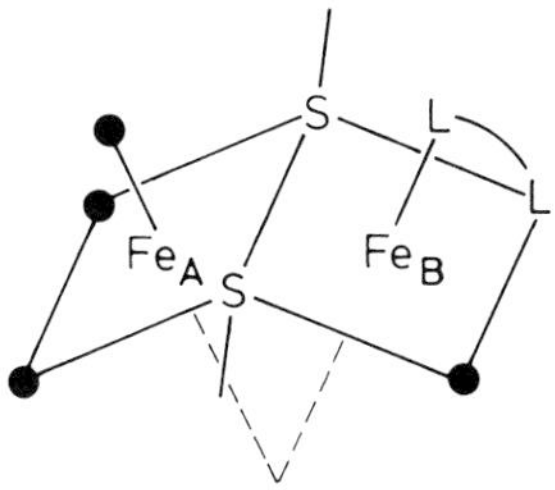

8

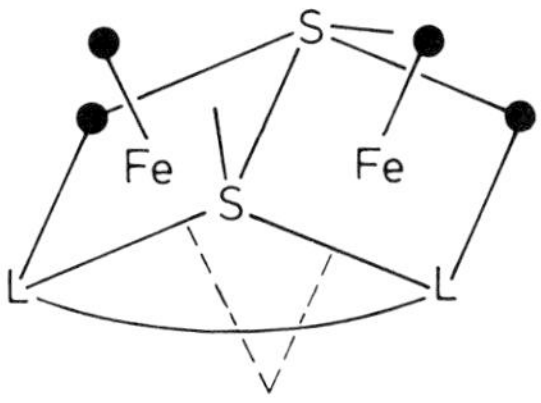

9

Table 4: Derivatives of $(OC)_3Fe(SMe)_2Fe(CO)_3$

	Fe_A δ_{Fe} [mm s^{-1}]	Fe_A Δ [mm s^{-1}]	Fe_B δ_{Fe} [mm s^{-1}]	Fe_B Δ [mm s^{-1}]
syn-$(OC)_3Fe(SMe)_2Fe(CO)_2L$ (6)				
L = $PhMe_2P$	0.03	1.29	0.04	0.78
$Ph_2PCH_2PPh_2$	0.03	1.13	0.05	0.74
Ph_3P	0.04	1.13	0.06	0.73
$Ph_2AsCH_2AsPh_2$	0.03	1.12	0.10	0.53
Ph_3As	0.05	1.05	0.11	0.53
Ph_3Sb	0.03	1.11	0.10	0.47
syn-$L(OC)_2Fe(SMe)_2Fe(CO)_2L$ (7)				
L = $(MeO)_3P$			0.04	1.00
$PhMe_2P$			0.02	1.00
$Ph_2PCH_2PPh_2$			0.05	0.84
Ph_3P			0.07	0.74
Ph_3As			0.10	0.57
Ph_3Sb			0.13	0.45
anti-$(OC)_3Fe(SMe)_2Fe(CO)(L-L)$ (8)				
L-L = f_8fos	0.01	1.06	0.16	0.76
$Ph_2PCH{=}CHPPh_2$	0.04	1.20	0.13	0.63
ffos	0.02	1.14	0.14	0.47
anti-$\overline{L(OC)_2Fe(SMe)_2Fe(CO)_2L}$ (9)				
L-L = $Ph_2AsCH_2AsPh_2$			0.09	1.22
ffars			0.10	1.08
$Ph_2PCH_2PPh_2$			0.05	1.08
$Ph_2PNEtPPh_2$			0.05	1.03
ffos			0.07	1.01

Data from Refs. *41* and *48*, all $\pm$ 0.01 mm s^{-1}.

which can be differentiated by their Mössbauer spectra. Data for derivatives of $(OC)_3Fe(SMe)_2Fe(CO)_3$ are collected in Table 4. In the unsymmetrical compounds, the two different iron atoms give well-resolved signals, the substituted atom showing an increase in isomer shift and a decrease in quadrupole splitting, both of which are attributable to decreased back-donation. Equatorial substitution appears to give a greater quadrupole splitting than axial substitution. Interestingly, the quadrupole splitting for the unsubstituted atom shows a slight increase, which may indicate a change in geometry or possibly an effect transmitted by the metal-metal bond. Similar effects are found in the parent compounds $(OC)_3FeQ_2Fe(CO)_3$ (Q = MeS, Me_2P, Me_2As) on one-electron reduction (Table 3). The ESR spectrum of $[(OC)_3Fe(PMe_2)_2Fe(CO)_3]^-$ shows that all the methyl groups are equivalent *(51)*, suggesting that the metal-metal bond is weakened sufficiently to allow a planar or fluxional structure.

Clearly non-equivalent iron atoms are also found in the compounds (chelate)$Fe_2(CO)_6$ [chelate = ffars, ffos, $f_6$fos, (10)] *(44)*, the diazepine derivatives $[-N{=}C(Ph)-CH_2-C(Ph){=}CH-C(Ph){=}N-]Fe_2(CO)_6$ and $[-N{=}C(Ph)-CH_2-C(Ph)-CH_2-C(Ph){=}N-]Fe_2(CO)_6$ *(102)*, and $L(CO)_2Fe(PhCO)_2Fe(CO)_3$ (L = CO, Et_2NH, $C_5H_{10}NH$) *(60)*, showing asymmetric structures which have been confirmed by X-ray measurement in three cases *(55,100,102)*.

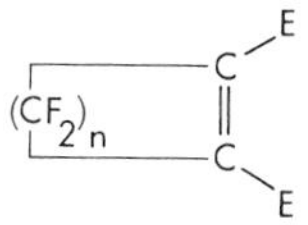

10

ffos : n = 2, E = PPh_2
$f_6$fos : n = 3, E = PPh_2
$f_8$fos : n = 4, E = PPh_2
ffars : n = 2, E = $AsMe_2$

However, such differentiation cannot be guaranteed. The compounds (triene)$Fe_2(CO)_6$ (triene = cyclo-octatriene, cycloheptatriene) and (cyclo-octatetraene)$Fe_2(CO)_6$ all give Mössbauer spectra consisting of sharp doublets, suggesting identical environments for the two iron atoms. The room-temperature NMR spectra also suggested symmetrical structures, and di-π-allylic forms (11) were proposed *(56,86)*. Subsequent X-ray examination has shown this structure to be correct for the cycloheptatriene compound, but the cyclooctatetraene deriva-

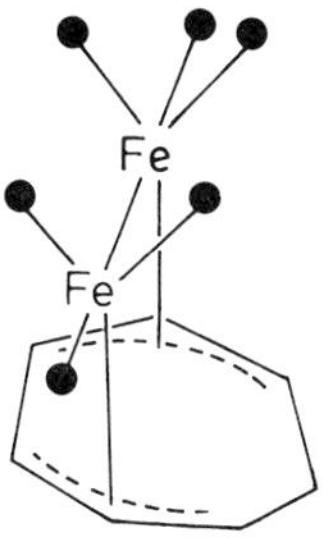

11

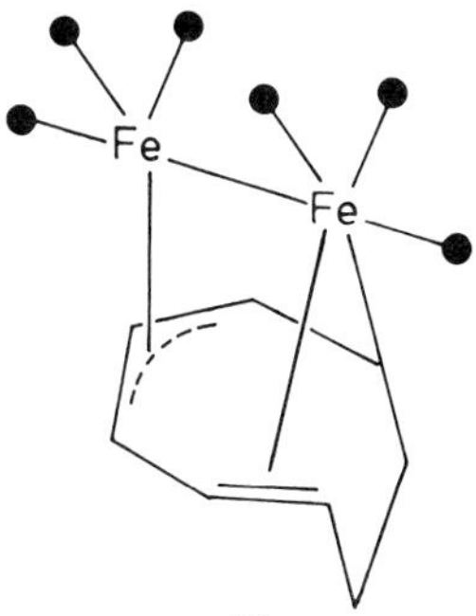

12

tive has an unsymmetrical structure with no plane of symmetry and rather different environments for the iron atoms (12). The low-temperature NMR spectra are typical of fluxional behaviour in the latter case only *(36,37)*.

3. Tri- and Tetra-nuclear Derivatives

The Mössbauer spectra of $Fe_3(CO)_{12}$ and $Fe_3(CO)_{11}H^-$ (Table 5) clearly show the presence of two types of iron atom in the ratio 2:1, which observations were instrumental in establishing the correct structure of the former *(57)*. (A fascinating case history of this investigation has been given *(134)*.) The structure may be regarded as being derived from that of $Fe_2(CO)_9$ by replacing one bridging carbonyl group by a bridging $Fe(CO)_4$-group (13). The isomer shift of the two bridged atoms is very similar to that of $Fe_2(CO)_9$ but the quadrupole splitting is greater, reflecting the greater asym-

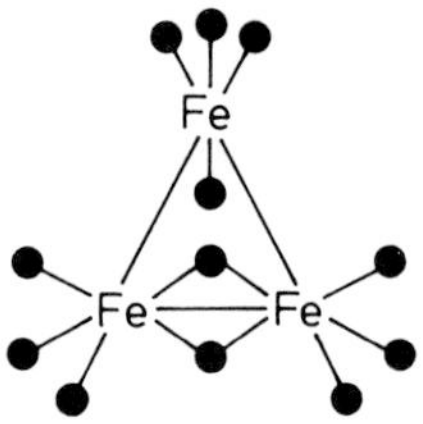

13

metry of the iron environment. The other iron atom has a lower isomer shift and a very small quadrupole splitting, consistent with the almost regular octahedral coordination. Substitution of carbonyl groups by tertiary phosphines and arsines leads to an increase in isomer shift and decrease in quadrupole splitting *(28,43,66)*, as in other carbonyl systems (*cf.* pp.178,183,194). Recently several mixed trinuclear carbonyls have been prepared, the Mössbauer spectra of which show clearly that the structures are related to that of $Fe_3(CO)_{12}$, and allow the positions of the iron atoms to be identified. For instance, the compounds $CpMFe_2(CO)_9$ (M = Co, Rh) have quite different spectra, the cobalt compound showing two distinct iron environments while the rhodium derivative gives only a simple doublet. Comparison with the $Fe_3(CO)_{12}$-spectrum, in conjunction with IR data, strongly suggests the structures shown in Table 5.

The parameters for the iron atom in $Co_2Fe(CO)_9S$ are similar to those of the unique atom in $Fe_3(CO)_{12}$, showing again

Table 5: Trinuclear carbonyl compounds

	Basal Fe		Apical Fe		
	δ_{Fe}[mm s^{-1}]	Δ[mm s^{-1}]	δ_{Fe}[mm s^{-1}]	Δ[mm s^{-1}]	*(Ref.)*
14	+0.11(1)	1.13(1)	0.05(1)	0.13(2)	*(58)*
15	+0.16(1)	1.52(1)	+0.02(1)	0[a]	*(43)*
16	+0.35(2)	0.57(2)	+0.28(2)	0	*(107)*
17	+0.04(1)	1.41(1)	+0.02(1)	0.16(2)	*(58)*
18	+0.08(1)[b]	0.91(1)			*(99)*
19	+0.09(1)	1.01(1)	+0.01(1)	0.43(1)	*(89)*
20	-0.01(1)	1.04(1)			*(89)*
21	-0.02(1)	1.05(1)			*(89)*
22	+0.04(2)	1.10(2)			*(78)*

[a]Half-width = 0.38mm s^{-1}; [b]At 293 K.

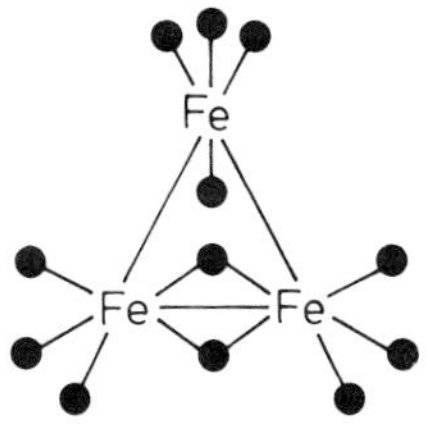

14

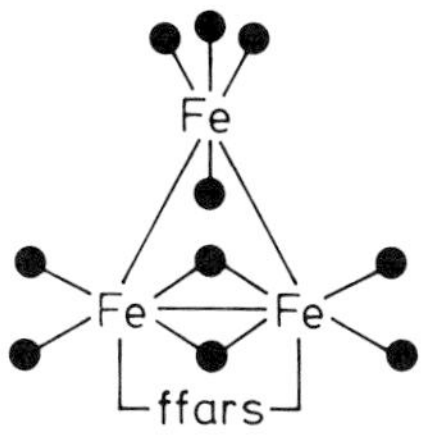

15

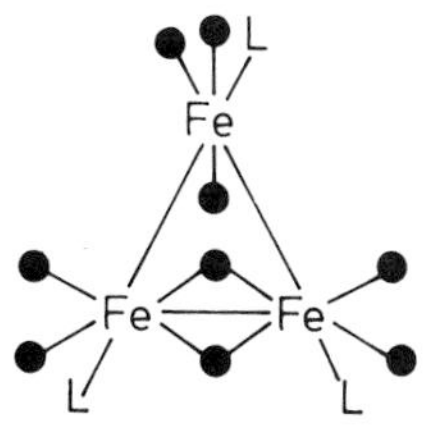

16

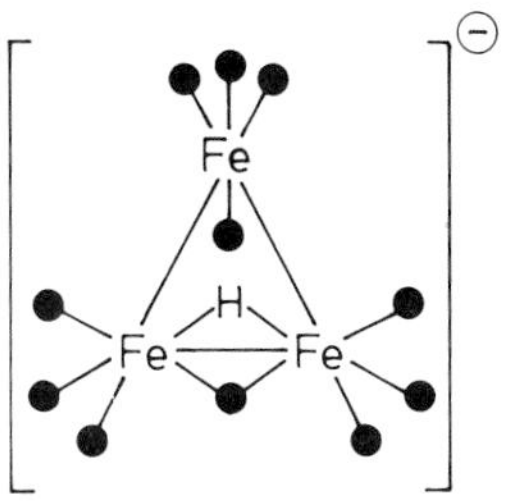

17

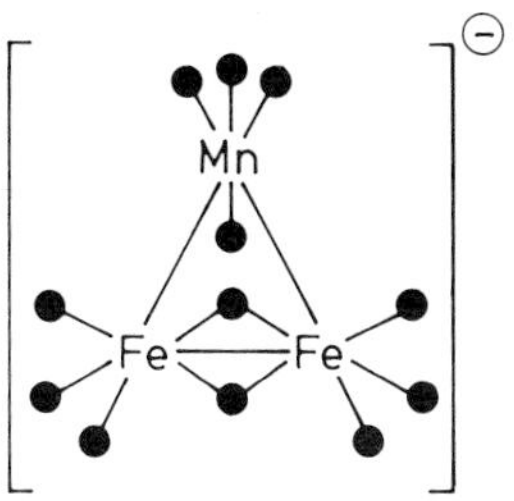

18

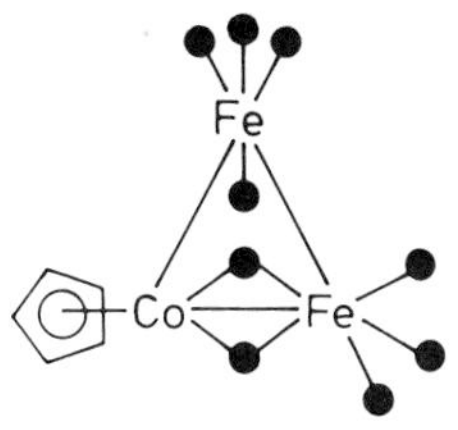

19

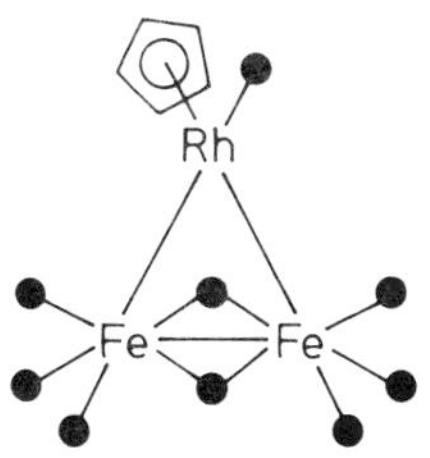

20

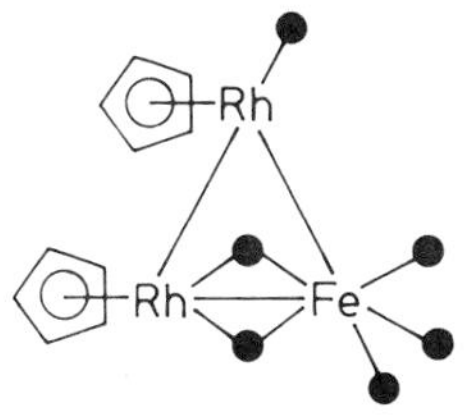

21

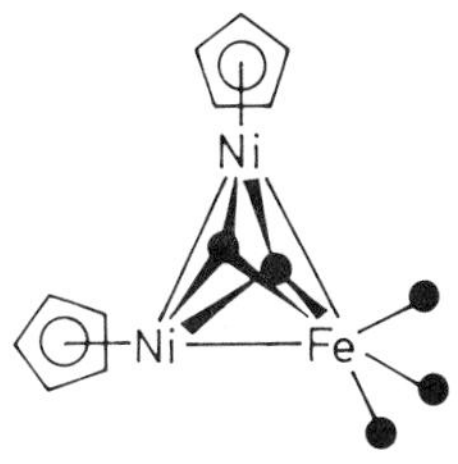

22

the essential six-coordination. The insensitivity of these parameters to substitution of one or two carbonyl groups by triphenylphosphine suggests that substitution occurs at cobalt rather than iron *(22)*. In contrast, the spectra of the compounds $Fe_3(CO)_8LS_2$ (L = CO, PPh_3, $P(n\text{-}Bu)_3$) are very similar *(41)* and do not reflect the inequivalence between the iron atoms which the X-ray data indicate *(47)*.

The spectrum of the anion $Fe_4(CO)_{13}^{2-}$ also appears as a simple doublet, but it is likely that the signal for the unique iron atom is masked by that of the three equivalent atoms *(58)*. Doublets with somewhat broadened lines are found for $Cp_2Rh_2Fe_2(CO)_8$ and $CpRhFe_3(CO)_{11}$ suggesting non-equivalence of the iron atoms *(89)* subsequently confirmed by X-ray investigation *(29)*.

B. CYCLOPENTADIENYL-CARBONYL COMPOUNDS

The parent compound for this series, $[CpFe(CO)_2]_2$, and its derivatives in which the bridging groups are PMe_2 or SPh display *cis-trans* isomerism (23), but the Mössbauer parameters of the two isomers are not significantly different (Table 6). The crystal structures show that for $[CpFe(CO)_2]_2$ the geometry

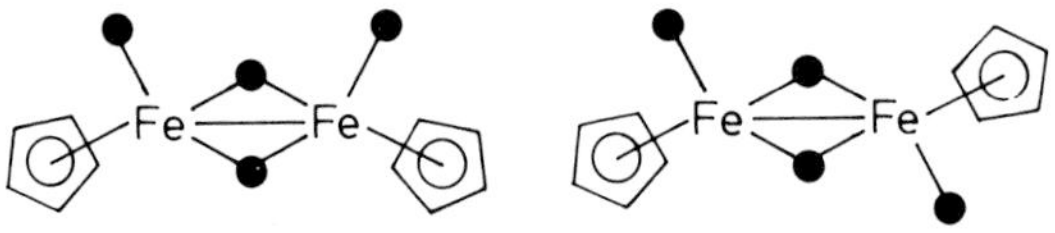

23

about the iron and the iron-iron bond distance are the same for both isomers *(21)*. The isomer shift trend, Q = R_2P < $R_2As \sim RS$, is similar to that for $[(CO)_3FeQ]_2$ (see above).

The ligand $Ph_2PC{\equiv}CPPh_2$ replaces one terminal carbonyl group on each of two molecules of the dimeric carbonyl compound, giving the unsymmetrical complex $Cp(CO)Fe(CO)_2Fe(Cp)$-$Ph_2PC{\equiv}CPPh_2Fe(Cp)(CO)_2Fe(CO)Cp$ the Mössbauer spectrum of which displays only one sharp doublet *(27)*. It has been proposed that a partial isomer shift may be assigned to each ligand in certain carbonyl complexes *(73,77)*, and application of this method suggested that the isomer shifts of the two iron atoms would be expected to differ by only 0.017 mm s^{-1} *(26)*. It is not possible to predict the quadrupole splitting in this type of compound.

This complex undergoes unsymmetrical cleavage by halogens to give the cation $[Cp(CO)_2FePh_2PC{\equiv}CPPh_2Fe(CO)_2Cp]^{2+}$ which is clearly differentiated by its isomer shift from the halogen-substituted derivatives $Cp(CO)FeXPh_2PC{\equiv}CPPh_2XFe(CO)Cp$ (X = Cl,

Table 6: Cyclopentadienyltricarbonyliron derivatives

	δ_{Fe}[mm s^{-1}]	Δ[mm s^{-1}]	*(Ref.)*
cis-$[CpFe(CO)_2]_2$	+0.21(2)	1.92(2)	*(21)*
trans-$[CpFe(CO)_2]_2$	+0.21(2)	1.90(2)	*(21)*
cis-$[CpFe(CO)PMe_2]_2$	+0.14(1)	1.61(1)	*(63)*
trans-$[CpFe(CO)PMe_2]_2$	+0.16(1)	1.64(1)	*(63)*
cis-$[CpFe(CO)PPh_2]_2$	+0.17(5)	1.60(5)	*(76)*
trans-$[CpFe(CO)PPh_2]_2$	+0.17(5)	1.66(5)	*(76)*
cis-$[CpFe(CO)AsMe_2]_2$	+0.26(5)	1.42(5)	*(76)*
trans-$[CpFe(CO)AsMe_2]_2$	+0.26(5)	1.57(5)	*(76)*
'stable' $[CpFe(CO)SPh]_2$	+0.34(1)	1.67(1)	*(63)*
'unstable' $[CpFe(CO)SPh]_2$	+0.35(1)	1.72(1)	*(63)*
$[Cp(CO)Fe(CO)_2FeCp]_2$DPPA	+0.27(1)	1.94(1)	*(26)*
$\{[CpFe(CO)_2]_2$DPPA$\}(FeCl_4)_2$[a]	+0.09(1)	1.80(1)	*(26)*
$\{[CpFe(CO)_2]_2$DPPA$\}(FeBr_4)_2$[b]	+0.10(1)	1.79(1)	*(26)*
$\{[CpFe(CO)_2]_2$DPPA$\} I_3$	+0.10(1)	1.80(1)	*(26)*
$[CpFe(CO)Cl]_2$DPPA	+0.31(1)	1.91(1)	*(26)*
$[CpFe(CO)Br]_2$DPPA	+0.32(1)	1.96(1)	*(26)*
$[CpFe(CO)I]_2$DPPA	+0.30(1)	1.88(1)	*(26)*
$[CpFe(CO)_2CS]PF_6$	-0.05(2)	1.89(2)	*(23)*
$[CpFe(CO)_3]PF_6$	+0.01(2)	1.78(2)	*(23)*
$CpFe(CO)_2CN$	+0.04(2)	1.96(2)	*(23)*
$CpFe(CO)_2CO\text{-}CH_3$	+0.04	1.68	*(77)*
$[CpFe(CO)_2PPh_3]Cl$	+0.05(2)	1.92(2)	*(23)*
$[CpFe(CO)_2P(NMe_2)_3]I$	+0.05	1.66	*(77)*
$CpFe(CO)_2CO\text{-}NEt_2$	+0.06	1.73	*(77)*
$CpFe(CO)_2(CH_2)_3Fe(CO)_2Cp$	+0.08	1.67	*(77)*
$CpFe(CO)_2SnPh_3$	+0.10(2)	1.75(2)	*(118)*
$CpFe(CO)_2SnBr_3$	+0.13(2)	1.80(2)	*(118)*
$CpFe(CO)_2SnCl_3$	+0.15(2)	+1.80(2)	*(118)*
$[CpFe(CO)_2(py)]PF_6$	0.15(1)	1.86(1)	*(7)*
$[CpFe(CO)_2(C_2H_4)]PF_6$	0.17(1)	1.77(1)	*(7)*
$CpFe(CO)_2SnI_3$	+0.18(2)	1.74(2)	*(118)*
$[CpFe(CO)_2(NCMe)]PF_6$	0.18(1)	1.95	*(7)*
$CpFe(CO)_2I$	+0.22(2)	1.86(2)	*(118)*
$CpFe(CO)_2Br$	+0.24(2)	1.77(2)	*(118)*
$CpFe(CO)_2Cl$	+0.24(2)	1.89(2)	*(118)*

[a] δ(anion) = +0.24 mm s^{-1}; [b] δ(anion) = +0.36 mm s^{-1}

Br, I). The isomer shift of the former is similar to that found for $[CpFe(CO)_2PR_3]^+$ (R = Ph, NMe_2), while those of the halogen-substituted compounds are intermediate between those of $CpFe(CO)_2X$ and $\{CpFe[P(OMe)_3]_2\}Br$ *(26)*. As in other systems, (*cf.* pp.178,183,190), replacement of a carbonyl group by a phosphine gives an increase in isomer shift.

Several series of compounds of the type $CpFe(CO)_2Y$ have been examined, in which the isomer shifts represent the σ/π-character of the ligand Y, good donor/acceptors giving the lower values. Thus, the lowest values are found for CS and CO, which are both good acceptor ligands. Good σ-donor groups such as alkyl or SnR_3 give intermediate values, while poorer donors, such as the halogens, give the highest shifts of all. On this basis, it might be argued that the acetyl group is a modest π-acceptor.

The whole group of complexes $CpFe(CO)_2Y$ is unusual among organoiron compounds in that the isomer shift shows systematic variations while the quadrupole splitting varies irregularly. The electric field gradient is presumably dominated by the contribution from the cyclopentadienyl group, and the sign of the gradient in $CpFe(CO)_2SnCl_3$ has been found to be positive *(15)*, as in ferrocene. The variations in quadrupole splitting probably reflect changes in geometry rather than the nature of the Fe-Y bond.

C. BIS-CYCLOPENTADIENYL AND RELATED COMPLEXES

Large numbers of ferrocene derivatives have been studied by the Mössbauer technique, but the parameters obtained are almost independent of the nature of the substituents on the rings, e.g. isomer shifts occur in the range 0.40 - 0.55 mm s^{-1} and quadrupole splittings in the range 2.05 - 2.42 mm s^{-1}, with only isolated values outside these ranges. Closer inspection of the data shows that low quadrupole splitting values (Δ < 2.30 mm s^{-1}) are associated with electron-withdrawing substituents, e.g. $-COCH_3$, $-NO_2$, $-CO_2H$ (Table 7). Particularly interesting in this connection are the ferrocenyl carbene complexes, for which other physical data suggest that the ferrocenyl group acts as a strong π-donor to the trigonal carbon atom in a similar way to that thought to occur in the ferrocenyl carbonium ions *(34)*. The Mössbauer parameters for the ferrocenyl carbenes and the carbonium ions are very similar.

Analogous results are found for the series of substituted benzene complexes $[(C_6H_5R)Fe(C_5H_5)]^+$. Although there is some variation in the quadrupole splitting with change of anion, the same trend is found as for the ferrocenes; the quadrupole splitting decreases as R becomes more electron-withdrawing, π-interaction again seeming to be the most important factor *(132)*.

In a classic paper, Collins has shown that the electric field gradient in ferrocene is positive and that this result

Table 7: Substituted cyclopentadienyl and benzene complexes

	δ_{Fe} [mm s^{-1}]	Δ [mm s^{-1}]	*(Ref.)*
$C_5H_5FeC_5H_4R$			
R =			
Cl	0.51(5)	2.42(5)	*(127)*
H	0.51(1)	+2.40(3)	*(97)*
CH_2OH	0.51(2)	3.39(5)	*(97)*
$[CH_2NMe_3]^+I^-$	0.51(5)	2.38(5)	*(127)*
C_6H_5	0.42(5)	2.34(5)	*(127)*
CO_2Na	0.48(5)	2.34(5)	*(127)*
$C(CH_3)_3$ [a]	0.44	2.32	*(136)*
CN	0.41(1)	2.30(3)	*(97)*
$COCH_3$	0.45(5)	2.27(5)	*(127)*
NO_2 [a]	0.43	2.27	*(136)*
$COC_{15}H_{13}$	0.54(2)	2.25(5)	*(97)*
$C(OCH_3)Mn(CO)_2C_5H_4CH_3$	0.50(2)	2.24(2)	*(117)*
$CH_2CH_2CO_2H$	0.54(1)	2.22(3)	*(97)*
CO_2H [a]	0.44	2.16	*(136)*
$C(OCH_3)Cr(CO)_5$	0.52(2)	2.15(2)	*(117)*
$[CH(C_5H_4FeC_5H_5)]^+BF_4^-$ [b]	0.52(2)	2.11(2)	*(74)*
$[CH(C_5H_4FeC_5H_5)]^+ClO_4^-$	0.41	2.05	*(140)*
$[C(C_5H_4FeC_5H_5)_2]^+ClO_4^-$	0.48	2.05	*(140)*
$[C_5H_5FeC_6H_5R]^+PF_6^-$			
R =			
NH_2	0.49(4)	1.83(4)	*(132)*
F	0.51(4)	1.78(4)	*(132)*
CH_3	0.47(4)	1.78(4)	*(132)*
$NHCOCH_3$	0.53(4)	1.76(4)	*(132)*
OCH_3	0.51(4)	1.76(4)	*(132)*
H	0.49(4)	1.67(4)	*(132)*
$CONH_2$	0.53(4)	1.62(4)	*(132)*
CO_2H	0.50(4)	1.51(4)	*(132)*
CN	0.52(4)	1.44(4)	*(132)*

[a] At 298 K; [b] At 100 K.

is in accord with the molecular orbital treatments of the system *(31)*. The results of two such treatments are summarized in Table 8 *(46)*. The "metal" 3*d*-electrons are disposed as follows: d_{xz}, d_{yz}, populated by donation from the rings; d_{z^2}, non-bonding; d_{xy}, $d_{x^2-y^2}$, formally non-bonding, depopulated by back-donation to the ring anti-bonding orbitals. There is

Table 8: Molecular orbital results for ferrocene *(46)*

Orbital		Occupancy	3*d*-population		Contribution to e.f.g.[a]	
d_{xy}, $d_{x^2-y^2}$	(e_{2g})	4	2.888[b]	(3.2256)[c]	+2.888[b]	(+3.2256)[c]
d_{z^2}	(a_{1g})	2	2.000[b]	(2.0000)[c]	-2.000[b]	(-2.0000)[c]
d_{xz}, d_{yz}	(e_{1g})	4	0.548[b]	(0.8244)[c]	-0.274[b]	(-0.4122)[c]
				Total[d]	+0.614	(+0.8134)

[a] Relative to -1 as the contribution of one $3d_{z^2}$-electron; e.f.g. = electric field gradient; [b] Results of Shustorovich and Dyatkina; [c] Results of Ballhausen and Dahl; [d] After removal of one e_{2g}-electron, the total contributions are -0.108 (+0.0070).

thus an excess of electron density localized in an equatorial belt between the rings, giving a positive electric field gradient along the molecular axis. (There is a smaller contribution, also positive, from the 4*p*-electrons). A similar description applies to the arene complexes *(125)*, and these treatments can be extended to explain the effects of ring-substitution. The introduction of π-withdrawing substituents would lower the energy of the ring orbitals and favour back-donation, thus reducing the field gradient.

Complete removal of one electron, to give the ferricenium cation results in collapse of the quadrupole doublet to a single, rather broad line (Table 9), again in agreement with theory. The broadening may be due to an unresolved quadrupole splitting together with magnetic relaxation effects; the latter suggestion is supported by the observation that the narrowest line is obtained with a paramagnetic anion *(14)*. Consistently with loss of a 3*d*-electron, the isomer shift decreases.

The spectrum of ferricenium ferrichloride consists of a single broad line, which would not be expected for the obvious formulation $[Cp_2Fe][FeCl_4]$, and the chlorine-bridged structure $[CpFeCl(\mu\text{-}Cl)]_2$ was suggested, since the two iron atoms would then be equivalent *(128)*. The presence of two different iron sites was demonstrated by tracer experiments *(129)*, and re-examination of the Mössbauer spectrum *(14)* revealed that the spectrum was asymmetric and contained two lines characteristic of $[Cp_2Fe]^+$ and $[FeCl_4]^-$.

Biferrocenyl has the *trans*-configuration (<u>24</u>) *(84)*, so that interaction between the two iron atoms is unlikely, and the Mössbauer parameters are very similar to those of ferrocene. One-electron oxidation gives a species containing one

Table 9: Ferricenium salts and related compounds

	Fe(III) δ_{Fe} [mm s^{-1}]	Fe(III) Δ [mm s^{-1}]	Fe(II) δ_{Fe} [mm s^{-1}]	Fe(II) Δ [mm s^{-1}]	*(Ref.)*
$[(C_5H_5)_2Fe]FeCl_4$ [a]	0.55(1)	0.00[b]			*(14)*
$[(C_5H_5)_2Fe]BF_4$	0.58(1)	0.00[c]			*(14)*
$[(C_5H_5)_2Fe]Cl$	0.54(1)	0.00[d]			*(14)*
$[(C_5H_5)_2Fe]BPh_4$	0.58(1)	0.00[e]			*(14)*
$C_5H_5Fe(II)C_5H_4$-$C_5H_4Fe(II)$-C_5H_5 (<u>24</u>)			0.48(3)	2.30(3)	*(38)*
$[C_5H_5Fe(III)C_5H_4$-C_5H_4Fe-$(II)C_5H_5][OC_6H_2(NO_2)_3]$	0.52(3)	0.29(3)	0.51(3)	2.14(3)	*(38)*
$[C_5H_5Fe(III)C_5H_4$-C_5H_4-$Fe(III)C_5H_5][BF_4]_2$	0.50(3)	0.16(3)			*(38)*
$[C_5H_5FeC_5H_3$-$C(CH_3)C_2H_5]_n$	0.54(4)	0.92(4)	0.57(4)	2.40(4)	*(3)*
$[C_5H_5FeC_5H_3$-CO-C_6H_4-$CO]_n$	0.51(4)	0.75(4)	0.57(4)	2.45(4)	*(3)*
$[C_5H_5Fe(III)C_5H_4$-C_5H_4-$Fe(II)C_5H_5]I_3$ [f]	0.53(0.3)	0.38(0.3)	0.52(0.3)	2.12(0.3)	*(109)*
$[C_5H_4Fe(II)C_5H_4$-C_5H_4-$Fe(II)C_5H_4]$ [f] (<u>25</u>)			0.53(0.1)	2.40(0.1)	*(109)*
$[C_5H_4Fe(III)C_5H_4$-C_5H_4-$Fe(II)C_5H_4]I_3$ [f]	0.54(0.2)	1.76(0.2)	0.54(0.2)	1.76(0.2)	*(109)*
$[C_5H_4Fe(III)C_5H_4$-C_5H_4-$Fe(III)C_5H_4][PF_6]_2$ [f]	0.57(0.1)	2.95(0.1)			*(110)*
$C_5H_5FeC_2B_9H_{11}$ [g]	0.35(3)	0.53(3)			*(74)*
$Me_4N[Fe(C_2B_9H_{11})_2]$	0.24(1)	0.64(1)			*(14)*
$(Me_4N)_2[Fe(C_2B_9H_{11})_2]$			0.30(1)	2.80(1)	*(14)*

[a] δ(anion) = 0.32 mm s^{-1}, half-width (anion) = 0.35 mm s^{-1}; [b] half-width = 0.43 mm s^{-1}; [c] half-width = 0.61 mm s^{-1}; [d] half-width = 0.70 mm s^{-1}; [e] half-width = 0.77 mm s^{-1}; [f] at 4.2 K; [g] at 140 K.

iron(II) atom and one iron(III) atom and the spectra show that electron-exchange between them is slower than $10^7 s^{-1}$ *(2,38, 109)*. A similar conclusion was reached in a study of the X-ray photo-electron spectra, in which separate signals for the iron atoms could be seen. That for the paramagnetic atom showed characteristic broadening, and the line-widths correlated with the quadrupole splittings *(40)*. In these derivatives, the splitting is just resolved, and further splitting is found in oxidized ferrocene polymers *(3)*. It seems possible that the lack of interaction between the iron atoms is due to the retention of the *trans*-configuration. A *cis*-system is achieved in biferrocenylene (<u>25</u>), the Mössbauer parameters of which are again similar to those of ferrocene. After one-electron oxidation, however, only one quadrupole doublet is observed, with splitting intermediate between those of ferrocene and the ferricenium ion, showing that electron-exchange is fast, i.e. only an average spectrum is seen *(39,109)*. Further

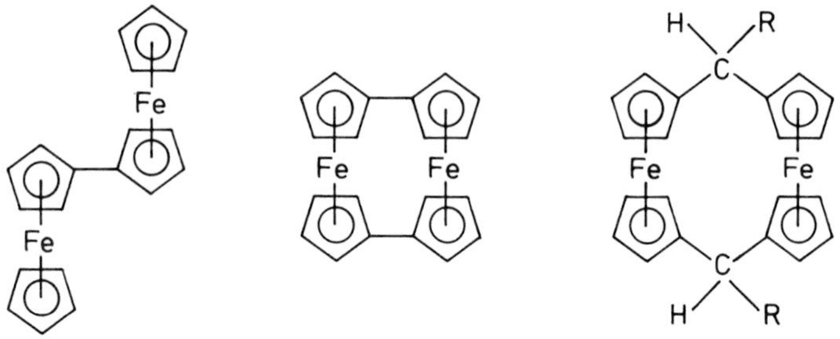

24 25 26

one-electron oxidation might be expected to give a product with a small quadrupole splitting, as in the ferricenium ion. However, the observed splitting is larger than that in ferrocene itself. It has been suggested that this results from interaction between the iron atoms allowing extensive delocalization of the e_{2g}-electrons, thus diminishing their contribution to the electric field gradient *(110)*. In the ferrocenophane derivatives (26, R = H, Me) it seems likely that both mixed-valence and average-valence forms may co-exist *(110)*.

The carbollyl ligand, $C_2B_9H_{11}^{2-}$, forms complexes similar to the cyclopentadienyls in that the metal is coordinated to an open C_2B_3-face of the cage *(141)*. The Mössbauer spectra of the two types of complex are similar, the ferrous complexes showing large quadrupole splittings which are drastically reduced on oxidation. The splittings are greater than for the corresponding bis-cyclopentadienyl complexes, and this has been attributed to the greater asymmetry in the ring *(14)*. However, the NQR spectrum of $Cs[Co(C_2B_9H_{11})_2]$ shows that the asymmetry parameter is negligible (η = 0.03 ± 0.01) *(71)*. The increase in splitting must therefore be an electronic effect, probably decreased back-donation to the doubly-negatively charged ligand. The lower isomer shift for the carbollyl complexes suggests that this ligand is also a better donor than cyclopentadienyl. The quadrupole doublet for the iron(III) complex is markedly asymmetric, and this is thought to be due to an increased magnetic relaxation time (relative to the ferricenium cation or the mixed complex), allowed by the "insulating" effect of the large carbollyl ligands *(14)*.

D. CYANIDE AND PHOSPHINE COMPLEXES

Many other low-spin systems have been studied, of which space permits the mention of only two types, the cyano-complexes and the Group V donor complexes.

1. Cyano-complexes

The cyano-complexes of iron have been widely studied. For hexacyanoferrate(II) salts, the isomer shift appears to decrease with increasing polarizing power of the cation (Table 10), which suggest polarization of the cyanide ligands resulting in enhanced delocalization of the non-bonding 3*d*-electrons. Coordination of BF_3 or a carbonium ion to the nitrogen

Table 10: Iron(II) cyano-complexes (at 293 K)

	δ_{Fe} [mm s^{-1}]	*(Ref.)*
$H_4Fe(CN)_6$	-0.14(1)	*(18)*
$Mg_2Fe(CN)_6$	-0.10(1)	*(18)*
$Al_4[Fe(CN)_6]_3$	-0.16(1)	*(18)*
$ZrFe(CN)_6$	-0.13(1)	*(18)*
$H_4Fe(CN)_6{\cdot}nEt_2O$	-0.17(1)	*(18)*
$K_4Fe(CN)_6$	-0.03(3)	*(70)*
$Fe(CNH)_4(CN)_2$	-0.10(3)	*(70)*
$K_4Fe(CNBF_3)_6$	-0.11(3)	*(70)*
$Fe(CNH)_4(CNBF_3)_2$	-0.12(3)	*(70)*
trans-$Fe(CNMe_4)(CN)_2$ [a]	-0.14(3)	*(70)*
$Fe(CNMe)_4(CNBF_3)_2$	-0.08(3)	*(70)*
$[Fe(CNMe)_6](HSO_4)_2$	-0.11(5)	*(13)*
$[Fe(CNEt)_6](ClO_4)_2$	-0.09(5)	*(13)*
$[Fe(CNCH_2Ph)_6](ClO_4)_2$	-0.05(5)	*(13)*

[a] Δ = 0.44 mm s^{-1}

atom gives similar decreases in isomer shift *(13,70)*. None of these complexes shows a quadrupole splitting. When one cyanide ligand is replaced by another ligand, both the isomer shift and the quadrupole splitting reflect the varying degree of back-donation possible. That π-delocalization is the dominant mechanism in producing the quadrupole splitting is shown by the positive sign of the electric field gradient for $[Fe(CN)_5NO]^{2-}$ *(87)*. The corresponding iron(III) complexes show similar trends, although interpretation of the quadrupole splitting data is complicated by the effects of the low-spin d^5-configuration.

Prussian Blue and related complexes have also received much attention, culminating in some elegant work using various iron isotopes to enhance or diminish the signals from the different iron sites *(4,17,104)*. As is now wellknown, Prussian Blue (Fe^{3+} + $Fe(CN)_6{}^{4-}$) and Turnbull's Blue (Fe^{2+} + $Fe(CN)_6{}^{3-}$) are identical, containing the (presumably hydrated) high-spin ferric and low-spin ferrocyanide ions. Dehydration of Prussian Blue at 300-400°C does give the alternative material, ferrous ferricyanide, which reverts to ferric ferrocyanide on contact with moisture *(35)*. It is postulated that the redox reaction occurs as a result of depressurization of the lattice on removal of water.

In mixed cyanides, involving iron and another metal, linkage isomerism has been detected *(20)*.

Isonitrile complexes have also been investigated *(6,8, 13)*. A recent correlation of iron 2*p*- and 3*p*-binding energies (by ESCA) with isomer shifts suggests that the binding energies are more sensitive to change in π-acceptance by the ligands than σ-donation *(1)*.

2. Group V Donor Complexes

A wide variety of complexes of the type (diphosphine)FeXY has been examined, and attempts have been made to assign partial isomer shift and partial quadrupole splitting values to individual ligands *(8,9)*. Since the isomer shift responds to the ($\sigma + \pi$) properties of the ligands and the quadrupole splitting to the ($\pi - \sigma$) properties, the relative magnitude of the donor and acceptor properties should be assessable. Data for complexes $[(depe)_2FeHX]^+$ are shown in Figure 4, in which the quadrupole splitting is plotted against the isomer shift. If H^-, RCN, and Cl are assumed to be purely σ-bonding ligands, the distance at which the points for other ligands lie above this line may represent the π-acceptor ability. Thus, CO emerges as a powerful donor and a good acceptor, isonitriles are slightly less so, whereas N_2 is relatively poor in both respects *(5,10)*. Similar arguments have been used to suggest that the ligand $P(OMe)_3$ has bonding properties close to those of isocyanides, and is appreciably stronger both in σ-donation and π-acceptance than other tertiary phosphines and phosphites *(9,98)*.

Data for the unique iron(IV) complexes $[Fe(diars)_2X_2]$-$(BF_4)_2$ (X = Cl, Br) *(72)* and the analogous iron(III) and iron(II) complexes have been obtained using iron enriched in ^{57}Fe to overcome the large absorption of the soft γ-rays by the arsenic atoms *(59,114,115)*. The quadrupole splitting of the ferrous compound (low-spin d^6) is small, reflecting only the difference between Fe-Cl and Fe-As bonds. On oxidation, electrons are removed from the d_{xz}- and d_{yz}- orbitals, giving larger splittings. For the iron(III) compound, analysis of the magnetically-perturbed spectrum, together with the tem-

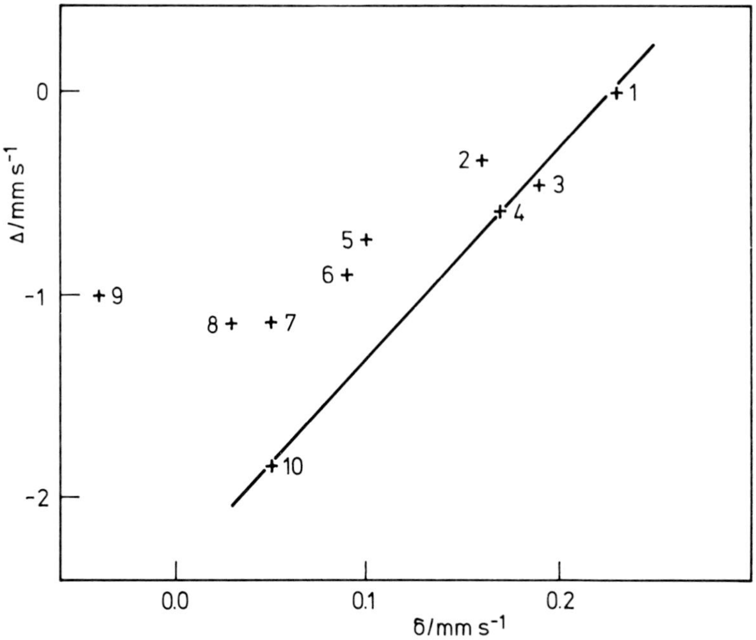

Fig. 4: Isomer shift vs. quadrupole splitting for complexes of the type (depe)$_2$FeHX [depe = 1,2-bisdiethylphosphinoethane; X = Cl, I(1), N_2 (2), MeCN (3), PhCN (4), (PhO)$_3$P (5), (MeO)$_3$P (6), t-Bu-NC (7), p-MeO-C_6H_4NC (8), CO (9)]. Point 10 corresponds to (depb)$_2$FeH$_2$ (depb = o-phenylenebisdiethylphosphine). The electric field gradient is assumed to be negative in all cases. Data from Ref. 10.*

perature dependence of the quadrupole splitting, showed removal of the degeneracy of d_{xz} and d_{yz} by a small amount (ca. 18 cm^{-1}), while the separation from the third t_{2g}-orbital was much greater (2700 - 3000 cm^{-1}).

E. BIOLOGICAL SYSTEMS

Iron-containing proteins are very important in biological systems, in transporting oxygen, performing a variety of oxidations and reductions (including nitrogen fixation), and various catalyses, and the application of the Mössbauer technique to the study of these materials is growing rapidly. The metal atom can be probed specifically and in this respect the technique is similar to ESR spectroscopy, with the important advantage that it is not restricted to paramagnetic samples. A major disadvantage is the low iron content of these high molecular weight materials, but this can often be offset by ^{57}Fe-enrichment. The isotope is not radioactive, and may safely be included in the diet of suitable organisms. Another factor which results from the high dilution of the metal is the frequent appearance of magnetic relaxation effects, especially in high-spin iron(III) systems. Antiferromagnetic coupling between pairs of iron atoms often occurs, giving a spectrum very different from that of single-iron species, and this type of coupling is usually not easy to observe directly by other methods. Although these effects complicate the spectrum and usually require low temperatures (liquid helium) and the application of magnetic fields, they also, in principle, allow the derivation of much information about the interaction of the iron atom with its immediate environment. It is usually necessary to compare the observed spectrum with a calculated spectrum. A proper treatment of this subject is beyond the scope of this chapter, but good qualitative descriptions of the technique may be found in Refs. *80* and *91*.

The haem proteins have been studied extensively and the effect of systematic changes in the axial ligands are now well-documented *(137)*. Detailed information is available on the effect which spin-orbit coupling and the ligand fields of the protein and the axial ligands have on the energy levels of the iron atom. These (relatively) simple systems are also of value in that they establish reference spectra for the four major states in which iron is found (high- and low- spin ferrous and ferric). Data for these systems have been reviewed *(92)*.

Systems in which information may be derived relatively easily are those containing more than one iron atom and those for which redox reactions involve the metal atom. For instance, in the non-heme protein hemerythrin (an oxygen-carrier in molluscs) iron atoms occur in pairs. In the reduced form (deoxyhemerythrin) both iron atoms are in the high-spin ferrous state and do not interact with each other. After chemical oxidation (methemerythrin), the metal atoms are in the high-spin ferric state, but the spins are coupled antiferromagnetically, probably *via* an oxo (O^{2-}) bridge. In the oxygenated complex (oxyhemerythrin), a similar structure is thought to occur with the O_2 molecule acting as an additional bridge between the iron atoms *(61)*.

Iron-sulphur proteins are also very important (*e.g.* rubredoxin, ferridoxin, putidaredoxin, xanthine oxidase). Many contain pairs or higher clusters of iron atoms, but reference spectra may be established by studying the single-iron species such as rubredoxin *(120,121)*. Two-iron ferredoxins from a variety of sources (*e.g.* spinach, *azobacter vinelandii*, pig adrenal cortex) all give similar spectra and are probably identical *(53)*. The iron site is probably a slightly distorted S_4-tetrahedron, with two sulphur atoms bridging between the two iron atoms. The oxidized form contains two high-spin iron(III) atoms although antiferromagnetic spin-coupling between them results in diamagnetism, at least at low temperatures. After one-electron reduction, signals characteristic of both ferrous and ferric iron are found, the latter being very similar to that of the unreduced material. Again, the two iron atoms are antiferromagnetically coupled. This case in which ^{57}Fe-enrichment was valuable as data on natural material was originally interpreted in terms of the reduction-electron being delocalized over both iron atoms *(82)*.

Horse-radish peroxidase (HRP) contains high-spin ferric iron, and reacts with peroxide to give a derivative (HRP-I) capable of two-electron oxidations. In the absence of a reducing substrate this compound decomposes to a second compound (HRP-II), which is only one oxidizing-equivalent above the starting material. The Mössbauer spectra of the two oxidized compounds are very similar (but not identical), suggesting that the second oxidizing-equivalent of HRP-I is associated with orbitals not localized on the iron atom. For both compounds, the isomer shift is lower than that for HRP, which is consistent with an iron(IV)-like configuration for the metal *(112)*.

Application of the Mössbauer technique to biological systems has been reviewed recently *(12,67,80,81,91,92,105,130)*; references to other recent work are given in Table 11.

Table 11: Recent studies on biological systems.

	Ref.
Haem derivatives	*92,95,103,137*
Haemoglobin	*131,137,138*
Myoglobin	*65,94*
Cytochrome	*49,93,96,124*
Ferredoxin	*24,53,82,121,123*
Xanthine oxidase	*82*
Nitrogenase	*54,126*
Hemerythrin	*61,139*
Ferritin	*16*
Rubredoxin	*120,122*

Abbreviations

δ_{Fe}	= isomer shift, relative to iron metal
Δ	= quadrupole splitting
η	= asymmetry parameter
Me	= CH_3
Et	= C_2H_5
Bu	= C_4H_9
Ph	= C_6H_5
py	= pyridine
bipy	= 2,2'-bipyridyl
en	= $H_2N-CH_2-CH_2NH_2$
Cp	= $\eta^5-C_5H_5$
R	= alkyl
depe	= $Et_2P-CH_2-CH_2-PEt_2$
depb	= $o-C_6H_4(PEt_2)_2$
diars	= $o-C_6H_4(AsMe_2)_2$
DPPA	= $Ph_2P-C{\equiv}C-PPh_2$

REFERENCES

1. Adams, I., Thomas, J.M., Bancroft, G.M., Butler, K.D., and Barber, M., *J. Chem. Soc. Chem. Commun.*, *1972*, 751.
2. Alekseev, V.P., Stukan, R.A., and Koridze, A.A., *Izv. Akad. Nauk SSSR, Ser. Khim.*, *1973*, 132; *Bull. Acad. Sci. USSR, Div. Chem. Sci.*, *1973*, 129.
3. Aliev, L.A., Vishnyakova, T.P., Paushkin, Ya.M., Pendin, A.A., Sokolinskaya, T.A., and Stukan, R.A., *Izv. Akad. Nauk SSSR, Ser. Khim.*, *1970*, 306; *Bull. Acad. Sci. USSR, Div. Chem. Sci.*, *1970*, 256.
4. Allen, J.F., Edwards, B.R., and Bonnette, A.K., *J. Inorg. Nucl. Chem.*, *35*, 3547 (1973).
5. Bancroft, G.M., *Coord. Chem. Rev.*, *11*, 247 (1973).
6. Bancroft, G.M., and Butler, K.D., *J. Chem. Soc. Dalton Trans.*, *1972*, 1209.
7. Bancroft, G.M., Butler, K.D., Manzer, L.E., Shaver, A., and Ward, J.E.H., *Canad. J. Chem.*, *52*, 782 (1974).
8. Bancroft, G.M., Mays, M.J., and Prater, B.E., *J. Chem. Soc. A*, *1970*, 956.
9. Bancroft, G.M., and Libbey, E.T., *J. Chem. Soc. Dalton Trans.*, *1973*, 2103.
10. Bancroft, G.M., Mays, M.J., Prater, B.E., and Stefanini, F.P., *J. Chem. Soc. A*, *1970*, 2146; Bancroft, G.M., Garrod, R.E.B., Maddock, A.G., Mays, M.J., and Prater, B.E., *J. Amer. Chem. Soc.*, *94*, 647 (1972).
11. Bancroft, G.M., and Platt, R.H., *Advan. Inorg. Chem. Radiochem.*, *15*, 59 (1972).
12. Bearden, A.J., and Dunham, W.R., *Struct. Bonding (Berlin)*, *8*, 1 (1970).
13. Berrett, R.R., and Fitzsimmons, B.W., *J. Chem. Soc. A*, *1967*, 525.
14. Birchall, T., and Drummond, I., *Inorg. Chem.*, *10*, 399 (1971).
15. Bird, S.R.A., Donaldson, J.D., Holding, A.F.LeC., Senior, B.J., and Tricker, M.J., *J. Chem. Soc. A*, *1971*, 1616.
16. Boas, J.F., and Troup, G.J., *Biochim. Biophys. Acta*, *229*, 68 (1971).
17. Bonnette, A.K., Jr., and Allen, J.F., *Inorg. Chem.*, *10*, 1613 (1971).
18. Borshagovskii, B.V., Gol'danskii, V.I., Seifer, G.B., and Stukan, R.A., *Izv. Akad. Nauk SSSR, Ser. Khim.*, *1968*, 87; *Bull. Acad. Sci. USSR, Div. Chem. Sci.*, *1968*, 81.
19. Bowen, L.H., Garrou, P.E., and Long, G.G., *Inorg. Chem.*, *11*, 182 (1972).

20. Brown, D.B., Shriver, D.F., and Schwartz, L.H., *Inorg. Chem., 7,* 77 (1968); Brown, D.B., and Shriver, D.F., *Inorg. Chem., 8,* 37 (1969).
21. Bryan, R.F., Greene, P.T., *J. Chem. Soc. A, 1970,* 3064; Bryan, R.F., Greene, P.T., Newlands, M.J., and Field, D.S., *J. Chem. Soc. A, 1970,* 3068.
22. Burger, K., Korecz, L., and Bor, G., *J. Inorg. Nucl. Chem., 31,* 1527 (1969).
23. Burger, K., Korecz, L., Mag, P., Belluco, U., and Busetto, L., *Inorg. Chim. Acta, 5,* 362 (1971).
24. Cammack, R., Johnson, C.E., Hall, D.O., and Rao, K.K., *Biochem. J., 125,* 18, 849 (1971).
25. Carroll, W.E., Deeney, F.A., Delaney, J.A., and Lalor, F., *J. Chem. Soc. Dalton Trans., 1973,* 718.
26. Carty, A.J., Efraty, A., Ng, T.W., and Birchall, T., *Inorg. Chem., 9,* 1263 (1970).
27. Carty, A.J., Ng, T.W., Carter, W., Palenik, G.J., and Birchall, T., *Chem. Commun., 1969,* 1101.
28. Chia, L.S., Cullen, W.R., Sams, J.R., and Ward, J.E.H., *Canad. J. Chem., 51,* 3223 (1973).
29. Churchill, M.R., and Veidis, M.V., *J. Chem. Soc. A, 1971,* 2170, 2995.
30. Clark, M.G., Cullen, W.R., Garrod, R.E.B., Maddock, A. G., and Sams, J.R., *Inorg. Chem., 12,* 1045 (1973).
31. Collins, R.L., *J. Chem. Phys., 42,* 1072 (1965).
32. Collins, R.L., and Pettit, R., *J. Amer. Chem. Soc., 85,* 2332 (1963).
33. Collins, R.L., and Pettit, R., *J. Chem. Phys., 39,* 3433 (1963).
34. Connor, J.A., and Lloyd, J.P., *J. Chem. Soc. Dalton Trans., 1972,* 1470.
35. Cosgrove, J.G., Collins, R.L., and Murty, D.S., *J. Amer. Chem. Soc., 95,* 1083 (1973).
36. Cotton, F.A., De Boer, B.G., and Marks, T.J., *J. Amer. Chem. Soc., 93,* 5069 (1971).
37. Cotton, F.A., and Edwards, W.T., *J. Amer. Chem. Soc., 91,* 843 (1969).
38. Cowan, D.O., Collins, R.L., and Kaufman, F., *J. Phys. Chem., 75,* 2025 (1971).
39. Cowan, D.O., Le Vanda, C., Collins, R.L., Candela, G.A., Mueller-Westerhoff, U.T., and Eilbracht, P., *J. Chem. Soc. Chem. Commun., 1973,* 329.
40. Cowan, D.O., Park, J., Barber, M., and Swift, P., *Chem. Commun., 1971,* 1444.
41. Crow, J.P., and Cullen, W.R., *Can. J. Chem., 49,* 2948 (1971).
42. Crow, J.P., Cullen, W.R., Herring, F.G., Sams, J.R., and Tapping, R.L., *Inorg. Chem., 10,* 1616 (1971).

43. Cullen, W.R., Harbourne, D.A., Liengme, B.V., and Sams, J.R., *J. Amer. Chem. Soc., 90,* 3293 (1968).
44. Cullen, W.R., Harbourne, D.A., Liengme, B.V., and Sams, J.R., *Inorg. Chem., 8,* 95 (1969).
45. Cullen, W.R., Harbourne, D.A., Liengme, B.V., and Sams, J.R., *Inorg. Chem., 8,* 1464 (1969).
46. Dahl, J.P., and Ballhausen, C.F., *Mat. Fys. Medd. Dan. Vid. Selsk., 33,* No. 5 (1961); Shustorovich, E.M., and Dyatkina, M.E., *Dokl. Akad. Nauk SSSR, 128,* 1234 (1959).
47. Dahl, L.F., and Wei, C.-H., *Inorg. Chem., 2,* 328 (1963).
48. de Beer, J.A., Haines, R.J., Greatrex, R., and Greenwood, N.N., *J. Chem. Soc. A, 1971,* 3271.
49. Debrunner, P.G., in S.G. Cohen and M. Pasternak (Eds.), *Perspectives in Mössbauer Spectroscopy,* Plenum Press, New York, 1973, p. 89.
50. Dessey, R., Charkordian, J.C., Abeles, T.P., and Rheingold, A.L., *J. Amer. Chem. Soc., 92,* 3947 (1970).
51. Dessey, R.E., and Wieczorek, L., *J. Amer. Chem. Soc., 91,* 4963 (1969).
52. Dominelli, N., Wood, E., Vasudev, P., and Jones, C.H.W., *Inorg. Nucl. Chem. Lett., 8,* 1077 (1972).
53. Dunham, W.R., Bearden, A.J., and Salmeen, I.T., *Biochim. Biophys. Acta, 253,* 134 (1971).
54. Eady, R.R., Smith, B.E., Cook, K.A., and Postgate, J.R., *Biochem. J., 128,* 655 (1972).
55. Einstein, F.W.B., and Trotter, J., *J. Chem. Soc. A, 1967,* 824.
56. Emerson, G.E., Mahler, J.E., Pettit, R., and Collins, R.L., *J. Amer. Chem. Soc., 86,* 3590 (1964).
57. Erickson, N.E., and Fairhall, A.W., *Inorg. Chem., 4,* 1320 (1965).
58. Farmery, K., Kilner, M., Greatrex, R., and Greenwood, N.N., *J. Chem. Soc. A, 1969,* 2339.
59. Feltham, R.D., Silverthorn, W., Wickman, H., and Wesolowski, W., *Inorg. Chem. 11,* 676 (1972).
60. Frank, E., and Bunbury, D. St. P., *J. Chem. Soc. A, 1970,* 2143.
61. Garbett, K., Johnson, C.E., Klotz, I.M., Okamura, M.Y., and Williams, R.J.P., *Arch. Biochem. Biophys., 142,* 574 (1971); Garbett, K., Darnall, D.W., Klotz, I.M., and Williams, R.J.P., *Arch. Biochem. Biophys., 135,* 419 (1969).
62. Gibb, T.C., Greatrex, R., and Greenwood, N.N., *J. Chem. Soc. A, 1968,* 890.
63. Gibb, T.C., Greatrex, R., Greenwood, N.N., and Thompson, D.T., *J. Chem. Soc. A, 1967,* 1663.

64. Gol'danskii, V.I., and Herber, R.H., (Eds.), *Chemical Applications of Mössbauer Spectroscopy, Academic* Press, New York and London, 1968.
65. Grant, R.W., and Topol, L.E., *Biophys. J., 9,* 1446 (1969).
66. Greatrex, R., and Greenwood, N.N., *Discuss. Faraday Soc., 47,* 126 (1969).
67. Greenwood, N.N., and Gibb, T.C., *Mössbauer Spectroscopy,* Chapman Hall, London, 1971.
68. Grubbs, R., Breslow, R., Herber, R.H., and Lippard, S.J., *J. Amer. Chem. Soc., 89,* 6864 (1967).
69. Gütlich, P., and Prange, H., *Chem.-Ing.-Tech., 43,* 1049 (1971).
70. Hall, D., Slater, J.H., Fitzsimmons, B.W., and Wade, K., *J. Chem. Soc. A, 1971,* 800.
71. Harris, C.B., *Inorg. Chem., 7,* 1517 (1968).
72. Hazeldean, G.S.F., Parish, R.V., and Nyholm, R.S., *J. Chem. Soc. A, 1966,* 162.
73. Herber, R.H., *Progr. Inorg. Chem., 8,* 1 (1967).
74. Herber, R.H., *Inorg. Chem., 8,* 174 (1969).
75. Herber, R.H., *Sci. American, 225,* No. 4, 86 (1971).
76. Herber, R.H., and Hayter, R.G., *J. Amer. Chem. Soc., 86,* 301 (1964).
77. Herber, R.H., King, R.B., and Wertheim, G.K., *Inorg. Chem., 3,* 101 (1964).
78. Hsieh, A.T.T., and Knight, J., *J. Organometal. Chem., 26,* 125 (1971).
79. Hsieh, A.T.T., Mays, M.J., and Platt, R.H., *J. Chem. Soc. A, 1971,* 3296.
80. Johnson, C.E., *J. Appl. Phys., 42,* 1325 (1971).
81. Johnson, C.E., in S.G. Cohen and M. Pasternak (Eds.), *Perspectives in Mössbauer Spectroscopy,* Plenum Press, New York, 1973, p. 79.
82. Johnson, C.E., Bray, R.C., Cammack, R., and Hall, D. O., *Proc. Nat. Acad. Sci. U.S., 63,* 1234 (1969).
83. Jones, M.T., *Inorg. Chem., 6,* 1249 (1967).
84. Kaluski, Z.L., Struchkov, Yu.T., and Avoyan, R.L., *Zh. Strukt. Khim., 5,* 743 (1964); *J. Struct. Chem., 5,* 683 (1964).
85. Kalvius, M., Zahn, U., Kienle, P., and Eicher, H., *Z. Naturforsch., 17a,* 494 (1962).
86. Keller, C.E., Emerson, G.E., and Pettit, R., *J. Amer. Chem. Soc., 87,* 1389 (1965).
87. Kerler, W., *Z. Phys., 167,* 194 (1962).
88. King, R.B., Epstein, L.M., and Gowling, E.W., *J. Inorg. Nucl. Chem., 32,* 441 (1970).
89. Knight, J., and Mays, M.J., *J. Chem. Soc. A, 1970,* 654.

90. Kostiner, E., and Massey, A.G., *J. Organometal. Chem.*, *19*, 233 (1969).

91. Lang, G., in I. Deszi (Ed.), *Proc. Conf. Appl. Mössbauer Effect, Tihany, 1969*, Akadêmiai Kiadó, Budapest, 1971.

92. Lang, G., *Quart. Rev. Biophys.*, *3*, 1 (1970).

93. Lang, G., Asakura, T., and Yonetani, T., *J. Phys. C*, *2*, 2246 (1969).

94. Lang, G., Asakura, T., and Yonetani, T., *Biochim. Biophys. Acta*, *214*, 381 (1970).

95. Lang, G., Asakura, T., and Yonetani, T., *Phys. Rev. Lett.*, *24*, 981 (1970).

96. Lang, G., Lippard, S.J., and Rosen, S., *Biochim. Biophys. Acta*, *336*, 6 (1974).

97. Lesikar, A.V., *J. Chem. Phys.*, *40*, 2746 (1964).

98. Libbey, E.T., and Bancroft, G.M., *J. Chem. Soc. Dalton Trans.*, *1974*, 87.

99. Lindauer, M.W., Spiess, H.W., and Sheline, R.K., *Inorg. Chem.*, *9*, 1694 (1970).

100. Lindley, P.F., and Mills, O.S., *J. Chem. Soc. A*, *1969*, 1279.

101. Loew, G.H., and Lo, D., *Theor. Chim. Acta*, *33*, 137 (1974).

102. Madden, D.P., Carty, A.J., and Birchall, T., *Inorg. Chem.*, *11*, 1453 (1972).

103. Maeda, Y., Trautwein, A., Gonser, U., Yoshida, K., Kikuchi-Torii, K., Homma, T., and Ogura, Y., *Biochim. Biophys. Acta*, *303*, 230 (1973).

104. Maer, K., Jr., Beasley, M.L., Collins, R.L., and Milligan, W.O., *J. Amer. Chem. Soc.*, *90*, 3201 (1968).

105. May, L., (Ed.), *Introduction to Mössbauer Spectroscopy*, Plenum Press, New York, 1971.

106. Mazak, R.A., and Collins, R.L., *J. Chem. Phys.*, *51*, 3220 (1969).

107. McDonald, W.S., Moss, J.R., Raper, G., Shaw, B.L., Greatrex, R., and Greenwood, N.N., *Chem. Commun.*, *1969*, 1295.

108. Mills, O.S., and Palmer, A., quoted by J. Lewis in *Pure Appl. Chem.*, *10*, 11 (1965).

109. Morrison, W.H., Jr., and Hendrickson, D.N., *J. Chem. Phys.*, *59*, 380 (1973).

110. Morrison, W.H., Jr., and Hendrickson, D.N., *Chem. Phys. Letters*, *22*, 119 (1973).

111. Moss, T.H., *Methods in Enzymology*, *27*, 912 (1973).

112. Moss, T.H., Ehrenberg, A., and Bearden, A.J., *Biochem.*, *8*, 4159 (1969).

113. Mößbauer, R.L., *Angew. Chem.*, *83*, 524 (1971); *Angew. Chem. Int. Ed. Engl.*, *10*, 462 (1971).

114. Oosterhuis, W.T., Weaver, D.L., and Paez, E.A., *J. Chem. Phys.*, *60*, 1018 (1974).
115. Paez, E.A., Weaver, D.L., and Oosterhuis, W.T., *J. Chem. Phys.*, *57*, 3709 (1972).
116. Parish, R.V., *Progr. Inorg. Chem.*, *15*, 101 (1972).
117. Parish, R.V., unpublished results.
118. Parish, R.V., and Rowbotham, P.J., unpublished results.
119. Pedone, C., and Sirigu, A., *Inorg. Chem.*, *7*, 2614 (1968); Luxmoore, A.R., and Truter, M.R., *Acta Crystallogr.*, *15*, 1117 (1962).
120. Phillips, W.D., Poe, M., Weiher, J.F., McDonald, C.C., and Lovenberg, W., *Nature*, *227*, 574 (1970).
121. Rao, K.K., Cammack, R., Hall, D.O., and Johnson, C.E., *Biochem. J.*, *122*, 257 (1971).
122. Rao, K.K., Evans, M.C.W., Cammack, R., Hall, D.O., Tompson, C.L., Jackson, P.J., and Johnson, C.E., *Biochem. J.*, *129*, 1063 (1972).
123. Rao, K.K., Smith, R.V., Cammack, R., Evans, M.C.W., Hall, D.O., and Johnson, C.E., *Biochem. J.*, *129*, 1159 (1972).
124. Sharrock, M., Münck, E., Debrunner, P.G., Marshall, V., Lipscomb, J.D., and Gonsalas, I.C., *Biochem.*, *12*, 258 (1973).
125. Shustorovich, E.M., and Dyatkina, M.E., *Dokl. Akad. Nauk SSSR*, *131*, 113 (1960); *Proc. Acad. Sci. USSR, Chem. Sect.*, *131*, 215 (1960).
126. Smith, B.E., and Lang, G., *Biochem. J. 137*, 169 (1974).
127. Stukan, R.A., Gubin, S.P., Nesmeyanov, A.N., Gol'danskii, V.I., and Makarov, E.F., *Theor. Eksp. Khim.*, *2*, 805 (1966); *Theor. Exper. Chem.*, *2*, 581 (1966).
128. Stukan, R.A., and Yurieva, L.P., *Dokl. Akad. Nauk SSSR*, *167*, 1311 (1966); *Doklady Chemistry*, *167*, 448 (1966).
129. Stukan, R.A., quoted in *Ref. 64*, p. 296.
130. Trautwein, A., in S.G. Gohen and M. Pasternak (Eds.), *Perspectives in Mössbauer Spectroscopy*, Plenum Press, New York, 1973, p. 101.
131. Trautwein, A., Eicher, H., Mayer, A., Alfsen, A., Waks, M., Rosa, J., and Benzard, Y., *J. Chem. Phys.*, *53*, 963 (1970).
132. Turta, K.I., Stukan, R.A., Gol'danskii, V.I., Vol'kenau, N.A., Sirotkina, E.I., Bolesova, I.N., Isaeva, L.S., and Nesmeyanov, A.N., *Teor. Eksp. Khim.*, *7*, 486 (1971); *Theor. Exper. Chem.*, *7*, 401 (1971).

133. Vasudev, P., and Jones, C.H.W., *Canad. J. Chem.*, *51*, 405 (1973).
134. Wei, C.H., and Dahl, L.F., *J. Amer. Chem. Soc.*, *88*, 1821 (1966).
135. Wertheim, G.K., *Mossbauer Effect: Principles and Applications*, Academic Press, New York and London, 1964.
136. Wertheim, G.K., and Herber, R.H., *J. Chem. Phys.*, *38*, 2106 (1963).
137. Winter, M.R.C., Johnson, C.E., Lang, G., and Williams, R.J.P., *Biochim. Biophys. Acta*, *263*, 515 (1972).
138. Winterhalter, K.H., Di Iorio, E.E., Beetlestone, J.G., Kushimo, J.B., Uebelhack, H., Eicher, H., and Mayer, E., *J. Mol. Biol.*, *70*, 665 (1972).
139. York, J.L., and Bearden, A.J., *Biochem.*, *9*, 4549 (1970).
140. Zahn, U., Kienle, P., and Eicher, H., *Z. Phys.*, *166*, 220 (1962).
141. Zalkin, A., Templeton, D.H., and Hopkins, T.E., *J. Amer. Chem. Soc.*, *87*, 3988 (1965).
142. Zuckerman, J.J., *Adv. Organometal. Chem.*, *9*, 21 (1970).

Magnetic Properties

By EDGAR KÖNIG

Institut für Physikalische Chemie II,
Universität Erlangen-Nürnberg, 8520 Erlangen, Germany

TABLE OF CONTENTS

ISBN 0-12-417101-X (v.1)

I. INTRODUCTION

The measurement of magnetic susceptibility provides results which are related and to some extent complementary to those obtained by electron paramagnetic resonance (EPR) to be discussed in the next chapter. While any substance yields at least a diamagnetic contribution to the susceptibility, diamagnetic materials do not produce EPR signals and, consequently, neither affect the EPR of a paramagnetic probe. With respect to paramagnets, the susceptibility of any substance can be measured, whereas an EPR study may fail if the zero-field splitting of the lowest doublet state is too large, if the spin-lattice relaxation time is unfavourable, or if some other reason appears.

A recent compilation of magnetic data *(19)* shows that the major part of all magnetic measurements is performed on polycrystalline samples at room temperature. Since this type of result is not particularly useful - except to determine the overall electron configuration - the impression arises that magnetic studies provide but little information on the molecular and electronic structure. This assertion is by no means correct. On the other hand, it is certainly true that the handling of a sensitive magnetic balance may require more attention than the operation of a modern push-button spectrometer. None the less, the complementary nature of the magnetic susceptibility demands that full advantage of the information content of this method be utilized. This may be achieved, in general, by extending magnetic measurements down to the cryogenic temperature range and by carrying out a detailed comparison with appropriate theoretical calculations. Although suitable equipment for this type of studies is available in many laboratories, the average chemist may experience difficulties in familiarizing himself with theory. After a brief introduction into the physical basis of magnetism we will concentrate, therefore, on the theoretical background. Only a minimum workable knowledge of quantum mechanics is required to follow the outline of the electronic structure of iron and the so-called Van Vleck approach to the calculation of paramagnetic susceptibilities. The high-spin iron(II) problem has been chosen as a typical example for the detailed treatment. The results of recent more advanced theoretical studies are likewise presented. Space does not permit an extensive discussion of experimental data. However, the provided results of theoretical studies demonstrate clearly the type of experimental data to be expected as well as the information which may be extracted from a detailed comparison of such data with theory.

The available experimental data are listed in a compilation *(19)* which covers all literature until the beginning of 1965. A supplement *(20)* including the magnetic studies published between 1964 and 1968 has been published, additional supplemental volumes being in preparation. The excellent reviews of Figgis and Lewis *(8,9)* might likewise be consulted in all experimentally based problems of the magnetism of transition metal ions and the rare earths. In addition, one of the latter references *(9)* gives many details concerning experimental set-ups for magnetic susceptibility measurements.

II. BASIC MAGNETISM

A. *DEFINITIONS AND FUNDAMENTAL RELATIONS*

The magnetic field strength $\mathbf{H}^{int}$ inside a substance differs from the field strength of the applied field $\mathbf{H}$ (measured in vacuo) according to

$$\mathbf{H}^{int} = \mathbf{H} + \Delta \mathbf{H} \qquad [1]$$

Conventionally, the relationship of eq. [1] is written as follows

$$\mathbf{B} = \mathbf{H} + 4\pi\mathbf{M} \qquad [2]$$

where $\mathbf{B}$, called the magnetic induction, is equivalent to $\mathbf{H}^{int}$ in eq. [1] and $\mathbf{M}$ is the (induced) magnetic moment per unit volume (often called intensity of magnetization and denoted then by $\mathbf{I}$). Here $\mathbf{B}$, $\mathbf{H}$, and $\mathbf{M}$ are vector quantities. If we consider an isotropic substance, $\mathbf{M}$ depends only on $\mathbf{H}$ and is independent of direction

$$\mathbf{M} = \kappa \mathbf{H} \qquad [3]$$

where κ is a scalar, the volume susceptibility, which is a measure of the ease of magnetic polarization of the substance. Eq. [2] may now be rewritten

$$\mathbf{B} = (1 + 4\pi\kappa)\mathbf{H} \qquad [4]$$

Often, the quantity that is actually measured is the gram susceptibility (also specific susceptibility) χ_g defined by

$$\chi_g = \kappa/\rho \qquad [5]$$

and therefrom the molar susceptibility χ_M is derived where

$$\chi_M = \chi_g \times M = \kappa \times V \qquad [6]$$

Here, ρ is the density, M the molecular weight and V the molecular volume. Paramagnetic substances have $\chi_g > 0$, whereas diamagnetic substances have $\chi_g < 0$. Table 1 shows the most common types of magnetic behaviour and their properties. The units commonly employed are in case of χ_g, 10^{-6} cgs/g (cgs/g = emu/g = cm^3/g) and in case of χ_M, 10^{-6} cgs/mole.

Table 1: Types of magnetic behaviour.

Type	Sign of Susceptibility	Magnitude of χ_g at 20°C	Dependence on H	Origin
Diamagnetism	negative	1×10^{-6}	no	electron charge
Paramagnetism	positive	$0 - 100 \times 10^{-6}$	no	electron spin
Ferromagnetism	positive	$10^{-2} - 10^{4}$	yes	electron spin exchange
Antiferromagnetism	positive	$0 - 1000 \times 10^{-6}$	yes, no	

In anisotropic substances, *e.g.* a noncubic single crystal, **M** depends on the direction and on the magnitude of **H**, and eq. [3] has to be replaced by

$$\mathbf{M} = (\kappa)\mathbf{H} \qquad [7]$$

where (κ) is a symmetric tensor according to

$$\begin{bmatrix} M_1 \\ M_2 \\ M_3 \end{bmatrix} = \begin{bmatrix} \kappa_{11} & \kappa_{12} & \kappa_{13} \\ \kappa_{21} & \kappa_{22} & \kappa_{23} \\ \kappa_{31} & \kappa_{32} & \kappa_{33} \end{bmatrix} \begin{bmatrix} H_1 \\ H_2 \\ H_3 \end{bmatrix} \qquad [8]$$

In eq. [8], the vector components of **M** and **H** refer to an orthogonal coordinate system fixed in the crystal.

B. DIAMAGNETISM

The effect of a magnetic field on the motion of electrons is to cause a precession of each electronic orbit about the direction of **H** . As a consequence, the atom to which the electrons belong acquires a magnetic moment proportional in magnitude but opposite in direction to the field. It may be shown *(33)* that the resulting susceptibility per gram-atom is given by

$$\chi_A = -\frac{Ne^2}{6mc^2} \sum_i \langle r_i^2 \rangle \qquad [9]$$

where e and m refer to the charge and mass of the electron, respectively, c is the velocity of light, N the Avogadro number, and $\langle r_i^2 \rangle$ the expectation value (mean value) of the square distance of the electron from the nucleus

$$\langle r^2\rangle = \langle x^2\rangle + \langle y^2\rangle + \langle z^2\rangle \quad [10]$$

According to eq. [9], diamagnetic susceptibilities should be independent of temperature which is observed in practice to a good approximation. Since all substances contain electrons, diamagnetism is an inherent property of all atomic and molecular systems. In simple cases eq. [9] remains valid for the molar susceptibility, χ_M. Alternatively, an additional term describing the so-called temperature-independent paramagnetism has to be added *(42)*. Although diamagnetism is thus amenable to quantum chemical calculations, at present theoretical results are of no significance in experimental work.

Instead of direct calculations use is often made of the law of additivity of atomic diamagnetic susceptibilities. According to Pascal, the molar diamagnetic susceptibility of a pure compound is approximately determined by

$$\chi_M = \sum_i n_i \chi_{Ai} + \sum_j n_j \chi_{Bj} \quad [11]$$

where n_i is the number of atoms i, χ_{Ai} their susceptibility, per gram-atom, n_j the number of certain structural elements j and χ_{Bj} their contribution to the susceptibility. The quantities χ_{Ai} and χ_{Bj} are called the Pascal constants and the constitutive corrections, respectively. Tabulations of these values for various atoms and bonds are available *(9,10,19)*. In general, the susceptibility of an organic substance may be estimated on the basis of the tabulated values with an accuracy of a few percent. We will be interested, in particular, in the susceptibility χ_A of a paramagnetic ion within a compound containing large organic constituents. This value may be obtained by subtracting, from the observed molar susceptibility χ_M, the susceptibilities of all diamagnetic groups present including the diamagnetic contribution of the paramagnetic ion,

$$\chi_A = \chi'_M = \chi_M - \sum_i \chi_{Ai}^{dia} \quad [12]$$

Here χ_A is often referred to as the "corrected" molar susceptibility and is denoted then by χ'_M. Since $\Sigma\chi_{Ai}^{dia}$ is considerably smaller than χ_M, this quantity is usually estimated on the basis of eq. [11]. However, it should be observed that in all complicated organic moieties it is, in general, more accurate to measure $\Sigma\chi_{Ai}^{dia}$ directly.

Except for the purpose of "diamagnetic corrections", the susceptibilities of diamagnetic systems are not often measured. A review concerned with diamagnetic susceptibilities of organometallics has recently appeared *(34)*.

C. PARAMAGNETISM

Paramagnetic properties are conferred upon an atomic or molecular system, if there are partially filled electron shells. It may be assumed that, in general, this condition should be met in compounds of iron due to the incompletely filled 3*d* shell. Another condition for simple paramagnetic behaviour is magnetic dilution. With this we describe the situation where, *e.g.*, the presence of coordinated large organic ligands keeps the paramagnetic centers separated, thus reducing the influence of ion-ion coupling and avoiding co-operative interactions like ferro- and antiferromagnetism. It may be shown that, at sufficiently high temperatures, the susceptibility of the paramagnetic atom should follow the Curie law

$$\chi_A = \frac{C}{T} = \frac{N\mu_{eff}^2\beta^2}{3kT} \qquad [13]$$

where C is the Curie constant, N the Avogadro number, and k the Boltzmann constant. In the chemical literature, it is customary to describe the magnetic behaviour in terms of the effective magnetic moment μ_{eff}. Following eq. [13] it is

$$\mu_{eff} = \left(\frac{3k}{N\beta^2}\right)^{\frac{1}{2}} (\chi_A T)^{\frac{1}{2}} = 2.828\ (\chi_A T)^{\frac{1}{2}} \qquad [14]$$

where μ_{eff} is given in units of the Bohr magneton β (also denoted μ_B or B.M.) and

$$\beta = \frac{eh}{4\pi mc} = 0.927314 \cdot 10^{-20}\ \text{erg Gauss}^{-1}$$

Sometimes the temperature independent paramagnetism Nα is added to eq. [13]. This term arises from second order Zeeman effect contributions of states separated from the ground state by an energy E >> kT.

The temperature dependence of the susceptibility for substances which are not magnetically dilute often follows the Curie-Weiss law

$$\chi_A = \frac{C}{T-\Theta} = \frac{N\mu_{eff}^2\beta^2}{3k(T-\Theta)} \qquad [15]$$

where Θ is the Weiss constant or paramagnetic Curie temperature (Θ_p). The relation eq. [15] holds especially for antiferromagnetic (Θ<0) and ferromagnetic (Θ>0) substances far above the Néel and Curie temperature, respectively. In this case the effective magnetic moment is given by

$$\mu_{eff} = 2.828\{\chi_A(T-\Theta)\}^{\frac{1}{2}} \qquad [16]$$

in units of β. However, the majority of paramagnetic substances also obeys the Curie-Weiss rather than the Curie law in a limited temperature range. Here eq. [15] is employed solely in an empirical manner and no particular significance can be assigned to Θ. In addition, the magnetic moment becomes a function of temperature and should be calculated on the basis of eq. [14].

The magnetic susceptibility ($\boldsymbol{\chi}$) of single crystals of lower than cubic symmetry has the character of a symmetric tensor of rank 2, *cf* eq. [8]. If the six unique elements of the tensor ($\boldsymbol{\chi}$) are known, it is always possible to find a transformation to a new coordinate system in which ($\boldsymbol{\chi}$) is diagonal,

$$(\boldsymbol{\chi})\,\mathbf{C} = \mathbf{C}\,(\boldsymbol{\chi}_p) \qquad [17]$$

$$\boldsymbol{\chi}_p = \begin{bmatrix} \chi_1 & 0 & 0 \\ 0 & \chi_2 & 0 \\ 0 & 0 & \chi_3 \end{bmatrix} \qquad [18]$$

where $\mathbf{C}$ is the transformation matrix. The diagonal elements χ_1, χ_2, χ_3 are called principal susceptibilities of the crystal and their directions are the principal magnetic axes. The magnetic anisotropies of the crystal are the differences $(\chi_1-\chi_2)$, $(\chi_1-\chi_3)$, $(\chi_2-\chi_3)$. The magnetic properties of the individual molecules or ions in the crystal are defined in terms of three principal molecular susceptibilities K_1, K_2, and K_3 which form an orthogonal set. In compounds where the orientations of the molecules in the unit cell are known from X-ray diffraction studies, K_1, K_2, and K_3 may be determined from the principal molar susceptibilities of the crystal and the molecular direction cosines. The relationships between these quantities for the various crystal classes have been discussed by Lonsdale and Krishnan *(30)*. A more recent review concerned with paramagnetic anisotropy is available *(14)*. So far only few iron-organic compounds have been studied as single crystals.

D. MAGNETIC EXCHANGE INTERACTIONS

A serious limitation to the accurate determination of susceptibilities in paramagnetic metal ions is sometimes the existence of magnetic dipolar coupling and exchange effects. When the magnetic dilution is not adequate, the magnetic moments of different paramagnetic ions start to interact thus giving rise to two associated types of cooperative phenomena: ferromagnetism and antiferromagnetism. Experimentally, ferromagnetic substances are characterized by high values of the susceptibility which is dependent on the strength of the applied field and on the previous magnetic history of the specimen. At infinite field strength, the extrapolated susceptibility decreases toward a normal paramagnetic value, the corresponding magnetic moment being known as the saturation magnetization σ_s. If the temperature is raised sufficiently, the saturation magnetization vanishes and the substance becomes

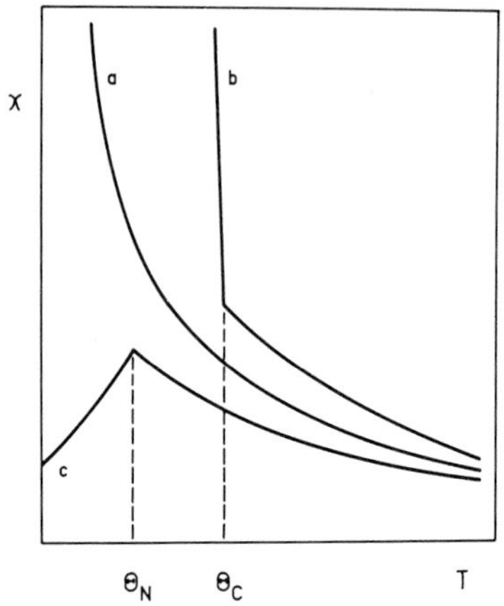

Fig. 1: The temperature dependence of the magnetic susceptibility of typical paramagnetic (a), ferromagnetic (b), and antiferromagnetic (c) substances. Θ_C and Θ_N are the Curie and Néel temperature, respectively.

paramagnetic. The transition temperature is called the ferromagnetic Curie temperature Θ_C, *viz.* Fig. 1. At $T>\Theta_C$, the susceptibility follows the Curie-Weiss law eq. [15] with positive Θ. Although often Θ approximates Θ_C, there are deviations from eq. [15] in the transition region. Iron-organic compounds are not infrequently contaminated with traces of a ferromagnetic impurity. The presence of such impurity shows up in the dependence of the susceptibility on **H**. Consequently, a plot of susceptibility *versus* 1/H is extrapolated to zero to determine the limiting value of the susceptibility.

Antiferromagnetic substances which are quite common follow the Curie-Weiss law eq. [15] with negative values of Θ at temperatures $T>\Theta_N$. At the Néel temperature Θ_N, a characteristic maximum in the susceptibility is observed, *viz.* Fig. 1, indicating the transition to an antiferromagnetic state. The susceptibility decreases below Θ_N and, for lattice antiferromagnets, becomes field dependent in the direction of orientation of the spin dipoles. For antiferromagnetic compounds, the effective magnetic moment should be determined according to eq. [16]. The coupling between the paramagnetic centers may take place either directly (direct exchange) or, more commonly, by intervening atoms of a different kind, *e.g.* oxygen or halogen atoms, a phenomenon referred to as superexchange.

Intermolecular antiferromagnetism may be considered as a special type of exchange interaction encountered in molecules which contain more than one metal atom. Entities of this type are often termed 'clusters', each cluster being effectively shielded from its neighbours. The interaction may be described by the Hamiltonian

$$\mathcal{H}_{ex} = \sum_{i,j} J_{ij}\, \mathbf{s}_i \cdot \mathbf{s}_j \qquad [19]$$

where J_{ij} is the exchange integral and $\mathbf{s}_i$ and $\mathbf{s}_j$ are the spin operators for the respective electrons. Probably the simplest antiferromagnetic system is that of a pair of ions each having a spin s = 1/2, *e.g.* in copper(II) acetate monohydrate and its higher homologues. In this case, the pair interaction produces a singlet and a triplet state separated in energy by J_{12}, the resulting susceptibility expression being

$$\chi_A = \frac{Ng^2\beta^2}{3kT}\left\{ 1 + \frac{1}{3}\exp(J_{12}/kT) \right\}^{-1} + N\alpha \qquad [20]$$

Antiferromagnetic coupling within a pair of iron(III) ions each having s = 5/2 is involved, *e.g.* in $[(phen)_2ClFeOFeCl(phen)_2]^{2+}$ ion *(6,7)*. Trinuclear clusters have been observed in basic carboxylates of iron(III), *viz.* $[Fe_3O(RCOO)_6(H_2O)_3]^+$ *(5)* as well as in iron(III) alkoxides, $[Fe_3(OR)_9]$ *(1)*. An introductory survey *(32)* of metal-metal interactions and a more specialized review *(38)* recently became available.

III. ELECTRONIC STATES OF IRON

A. *FREE IONS AND FREE ION TERMS*

The neutral iron atom corresponds to the electron configuration $[Ar]3d^6 4s^2$, 18 of the total of 26 electrons being assigned to the argon core. The most common oxidation states of iron are the trivalent (Fe^{3+}) and the divalent (Fe^{2+}) state characterized by the configurations outside of closed shells $3d^5$ and $3d^6$, respectively. In iron organic compounds, the oxidation states of Fe^+ ($3d^7$), Fe^0 ($3d^8$), and Fe^- ($3d^9$) may occur, in addition.

In the 3*d* shell, a maximum of 10 electrons may be accommodated if two electrons with spins paired are placed in each of the five orbitals. When there are less than 10 electrons, in general several arrangements are possible. Suppose that six electrons should be distributed on five *d*-orbitals in agreement with the Pauli exclusion principle. Then the maximum value of the total spin S = 2 results if only two electrons have their spins aligned antiparallel. Other possible arrangements give S = 1 and S = 0, *viz.* Fig. 2b. According to Hund's rule the state of lowest energy is one of the highest possible value of S, *i.e.* in Fe^{2+}, ^5D. The possible spin alignments in the various oxidation states of iron considered here are displayed in Fig. 2.

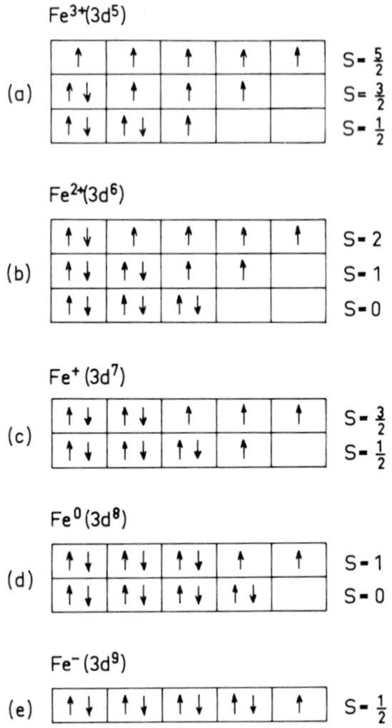

Fig. 2: Possible spin alignments in the various ions of iron.

In a free ion, the five 3*d* orbitals are degenerate and may be written according to

$$\psi_{nlm}(r,\theta,\varphi) = R_{nl}(r)Y_l^m(\theta,\varphi) \qquad [21]$$

where n = 3 and l = 2. Since the atomic Hamiltonian is spherically symmetric, the functions form bases of the representation $D^{(2)}$ of the three-dimensional rotation group, R_3. We will use for convenience the ket notation $|nlm_l m_s\rangle = |m_l^{m_s}\rangle$ where we suppress, in the present discussion, the quantum numbers n and l and where we leave only the sign + or − to indicate the spin $m_s = +1/2$ or $m_s = -1/2$. The ten possible spin functions of 3*d* electrons are thus $|2^+\rangle$, $|1^+\rangle$, $|0^+\rangle$, ..., $|-2^-\rangle$.

The effect of the electronic interactions is to split a particular configuration into a number of terms. In the $3d^6$ configuration, the terms produced are as follows:

Quintet: ^5D
Triplets: ^3H, ^3G, 2^3F, ^3D, 2^3P
Singlets: ^1I, 2^1G, ^1F, 2^1D, 2^1S

The energies of these terms may be expressed by the Slater-Condon parameters F_0, F_2, F_4 or by the related Racah parameters A, B, C, where

$$\begin{aligned} A &= F_0 - 49\ F_4 \\ B &= F_2 - 5\ F_4 \\ C &= 35\ F_4 \end{aligned} \qquad [22]$$

We will be particularly interested in ^5D, ^3H, and ^1I terms, the energies of which are

$$\begin{aligned} E(^5D) &= 6\ A - 21\ B \\ E(^3H) &= 6\ A - 17\ B + 4\ C \\ E(^1I) &= 6\ A - 15\ B + 6\ C \end{aligned} \qquad [23]$$

From the emission spectrum of the Fe^{2+} ion, B = 917 cm^{-1} and C = 4040 cm^{-1} *(24)*.

The total degeneracy of a term is (2S + 1)(2L + 1), thus amounting to 25 for ^5D, 33 for ^3H, and 13 for ^1I. Each component may be expressed as a normalized and antisymmetrized six-electron wavefunction (Heisenberg-Slater determinant). If we employ the ket notation introduced above, the component of ^3H with $M_L = 5$ and $M_S = 1$ may be abbreviated

$$|^3H\ 5\ 1\rangle = |2^+2^-1^+1^-0^+-1^+\rangle \qquad [24]$$

This ket being the only possible with the specified values of $M_L = \Sigma m_{li}$ and $M_S = \Sigma m_{si}$, other wavefunctions may be obtained by application of the step-up and step-down operators L_+ and L_- (and similarly S_+ and S_-) where

$$L_\pm|LSM_LM_S\rangle = \{L(L + 1) - M_L(M_L \pm 1)\}^{\frac{1}{2}}|LSM_L \pm 1\ M_S\rangle \qquad [25]$$

with a similar expression for $S_\pm$. Operating with L_- according to eq. [25] on $|^3H\ 5\ 1\rangle$ we obtain

$$L_-|^3H\ 5\ 1\rangle = \sqrt{10}\ |^3H\ 4\ 1\rangle \qquad [26]$$

and, since $L_\pm = \Sigma\ l_{i\pm}$ where $l_{i\pm}$ operates on the single electron function of electron i, application to the right hand ket of eq. [24] gives

$$\Sigma l_{i-}|2^+2^-1^+1^-0^+-1^+\rangle = \sqrt{6}|2^+2^-1^+0^-0^+-1^+\rangle + \sqrt{4}|2^+2^-1^+1^-0^+-2^+\rangle \qquad [27]$$

All other kets generated in eq. [27] vanish because of violation of the Pauli principle. It follows

$$|^3H\ 4\ 1> = -\sqrt{\frac{3}{5}}|2^+2^-1^+0^+0^--1^+> + \sqrt{\frac{2}{5}}|2^+2^-1^+1^-0^+-2^+> \qquad [28]$$

Continuous application of L_- produces all functions $|^3HM_L\ 1>$ and therefrom the remaining kets $|^3HM_LM_S>$ of the ^3H term may be obtained using S_-. In addition, eq. [28] may be employed to construct $|^3G\ 4\ 1>$ which must be orthogonal to $|^3H\ 4\ 1>$

$$|^3G\ 4\ 1> = \sqrt{\frac{2}{5}}|2^+2^-1^+0^+0^--1^+> - \sqrt{\frac{3}{5}}|2^+2^-1^+1^-0^+-2^+> \qquad [29]$$

In this way, all required wavefunctions of the terms within the $3d^6$ configuration are obtained. For the ^5D term, the component of highest M_L and $M_S = 2$, *e.g.* is

$$|^5D\ 2\ 2> = |2^+2^-1^+0^+-1^+-2^+> \qquad [30]$$

A splitting of each term may arise by way of the spin-orbit interaction described by the Hamiltonian

$$\mathscr{H}_{so} = \sum_i \zeta_i\ \mathbf{l}_i \cdot \mathbf{s}_i \qquad [31]$$

Since $\mathscr{H}_{so}$ is the sum of one-electron operators, it is straightforward to calculate matrix elements of eq. [31] provided the required wavefunctions are known in terms of one-electron kets as in eqs. [28], [29], and [30]. Here we are particularly interested in the spin-orbit interaction within the ^5D ground term. However, due to the 25-fold degeneracy, the direct calculation of spin-orbit matrix elements is at most time consuming. In general, therefore, the application of the Landé interval rule is preferred. According to the rule, the energy of a spin-orbit level is determined according to

$$E(^{2S+1}L_J) = \frac{\lambda}{2}\{J(J+1) - L(L+1) - S(S+1)\} \qquad [32]$$

where J assumes one of the values $J = L + S,\ L + S - 1,\ \ldots\ |L - S|$ and

$$\lambda = \pm\zeta/\ 2\ S \qquad [33]$$

the negative sign being valid if the shell is more than half filled. The total energies of the ^5D levels thus result to

$$E(^5D_0) = 6\ A - 21\ B + \frac{3}{2}\zeta$$
$$E(^5D_1) = 6\ A - 21\ B + \frac{5}{4}\zeta$$
$$E(^5D_2) = 6\ A - 21\ B + \frac{3}{4}\zeta \qquad [34]$$

$$E(^5D_3) = 6A - 21B$$

$$E(^5D_4) = 6A - 21B - \zeta$$

Another very powerful technique of general applicability has been developed by Racah *(36)*. The method is based on the theory of irreducible tensor operators and employs the Wigner-Eckart theorem *(18,39)*. We demonstrate this technique by the calculation of the matrix element $\langle ^5D_4 | \mathscr{H}_{so} | ^5D_4 \rangle$. Within the configuration l^n it is, according to Slater *(39)*

$$\langle \alpha\, S\, L\, J | \mathscr{H}_{so} | \alpha' S' L' J \rangle = (-1)^{S+L'-J} [l(l+1)(2l+1)]^{\frac{1}{2}} \times \langle \alpha\, S\, L \| V^{11} \| \alpha' S' L' \rangle\, W(SLS'L';J1) \qquad [35]$$

where α stands for additional quantum numbers that may be required and W(SLS'L';J1) is a Racah W coefficient. In terms of the equivalent 6j symbol

$$\begin{Bmatrix} S & L & J \\ L' & S' & 1 \end{Bmatrix} = (-1)^{S+L+S'+L'}\, W(SLS'L';J1) \qquad [36]$$

the matrix element of eq. [35] may be expressed, in units of $-\zeta$, (for $n > 2l+1$) as

$$\langle \alpha\, S\, L\, J | \mathscr{H}_{so} | \alpha' S' L' J \rangle = (-1)^{L+S'+J} [l(l+1)(2l+1)]^{\frac{1}{2}} \times \langle \alpha\, S\, L \| V^{11} \| \alpha' S' L' \rangle \begin{Bmatrix} S & L & J \\ L' & S' & 1 \end{Bmatrix} \qquad [37]$$

The reduced matrix elements $\langle \alpha\, S\, L \| V^{11} \| \alpha' S' L' \rangle$ have been tabulated by Slater *(39)*. Thus (from *(39)*, Vol. II, Appendix 26)

$$\langle ^5D_4 \| V^{11} \| ^5D_4 \rangle = \frac{1}{4}\sqrt{30}$$

and with $l = 2$, $S = S' = 2$, $L = L' = 2$, $J = 4$ it is

$$(-1)^{L+S'+J} = 1$$

$$[l(l+1)(2l+1)]^{\frac{1}{2}} = \sqrt{30}$$

The 6j coefficient is taken from tables *(37)* as

$$\begin{Bmatrix} S & L & J \\ L' & S' & 1 \end{Bmatrix} = \begin{Bmatrix} 2 & 2 & 4 \\ 2 & 2 & 1 \end{Bmatrix} = \frac{2}{15}$$

finally giving thus

$$\langle ^5D_4 | \mathscr{H}_{so} | ^5D_4 \rangle = 1 \cdot \sqrt{30} \cdot \frac{1}{4}\sqrt{30} \cdot \frac{2}{15}\,(-\zeta) = -\zeta \qquad [38]$$

in agreement with eq. [34].

B. *CONCEPT AND BASIC FACTS OF LIGAND FIELD THEORY*

So far we have considered the effect of electronic interaction and of spin-orbit coupling on a specific electronic configuration of the free iron ion. In actual compounds, the iron is surrounded either by ions functioning directly as ligands, by atoms forming the coordinating part of a ligand molecule, or by atoms bonded to the adjacent iron similar to the organic type of compounds. In any one of these cases, the most general approach to the theoretical description involves the use of molecular orbitals (MO). The iron and the ligands are then treated as of equivalent standing and the character of the bonds rests with the relative magnitude of the LCAO coefficients of metal *versus* ligand orbitals. The difficulty with MO calculations is, however, that if carried out to completeness, each molecule has to be considered as a problem on its own and results of general applicability cannot be expected. Needless to say that the computational effort required is considerable. On the other hand, if rigorous simplifications are introduced, the complexity of the calculations is reduced at the expense of reliability.

We turn therefore, in the present context, to ligand field theory which, in its semi-empirical form, offers high accuracy of calculated relative energies in conjunction with predictive capability. This approach is based on the fact that the iron atom may be considered, in the majority of its compounds, as being subject to an electric field set up by the ligands. Consequently, the ligands are assumed to be charged or if they are of dipolar nature then with the negatively charged end directed at the metal.

Assuming, for convenience, an octahedral disposition of ligands, the corresponding potential which produces non-zero matrix elements with *d* orbitals may be written, in terms of spherical harmonics,

$$V_{oct} = A\, r^4 \{Y_4^0(\theta,\varphi) + \sqrt{\frac{5}{14}}\,[Y_4^4(\theta,\varphi) + Y_4^{-4}(\theta,\varphi)]\} \qquad [39]$$

or, equivalently, in cartesian coordinates

$$V_{oct} = D\,[x^4 + y^4 + z^4 - \frac{3}{5}r^4] \qquad [40]$$

If six ligands of charge Ze are explicitly considered at the distance a from the origin along the three cartesian axes (crystal field model) it is

$$A = \frac{7\sqrt{\pi}}{3}\frac{Ze}{a^5} \quad \text{and} \quad D = \frac{35}{4}\frac{Ze}{a^5} \qquad [41]$$

The free ion *d* orbitals $|2\rangle, |1\rangle, |0\rangle, |-1\rangle$, and $|-2\rangle$ are no longer eigenfunctions within a field of octahedral symmetry. It may be shown, on the basis of group theory, that the original representation $D^{(2)}$ spanned by the *d* orbitals is split according to

$$D^{(2)} \longrightarrow E_g + T_{2g} \qquad [42]$$

We require, therefore, linear combinations of *d* orbitals transforming as the irreducible representations E_g and T_{2g} of O_h. These functions are listed below employing various ways of writing commonly encountered in literature *(12)*.

$$\begin{aligned}
e\theta &= |0\rangle &&= d_{z^2} &&= \tfrac{1}{2}(3z^2 - r^2)\\
e\varepsilon &= \tfrac{1}{\sqrt{2}}\,[|2\rangle + |-2\rangle] &&= d_{x^2-y^2} &&= \tfrac{\sqrt{3}}{2}\,(x^2-y^2)\\
t_2\xi &= \tfrac{i}{\sqrt{2}}\,[|1\rangle + |-1\rangle] &&= d_{yz} &&= \sqrt{3}\,(yz)\\
t_2\eta &= -\tfrac{1}{\sqrt{2}}[|1\rangle - |-1\rangle] &&= d_{zx} &&= \sqrt{3}\,(zx)\\
t_2\zeta &= \tfrac{1}{i\sqrt{2}}\,[|2\rangle - |-2\rangle] &&= d_{xy} &&= \sqrt{3}\,(xy)
\end{aligned} \qquad [43]$$

All the functions eq. [43] are *g* by parity. It is instructive to study the effect of the field first in a more qualitative way. The disposition of the orbitals in an octahedral field (*cf* Fig. 3) is such that the orbitals $e\theta$ and $e\varepsilon$ have their lobes of electron density pointing towards the negatively charged ligands, whereas $t_2\xi$, $t_2\eta$, and $t_2\zeta$ are concentrated between the x, y, and z axes. It follows that the electrostatic interaction of *e* orbitals with the ligands will be larger than the interaction of t_2 orbitals. The octahedral potential V_{oct} thus removes the five-fold degeneracy of *d* orbitals in the free ion producing an orbital triplet (t_{2g}) at lower energy and an orbital doublet (e_g) at higher energy. It is now a simple matter to calculate the energy of these orbitals according to

$$\begin{aligned}
E(t_{2g}) &= \langle t_2\xi|V_{oct}|t_2\xi\rangle\\
&= \int (t_2\xi)^* V_{oct} (t_2\xi)\, d\tau = -4\ Dq\\
E(e_g) &= \langle e\theta|V_{oct}|e\theta\rangle\\
&= \int (e\theta)^* V_{oct}\, (e\theta)\, d\tau = 6\ Dq
\end{aligned} \qquad [44]$$

The matrix elements eq. [44] may be promptly separated into the radial integral

$$q = \frac{2}{105} \int r^4 [R_{nd}(r)]^2 r^2 dr = \frac{2}{105} \langle r^4\rangle \qquad [45]$$

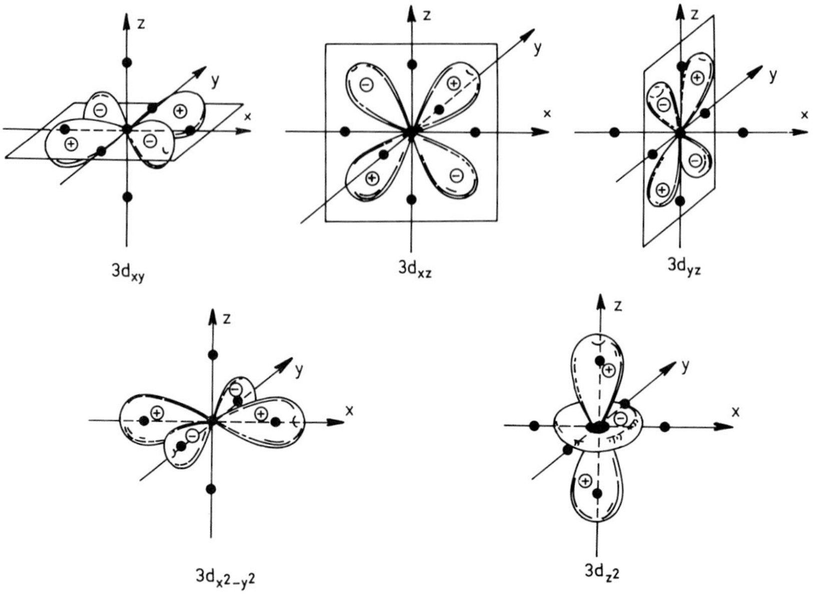

Fig. 3: Disposition of d orbitals (angular parts) within a field of octahedral symmetry.

where $R_{nd}(r)$ is the radial part of the *d* electron (*cf* eq.[21]) and into an angular integral. The latter is just a number multiplied by D which, in turn, originates in V_{oct}, *viz.* eq. [41]. The separation between the two sets of levels then results to

$$\Delta = E(e_g) - E(t_{2g}) = 10\ Dq = \frac{5}{3}\frac{Ze^2}{a^5}\langle r^4\rangle \qquad [46]$$

In principle, the quantity 10 Dq may be calculated according to its definition in the above equations. Computations of 10 Dq have indeed been successful in a number of cases *(35, 40)*. More often, however, 10 Dq is employed on a purely empirical basis and determined from suitable experimental data, *e.g.*, absorption spectra. In what follows we reserve the symbol 10 Dq to this sort of an empirical parameter. Thus 10 Dq should be a measure of the field strength originating in the ligands. In fact it has been demonstrated that the experimental 10 Dq values obey the formula

$$10\ Dq = f \cdot g \qquad [47]$$

where f depends only on the ligands and g only on the metal *(17)*.

C. *THE Fe^{2+} ION IN A FIELD OF OCTAHEDRAL SYMMETRY*

We recall that the Fe^{2+} ion is characterized by the electron configuration [Ar]$3d^6$. If the ion is subject to a ligand field of octahedral symmetry, the six *d* electrons have to be distributed between the symmetry orbitals eq.[43]. Of these, the t_2 orbitals are lower in energy by 10 Dq than the *e* orbitals and, therefore, the occupation of the t_2 orbitals will be favoured. This stabilization by the ligand field is counteracted by the interelectronic repulsion. In particular, electrons in the same orbital have higher repulsion energy than if they occupy different orbitals. To this comes a contribution from exchange energy again favouring states of high spin. In summary then, the relative magnitude of 10 Dq *versus* interelectronic repulsion determines the electron distribution between t_2 and *e* orbitals. A more quantitative treatment will be given below.

With respect to the ground state, two limiting situations may be visualized. If the relative magnitude of 10 Dq is small, the electrons tend to achieve a distribution with maximum spin which will correspond to the free ion ground state (high-spin state). In the Fe^{2+} ion, it is the configuration $t_2{}^4e^2$ with S = 2 that has the lowest energy. On the contrary, if 10 Dq is much larger than the interelectronic repulsion energy, the electrons will prefer to fill the t_2 orbitals thus producing a state with minimum spin (low-spin state). In this case, the $t_2{}^6$ configuration with S = 0 is lowest in Fe^{2+}. It is obvious that by large the value of 10 Dq will determine the magnetic properties of an iron compound. In iron(II), *e.g.*, a weak field ligand will produce a compound showing a paramagnetism of similar magnitude as a free Fe^{2+} ion, whereas a diamagnetic ground state will result with a strong field ligand.

The possible electron distributions originating in the $3d^6$ configuration of Fe^{2+} are illustrated in Fig. 4. For convenience of presentation, we have assumed a spin orientation corresponding to the maximum values of S. Lower values of S are, in general, possible and thus each of the electron

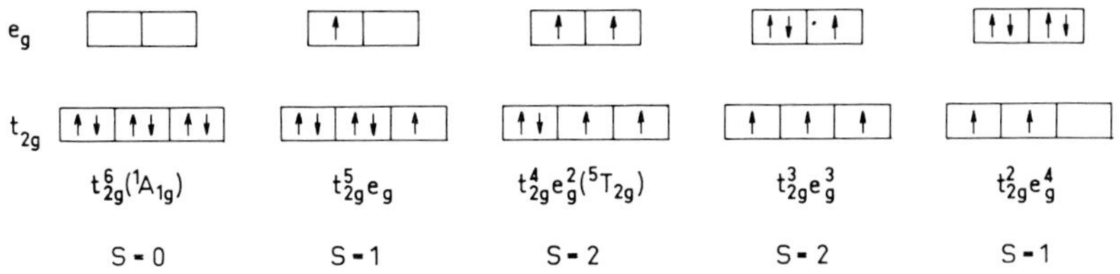

Fig. 4: Possible electron distributions and resulting maximum total spin of the $3d^6$ configuration of iron(II) within a field of octahedral symmetry.

configurations $t_{2g}^{m}e_{g}^{n}$ (m + n = 6) may give rise to more than one term consistent with the Pauli principle. A complete listing of these terms is provided in Table 2.

Table 2: Terms arising from the configuration $3d^6$ (Fe^{2+}) in a field of octahedral symmetry.

Electron Configuration	Terms S = 2	S = 1	S = 0
t_2^6			$^1A_1$
t_2^5e		$^3T_1$, $^3T_2$	$^1T_1$, $^1T_2$
$t_2^4e^2$	$^5T_2$	$^3A_2$, ^3E, 3^3T_1, 2^3T_2	2^1A_1, $^1A_2$, 3^1E, $^1T_1$, 3^1T_2
$t_2^3e^3$	^5E	$^3A_1$, $^3A_2$, 2^3E, 2^3T_1, 2^3T_2	$^1A_1$, $^1A_2$, ^1E, 2^1T_1, 2^1T_2
$t_2^2e^4$		$^3T_1$	$^1A_1$, ^1E, $^1T_2$

The energies of the various terms may be expressed *(12)* as functions of the octahedral field splitting 10 Dq and the Racah parameters A, B, and C (alternatively, the parameters F_0, F_2, and F_4 may be used, *cf* eq. [22]). We list below the energies of some of the lowest states of Fe^{2+} which are of interest to us:

$$\begin{aligned} E[^1A_1(t_2{}^6)] &= -24\ Dq + 15\ A - 30\ B + 15\ C \\ E[^3T_1(t_2{}^5e)] &= -14\ Dq + 15\ A - 30\ B + 12\ C \\ E[^3T_2(t_2{}^5e)] &= -14\ Dq + 15\ A - 22\ B + 12\ C \\ E[^5T_2(t_2{}^4e^2)] &= -\ 4\ Dq + 15\ A - 35\ B + 7\ C \end{aligned} \qquad [48]$$

These terms are plotted in Fig. 5 as functions of Dq employing the free ion values B = 917 cm^{-1} and C = 4040 cm^{-1}. At small values of 10 Dq, the ground state is $^5T_2$, whereas for large values of 10 Dq, $^1A_1$ is the state of lowest energy. The condition for the crossover between $^5T_2$ and $^1A_1$ states follows from eq. [48] to

$$(10\ Dq)_{^1A_1-^5T_2} = \frac{5}{2}B + 4\ C = 18{,}452\ cm^{-1} \qquad [49]$$

The critical value of 10 Dq = Π is known as spin pairing energy *(25)*. Similarly, $^3T_1$ crosses $^1A_1$ at

$$(10\ Dq)_{^1A_1-^3T_1} = 3\ C = 12{,}120\ cm^{-1} \qquad [50]$$

It should be observed that, however, the $^3T_1$ cannot become ground state within octahedral symmetry, since either $^5T_2$ or $^1A_1$ always is lower in energy than $^3T_1$. In addition, the same applies to $^3T_2$, *cf* Fig. 5.

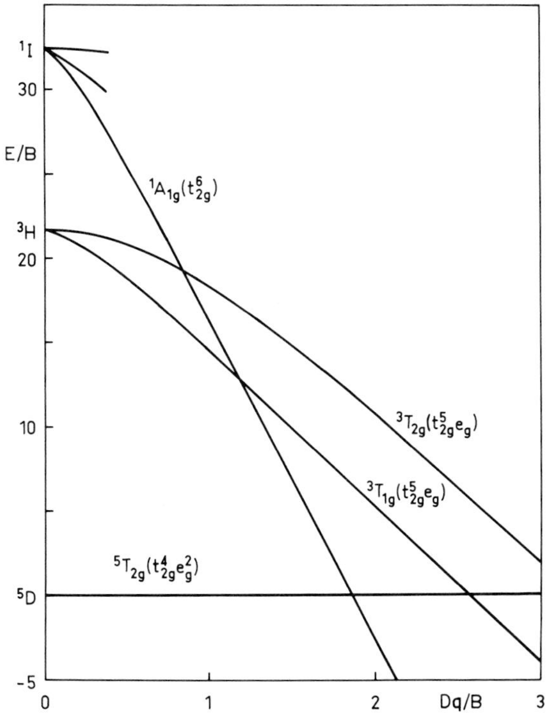

Fig. 5: Energies E of $^5T_2$, $^3T_1$, $^3T_2$, and $^1A_1$ terms as function of Dq, both in units of B, for the octahedral $3d^6$ configuration.

The wavefunctions corresponding to the terms discussed above may be written in a ket notation as $|S\Gamma M\gamma\rangle$ or $|^{2S+1}\Gamma M\gamma\rangle$. Here S is the total spin quantum number, whereas M denotes the z component of S, Γ is an irreducible representation of O_h and γ a component of Γ. Thus the only wavefunction of the non-degenerate $^1A_1$ term is $|^1A_1\ 0\ a_1\rangle$. On the other hand, the $^5T_2$ term has a total degeneracy of 15, *i.e.* a 5-fold spin and a 3-fold orbital degeneracy. The wavefunctions may be written

$$\begin{matrix} |^5T_2 2\xi\rangle & |^5T_2 1\xi\rangle & |^5T_2 0\xi\rangle & |^5T_2 {-1}\xi\rangle & |^5T_2 {-2}\xi\rangle \\ |^5T_2 2\eta\rangle & |^5T_2 1\eta\rangle & |^5T_2 0\eta\rangle & |^5T_2 {-1}\eta\rangle & |^5T_2 {-2}\eta\rangle \\ |^5T_2 2\zeta\rangle & |^5T_2 1\zeta\rangle & |^5T_2 0\zeta\rangle & |^5T_2 {-1}\zeta\rangle & |^5T_2 {-2}\zeta\rangle \end{matrix} \qquad [51]$$

where ξ, η, ζ represent real components of the representation

T_2 *(12)*. Alternatively, complex components of T_2 may be used. In this case, the functions with M = 2 would be

$$|{}^5T_2\ 2\ 1>\qquad |{}^5T_2\ 2\ 0>\qquad |{}^5T_2\ 2\ -1> \tag{52}$$

with a corresponding notation of the other spin components.

D. THE EFFECT OF A TETRAGONAL DISTORTION

Let us consider a distortion of the octahedron along the z axis (elongation or compression). The corresponding ligand field potential is

$$V_{tet} = -4\sqrt{\frac{\pi}{5}}\,B_2r^2Y_2^0(\theta,\varphi) + 4\sqrt{\pi}\,B_4r^4\,\{Y_4^0(\theta,\varphi) + \frac{1}{3}\sqrt{\frac{35}{2}}\,[Y_4^4(\theta,\varphi) - Y_4^{-4}(\theta,\varphi)]\} \tag{53}$$

V_{tet} is invariant with respect to operations of the group D_{4h}. The representations *E* and T_2 of O_h are now reducible and may be decomposed into irreducible representations within D_{4h} according to

$$\begin{aligned} E &\rightarrow A_1 + B_1 \\ T_2 &\rightarrow B_2 + E \end{aligned} \tag{54}$$

Consequently, levels which were degenerate within an octahedral field will be split, the eigenfunctions of a D_{4h} field being now $e\theta = a_1$, $e\varepsilon = b_1$, $t_2\zeta = b_2$, and $(t_2\xi,\ t_2\eta) = e$. The energies of the corresponding levels result to

$$\begin{aligned} E(b_1) &= 6\ Dq + 2\ Ds - Dt \\ E(a_1) &= 6\ Dq - 2\ Ds - 6\ Dt \\ E(b_2) &= -4\ Dq + 2\ Ds - Dt \\ E(e) &= -4\ Dq - Ds + 4\ Dt \end{aligned} \tag{55}$$

In these expressions, integrals over the radial part have been denoted by s and t similar to eq. [45] *(2)*. The quantities Ds and Dt may then be considered as empirical parameters specifying the tetragonal field.

Within the $3d^6$ configuration, the energies of the tetragonal terms resulting from the octahedral terms $^1A_1$, $^3T_1$, and $^5T_2$ are

$$\begin{aligned} E(^1A_1) &= -24\ Dq + 14\ Dt + 15\ A - 30\ B + 15\ C \\ E(^3A_2) &= -14\ Dq + 14\ Dt + 15\ A - 30\ B + 12\ C \\ E(^3E) &= -14\ Dq + \frac{21}{4}\ Dt + 15\ A - 30\ B + 12\ C \\ E(^5B_2) &= -\ 4\ Dq + 2\ Ds - Dt + 15\ A - 35\ B + 7\ C \\ E(^5E) &= -\ 4\ Dq - Ds + 4\ Dt + 15\ A - 35\ B + 7\ C \end{aligned} \tag{56}$$

It follows that, in D_{4h} symmetry, a $^3A_2$ term may likewise become ground term. The conditions required are easily deduced from eq. [56] as

$$\begin{aligned} 10\,Dq &\leq 3\,C \\ 10\,Dq - Ds - 10\,Dt &\geq 5\,B + 5\,C \end{aligned} \qquad [57]$$

Often, particularly in magnetic studies, the separation δ of the two lowest levels (^5E and $^5B_2$) is of interest where from eq. [56]

$$\delta = 3\,Ds - 5\,Dt \qquad [58]$$

To the first approximation, this splitting is the same if a distortion along the C_3 axis of the octahedron (of symmetry D_3) is assumed. Complete ligand field energy diagrams for all electron configurations d^n and for the most important symmetries have been recently published (26a), and these will be often useful to study the effect of a particular distortion.

IV. MAGNETIC SUSCEPTIBILITY

A. *THE EQUATION OF VAN VLECK*

In his well-known book Van Vleck *(42)* developed an expression for the susceptibility which is generally applicable to almost any paramagnetic system. Since the derivation is repeated in most textbooks of magnetism, we confine this section to a brief account of the most essential relations employed thereby and to a discussion of the result.

Consider the energy E_n of state n which is supposed to be of degeneracy j_n. The magnetic field removes, in general, the degeneracy and, if the sublevels are specified by the suffix m, the energy of any magnetic level E_{nm} may be expanded according to

$$E_{nm} = E_n^{(0)} + E_{nm}^{(1)}\,H + E_{nm}^{(2)}\,H^2 + \ldots \qquad [59]$$

The magnetic moment in direction of **H** is then

$$\mu_{nm} = -\frac{\partial E_{nm}}{\partial H} = -E_{nm}^{(1)} - 2\,E_{nm}^{(2)}\,H \qquad [60]$$

and the total magnetic moment M results by taking the statistical average over the thermal distribution of the μ_{nm}. Therefrom and using eq. [3] as well as eq. [6] the molar susceptibility obtains as

$$\chi_m = N \frac{\sum_{nm} \left\{ \frac{[E_{nm}^{(1)}]^2}{kT} - 2\, E_{nm}^{(2)} \right\} e^{-E_n^{(0)}/kT}}{\sum_n j_n \cdot e^{-E_n^{(0)}/kT}} \qquad [61]$$

where $E_n^{(0)}$ is the energy in zero field and N the Avogadro number. In the derivation, it has been assumed that there is no net magnetization in absence of the field, *i.e.* M = 0 at H = 0, and that $H\,E_{nm}^{(1)} << kT$.

The interaction with a magnetic field of strength **H** is described by the Hamiltonian

$$\mathcal{H}_m = \beta(\kappa \mathbf{L} + 2\mathbf{S})\mathbf{H} \qquad [62]$$

where β is the Bohr magneton and κ will be explained below. The first and second order Zeeman coefficients in eq. [59] are then determined by *(12)*

$$E_{nm}^{(1)} = \beta \langle nm|\kappa L_z + 2S_z|nm\rangle$$
$$E_{nm}^{(2)} = \sum_{n',m'} \frac{\beta^2 |\langle nm|\kappa L_z + 2S_z|n'm'\rangle|^2}{E_n^{(0)} - E_m^{(0)}}, \quad n \neq m \qquad [63]$$

where $|nm\rangle$ denotes an eigenfunction of the level with energy E_{nm}. In eq. [63], the use of L_z and S_z instead of **L** and **S** is restricted to octahedral symmetry.

For convenience, our treatment has been limited so far to the application of ligand field theory. However, matrix elements of type $\langle n|L_z + 2S_z|m\rangle$ should be calculated rather between functions $|n\rangle$, $|m\rangle$ corresponding to molecular orbitals. This may be adequately taken into account by the introduction of the orbital reduction factor κ *(11)* where

$$\kappa_\alpha = \frac{\langle\psi|L_\alpha|\psi\rangle}{\langle\varphi|L_\alpha|\varphi\rangle} \qquad [64]$$

where ψ represents an MO and φ is the metal *d* orbital contributing to ψ. Obviously, κ = 1.0 if the magnetic electrons are confined to the metal orbitals ("ionic" compound) and κ< 1.0 otherwise.

B. *AN EXAMPLE: MAGNETISM OF THE OCTAHEDRAL $^5T_2$ GROUND STATE IN Fe^{2+}*

In the calculation of magnetic susceptibility, a reasonable approximation may be achieved by considering only the contribution of the ground term. The starting point should consist of setting up the wavefunctions for the term in question as function of one-electron kets. For the $^5T_2$ term this would involve the 15 functions eq. [51]. However, a considerable simplification is possible since a $^5T_2$ term behaves analogously to a ^5P term of the free ion. The ^5P functions may be written down immediately in the $|SLM_SM_L\rangle \equiv |M_SM_L\rangle$ notation where $M_S = 2,1,0,-1,-2$ and $M_L = 1,0,-1$. It should be observed, however, that the operator L within the $^5T_2$ manifold has to be replaced by -L within ^5P. The subsequent calculation consists of three steps: (1) determination of the spin-orbit interaction energies and corresponding eigenfunctions; (2) calculation of the magnetic field energy; (3) insertion into eq. [61] to obtain the magnetic susceptibility.

The spin-orbit energies may be calculated by the formulae eqs. [32] or [35] as shown above. Since we require the resulting eigenfunctions, a direct calculation is preferred here.

Table 3: Matrices of the operator $-\lambda\,\mathbf{L}\cdot\mathbf{S}$ within the $|M_SM_L\rangle$ functions of a ^5P term (isomorphous to $^5T_2$), resulting eigenvalues and eigenfunctions.

$M_J = 3$:

	$\lvert 21\rangle$		
$\langle 21\rvert$	-2λ	$E_1 = -2\lambda$	$\psi_1 = \lvert 21\rangle$

$M_J = 2$:

	$\lvert 20\rangle$	$\lvert 11\rangle$		
$\langle 20\rvert$	0	$-\sqrt{2}\lambda$	$E_2 = \lambda$	$\psi_2 = \sqrt{\frac{2}{3}}\lvert 20\rangle - \sqrt{\frac{1}{3}}\lvert 11\rangle$
$\langle 11\rvert$	$-\sqrt{2}\lambda$	$-\lambda$	$E_3 = -2\lambda$	$\psi_3 = \sqrt{\frac{1}{3}}\lvert 20\rangle + \sqrt{\frac{2}{3}}\lvert 11\rangle$

$M_J = 1$:

	$\lvert 2{-1}\rangle$	$\lvert 10\rangle$	$\lvert 01\rangle$		
$\langle 2{-1}\rvert$	2λ	$-\sqrt{2}\lambda$	0	$E_4 = 3\lambda$	$\psi_4 = \sqrt{\frac{3}{5}}\lvert 2{-1}\rangle - \sqrt{\frac{3}{10}}\lvert 10\rangle + \sqrt{\frac{1}{10}}\lvert 01\rangle$
$\langle 10\rvert$	$-\sqrt{2}\lambda$	0	$-\sqrt{3}\lambda$	$E_5 = \lambda$	$\psi_5 = \sqrt{\frac{1}{3}}\lvert 2{-1}\rangle + \sqrt{\frac{1}{6}}\lvert 10\rangle - \sqrt{\frac{1}{2}}\lvert 01\rangle$
$\langle 01\rvert$	0	$-\sqrt{3}\lambda$	0	$E_6 = -2\lambda$	$\psi_6 = \sqrt{\frac{1}{15}}\lvert 2{-1}\rangle + \sqrt{\frac{8}{15}}\lvert 10\rangle + \sqrt{\frac{6}{15}}\lvert 01\rangle$

$M_J = 0$:

	$\lvert 1{-1}\rangle$	$\lvert 00\rangle$	$\lvert {-1}1\rangle$		
$\langle 1{-1}\rvert$	λ	$-\sqrt{3}\lambda$	0	$E_7 = -2\lambda$	$\psi_7 = \sqrt{\frac{1}{5}}\lvert 1{-1}\rangle + \sqrt{\frac{3}{5}}\lvert 00\rangle + \sqrt{\frac{1}{5}}\lvert {-1}1\rangle$
$\langle 00\rvert$	$-\sqrt{3}\lambda$	0	$-\sqrt{3}\lambda$	$E_8 = \lambda$	$\psi_8 = \sqrt{\frac{1}{2}}\lvert 1{-1}\rangle - \sqrt{\frac{1}{2}}\lvert {-1}1\rangle$
$\langle {-1}1\rvert$	0	$-\sqrt{3}\lambda$	λ	$E_9 = 3\lambda$	$\psi_9 = \sqrt{\frac{3}{10}}\lvert 1{-1}\rangle - \sqrt{\frac{2}{5}}\lvert 00\rangle + \sqrt{\frac{3}{10}}\lvert {-1}1\rangle$

We use the operator in the form

$$-\lambda\,\mathbf{L}\cdot\mathbf{S} = -\lambda(L_zS_z + \frac{1}{2}L_+S_- + \frac{1}{2}L_-S_+) \qquad [65]$$

where $L_\pm$ is the operator eq. [25] and similarly $S_\pm$. Application to the ket $|21>$ yields

$$-\lambda\,\mathbf{L}\cdot\mathbf{S}\;|21> = -\lambda(2|21> + \frac{1}{2}\cdot 0 + \frac{1}{2}\cdot 0)$$
$$= -2\lambda|21>$$

and thus

$$<21|-\lambda\,\mathbf{L}\cdot\mathbf{S}\,|21> = -2\lambda$$

Correspondingly, using the ket $|20>$ we obtain

$$-\lambda\,\mathbf{L}\cdot\mathbf{S}\,|20> = -\lambda(0|20> + \sqrt{2}|11> + \frac{1}{2}\cdot 0)$$
$$= -\sqrt{2}\lambda|11>$$

Here, only an off-diagonal matrix element is non-zero, since

$$<20|-\lambda\,\mathbf{L}\cdot\mathbf{S}\,|20> = 0$$

$$<11|-\lambda\,\mathbf{L}\cdot\mathbf{S}\,|20> = -\sqrt{2}\lambda$$

Proceeding along these lines, all matrix elements of $-\lambda\,\mathbf{L}\cdot\mathbf{S}$ may be calculated. It is found that the resulting matrix is diagonal in $M_J = M_S + M_L = 3,2,1,\ldots,-3$. Stated in a different way, the overall matrix is decomposed into seven smaller non-zero matrices arranged along the diagonal, each being characterized by one of the values of M_J. In Table 3, we list these matrices of the operator eq. [65] together with the eigenvalues and eigenfunctions resulting from diagonalization of the corresponding secular equations

$$|H_{ij} - E\,\delta_{ij}| = 0 \qquad [66]$$

where

$$H_{ij} = <M_SM_L|-\lambda\,\mathbf{L}\cdot\mathbf{S}\,|M_S'M_L'> \qquad [67]$$

In actual fact, details concerning $M_J = -1,-2$, and -3 are not included in Table 3 since all results are identical to those with positive values of M_J if only the $|M_SM_L>$ functions are replaced as follows

$M_J = -1$: $|-21>$, $|-10>$, $|0-1>$ instead of $|2-1>$, $|10>$, $|01>$

$M_J = -2$: $|-20>$, $|-1-1>$ instead of $|20>$, $|11>$

$M_J = -3$: $|-2-1>$ instead of $|21>$

Due to the replacement of L_z by $-L_z$, the operator of the magnetic field interaction is now

$$\mathscr{H}_m = \beta(-L_z + 2S_z)\ H \qquad [68]$$

Application to the kets $|21\rangle$, $|20\rangle$, and $|11\rangle$ yields immediately

$$\begin{aligned} \langle 21|(-L_z + 2S_z)\beta H|21\rangle &= 3\beta H \\ \langle 20|(-L_z + 2S_z)\beta H|20\rangle &= 4\beta H \\ \langle 11|(-L_z + 2S_z)\beta H|11\rangle &= \beta H \end{aligned} \qquad [69]$$

The first of the matrix elements of eq. [69] is identical to that of ψ_1. The matrix elements of the functions ψ_2 and ψ_3 of Table 3 result as

$$\begin{aligned} \langle \psi_2|(-L_z + 2S_z)\beta H|\psi_2\rangle &= 3\beta H \\ \langle \psi_3|(-L_z + 2S_z)\beta H|\psi_3\rangle &= 2\beta H \\ \langle \psi_2|(-L_z + 2S_z)\beta H|\psi_3\rangle &= \sqrt{2}\beta H \end{aligned} \qquad [70]$$

Table 4 lists the complete matrices of the spin-orbit and mag-

Table 4: Matrices of the combined spin-orbit and magnetic operators $-\lambda\,\mathbf{L}\cdot\mathbf{S} + (-L_z + 2S_z)\beta H$ within the functions ψ_j of Table 3.

$M_J = 3$:

	ψ_1
ψ_1	$-2\lambda + 3\beta H$

$M_J = 2$:

	ψ_2	ψ_3
ψ_2	$\lambda + 3\beta H$	$\sqrt{2}\beta H$
ψ_3	$\sqrt{2}\beta H$	$-2\lambda + 2\beta H$

$M_J = 1$:

	ψ_4	ψ_5	ψ_6
ψ_4	$3\lambda + \frac{7}{2}\beta H$	$\frac{9\sqrt{5}}{10}\beta H$	0
ψ_5	$\frac{9\sqrt{5}}{10}\beta H$	$\lambda + \frac{3}{2}\beta H$	$\frac{4\sqrt{5}}{5}\beta H$.
ψ_6	0	$\frac{4\sqrt{5}}{5}\beta H$	$-2\lambda + \beta H$

$M_J = 0$:

	ψ_7	ψ_8	ψ_9
ψ_7	-2λ	$\frac{3\sqrt{10}}{5}\beta H$	0
ψ_8	$\frac{3\sqrt{10}}{5}\beta H$	λ	$\frac{3\sqrt{15}}{5}\beta H$
ψ_9	0	$\frac{3\sqrt{15}}{5}\beta H$	3λ

netic field interaction. Again results for $M_J = -1, -2$, and -3 are not given in detail since these are simply related by a sign change of all matrix elements of the magnetic interaction to those of positive M_J. For the present case, eq. [61] may be rewritten as

$$\mu_{eff}^2 = \frac{\sum_{i=1}^{15}\left[\langle\psi_i|\mathcal{H}_m|\psi_i\rangle^2 + 2kT\sum_{j\neq i}^{15}\frac{|\langle\psi_j|\mathcal{H}_m|\psi_i\rangle|^2}{E_j - E_i}\right]\exp(-E_i/kT)}{\sum_{i=1}^{15}\exp(-E_i/kT)} \quad [71]$$

Inserting the required matrix elements of $\mathcal{H}_m$ from Table 4 finally produces

$$\mu_{eff}^2 = 3\frac{\left(\frac{49}{2} - \frac{1}{x}\frac{27}{2}\right)e^{-3x} + \left(\frac{45}{2} + \frac{1}{x}\frac{25}{6}\right)e^{-x} + \left(28 + \frac{1}{x}\frac{28}{3}\right)e^{2x}}{3e^{-3x} + 5e^{-x} + 7e^{2x}} \quad [72]$$

in units of β and where $x = \lambda/kT$ *(12)*. This equation then gives the temperature dependence of the square of the effective magnetic moment for a $^5T_2$ $(t_2{}^4e^2)$ ground state within octahedral symmetry.

In Fig. 6, the curve marked zero gives μ_{eff} of eq. [72] as a function of kT/λ. The remaining curves refer to the corresponding dependence of μ_{eff} if an axial field distortion is present. The figures marked on the curves refer to values of δ/λ where δ has been defined in eq. [58] and is assumed to be positive if the ^5E term is lower in energy than the orbitally non-degenerate state ($^5B_2$ in D_{4h} symmetry) *(21,22)*.

Within octahedral symmetry then, the magnetic moment increases with lowering of temperature from about 5.38 BM (assuming $\lambda = -80$ cm^{-1}) at 575 K to 5.72 BM at 115 K. To lower temperatures a decrease of μ_{eff} follows assuming μ_{eff} = 5.08 BM at 11.5 K. An increase of covalency (smaller values of κ) causes, in general, decreasing values of μ_{eff} *(22)*. An axial distortion of the octahedron likewise has the effect to lowering the magnetic moment, *viz.* Fig. 6. However, there are significant differences which depend on the sign of δ. If δ is positive, the ground state in D_{4h} symmetry is ^5E and the magnetic moment is lowered, although its temperature dependence is preserved. If, on the other hand, δ is negative, the lowest state is $^5B_2$ having the consequence of almost temperature independent magnetic moment values.

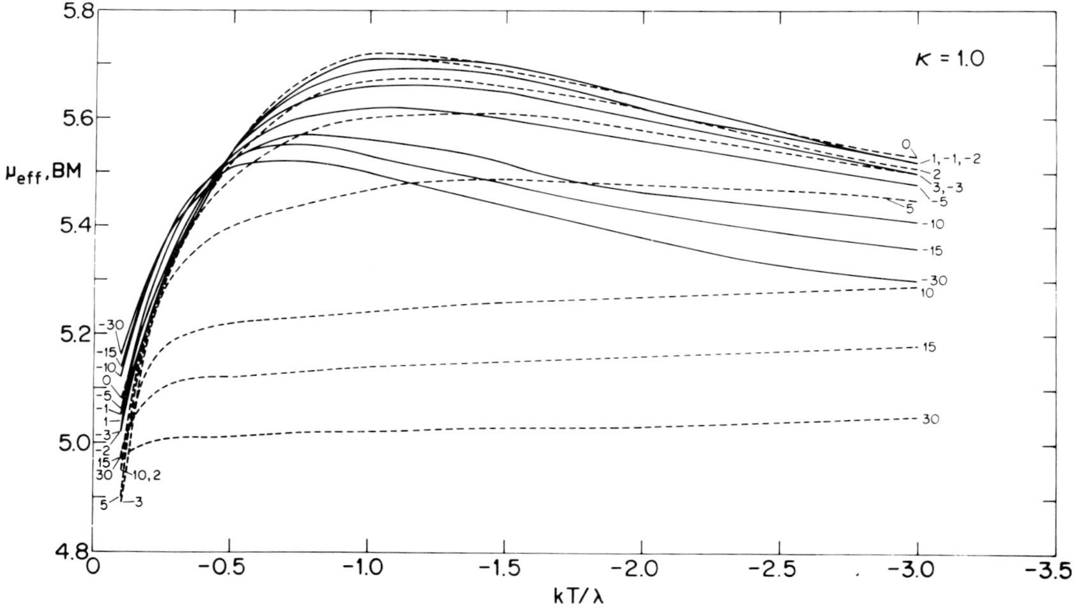

Fig. 6: Magnetic moment μ_{eff} in units of β (BM) for a $^5T_2$ term as function of kT/λ and δ/λ (marked on the curves). Full curves for δ>0, broken curves for δ<0 and κ = 1.0. Reproduced by permission of the publisher from reference (22).

C. DETAILED THEORY OF MAGNETISM IN OCTAHEDRAL Fe^{2+}

The terms arising from the configuration $3d^6$ under the influence of an octahedral field of ligands have been listed in Table 2. We studied alone the magnetism of a pure $^5T_2$ $(t_2{}^4e^2)$ term so far which produces the result of eq. [72]. It has been pointed out, however, that if 10 Dq > Π(*cf* eq. [49]), the $^1A_1(t_2{}^6)$ term becomes the ground state. Thus the magnetism represented by eq. [72] covers but a limited range of 10 Dq values, *viz.* high-spin iron(II). On the other hand, diamagnetism should be expected for a spin singlet ground state, *e.g.* the $^1A_1$ term. Experience shows, however, that weak paramagnetism is always encountered in low-spin iron(II). This paramagnetism is a result of contributions from higher energy terms of different total spin. In general, there are connecting non-zero matrix elements of both the operators of spin-orbit and of the magnetic interaction between different terms. The resulting mixing of states may affect the magnetism considerably, at least under certain circumstances.

A detailed theory of magnetism should account for the complete electron repulsion, ligand field, spin-orbit, and magnetic field interactions between all terms arising from the $3d^6$ configuration within octahedral symmetry. Although it is straight forward to perform such calculations with the methods outlined above, the amount of work required is surprisingly large. Thus there are 91 spin-orbit levels and, if the magnetic field interaction is included, 210 separate states result corresponding to the total degeneracy of the problem. Therefore, more powerful methods like those used in setting up eq. [35] are used.

An additional advantage of complete calculations is that magnetic properties are obtained as functions of the parameters of semi-empirical ligand theory, *viz.* 10 Dq, B, C, and ζ or λ. In Fig. 7 and Fig. 8 we display the results of complete magnetic calculations *(26)* on the octahedral $3d^6$ configuration in terms of $1/\chi_M$ and μ_{eff}, respectively. These plots may be employed in a direct comparison with spectroscopic measurements.

In particular, reasonable agreement with magnetic moment values obtained on the basis of an isolated $^5T_2(t_2{}^4e^2)$ multiplet *(21,22)* is achieved as long as low values of 10 Dq up to about 13,000 cm^{-1} are considered. If 10 Dq is significantly higher, say 10 Dq > 14,500 cm^{-1}, the $^1A_1(t_2{}^6)$ ground state is well separated from all excited states. The non-negligible temperature dependence of μ_{eff} is caused by mixing with the excited states. In the intermediate region of 10 Dq, a typical crossover behaviour of magnetism is encountered. In fact, the crossover between the ground states $^5T_2(t_2{}^4e^2)$ and

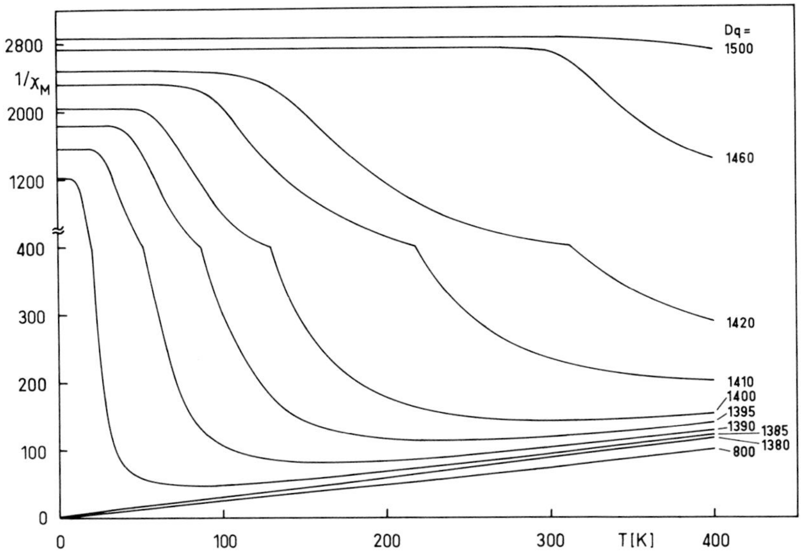

Fig. 7: Results of complete magnetic calculations in terms of $1/\chi_M$ for the octahedral $3d^6$ configuration vs T and Dq (B = 806 cm^{-1}, C = 4B, ζ = 420 cm^{-1}, κ = 1.0).

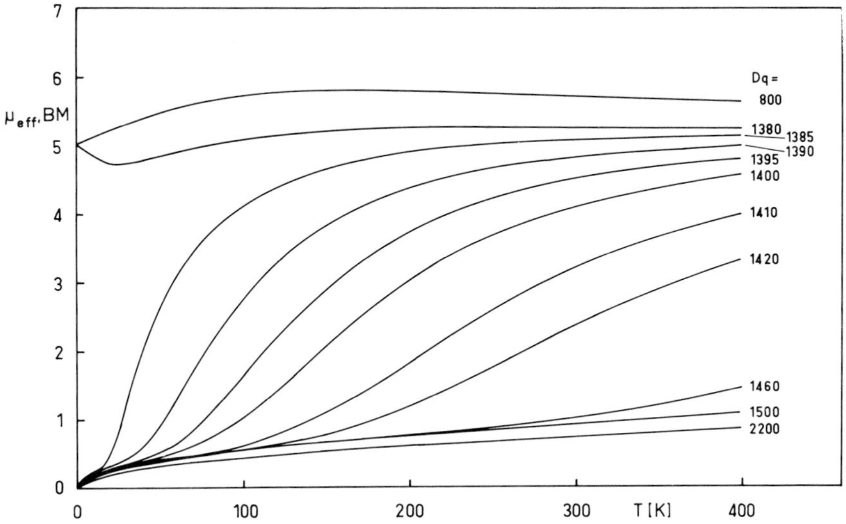

Fig. 8: Results of complete magnetic calculations in terms of μ_{eff} for the octahedral $3d^6$ configuration vs T and Dq (B = 806 cm^{-1}, C = 4B, ζ = 420 cm^{-1}, κ = 1.0).

$^1A_1(t_2{}^6)$ is found, with the parameter values of Fig. 8 at about Π = 13,850 cm^{-1}. The s shaped μ_{eff} *vs* T curves extend over a considerable range of Dq values, the direct contribution from the $^5T_2$ term being still discernible at 10 Dq = 14,400 cm^{-1}. There is now a large number of iron(II) organic compounds showing crossover magnetic effects *(23)*.

D. MAGNETISM OF TETRAHEDRAL AND TRIPLET GROUND STATE Fe^{2+}

In tetrahedral symmetry, the ground term within the $3d^6$ configuration of Fe^{2+} is $^5E(t_2{}^3(^4A_2)e^3)$. A complete calculation of the temperature dependence of magnetism has been performed *(26)*. The magnetic moment is 4.95 - 5.03 BM over almost the whole range of temperature if values of 10 Dq characteristic for a weak-field ligand complex (10 Dq = -8000 to -12,000 cm^{-1}) are assumed. It should be pointed out that, in tetrahedral symmetry, the ligand field splitting is smaller and opposite to that in an octahedron, *viz.* $(10\ Dq)_{tet} = -(4/9)(10\ Dq)_{oct}$. The magnetic moment values show that the ground term ^5E is primarily responsible for these, the orbital contribution being essentially suppressed. Below 50 K, μ_{eff} starts to decrease and falls off rapidly below 20 K. This is caused by progressive depopulation of all spin-orbit levels except the lowest in energy, $\Gamma_1(^5E)$, the magnetic moment of which is zero.

Subject to the condition eq. [57], a $^3A_2$ term may become ground state in D_{4h} symmetry and, in specific circumstances, $^3B_2$ and ^3E ground states may be formed *(27)*. Since some spin-mixing is nearly always associated with spin triplet ground states, magnetic moments between 3.0 and 4.5 BM are encountered, practically independent of temperature.

E. MAGNETISM OF IRON ORGANIC COMPOUNDS WITH IRON OXIDATION STATE DIFFERENT FROM +II

The oxidation states +VI ($3d^2$) and +V ($3d^3$) of iron are encountered in tetrahedral iron oxygen anions, whereas apparently only a single iron(IV) compound ($3d^4$) has been reported. The magnetic properties of these configurations are of no particular relevance in this context and will not be discussed here. Instead reference to other sources *(8,19,20)* is made.

The oxidation state +III ($3d^5$) of iron is very common and its magnetism has received due attention. The ground term of the free Fe^{3+} ion is ^6S which, in a field of octahedral symmetry and at small values of 10 Dq, becomes $^6A_1(t_2{}^3e^2)$. At high values of 10 Dq, the state $^2T_2(t_2{}^5)$ becomes the ground state, the $^6A_1$ - $^2T_2$ crossover occuring if

$$\Pi = 7\frac{1}{2}B + 5C \qquad [73]$$

within the usual approximation (see, however, a recent exact treatment *(25)*).

Since there is no orbital angular momentum in a $^6A_1$ state, the Hamiltonian eq. [62] becomes

$$\mathscr{H}_m = 2\beta S_z H_z \qquad [74]$$

and the corresponding magnetic energy is simply

$$E_m = 2\beta H_z \langle n|S_z|n\rangle = 2\beta M_S H_z \qquad [75]$$

where $M_S = S, S-1, \ldots, -S$. Using eq. [61] the magnetic susceptibility obtains as

$$\chi_M = \frac{4N\beta^2}{3kT} S(S+1) \qquad [76]$$

This is the Curie law within an effective magnetic moment

$$\mu_{eff} = 2[S(S+1)]^{\frac{1}{2}} \beta = 5.92 \text{ BM} \qquad [77]$$

if S = 5/2 ("spin-only" magnetic moment). Consequently, octahedrally coordinated high-spin iron(III) is expected to possess moments very close to eq. [77] and to be independent of temperature.

In a field of D_{4h} symmetry, the $^6A_1$ state is split into three doublets via ligand field and spin-orbit interactions with higher energy states, *e.g.* the $^4T_1(t_2{}^4e)$ term, *viz.* the EPR chapter. The doublets may be denoted by $|\pm 1/2\rangle$, $|\pm 3/2\rangle$, $|\pm 5/2\rangle$. If a magnetic field parallel to the z axis is applied, the energies of the magnetic levels result as shown below (*cf* eq. [32]),

$$\begin{aligned} E(\pm 1/2) &= 0 \pm \beta H_z \\ E(\pm 3/2) &= 2D \pm 3\beta H_z \\ E(\pm 5/2) &= 6D \pm 5\beta H_z \end{aligned} \qquad [78]$$

where D is the so-called zero field splitting parameter. From eq. [78], the susceptibility obtains *via* eq. [61] as

$$\chi_M^{\parallel} = \frac{N\beta^2}{kT} \frac{1 + 9e^{-2\xi} + 25e^{-6\xi}}{1 + e^{-2\xi} + e^{-6\xi}} \qquad [79]$$

On the other hand, a magnetic field perpendicular to the z axis produces energies of the magnetic levels (*cf* eq. [36])

$$E(\pm 1/2) = 0 \pm 3\beta H_{\perp} - \frac{4\beta^2 H_{\perp}^2}{D}$$
$$E(\pm 3/2) = 2D + \frac{11}{4}\frac{\beta^2 H_{\perp}^2}{D} \qquad [80]$$
$$E(\pm 5/2) = 6D + \frac{5}{4}\frac{\beta^2 H_{\perp}^2}{D}$$

and therefrom

$$\chi_M^{\perp} = \frac{N\beta^2}{kT}\,\frac{9 + \frac{8}{\xi} - \frac{11}{2\xi}e^{-2\xi} - \frac{5}{2\xi}e^{-6\xi}}{1 + e^{-2\xi} + e^{-6\xi}} \qquad [81]$$

Finally, the averaged susceptibility is calculated from eq. [79] and eq. [81] according to

$$\chi_M^{av} = \frac{1}{3}(\chi_M^{\parallel} + 2\chi_M^{\perp}) \qquad [82]$$

and the squared moment results as

$$\mu_{eff}^2 = \frac{19 + \frac{16}{\xi} + (9 - \frac{11}{\xi})e^{-2\xi} + (25 - \frac{5}{\xi})e^{-6\xi}}{1 + e^{-2\xi} + e^{-6\xi}} \qquad [83]$$

in units of β^2. In the eqs. [79], [81], and [83] it is $\xi = D/kT$. Limiting values of eq. [83] are $\mu_{eff} = 5.92$ BM if $T \rightarrow \infty$ and $\mu_{eff} = 4.36$ BM if $T \rightarrow 0$.

The treatment of the $^2T_2(t_2{}^5)$ term of low-spin iron(III) may be much simplified if it is realized that the $t_2{}^5$ configuration corresponds to a single hole in the filled t_2 orbital. Spin-orbit coupling produces a doublet and a quadruplet which are further split on application of a magnetic field. If the field is of octahedral symmetry, the detailed calculation yields *(28)*

$$\mu_{eff}^2 = \frac{8 + 3x - 8e^{-3x/2}}{x(1 + 2e^{-3x/2})} \qquad [84]$$

in units of β^2. In eq. [84], it is $x = \lambda/kT$ as above. Thus the magnetic moment is expected to be about 2.4 BM at room temperature and to decrease if the temperature is lowered approaching $\mu_{eff} = 1.73$ BM as $T \rightarrow 0$. A calculation of μ_{eff} based on tetragonal symmetry has been provided by Kotani *(29)*.

In the oxidation state +I ($3d^7$) of iron, the majority of compounds is polynuclear and thus diamagnetic *(19,20)*. The octahedral $3d^7$ configuration again produces two possible ground states, *viz.* $^4T_1(t_2{}^5e^2)$ and $^2E(t_2{}^6e)$, depending on the value of 10 Dq. The high covalency expected in iron(I) organic compounds invariably leads to an ^2E ground state if complete spin pairing does not occur. In such case, the moment is expected to be about 1.9 BM with little temperature dependence. Formal oxidation states of 0 ($3d^8$) and -I ($3d^9$) and sometimes even lower values are encountered in certain iron organic systems. As a rule, due to complete spin pairing between different metal centers, these compounds are diamagnetic.

V. ILLUSTRATED EXAMPLES

This section serves to illustrate the theoretical basis of section IV by examples chosen from the broad area of complex and organometallic compounds of iron. It should be kept in mind that the preceeding section was dictated by the requirement of providing a general introduction to the field rather than by the necessity of immediate application to specific problems. Consequently, in what follows, only limited use can be made of the specific results obtained above. In reality, often a "theory" tailored to measure is used for a particular compound, the possible reason being an unusual geometry of or the specific bonding properties within the molecule considered. Such theory may be at least of two different kinds: (1) if a general theory of bonding is available for the molecule in question, the resulting wavefunctions may be employed to calculate the magnetic properties; (2) alternatively, if such is not the case, a simple empirical approach may be chosen (corresponding, *e.g.*, to the spin Hamiltonian formalism considered in the EPR chapter) to interpret the magnetism. Examples for both treatments will be presented below.

A. *IRON(II) HYDRIDOTRIS(1-PYRAZOLYL)BORATE*

The methyl 3,5-disubstituted hydridotris(1-pyrazolyl)borate of iron(II), usually abbreviated as $\{HB[3,5\text{-}(CH_3)_2pz]_3\}_2$-Fe *(41)*, is paramagnetic at room temperature ($\mu_{eff} \sim 5.16$ BM) but becomes diamagnetic at 147 K *(15)*. Various physical techniques such as the ^{57}Fe Mössbauer effect were utilized to show *(15,16)* that a reversible spin transition $^5T_{2g}(t_{2g}^4e_g^2) \rightleftarrows {}^1A_{1g}(t_{2g}^6)$ takes place in this compound between about

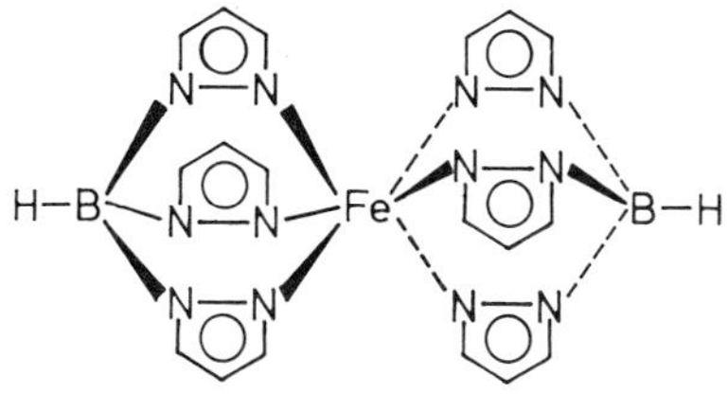

245 K and 147 K. A schematic of the structure of the molecule is given above.

We might like to compare the experimental values of the magnetic moment for this compound with Fig. 8. However, it should be made completely clear that we cannot expect complete agreement between the experimental data on this compound and the theory presented above for the $3d^6$ configuration of iron(II). The reason is simply that during the spin transition, the value of 10 Dq should change from a low value characteristic of the $^5T_{2g}$ ground state to a higher value characteristic of the $^1A_{1g}$ state or *vice versa*. On the other hand, the calculations are normally performed for fixed values of Dq as well as of the other parameters involved. We will resort, therefore, to an alternative approach *(24)*. In

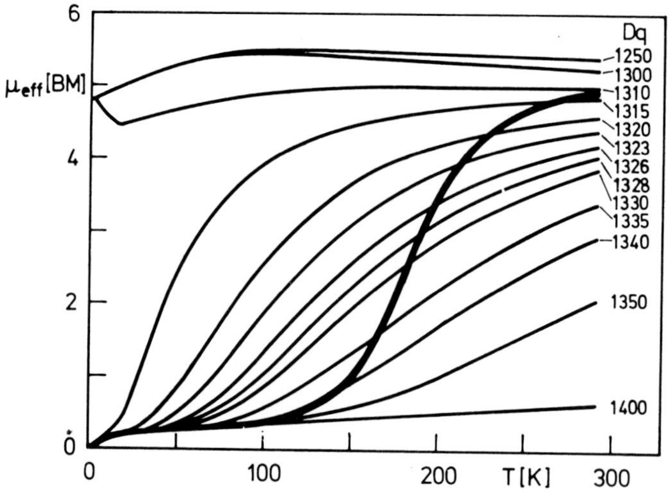

Fig. 9: Comparison of results from complete magnetic calculations on the octahedral $3d^6$ configuration (B = 765 cm^{-1}, C = 4.0 B, ζ = 420 cm^{-1}, κ = 0.80) with experimental data on the iron(II)hydridotris(1-pyrazolyl)borate (heavy line) (24).

Fig. 9, the heavy curve refers to experimental μ_{eff}-values of the compound, whereas the light curves are the results of a calculation using the specific parameter values appropriate to $\{HB[3,5-(CH_3)_2pz]_3\}_2Fe$ at room temperature and the Dq-values marked on the curves. From the points of intersection, the values of Dq at a number of temperatures may then be extracted and these were plotted in Fig. 10 *versus* temperature. Indeed, the expected temperature variation of Dq is clearly evident. The results are not quantitative, however, owing to the neglect of an axial field distortion of $\delta \sim -1000$ cm^{-1} *(16)*.

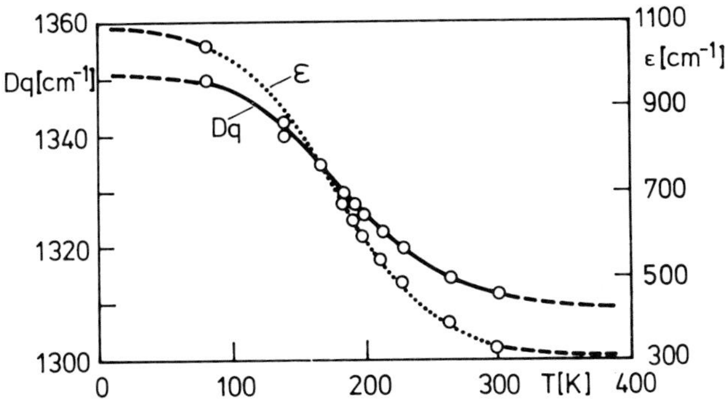

Fig. 10: Octahedral ligand field splitting parameter Dq and separation of centers of gravity of $^5T_{2g}$ and $^1A_{1g}$ terms ε for the iron(II)hydridotris(1-pyrazolyl)borate. Broken lines are extrapolated (24).

B. IRON(II) PHTHALOCYANINE, Fe(pc)

This is one of the few compounds containing an iron atom in an almost square planar environment. In addition, the value μ_{eff} = 3.89 BM at 296 K *(3)* suggests a spin triplet (S = 1) ground state. From the various triplet ground states which may become stabilized within the $3d^6$ configuration *(27)*, the assumption of a $^3B_2$ ground state seems plausible. The corresponding electron configuration is shown in Fig. 11. A complete calculation being somewhat complicated, an empirical model has been suggested by both Barraclough *et al.* *(3)* and Dale *et al.* *(4)* to account for the observed magnetism within the temperature range between 296 K and 1.57 K. If a zero-field splitting of the $^3B_2(\xi^2\eta^2\zeta\theta)$ ground state is assumed into the levels $|M_S = 0\rangle$ and $|M_S = \pm 1\rangle$ with separation D, the Zeeman energy in z direction is obtained by application of the operator

$$\mathcal{H}_m^z = g_z \beta S_z H_z \qquad [85]$$

to the levels in question. Inserting the resulting energy terms into the Van Vleck equation, eq. [61], produces the magnetic susceptibility in z direction,

$$\chi_{\parallel} = 2 g_{\parallel}^2 \frac{N_L \beta^2}{kT} \frac{\exp(-x)}{1 + 2\exp(-x)} \qquad [86]$$

where x = D/kT. On the other hand, if H is in x direction, the operator

$$\mathscr{H}_m^x = \frac{1}{2} g_x \beta H_x (S_+ + S_-) \qquad [87]$$

has to be applied to the levels of Fig. 11. Therefrom the ma-

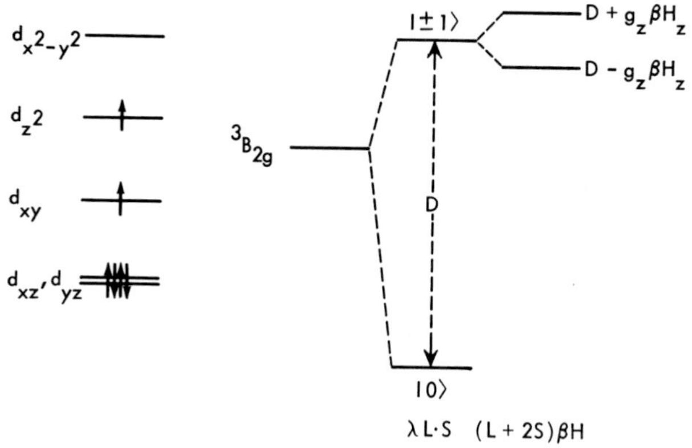

Fig. 11: Single-electron orbitals and schematic splitting of $^3B_2$ state by spin-orbit coupling and external magnetic field.

trix

$$\begin{pmatrix} D & 0 & g_x \beta H_x/\sqrt{2} \\ 0 & D & g_x \beta H_x/\sqrt{2} \\ g_x\beta H_x/\sqrt{2} & g_x\beta H_x/\sqrt{2} & 0 \end{pmatrix} \qquad [88]$$

results which has the eigenvalues

$$\begin{aligned} E_1 &= \frac{1}{2} [D - (D^2 + 4g_x^2 \beta^2 H_x^2)^{\frac{1}{2}}] \sim -(g_x \beta H_x)^2/D \\ E_2 &= D \qquad [89] \\ E_3 &= \frac{1}{2} [D + (D^2 + 4g_x^2 \beta^2 H_x^2)^{\frac{1}{2}}] \sim D + (g_x\beta H_x)^2/D \end{aligned}$$

where it has been assumed that $2g_x\beta H_x << D$. Application of eq. [61] as above gives

$$\chi_\perp = 2 g_\perp^2 \frac{N_L\beta^2}{D} \frac{1 - \exp(-x)}{1 + 2\exp(-x)} \qquad [90]$$

To obtain the average susceptibility, eq. [86] and eq. [90] are appropriately averaged assuming, in addition, $g_\parallel = g_\perp = g$. The result is

$$\chi_{av} = 2 g^2 \frac{N_L\beta^2}{3kT} \frac{x - 2 + 2\exp x}{x[2 + \exp x]} \qquad [91]$$

Fig. 12 shows experimental values of μ_{eff} for Fe(pc) together with a curve obtained by fitting eq. [91] to these data. The parameter values employed are D = 64 cm^{-1} and g = 2.74. The agreement between theory and experimental data is fairly good thus giving weight to the derived values of the zero-field splitting D and the spectroscopic splitting factor g.

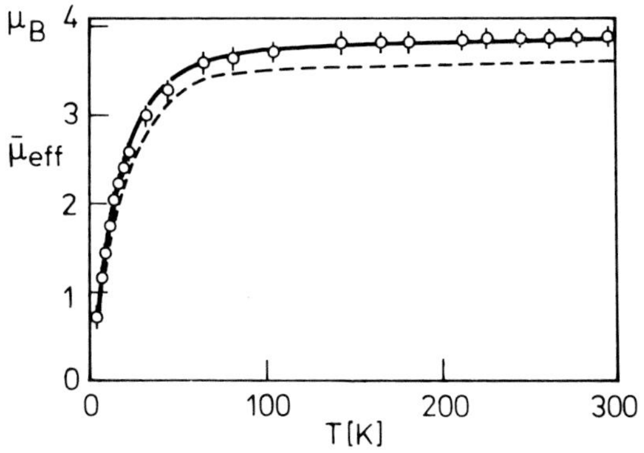

Fig. 12: Temperature dependence of μ_{eff} for Fe(pc). Experimental points marked with circles, solid line calculated using D = 64 cm^{-1} and g = 2.74. Broken line refers to a theoretical fit proposed by Dale et al. (4). Reproduced by permission from reference (3).

C. FERRICENIUM CATION AND IRON(III) DICARBOLLIDE COMPOUNDS

In the case of ferricenium compounds and their analogues, the extensive investigations on metal sandwich compounds by MO theory may be exploited. With particular reference to ferrocene, these studies suggest the MO configuration, in D_{5d} symmetry, $(a'_{1g})^2(e^{\pm}_{2g})^4$. Consequently, the configuration of ferricenium should be either $(a'_{1g})^2(e^{\pm}_{2g})^3$ or $(e^{\pm}_{2g})^4(a'_{1g})^1$ giving rise to the total states $^2E_{2g}$ and $^2A'_{1g}$, respectively. The five-electron problem may then be replaced by the equivalent problem of a single conjugate hole, *i.e.* $(e^{\pm}_{2g})$ or (a'_{1g}).

We start by assuming that $(e^{\pm}_{2g})$ is lowest in energy and consider the effect of spin-orbit coupling, $\mathcal{H}_{SO} = \xi(r)\ \mathbf{l}\cdot\mathbf{s}$. The assumption of a lowest (a'_{1g}) state may be ruled out as demonstrated by Maki and Berry *(31)* whose treatment we essentially follow. The spin states $\{e^{+}_{2g}\alpha,\ e^{-}_{2g}\beta\}$ remain degenerate and span the irreducible representation E'' of the double group D_5'. The states $\{e_{2g}^{+}\beta, e_{2g}^{-}\alpha\}$ and $\{a_{1g}'\alpha, a_{1g}'\beta\}$ transform according to A', A'', and E', respectively, and likewise form degenerate (Kramers) doublets. Putting

$$\langle e_{2g}^{+}|\xi(r)|e_{2g}^{+}\rangle = \zeta \sim -\kappa^2\zeta_0 \qquad [92]$$

the energies of the three lowest doublets result as

$$\begin{aligned} E(E'') &= +\zeta \\ E(A',A'') &= -\zeta \\ E(E') &= \Delta \end{aligned} \qquad [93]$$

where Δ is the excitation energy to the hole configuration (a_{1g}'). In eq. [92], ζ_0 is the one-electron spin-orbit coupling constant of the iron atom, and, since ζ is negative, $\psi_+(E'')$ is the ground doublet. However, with $\zeta_0 \sim 400\ \text{cm}^{-1}$, there may be significant mixing between $\psi(A',A'')$ and $\psi(E')$ by low-symmetry ligand fields. By solving the appropriate perturbation Hamiltonian, the lowest pair of Kramers' doublets follows as

$$\psi_{\pm}^{(a)} = \begin{cases} N(e_{2g}^{+} + \lambda e_{2g}^{-})\,\alpha \\ N(e_{2g}^{-} + \lambda e_{2g}^{+})\,\beta \end{cases}$$

$$\psi_{\pm}^{(b)} = \begin{cases} N(e_{2g}^{-} - \lambda e_{2g}^{-})\,\alpha \\ N(e_{2g}^{+} - \lambda e_{2g}^{-})\,\beta \end{cases} \qquad [94]$$

with corresponding energies

$$\begin{aligned} E^{(a)} &= -(\zeta^2 + \delta^2)^{\frac{1}{2}} \\ E^{(b)} &= +(\zeta^2 + \delta^2)^{\frac{1}{2}} \end{aligned} \qquad [95]$$

In eqs. [94] and [95], it is

$$\begin{aligned} \delta &= \langle e_{2g}^{+}|\mathscr{H}'_{\text{eff}}|e_{2g}^{-}\rangle \\ N &= (1+\lambda)^{-\frac{1}{2}} \\ \lambda &= x/[1 + (1 + x^2)^{\frac{1}{2}}] \end{aligned} \qquad [96]$$

where $x = \delta/\xi$ and $\mathscr{H}'_{\text{eff}}$ the effective perturbation Hamiltonian.

The wavefunctions eq. [94] may now be employed to calculate the paramagnetic susceptibility *(13)*. As outlined in section IV, matrix elements of the Zeeman operators

$$\beta(\kappa L_z + 2S_z)H_z \quad \text{and} \quad \beta(\kappa L_x + 2S_x)H_x \qquad [97]$$

are derived and the eigenvalues of the resulting secular equations are used to calculate $\chi_{\parallel}$ and $\chi_{\perp}$ separately. This procedure employs the Van Vleck equation, eq. [61]. The lower doublet $\psi_{\pm}^{(a)}$ thus yields

$$\chi_{\parallel}(\psi_{\pm}^{(a)}) = \frac{N_L\beta^2}{kT}\left[\left(1 + \frac{2\kappa(1-\lambda^2)}{(1+\lambda^2)}\right)^2 + \frac{16\lambda^2\kappa^2(kT)}{(\zeta^2+\delta^2)^{\frac{1}{2}}(1+\lambda^2)^2}\right]$$

$$\chi_{\perp}(\psi_{\pm}^{(a)}) = \frac{N_L\beta^2}{kT}\left[\frac{4\lambda^2}{(1+\lambda^2)^2} + \frac{kT(1-\lambda^2)^2}{(1+\lambda^2)^2(\zeta^2+\delta^2)^{\frac{1}{2}}}\right] \qquad [98]$$

where the quantities are the same as above and κ is the orbital reduction factor. In addition, it may be shown *(13)* that incorporation of the upper doublet $\psi_{\pm}^{(b)}$ produces only a minor effect on the susceptibility. The average susceptibility then simply follows as

$$\chi_{av} = \frac{1}{3}(\chi_{\parallel} + 2\chi_{\perp}) \qquad [99]$$

and μ_{eff} obtains from eq. [14].

As a correction to these results, thermal population of the $^2A'_{1g}$ $(e^{\pm}_{2g})^4(a'_{1g})^1$ state turns out to be important. The susceptibility expressions then have to be modified according to

$$\chi_{\alpha} = \frac{\chi_{\alpha}[\psi_{\pm}^{(a)}(^2E_{2g})] + e^{-\Delta E/kT}\chi_{\alpha}(^2A'_{1g})}{1 - e^{-\Delta E/kT}} \qquad [100]$$

where χ_{α} is either $\chi_{\parallel}$ or $\chi_{\perp}$ from eq. [98],

$$\Delta E = \Delta E[^2A'_{1g} - \psi_{\pm}^{(a)}(^2E_{2g})] \qquad [101]$$

and

$$\chi_{\parallel}(^2A'_{1g}) = \chi_{\perp}(^2A'_{1g}) = \frac{N_L\beta^2}{kT} \qquad [102]$$

Experimental data *(13)* are plotted in Fig. 13 as μ_{eff} *vs* T for three ferricenium salts. The solid curves were calculated on the basis of eq. [100] using values of δ and ΔE as listed in the caption. Fig. 14 shows similar results for two iron(III) dicarbollide compounds where dicarbollide denotes the anion $(B_9C_2H_{11})^{2-}$ which is formally related to the highly symmetric icosahedral $(B_{12}H_{12})^{2-}$ ion. The ion $(B_9C_2H_{11})^{2-}$ offers a planar face of five boron atoms which is used in the sandwich mode of bonding to the iron atom. Thus the experimental data may be approximately fitted by calculated curves with values of δ of 200 - 330 cm^{-1} and ΔE between 380 and 520 cm^{-1}. These results demonstrate the appreciable effect of both lower than D_5 symmetry ligand field distortion (δ)

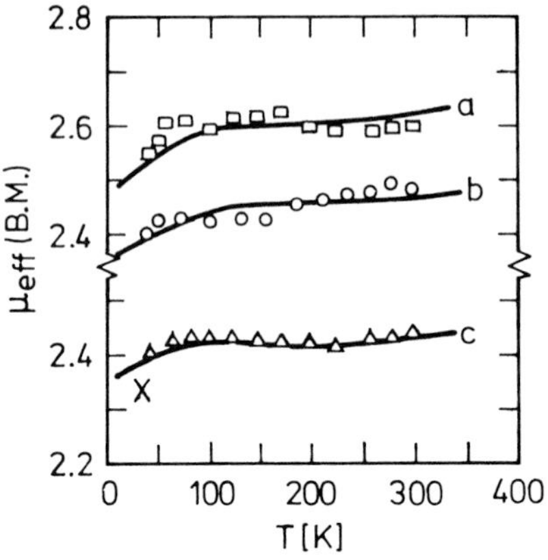

Fig. 13: Temperature dependence of μ_{eff} for three ferricenium salts: □, *$[Fe(C_5H_5)_2]$picrate;* O, *$[Fe(n\text{-}C_4H_9C_5H_4)(C_5H_5)]$picrate;* Δ, *$[Fe(C_5H_5)_2]BF_4$.*
The solid lines are calculated: a, δ = 200 cm^{-1}, ΔE = 460 cm^{-1}; b, δ = 330 cm^{-1}, ΔE = 400 cm^{-1}; c, δ = 330 cm^{-1}, ΔE = 380 cm^{-1}. Reproduced by permission from reference (13).

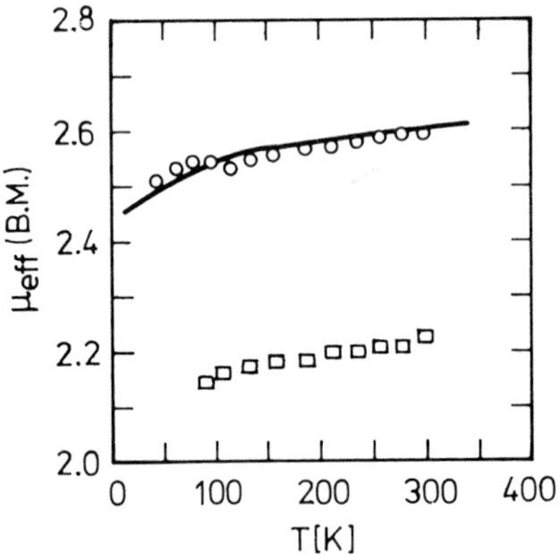

Fig. 14: Temperature dependence of μ_{eff} for two iron(III) dicarbollide compounds: O, *$[(CH_3)_4N][Fe(DCB)_2]$;* □, *Fe(cp)(DCB), where DCB = $(1,2\text{-}B_9C_2H_{11})^{2-}$. The solid line is calculated for δ = 240 cm^{-1} and ΔE = 520 cm^{-1}. Reproduced by permission from reference (13).*

and thermal population of the ${}^2A'_{1g}(e^{\pm}_{2g})^4(a'_{1g})^1$ state in ferricenium and iron(III) dicarbollide compounds.

VI. CONCLUSIONS

It is obvious that the most simple approach based on an octahedral field of ligands, *viz.* eqs. [72], [77], and [84], does not provide any additional information except to confirm the assumed symmetry if the calculational results agree with experiment. In reality, however, a strictly octahedral field is rarely encountered. If the ligand field is of lower than octahedral symmetry, the fit of magnetic data to the theoretical curves may provide both low symmetry field splittings and the covalency parameter, *viz.*, *e.g.*, δ and κ in Fig. 6. It is likewise evident from Fig. 6 that measurements have to be extended to as low temperatures as possible to provide a unique fit of the data. Another relevant case presented above is that of a low symmetry field in high-spin iron(III). Thus eq. [83] provides a value of the zero-field splitting parameter D which value may be checked by the results of electron paramagnetic resonance, *cf* the EPR chapter. Another approach to the same problem is exemplified in section V.C.. If complete calculations are available, *viz.* Fig. 8, the well-known spectroscopic parameters B, C, and 10 Dq in an O_h field, *e.g.*, may be determined or, preferably, a combined fit of optical spectra and magnetism may be performed.

REFERENCES

1. Adams, R.W., Barraclough, C.G., Martin, R.L., and Winter, G., *Inorg. Chem.*, *5*, 346 (1966).
2. Ballhausen, C.J., *Introduction to Ligand Field Theory*, McGraw-Hill, New York, 1962.
3. Barraclough, C.G., Martin, R.L., Mitra, S., and Sherwood, R.C., *J. Chem. Phys.*, *53*, 1643 (1970).
4. Dale, B.W., Williams, R.J.P., Johnson, C.E., and Thorp, T.L., *J. Chem. Phys.*, *49*, 3441 (1968).
5. Earnshaw, A., Figgis, B.N., and Lewis, J., *J. Chem. Soc. A*, *1966*, 1656.
6. Earnshaw, A., and Lewis, J., *J. Chem. Soc*, *1961*, 396.
7. Elliott, N., *J. Chem. Phys.*, *35*, 1273 (1961).
8. Figgis, B.N., and Lewis, J., *Progr. Inorg. Chem.*, *6*, 37 (1964).
9. Figgis, B.N., and Lewis, J. in H.B. Jonassen and A. Weissberger (Eds.), *Technique of Inorganic Chemistry*, *Vol. IV*, Interscience, New York, 1965, p. 137.
10. Foex, G., *Constantes Sélectionnées:Diamagnetisme et Paramagnetisme*, Masson et Cie., Paris, 1957.
11. Gerloch, M., and Miller, J.R., *Progr. Inorg. Chem.*, *10*, 1 (1968).
12. Griffith, J.S., *The Theory of Transition Metal Ions*, Cambridge University Press, 1961.
13. Hendrickson, D.N., Sohn, Y.S., and Gray, H.B., *Inorg. Chem.*, *10*, 1559 (1971).
14. Horrocks, W.DeW. Jr., and Hall, D.DeW., *Coord. Chem. Rev.*, *6*, 147 (1971).
15. Jesson, J.P., and Weiher, J.F., *J. Chem. Phys.*, *46*, 1995 (1967).
16. Jesson, J.P., Weiher, J.F., and Trofimenko, S., *J. Chem. Phys.*, *48*, 2058 (1968).
17. Jørgensen, C.K., *Absorption Spectra and Chemical Bonding in Complexes*, Pergamon Press, London, 1962.
18. Judd, B.R., *Operator Techniques in Atomic Spectroscopy*, McGraw-Hill, New York, 1963.
19. König, E., *Magnetic Properties of Coordination and Organo-metallic Transition Metal Compounds*, Landolt-Börnstein, New Series, Vol.II/2, Springer, Berlin 1966.
20. König, E. and König, G., *Magnetic Properties of Coordination and Organo-metallic Transition Metal Compounds*, Supplement 1 (1964-1968), Landolt-Börnstein, New Series Vol. II/8, Springer, Berlin 1976.
21. König, E., and Chakravarty, A.S., *Theor. Chim. Acta*, *9*, 151 (1967).
22. König, E., Chakravarty, A.S., and Madeja, K., *Theor.*

Chim. Acta, 9, 171 (1967).

23. König, E., and Kremer, S., *Theor. Chim. Acta, 20,* 143 (1971).
24. König, E., and Kremer, S., *Theor. Chim. Acta, 22,* 45 (1971).
25. König, E., and Kremer, S., *Theor. Chim. Acta, 23,* 12 (1971).
26. König, E., and Kremer, S., *Ber. Bunsenges. Phys. Chem., 76,* 870 (1972).
26a. König, E., and Kremer, S., *Ligand Field Energy Diagrams,* Plenum Press, New York 1976.
27. König, E., and Schnakig, R., *Theor. Chim. Acta, 30,* 205 (1973); *Inorg. Chim. Acta, 7,* 383 (1973).
28. Kotani, M., *J. Phys. Soc. Japan, 4,* 293 (1949).
29. Kotani, M., *Progr. Theor. Phys. Suppl., 17,* 4 (1961).
30. Lonsdale, K., and Krishnan, K.S., *Proc. Roy. Soc. (London) Ser. A, 156,* 597 (1936).
31. Maki, A.H., and Berry, T.E., *J. Amer. Chem. Soc., 87,* 4437 (1965).
32. Martin, R.L., in E.A.V. Ebsworth, A.G. Maddock, and A.G. Sharpe (Eds.), *New Pathways in Inorganic Chemistry,* Cambridge University Press, 1968, p. 175.
33. McMillan, J.A., *Electron Paramagnetism,* Reinhold, New York, 1968.
34. Mulay, L.N., and Dehn, J.T. in M. Tsutsui (Ed.), *Characterization of Organometallic Compounds, Vol. II,* Wiley, London, 1971, p. 439.
35. Offenhartz, P.O'D., *J. Amer. Chem. Soc., 91,* 5699 (1969).
36. Racah, G., *Phys. Rev., 63,* 367 (1943).
37. Rotenberg, M., Bivins, R., Metropolis, N., and Wooten, J.K., *The 3-j and 6-j Symbols,* MIT Press, Cambridge (Mass.), 1960.
38. Sinn, E., *Coord. Chem. Rev., 5,* 313 (1970).
39. Slater, J.C., *Quantum Theory of Atomic Structure,* McGraw-Hill, New York, 1960.
40. Soules, T.F., Richardson, J.W., and Vaight, D.M., *Phys. Rev., B3,* 2186 (1971).
41. Trofimenko, S., *Accounts Chem. Res., 4,* 17 (1971).
42. Van Vleck, J.H., *The Theory of Electric and Magnetic Susceptibilities,* Oxford University Press, 1932.

ELECTRON PARAMAGNETIC RESONANCE

By EDGAR KÖNIG

Institut für Physikalische Chemie II,
Universität Erlangen-Nürnberg, 8520 Erlangen, Germany

TABLE OF CONTENTS

ISBN 0-12-417101-X (v.1)

I. INTRODUCTION

Magnetic resonance techniques are between the most powerful methods which are available to the detailed study of the electronic structure of molecules. Electron paramagnetic resonance (EPR) which is the subject of this chapter, is somewhat limited in its applicability, since only systems containing unpaired electron spins will show a resonance signal. However, it was essentially on the basis of this method that our present knowledge about the distribution of the unpaired electrons in transition metal compounds and various other molecules was achieved. It is without any question that the full potentialities of the EPR should be exploited whenever possible.

The analysis of the EPR spectrum of an isolated (sufficiently diluted) transition metal ion provides three different types of parameters. The spectroscopic splitting parameter g determines the field strength required to produce transitions between the individual sublevels originating in the effect of the magnetic field. The hyperfine coupling parameter A is a measure of the electron spin - nuclear spin interaction which gives rise to the hyperfine structure of the spectrum. Furtheron, ions containing more than one unpaired electron will show zero-field splitting (fine structure) described by parameters D and E. All of these quantities will be affected by the presence of ligands around the metal ion. It may be shown that this observation is indirect evidence for the breakdown of the simple crystal field theory of transition metal complexes. Contrary to the treatment of magnetic susceptibility in the preceding chapter, it is not possible to account for these effects by a different parametrization. Rather the so-called super hyperfine splitting may be observed which is due to the direct interaction with ligand nuclei. This SHF splitting provides a measure of the probability density of the unpaired electron at the ligand nuclei. Under these circumstances, the orbitals centered on the metal ion have to be replaced by molecular orbitals explicitly containing ligand atomic orbitals. It is then possible to derive from the original parameters g and A a set of secondary parameters (α, β, etc.) which are LCAO coefficients in the underlying MO of the unpaired electron. It is these derived parameters which provide a detailed "mapping" of the unpaired electron density within the molecule and which are therefore of utmost interest to the chemist.

The complementary nature of the studies of magnetic susceptibility and EPR has been pointed out in the introduction to the preceding chapter on magnetic properties. Another complementary relationship exists between EPR and nuclear magnetic resonance (NMR). In both methods, interactions between electron and nuclear spins are considered. However, in EPR, attention is focused on the changes of the metal ion properties introduced by the ligands, whereas, in NMR, the focus is centered on the modification of ligand electronic structure by the metal ion. Thus, similar to the preceding chapter, our emphasis when dealing with the EPR of organic compounds of iron will be on *iron* rather than on the organic constituent. There are situations, however, where the unpaired electron is mainly concentrated on the ligand. Compounds of this sort should be considered as metal salts of free radicals, the metal usually causing only a minor modification to the unpaired electron density. The EPR in this situation will be of no concern to us and reference to the relevant literature is made here *(24,26)*.

Similar to the preceding chapter, we will again concentrate here on the underlying theory in the various situations encountered with organic compounds of iron. There are various general texts on EPR *(3,4,9)* as well as those dealing with transition metal compounds in particular *(1,27,30,37)* which may be consulted for details not covered in this chapter. Experimental arrangements of EPR spectrometers, associated equipment, and the required techniques are dealt with in two recent volumes *(3,40)*. A compilation of relevant experimental data covering the period until the end of 1965 *(28)* as well as a supplement for the years 1965 to 1968 *(29)* are likewise available and additional supplemental volumes are being prepared for publication.

II. RESONANCE CONDITION AND THE BASIC PRINCIPLES OF EPR

If an atomic or molecular system containing unpaired electron spins is placed in a magnetic field of strength $\mathbf{H}$, the essential part of the interaction energy may be described by the Hamiltonian (*cf* eq. [62] of the preceding chapter (p. 234)

$$\mathscr{H}_m = \beta \mathbf{H}(\mathbf{L} + g\mathbf{S}) \qquad [1]$$

where g is the spectroscopic splitting parameter and $\mathbf{L}$ and $\mathbf{S}$ are the orbital and spin angular momentum, respectively. In addition,

$$\beta = \frac{eh}{4\pi mc} = 0.92731 \times 10^{-20} \text{ erg Gauss}^{-1} \qquad [2]$$

is the Bohr magneton. The resulting magnetic sub-levels may be characterized by the quantum number M of the z component of the total angular momentum

$$\mathbf{J} = \mathbf{L} + \mathbf{S} \qquad [3]$$

and, in spherical and cubic symmetries, the z direction may be taken as the direction of the external magnetic field. The most simple case, of course, arises in an S = 1/2 system. In general, however, the effect of the Hamiltonian $\mathscr{H}_m$ of eq. [1] may be complicated by various additional interactions, the fine structure and hyperfine structure effects being the most important ones.

If an oscillating electromagnetic field oriented perpendicular to **H** is applied, transitions between the resulting levels are introduced. In a spin-only system (**L** = 0), $\mathscr{H} = g\beta HS_z$, and the transitions occur at a frequency ν determined according to

$$\Delta E = h\nu = g\beta H \qquad [4]$$

(resonance condition). For a completely free electron, g = 2.00232 and somewhat different otherwise. Similar to NMR, measurements of EPR are commonly carried out at fixed frequency. Most often employed is the frequency $\nu \simeq 9500$ MHz (X band), the resonance field H required for a free electron being about 3400 Gauss. Another useful frequency is $\nu \simeq 35{,}000$ MHz (Q band) where $H \simeq 12{,}500$ Gauss. For illustration purposes we will consider below three cases of spin systems with different type of interaction.

(a) The isolated spin S = 1/2: Fig. 1 shows the result-

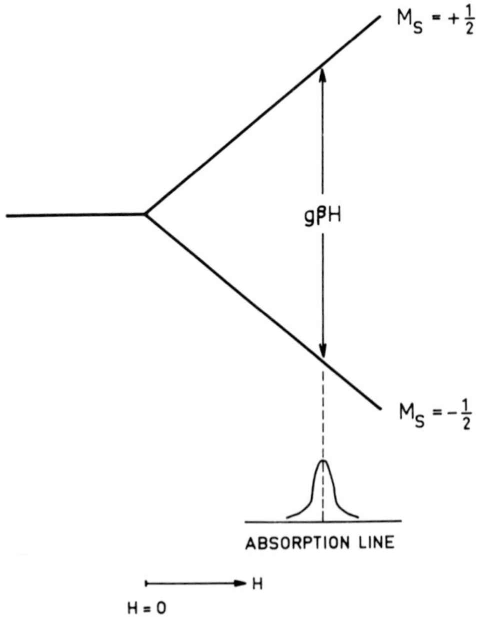

Fig. 1: Energy level diagram of an isolated electron spin S = 1/2 in an external magnetic field H.

ing energy levels as function of the external magnetic field H. The energies of the levels are $\pm 1/2\ g\beta H$. On application of the spectroscopic selection rule for magnetic dipole transitions, *viz.* $\Delta M_s = \pm 1$, the condition eq. [4] is obtained. Therefore, a single EPR line without any structure is observed.

(b) Spin S = 1 and zero field splitting (fine structure): A system with a total spin of S = 1 contains two electrons with their spins aligned parallel and $M_s = +1, 0, -1$. In transition metal ions, the original degeneracy of the spin levels may be very often lifted by the action of a lower symmetry field and/or spin-orbit coupling (at H = 0). This so-called zero field splitting may be accounted for using the parameters D and E. Fig. 2 shows a typical energy level diagram as function of H. Two types of transition are possible,

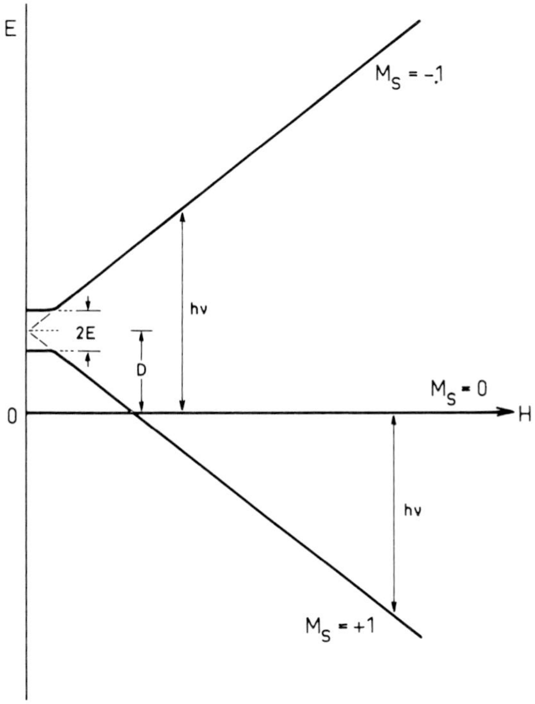

Fig. 2: Energy level diagram of a spin S = 1 with zero-field splitting described by the parameters D and E. Allowed transitions correspond to $\Delta M_s = \pm 1$.

viz. $\Delta M_s = \pm 1$ and $\Delta M_s = \pm 2$. If a fixed frequency is applied, the $\Delta M_s = \pm 2$ transitions will occur at about half the average field $\bar{H}$ for $\Delta M_s = \pm 1$ transitions. The polarization of the radiation field $H_1 \perp H$ is required for $\Delta M_S = \pm 1$ as in (a), whereas $H_1 \parallel H$ for $\Delta M_S = \pm 2$.

(c) Spin S = 1/2 and nuclear spin I = 1/2 (hyperfine structure): Very often the environment of an unpaired electron contains nuclei with non-zero nuclear magnetic moment and nuclear spin I. These nuclei may involve the metal ion originally providing the unpaired electron and/or any other suitable nuclei with non-negligible probability density of the electron. Fig. 3 demonstrates the disposition of energy levels as

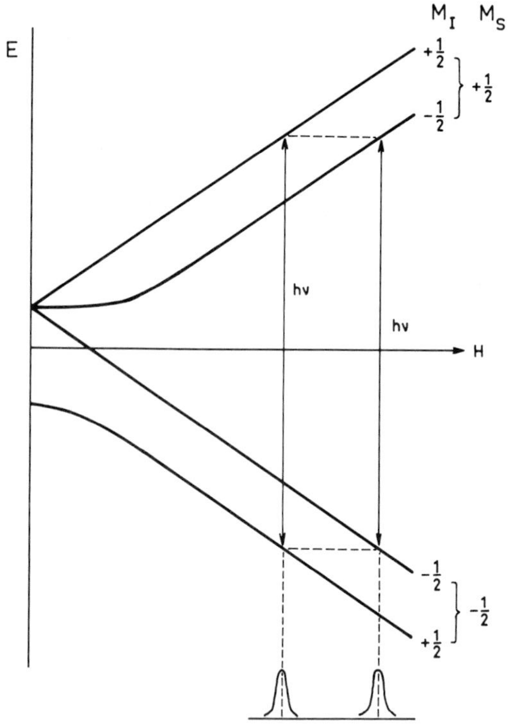

Fig. 3: Energy level diagram of a system with S = 1/2 and I = 1/2. Allowed transitions correspond to $\Delta M_S = \pm 1$, $\Delta M_I = 0$.

function of H for the interaction with a nuclear spin I = 1/2. Each level specified by M_S shows a 2I + 1 = 4-fold degeneracy with respect to the nuclear spin. The external magnetic field is, in general, much more intense than the field due to the nuclear spin. Consequently, the splitting is independent of H, the transition being allowed according to $\Delta M_S = \pm 1$, $\Delta M_I = 0$. The polarization of the radiation field is again $H_1 \perp H$ and the resonance condition eq. [4] has to be generalized to

$$h\nu = g\beta H + \sum_i A_i M_{Ii} \qquad [5]$$

Deviations from this simple behaviour occur if the applied field H is extremely weak (*cf* Fig. 3).

It is important to realize that EPR absorption can be detected only if there is a population difference, *e.g.* between the two spin levels of Fig. 1. Denoting the respective populations of the $M_S = +1/2$ and $M_S = -1/2$ levels as N^+ and N^- as shown in Fig. 4, a thermal equilibrium will be characterized by

$$N^-/N^+ = W_e/W_a = \exp(\varepsilon/kT) \qquad [6]$$

where W_e and W_a are the emission and absorption transition

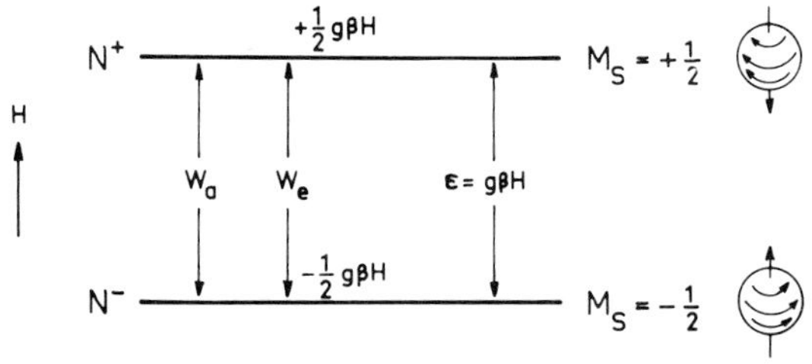

Fig. 4: Spin levels for a simple S = 1/2 system. The populations of the $M_S = +1/2$ and $M_S = -1/2$ levels are denoted N^+ and N^-, respectively. W_e and W_a are the emission and absorption probabilities and ε the energy difference between the levels.

probabilities and $\varepsilon = g\beta H$. The radiation field available in an usual EPR experiment is not sufficient to appreciably upset the thermal distribution of electrons. Consequently, there are two possibilities according to eq. [6] to enhance the N^-/N^+ ratio and thus to increase the signal intensity: (1) by minimizing the temperature T; (2) by maximizing the magnetic field strength H and thus by maximizing ε. If the radiation field is increased until $N^- = N^+$, there is no net energy absorption (saturation).

III. CONCEPT OF THE SPIN HAMILTONIAN

In the real physical situation of an unpaired electron, the simple postulates underlying Figs. 1, 2, and 3 are rarely applicable. Most often, various sorts of interaction of the electron with a number of nuclei and/or electrons have to be accounted for. Consequently, additional terms arise on the right hand side of eq. [4] or, stated in a more precise form, the energy of the levels involved in EPR results from the solution of a complicated eigenvalue problem. It is obvious that the general (theoretical) Hamiltonian $\mathcal{H}$ of the system will reflect this situation. We will present here a schematic representation of $\mathcal{H}$ and indicate the order of magnitude of the terms involved. According to Abragam and Pryce *(2)*

$$\mathcal{H} = \mathcal{H}_O + \mathcal{H}_{LF} + \mathcal{H}_{LS} + \mathcal{H}_{SS} + \mathcal{H}_N + \mathcal{H}_Q + \mathcal{H}_H + \mathcal{H}_h + \mathcal{H}_e \qquad [7]$$

where the individual terms have the significance shown in Table 1. For more details the reader should consult one of the

Table 1: Significance and magnitude of individual terms in the Hamiltonian eq. [7].

			Energy [cm^{-1}]	Energy [Gauss]
$\mathcal{H}_O$	=	Free ion and Coulomb energy	10^5	
$\mathcal{H}_{LF}$	=	Ligand field energy	10^4	
$\mathcal{H}_{LS}$	=	Spin-orbit coupling energy	10^2-10^3	
$\mathcal{H}_{SS}$	=	Spin-spin interaction energy	1	10^4
$\mathcal{H}_N$	=	Electron spin - nuclear spin interaction energy (hyperfine splitting energy)	10^{-1}-10^{-3}	10^3-10
$\mathcal{H}_Q$	=	Electron-nuclear quadrupole moment interaction energy	10^{-3}	10
$\mathcal{H}_H$	=	Electron interaction with external magnetic field (Zeeman energy)	1	10^4
$\mathcal{H}_h$	=	Nucleus interaction with external magnetic field (nuclear Zeeman energy)	10^{-4}	1
$\mathcal{H}_e$	=	Electron exchange interaction energy		

available reviews on the EPR of transition metal ions *(1,27, 30,37)*. In section IV below we will consider the theoretical treatment of the EPR energy with particular emphasis on the subject of this volume.

The listing of Table 1 shows that the various energy contributions cover nine decades varying between 10^5 and 10^{-4} cm^{-1}. It is obvious that only part of these interactions will be explicitly observed in an EPR spectrum. The most important contributions thus arise from the Zeeman splitting, the fine structure, and the hyperfine structure, *i.e.* from the terms $\mathcal{H}_H$, $\mathcal{H}_{SS}$, and $\mathcal{H}_N$ in eq. [7]. The larger terms usually require too high an energy to be excited by EPR, whereas the smaller terms may give an observable effect under favourable conditions. A schematic of the various splittings discussed is displayed in Fig. 5.

Due to the inherent complexity of the theoretical Hamil-

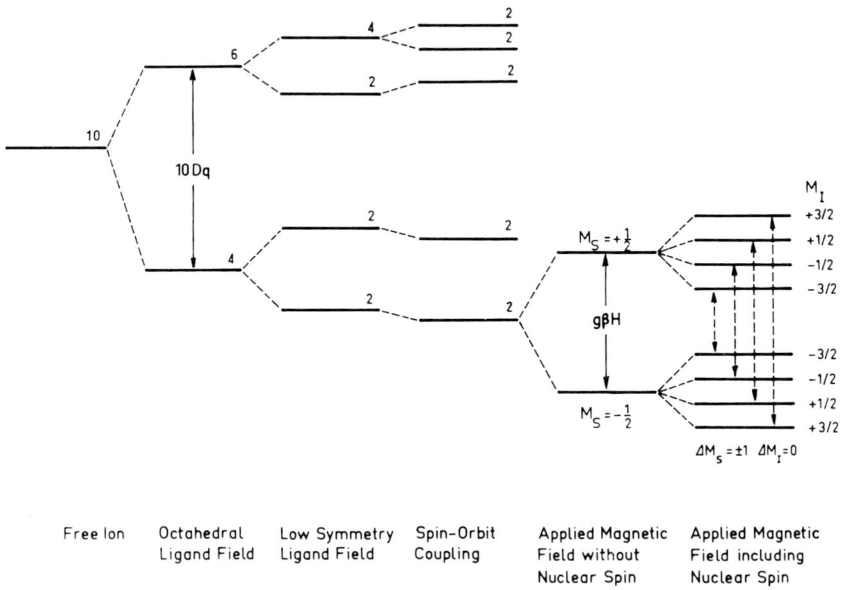

Fig. 5: Energy level diagram showing, for convenience, the various interactions and splittings of the ^2D term of a Cu^{2+} ion in ligand electric and external magnetic fields.

tonian another, more practical, approach has been widely accepted from the beginning of EPR spectroscopy. This so-called spin Hamiltonian contains, in a parametric form, those interactions directly evident from an EPR spectrum. The spin Hamiltonian may be considered thus as a shorthand description of the experimental results. Referred to a principal axes system, the usual spin Hamiltonian may be written

$$\begin{aligned}\mathcal{H} = {} & \beta(g_x S_x H_x + g_y S_y H_y + g_z S_z H_z) \\ & + D[S_z^2 - \frac{1}{3} S(S+1)] + E(S_x^2 - S_y^2) + A_x I_x S_x \\ & + A_y S_y I_y + A_z S_z I_z + Q[I_z^2 - \frac{1}{3} I(I+1)] \\ & - g_I \beta_I H \cdot I \end{aligned} \qquad [8]$$

Here, S is the effective electronic spin, I the nuclear spin, g_i and A_i the spectroscopic and hyperfine structure parameters along the respective axes, D and E the fine structure parameters, and Q the coefficient of the nuclear quadrupole interaction. By convention, if EPR transitions between 2S + 1 levels are observed, the effective spin is assigned the value S. This treatment obviously parallels that of a free ion where a state of quantum number J is split into 2J + 1 components by an external magnetic field. The inherent assumption is always that the lowest energy levels are well separated from all higher lying levels. If this is correct, the effective spin may equal the true spin, *e.g.* in the $[Fe(CN)_6]^{3-}$ ion where the lowest orbital state $^2T_2$ gives $S = S_{eff} = 1/2$. On the other hand, the lowest state in FeF_2 is $^5T_2$ characterized by S = 2. Lower symmetry fields and spin-orbit interaction lift the fifteen-fold degeneracy of the $^5T_2$ state. Consequently, EPR within the lowest level has been fitted to an $S_{eff} = 1/2$ spin Hamiltonian. In this as well as in similar situations, the g value is a function of the spin Hamiltonian employed and thus may deviate considerably from the Landé g factor. In actual fact, higher energy levels frequently affect the ground state significantly. In the spin Hamiltonian formalism, this interaction is commonly absorbed into the parameters of eq. [8] or similar expressions. It should be remembered that spin Hamiltonian parameters are therefore, in general, empirical quantities.

IV. LIGAND FIELD EFFECTS IN IRON(III)

We recall from the preceding chapter that the free Fe^{3+} ion with an electronic configuration described by [Ar] $3d^5$ gives rise to a ^6S ground state. This being the only term of spin multiplicity six, excited states are characterized as spin quartets or spin doublets. The splittings resulting from the effect of a ligand field have been discussed in section III of the preceding chapter. Accordingly, the free ion terms are, in general, split within a field of octahedral symmetry, whereas the ground state transforms into $^6A_1(t_2^3e^2)$. The energies of the resulting terms are again determined by the octahedral splitting parameter 10 Dq and the Racah parameters A, B, and C. Below we list energies of some states of the octahedral Fe^{3+} ion:

$$E[^6A_1(t_2^3e^2)] = 10\ A - 35\ B$$

$$E[^4T_1(t_2^4e)] = -10\ Dq + 10\ A - 25\ B + 6\ C \qquad [9]$$

$$E[^2T_2(t_2^5)] = -20\ Dq + 10\ A - 20\ B + 10\ C$$

In Fig. 6 these terms are plotted as function of Dq assuming the relation between the free ion Racah parameters C = 4B. If small values of 10 Dq are assumed, the ground state is $^6A_1$ (*viz.* high-spin iron(III)), whereas for large values of 10 Dq, $^2T_2$ is the lowest energy state (*viz.* low-spin iron(III)). The crossover between the terms $^6A_1$ and $^2T_2$ occurs subject to the

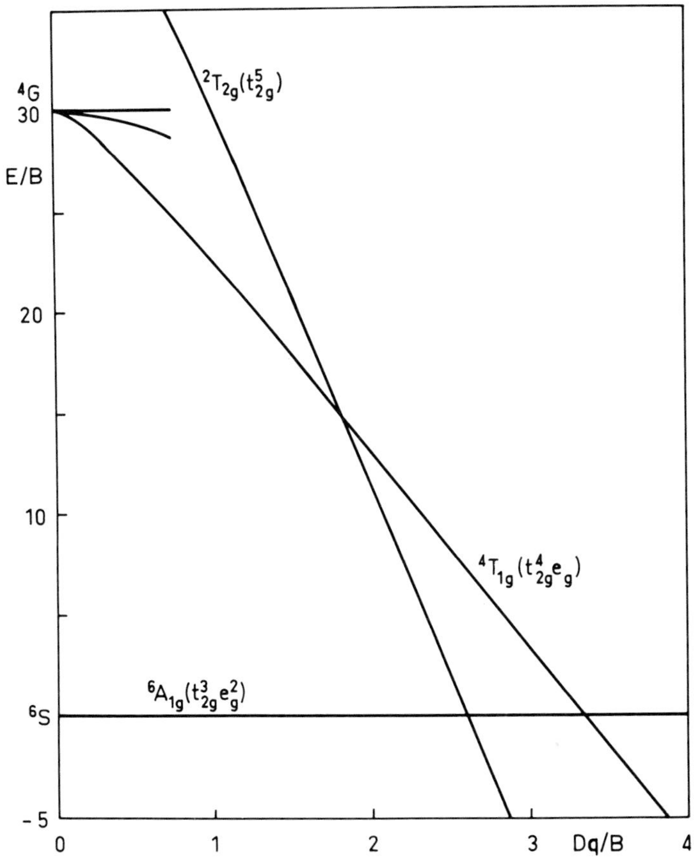

Fig. 6: Energies of the $^6A_1$, $^4T_1$, and $^2T_2$ terms as function of Dq, in units of B (assuming C = 4B).

condition

$$(10\ Dq)_{^2T_2-^6A_1} = 7\frac{1}{2}B + 5C = 27{,}900\ cm^{-1} \quad [10]$$

which follows directly from eq. [9]. The critical value of eq. [10], *cf* 10 Dq = Π, is usually referred to as spin pairing energy *(32)*. In addition, the terms $^4T_1$ and $^6A_1$ cross at

$$(10\ Dq)_{^4T_1-^6A_1} = 10B + 6C = 34{,}630\ cm^{-1} \quad [11]$$

However, similar to the $^3T_1$ term in the $3d^6$ configuration, the $^4T_1$ can never become ground state within regular octahedral symmetry. It is always $^6A_1$ or $^2T_2$ that is lower in energy than $^4T_1$ for any value of 10 Dq.

The wavefunctions corresponding to the above terms may again be written in the convenient ket notation as

$$|^{2S+1}\Gamma M\gamma>.$$

Since the $^6A_1$ term is orbitally non-degenerate but has a six-fold degeneracy with respect to spin, the appropriate kets are

$$|^6A_1\ \frac{5}{2}\ a_1> ,\ |^6A_1\ \frac{3}{2}\ a_1> ,\ldots,\ |^6A_1\ -\ \frac{5}{2}\ a_1> \qquad [12]$$

one function for each of the values M = 5/2, 3/2, 1/2, -1/2, -3/2, -5/2. Similarly, the $^4T_1$ term has a three-fold orbital degeneracy and a four-fold degeneracy with respect to spin. Here we have the choice to use the real components of the T_1 representation, *viz.*

$$|^4T_1\ \frac{3}{2}\ x > ,\ |^4T_1\ \frac{3}{2}\ y > ,\ |^4T_1\ \frac{3}{2}\ z > \qquad [13]$$

for M = 3/2 with similar kets if M = 1/2, -1/2, -3/2. Alternatively, the complex form of the T_1 components may be employed, *viz.*

$$|^4T_1\ \frac{3}{2}\ 1 > ,\ |^4T_1\ \frac{3}{2}\ 0 > ,\ |^4T_1\ \frac{3}{2}\ -1 > \qquad [14]$$

if again M = 3/2. The real and complex components of T_1 are simply related *(18)*.

If spin-orbit coupling is to be considered, due regard to transformation according to the double groups should be paid. Thus for the $^6A_1$ term, S = 5/2 and the corresponding spin part of the wavefunction may be decomposed, within the octahedral double group O* according to *(5,18)*

$$D^{(5/2)} \rightarrow E'' + U' \qquad [15]$$

Since the orbital part of the wavefunction is totally symmetric, *viz.* A_1, the result of eq. [15] is preserved even if the spin and orbital parts are coupled. Spin-orbit interaction thus produces a decomposition of the $^6A_1$ term into two levels with respective degeneracies of two and four,

$$^6A_1 \rightarrow E'' + U' \qquad [16]$$

For the corresponding basis functions within the O* group we will use the notation $|S\Gamma\Gamma^*\gamma^*>$ or the equivalent writing

$$|^{2S+1}\Gamma\Gamma^*\gamma^*>$$

where Γ^* is a representation in O* and γ^* a component of Γ^*. Following Griffith *(18)*, the two components of E" are denoted as α'' and β'', whereas the four components of U' are κ, λ, μ, ν.

For the $^4T_1$ term, it is S = 3/2 and $D^{(3/2)}$ = U'. Consequently, the product of the orbital and spin representation may be decomposed into

$$U' \times T_1 = E' + E'' + 2U' \quad [17]$$

thus producing two representations of each of the degeneracies of two and four. Finally, we consider the $^2T_2$ term where $S = 1/2$ and according to the O* group $D^{(1/2)} = E'$. The spin and orbital product representation then expands according to

$$E' \times T_2 = E'' + U' \quad [18]$$

which is equivalent to the transformation of the $^6A_1$ term, *cf* eq. [16]. The basis functions corresponding to the irreducible representations of the group O* obtained above may be set up from a table of coupling coefficients *(18)*.

The effect of a reduction of the symmetry from octahedral to tetragonal on the transformation property of the orbitals has been discussed in the preceding chapter. Terms are changed or decomposed into those labelled by the irreducible representations in D_{4h}, whereas the spin multiplicity is not affected. Application of these same arguments to the lowest terms of the $3d^5$ configuration produces

$$\begin{array}{lll} ^6A_1 & \rightarrow & ^6A_1 \\ ^4T_1 & \rightarrow & ^4A_2 + {}^4E \\ ^2T_2 & \rightarrow & ^2B_2 + {}^2E \end{array} \quad [19]$$

If spin-orbit coupling is introduced, the classification of states has to be based on the double group D_4*. There is no effect of a lowering of symmetry from cubic to tetragonal on the levels E' and E" while the U' level splits into an E' and an E" level. A correspondence between basis functions in O* and in D_4* may then be set up. If there are two sets of functions transforming according to the same representation in D_4* an additional label has to be introduced in order to distinguish between them. This label is often suggested by the p isomorphism and is called therefore a J value. The detailed results on the spin-orbit interaction will be discussed in relation to the expected EPR transitions in the following section.

V. EPR SPECTRA IN COMPOUNDS OF IRON

A. *HIGH-SPIN IRON(III)*

In the previous section, we have shown that the ^6S ground state of an Fe^{3+} ion is not split within an octahedral or even a lower symmetry ligand field. In addition, there can be no splitting by spin-orbit interaction or by the combined action of the ligand field and spin-orbit coupling. It is only by application of a magnetic field that the six-fold degeneracy is lifted to yield a set of levels with an equal separation of $2\beta H$. Due to the selection rule $\Delta M_s = \pm 1$, transitions are allowed between adjacent levels only resulting in a spectroscopic splitting factor $g = 2$. This result is independent of the orientation of the magnetic field and, therefore, an isotropic resonance line at $g = 2$ should be always observed, *cf* Fig. 7. Although there are examples of this type of

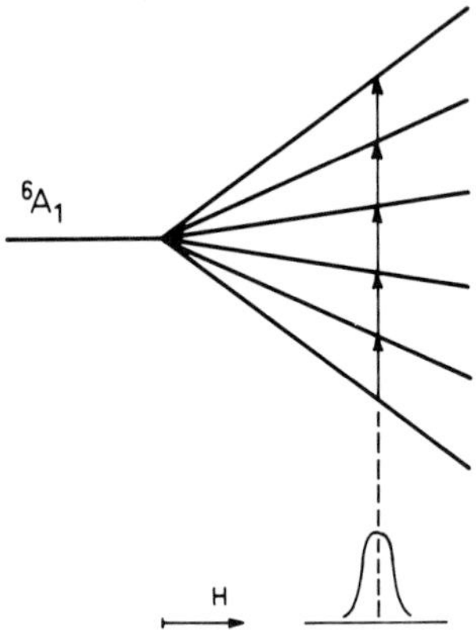

Fig. 7: Splitting of a $^6A_1$ ground state in an external magnetic field assuming perfect octahedral symmetry.

behaviour *(28,39)* it is the exception rather than the rule. To account for the frequently encountered large anisotropy of g values the interaction of the excited states with the ground state through spin-orbit coupling will therefore be investigated.

If we assume initially cubic (O_h) symmetry, the ground state $^6A_1(t_2^3e^2)$ possesses, according to standard selection rules, non-zero matrix elements of spin-orbit coupling with the $^4T_1(t_2^4e)$ state only. Matrix elements with all other excited states vanish. The evaluation of matrix elements of spin-orbit coupling may be accomplished employing methods similar to those used for free ions in the preceding chapter. For details the reader is referred to the treatment by Griffith *(19)*. There are six states within $^6A_1$ and twelve states within $^4T_1$ which will give rise to an overall matrix of the spin-orbit interaction of dimension 18. If the substates of $^6A_1$ and $^4T_1$ are classified according to representations of the group D_4^*, the matrix may be decomposed into four blocks arranged along the diagonal. These blocks consist of matrix elements belonging to one specific component of a particular representation and may be labelled accordingly $E'\alpha'$, $E'\beta'$, $E''\alpha''$, $E''\beta''$. The matrices $E'\alpha'$ and $E'\beta'$ are 4×4 and identical, the matrices $E''\alpha''$ and $E''\beta''$ are 5×5

Table 2: Matrix of spin-orbit interaction between the components of $^6A_1$ and $^4T_1$ terms belonging to E'α' within the group D_4* (in units of ζ). The matrix is symmetric to the main diagonal.

E'α'	$^6A_1\ \frac{1}{2}\ a_1$	$^4T_1\ \frac{1}{2}\ 0$	$^4T_1\ -\frac{1}{2}\ 1$	$^4T_1\ \frac{3}{2}\ -1$
$^6A_1\ \frac{1}{2}\ a_1$	0	$-\frac{\sqrt{6}}{\sqrt{5}}$	$-\frac{\sqrt{3}}{\sqrt{5}}$	$-\frac{1}{\sqrt{5}}$
$^4T_1\ \frac{1}{2}\ 0$		0	$-\frac{1}{3\sqrt{2}}$	$-\frac{1}{2\sqrt{6}}$
$^4T_1\ -\frac{1}{2}\ 1$			$\frac{1}{12}$	0
$^4T_1\ \frac{3}{2}\ -1$				$\frac{1}{4}$

Table 3: Matrix of spin-orbit interaction between the components of $^6A_1$ and $^4T_1$ terms belonging to E"α" within the group D_4* (in units of ζ). The matrix is symmetric to the main diagonal.

E"α"	$^6A_1\ \frac{5}{2}\ a_1$	$^6A_1\ -\frac{3}{2}\ a_1$	$^4T_1\ -\frac{3}{2}\ 0$	$^4T_1\ \frac{3}{2}\ 1$	$^4T_1\ -\frac{1}{2}\ -1$
$^6A_1\ \frac{5}{2}\ a_1$	0	0	0	$-\sqrt{2}$	0
$^6A_1\ -\frac{3}{2}\ a_1$		0	$-\frac{2}{\sqrt{5}}$	0	$-\frac{\sqrt{6}}{\sqrt{5}}$
$^4T_1\ -\frac{3}{2}\ 0$			0	0	$-\frac{1}{2\sqrt{6}}$
$^4T_1\ \frac{3}{2}\ 1$				$-\frac{1}{4}$	0
$^4T_1\ -\frac{1}{2}\ -1$					$-\frac{1}{12}$

and likewise identical. The matrix elements are listed in Table 2 and Table 3, respectively, where the state functions are written in the form

$$|^{2S+1}\Gamma M\gamma>$$

(43). To obtain matrix E'β' from E'α' the basis functions in the top line of Table 2 have to be replaced (from left to right) by $|^6A_1 -1/2\ a_1>$, $|^4T_1 -1/2\ 0>$, $|^4T_1\ 1/2\ -1>$, $|^4T_1\ -3/2\ 1>$. Similarly, to obtain the matrix E"β" from E"α" the basis functions in the top line of Table 3 should be replaced by $|^6A_1\ -5/2\ a_1>$, $|^6A_1\ 3/2\ a_1>$, $|^4T_1\ 3/2\ 0>$, $|^4T_1\ -3/2\ -1>$, $|^4T_1\ 1/2\ 1>$. In both Tables, the first column should be changed accordingly.

We have shown above that there is no first-order spin-orbit interaction energy within the $^6A_1$ term. The second-order contribution from the interaction between $^6A_1$ and $^4T_1$ may be extracted directly from Table 2 and Table 3. To this end we denote the zero-order energy difference between the terms in question as

$$\Delta E = E(^4T_1) - E(^6A_1) \qquad [20]$$

and obtain

$$\begin{aligned} E[^6A_1 \tfrac{1}{2} a_1] &= E[^6A_1 - \tfrac{1}{2} a_1] = \\ &= -\left[\frac{6\zeta^2}{5\Delta E} + \frac{3\zeta^2}{5\Delta E} + \frac{\zeta^2}{5\Delta E}\right] = -\frac{2\zeta^2}{\Delta E} \\ E[^6A_1 \tfrac{3}{2} a_1] &= E[^6A_1 - \tfrac{3}{2} a_1] = \\ &= -\left[\frac{4\zeta^2}{5\Delta E} + \frac{6\zeta^2}{5\Delta E}\right] = -\frac{2\zeta^2}{\Delta E} \\ E[^6A_1 \tfrac{5}{2} a_1] &= E[^6A_1 - \tfrac{5}{2} a_1] = -\frac{2\zeta^2}{\Delta E} \end{aligned} \qquad [21]$$

Thus each of the six components of the $^6A_1$ term is shifted by exactly the same amount, *viz.* $-2\zeta^2/\Delta E$, and the term remains six-fold degenerate. It can be shown that this result is valid to any order as long as the $^4T_1$ term is assumed to be 3-fold orbitally degenerate.

Next let us assume that the cubic symmetry at the site of the Fe^{3+} ion is reduced to tetragonal. We have shown above that, in this case, the $^4T_1$ term is split into $^4A_2$ and ^4E, whereas the $^6A_1$ is not affected. Let us follow the treatment due to Griffith *(18)* and let us assume that $^4A_2$ is lower in energy than ^4E, the energy separation between these states being sufficiently large. Then the essential contribution to the spin-orbit interaction energy will derive from that between $^4A_2$ and $^6A_1$ terms. This contribution may be obtained again from Table 2 and Table 3. It should be observed that $^4A_2$ in D_4 symmetry corresponds to ^4T_1z in O and

$$|^4T_1z\rangle = -i|^4T_1O\rangle \qquad [22]$$

Therefore, if we introduce

$$\Delta E' = E(^4T_1O) - E(^6A_1) \qquad [23]$$

it follows

$$\begin{aligned} E[^6A_1 \tfrac{1}{2} a_1] &= E[^6A_1 - \tfrac{1}{2} a_1] = -\frac{6\zeta^2}{5\Delta E'} \\ E[^6A_1 \tfrac{3}{2} a_1] &= E[^6A_1 - \tfrac{3}{2} a_1] = -\frac{4\zeta^2}{5\Delta E'} \\ E[^6A_1 \tfrac{5}{2} a_1] &= E[^6A_1 - \tfrac{5}{2} a_1] = 0 \end{aligned} \qquad [24]$$

In tetragonal symmetry then, the $^6A_1$ term is split into three doublets, the states having $+M_S$ and $-M_S$ remaining degenerate (Kramers doublets). If we introduce the quantity

$$D = \frac{\zeta^2}{5\Delta E'} \qquad [25]$$

the separations between the components of the $^6A_1$ term may be written

$$E[^6A_1 \pm \frac{3}{2} a_1] - E[^6A_1 \pm \frac{1}{2} a_1] = 2D$$
$$E[^6A_1 \pm \frac{5}{2} a_1] - E[^6A_1 \pm \frac{1}{2} a_1] = 6D \qquad [26]$$
$$E[^6A_1 \pm \frac{5}{2} a_1] - E[^6A_1 \pm \frac{3}{2} a_1] = 4D$$

The resulting energy levels are illustrated in Fig. 8.

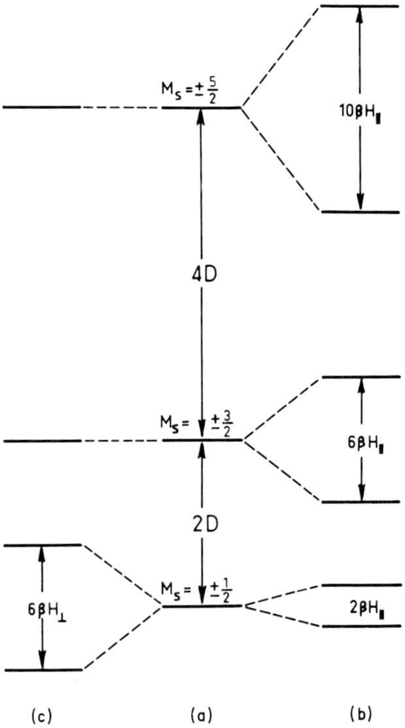

Fig. 8: Splitting of a $^6A_1$ ground state in (a) zero magnetic field, (b) magnetic field parallel to the 4-fold axis, (c) magnetic field perpendicular to the 4-fold axis, assuming tetragonal symmetry (E = O).

The results deduced above may be described in somewhat different terms, employing a spin Hamiltonian (*cf* section III)

$$\mathcal{H} = D[S_z^2 - \frac{1}{3} S (S + 1)] \qquad [27]$$

where, in the present case, since S = 5/2

$$\frac{1}{3} S (S + 1) = \frac{35}{12}$$

The energy results therefore as

$$E = D(M_S^2 - \frac{35}{12})$$

giving, in more detail,

$$\begin{aligned} E &= \frac{10}{3} D && \text{if } M_S = \pm \frac{5}{2} \\ E &= -\frac{2}{3} D && \text{if } M_S = \pm \frac{3}{2} \\ E &= -\frac{8}{3} D && \text{if } M_S = \pm \frac{1}{2} \end{aligned} \qquad [28]$$

Here, the energy zero is at the center of gravity of the three levels. If we choose instead the zero of energy at the lowest doublet, we obtain the same result as from the detailed calculation above

$$\begin{aligned} E\left(\pm \frac{5}{2}\right) &= 6D \\ E\left(\pm \frac{3}{2}\right) &= 2D \\ E\left(\pm \frac{1}{2}\right) &= 0 \end{aligned} \qquad [29]$$

The three eigenstates are frequently denoted as $|\pm 5/2>$, $|\pm 3/2>$, and $|\pm 1/2>$. If the ligand field contains a rhombic component, a more general spin Hamiltonian, *viz.*

$$\mathscr{H} = D[S_z^2 - \frac{1}{3} S (S + 1)] + E(S_x^2 - S_y^2) \qquad [30]$$

should be applied where $E \neq 0$. There will be again three doublets originating in the $^6A_1$ term, however, these doublets will be no pure eigenstates of S_z. If the ligand field is of strictly cubic symmetry, $D = E = 0$ and there is no zero field splitting.

In order to calculate g values the interaction with the external magnetic field needs to be calculated. Now, since $L = 0$, the corresponding Hamiltonian eq. [1] is simply $2\beta\,\mathbf{H}\cdot\mathbf{S}$ or

$$\begin{aligned} \mathscr{H}_m = 2\beta\,\mathbf{H}\cdot\mathbf{S} &= 2\beta[H_zS_z + H_xS_x + H_yS_y] \\ &= 2\beta[H_zS_z + \frac{1}{2}(H_+S_- + H_-S_+)] \end{aligned} \qquad [31]$$

The resulting matrix elements of $\mathscr{H}_m$ within the substates of the $^6A_1$ term are listed in Table 4. If the magnetic field is assumed in z direction, the appropriate matrix elements are on the diagonal and the corresponding total energies are

$$\begin{aligned} E\left(\pm \frac{1}{2}\right) &= 0 \pm \beta H_z \\ E\left(\pm \frac{3}{2}\right) &= 2D \pm 3\beta H_z \end{aligned} \qquad [32]$$

$$E\left(\pm \frac{5}{2}\right) = 6D \pm 5\beta H_z$$

Table 4: Matrix of the magnetic field interaction operator $2\beta \mathbf{H} \cdot \mathbf{S}$ within the substates of the $^6A_1$ term. Upper and lower entries for each value of M_S refer to the x and y components, respectively, single entries refer to the z component of H.

M_S	$\frac{1}{2}$	$-\frac{1}{2}$	$\frac{3}{2}$	$-\frac{3}{2}$	$\frac{5}{2}$	$-\frac{5}{2}$
$\frac{1}{2}$	βH_z	$3\beta H_x$ $-3i\beta H_y$	$2\sqrt{2}\beta H_x$ $2\sqrt{2}i\beta H_y$			
$-\frac{1}{2}$	$3\beta H_x$ $3i\beta H_y$	$-\beta H_z$		$2\sqrt{2}\beta H_x$ $-2\varphi i\varphi\sqrt{2}\beta H_y$		
$\frac{3}{2}$	$2\sqrt{2}\beta H_x$ $-2\varphi i\varphi\sqrt{2}\beta H_y$		$3\beta H_z$		$\sqrt{5}\beta H_x$ $\sqrt{5}i\beta H_y$	
$-\frac{3}{2}$		$2\sqrt{2}\beta H_x$ $2\sqrt{2}i\beta H_y$		$-3\beta H_z$		$\sqrt{5}\beta H_x$ $-\sqrt{5}i\beta H_y$
$\frac{5}{2}$			$\sqrt{5}\beta H_x$ $-\sqrt{5}i\beta H_y$		$5\beta H_z$	
$-\frac{5}{2}$				$\sqrt{5}\beta H_x$ $\sqrt{5}i\beta H_y$		$-5\beta H_z$

Thus the degeneracy of the $^6A_1$ term is completely removed by a magnetic field in z direction, each of the doublets $|\pm 1/2\rangle$, $|\pm 3/2\rangle$, $|\pm 5/2\rangle$ being split into two components, *viz*. Fig. 8. According to the selection rule $\Delta M_S = \pm 1$ only transitions between the components of the lowest doublet are allowed and, since $\Delta E = h\nu = 2\beta H_z$,

$$g_z = g_{\parallel} = 2 \qquad [33]$$

Of course, on the basis of the selection rule, transitions like $|+3/2\rangle \leftrightarrow |+1/2\rangle$ are likewise allowed. However, these transitions are expected to be weak.

If the magnetic field is oriented in the x or y direction, due to the application of the shift operators S_+ and S_- in eq. [31] only off-diagonal non-zero matrix elements appear. According to Table 4 the splitting of the $|\pm 1/2\rangle$ doublet is determined by either one of the matrices

$$\begin{pmatrix} 0 & 3\beta H_x \\ 3\beta H_x & 0 \end{pmatrix} \qquad \begin{pmatrix} 0 & -3i\beta H_y \\ 3i\beta H_y & 0 \end{pmatrix} \quad [34]$$

Diagonalization produces an energy separation of $6\beta H_\perp$ and thus (*cf* Fig. 8),

$$g_\perp = 6 \quad [35]$$

To first order, there is no splitting of the $|\pm 3/2>$ and $|\pm 5/2>$ doublets, while to second order some mixing between the doublets occurs giving the energies

$$\begin{aligned} E\left(\pm\frac{1}{2}\right) &= 0 \pm 3\beta H_\perp - \frac{4\beta^2 H_\perp^2}{D} \\ E\left(\pm\frac{3}{2}\right) &= 2D + \frac{11}{4}\frac{\beta^2 H_\perp^2}{D} \\ E\left(\pm\frac{5}{2}\right) &= 6D + \frac{5}{4}\frac{\beta^2 H_\perp^2}{D} \end{aligned} \quad [36]$$

We have thus demonstrated above that the highly anisotropic g values, *viz.*

$$g_\parallel = 2, \quad g_\perp = 6$$

encountered in some high-spin compounds of iron(III) may be explained on the basis of the spin Hamiltonian eq. [27] by assuming a zero field splitting D large compared to magnetic field energies and E = 0. It has likewise been shown that the zero field splitting arises *via* spin-orbit interaction with the excited $^4A_2(^4T_1)$ state in tetragonal (or lower) symmetry. Another limiting case of interest arises with the spin Hamiltonian eq. [30] if D = 0 and E ≠ 0. Here the middle doublet shows an effective g = 4.29. It has been clearly shown *(7)* that this case does not represent the maximum possible rhombic field, as might have been thought at the outset. In fact, if the ratio

$$\lambda = E/D \quad [37]$$

is introduced, then λ = 0 determines axial symmetry and an increase of λ above this value indicates a departure towards rhombic symmetry. It then follows that λ = 1/3 represents maximum rhombic symmetry with equally spaced Kramers doublets and values of λ larger than 1/3 indicate convergence toward axial symmetry again. Finally, λ = 1 represents entirely axial symmetry.

It is evident from the discussion above that the most general EPR spectrum of a high-spin iron(III) compound which would be characterized by D ≠ 0, E ≠ 0 will be rather complex. In fact, there has been a long controversy in the literature with respect to interpretation of the EPR spectra of iron(III). We will confine our discussion to the results of the most general treatment of the problem currently available. Thus Dowsing and Gibson *(13)* determined recently, by numerical methods, the eigenvalues and eigenfunctions of the spin Hamiltonian

$$\mathscr{H} = \beta \mathbf{H} \cdot (g) \cdot \mathbf{S} + D[S_z^2 - \frac{1}{3} S(S+1)] + E(S_x^2 - S_y^2) \qquad [38]$$

avoiding any assumptions about the size of D and E. The calculations were performed for various values of the parameter λ, and for the field H parallel to the x, y, and z directions, always assuming $g_x = g_y = g_z = 2.0$. Fig. 9 shows the predicted EPR transitions in terms of the magnetic field H as

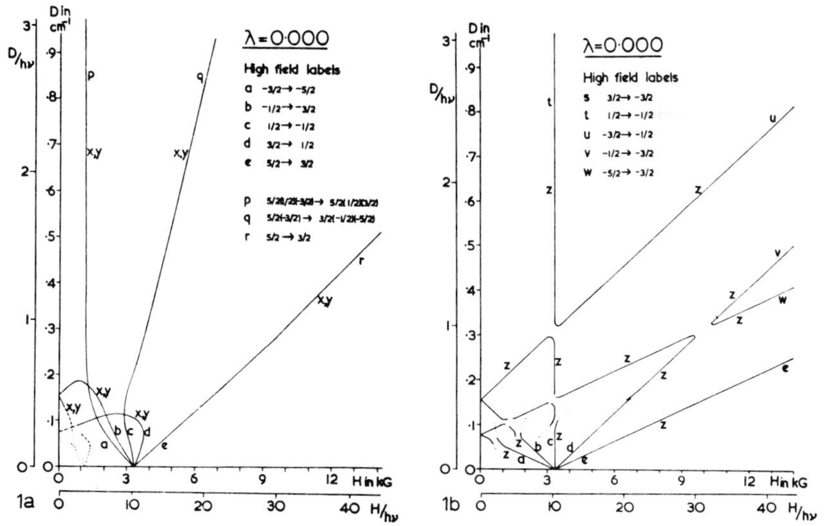

Fig. 9: Calculated diagrams of EPR transitions for S = 5/2 in terms of the magnetic field strength H assuming λ = 0. Full, dashed, and dotted lines represent transitions with high, low, and zero transition probability, respectively. The letter x, y, z close to a line gives the axis to which H is parallel. Reproduced by permission from ref. 13.

function of D for λ = 0. The eigenfunctions were employed to calculate relative transition probabilities, and their large, small, and zero values are indicated, in the diagrams, by full, dashed, and dotted lines, repectively. The inner set of axes applies to D *vs* H for a quantum of 0.310 cm^{-1} (X band), whereas the outer set of axes refers to any microwave frequency in terms of D/hν *vs* H/hν. The letter x, y, or z close to a line gives the axis to which H is parallel for that particular line. The remaining letters are used to identify the transitions to which the lines refer. Fig. 10 shows the cor-

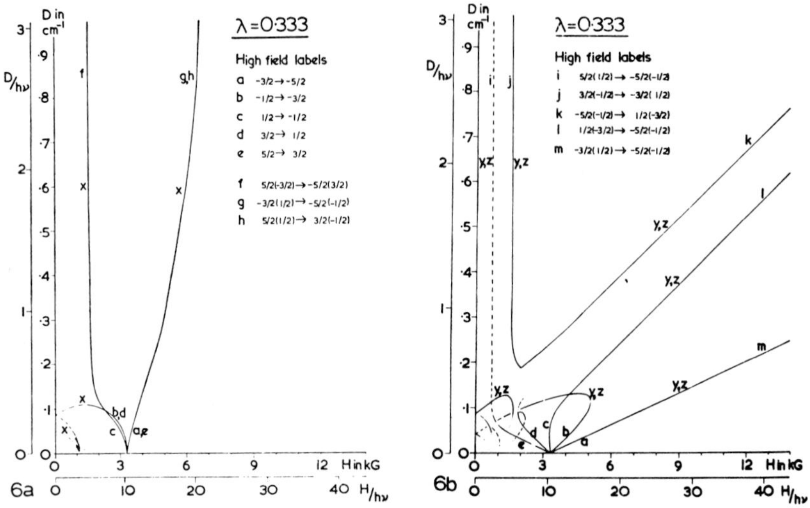

Fig. 10: Calculated diagrams of EPR transitions for $S = 5/2$ in terms of the magnetic field strength H assuming $\lambda = 0.333$. For the nomenclature refer to caption of Fig. 9. Reproduced by permission from ref. 13.

responding diagrams for $\lambda = 0.333$. Additional diagrams may be found in the original paper *(13)*. We add here some comments of the authors *(13)* on the general appearance of the spectra which apply to both single-crystal and to polycrystalline or rigid-solution spectra at X band: (1) For $D > 0.1$ cm^{-1} an absorption is expected between 650 and 700 Gauss if λ is close to 1/3. The intensity is always expected to be weak. (2) A line at H lower than ∿ 650 Gauss which is an allowed transition indicates that the zero field splitting between a pair of Kramers doublets is of the order of $h\nu$. (3) An absorption at ∿ 1500 Gauss is expected if $\lambda \simeq 1/3$ and $D > 0.23$ cm^{-1}. This line broadens or splits as λ departs from 1/3. (4) The observation of several lines at H between ∿ 5000 and 15 000 Gauss indicates D lying between 0.1 and 0.6 cm^{-1}. (5) In almost axial symmetry (λ∿0), observation of the $g_{\parallel}$ band at ∿ 1100 Gauss and the $g_{\perp}$ band at 3300 Gauss indicates $D \geq 0.2$ cm^{-1}. (6) In axial symmetry a strong single line at 1100 Gauss indicates $\lambda = 0$, this band will split or broaden as λ increases, and may still be discerned at $\lambda = 0.1$. The band moves up in field as D decreases below 0.2 cm^{-1}.

B. LOW-SPIN IRON(III)

We have shown in section III above that a $^2T_2(t_2^5)$ ground state is formed within the $3d^5$ configuration of iron(III) if 10 Dq exceeds the value determined by eq. [10]. Since in this situation all excited states are considerably higher in energy, configurational mixing in the ground state may be ignored. The t_2^5 configuration may then be treated as a single t_2 hole within the filled t_2 subshell. This well-known relationship between holes and particles is generally applicable provided appropriate changes in the respective Hamiltonian are introduced *(5,18)*. Below we list the $^2T_2$ ground state wavefunctions and the corresponding functions of an equivalent hole *(16)*.

$$\begin{aligned}
|^2T_2\ \tfrac{1}{2}\ \xi > &= |\xi^+\eta^2\zeta^2 > \ \sim\ |\ \xi^- > \\
|^2T_2\ \tfrac{1}{2}\ \eta > &= |\xi^2\eta^+\zeta^2 > \ \sim\ |\ \eta^- > \\
|^2T_2\ \tfrac{1}{2}\ \zeta > &= |\xi^2\eta^2\zeta^+ > \ \sim\ |\ \zeta^- > \\
|^2T_2\ \tfrac{1}{2}\ \xi > &= |\xi^-\eta^2\zeta^2 > \ \sim\ |\ \xi^+ > \\
|^2T_2\ \tfrac{1}{2}\ \eta > &= |\xi^2\eta^-\zeta^2 > \ \sim\ |\ \eta^+ > \\
|^2T_2\ \tfrac{1}{2}\ \zeta > &= |\xi^2\eta^2\zeta^- > \ \sim\ |\ \zeta^+ >
\end{aligned} \qquad [39]$$

To account for the effect of spin-orbit interaction the matrix elements of the operator

$$-\zeta\,\mathbf{l}\cdot\mathbf{s} = -\zeta[l_z s_z + \tfrac{1}{2}(l_+ s_- + l_- s_+)] \qquad [40]$$

have to be calculated within the subspace of the functions eq. [39]. In eq. [40], ζ is the spin-orbit coupling parameter, and $l_\pm$, $s_\pm$ are the step-up and step-down operators defined according to

$$l_\pm = l_x \pm i l_y, \qquad s_\pm = s_x \pm i s_y \qquad [41]$$

The effect of the operators eq. [41] on the functions $|nlm_l m_s>$ has been discussed in the previous chapter (*cf* eq. [25]). Since the functions eq. [39] are certain linear combinations of the kets $|m_l m_s>$ where $m_l = 2,1,\ldots,-2$ and $m_s = +1/2$ or $-1/2$ *(18)*, it is straight forward to set up the required matrix

$$\begin{array}{c} \\ <\xi^+| \\ <\eta^+| \\ <\zeta^-| \end{array}
\begin{array}{c} \begin{array}{ccc} |\xi^+> & |\eta^+> & |\zeta^-> \end{array} \\
\begin{bmatrix} 0 & i\zeta/2 & -\zeta/2 \\ -i\zeta/2 & 0 & i\zeta/2 \\ -\zeta/2 & -i\zeta/2 & 0 \end{bmatrix} \end{array} \qquad [42]$$

$$\begin{array}{c|ccc} & |\xi^- > & |\eta^- > & |\zeta^+ > \\ \hline <\xi^-| & 0 & -i\zeta/2 & \zeta/2 \\ <\eta^-| & i\zeta/2 & 0 & i\zeta/2 \\ <\zeta^+| & \zeta/2 & -i\zeta/2 & 0 \end{array} \qquad [42]$$

Evidently, the matrix of eq. [40] within the functions eq.[39] is split into the matrices eq. [42] with no non-zero interconnecting elements. Diagonalization yields the eigenvalues 1 and -1/2 which are associated with the representations E" and U' of respective degeneracies two and four, *cf* eq. [18]. Consequently, spin-orbit interaction splits the $^2T_2$ ground state into a Kramers doublet E" and a quadruplet U'. Of these, the Kramers doublet is at lower energy and will therefore account for the observed EPR spectrum. If the symmetry is lower than octahedral, the U' level is likewise split into two Kramers doublets. From eq. [42] the general expression for the lowest Kramers doublet will thus be

$$\begin{aligned} \psi_1 &= a_1|\xi^+ > + b_1|\eta^+ > + c_1|\zeta^- > \\ \psi_2 &= a_2|\xi^- > + b_2|\eta^- > + c_2|\zeta^+ > \end{aligned} \qquad [43]$$

where ψ_1 and ψ_2 are related on the basis of the Kramers theorem *(18)* by

$$\psi_2 = i\psi_1^* \qquad [44]$$

With real coefficients A_1, B_1, C_1, the lowest Kramers doublet may thus be written

$$\begin{aligned} \psi_1^+ &= A_1|\xi^+ > + iB_1|\eta^+ > + C_1|\zeta^- > \\ \psi_1^- &= -A_1|\xi^- > + iB_1|\eta^- > + C_1|\zeta^+ > \end{aligned} \qquad [45]$$

Next we consider the interaction with a magnetic field described by the Hamiltonian

$$\mathscr{H}_m = \beta \mathbf{H}(\mathbf{l} + 2\mathbf{s}) \qquad [46]$$

To this end matrix elements of $\mathbf{l} + 2\mathbf{s}$ within the orbitals eq. [39] are calculated by way of the corresponding elements within the functions $|m_l m_s>$. Whereas matrix elements of $l_z + 2s_z$ are easily obtained, it is most convenient to rewrite $l_x + 2s_x$ and $l_y + 2s_y$ in terms of the operators eq. [41] (*cf* also eq. [31]). The results are listed in Table 5. The conversion to the basis set ψ_1^+, ψ_1^- is then accomplished by means of eq. [45], the results being presented in Table 6.

If the magnetic field is in the z direction, the energies corresponding to eq. [46] are determined directly from Table 6 as

Table 5: Matrices of the operator **l** + 2 **s** within the set of functions ξ, η, ζ of a single *d* electron, *cf* eq. [39]. Upper and lower entries refer to matrix elements of the x and y components, respectively, single entries refer to the z components of **l** + 2 **s**.

l + 2 **s**	ξ^+	η^+	ζ^+	ξ^-	η^-	ζ^-
ξ^+	1	i	-i	1 -i		
η^+	-i	1	i		1 -i	
ζ^+	i	-i	1			1 -i
ξ^-	1 i			-1	i	-i
η^-		1 i		-i	-1	i
ζ^-			1 i	i	-i	-1

Table 6: Matrices of the operator **l** + 2 **s** for the lowest Kramers doublet, *cf* eq. [45].

$l_z + 2s_z$	ψ_1^+	ψ_1^-
ψ_1^+	$(A_1 - B_1)^2 - C_1^2$	0
ψ_1^-	0	$-[(A_1 - B_1)^2 - C_1^2]$

$l_x + 2s_x$	ψ_1^+	ψ_1^-
ψ_1^+	0	$(B_1 + C_1)^2 - A_1^2$
ψ_1^-	$(B_1 + C_1)^2 - A_1^2$	0

$l_y + 2s_y$	ψ_1^+	ψ_1^-
ψ_1^+	0	$i[(A_1 - C_1)^2 - B_1^2]$
ψ_1^-	$-i[(A_1 - C_1)^2 - B_1^2]$	0

$$E_1^z = \beta H_z[(A_1 - B_1)^2 - C_1^2]$$
$$E_2^z = -\beta H_z[(A_1 - B_1)^2 - C_1^2] \qquad [47]$$

As expected, the two-fold degeneracy of the Kramers doublet eq. [45] is lifted by application of a magnetic field, the energy separation being

$$\Delta E^z = 2\beta H_z[(A_1 - B_1)^2 - C_1^2] \qquad [48]$$

From the resonance condition

$$\Delta E = h\nu = g_z \beta H_z \qquad [49]$$

it then follows

$$g_z = 2|(A_1 - B_1)^2 - C_1^2| \qquad [50]$$

The eigenvalues of $l_x + 2s_x$ and $l_y + 2s_y$ obtain from solution of the secular determinants, *cf* Table 6 as

$$E^x = \pm\beta H_x[(B_1 + C_1)^2 - A_1^2]$$
$$E^y = \pm\beta H_y[(A_1 - C_1)^2 - B_1^2] \qquad [51]$$

The principal components of the g tensor are therefore determined by

$$g_x = 2|(B_1 + C_1)^2 - A_1^2|$$
$$g_y = 2|(A_1 - C_1)^2 - B_1^2| \qquad [52]$$
$$g_z = 2|(A_1 - B_1)^2 - C_1^2|$$

Finally, the orbital separations between the Kramers doublets may be calculated. Let us assume that we are dealing with three separate Kramers doublets, subject to the combined effect of spin-orbit coupling and a low symmetry ligand field described by

$$\mathcal{H}' = -\lambda\, \mathbf{l} \cdot \mathbf{s} + V \qquad [53]$$

If ξ, η, ζ, are eigenfunctions of the operator V with eigenvalues ε_ξ, ε_η, ε_ζ, the energies E resulting from

$$\mathcal{H}'\psi_1^{\pm} = E\psi_1^{\pm} \qquad [54]$$

are required. Inserting eq. [45] the secular equations given below follow *(43)*

$$A_1[\langle\xi^+|\mathcal{H}|\xi^+\rangle - E] + iB_1\langle\xi^+|\mathcal{H}|\eta^+\rangle + C_1\langle\xi^+|\mathcal{H}|\zeta^-\rangle = 0$$
$$A_1\langle\eta^+|\mathcal{H}|\xi^+\rangle + iB_1[\langle\eta^+|\mathcal{H}|\eta^+\rangle - E] + C_1\langle\eta^+|\mathcal{H}|\zeta^-\rangle = 0 \qquad [55]$$
$$A_1\langle\zeta^-|\mathcal{H}|\xi^+\rangle + iB_1\langle\zeta^-|\mathcal{H}|\eta^+\rangle + C_1[\langle\zeta^-|\mathcal{H}|\zeta^-\rangle - E] = 0$$

and using the matrix elements of spin-orbit coupling from eq. [42] one obtains

$$\begin{aligned} A_1(\varepsilon_\xi - E) - iB_1 \frac{i}{2}\lambda + C_1 \frac{\lambda}{2} &= 0 \\ A_1 \frac{i}{2}\lambda + iB_1(\varepsilon_\eta - E) - C_1 \frac{i}{2}\lambda &= 0 \\ A_1 \frac{\lambda}{2} + iB_1 \frac{i}{2}\lambda + C_1(\varepsilon_\zeta - E) &= 0 \end{aligned} \qquad [56]$$

In eq. [56] A_1, B_1, C_1 are the coefficients of the lowest Kramers doublet and E its energy. Since we are interested in energy differences, we may set E = O and thus it is

$$\begin{aligned} \varepsilon_\xi &= -\frac{B_1 + C_1}{A_1}\frac{\lambda}{2} \\ \varepsilon_\eta &= \frac{C_1 - A_1}{B_1}\frac{\lambda}{2} \\ \varepsilon_\zeta &= \frac{B_1 - A_1}{C_1}\frac{\lambda}{2} \end{aligned} \qquad [57]$$

Provided we know the principal g values, the coefficients A_1, B_1, C_1 may be obtained from eq. [52] and therefrom the energy separations $\varepsilon_\xi - \varepsilon_\eta$ and $\varepsilon_\xi - \varepsilon_\zeta$ may be calculated using eq. [57]. Finally, the coefficients A_i, B_i, C_i of the remaining two doublets where i = 2,3 may be obtained by a procedure analogous to that applied above. In addition, covalency effects may be taken into account by a simple modification of the above expressions *(16)*.

C. INTERMEDIATE SPIN (S = 3/2) IRON(III)

Spin quartet ground states have been encountered in a few compounds of iron(III) *(36)*. In particular, EPR has been observed in bis(N,N-diisopropyldithiocarbamato)iron(III) chloride and a detailed analysis has been made. For details we refer to the original literature *(36, 44)*.

D. IRON ORGANIC COMPOUNDS WITH IRON OXIDATION STATE DIFFERENT FROM +III

The electronic states arising from the $3d^6$ configuration of high-spin iron(II) compounds have been discussed in the previous chapter. In octahedral symmetry, the fifteen-fold degeneracy of the $^5T_2(t_2^4e^2)$ ground state is split by spin-orbit interaction producing a lowest triplet with energy 3λ, followed by a quintet at λ and a septet at -2λ. In most cases of interest, lower symmetry fields further reduce the degeneracy. Since Kramers theorem does not apply to systems with an even number of electrons, in fact all degeneracies may be removed. The application of a magnetic field can only shift the resulting energy levels. In such a system, EPR is not likely to be observed except when two levels accidentally come sufficiently close for a microwave photon to induce a transition. Thus there are very few reported EPR measurements on compounds containing high-spin iron(II) *(28,29)*. A well studied example is that of Fe^{2+} ion in ZnF_2 where the forbidden $\Delta M_s = 4$ transition was measured and the hyperfine structure from the six fluoride ligands was observed *(42)*. Although this example does not fall within the scope of this volume, it might be useful for future reference.

No EPR is expected, of course, from the $^1A_1(t_2^6)$ ground state of low-spin iron(II) compounds.

Intermediate spin (S = 1) compounds of iron(II) definitely exist *(6,31,33)*, however, their EPR spectra have not been well characterized.

To the author's knowledge, no detailed EPR in iron-organic compounds containing iron in oxidation states definitely higher than +III or definitely lower than +II has been reported. In principle, there is no reason why, *e.g.*, the EPR in the $3d^7$ configuration of iron(I) should not be observed, at least at cryogenic temperatures. There is, however, a considerable amount of resonance data on NO-containing compounds of iron and hyperfine structure from ^{14}N has frequently been observed. An example is presented in section VI.D. Additional examples of such systems are the ion $[Fe(CN)_5NO]^{3-}$ *(23)* and $\{Fe[(CH_3)_2NC(S)S-]_2NO\}$ *(14)*. For details the original literature should be consulted.

VI. ILLUSTRATED EXAMPLES

In this section, we will demonstrate the application of the theoretical treatment of section V to a few specific systems. The examples were selected from the literature in such a way as to provide a sufficiently detailed account of the analysis of the data. In general, therefore, the EPR spec-

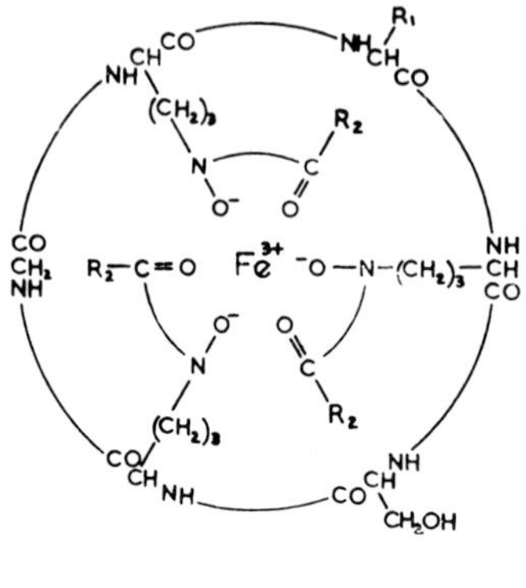

Fig. 11: Probable structure of ferrichrysin (R_1 = $-CH_2OH$, R_2 = $-CH_3$). Reproduced by permission from ref. 13.

tra and the interpretation furnished will be representative for a small number of similar systems only. It should not be expected that these examples would cover all the variety of EPR which may be obtained in the organic compounds of iron.

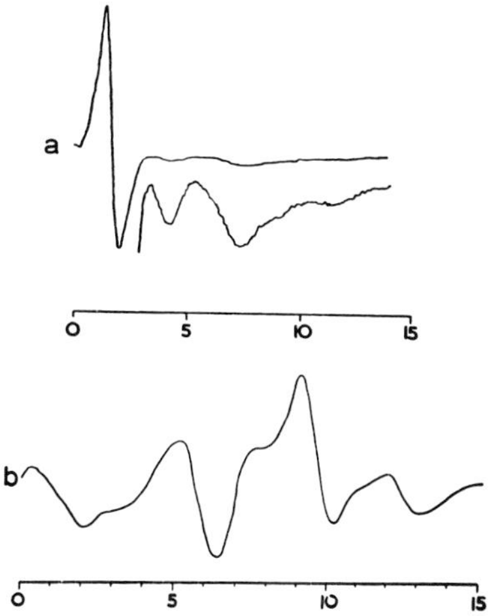

Fig. 12: EPR spectrum of solid ferrichrysin at 93 K (a) X band spectrum (9300 MHz); (b) Q band spectrum (36,000 MHz); field in kGauss; reproduced by permission from ref. 13.

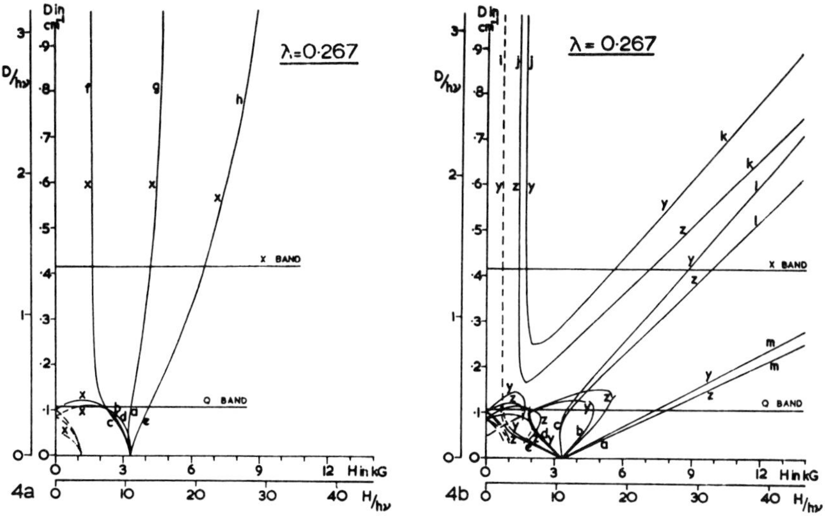

Fig. 13: Calculated diagrams of EPR transitions for S = 5/2 in terms of the magnetic field strength H assuming λ = 0.267. For the nomenclature refer to caption of Fig. 9. Reproduced by permission from ref. 13.

A. FERRICHRYSIN

This compound plays an important biological role as one of the iron-containing growth factors usually called sideramines. The probable structure of ferrichrysin is illustrated in Fig. 11. The EPR spectrum has been analysed in detail by Dowsing and Gibson *(13)* and is shown, both at X and Q band frequencies, in Fig. 12. Our discussion below essentially follows that by the above authors. It is important to realize at the outset that the intense line at g = 4 is indicative of high D and a value of λ close to 1/3. More accurate values for D and λ result from the higher field lines in conjunction with line positions in the Q band spectra. For this purpose, the

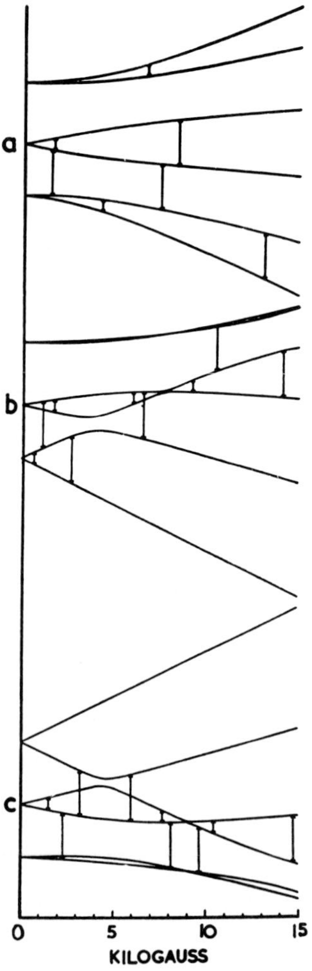

Fig. 14: Energy versus magnetic field strength for the spin Hamiltonian eq. [38], $g_x = g_y = g_z = 2.0$*,* $D = 0.42$ cm^{-1}*, and* $\lambda \cong E/D = 0.267$*. All transitions with finite probability are plotted as vertical arrows. Short arrows represent X band, long arrows represent Q band transitions. Reproduced by permission from ref. 13.*

plots displayed in Fig. 13 may be employed. From the resulting data a graph giving energy *vs* field strength, *viz.* Fig. 14, may be constructed. Here, the short arrows show X band transitions, whereas long arrows represent Q band transitions. Table 7 lists in detail the wavefunctions of those states connected by the short vertical arrows of Fig. 14. These functions were calculated using the spin Hamiltonian of eq. [38] where $g_x = g_y = g_z = 2.0$, $D = 0.42\ cm^{-1}$, and $\lambda = E/D = 0.267$. Finally, Table 8 gives the experimental fields corresponding to observed transitions and the calculated field values providing the best fit to the EPR spectrum. In ferrichrysin then, the $^6A_1$ ground state of the Fe^{3+} ion is split into three Kra-

Table 7: Wavefunctions of the upper and lower levels involved in the transitions labelled in Fig. 13 and represented by the short vertical arrows in Fig. 14. Reproduced by permission from ref. *13*.

Transition		Wavefunctions					
		$\frac{5}{2}$	$\frac{3}{2}$	$\frac{1}{2}$	$-\frac{1}{2}$	$-\frac{3}{2}$	$-\frac{5}{2}$
Field parallel to *x*							
f	upper level	0.799	0.000	0.148	-0.000	-0.582	0.000
	lower level	0.000	-0.615	-0.000	-0.262	-0.000	0.744
g	upper level	0.219	-0.000	0.599	0.000	0.770	0.000
	lower level	0.000	-0.272	-0.000	-0.573	0.000	-0.773
h	upper level	0.789	0.000	-0.573	0.000	0.222	-0.000
	lower level	0.000	-0.825	-0.000	0.549	-0.000	-0.136
Field parallel to *y*							
i	upper level	0.964	0.000	-0.241	-0.000	0.115	-0.000
	lower level	0.000	0.072	-0.000	-0.208	-0.000	0.975
j	upper level	0.000	-0.826	0.000	0.543	-0.000	0.152
	lower level	0.273	0.000	0.341	0.000	-0.900	0.000
k	upper level	0.000	-0.638	0.000	0.760	0.000	0.129
	lower level	0.939	0.000	-0.213	0.000	-0.271	0.000
l	upper level	0.899	0.000	0.406	0.000	-0.164	0.000
	lower level	0.000	-0.473	0.000	0.873	-0.000	0.116
Field parallel to *z*							
i	upper level	0.990	0.000	0.136	0.000	0.033	0.000
	lower level	0.000	0.049	-0.000	0.152	-0.000	0.987
j	upper level	-0.000	0.940	0.000	0.320	-0.000	-0.116
	lower level	-0.085	0.000	0.470	-0.000	0.879	0.000
k	upper level	0.000	-0.115	0.000	0.231	0.000	0.966
	lower level	0.083	0.000	-0.849	0.000	-0.521	0.000
l	upper level	0.078	-0.000	-0.926	0.000	-0.369	0.000
	lower level	0.000	-0.090	0.000	0.364	0.000	0.927

mers doublets $|\pm 1/2>$, $|\pm 3/2>$, and $|\pm 5/2>$ as in (a) of Fig. 8. However, the separations which are listed in the figure as 2D and 4D and where now, in ferrichrysin, D = 0.42 cm^{-1}, are further modified and the doublets split again according to the value of E = 0.224 cm^{-1}. The obtained values of the zero-field splitting parameters D and E provide a measure of the rhombic field distortion at the site of the Fe^{3+} ion. The interested reader is referred to similar analyses of EPR spectra of iron(III) complexes with sulphur- *(10)*, oxygen- *(11)*, and nitrogen-containing *(12)* ligands.

Table 8: Experimental values of magnetic field corresponding to some turning points in the EPR derivative spectrum of ferrichrysin and values of the magnetic field calculated on the basis of eq. [38] employing $g_x = g_y = g_z = 2.0$, D = 0.42 ± 0.04 cm^{-1}, and λ = 0.267 ± 0.02. Reproduced by permission from ref. *13*.

Frequency (band)	Experimental field [G]	Calculated field [G]
X	1720	1438, 1634, 1758
	3710	4260
	6525	5722, 6619
		also 702, 7269, 8959, 10 027
Q	900	705, 829, 1157
	5900	4632, 6090, 6764
	10 000	9445
	12 650	13 084, 13 926, 14 422
		also 2542, 7462, 8176, 8658

B. FERRIC HEMOGLOBIN AZIDE

The hemoglobin molecule contains, in its active form, high-spin iron (II) and, for reasons discussed in section V.D above, cannot be studied by EPR. However, the corresponding ferric molecule and its derivatives may be easily produced by oxidation and their resonance spectra are well known by large. Depending on the nature of the sixth ligand, ferric heme compounds exist in two forms: high-spin compounds with S = 5/2 and low-spin compounds with S = 1/2. The EPR of these ground states is very distinctive as evidenced by Fig. 15. Ferric hemoglobin azide is low-spin and its g values were measured as shown in the lower part of Fig. 15, using a polycrystalline sample, giving the result *(15)*

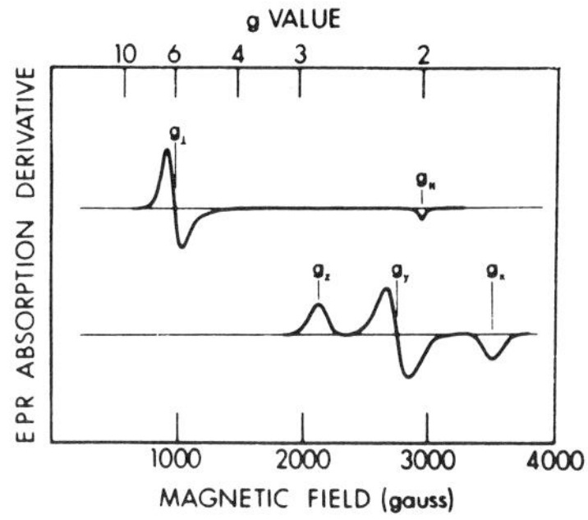

Fig. 15: Typical X band EPR spectra of high-spin (upper) and low-spin (lower) ferric heme compounds as examined in frozen solutions. Reproduced by permission from ref. 8.

$$g_x = 1.72 \qquad g_y = 2.22 \qquad g_z = 2.80$$

Here, the x and y axes refer to the plane of the heme and the z axis is oriented perpendicular to it. Since the principal g-values are all different, we may conclude that the symmetry at the iron site cannot be higher than rhombic. We may now proceed according to section V.B and if we insert the above g-values into eq. [51] we obtain $A_1 = 0.973$, $B_1 = 0.209$, and $C = -0.097$. Of course, we have tacitly assumed that the observed g-values are associated with the lowest Kramers doublet which may now be written using eq. [45] as

$$\begin{aligned} \psi_1^+ &= 0.973\,|\xi^+\rangle - 0.209\,i\,|\eta^+\rangle - 0.097\,|\zeta^-\rangle \\ \psi_{\bar{1}}^- &= -0.973\,|\xi^-\rangle - 0.209\,i\,|\eta^-\rangle - 0.097\,|\zeta^+\rangle \end{aligned} \qquad [58]$$

It is now an easy matter to determine the energy differences between the Kramers doublets from eq. [57]. The ξ orbital results as lowest in energy with the η orbital higher by 2.403λ and the ζ orbital still above by 3.533λ, *viz.* Fig. 16. With a spin-orbit coupling parameter $\lambda \sim 435$ cm^{-1}, the orbital separations are $\varepsilon_\xi - \varepsilon_\eta = 1040$ cm^{-1} and $\varepsilon_\xi - \varepsilon_\zeta = 2580$ cm^{-1}. Finally, the two higher Kramers doublets may be determined. If we

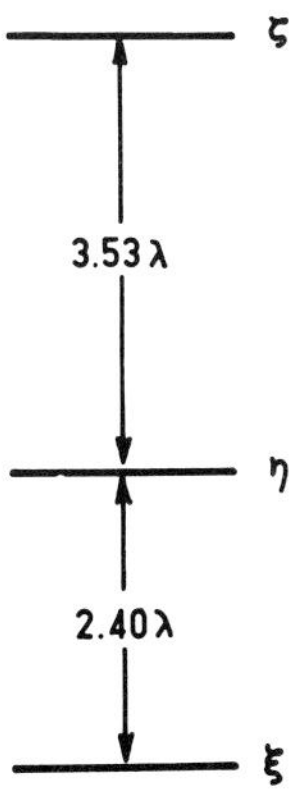

Fig. 16: Orbital energies for ferric hemoglobin azide.

describe a general doublet by an equation similar to eq. [45] with coefficients A_i, B_i, C_i and i = 1,2,3, we shall arrive at a secular equation of the form of eq. [56]. Since ε_ξ, ε_η, and ε_ζ are known, the coefficients may be determined to

i	A_i	B_i	C_i
1	0.973	-0.209	-0.097
2	0.219	0.970	0.108
3	0.071	-0.126	0.990

[59]

Thus each of the Kramers doublets approximates quite closely to one of the original orbitals ξ, η, ζ.

Whereas in ferric hemoglobin the g-tensor is of D_{4h} symmetry, *viz.* g = (2,6,6) *(25)*, the addition of the N_3^- ion produces a lower symmetry such as D_{2h}, *cf* g = (1.72, 2.22, 2.80). The immediate consequence is that the orbitals ξ and η which are degenerate in D_{4h} are now separated, in fact, by as much as ∿1000 cm^{-1}. However, the orientation of the principal g-axes is the same as in the D_{4h} symmetry of ferric hemoglobin. This shows that conversion to hemoglobin azide does not influence the relative orientation of the hemes and the peptide chains. There has been much speculation *(20,21,35)* about the origin of the distortion. An X-ray structure determination *(41)* finally has shown that the N_3^- ion is inclined at 21° to the heme plane. Presumably then, the orientation of the azide ion is the major cause of the anisotropy.

C. *TRIS(3-METHYLPYRAZOLE)IRON(III) CHLORIDE, Fe(Mepz)$_3$Cl$_3$*

According to Cotton and Gibson *(12)* the undiluted complex $Fe(Mepz)_3Cl_3$ gives resonances close to g_{eff} = 4.3 and g_{eff} = 2.0. Here g_{eff} is always taken as $g_{eff} = h\nu/\beta H$, whereas the real g is taken as isotropic and equals 2.00, *cf* section V.A. Since exchange interactions between the iron atoms which might broaden the spectrum can be eliminated by dilution, the EPR of the complex diluted by its indium analogue was investigated at both X and Q band frequencies. In fact, sharp resonances were now obtained at 1550 G at X band and at 5800 and 6500 G at Q band as well as other features up to 12.8 kG, *cf* Fig. 17. The complete EPR spectra are listed in Table 9. The spin Hamiltonian parameters D and λ were determined by the method of Dowsing and Gibson *(13)* outlined in section V.A. In fact, the plots of Fig 10 may be applied approximately since the best fit of both X and Q band spectra is provided by D = 0.90 cm^{-1} and λ = 0.31 (*i.e.* E = 0.28 cm^{-1}). The situation where λ = 1/3 has been termed *(7)* the 'completely rhombic field' and the three Kramers doublets are then equally spaced. Cotton and

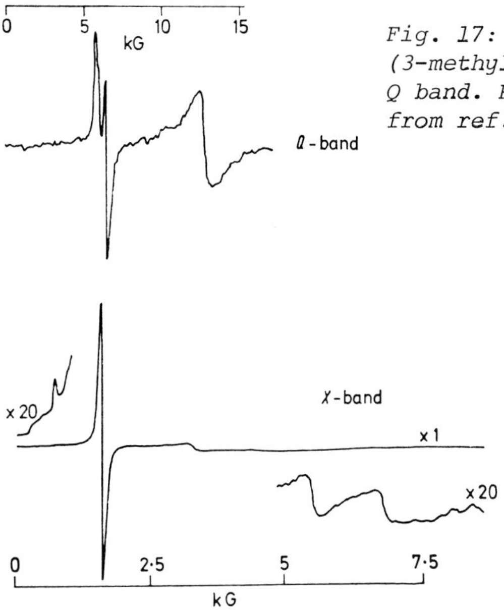

Fig. 17: EPR spectra of In(Fe)-(3-methylpyrazole)$_3$Cl$_3$ at X and Q band. Reproduced by permission from ref. 12.

Table 9: Experimental values of the magnetic field in the EPR spectra of Fe(3-methylpyrazole)$_3$Cl$_3$ and the complex diluted by the indium analogue (in G). Reproduced by permission from ref. *12*.

Compound	Frequency (band)	Experimental field [G]
Fe(Mepz)$_3$Cl$_3$	X-band:	685w; 1550s, br
	Q-band:	6350s, br; 12 800m, br
In(Fe)(Mepz)$_3$Cl$_3$	X-band:	685w; 1550s; 3300w; 5800w; 7100w
	Q-band:	2700w; 5800s; 6500s; 12 800m

Gibson *(12)* likewise studied Fe(pz)$_3$Cl$_3$ and, in this compound, D = 0.24 cm^{-1} and λ = 0.133. The large difference between the two compounds has been accounted for by the assumption that the pyrazole complex is *fac*-ML$_3$X$_3$ (C_{3v} symmetry), whereas the methylpyrazole complex is *mer*-ML$_3$'X$_3$ (C_{2v} symmetry) consistent with $\lambda \sim 1/3$.

D. BIS(N,N-DIETHYLDITHIOCARBAMATO)NITROSYLIRON

In the theoretical outline of section V we have completely neglected the effects due to nuclear hyperfine interactions. In order to show how the nuclear hyperfine structure (HFS) may be exploited to derive information concerning the electronic structure, we consider the EPR of $Fe(NO)[-S_2CN-(C_2H_5)_2]_2$. Our discussion closely follows the study by Goodman, Raynor, and Symons *(17)*. Fig. 18 shows the coordinate

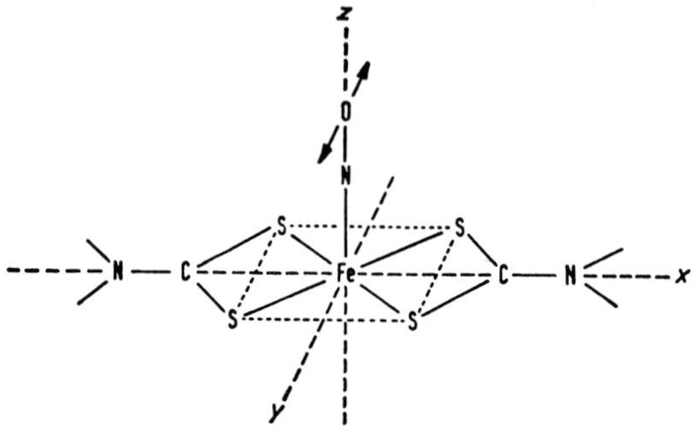

Fig. 18: Coordinate system and principal directions of the g-tensor in $Fe(NO)[-S_2CN(C_2H_5)_2]_2$. The arrows indicate the direction of vibration of the oxygen atom. Reproduced by permission from ref. 17.

system which we will adopt. The EPR spectrum of the compound in EPA solvent (2:5:5 ethanol-isopentane-diethyl ether, often used to form a glass for low-temperature studies) at room temperature exhibits three lines with a spacing of 12.6 G. The splitting is attributed to the hyperfine interaction of the unpaired electron with the ^{14}N nucleus (I = 1) of the NO^+ group. Fig. 19 shows the spectrum of the complex enriched with ^{57}Fe in 90% abundance. The ^{14}N-HFS and the additional splitting due to interaction with ^{57}Fe (I = 1/2) are indicated.

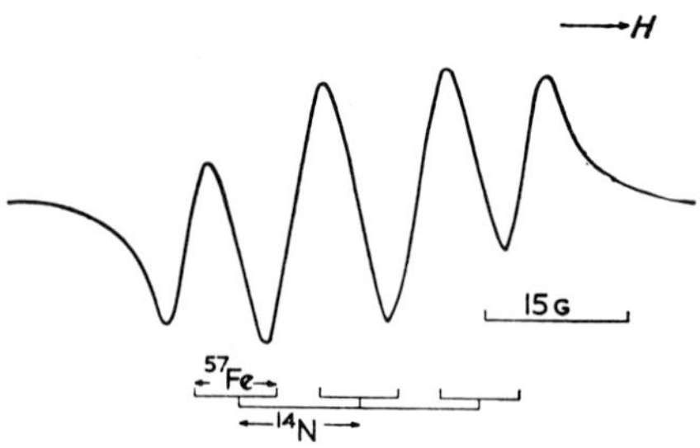

Fig. 19: EPR spectrum of 90% enriched $Fe(NO)[-S_2CN(C_5H_5)_2]_2$ in a fluid solution of EPA. Reproduced by permission from ref. 17.

These data are sufficient to determine the isotropic parameters on the basis of the spin Hamiltonian

$$\mathcal{H} = g\beta H \cdot S + A^N S \cdot I^N + A^{Fe} S \cdot I^{Fe} \qquad [60]$$

The resulting values of g and of the splitting parameters A^N and A^{Fe} are listed in Table 10.

Table 10: Experimental results of the EPR spectra for Fe(NO)-[$-S_2CN(C_2H_5)_2$]$_2$ at X, Q, and S band. Reproduced by permission from ref. *17*.

	g_{av} (300 K)	g_{av} (100 K)	g: g_z	g_x	g_y
X-band	2.040	2.0368	2.025	2.039	2.035
Q-band			2.027	2.042	2.038
S-band					

	$A(^{14}N)$, [G]: A_{iso} (300 K)	A_{iso} (100 K)	A_z	A_x	A_y	$A(^{57}Fe)$, [G]: A_{iso}	A_z	A_x	A_y
X-band	12.6	13.4	15.5	13.4	12.1	8.6	2±4	14±4	
Q-band			14.5	12.8	11.1				
S-band			16.0	13.2				14	

However, from the study of frozen solution spectra, anisotropic EPR parameters directly relating to properties of the molecule may be derived in addition. The corresponding X band spectra are displayed in Fig. 20. Seven out of the total of nine possible ^{14}N-HFS lines centered on three different g-values were observed. This interpretation indicated in Fig. 20

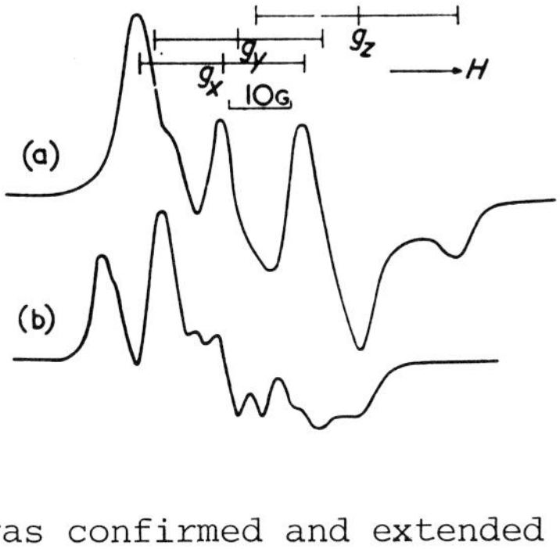

Fig. 20: Rigid solution X band EPR spectrum of Fe(NO)[$-S_2CN(C_2H_5)_2$]$_2$ (a) unenriched, (b) enriched with 90% ^{57}Fe. Reproduced by permission from ref. 17.

was confirmed and extended by running the spectra at Q and at S band (∿3 GHz) frequencies, *cf* Figs. 21 and 22. The spectra of the ^{57}Fe-enriched complex (*cf* Figs. 20 and 22) show additional lines due to ^{57}Fe-HFS and were analysed accordingly. The parameter values determined by application of the spin

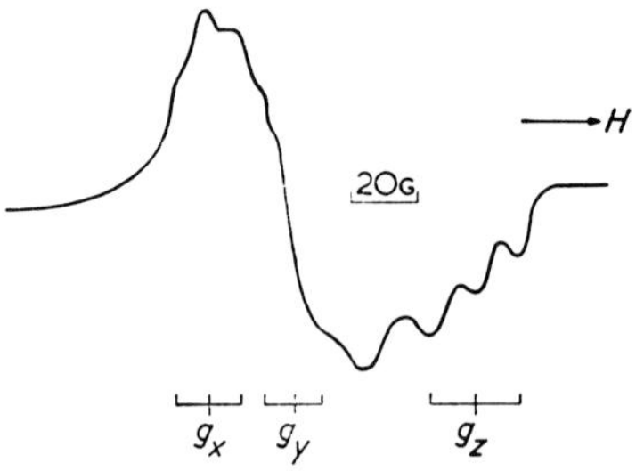

Fig. 21: The rigid solution EPR spectrum of $Fe(NO)[-S_2CN(C_2H_5)_2]_2$ at Q band frequencies. Reproduced by permission from ref. 17.

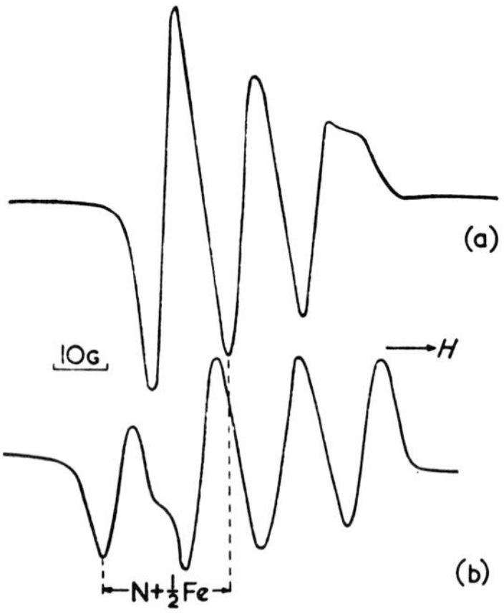

Fig. 22: The rigid solution EPR spectrum of $Fe(NO)[-S_2CN(C_2H_5)_2]_2$ at S band frequencies (a) unenriched, (b) enriched with 90% ^{57}Fe. Reproduced by permission from ref. 17.

Hamiltonian

$$\mathcal{H} = \beta \mathbf{H} \cdot (\mathbf{g}) \cdot \mathbf{S} + \mathbf{S} \cdot (\mathbf{A}^{N}) \cdot \mathbf{I}^{N} + \mathbf{S} \cdot (\mathbf{A}^{Fe}) \cdot \mathbf{I}^{Fe} \quad [61]$$

are compiled in Table 10. In eq. [61], $(\mathbf{g})$, $(\mathbf{A}^{N})$, and $(\mathbf{A}^{Fe})$ are tensor quantities and may be decomposed into the components along the directions of the x, y, and z axes.

If the nitrosyl ligand in $Fe(NO)[S_2CN(C_2H_5)_2]_2$ is considered as NO^+, the formal electronic configuration of the iron atom results as $3d^7$. It may now be shown that the above results are best accounted for if the unpaired electron is accomodated in an $a_1'(d_{z^2})$ orbital and the order of energy levels is

$$a_2 < b_1 < b_2 < a_1' < a_1'' \quad [62]$$

Here, we have employed the notation of C_{2v} symmetry. Adding a more common designation, the electron configuration of the complex is obtained as

$$[a_2(d_{xy})]^2[b_1(d_{xz})]^2[b_2(d_{yz})]^2[a_1'(d_{z^2})]^1[a_1''(d_{x^2-y^2})]^0 \quad [63]$$

This is in complete agreement with the $(\mathbf{g})$, $(\mathbf{A}^{Fe})$ and $(\mathbf{A}^{N})$

tensors as discussed in detail by Goodman *et al.* *(17)*. The ^{14}N hyperfine tensor, when corrected for dipolar coupling, may be written as $(\mathbf{A}^N) = (1.3, -0.6, -0.6)$ Gauss. With these values, the ^{14}N *s*-and *p*-character of the unpaired electron may be calculated from *(38)*

$$A_s = A_{iso}/550 \qquad [64]$$

$$A_p = A_{aniso}/34.1$$

The resulting values are $A_s = 2.6\%$ and $A_p = 4\%$. In summary, a careful EPR study of $Fe(NO)[-S_2CN(C_2H_5)_2]_2$ provides very valuable insight into the electronic structure, particularly with regard to the unpaired electron. Further details may be inferred from a comparison with similar compounds of iron *(17)* and from, *e.g.*, a study of the solvent interaction with the complex *(22)*.

VII. CONCLUSIONS

There is obvious reason that the effect of hyperfine splitting has been discussed only in the sample of section VI.D. For one, the most abundant isotope of iron, ^{56}Fe, has $I = 0$ and the narrow hyperfine structure due to ^{57}Fe (natural abundance 2.21%) is rarely observed. Thus, in general, HFS due to ^{57}Fe may be encountered only if samples enriched in ^{57}Fe are studied. In addition, as a rule, ligand hyperfine structure (SHFS) in compounds of iron(III) is likewise not observed. Since, in this case, the EPR parameters are determined predominantly by the geometry and spin-orbit coupling of the compound, the information on metal-ligand interactions is, at best, qualitative. However, the symmetry of the complex may be studied in detail, sometimes more accurately than by X-ray crystal structure methods.

The restriction mentioned above is generally true for orbitally degenerate ground states, Therefore, the distribution of the unpaired electron has been deduced most frequently from systems containing a single unpaired electron, *viz.*, *e.g.*, compounds of Cu^{2+}, Ag^{2+}, Mo^{5+}, and W^{5+}. Within the subject of this volume, in particular, it is only in organic iron compounds containing the NO molecule that more detailed information on the unpaired electron density may be obtained.

REFERENCES

1. Abragam, A., and Bleany, B., *Electron Paramagnetic Resonance of Transition Ions*, Oxford University Press, 1970.
2. Abragam, A., and Pryce, M.H.L., *Proc. Roy. Soc. (London), A 205*, 135 (1951).
3. Alger, R.S., *Electron Paramagnetic Resonance*, Interscience, New York, 1968.
4. Assenheim, H.M., *Introduction to Electron Spin Resonance*, Hilger, London, 1966.
5. Ballhausen, C.J., *Introduction to Ligand Field Theory*, McGraw-Hill, New York, 1962.
6. Barraclough, C.G., Martin, R.L., Mitra, S., and Sherwood, R.C., *J. Chem. Phys., 53*, 1643 (1970).
7. Blumberg, W.E., in A. Ehrenberg, B.E. Malmström, and T. Vänngard (Eds.) *Magnetic Resonance in Biological Systems*, Pergamon Press, London, 1967, p. 119.
8. Blumberg, W.E., and Peisach, J., *Advan. Chem. Ser., 100*, 271 (1971).
9. Carrington, A., and McLachlan, A.D., *Introduction to Magnetic Resonance*, Harper and Row, New York, 1967.
10. Cotton, S.A., and Gibson, J.F., *J. Chem. Soc. A, 1971*, 803.
11. Cotton, S.A., and Gibson, J.F., *J. Chem. Soc. A, 1971*, 1690.
12. Cotton, S.A., and Gibson, J.F., *J. Chem. Soc. A, 1971*, 1696.
13. Dowsing, R.D., and Gibson, J.F., *J. Chem. Phys., 50*, 294 (1969).
14. Gibson, J.F., *Nature, 196*, 64 (1962).
15. Gibson, J.F., and Ingram, D.J.E., *Nature, 180*, 29 (1957).
16. Golding, R.M., *Applied Wave Mechanics*, Van Nostrand, London, 1969.
17. Goodman, B.A., Raynor, J.B., and Symons, M.C.R., *J. Chem. Soc. A, 1969*, 2572.
18. Griffith, J.S., *The Theory of Transition Metal Ions*, Cambridge University Press, 1961.
19. Griffith, J.S., *The Irreducible Tensor Method for Molecular Symmetry Groups*, Prentice-Hall, Englewood Cliffs (N.J.), 1962.
20. Griffith, J.S., *Nature, 180*, 30 (1957).
21. Griffith, J.S., *Biopolymers Symp., 1*, 35 (1964).
22. Guzy, C.M., Raynor, J.B., and Symons, M.C.R., *J. Chem. Soc. A, 1969*, 2987.
23. Hayes, R.G., *J. Chem. Phys., 48*, 4806 (1968).
24. Ingram, D.J.E., *Free Radicals*, Butterworths, London, 1958.
25. Ingram, D.J.E., Gibson, J.F., and Perutz, M.F., *Nature, 178*, 906 (1956).

26. Kaiser, E.T., and Kevan, L., (Eds.), *Radical Ions*, Interscience, New York, 1968.
27. Kokoszka, G.F., and Gordon, G., in H.B. Jonassen and A. Weissberger (Eds.), *Technique of Inorganic Chemistry, Vol. 7*, Interscience, London, 1968, p. 151.
28. König, E., *Magnetic Properties of Coordination and Organometallic Transition Metal Compounds*, Landolt-Börnstein, New Series, Vol. II/2, Springer, Berlin, 1966.
29. König, E., and König, G., *Magnetic Properties of Coordination and Organometallic Transition Metal Compounds*, Landolt-Börnstein, New Series, Vol. II/8, Supplement 1 (1964-1968), Springer, Berlin, 1976.
30. König, E., in H.A.O. Hill and P. Day (Eds.), *Physical Methods in Advanced Inorganic Chemistry*, Interscience, London, 1968, p. 266.
31. König, E., and Kanellakopulos, B., *Chem. Phys. Lett., 12*, 485 (1972).
32. König, E., and Kremer, S., *Theor. Chim. Acta, 23*, 12 (1971).
33. König, E., and Madeja, K., *Inorg. Chem., 7*, 1848 (1968).
34. Kotani, M., *Rev. Mod. Phys., 35*, 717 (1963).
35. Kotani, M., *Biopolymers Symp., 1*, 67 (1964).
36. Martin, R.L., and White, A.H., *Inorg. Chem., 6*, 712 (1967).
37. McGarvey, B.R., *Transition Metal Chem., 3*, 90 (1966).
38. McNeil, D.A.C., Raynor, J.B., and Symons, M.C.R., *J. Chem. Soc., 1965*, 410.
39. Orton, J.W., *Rep. Progr. Phys., 22*, 204 (1959).
40. Poole, C.P., *Electron Spin Resonance*, Interscience, London, 1967.
41. Stryer, L., Kendrew, J.C., and Watson, H.C., *J. Mol. Biol., 8*, 96 (1964).
42. Tinkham, M., *Proc. Roy. Soc. (London), A 236*, 535 (1956).
43. Weissbluth, M., *Struct. Bonding (Berlin), 2*, 1 (1967).
44. Wickman, H.H., and Merritt, F.R., *Chem. Phys. Lett., 1*, 117 (1967).

OPTICAL ACTIVITY

By HENRI BRUNNER

Chemisches Institut der Universität,
Regensburg, Germany

TABLE OF CONTENTS

ISBN 0-12-417101-X (v.1)

The review is divided into two parts: Compounds containing and compounds not containing direct iron-carbon bonds. The complexes in which the iron atom is the centre of chirality are treated more extensively than the complexes in which the optical activity is only associated with the ligands.

Only compounds are included the optical activity of which was demonstrated experimentally. Complexes which according to NMR and other physical measurements should have chiral structures and hence are supposed to be optically active are not considered. Also compounds are excluded which occur in diastereoisomeric forms if they are not optically active, even if the diastereoisomerism arises from asymmetric centres.

I. COMPOUNDS WITHOUT Fe-C BONDS

A. THE IRON ATOM BEING THE CENTRE OF CHIRALITY

Following the resolution of the octahedral cation $[Co(en)_2(NH_3)Cl]^{2+}$ by A. Werner in 1911 *(178)* many optically active metal complexes were isolated *(14)*. In the early investigations the complexes of Co(III) and Cr(III) played an important part, as these compounds in most cases were configurationally stable and did not lose their optical activity. They could be used for the elucidation of the stereochemistry of substitution reactions.

The first resolution of an iron complex was reported by

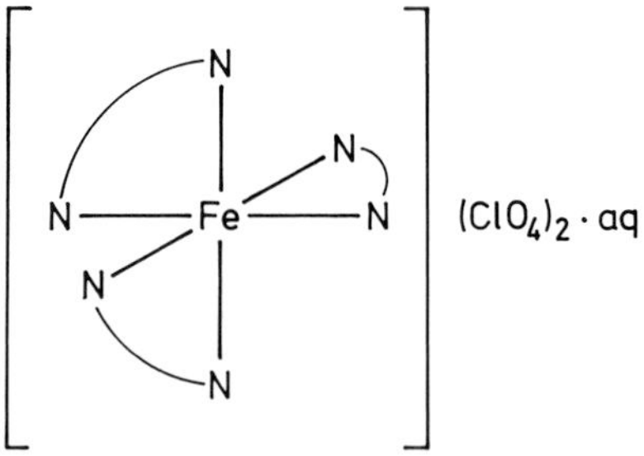

aq = $2H_2O$

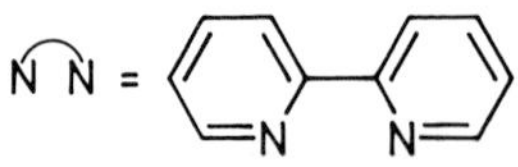

$[\alpha]^{4}_{D}$ +4800°; -4100°

1a, 1b

aq = $3H_2O$

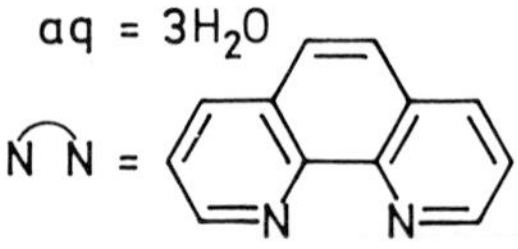

$[\alpha]^{6}_{D}$ +1432°; -1416°

2a, 2b

A. Werner in 1912. He combined the racemic cations of $[Fe(dipy)_3]^{2+}$ (dipy = α,α'-dipyridyl) with the optically active anion of tartaric acid, separated the diastereoisomers and converted the optically active cations into their bromide and iodide salts *(179)*. Contrary to most of the Co(III) and Cr(III) compounds these iron complexes racemize in solution with half lives of about 1/2 hour at room temperature. Higher rotations could be achieved by resolving $[Fe(dipy)_3]^{2+}$ through the iodide antimonyl tartrate followed by isolation in the form of the enantiomeric perchlorates 1a, 1b *(66,67,202)*. Similarly the corresponding o-phenanthroline complex 2a, 2b was resolved *(65)*. (-)-$[Fe(dipy)_3]^{2+}$ was also obtained by interaction of dipy with ferredoxin in a stereoselective reaction which demonstrated that in the protein the iron atom is located at a highly asymmetric site *(80)*. The toxicity of the optically active cation $[Fe(dipy)_3]^{2+}$ was tested *(69,162)*.

The absolute configuration of 2b at the iron atom was determined by an X-ray structure analysis *(167,189)*. Λ-configuration *(98)* was assigned to the octahedral cation $[Fe(phen)_3]^{2+}$ in $(-)_{589}$-Tris(1,10-phenanthroline)iron(II)-bis-[antimony(III)-*d*-tartrate]octahydrate. See also refs. *221-224,244*.

The electronic structure of $[M(dipy)_3]^{n+}$ and $[M(phen)_3]^{n+}$ is a matter of continuing interest. The contributions which can be made to this topic by the interpretation of the CD spectra have been discussed extensively. For the corresponding iron complexes some recent literature references are refs. *90,105,117,118,149,226*.

The tris-chelate complexes of iron(III) with the ligands dipy and phen are even more labile than the corresponding iron(II) compounds 1a, 1b and 2a, 2b. Thus $[Fe(dipy)_3]^{3+}$ cannot be resolved directly. (+)-$[Fe(dipy)_3](ClO_4)_3$ can be prepared by the rapid oxidation of (+)-$[Fe(dipy)_3](ClO_4)_2$ with Ce^{4+} salts which occurs with retention of configuration. The iron(III) complex racemizes very fast. At 15°C no rotation is left after 5 minutes *(64,54)*.

The resolution of $[Fe(C_2O_4)_3]^{3-}$ was claimed *(168)* but could not be reproduced by several groups *(99,101,70)*. The anion seems to be so labile that racemization cannot be observed experimentally at room temperature. At -40°C, however, $[Fe(C_2O_4)_3]^{3-}$ could partly be resolved by chromatography on a starch column *(106)*. As many coordination compounds of other metals $[Fe(C_2O_4)_3]^{3-}$ shows the Pfeiffer effect *(136)*, *i.e.* the rotation of an optically active substance is significantly changed upon addition of racemic $[Fe(C_2O_4)_3]^{3-}$ *(104,190)*, indicating a rapid equilibrium between the two enantiomers differing in the configuration at the iron atom *(87)*. $Fe(acac)_3$ (acac = acetylacetonate) also is configurationally so unstable that it could not be resolved by different techniques *(68,121, 216)*.

Besides complexes with bidentate ligands iron complexes with tridentate and polydentate ligands were obtained in optically active form. The bis-tridentate iron(II) cation 3a, 3b was resolved through the (+)-antimonyl tartrate *(63)*. The enantiomeric iodides 3a and 3b were configurationally stable at room temperature but racemized in 2 minutes at 100°C.

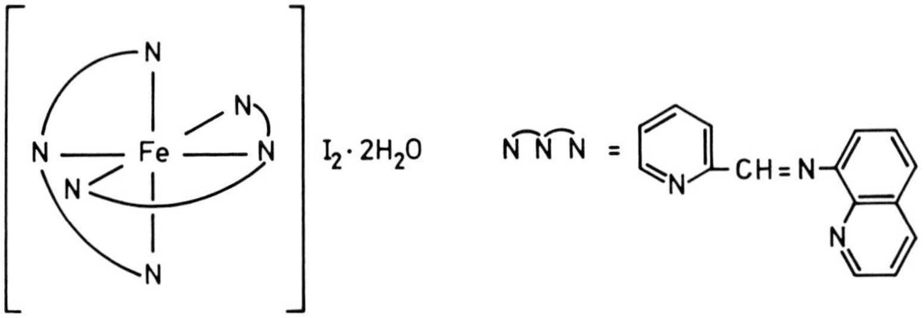

$[\alpha]_D$ +2000°; −2000°

3a 3b

With the same method the resolution of the iron(III) complex 4a, 4b with a hexadentate ligand was accomplished *(150)*. This resolution is an exception to the rule that optically active iron(III) complexes cannot be prepared directly. The complexes 4a and 4b do not racemize.

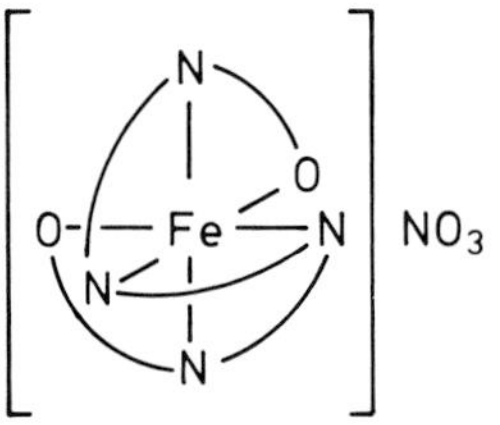

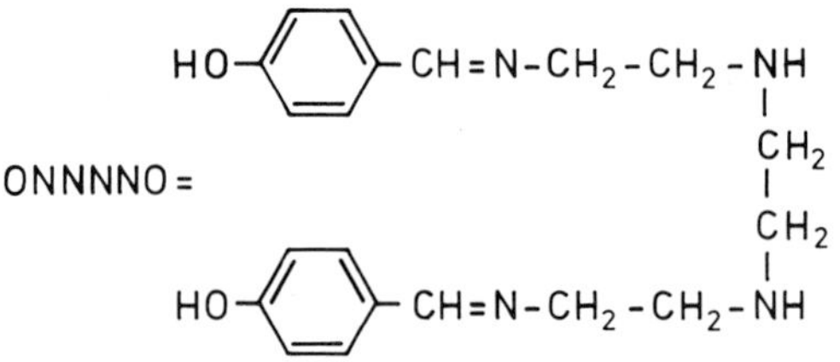

$[\alpha]_D$ +545°; −515°

4a, 4b

For attempts to prepare optically active iron complexes with tetradentate Schiff bases of salicyl aldehyde see ref. *299*.

At the outset only rough estimates for the half lives of the optically active iron compounds were obtained. Later on the kinetics of their racemization and epimerization were measured accurately and these data were used for the elucidation of the mechanisms of the change in configuration associated with the racemization and epimerization.

In the complexes $[Ni(phen)_3]^{2+}$ and $[Ni(dipy)_3]^{2+}$ the rates of ligand exchange are the same as the rates of racemization *(12,183,21)*. Therefore the loss of optical activity is due to an intermolecular process. Each dissociation step leads to loss of optical activity. No intramolecular racemization is involved (scheme [1]; path A).

The rates of racemization as well as the rates of ligand dissociation of $[Fe(phen)_3]^{2+}$ and $[Fe(dipy)_3]^{2+}$ were measured under a variety of conditions *(53,13,62)*. A comparison showed that the racemization of $[Fe(phen)_3]^{2+}$ and $[Fe(dipy)_3]^{2+}$ is much more rapid than ligand dissociation.

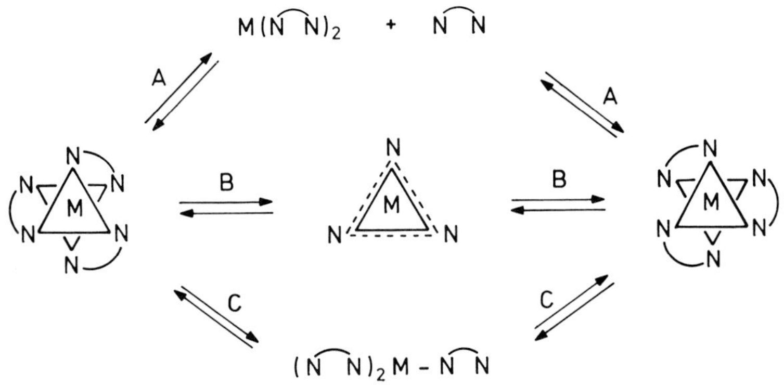

Scheme [1]

M = Ni^{2+}; N⌒N = phen, dipy; path A

M = Fe^{2+}; N⌒N = phen; paths A, and B

M = Fe^{2+}; N⌒N = dipy; paths A, B, and C

This implies, that contrary to $[Ni(phen)_3]^{2+}$ and $[Ni(dipy)_3]^{2+}$ racemization of $[Fe(phen)_3]^{2+}$ and $[Fe(dipy)_3]^{2+}$ also takes place by intramolecular mechanisms, the rate of which may be obtained by the difference in the rates of racemization and dissociation *(13)*. Intramolecular racemization of octahedral chelate complexes may occur either by a twist mechanism *(140, 11)* as shown in path B or by chelate ring opening as shown in path C of scheme [1].

The rigid ligand phen cannot open up at one end to the same extent as the flexible ligand dipy. Therefore the intramolecular contribution to the racemization of $[Fe(phen)_3]^{2+}$ must involve a twist mechanism (path B) *(13,62)*. For $[Fe(dipy)_3]^{2+}$, however, besides the twist mechanism an intramolecular racemization by chelate ring opening (path C) was verified by kinetic measurements at different acid concentrations *(13)*. In addition to variation in p_H-value the racemization of $[Fe(dipy)_3]^{2+}$ and $[Fe(phen)_3]^{2+}$ was studied in the presence of cations and anions *(59,100)* and in nonaqueous solvents *(55,160)*.

The differences in the racemization mechanism of Ni(II) and Fe(II) compounds have been rationalized in terms of crystal field theory. The low spin d^6 ground state of the diamagnetic Fe(II) complexes may be excited to a high spin d^6 state which either dissociates or returns to the stable ground state. Rearrangement or chelate ring opening in the expanded excited state would account for the intramolecular racemization. As no such excited state is possible for Ni(II) only the dissociation mechanism seems to be available to it *(52)*.

Contrary to the solution behaviour the optically active salts $[Fe(phen)_3]X_2$ and $[Fe(dipy)_3]X_2$ are configurationally stable in the solid state *(65)*. Pressure of 20 000 to 50 000 atm, however, leads to a racemization of (-)-$[Fe(phen)_3](ClO_4)_2$•aq *(158)*. An investigation of the effect of temperature on the racemization of solid (+)-$[Fe(phen)_3](ClO_4)_2$•aq and (+)-$[Fe(dipy)_3](ClO_4)_2$•aq showed that between 50 -120°C the water of crystallization is lost. Above this temperature range racemization was observed without decomposition which starts only above 250°C *(188)*. An exothermic peak on the DTA curve of (-)-$[Fe(phen)_3]I_2$•aq and (-)-$[Fe(phen)_3](ClO_4)_2$•aq between dehydration and decomposition at around 190 - 200°C was attributed to the racemization process *(166)*. The results of these solid state studies are consistent with the trigonal twist mechanism proposed in scheme [1].

The reaction of Δ-$[Fe(phen)_3](ClO_4)_2$•aq with CN^- in water yields the optically active substitution product $Fe(phen)_2(CN)_2$ which contrary to the starting material is almost configurationally stable at room temperature in water solution *(7, 227)*. Based on the CD spectra it was concluded that in this reaction an inversion of configuration takes place *(6,7)*. On the other hand the corresponding dipy complex Δ-$[Fe(dipy)_3](ClO_4)_2$•aq reacts with aqueous CN^- with predominant retention of configuration at the iron atom *(207,227)*. The stereochemical differences in the same substitution reaction of the two closely related systems is rationalized in terms of the different ligand flexibilities of dipy and phen as has been done to account for the results of the acid hydrolysis reaction *(13,62)*, discussed before. Whereas in the dissociation reaction of the rigid ligand phen from tris-chelate complexes both donors must leave simultaneously the increased flexibility of the ligand dipy allows chelate ring opening and the formation of complexes with unidentate dipy, supposed to favour the retentive path *(207)*.

In the last decade the NMR method was used increasingly as a tool for the study of reaction mechanisms which earlier had been investigated by polarimetric kinetics of racemization and epimerization or by kinetics of ligand exchange with isotopically labelled compounds and similar methods. With NMR techniques, especially with the help of diastereotopic groups, the course of reactions including isomerizations such as the interconversion of enantiomers can be elucidated without resolving the compounds under investigation. For recent reviews see refs. *96,137*.

B. THE IRON ATOM BEING ASSOCIATED WITH OPTICALLY ACTIVE LIGANDS

Besides resolving compounds with a centre of chirality at the metal atom optically active complexes result from a combination of a central atom with one or more optically active ligands. Depending on the type of complex formed, asymmetric centres in the ligands may or may not lead to a centre of chirality at the metal atom. In this chapter the iron complexes with optically active amino acids, peptides and similar systems would have to be treated. But as in these compounds the main problems are not those of optical activity they are not further discussed here.

II. COMPOUNDS WITH Fe-C BONDS

A. THE IRON ATOM BEING THE CENTRE OF CHIRALITY

The magnetic nonequivalence of diastereotopic groups *(120)* demonstrated, that iron complexes of the type $C_5H_5Fe(CO)(L)(X)$ are configurationally stable at room temperature as well as at higher temperatures *(32,31,23,209,228,241,237,238)*. L means a two-electron ligand, for instance $P(CH_3)_2C_6H_5$ or $P(C_6H_5)_3$ and X a one-electron ligand, for instance COOR, $COCH_3$ or I. On the basis of these NMR results the synthesis of optically active organometallic compounds with four different ligands at the iron atom appeared possible. The subject of optically active organometallic compounds containing an asymmetric transition metal atom has been reviewed previously *(23-26)*.

To resolve compounds of the type $C_5H_5Fe(CO)[P(C_6H_5)_3]$-COOR the salt $[C_5H_5Fe(CO)_2P(C_6H_5)_3][PF_6]$ (5) *(170)* was treated with the sodium salt of the optically active alcohol menthol, $NaOC_{10}H_{19}$, according to equation [2]:

$$[C_5H_5Fe(CO)_2P(C_6H_5)_3][PF_6] + NaOC_{10}H_{19} \longrightarrow \quad [2]$$

5

$$C_5H_5Fe(CO)[P(C_6H_5)_3]COOC_{10}H_{19} + NaPF_6$$

6a,6b

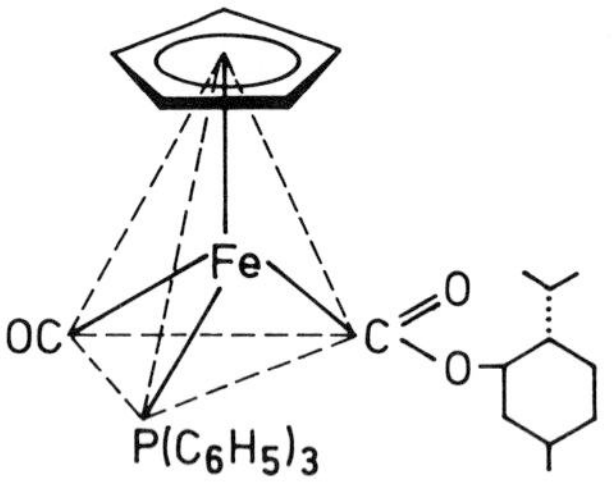

$[\alpha]_{546}$ + 70°; -120°

6a, 6b

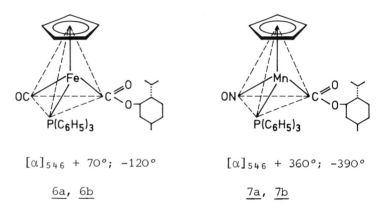

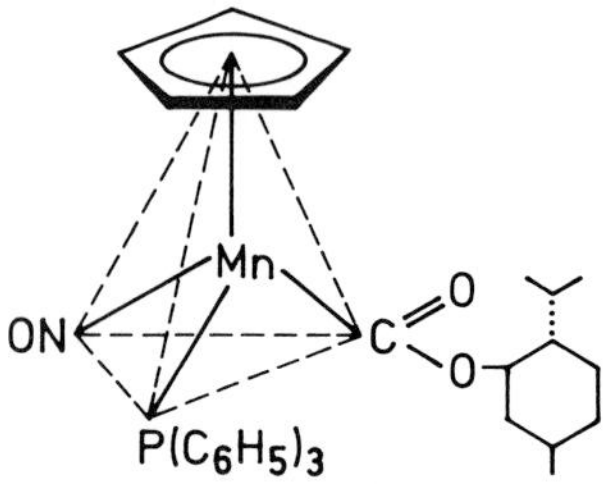

$[\alpha]_{546}$ + 360°; -390°

7a, 7b

In this reaction the alkoxide anion adds nucleophilically to the carbon atom of one of the enantiotopic carbonyl ligands of the cation of 5 with formation of an ester group. The resulting diastereoisomeric complexes (+)- and (-)-$C_5H_5Fe(CO)$-$[P(C_6H_5)_3]COOC_{10}H_{19}$ (6a and 6b) can be separated on the basis of their different solubilities in aliphatic hydrocarbons *(33, 35)*. They correspond to the previously described isoelectronic manganese complexes (+)- and (-)-$C_5H_5Mn(NO)[P(C_6H_5)_3]COOC_{10}H_{19}$ (7a and 7b) *(22,28,30)* which epimerize in benzene solution at room temperature in the course of hours by dissociation of triphenylphosphine *(30,27)*. The corresponding iron complexes 6a and 6b on the other hand are configurationally stable *(33)*. Their benzene solutions do not lose their optical rotations.

The diastereoisomeric manganese compounds (+)- and (-)-$C_5H_5Mn(NO)[P(C_6H_5)_3]COOC_{10}H_{19}$ (7a and 7b) could be converted by HCl to the enantiomeric salts (+)- and (-)-$\{C_5H_5Mn(NO)[P(C_6H_5)_3]CO\}^+[X]^-$ (X = Cl, PF_6) *(29)*. A similar reaction with the iron compounds (+)- and (-)-$C_5H_5Fe(CO)[P(C_6H_5)_3]COOC_{10}H_{19}$ (6a and 6b) would lead to the salts $[C_5H_5Fe(CO)_2P(C_6H_5)_3]^+[X]$ *(170)*, which contain only achiral cations. Therefore in order to come from the diastereoisomeric compounds 6a and 6b to enantiomeric complexes the transesterification to the corresponding methyl esters was studied.

In the manganese system the menthylesters (+)- and (-)-$C_5H_5Mn(NO)[P(C_6H_5)_3]COOC_{10}H_{19}$ (7a and 7b) react with CH_3OH/$NaOCH_3$ to give the methylesters (+)- and (-)-$C_5H_5Mn(NO)[P(C_6H_5)_3]COOCH_3$. As only the ligands are changed the reactions occur with retention of configuration at the asymmetric manganese atom *(28,30)*.

In contrast to the manganese system the transesterification of (+)- and (-)-$C_5H_5Fe(CO)[P(C_6H_5)_3]COOC_{10}H_{19}$ (6a and 6b) in CH_3OH leads to racemic $C_5H_5Fe(CO)[P(C_6H_5)_3]COOCH_3$ (8) according to equation [3].

$$\left.\begin{array}{l} (+)\text{-}C_5H_5Fe(CO)[P(C_6H_5)_3]COOC_{10}H_{19} + CH_3OH \\ \qquad\qquad\qquad\qquad \underline{6a} \\ (-)\text{-}C_5H_5Fe(CO)[P(C_6H_5)_3]COOC_{10}H_{19} + CH_3OH \\ \qquad\qquad\qquad\qquad \underline{6b} \end{array}\right\} \longrightarrow \qquad [3]$$

$$\begin{array}{c} C_5H_5Fe(CO)[P(C_6H_5)_3COOCH_3 + C_{10}H_{19}OH \\ \text{racemic } \underline{8} \end{array}$$

As iron esters of type 6 and 8 were shown to be configurationally stable *(33)* the loss of optical activity during the transesterification must be connected with the mechanism of this reaction *(35)*. An explanation for the differences in the stereochemical course of the transesterification of the isoelectronic manganese and iron compounds 6a, 6b and 7a, 7b is offered after consideration of the following reaction.

On treatment of the ester derivative (+)-$C_5H_5Fe(CO)[P(C_6H_5)_3]COOC_{10}H_{19}$ (6a) with $LiCH_3$ the acetyl derivative (-)-$C_5H_5Fe(CO)[P(C_6H_5)_3]COCH_3$ (9b) is formed *(34,25)* according to equation [4].

In the same way (-)-$C_5H_5Fe(CO)[P(C_6H_5)_3]COOC_{10}H_{19}$ (6b) yields (+)-$C_5H_5Fe(CO)[P(C_6H_5)_3]COCH_3$ (9a) on reaction with $LiCH_3$ *(34)*. Contrary to the results in the manganese system the ORD- and CD-spectra of reactants and products in the iron series are opposite. This indicates that in the reactions of (+)- and (-)-$C_5H_5Fe(CO)[P(C_6H_5)_3]COOC_{10}H_{19}$ (6a and 6b) with $LiCH_3$ the configuration at the asymmetric iron atom must have been inverted. Obviously the nucleophile does not attack at the carbon atom of the ester function but at the carbon atom of the carbonyl ligand. The anion $OC_{10}H_{19}^-$ on the other hand

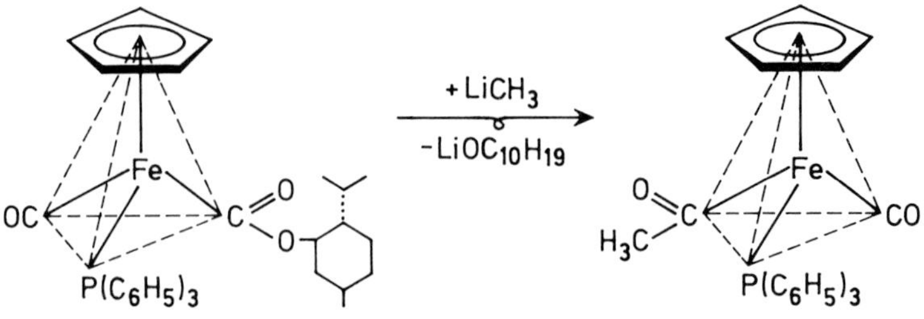

[4]

$[\alpha]_{546}$ + 70° $[\alpha]_{546}$ -228°

6a 9b

is split off from the carbon atom of the ester group. In this way the original carbonyl ligand is transformed into the new functional group and the original functional group into the new carbonyl ligand. Thus the configuration at the asymmetric iron atom in 6a and 6b is inverted without rupture of the bonds between the iron atom and its four substituents *(34)*.

In compounds of the type $C_5H_5Fe(CO)[P(C_6H_5)_3]COOR$ the carbon atom of the carbonyl ligand seems to be more reactive than the carbon atom of the ester group towards nucleophiles. With this reactivity sequence the results of the transesterification reaction [3] can be explained. Nucleophilic attack occurs easier on CO than on COOR. One reaction of this kind leads to an inversion of the configuration at the asymmetric iron atom as in the stoichiometric reaction of (+)- and (-)-$C_5H_5Fe(CO)[P(C_6H_5)_3]COOC_{10}H_{19}$ with $LiCH_3$ (equation [4]). Successive nucleophilic attacks on the other hand as in the transesterification of (+)- and (-)-$C_5H_5Fe(CO)[P(C_6H_5)_3]COO$-$C_{10}H_{19}$ in methanol (equation [3]) lead to a series of inversion steps and ultimately to racemization *(34,35)*.

In the manganese compounds (+)- and (-)-$C_5H_5Mn(NO)[P(C_6$-$H_5)_3]COOC_{10}H_{19}$ (7a and 7b) the attack of the nucleophile occurs at the more reactive carbon atom of the ester group and not at the nitrogen atom of the nitrosyl ligand proceeding with retention of configuration *(28,30)*. These experiments establish the following reactivity sequence against nucleophilic attack in compounds of type 6 and 7 *(34,35)*.

$$CO > COOR > NO$$

This series explains the inversion of configuration in the reactions of (+)- and (-)-$C_5H_5Fe(CO)[P(C_6H_5)_3]COOC_{10}H_{19}$ with $LiCH_3$ (equation [4]) *(34)*, the racemization in the transesterification of (+)- and (-)-$C_5H_5Fe(CO)[P(C_6H_5)_3]COOC_{10}H_{19}$ with CH_3OH (equation [3]) *(35)* and the retention of configuration in the transesterification of (+)- and (-)-$C_5H_5Mn(NO)[P(C_6H_5)_3]COOC_{10}H_{19}$ (7a and 7b) with CH_3OH/ $NaOCH_3$ *(28,30)*. It explains also why in the transesterification of (+)- and (-)-$C_5H_5Fe(CO)[P(C_6H_5)_3]COOC_{10}H_{19}$ (equation [3]) the nucleophilic attack on the more reactive CO group occurs already in pure methanol whereas in the transesterification of (+)- and (-)-$C_5H_5Mn(NO)[P(C_6H_5)_3]COOC_{10}H_{19}$ (7a and 7b) for the nucleophilic attack on the less reactive COOR group a higher concentration of alkoxide ions is necessary *(35,28,30)*.

The optically active iron acyls 9a and 11a, accessible by reaction [4], proved to be suitable starting materials for the investigation of the stereochemical course of the decarbonylation reaction *(184)*. Whereas the stereochemistry of the decarbonylation of compounds of the type $C_5H_5Fe(CO)_2COR$ with respect to the α-carbon atom of the R group (discussed in chapter II.B.5) was studied some time ago, the stereochemistry at the iron center was investigated only recently. With a number of diastereomerically related pairs of enantiomers containing an asymmetric center at the iron atom (for instance $C_5H_5Fe(CO)[P(C_6H_5)_3]COCH_2CH(CH_3)(C_6H_5)$ *R,R-S,S* and *R,S-S,R* as well as 1-CH_3-3-C_6H_5-$C_5H_3Fe(CO)[P(C_6H_5)_3]COCH_3$ *R,R-S,S* and *R,S-S,R*) it was shown that the photochemical decarbonylation of the acyl to the corresponding alkyl compounds proceeds with a high degree of stereospecificity at the iron atom, the partial epimerization being due to a reaction subsequent to the decarbonylation step *(8,10,142)*.

As the optically active acetyl and propanoyl complexes 9a and 11a in equations [5] and [6] were prepared from $(+)_{436}$-$C_5H_5Fe(CO)[P(C_6H_5)_3]COOC_{10}H_{19}$ (6b), both have the same configuration at the iron atom *(57)*. The acetyl compound 9a was reduced with $NaBH_4$ *via* the intermediate salt $(-)_{436}$-$\{C_5H_5Fe(CO)[P(C_6H_5)_3]C(OC_2H_5)CH_3\}^+[BF_4]^-$ to the iron ethyl compound 10a.

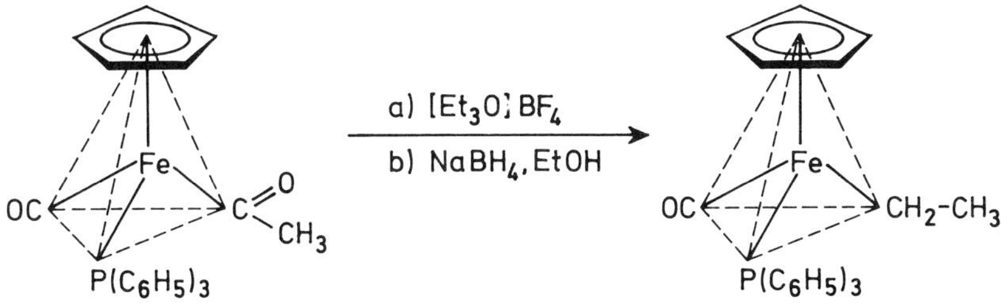

$[\alpha]^{27}_{436}$ -1550° 9a

$[\alpha]^{27}_{436}$ -256° 10a

Both complexes of equation [5] have the same configuration because the asymmetric center at iron is untouched *(57)*.

The propanoyl compound 11a was photochemically decarbonylated to give the ethyl derivative 10b (equation [6]). The optical rotations and CD spectra showed that it has opposite configuration to the ethyl derivative 10a of equation [5] *(57)*. The observed inversion of configuration at the iron atom in equation [6] is consistent with the proposed decarbonylation mechanism, in which the alkyl group migrates into

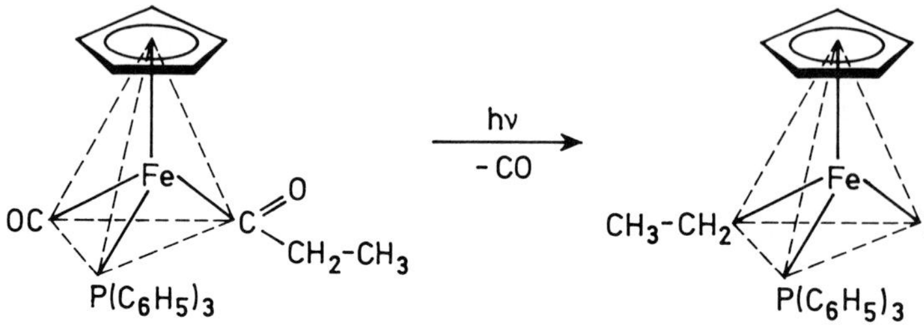

[6]

$[\alpha]^{27}_{436}$ -800° 11a

$[\alpha]^{27}_{436}$ +110° 10b

the site vacated by the leaving CO group *(184)*. The same conclusion was drawn from a study of the photochemical decarbonylation of (+)-$C_5H_5Fe(CO)[P(C_6H_5)_3]COCH_3$ *(36)*. Long irradiation times in the photochemical decarbonylation lead to racemization at the iron atom *(36,184)*. Racemization was also observed in the thermal decarbonylation of (−)-$C_5H_5Fe(CO)[P(C_6H_5)_3]COCH_3$ *(36)*.

To investigate the stereochemistry at the iron atom in the SO_2 insertion reaction *(185)*, which from NMR measurements is known to be highly stereospecific *(9,142)*, compound

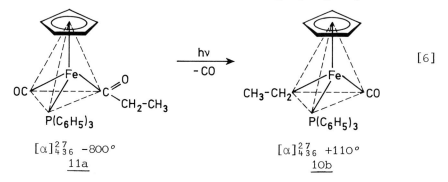

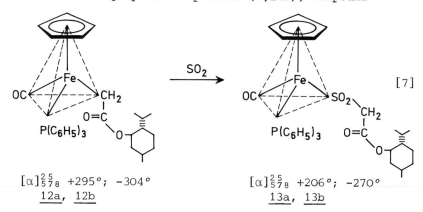

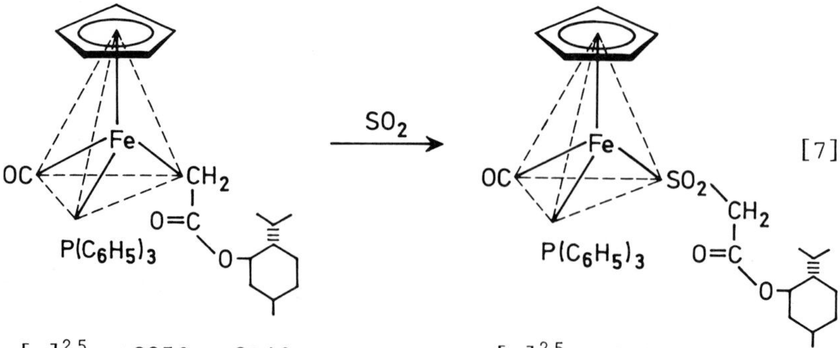

[7]

$[\alpha]^{25}_{578}$ +295°; -304° 12a, 12b

$[\alpha]^{25}_{578}$ +206°; -270° 13a, 13b

12a, 12b was separated into its two diastereoisomeric components which differ in their NMR spectra *(79)*. According to equation [7] liquid SO_2 inserts into the metal carbon bond. From the similarity of the CD spectra of the starting material 12a and 12b and the product 13a and 13b it was concluded that the SO_2 insertion takes place with retention of configuration at iron *(79)*. The stereochemistry at the α-carbon atom of the alkyl chain during the SO_2 insertion and the implications concerning the reaction mechanism are discussed in section II.B.5.

The two diastereoisomeric iron menthoxymethyl compounds 14a and 14b were separated by fractional crystallization from pentane *(56)*. These compounds were used as carbene transfer reagents to olefins. In the reaction with trans-1-phenylpropene according to scheme [8] an asymmetric induction of 26% and 38.5%, respectively, in the cyclopropanes formed was observed *(56)*. By reaction with HI the $OC_{10}H_{19}$ group in the compounds 14a and 14b could be replaced by I without change in configuration at the metal atom *(56)*.

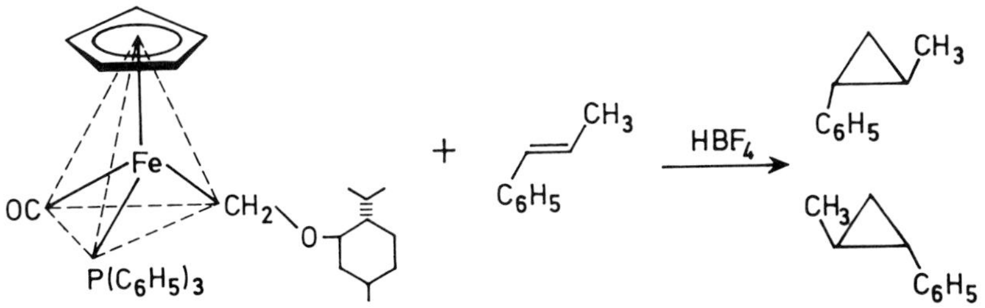

[8]

14a, 14b

A versatile, high yield synthesis of enantiomerically pure primary alkyl iron compounds is based on the easy diastereoisomer separation of (+)- and (-)-$C_5H_5Fe(CO)[P(C_6H_5)_3]CH_2O$-(-)-menthyl *(211)*. After conversion into (+)- or (-)-$C_5H_5Fe(CO)[P(C_6H_5)_3]CH_2Cl$ with anhydrous HCl alkylation with Li organic or Grignard reagents leads to a variety of compounds $C_5H_5Fe(CO)[P(C_6H_5)_3]CH_2R$ the optical purity of which can be demonstrated by the interaction of their SO_2-insertion products with optically active shift reagents *(211)*.

All the resolutions of organometallic iron compounds described in this section up to this point were achieved with the menthyl group. Other methods for converting racemic chiral or prochiral organometallic compounds into diastereoisomers have been described using the optically active acid chloride *R*-(-)-$C_6H_5CH(CH_3)CH_2COCl$, the optically active amine *S*-(-)-$H_2NCH(CH_3)(C_6H_5)$ and the optically active isonitriles *R*-(+)- and *S*-(-)-$CNCH(CH_3)(C_6H_5)$ *(175)*. As isonitriles are good ligands in organometallic systems *(110,177)*, (+)- and (-)-CN-$CH(CH_3)(C_6H_5)$ should be applicable in many cases.

In the reaction of $C_5H_5Fe(CO)_2I$ (15) with (+)-CNCH(CH_3)-(C_6H_5) a carbonyl group is replaced according to equation [9] *(37)*. The resulting diastereoisomers (+)- and (−)-$C_5H_5Fe(CO)$-

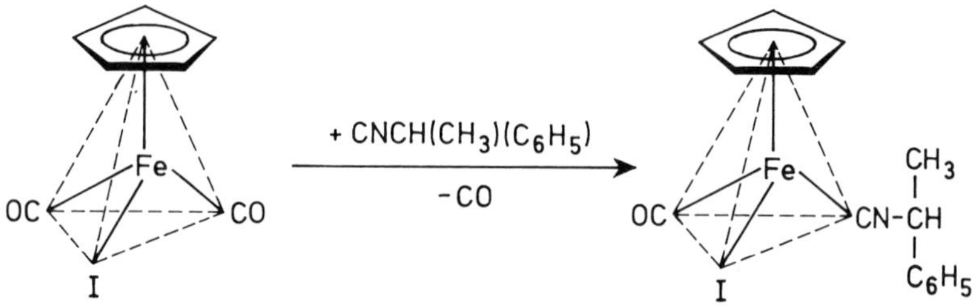

[9]

$[\alpha]_{546}$ +260°; −77°

15 16a, 16b

[CNCH(CH_3)(C_6H_5]I (16a and 16b) can be separated on the basis of solubility differences. The compounds do not epimerize in solution, but they are sensitive to light. In daylight the rotations decrease depending on the intensity of irradiation, whereas in the darkness the rotational values remain constant *(37)*.

If *S*-(−)-α-phenylethylamine is reacted with the enantiomers of complex 17 the two diastereoisomers 18a and 18b are formed (equation [10]), which could be separated by fractionation from ethanol *(58)*. The isomers, configurationally sta-

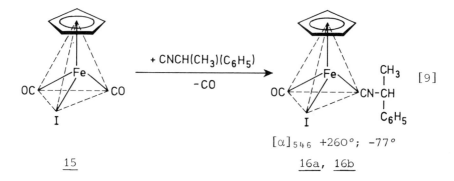

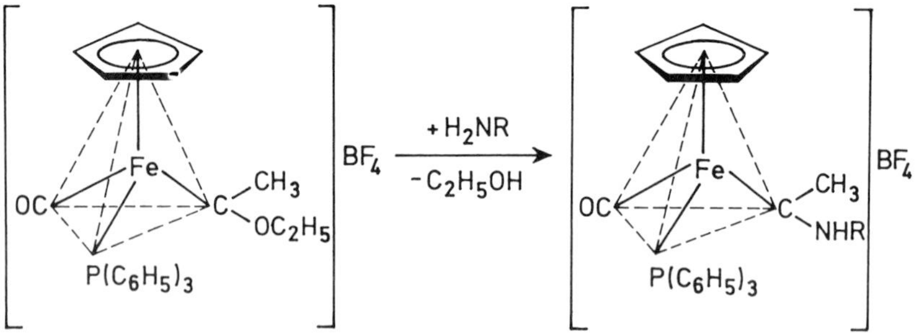

17 [10]

H_2NR = *S*-(−)-$H_2NCH(CH_3)(C_6H_5)$

18a, 18b

ble in refluxing acetone, differ in their NMR spectra and have opposite CD spectra *(58)*.

$(-)_D$-$C_5H_5Fe(CO)_2COCH_2CH(CH_3)(C_6H_5)$ (19) was prepared by the reaction of Na[$C_5H_5Fe(CO)_2$] with *R*-$C_6H_5CH(CH_3)CH_2COCl$ and decarbonylated to give $(-)_D$-$C_5H_5Fe(CO)_2CH_2CH(CH_3)(C_6H_5)$ (20) *(142)*. Compounds 19 and 20 can be transformed into the com-

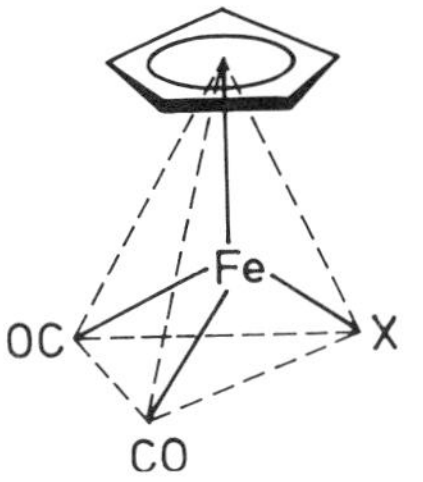

19 X = $COCH_2CH(CH_3)(C_6H_5)$
$[\alpha]_D^{23}$ -31°

20 X = $CH_2CH(CH_3)(C_6H_5)$
$[\alpha]_D^{23}$ -78°

plexes 21, 22, and 23 by carbonylation, decarbonylation or sulphur dioxide insertion *(142)*. As in 21, 22, and 23 the iron atom is an asymmetric center, all the compounds consist of two diastereoisomers a and b differing in their NMR spectra. For 21, 22, and 23 both isomers with *R* and *S* configuration at the iron could be enriched by a combination of column chromatography and fractional crystallization *(142)*. Configurational correlations based on CD spectra between the different isomers were not possible due to the instability of the alkyl derivatives 22a and 22b in solution *(142)*.

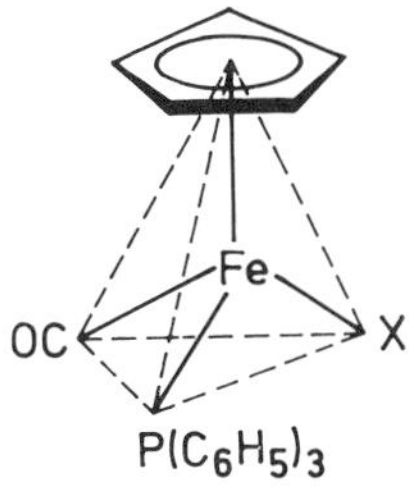

X = $COCH_2CH(CH_3)(C_6H_5)$
$[\alpha]_D^{23}$ +49°; -127°
21a 21b

X = $CH_2CH(CH_3)(C_6H_5)$
22a,22b

X = $SO_2CH_2CH(CH_3)(C_6H_5)$
23a,23b

Recently the amino phosphine *S*-(+)-$(C_6H_5)_2PN(CH_3)CH(CH_3)(C_6H_5)$, abbreviated with $P\emptyset_2R^*$, was applied as an optically active resolving agent in some cases *(194,195)*. In the reaction of $C_5H_5Fe(CO)_2I$ with $P\emptyset_2R^*$ the two diastereoisomers (+)- and (-)-$C_5H_5Fe(CO)(P\emptyset_2R^*)I$ are formed. The less soluble $(-)_{365}$-isomer, which can easily be obtained in larger quantities, was used as the starting material for the preparation of the compounds $(-)_{365}$-$C_5H_5Fe(CO)(P\emptyset_2R^*)X$ with X = CH_3, Br, and Cl by reaction with $LiCH_3$, Br_2, and Cl_2. In the same way $(-)_{365}$-$C_5H_5Fe(CO)(P\emptyset_2R^*)CH_3$ could be transformed into $(-)_{365}$-$C_5H_5Fe(CO)(P\emptyset_2R^*)$Hal with Hal = J, Br, and Cl by halogen cleavage of the iron-methyl bond. Reaction cycles prove that all these reactions, which involve bonds between the asymmetric Fe atom and the halogen or methyl ligands, occur with predominant retention of configuration at iron *(196)*.

The loss of stereospecificity in these reactions was attributed to the formation and pseudorotation of square pyramidal intermediates as in the reaction of the iron-methyl bond in 1-CH_3-3-C_6H_5-$C_5H_3Fe(CO)[P(C_6H_5)_3]CH_3$ with I_2, HI, and HgI_2 *(9,196)*. In contrast to the methyl derivative $(-)_{365}$-C_5H_5Fe-(CO)($PØ_2R$*)CH_3 the halogen compounds $(-)_{365}$-$C_5H_5Fe(CO)$($PØ_2R$*)-Hal in solution epimerize by a change in configuration at the Fe atom with the half lives $\tau_{\frac{1}{2}}$ (70°C) = 51 min for Hal = J, $\tau_{\frac{1}{2}}$ (50°C) = 20 min for Hal = Br, and $\tau_{\frac{1}{2}}$ (30°C) = 30 min for Hal = Cl *(196)*. Cleavage of the metal acyl and metal ester bonds in (-)-$C_5H_5Fe(CO)[P(C_6H_5)_3]COCH_3$ and (-)-$C_5H_5Fe(CO)[P$-$(C_6H_5)_3]COOC_{10}H_{19}$ with I_2 leads to optically inactive C_5H_5-$Fe(CO)[P(C_6H_5)_3]I$ *(36)*. Stereochemical changes at the α-carbon atom during the cleavage of metal carbon bonds with halogens are described in section II.B.5.

If $C_5H_5Fe(CO)_2Si(CH_3)(C_6H_5)(1$-$C_{10}H_7$ *(71)* containing an asymmetric Si atom *(50)* is irradiated in the presence of P-$(C_6H_5)_3$ one of the CO groups is replaced and two diastereoisomers (+)- and (-)-$C_5H_5Fe(CO)[P(C_6H_5)_3]Si(CH_3)(C_6H_5)(1$-$C_{10}$-$H_7)$ differing in the configuration at the Fe atom are formed with an asymmetric induction of about 10 % *(197)*. The diastereoisomers, differentiable in their NMR spectra, were separated by fractional crystallization and used as starting material for the cleavage of the Fe-Si bond. Whereas cleavage by Cl_2 occurs with retention of configuration at silicon, in the presence of an excess $P(C_6H_5)_3$ cleavage by Cl_2 proceeds with predominant inversion at Si *(197)*. The stereochemistry at the Fe centre, which should be configurationally labile according to ref. *196*, was not investigated.

With the X-ray structure of $[C_5H_5Mo(CO)_2NN']PF_6$, NN' = Schiff base of pyridine aldehyde(2) and (*S*)-(-)-α-phenylethylamine, the first absolute configuration of an optically active organometallic compound of the transition series was determined *(220,211,57)*. An extension of the *R,S*-nomenclature *(39)* to include complexes with polyhapto ligands was proposed in connection with the asymmetric synthesis of only one diastereoisomer of a pair of two differing only in the configuration at the iron atom *(239)*. In the sequence rules *(39)* polyhapto ligands should be considered pseudo-atoms of atomic weight equal to the sum of the atomic weights of all the atoms bonded to the metal atom *(239)*, see also refs. *57,142,193*.

B. *THE IRON ATOM NOT BEING THE CENTRE OF CHIRALITY*

1. Ferrocene Derivatives

Shortly after the discovery of ferrocene *(103,119)* it was found, that it can easily be acetylated in a Friedel-Crafts type reaction *(186)*. Ferrocene also undergoes other electrophilic substitution reactions with greater ease than benzene *(138)*. These facts were the starting point for the development of the organic chemistry of ferrocene which has been intensely studied in the past two decades *(147,38,154)*. Other arene π-complexes are also susceptible to ring substitution reactions, especially $(C_5H_5)_2Ru$ (ruthenocene) *(139,93)*, $C_5H_5Mn(CO)_3$ (cymantrene) *(78,51)* and $C_6H_6Cr(CO)_3$ (benchrotrene) *(143,71)*.

Into a monosubstituted ferrocene derivative a second substituent can be introduced either homoannularly or heteroannularly. If it enters the unsubstituted ring only one kind of 1,1'-derivatives is formed. If it occupies a position in the substituted ring two series of geometrical isomers 1,2 (or α) and 1,3 (or β) are possible. With two different substituents the heteroannular disubstitution products are achiral whereas the homoannularly disubstituted derivatives are chiral, as shown by their Newman type projection formulae 24a, 24b and 25a, 25b. Thus π-complex formation leads to optical isomerism not encountered in the chemistry of uncomplexed planar arenes.

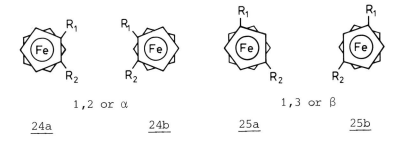

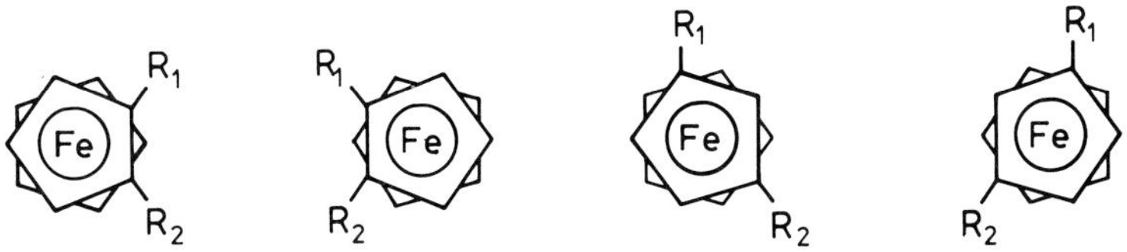

1,2 or α 1,3 or β

24a 24b 25a 25b

An increasing number of substituents gives rise to an increasing number of geometrical and optical isomers *(151-153)*. With only few exceptions the compounds do not contain any elements of symmetry. Homoannularly substituted ferrocene derivatives with two different substituents can be treated as possessing a plane of chirality *(155,39,115,151-153)*. Alternatively the 5 asymmetric carbon atoms in the disubstituted ring can be used for the specification of the configuration according to the sequence rule *(153)*.

The same situation holds for α- and β-disubstituted ruthenocene *(94,95)* and cymantrene *(144,153)* as well as for o- and m- disubstituted benchrotrene derivatives *(111,153)* if they contain different substituents. The chiral forms 26a and 26b of an α-derivative of cymantrene and 27a and 27b of an o-derivative of benchrotrene are shown *(151-153)*.

The differently substituted ferrocene compounds in most cases are synthesized by electrophilic substitution of monosubstituted products, preferentially by homoannular cycliza-

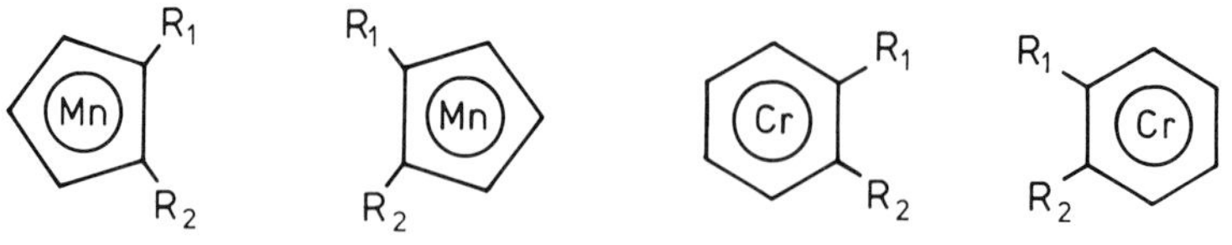

26a 26b 27a 27b

tion reactions in which only one isomer can be formed. Thus the intramolecular Friedel-Crafts reaction of γ-ferrocenyl butyric acid (28) yields 1,2-(α-ketotetramethylene)-ferrocene (29a,b) *(145)* (scheme [11]) which in 1959 was the first ferrocene derivative to be resolved into its antipodes 29a and 29b by use of its menthydrazone *(169)*.

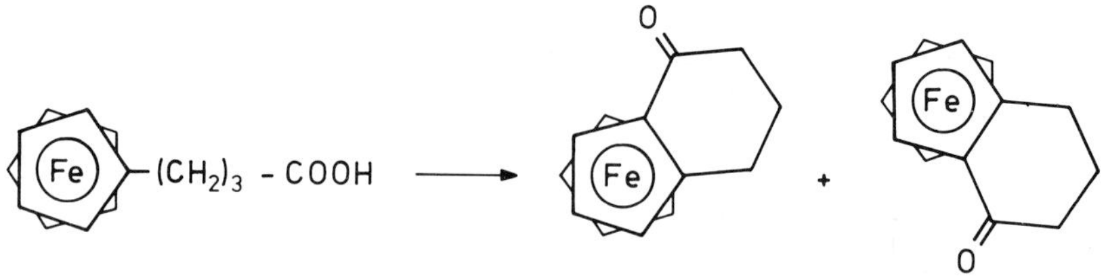

28 $[\alpha]_D$ +580° 29a $[\alpha]_D$ -580° 29b

Scheme [11]

In the meantime more than 200 optically active ferrocene derivatives have been synthesized. As the subject has been reviewed recently *(151-153)* only some examples in connection with general remarks will be given.

For the resolution of ferrocene derivatives the procedures of organic chemistry can be employed, for instance fractional crystallization of α-phenethylamine or alkaloid salts (of carboxylic acids), dibenzoyl tartrate salts (of amines) and menthydrazones (of carbonyl derivatives). Chromatography and kinetic methods were also used in some cases.

By the methods established in the organic chemistry of ferrocene *(147,38,154,212)* the functional groups of optically active ferrocene derivatives can be transformed in many ways. Reactions of this kind are used for chemical correlation of configuration. Besides the ketone 29a the methyl substituted ferrocenyl carboxylic acids *(88,89)* are key compounds of special importance. Thus methyl-ferrocenyl-α-carboxylic acid (30a) for instance can be converted into the trisubstituted derivative 32a by chain lengthening and cyclization of the corresponding butyric acid 31a *(89)* (scheme [12]).

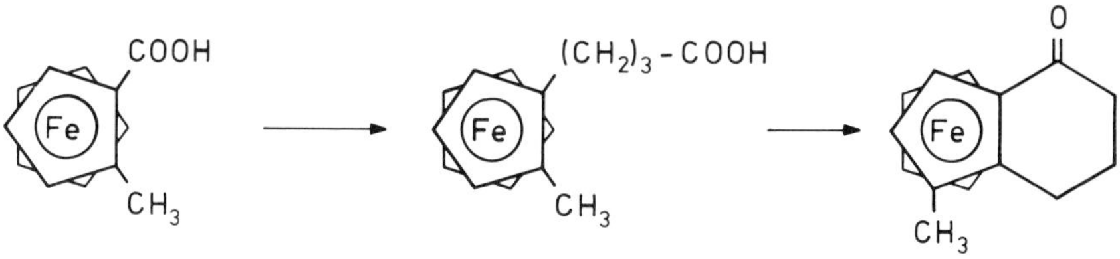

$[\alpha]_D$ +52° 30a — 31a — $[\alpha]_D$ +700° 32a

Scheme [12]

Extensive series of this kind as well as ORD- and CD-correlations link the configurations of most of the known optically active ferrocene derivatives *(151-153)*.

Many resolutions in the ferrocene series are assumed to give optically pure compounds. The optical purity of methylferrocenyl-α-carboxylic acid was examined by the isotope dilution method *(141,208)*. Because of the stereoselectivity of the transformations this optical purity is not lost in most of the reactions described.

The absolute configuration of 29a and its exo alcohol *(74,75)* was for the first time determined by the method of Horeau *(97)*. As a consequence the absolute configurations of all the compounds were known which were correlated with it. The applicability of Horeau's method for the determination of the absolute configuration of ferrocene derivatives was questioned *(115)* but an X-ray analysis of 1,1'-dimethylferrocenyl-β-carboxylic acid (33) confirmed the correct assignment of the absolute configurations *(41,76)*.

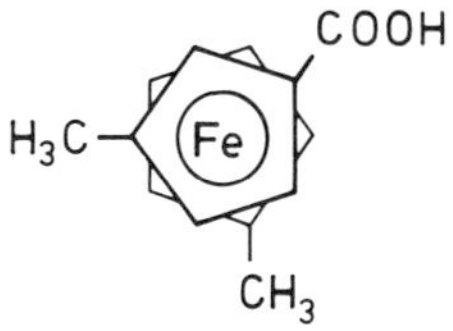

$[\alpha]_D$ -36°

33

The transformation of the optically active ketones 29a and 34a to the vinyl compounds 35a and 36a of the same configuration (scheme [13]) proceeds by reduction and subsequent dehydration. In both cases the vinyl compounds exhibit opposite signs of rotation as the parent ketones *(88)*. This significant change was extensively used for configurational correlations by optical comparison *(89)*.

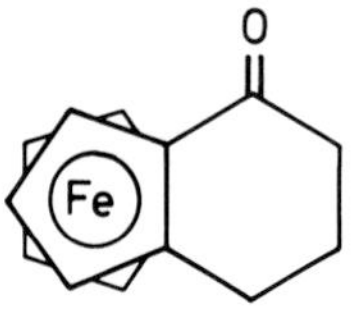

$[\alpha]_D$ +580°

29a

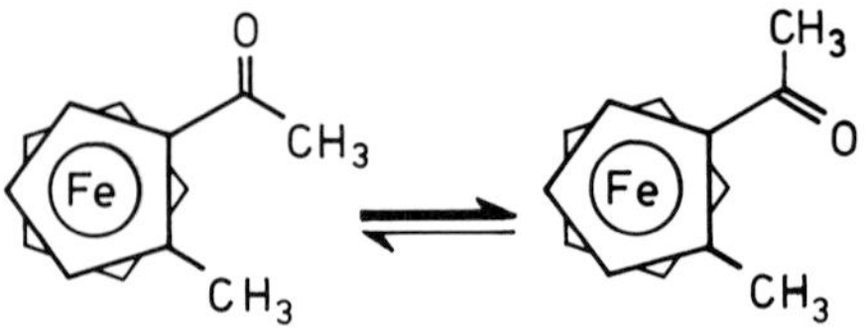

$[\alpha]_D$ -540°

34a

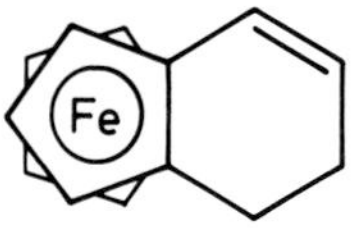

$[\alpha]_D$ -2090°

35a

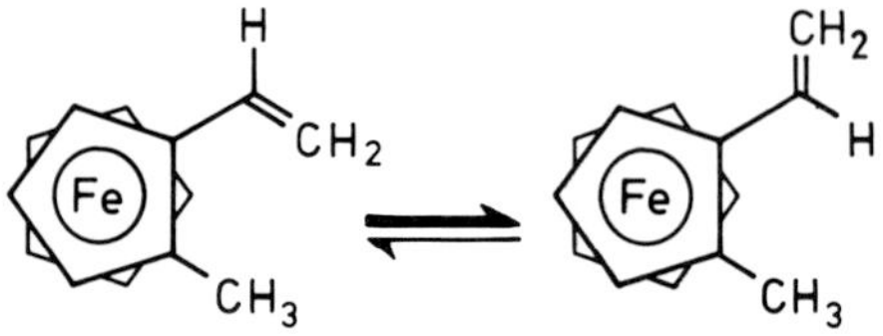

$[\alpha]_D$ +1130°

36a

Scheme [13]

It can also be applied to the elucidation of conformational problems. The ORD and the CD spectra of the open chain ketone 34a and its vinyl derivative 36a are opposite to the analogous cyclic derivatives 29a and 35a in which the chromophores are conformationally fixed. This fact was explained by assuming that the open chain compounds prefer the opposite conformations, imposed on the molecules by the methyl-methyl interactions in the acyclic complexes *(88,72)*.

An empirical rule concerning the relationship between configuration, preferred conformation and sign of rotation has been developed - valid for most of the optically active ferrocene derivatives *(153,77,82)*. The observer looks along the molecular axis onto the molecule with the differently substituted ring pointing towards him. Then the sign of $[\alpha]_D$ is (+) if the chromophor disturbing the ferrocene system is on the left side and (-) if the chromophor is on the right side of the plane bisecting the molecule as indicated in 29a and 35a.

Optically active [3]ferrocenophanes (1,1'-trimethyleneferrocene derivatives) were included in the stereochemical studies *(153)*. Correlation with known compounds could be achieved. New problems associated with ferrocenophanes are posed by the conformations of the bridge between the two cyclopentadienyl rings. For recent references see *(204-206, 218)*. Optically active ferrocenyl compounds containing an a-

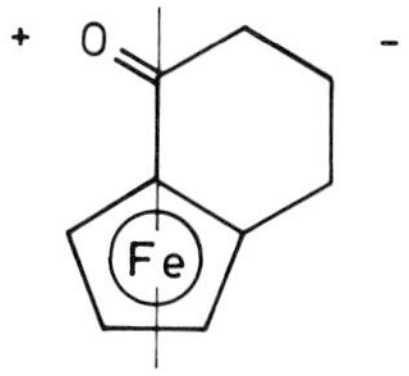

$[\alpha]_D$ +580°

29a

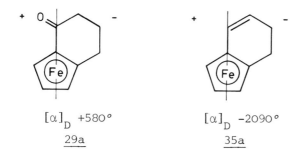

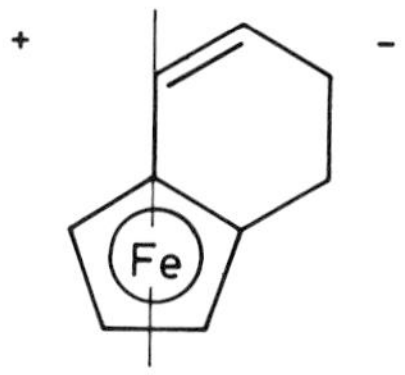

$[\alpha]_D$ -2090°

35a

symmetric Si atom in the side chain were prepared and studied with respect to the stereochemistry at silicon *(198-201)*. The asymmetry and pseudoasymmetry, which arises in ferrocene molecules when two homoannular sustituents are constitutionally identical, were described *(213-215)*.

For further details concerning optically active ferrocene derivatives see the reviews of Schlögl *(151-153)*, the last annual surveys *(146,116,225,233)*, and summaries *(135, 154,212)*. Only some recent development will be discussed extensively in the following paragraphs.

The lithiation of ferrocene derivatives has been carried out under asymmetric conditions *(4,5,77,114,115)*. In the metalation of N,N-dimethyl-1-ferrocenyl ethyl amine (37a) *(115,85)*, the optical activity of which is due to an asymmetric carbon atom in the α-position, with n-butyl-lithium the metal enters the 2-position of the ferrocene system because the nitrogen atom of the amine group interacts with the attacking metal. In 38a the methyl group is directed away from the iron atom and the unsubstituted ring, in 38a', however, it strongly interferes with the ferrocene system. Therefore the product ratio of the diastereoisomers 39a and 39a' is 96:4 in favour of 39a *(115)*. Starting from the organometallic derivatives 38 a variety of groups may be introduced. As the asymmetric group $-CH(CH_3)(NMe_2)$ can be transformed to achiral substituents, *i.e.* the vinyl group, the highly stereoselective lithiation of 37a represents a new synthesis of optically active ferrocene derivatives of known configuration *(115,176, 161,165)*.

These conclusions concerning the configuration were corroborated by an X-ray determination of the absolute configuration of 40a, the reaction product of 38a in scheme [14] with anisaldehyde *(15,16)*.

The assumption that no racemization takes place during reaction and work up of optically active ferrocene derivatives was the basis for the chemical correlation of their configurations and the assignment of minimum values for their optical

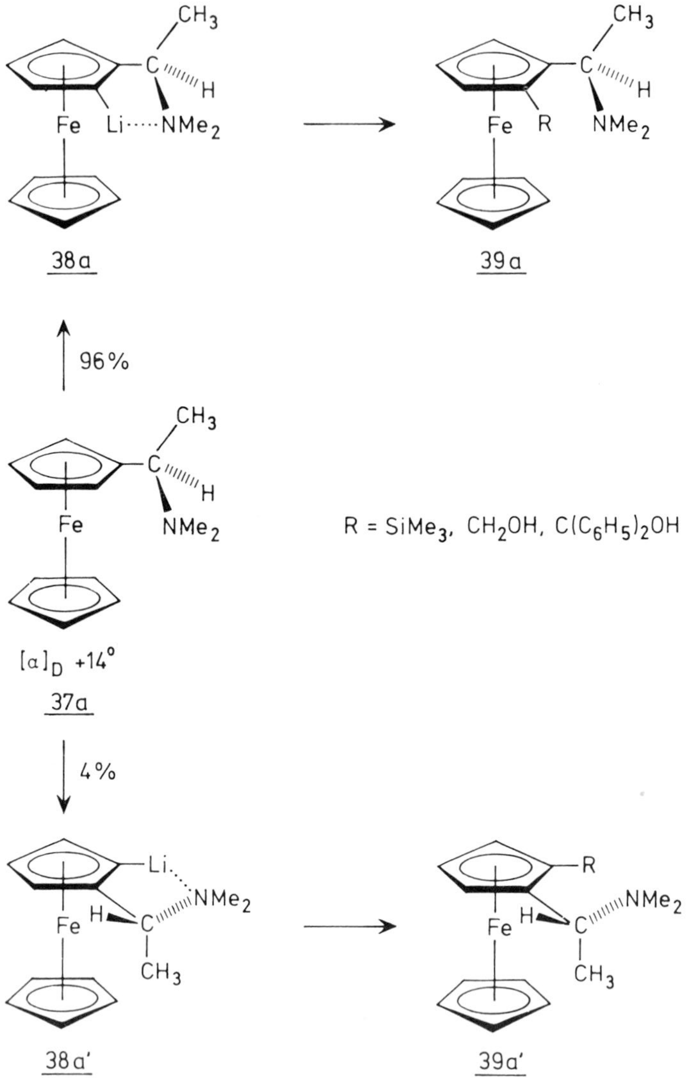

Scheme [14]

purities *(151-153)*. The more surprising was the discovery of Slocum *et al.* that (+)-1,2-(α-ketotetramethylene)ferrocene (29a) racemizes at 100°C in nitromethane in the presence of $AlCl_3$ *(163)*. After 1 hour only half of the rotation is left. Decomposition can be excluded as an explanation for the loss of optical activity. A racemization mechanism with cleavage

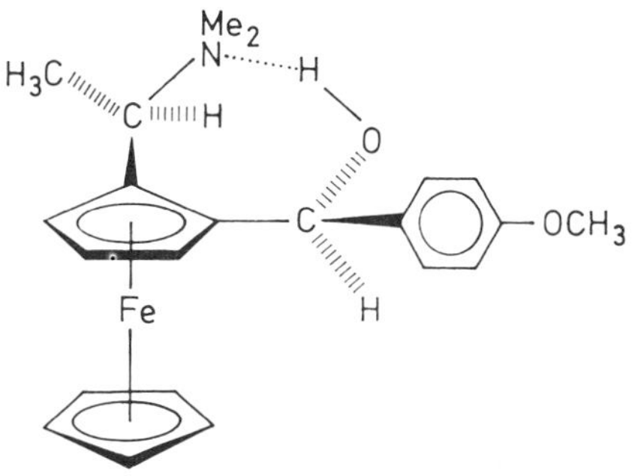

$[\alpha]_D^{22}$ -21°

40a

of the ring metal bond is suggested *(163)*. Acids such as $FeCl_3$, $BF_3 \cdot OEt_2$, H_3PO_4 or $HClO_4$ instead of $AlCl_3$ also catalyse the racemisation of optically active ferrocene derivatives *(17)*. According to kinetic measurements, deuteration experiments and solvent variation this reaction is an intramolecular process of first order *(73)*. These investigations show that the scope for this new racemization is so limited that no corrections of the values hitherto obtained for optical purity seem to be necessary. The same conclusions were drawn from an epimerization study of ketones similar to 29a, which in addition to the planar chirality contain a methyl phenyl substituted chiral carbon center in the six membered ring *(2)*. The epimerization is thermodynamically controlled favouring the isomers with the phenyl substituent in the exo position. In addition an unexpected change in configuration at the quaternary asymmetric carbon atom in the six membered ring was observed *(2)*.

Besides from the so called metallocene chirality optical activity of ferrocene derivatives may also arise from asymmetric carbon atoms in the functional groups. Because of the pronounced anchimeric effect of the metallocenyl substituent unexpected reactions at the carbon centres in the side chain of ferrocene derivatives may be encountered. Thus, whereas nucleophilic substitution normally is accompanied by racemisation or inversion depending on whether the reaction proceeds by an S_N1 or S_N2 mechanism, retention stereochemistry was observed for nucleophilic substitution of α-substituted alkylferrocenes. The optically active ferrocenyl ethyl ammonium ion (41a), prepared by quaternization of the corresponding N,N-dimethyl-1-ferrocenyl ethyl amine (37a), reacts with NaN_3, NH_3 or $NHMe_2$ with complete retention of configuration *(83)* (scheme [15]).

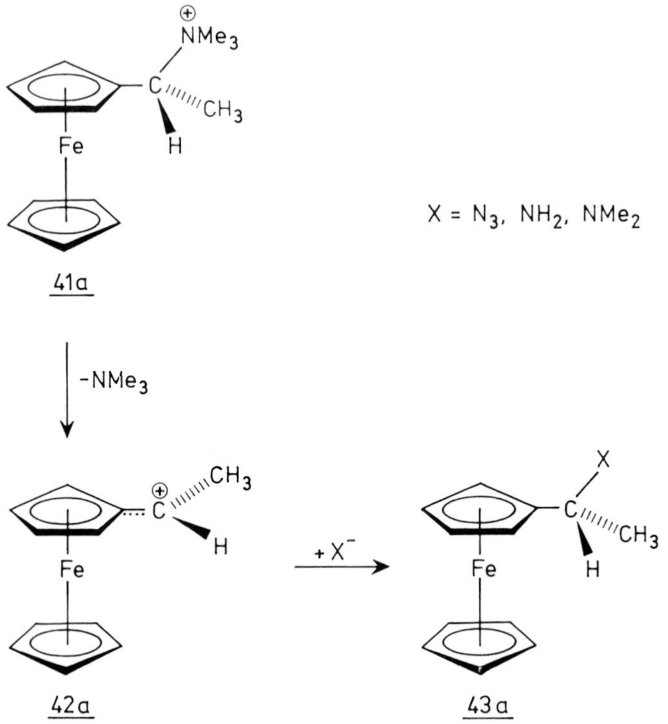

Scheme [15]

In this reaction NMe_3 departs from the exo position of the conformation 41a away from the iron atom and the stable α-ferrocenyl carbonium ion 42a is formed. The nucleophile X^- attacks the carbonium ion from the outside with formation of 43a before rotation around the bond between the ferrocene unit and the α-carbon atom occurs *(83)*. Both enantiomers of N,N-dimethyl-1-ferrocenyl ethyl amine are easy to obtain *(115,85)* and represent a convenient starting material for the synthesis of O and N substituted optically active α-ferrocenyl compounds *(83a,85)*. Extensive stereochemical investigations showed that most of the reactions similar to those in scheme [15] proceed with complete retention of configuration *(83a,60,61,84,85,115, 164,165,218,191,192,230)*. The optically active amine 37a was used in stereoselective peptide synthesis by four component condensations *(81,83,84,113,114,173,174,242)*.

The intermediates of scheme [15], the enantiomeric α-ferrocenyl methyl carbonium ions 42a and 42b, were observed on dissolution of the corresponding 1-ferrocenyl ethanols 44a

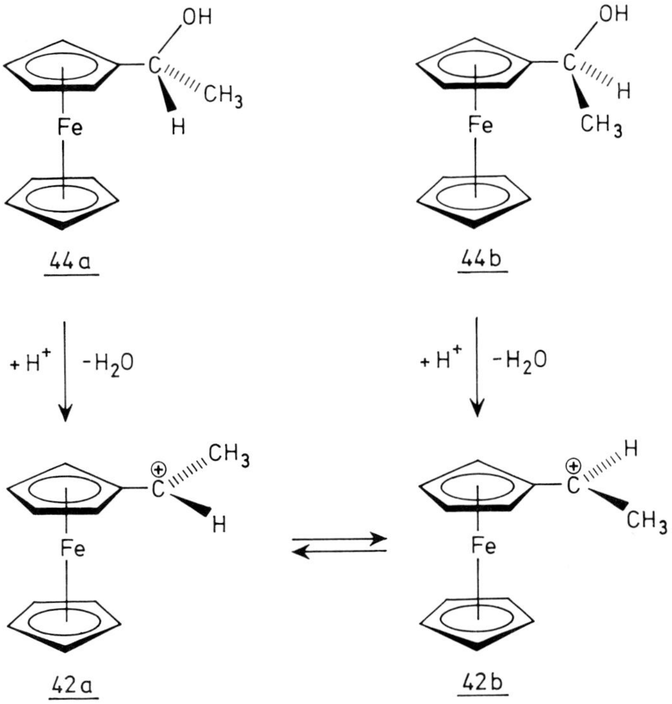

Scheme [16]

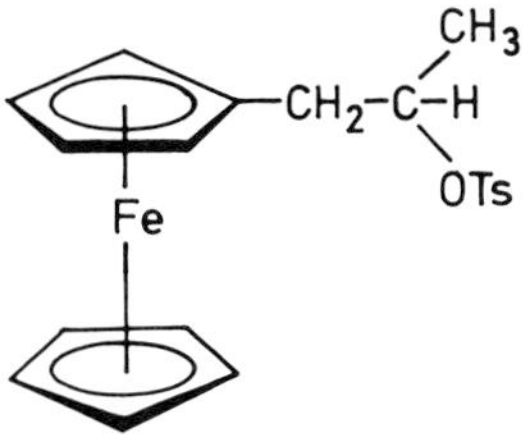

$[\alpha]_D$ -58°

45

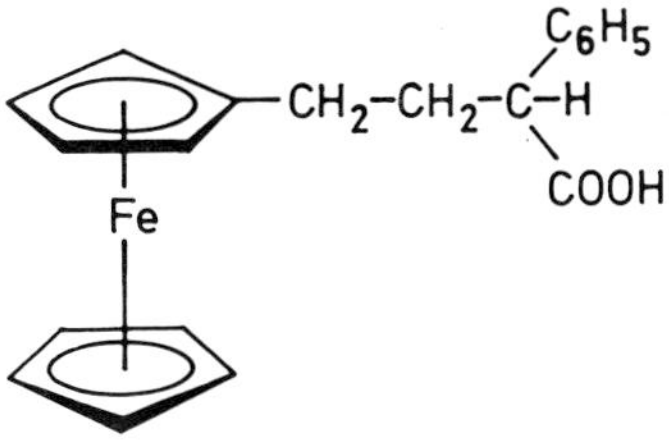

$[\alpha]_D$ +67°

46

and 44b in CF_3COOH (scheme [16]) and their optical rotations could be determined. On standing racemization took place because the two carbonium ions interconvert slowly by rotation around the exo-cyclic bond. In the temperature range 40-60°C a first order process was established by polarimetric kinetics, the free energy barrier to rotation being 19.5 kcal/mole in accord with earlier spectroscopic results *(171,172,192, 230)*. The corresponding α-ruthenocenyl methyl carbonium ions are configurationally stable and do not racemize during several days at 70°C *(172)*.

Retention of configuration was also observed in the solvolysis of optically active 1-ferrocenyl-2-propyl tosylate (45) with the asymmetric centre in the β-position with respect to the ferrocene system *(126)*. Optically active 1-ferrocenyl-2-amino-propane with an asymmetric centre in the β-position of the side chain was prepared *(231)*. Its optical purity could be determined by NMR spectroscopy *(232)*. High stereoselectivity induced by an asymmetric centre in the γ-position of the side chain was also found in the homoannular cyclization reaction of α-phenyl-γ-ferrocenyl-butyric acid (46) with trifluoroacetic acid anhydride *(75a,1)*. See also refs. *203,219, 243*.

2. Diene Complexes

If a substituent is introduced into the diene part of a diene tricarbonyliron complex (47) geometrical isomerism arises because it may be bonded to one of the inner carbon atoms 2,3 or to one of the outer carbon atoms 1,4 of the *cis*-diene moiety. Substitution at the carbon atoms 1 or 4 gives rise to *cis-trans* isomerism. Each of these geometrical isomers occurs in enantiomeric pairs. Whereas two different substituents in the same ring are necessary to cause metallocene chirality, in diene-$Fe(CO)_3$ compounds one substituent is sufficient to generate chiral forms.

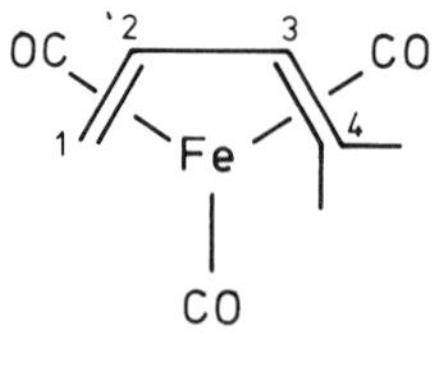

47

Much work in the field of diene-$Fe(CO)_3$ complexes has been carried out with diastereoisomerically related pairs of enantiomers but only few diene-$Fe(CO)_3$ complexes have been obtained in optically active form up to now: Acids of type 48a were resolved by formation of salts with brucine and α-phenethylamine and ketones of type 49a were isolated from the reaction of the acid chlorides of the optically active acids 48 with $Cd(CH_3)_2$ *(122-124)*.

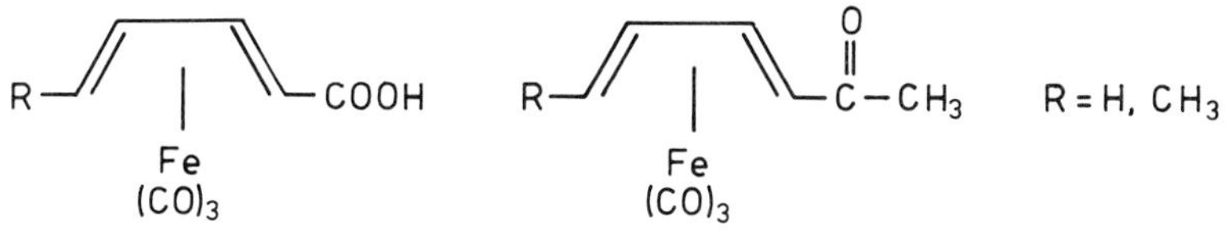

48a 49a

The configurations of the optically active diene-$Fe(CO)_3$ complexes 48 and 49 and similar compounds were correlated by use of the lowest energy transition at 390 mμ of the CD spectra which was assigned to a dissymmetrically perturbed *d-d* transition of the metal atom *(122,124)*.

Exo-attack of nucleophiles on the carbon atoms 2 and 6 of the dienyl tricarbonyliron cation 50, normally leading

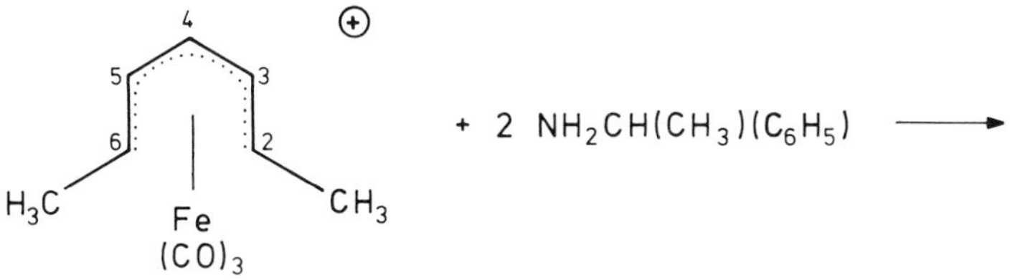

50 [17]

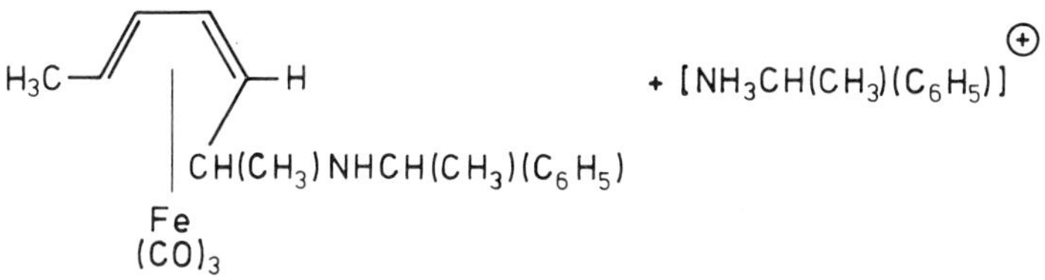

$[\alpha]_{589}$ -203°

51a

to enantiomeric pairs of substituted diene-$Fe(CO)_3$ complexes, in the case of the optically active nucleophile α-phenethylamine results in the formation of two diastereoisomers which could be separated by column chromatography *(107,109)*. The product of exo-addition to C_2, 51a, is shown in equation [17]. An X-ray analysis of (-)-2-(α-phenethylamine)-*cis,trans*-hepta-3,5-dienyl tricarbonyliron (51a) revealed that the two independent chiral centres C_2 and C_3 formed in reaction [17] have the same *R*-configuration *(107,108)*.

Whereas optically active ferrocene derivatives are configurationally stable in the absence of acids *(163,17,73,2)*, the optically active diene-$Fe(CO)_3$ complexes 52a and 53a racemize above 100°C according to first order kinetics.

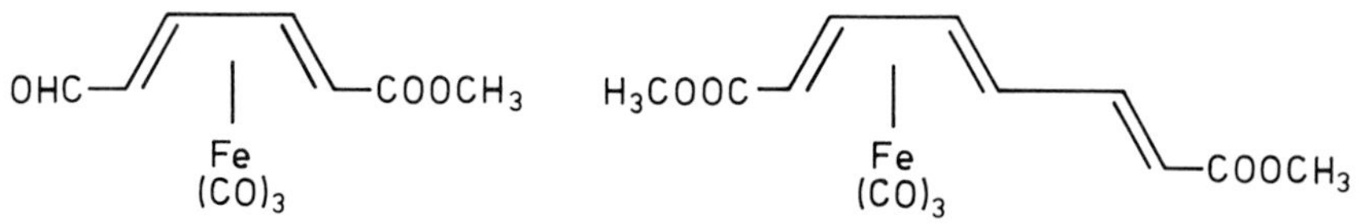

52a 53a

The rate of the racemization of 53a was found to be 2.6 times faster than the rate of $Fe(CO)_3$ shift from the 1,3- to the 3,5-position in the hexatriene system. This was explained by the mechanism shown in scheme [18] *(182)*.

From the η^4-diene-$Fe(CO)_3$ complexes 54a and 54b η^2-diene-

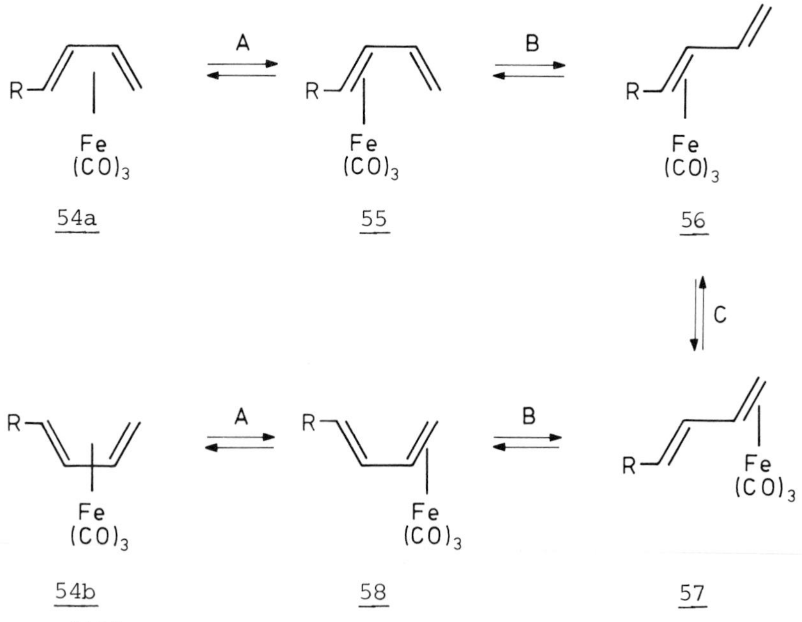

Scheme [18]

$Fe(CO)_3$ intermediates 55 and 58 are formed reversibly (steps A). Rotation about all single bonds not involved in bonding to the $Fe(CO)_3$ moiety is possible (steps B). The processes A and B account for the shift of the $Fe(CO)_3$ group in unsaturated systems but do not change the configuration. Racemization is only brought about by the shift of the $Fe(CO)_3$ group in a dihapto complex to a *trans* double bond (step C) *(182)*. These investigations were extended to octatetraene complexes with one *(182)* or two *(112)* $Fe(CO)_3$ groups.

The ferracyclopentadiene derivatives 59 were resolved through the menthoxy acetate ($R = C_2H_5$) or the camphor sulfonate methyl ether ($R = t\text{-}C_4H_9$) *(42)*. The optically active forms 59a and 59b obtained by alkaline hydrolysis of the separated diastereoisomers racemize at elevated temperatures by an intramolecular interchange of roles of the two $Fe(CO)_3$ groups. The optically active ferracyclopentadiene derivatives 59a and 59b are included in this section because in the transition state 60 of their racemization [19] a diene system symmetrically bonded between the two $Fe(CO)_3$ groups is postulated *(42,148)*.

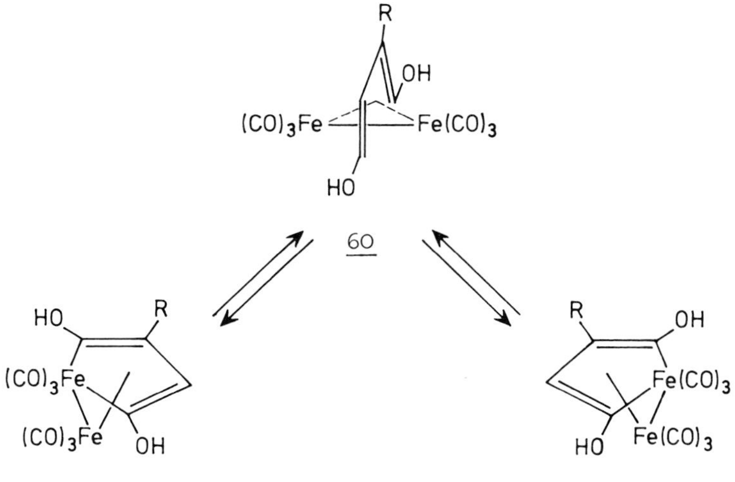

60

59a

59b

Cyclobutadiene complexes containing non-identical substituents in adjacent positions on the four membered ring are chiral. The first optical resolutions of 1,2-disubstituted cyclobutadiene derivatives were carried out with 1-methyl-2-(dimethylaminomethyl)cyclobutadiene tricarbonyliron. 61a was obtained in 30-40% optical purity by 2 fractional crystallizations of its (+)-camphor-10-sulfonate *(86)*. The resolved complex showed less than 5% decrease of optical rotation on being heated at 120°C for 48 hours *(86)*. Thus optically active cyclobutadiene complexes are much more configurationally stable than butadiene complexes. This was explained by assuming that the cyclobutadiene ligand must become completely detached from the metal for a racemization to occur, whereas the butadiene complex requires the decomplexation of only one bond in the racemization process *(86,182)*.

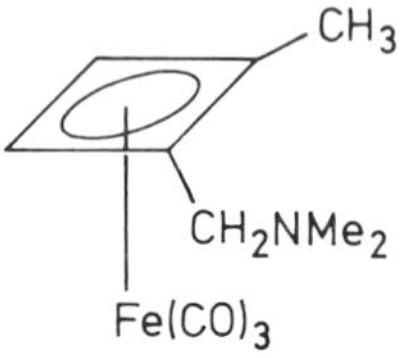

$[\alpha]_{578}$ 17.3°

61a

The (-)-1-acetyl-2-carboxy-cyclobutadiene derivative 62a was prepared by resolution with quinine in at least 99% optical purity *(156,157)*. Its optical rotation is strongly solvent dependent. 62a is reduced with B_2H_6/BF_3 to the optically active 1-ethyl-2-methyl-cyclobutadiene tricarbonyliron (63a), which is configurationally stable up to 200°C *(156,157)* (scheme [20]). The ring metal bond in 63a was cleaved with Ce^{4+} in the presence of the dienophiles dimethyl maleate, maleic anhydride or tetracyanoethylene. The formation of completely racemic Diels-Alder adducts demonstrates the intermediacy of free cyclobutadiene in these reactions *(156,217,235)*. On the other hand oxidative degradation of 63a to an optically active cyclobutene shows, that this process involves the cyclobutadiene system still attached to the metal atom *(234)*.

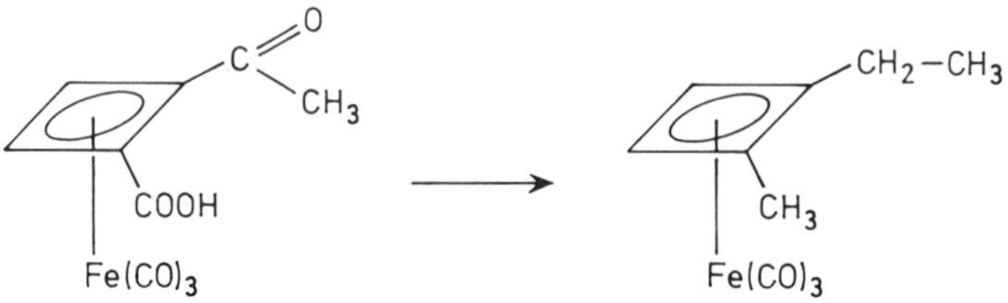

$[\alpha]^{24}_{578}$ -102° $[\alpha]^{24}_{578}$ -20.5°

62a 63a

Scheme [20]

3. Allyl Complexes

Three geometrical isomers are possible for monosubstituted π-allyl complexes depending on whether the substituent is bonded to the middle carbon atom C_2 or either *syn* or *anti* to the terminal carbon atoms C_1 and C_3. Whereas the first type of substitution leads to achiral compounds the terminally substituted derivatives should occur in enantiomeric pairs.

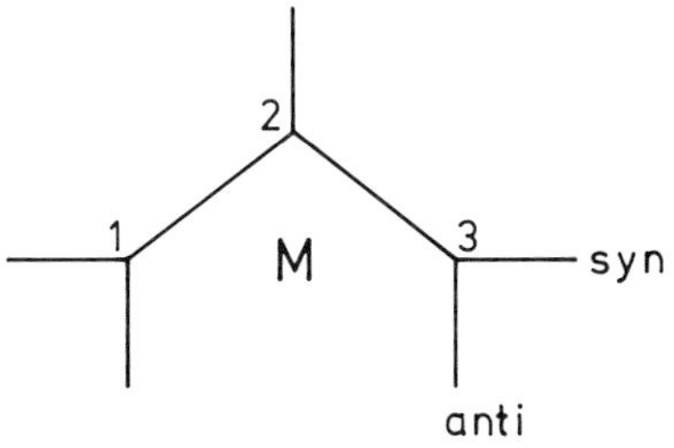

64

Optical activity arising from π-complexed substituted allylic systems has been demonstrated for Pd compounds *(47)*. It should also be observable in suitably substituted allylic complexes of iron, but no examples are known up to date.

4. Mono-olefin Complexes

The optical activity of olefin metal complexes arises from the metal olefin moiety, if the olefin is either chiral *(39)* or prochiral *(120,91)*. Chiral olefins contain at least one element of chirality, mostly chiral centres; they are characterized by diastereotopic faces. Prochiral olefins do not contain symmetry planes perpendicular to the plane of the olefin *(128,131)*; they are characterized by enantiotopic faces.

The subject of optically active olefin metal complexes has been reviewed recently *(127,92)*. The metals Pt and Pd, which form stable complexes with a variety of olefins *(92)* have been extensively studied. Resolution is achieved by introducing an optically active ligand L, preferentially (+)- or (-)-$H_2NCH(CH_3)(C_6H_5)$ into *cis* or *trans* platinum(II)-halide complexes with suitable olefins *(127)*. After separation of the diastereoisomers they can be converted into enantiomeric complexes by substitution of the optically active ligand L by an achiral ligand L' or the olefins can be split off from the complexes *(127)*. Thus cyclic olefins like cyclooctene *(45, 46)*, cyclononene *(44)*, and cyclodecene *(44)* were resolved.

Most of the optically active olefin complexes epimerize or racemize in solution, especially in the presence of olefins. This process occurs by dissociation of the metal-olefin bond; excess olefin leads to rapid exchange *(128,131,132)*.

With the absolute configuration of platinum-olefin complexes established by X-ray analyses *(18,133)*, the CD spectra can be used for the determination of the absolute configuration. The sign of the first *d-d* transition in the CD spectra was correlated to the absolute configuration *(48,187,159)*.

Whereas Pt and Pd form many complexes with ethylene and its alkyl and aryl derivatives, Fe favours bonding to activated olefins carrying electronegative substituents *(92)*.

The first mono-olefin $Fe(CO)_4$ complex prepared in optically active form was fumaric acid $Fe(CO)_4$ *(130)*. The resolution was accomplished *via* fractional crystallization of the mono brucine salt. Cleavage with HCl yielded the enantiomeric complexes (-)- and (+)-fumaric acid $Fe(CO)_4$ 65a and 65b *(130)*.

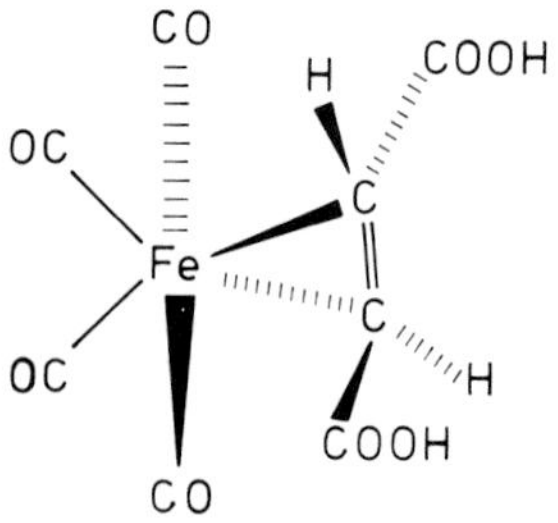

$[\alpha]_D$ -578°

65a

In the same way the resolution of acrylic acid $Fe(CO)_4$ was carried out *(129)*. As was to be expected maleic acid $Fe(CO)_4$, the olefinic ligand of which contains a plane of symmetry perpendicular to the plane of the olefin, could not be resolved *(130)*.

The crystal structure of (-)-fumaric acid $Fe(CO)_4$ (65a) was determined by an X-ray analysis *(49,134)*. The configuration around the iron atom is nearly trigonal bipyramidal the olefin being an equatorial substituent. Three different types of molecules occur in the crystal lattice in two of which the C=C axis is inclined to the equatorial plane. The 4 carbon atoms of the olefin ligand are not coplanar. The absolute configuration of (-)-fumaric acid $Fe(CO)_4$ is *R,R* *(49,127)*. The absolute configuration of (+)-acrylic acid $Fe(CO)_4$ is *S* at the unique asymmetric carbon atom *(122-124)*. The magnitude of the circular dichroism in the 350 mμ transition of (+)-acrylic acid $Fe(CO)_4$, containing one asymmetric carbon atom, is about half the value of (-)-fumaric acid $Fe(CO)_4$, containing two asymmetric carbon atoms. Therefore it is inferred that each asymmetric carbon atom contributes by the same amount to the optical activity of these compounds *(122-124)*.

5. Alkyl Complexes and Related Compounds

The stereochemical changes accompanying the formation, modification, and cleavage of iron alkyl bonds has been discussed in section II.A. as far as the asymmetric iron atom is concerned. The stereochemistry of these reactions with respect to the carbon atoms of the side chain is described in the following paragraphs.

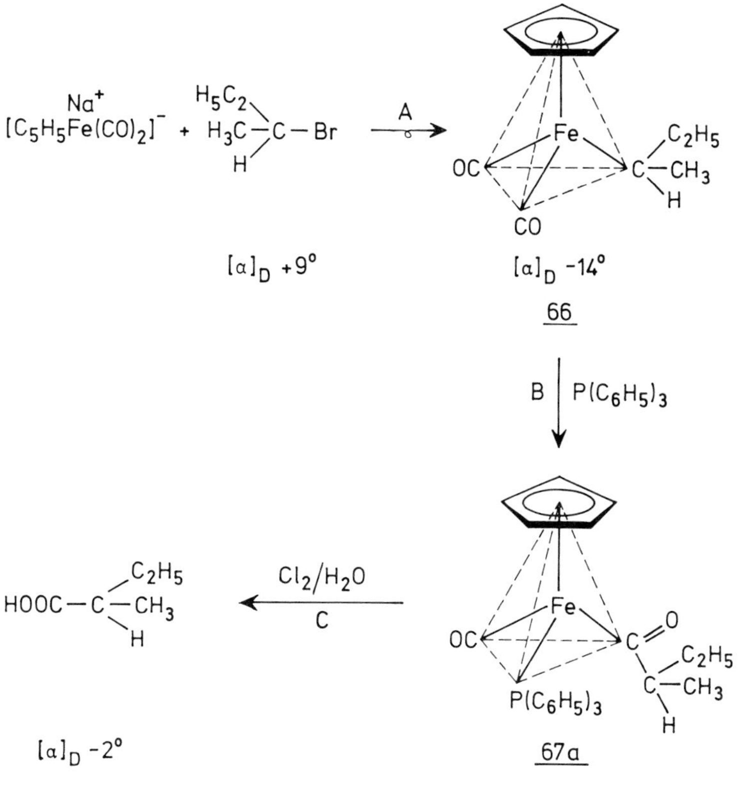

Scheme [21]

The optically active complex (-)-$C_5H_5Fe(CO)_2CH(CH_3)(C_2H_5)$ (66) was prepared by the reaction of $Na[C_5H_5Fe(CO)_2]$ with D-(+)-2-bromobutane (step A, scheme [21]) *(102)*. The attack of a transition metal nucleophile on an alkyl halide leading to displacement of the halide ion occurs with inversion of configuration at the carbon atom *(180,181,19,125)* as inferred from scheme [21] by assuming retention stereochemistry for the carbonylation step B and the cleavage step C *(102)*. For electrophilic cleavage of Fe-C σ-bonds similar to that in step C both retention and inversion of configuration at the α-C-atom have been encountered *(102,19,9,236,210)*.

As in the carbonylation step B of scheme [21] high stereospecificity at the chiral carbon atom has also been found in the decarbonylation of the acyl compound 68 *(3)*. This is in accord with the findings that alkyl migration during carbonyl insertion proceeds with retention of configuration *(184,40, 180,19)*.

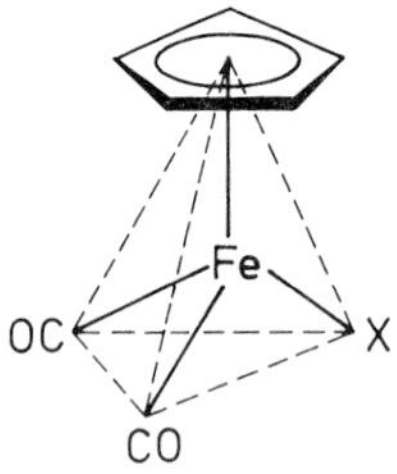

X = CO-CH(CH_3)(C_6H_5) 68 $[\alpha]_{546}^{27}$ -64°

X = CH(CH_3)(C_6H_5) 69 $[\alpha]_{546}^{27}$ +78°

X = SO_2CH(CH_3)(C_6H_5) 70 $[\alpha]_{546}^{27}$ -186°

X = SiMePhNp 71 $[\alpha]_D^{25}$ -25.5°

Similar to the reactions shown in scheme [21] the nucleophilic substitution in the first step of the ketone synthesis of scheme [22] occurs with inversion at the chiral carbon atom of *S*-(+)-2-octyl tosylate, whereas the carbonyl insertion takes place with retention of configuration *(43)*.

The SO_2 insertion was studied with (+)-$C_5H_5Fe(CO)_2CH(CH_3)$-(C_6H_5) (69). In the sulfonation reaction (-)-$C_5H_5Fe(CO)_2SO_2$-$CH(CH_3)(C_6H_5)$ (70) was obtained with slightly differing optical rotations depending on the reaction conditions, indicating high stereospecificity with respect to the α-carbon atom of the alkyl group *(3)*.

The stereochemistry at the α-carbon atom during carbonylation and sulphur dioxide insertion was demonstrated unambiguously by an elegant piece of work using deuterated *threo-erythro* isomers of the type $C_5H_5Fe(CO)_2CHDCHDCMe_3$. From the

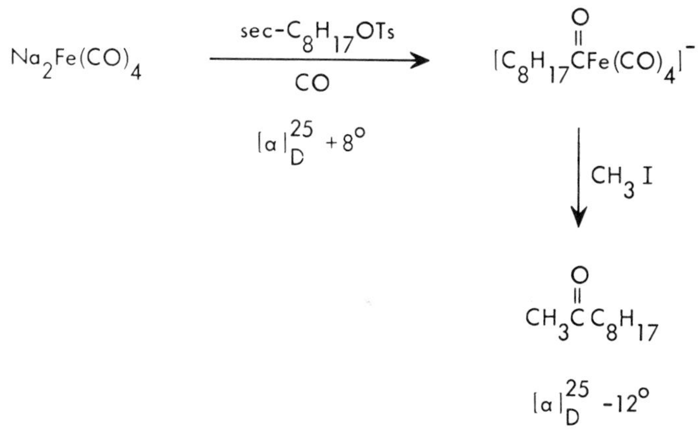

Scheme [22]

NMR analysis of the corresponding reaction products the stereochemistry with regard to the α-carbon atom of the alkyl group of reactions at the Fe-C bond can be inferred *(180,181,19,20)*. It could be shown that the carbonylation step proceeds with retention of configuration, whereas the SO_2 insertion occurs with inversion of configuration at the α-carbon atom.

As far as the SO_2 insertion is concerned the following suggestions for the mechanism can be made: Back side attack of SO_2 on the α-carbon atom of the alkyl group with inversion of configuration at the α-carbon atom affords an ion pair M^+-RSO_2^- from which the O and S bonded products MSO_2R are formed with retention of configuration at the metal atom *(185,79,19)*. In an extension of this mechanism an epimerization at the α-carbon atom of $C_5H_5Fe(CO)[P(C_6H_5)_3]CH(C_6H_5)Si(CH_3)_3$ in liquid SO_2 was reported *(239,240)*.

The preparation of the optically active cyclopentadienyl dicarbonyliron derivatives 17 and 18 and their transformations have been mentioned in section II.A. *(142)*. Also, optically active iron complexes like 71 containing an asymmetric silicon atom were synthesized and used to study the cleavage of the Fe-Si bond with respect to the stereochemistry at silicon *(50, 197-201)*.

REFERENCES

1. Abbayes, H.D., and Dabard, R., *C.R. Acad. Sci., Ser. C, 276,* 1763 (1973).
2. Abbayes, H.D., and Dabard, R., *J. Organometal. Chem., 61,* C51 (1973).
3. Alexander, J.J., and Wojcicki, A., *Inorg. Chim. Acta, 5,* 655 (1971).
4. Aratani, T., Gonda, T., and Nozaki, H., *Tetrahedron Lett., 1969,* 2265.
5. Aratani, T., Gonda, T., and Nozaki, H., *Tetrahedron, 26,* 5453 (1970).
6. Archer, R.D., and Dollberg, D.D., *Inorg. Chem., 13,* 1551 (1974).
7. Archer, R.D., Suydam, L.J., and Dollberg, D.D., *J. Amer. Chem. Soc., 93,* 6837 (1971).
8. Attig, T.G., Reich-Rohrwig, P., and Wojcicki, A., *J. Organometal. Chem., 51,* C21 (1973).
9. Attig, T.G., and Wojcicki, A., *J. Amer. Chem. Soc., 96,* 262 (1974).
10. Attig, T.G., and Wojcicki, A., *J. Organometal. Chem., 82,* 397 (1974).
11. Bailar, J.C., Jr., *J. Inorg. Nucl. Chem., 8,* 165 (1959).
12. Basolo, F., Hayes, J.C., and Neumann, H.M., *J. Amer. Chem. Soc., 75,* 5102 (1953).
13. Basolo, F., Hayes, J.C., and Neumann, H.M., *J. Amer. Chem. Soc., 76,* 3807 (1954).
14. Basolo, F., and Pearson, R.G., *Mechanisms of Inorganic Reactions,* Wiley, New York, 1967, p. 247.
15. Battelle, L.F., Bau, R., Gokel, G.W., Oyakawa, R.T., and Ugi, I., *Angew. Chem., 84,* 164 (1972); *Angew. Chem. Int. Ed. Engl., 11,* 138 (1972).
16. Battelle, L.F., Bau, R., Gokel, G.W., Oyakawa, R.T., and Ugi, I., *J. Amer. Chem. Soc., 95,* 482 (1973).
17. Bauer, K., Falk, H., Lehner, H., Schlögl, K., and Wagner, U., *Monatsh. Chem., 101,* 941 (1970).
18. Benedetti, E., Corradini, P., and Pedone, C., *J. Organometal. Chem., 18,* 203 (1969).
19. Bock, P.L., Boschetto, D.J., Rasmussen J.R., Demers, J.P., and Whitesides, G.M., *J. Amer. Chem. Soc., 96,* 2814 (1974).
20. Bock, P.L., and Whitesides, G.M., *J. Amer. Chem. Soc., 96,* 2826 (1974).
21. Broomhead, J.A., and Dwyer, F.P., *Aust. J. Chem., 16,* 51 (1963).
22. Brunner, H., *Angew. Chem., 81,* 395 (1969); *Angew. Chem. Int. Ed. Engl., 8,* 382 (1969).

23. Brunner, H., *Angew. Chem., 83,* 274 (1971); *Angew. Chem. Int. Ed. Engl., 10,* 249 (1971).
24. Brunner, H., *Ann. N.Y. Acad. Sci., 239,* 213 (1974).
25. Brunner, H., *Chimia, 25,* 284 (1971).
26. Brunner, H., *Top. Cur. Chem., 56,* 67 (1975).
27. Brunner, H., Aclasis, J., Langer, M., and Steger, W., *Angew. Chem., 86,* 864 (1974); *Angew. Chem. Int. Ed. Engl., 13,* 810 (1974).
28. Brunner, H., and Schindler, H.-D., *Chem. Ber., 104,* 2467 (1971).
29. Brunner, H., and Schindler H.-D., *J. Organometal. Chem., 24,* C7 (1970).
30. Brunner, H., and Schindler, H.-D., *Z. Naturforsch., 26b,* 1220 (1971).
31. Brunner, H., Schindler, H.-D., Schmidt, E., and Vogel, M., *J. Organometal. Chem., 24,* 515 (1970).
32. Brunner, H., and Schmidt, E., *Angew. Chem., 81,* 570 (1969); *Angew. Chem. Int. Ed. Engl., 8,* 616 (1969).
33. Brunner, H., and Schmidt, E., *J. Organometal. Chem., 21,* P53 (1970).
34. Brunner, H., and Schmidt, E., *J. Organometal. Chem., 36,* C18 (1972).
35. Brunner, H., and Schmidt, E., *J. Organometal. Chem., 50,* 219 (1973).
36. Brunner, H., and Strutz, J., *Z. Naturforsch., 29b,* 446 (1974).
37. Brunner, H., and Vogel, M., *J. Organometal. Chem., 35,* 169 (1972).
38. Bublitz, D.E., and Rinehart, K.L., jr., *Organic Reactions, 17,* 1 (1969).
39. Cahn, R.S., Ingold, C., and Prelog, V., *Angew. Chem., 78,* 413 (1966); *Angew. Chem. Int. Ed. Engl., 5,* 385 (1966).
40. Calderazzo, F., and Noack, K., *Coord. Chem. Rev., 1,* 118 (1966).
41. Carter, O.L., McPhail, A.T., and Sim, G.A., *J. Chem. Soc., A, 1967,* 365.
42. Case, R., Jones, E.R.H., Schwartz, N.V., and Whiting, M.C., *Proc. Chem. Soc., 1962,* 256.
43. Collman, J.P., Winter, S.R., and Clark, D.R., *J. Amer. Chem. Soc., 94,* 1788 (1972).
44. Cope, A.C., Banholzer, K., Keller, H., Pawson, B.A., Whang, J.J., and Winkler, H.J.S., *J. Amer. Chem. Soc., 87,* 3644 (1965).
45. Cope, A.C., Ganellin, C.R., Johnson, H.W., van Auken, T.V., and Winkler, H.J.S., *J. Amer. Chem. Soc., 85,* 3276 (1963).
46. Cope, A.C., and Pawson, B.A., *J. Amer. Chem. Soc., 87,*

3649 (1965).
47. Corradini, P., Maglio, G., Musco, A., and Paiaro, G., *Chem. Commun.*, *1966*, 618.
48. Corradini, P., Paiaro, G., Panunzi, A., Mason, S.F., and Searle, G.H., *J. Amer. Chem. Soc.*, *88*, 2863 (1966).
49. Corradini, P., Pedone, C., and Sirigu, A., *Chem. Commun.*, *1968*, 275.
50. Corriu, R.J.P., and Douglas, W.E., *J. Organometal. Chem.*, *51*, C3 (1973).
51. Cotton, F.A., and Leto, J.R., *Chem. Ind.*, *1958*, 1368.
52. Davies, N.R., *Rev. Pure Appl. Chem.*, *4*, 66 (1954).
53. Davies, N.R., and Dwyer, F.P., *Trans. Faraday Soc.*, *49*, 180 (1953).
54. Davies, N.R., and Dwyer, F.P., *Trans. Faraday Soc.*, *50*, 820 (1954).
55. Davies, N.R., and Dwyer, F.P., *Trans. Faraday Soc.*, *50*, 1325 (1954).
56. Davison, A., Krusell, W.C., and Michaelson, R.C., *J. Organometal. Chem.*, *72*, C7 (1974).
57. Davison, A., and Martinez, N., *J. Organometal. Chem.*, *74*, C17 (1974).
58. Davison, A., and Reger, D.L., *J. Amer. Chem. Soc.*, *94*, 9237 (1972).
59. Dickens, J.E., Basolo, F., and Neumann, H.M., *J. Amer. Chem. Soc.*, *79*, 1286 (1957).
60. Dixneuf, P., *Tetrahedron Lett.*, *1971*, 1561.
61. Dixneuf, P., and Dabard, R., *Bull. Soc. Chim. France*, *1972*, 2847.
62. Dowley, P., Garbett, K., and Gillard, R.D., *Inorg. Chim. Acta*, *1*, 278 (1967).
63. Dwyer, F.P., Gill, N.S., Gyarfas, E.C., and Lions, F., *J. Amer. Chem. Soc.*, *75*, 3834 (1953).
64. Dwyer, F.P., and Gyarfas, E.C., *J. Amer. Chem. Soc.*, *74*, 4699 (1952).
65. Dwyer, F.P., and Gyarfas, E.C., *J. Proc. Roy. Soc. N.S. Wales*, *83*, 263 (1949).
66. Dwyer, F.P., and Gyarfas, E.C., *J. Proc. Roy. Soc. N.S. Wales*, *85*, 126 (1951).
67. Dwyer, F.P., and Gyarfas, E.C., *J. Proc. Roy. Soc. N.S. Wales*, *85*, 135 (1951).
68. Dwyer, F.P., and Gyarfas, E.C., *Nature*, *168*, 29 (1951).
69. Dwyer, F.P., Gyarfas, E.C., Rogers, W.P., and Koch, J.H., *Nature*, *170*, 190 (1952).
70. Dwyer, F.P., and Sargeson, A.M., *J. Phys. Chem.*, *60*, 1331 (1956).
71. Ercoli, R., and Calderazzo, F., *Chim. Ind. (Milan)*, *41*, 404 (1959).

72. Falk, H., Haller, G., and Schlögl, K., *Monatsh. Chem.*, *98*, 2058 (1967).
73. Falk, H., Lehner, H., Paul, J., and Wagner, U., *J. Organometal. Chem.*, *28*, 115 (1971).
74. Falk, H., and Schlögl, K., *Angew. Chem.*, *76*, 570 (1964); *Angew. Chem. Int. Ed. Engl.*, *3*, 512 (1964).
75. Falk, H., and Schlögl, K., *Monatsh. Chem.*, *96*, 266 (1965).
75a. Falk, H., and Schlögl, K., *Monatsh. Chem.*, *96*, 1065 (1965).
76. Falk, H., and Schlögl, K., *Monatsh. Chem.*, *102*, 33 (1971).
77. Falk, H., and Schlögl, K., *Tetrahedron*, *22*, 3047 (1966).
78. Fischer, E.O., and Pleszke, K., *Chem. Ber.*, *91*, 2719 (1958).
79. Flood, T.C., and Miles, D.L., *J. Amer. Chem. Soc.*, *95*, 6460 (1973).
80. Gillard, R.D., McKenzie, E.D., Mason, R., Mayhew, S.G., Peel, J.L., and Stangroom, J.E., *Nature*, *208*, 769 (1965).
81. Gokel, G., Hoffmann, P., Kleimann, H., Klusacek, H., Lüdke, G., Marquarding, D., and Ugi, I. in I. Ugi, *Isonitrile Chemistry*, Academic Press, New York, 1971, *p. 201.*
82. Gokel, G., Hoffmann, P., Kleimann, H., Klusacek, H., Marquarding, D., and Ugi, I., *Tetrahedron Lett.*, *1970*, 1771.
83. Gokel, G., Hoffmann, P., Klusacek, H., Marquarding, D., Ruch, E., and Ugi, I., *Angew. Chem.*, *82*, 77 (1970); *Angew. Chem. Int. Ed. Engl.*, *9*, 64 (1970).
83a. Gokel, G.W., Marquarding, D., and Ugi, I., *J. Org. Chem.*, *37*, 3052 (1972).
84. Gokel, G.W., and Ugi, I.K., *Angew. Chem.*, *83*, 178 (1971); *Angew. Chem. Int. Ed. Engl.*, *10*, 191 (1971).
85. Gokel, G.W., and Ugi, I.K., *J. Chem. Educ.*, *49*, 294 (1972).
86. Grubbs, R.H., and Grey, R.A., *Chem. Commun.*, *1973*, 76.
87. Gyarfas, E.C., *Rev. Pure Appl. Chem.*, *4*, 73 (1954).
88. Haller, G., and Schlögl, K., *Monatsh. Chem.*, *98*, 603 (1967).
89. Haller, G., and Schlögl, K., *Monatsh. Chem.*, *98*, 2044 (1967).
90. Hanazaki, I., and Nagakura, S., *Inorg. Chem.*, *8*, 654 (1969).
91. Hanson, K.R., *J. Amer. Chem. Soc.*, *88*, 2731 (1966).
92. Herberhold, M., *Metal π-Complexes, Vol. II, Complexes with Monoolefinic Ligands, Part 1,*

Elsevier, Amsterdam, 1972.
93. Hofer, O., and Schlögl, K., *J. Organometal. Chem.*, *13*, 443 (1968).
94. Hofer, O., and Schlögl, K., *J. Organometal. Chem.*, *13*, 457 (1968).
95. Hofer, O., and Schlögl, K., *Tetrahedron Lett.*, *1967*, 3485.
96. Holm, R.H., *Accounts Chem. Res.*, *2*, 307 (1969).
97. Horeau, A., *Tetrahedron Lett.*, *1961*, 506.
98. *Inorg. Chem.* *9*, 1 (1970).
99. Jaeger, M.F.M., *Rec. Trav. Chim. Pays-Bas*, *38*, 171 (1919).
100. Jensen, A., Basolo, F., and Neumann, H.M., *J. Amer. Chem. Soc.*, *80*, 2354 (1958).
101. Johnson, C.H., *Trans. Faraday Soc.*, *28*, 845 (1932).
102. Johnson, R.W., and Pearson, R.G., *Chem. Commun.*, *1970*, 986.
103. Kealy, T.J., and Pauson, P.L., *Nature*, *168*, 1039 (1951).
104. Kirschner S., and Ahmad, N., *Coordination Chemistry*, Plenum Press, New York, 1969, p. 42.
105. Kral, M., *Collect. Czech. Chem. Commun.*, *37*, 3985 (1972).
106. Krebs, H., Diewald, J., Arlitt, H., and Wagner, J.A., *Z. Anorg. Allg. Chem.*, *287*, 98 (1956).
107. Maglio, G., Musco, A., and Palumbo, R., *J. Organometal. Chem.*, *32*, 127 (1971).
108. Maglio, G., Musco, A., Palumbo, R., and Sirigu, A., *Chem. Commun.*, *1971*, 100.
109. Maglio, G., and Palumbo, R., *J. Organometal. Chem.*, *76*, 367 (1964).
110. Malatesta, L., and Bonati, F., *Isocyanide Complexes of Metals*, *Wiley*, New York, 1969.
111. Mandelbaum, A., Neuwirth,Z., and Cais, M., *Inorg. Chem.*, *2*, 902 (1963).
112. Markezich, R.L., and Witlock, H.W., *J. Amer. Chem. Soc.*, *93*, 5291 (1971).
113. Marquarding, D., Hoffmann, P., Heitzer, H., and Ugi, I., *J. Amer. Chem. Soc.*, *92*, 1969 (1970).
114. Marquarding, D., Klusacek, H., Gokel, G., Hoffmann, P., and Ugi, I., *Angew. Chem.*, *82*, 360 (1970); *Angew. Chem. Int. Ed. Engl.*, *9*, 371 (1970).
115. Marquarding, D., Klusacek, H., Gokel, G., Hoffmann, P., and Ugi, I., *J. Amer. Chem. Soc.*, *92*, 5389 (1970).
116. Marr, G., and Rockett, B.W., *J. Organometal. Chem.*, *58*, 323 (1973).
117. Mason, S.F., Peart, B.J., and Waddell, R.E., *J. Chem. Soc., Dalton Trans.*, *1973*, 949.

118. McCaffery, A.J., Mason, S.F., and Norman, B.J., *J. Chem. Soc. A, 1969*, 1428 and references therein.
119. Miller, S.A., Tebboth, J.A., and Tremaine, J.F., *J. Chem. Soc., 1952*, 632.
120. Mislow, K., and Raban, M., *Top. Stereochem., 1*, 1 (1967).
121. Moeller, T., and Gulyas, E., *J. Inorg. Nucl. Chem. 5*, 245 (1957).
122. Musco, A., Paiaro, G., and Palumbo, R., *Chim. Ind. (Milan), 50*, 669 (1968).
123. Musco, A., Palumbo, R., and Paiaro, G., *Inorg. Chim. Acta, 5*, 157 (1971).
124. Musco, A., Paiaro, G., and Palumbo, R., *Ric. Sci., 39*, 417 (1969).
125. Nicholas, K.M., and Rosenblum, M., *J. Amer. Chem. Soc., 95*, 4449 (1973).
126. Nugent, M.J., Carter, R.E., and Richards, J.H., *J. Amer. Chem. Soc., 91*, 6145 (1969).
127. Paiaro, G., *Organometal. Chem. Rev. A, 6*, 319 (1970).
128. Paiaro, G., Corradini, P., Palumbo, R., and Panunzi, A., *Makromol. Chem. 71*, 184 (1964).
129. Paiaro, G., and Palumbo, R., *Gazz. Chim. Ital., 97*, 265 (1967).
130. Paiaro, G., Palumbo, R., Musco, A., and Panunzi, A., *Tetrahedron Lett., 1965*, 1067.
131. Paiaro, G., and Panunzi, A., *J. Amer. Chem. Soc., 86*, 5148 (1964).
132. Panunzi, A., and Paiaro, G., *J. Amer. Chem. Soc., 88*, 4843 (1966).
133. Pedone, C., and Benedetti, E., *J. Organometal. Chem., 31*, 403 (1971).
134. Pedone, C., and Sirigu, A., *Inorg. Chem., 7*, 2614 (1968).
135. Peet, J.H., and Rockett, B.W., *Rev. Pure Appl. Chem., 22*, 145 (1972).
136. Pfeiffer, P., and Quehl, K., *Ber. dtsch. chem. Ges., 64*, 2667 (1931).
137. Pignolet, L.H., *Top. Curr. Chem., 56*, 91 (1975).
138. Plesske, K., *Angew. Chem., 74*, 301, 347 (1962).
139. Rausch, M.D., Fischer, E.O., and Grubert, H., *Chem. Ind. [London], 1958*, 756.
140. Ray, P.C., and Dutt, N.K., *J. Indian Chem. Soc., 20*, 81 (1943).
141. Reich-Rohrwig, P., and Schlögl, K., *Monatsh. Chem., 99*, 1752 (1968).
142. Reich-Rohrwig, P., and Wojcicki, A., *Inorg. Chem., 13*, 2457 (1974).
143. Riemschneider, R., Becker, O., and Franz, K.,

Monatsh. Chem., *90*, 571 (1959).

144. Riemschneider, R., and Herrmann, W., *Justus Liebigs Ann. Chem.*, *648*, 68 (1961).

145. Rinehart, K.L., and Curby, R.J., *J. Amer. Chem. Soc.*, *79*, 3290 (1957).

146. Rockett, B.W., and Marr, G., *J. Organometal. Chem.*, *45*, 389 (1972).

147. Rosenblum, M., *Chemistry of the Iron Group Metallocenes; Ferrocene, Ruthenocene, Osmocene, Part I*, Interscience, New York, 1965.

148. Rosenblum, M., Giering, W.P., North, B., and Wells, D., *J. Organometal. Chem.*, *28*, C17 (1971).

149. Sanders, N., *J. Chem. Soc., Dalton Trans.*, *1972*, 345.

150. Sarma, B.D., and Bailar, J.C., Jr., *J. Amer. Chem. Soc.*, *77*, 5476 (1955).

151. Schlögl, K., *Fortschr. Chem. Forsch.*, *6*, 479 (1966).

152. Schlögl, K., *Pure Appl. Chem.*, *23*, 413 (1970).

153. Schlögl, K., *Top. Stereochem.*, *1*, 39 (1967).

154. Schlögl, K., and Falk, H., *Methodicum Chimicum*, Thieme-Verlag, Stuttgart, 1974, Vol. VIII, p. 433.

155. Schlögl, K., and Fried, M., *Monatsh. Chem.*, *95*, 558 (1964).

156. Schmidt, E.K.G., *Angew. Chem.*, *85*, 820 (1973); *Angew. Chem. Int. Ed. Engl.*, *12*, 777 (1973).

157. Schmidt, E.K.G., *Chem. Ber.*, *107*, 2440 (1974).

158. Schmulbach, C.D., Dachille, F., and Bunch, M.E., *Inorg. Chem.*, *3*, 808 (1964).

159. Scott, A.I., and Wrixon, A.D., *Chem. Commun.*, *1969*, 1184.

160. Seiden, L., Basolo, F., and Neumann, H.M., *J. Amer. Chem. Soc.*, *81*, 3809 (1959).

161. Shirafuji, T., Odaira, A., Yamamoto, Y., and Nozaki, H., *Bull. Chem. Soc. Jap.*, *45*, 2884 (1972).

162. Shulman, A., and Dwyer, F.P., in F.P. Dwyer and D.P. Mellor, *Chelating Agents and Metal Chelates*, Academic Press, New York, 1964, Chap. 9.

163. Slocum, D.W., Tucker, S.P., and Engelmann, T.R., *Tetrahedron Lett.*, *1970*, 621.

164. Sok, K.C.Y., Tainturier, G., and Gautheron, B., *C.R. Acad. Sci., Ser. C*, *278*, 1347 (1974).

165. Tainturier, G., Sok, K.C.Y., and Gautheron, B., *C.R. Acad. Sci., Ser. C*, *277*, 1269 (1973).

166. Tatehata, A., Kumamaru, T., and Yamamoto, Y., *J. Inorg. Nucl. Chem.*, *33*, 3427 (1971).

167. Templeton, D.H., Zalkin, A., and Ueki, T., *Acta Crystallogr.*, *21*, A 154 (supplement) (1966).

168. Thomas, W., *J. Chem. Soc.*, *119*, 1140 (1921).

169. Thomson, B., *Tetrahedron Lett., 1959* (6), 26.
170. Treichel, P.M., Shubkin, R.L., Barnett, K.W., and Reichard, D., *Inorg. Chem., 5,* 1177, (1966).
171. Turbitt, T.D., and Watts, W.E., *J. Chem. Soc. Chem. Commun., 1973,* 182.
172. Turbitt, T.D., and Watts, W.E., *J. Chem. Soc., Perkin Trans. II, 1974,* 177.
173. Ugi, I., *Intra. Sci. Chem. Rep., 5,* 229 (1971).
174. Ugi, I., *Rec. Chem. Progr., 30,* 289 (1969).
175. Ugi, I., Fetzer, U., Eholzer, U., Kupfer, H., and Offermann, K., *Neuere Methoden der präparativen organischen Chemie, Vol. IV,* Verlag Chemie, Weinheim, 1966, p. 37.
176. Valkovich, P.B., Gokel, G.W., and Ugi, I.K., *Tetrahedron Lett., 1973,* 2947.
177. Vogler, A., in I. Ugi, *Isonitrile Chemistry,* Academic Press, New York, 1971, p. 217.
178. Werner, A., *Ber. dtsch. chem. Ges., 44,* 1887 (1911).
179. Werner, A., *Ber. dtsch. chem. Ges., 45,* 433 (1912).
180. Whitesides, G.M., and Boschetto, D.J., *J. Amer. Chem. Soc., 91,* 4313 (1969).
181. Whitesides, G.M., and Boschetto, D.J., *J. Amer. Chem. Soc., 93,* 1529 (1971).
182. Whitlock, H.W., Jr., and Markezich, R.L., *J. Amer. Chem. Soc., 93,* 5290 (1971).
183. Wilkins, R.G., and Williams, M.J.G., *J. Chem. Soc., 1957,* 1763.
184. Wojcicki, A., *Adv. Organometal. Chem., 11,* 87 (1973).
185. Wojcicki, A., *Adv. Organometal. Chem., 12,* 31 (1974).
186. Woodward, R.B., Rosenblum, M., and Whiting, M.C., *J. Amer. Chem. Soc., 74,* 3458 (1952).
187. Wrixon, A.D., Premuzic, E., and Scott, A.I., *Chem. Commun., 1968,* 639.
188. Yamamoto, Y., Akabori, K., and Seno, T., *Inorg. Nucl. Chem. Lett., 9,* 195 (1973).
189. Zalkin, A., Templeton, D.H., and Ueki, T., *Inorg. Chem., 12,* 1641 (1973).
190. Ahmad, N., and Kirschner, S., *Inorg. Chim. Acta, 14,* 215 (1975).
191. Allenmark, S., and Kalen, K., *Tetrahedron Lett., 1975,* 3175.
192. Allenmark, S., Kalen, K., and Sandblom, A., *Chem. Scr., 7,* 97 (1975).
193. Alt, H., Herberhold, M., Kreiter, C.G., and Strack, H., *J. Organometal. Chem., 77,* 353 (1974).
194. Brunner, H., and Doppelberger, J., *Bull. Soc. Chim. Belg., 84,* 923 (1975).
195. Brunner, H., and Rambold, W., *Angew. Chem., 85,* 1118

(1973); *Angew. Chem. Int. Ed. Engl., 12,* 1013 (1973).
196. Brunner, H., and Wallner, G., *Chem. Ber., in press.*
197. Cerveau, G., Colomer, E., Corriu, R., and Douglas, W.E., *J. Chem. Soc. Chem. Commun., 1975,* 410.
198. Chauviere, G., Corriu, R., and Royo, G., *J. Organometal. Chem., 78,* C7 (1974).
199. Corriu, R.J.P., Larcher, F., and Royo, G., *J. Organometal. Chem., 104,* 161 (1976).
200. Corriu, R.J.P., Larcher, F., and Royo, G., *J. Organometal. Chem., 104,* 293 (1976).
201. Corriu, R.J.P., Larcher, F., and Royo, G., *J. Organometal. Chem., 92,* C18 (1975).
202. Davies, N.R., and Dwyer, F.P., *Trans. Faraday Soc., 48,* 244 (1952).
203. Des Abbayes, H., and Dabard, R., *Tetrahedron, 31,* 2111 (1975).
204. Dodey, P., and Gautheron, B., *J. Organometal. Chem., 94,* 441 (1975).
205. Dodey, P., and Gautheron, B., *C.R. Acad. Sci., Ser. C, 280,* 1113 (1975).
206. Dodey, P., and Gautheron, B., *C.R. Acad. Sci., Ser. C, 281,* 127 (1975).
207. Dollberg, D.D., and Archer, R.D., *Inorg. Chem., 14,* 1888 (1975).
208. Eberhardt, R., Glotzmann, C., Lehner, H., and Schlögl, K., *Tetrahedron Lett., 1974,* 4365.
209. Faller, J.W., and Anderson, A.S., *J. Amer. Chem. Soc., 91,* 1550 (1969).
210. Flood, T.C., and DiSanti, F.J., *J. Chem. Soc. Chem. Commun., 1975,* 18.
211. Flood, T.C., DiSanti, F.J., and Miles, D.L., *J. Chem. Soc. Chem. Commun., 1975,* 336.
212. Füssel, J., and Wagner, J., in U. Krüerke and A. Slawisch, *Gmelin, Vol. 14, Eisen-Organische Verbindungen, Part A, Ferrocen, 1,* 1974.
213. Goldberg, S.I., and Bailey, W.D., *J. Amer. Chem. Soc., 93,* 1046 (1971).
214. Goldberg, S.I., and Bailey, W.D., *J. Amer. Chem. Soc., 96,* 6381 (1974).
215. Goldberg, S.I., and Bailey, W.D., *Tetrahedron Lett., 1971,* 4087.
216. Gordon, J.G., O'Connor, M.J., and Holm, R.H., *Inorg. Chim. Acta, 5,* 381 (1971).
217. Grubbs, R.H., and Grey, R.A., *J. Amer. Chem. Soc., 95,* 5765 (1973).
218. Khay, C.Y.S., Tainturier, G., and Gautheron, B., *Tetrahedron Lett., 1974,* 2207.
219. Kimny, T., Moise, C., and Tainturier, G., *C.R. Acad.*

Sci., Ser. C, 278, 1157 (1974).

220. LaPlaca, S.J., Bernal, I., Brunner, H., and Herrmann, W.A., *Angew. Chem., 87,* 379 (1975); *Angew. Chem. Int. Ed. Engl., 14,* 353 (1975).

221. Leong, J., and Raymond, K.N., *Biochem. Biophys. Res. Commun., 60,* 1066 (1974).

222. Leong, J., and Raymond, K.N., *J. Amer. Chem. Soc., 96,* 1757 (1974).

223. Leong, J., and Raymond, K.N., *J. Amer. Chem. Soc., 96,* 6628 (1974).

224. Leong, J., and Raymond, K.N., *J. Amer. Chem. Soc., 97,* 293 (1975).

225. Marr, G., and Rockett, B.W., *J. Organometal. Chem., 106,* 259 (1976).

226. Mason, S.F., *Inorg. Chim. Acta Rev., 2,* 89 (1968).

227. Nord, G., *Acta Chem. Scand., 27,* 743 (1973).

228. Pannell, K.H., *Chem. Commun., 1969,* 1346.

229. Pfeiffer, P., Christeleit, W., Hesse, T., Pfitzner, H., and Thielert, H., *J. prakt. Chem., 150,* 261 (1938).

230. Ratajczak, A., and Misterkiewicz, B., *J. Organometal. Chem., 91,* 73 (1975).

231. Ratajczak, A., and Zmuda, H., *Bull. Acad. Pol. Sci., Ser. Sci. Chim., 22,* 261 (1974).

232. Ratajczak, A., and Zmuda, H., *Rocz. Chem., 49,* 215 (1975).

233. Rockett, B.W., and Marr, G., *J. Organometal. Chem., 79,* 223 (1974).

234. Schmidt, E.K.G., *Chem. Ber., 108,* 1598 (1975).

235. Schmidt, E.K.G., *Chem. Ber., 108,* 1609 (1975).

236. Slack, D., and Baird, M.C., *J. Chem. Soc. Chem. Commun., 1974,* 701.

237. Stanley, K., and Baird, M.C., *Inorg. Nucl. Chem. Lett., 10,* 1111 (1974).

238. Stanley, K., and Baird, M.C., *J. Amer. Chem. Soc., 97,* 4292 (1975).

239. Stanley, K., and Baird, M.C., *J. Amer. Chem. Soc., 97,* 6598 (1975).

240. Stanley, K., Groves, D., and Baird, M.C., *J. Amer. Chem. Soc., 97,* 6599 (1975).

241. Stanley, K., Zelonka, R.A., Thompson, J., Fiess, P., and Baird, M.C., *Can. J. Chem., 52,* 1781 (1974).

242. Urban, R., and Ugi, I., *Angew. Chem., 87,* 67 (1975); *Angew. Chem. Int. Ed. Engl., 14,* 61 (1975).

243. Uysal, H., and Gautheron, B., *C.R. Acad. Sci., Ser. C, 278,* 1297 (1974).

244. Zalkin, A., Forrester, J.D., and Templeton, D.H., *J. Amer. Chem. Soc., 88,* 1810 (1966).

THE ORGANIC CHEMISTRY OF IRON, VOLUME 1

Compounds with Iron-Carbon σ-Bonds

By F.L. BOWDEN and L.H. WOOD

Chemistry Department,
University of Manchester,
Institute of Science and Technology,
Manchester, England

TABLE OF CONTENTS

ISBN 0-12-417101-X (v.1)

I. INTRODUCTION

This chapter is a summary of the preparation, properties and reactions of compounds which contain an iron-carbon σ-bond. The emphasis is on preparative routes and reaction pathways common to a range of compounds although many individual variations occur.

For the purposes of this chapter an iron-carbon σ-bond is assumed to involve overlap between a filled sp^2 or sp^3 orbital on carbon and a suitable vacant orbital on the metal, metal-to-carbon π-bonding making at the most a very minor contribution. Thus, the complexes R-ML_n and R-CO-ML_n (R = alkyl, aryl, alkenyl, benzyl, allyl and propargyl; ML_n = iron-ancillary ligand combination) and their fluorocarbon analogues are included but metal carbonyls, nitriles, iso-nitriles and acetylides are not. Also excluded are the, mainly heterocyclic, iron-carbon compounds obtained from reactions between iron complexes and acetylenes. These have been reviewed elsewhere *(23)*. There are several excellent reviews which include aspects of the chemistry of iron-carbon σ-bonds *(35,75,107,113, 134,158)*.

Simple iron alkyls of the type R_2Fe or R_3Fe are unknown and only a few aryls Ar_2Fe (Ar = *p*-NH_2-C_6H_4, Ph-CH_2, *p*-Me-C_6H_4, and *p*-MeO-C_6H_4) have been reported *(123)*. The rarity of such compounds has been attributed to the intrinsic weakness of σ-bonds between transition metals and carbon *(37,38,54,84)*. However, what evidence there is points to comparable strengths for transition metal-carbon and main-group metal-carbon bonds *(25,55,126)*. Recently, it has been suggested that the apparent *instability* of transition metal-carbon σ-bonds relative to their main-group metal analogues is an increased *lability* due to the availability of lower activation energy pathways for decomposition; that is, the instability is kinetic rather than thermodynamic *(25,126)*. The implication is that it is the availability of non-bonding *d*-orbitals which in some way contributes to the lower activation energies and any means whereby these orbitals can be made less accessible for reaction is likely to decrease the lability of the metal-carbon bond *(37)*. Qualitative observations on the stabilizing effect on the metal-carbon bond of ancillary ligands such as η^5-C_5H_5(= Cp), CO, and Ph_3P support this view. Although the precise role of such ligands is not fully understood; nor are the quantitative data available which are necessary to distinguish between the various possibilities. Optimistically, if iron-carbon σ-bonds are thermodynamically stable, then it is to be anticipated that the range of compounds containing such bonds will be extended as the ability of new ancillary ligands to increase their kinetic stability is investigated.

II. PREPARATION OF IRON-CARBON σ-BONDS

A REACTIONS OF METALATE ANIONS WITH RX

The most widely used routes to iron-carbon σ-bonds are

$$\begin{aligned} NaFe(CO)_2Cp + RX &\longrightarrow RFe(CO)_2Cp \\ NaFe(CO)_2Cp + RCOX &\longrightarrow RCOFe(CO)_2Cp \end{aligned} \qquad [1]$$

A very wide range of R-X and R-CO-X has been used *(35,61,106, 107,108,111,113)*.

The $[CpFe(CO)_2]^-$ ion is a powerful nucleophile *(60)*, for example it displaces fluoride ion from C_6F_6 *(109)* and from C_2F_4 *(98)* neither of which react with weaker nucleophiles such as $[Mn(CO)_5]^-$ and $[CpMo(CO)_3]^-$. In addition, the π-bonding ancillary ligands CO and Cp stabilize the Fe-C bond in the R-$Fe(CO)_2Cp$ complexes. Bimolecular nucleophilic substitution reactions of the type represented by eq. [1] are expected to proceed with inversion of the configuration at the carbon atom. However, the formation of *threo*-Me_3C-CHD-CHD-$Fe(CO)_2Cp$ from *erythro*-Me_3C-CHD-CHD-SO_2-C_6H_4Br and $NaFe(CO)_2Cp$ provided the first direct evidence of such an inversion *(165)*. Despite the high nucleophilicity of the $[CpFe(CO)_2]^-$ ion it reacts sluggishly if at all with aryl or vinyl halides. The much greater reactivity of aroyl- and alkenoyl-halides allows good yields of the corresponding iron complexes to be obtained, *e.g.* *(108,113)*,

$$\begin{aligned} PhCOCl + NaFe(CO)_2Cp &\longrightarrow PhCOFe(CO)_2Cp \\ CH_2{=}CHCOCl + NaFe(CO)_2Cp &\longrightarrow CH_2{=}CHCOFe(CO)_2Cp \end{aligned} \qquad [2]$$

Photolytic decarbonylation of these complexes provides a convenient route to the aryl and vinyl compounds *(108,113)*. Alternatively, the aryl complexes may be obtained *via* the powerful arylating agents diphenyliodonium iodide *(36)*, aryldiazonium chlorides *(128)*, or triphenylsulphonium tetrafluoroborate *(128)*.

Substitution of cyano groups for the vinyl hydrogen atoms in vinyl halides causes a substantial increase in the reactivity of the carbon halogen bond (eq. [3]).
The $[CpFe(CO)_2]^-$ anion is sufficiently nucleophilic to displace both halogens from the 2,2-dicyanovinyldihalides forming as chief products analogues of the well known dimer $Cp_2Fe_2(CO)_4$ with a bridging CO group replaced by the $:C{=}C(CN)_2$ or $:C(CN)_2$ group *(114)*. The terminal CO frequencies in $Cp_2Fe_2(CO)_3[C{=}C(CN)_2]$ are about 30 cm^{-1} higher than those in $Cp_2Fe_2(CO)_4$ indicating that the dicyanovinylidine ligand is a better π-acceptor than CO *(114)*.

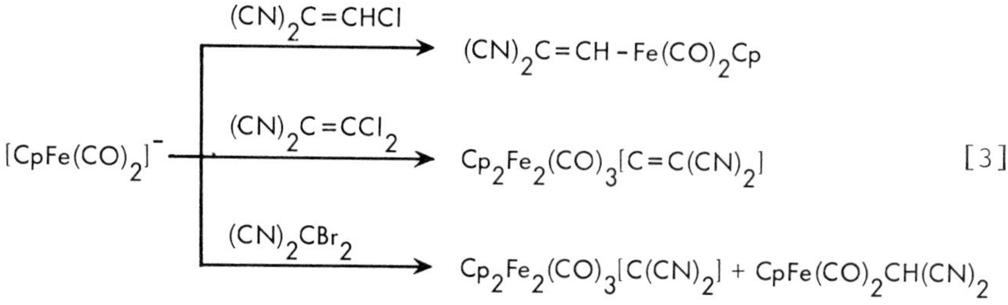

$$[CpFe(CO)_2]^- \quad \begin{cases} \xrightarrow{(CN)_2C=CHCl} (CN)_2C=CH-Fe(CO)_2Cp \\ \xrightarrow{(CN)_2C=CCl_2} Cp_2Fe_2(CO)_3[C=C(CN)_2] \\ \xrightarrow{(CN)_2CBr_2} Cp_2Fe_2(CO)_3[C(CN)_2] + CpFe(CO)_2CH(CN)_2 \end{cases} \qquad [3]$$

The nature of the products from reactions between the $[CpFe(CO)_2]^-$ ion and organic halides of the type $X-CH_2-CH_2-Cl$ depend on the nature of X *(113)*. When X = NMe$_2$ the cyclic acyl complex $Me_2\overset{\cdot\cdot}{N}-CH_2-CH_2-CO-\overline{Fe}(CO)Cp$ is formed by spontaneous carbon monoxide insertion into the M–C bond of an intermediate β-dimethylaminoethyliron complex. When X = SMe an analogous intermediate, $MeS-CH_2-CH_2-Fe(CO)_2Cp$ can be isolated. Irradiation of this intermediate affords a low yield of the acyl complex $Me\overline{S}-CH_2-CH_2-CO-\overline{Fe}(CO)Cp$ but the major photolysis product is $MeS-Fe(CO)_2Cp$ which arises *via* olefin elimination.

It seems that the product from the reaction between prop-2-ynyl bromide and the $[CpFe(CO)_2]^-$ anion is the σ-allenyl complex $CH_2=C=CH-Fe(CO)_2Cp$ rather than either the σ-prop-1-ynyl- *(100)* or the σ-prop-2-ynyl compound *(6)*. The ^{1}H-NMR data for the complex (τ 5.11, α-CH; τ 6.03, γ-CH$_2$; J(α,γ) 6.5 Hz) are characteristic of a σ-allenyl structure *(97)*.

Carbanion intermediates have been invoked to account for the outcome of the reaction between the $[CpFe(CO)_2]^-$ anion and the alkynes $CF_3-C\equiv CH$ and $CF_3-C\equiv C-CF_3$, *e.g.* *(29,73)* eq. [4].

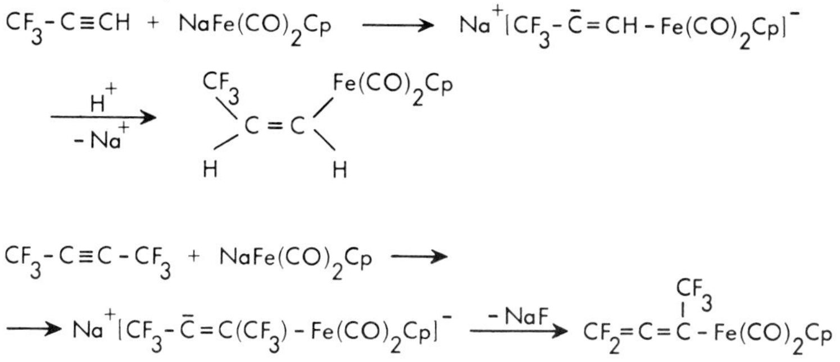

$$CF_3-C\equiv CH + NaFe(CO)_2Cp \longrightarrow Na^+[CF_3-\overset{\cdot\cdot}{C}=CH-Fe(CO)_2Cp]^-$$

$$\xrightarrow[-Na^+]{H^+} \quad \underset{H}{\overset{CF_3}{\diagdown}}C=C\underset{H}{\overset{Fe(CO)_2Cp}{\diagup}} \qquad [4]$$

$$CF_3-C\equiv C-CF_3 + NaFe(CO)_2Cp \longrightarrow$$

$$\longrightarrow Na^+[CF_3-\overset{\cdot\cdot}{C}=C(CF_3)-Fe(CO)_2Cp]^- \xrightarrow{-NaF} CF_2=C=\overset{\overset{\displaystyle CF_3}{|}}{C}-Fe(CO)_2Cp$$

Several reaction pathways have been distinguished for reactions between carbonylmetalate anions and polyfluoroaromatic compounds *(28)*. The products of these reactions were characterized by ^{19}F-NMR spectroscopy *(30)*, they depend on the nature and positions of groups other than fluorine on the aromatic ring (Figure 1).

Polarographic measurements show the $[Fe(CO)_4]^{2-}$ ion to be

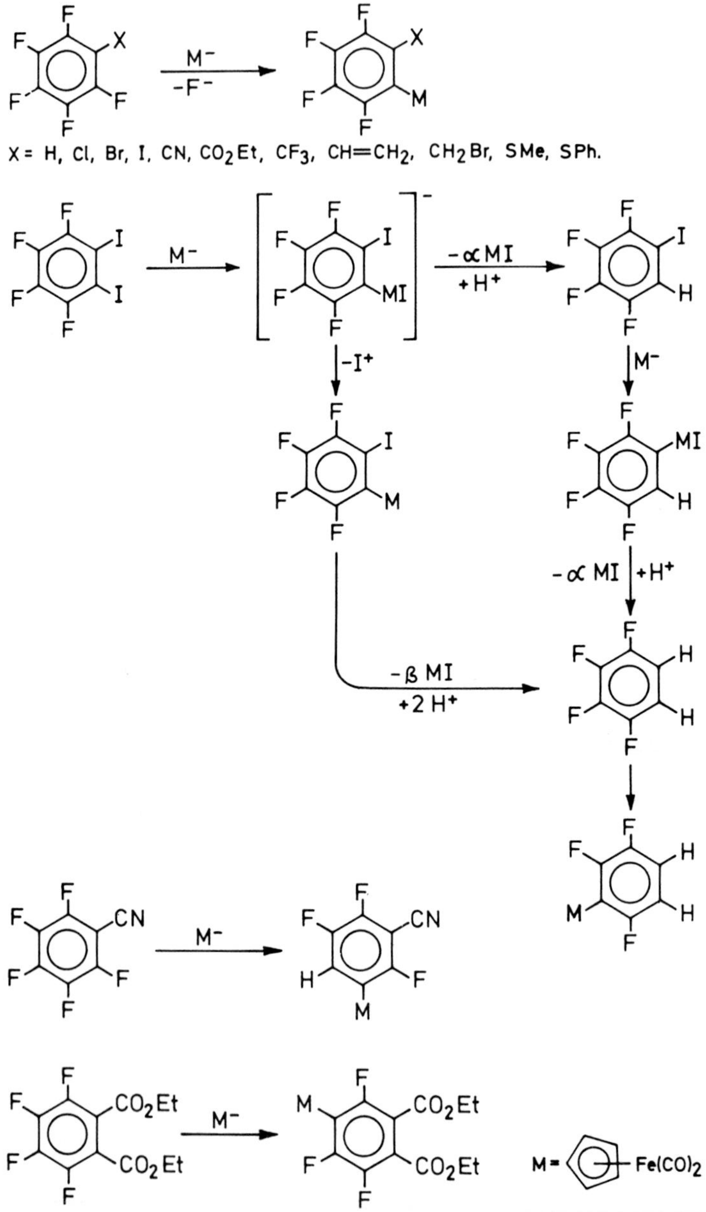

Fig. 1: The influence of substituents on the reactions between $NaFe(CO)_2Cp$ and polyfluoroaromatic compounds (ref. 28).

less nucleophilic than the $[CpFe(CO)_2]^-$ ion by a factor of ca. 10^7. Nevertheless it is sufficiently nucleophilic to react with alkyl halides and tosylates. The reaction can be viewed as an oxidative addition of R-X to the d^{10} saturated carbonyl anion or as an S_N2 displacement at carbon *(47)*. The primary intermediate $[R\text{-}Fe(CO)_4]^-$ may be in equilibrium with the unsaturated acyl derivative $[R\text{-}CO\text{-}Fe(CO)_3]^-$. Addition of triphenylphosphine and subsequent acidification of the reaction mixture produced the aldehydes R-CHO in 80-90 % yield *(51)*. The proposed acyl intermediates have been isolated as their bis(triphenylphosphine)iminium salts *(145)*.

Oxidation of the reaction mixture from R-X and $Na_2Fe(CO)_4$ produces carboxylic acids and acid derivatives. Oxidation either induces alkyl migration (see also section IV.D.) or traps the intermediate acyl derivative *(48)*.

Very large differences in the rates of reaction of $Na_2Fe(CO)_4$ and functional groups, *e.g.* Br reacts 10^4 times faster than Cl in R-X, should permit the selective reaction of multifunctional compounds. This coupled with the probable high stereospecificity of the reactions (see also section IV.D.) promises to make them very valuable in organic syntheses.

Other examples of the formation of iron-carbon σ-bonds *via* reactions of $Na_2Fe(CO)_4$ include the preparation of the first iron-formyl complex *(49)* and the formation of bis(perfluoroalkyl) compounds *via* CO elimination from the corresponding acyl complexes *(104,105)*. Attempts to prepare $(CF_3)_2Fe(CO)_4$ by this technique have been unsuccessful, although it has been made using $(CF_3\text{-}CO)_2O$ *(107)*.

There is only one example to date of an iron-metalate anion which does not have carbonyl groups as ancillary ligands. This is the $[Fe(salen)]^-$ ion produced by reduction of NN'-ethylenebis(salicylideneiminato)iron(II), Fe(salen), with metallic sodium *(68)*. It reacts with benzyl chloride to yield the stable *high-spin* σ-benzyl(salen)iron(III) complex 1. The square pyramidal structure was suggested by analogy with the [Fe(salen)]Cl monomer which has similar magnetic properties *(70)*.

The great majority of σ-organoiron complexes are diamagnetic. Rapid reaction occurred between the $[Fe(salen)]^-$ ion and acetyl chloride or methyl iodide but neither the acetyl nor the methyl complex could be isolated *(68)*; it is likely that a modification of the isolation technique will render these complexes accessible.

B. *REACTIONS OF ORGANOMETALLIC REAGENTS WITH IRON SALTS*

The reaction of an iron-X compound where X is a halide or similar leaving group with an organometallic reagent led to

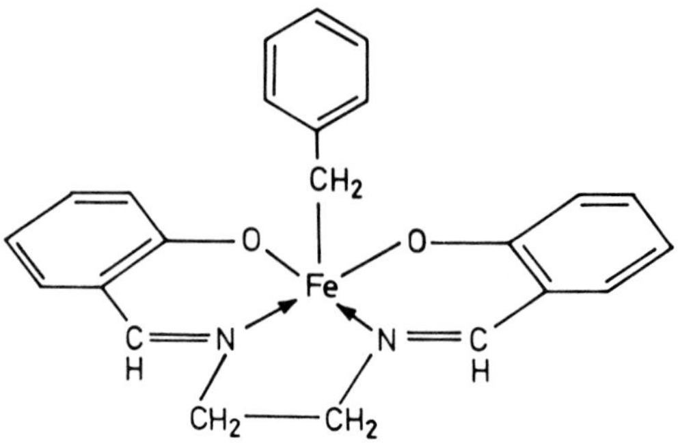

1

the first successful isolation of a stable σ-organoiron complex, *e.g.* *(136)* eq. [5].

$$CpFe(CO)_2Cl + RMgX \longrightarrow CpFe(CO)_2R + MgXCl \quad [5]$$

$$R = Me,\ Et,\ n\text{-}Pr,\ Ph$$

Other examples of this method, which has not been widely used in organoiron chemistry, are shown in eqs. [6] to [10].

$$FeCl_3 + 2\,NaCp \longrightarrow (\eta^1\text{-}C_5H_5)_2FeCl\cdot THF \quad (164) \quad [6]$$

$$Fe(CO)_4I_2 + C_6F_5Li \longrightarrow C_6F_5Fe(CO)_4I \quad (157) \quad [7]$$

$$Fe(acac)_3 + R_3Al + L\frown L \longrightarrow (L\frown L)_2FeR_2 \quad (24,169) \quad [8]$$
$$R = Me,\ Et; \quad L\frown L = bipy,\ o\text{-}phen$$

$$(Et_2PhP)_2FeCl_2 + C_6Cl_5MgCl \longrightarrow (Et_2PhP)_2Fe(C_6Cl_5)_2 \quad (39) \quad [9]$$

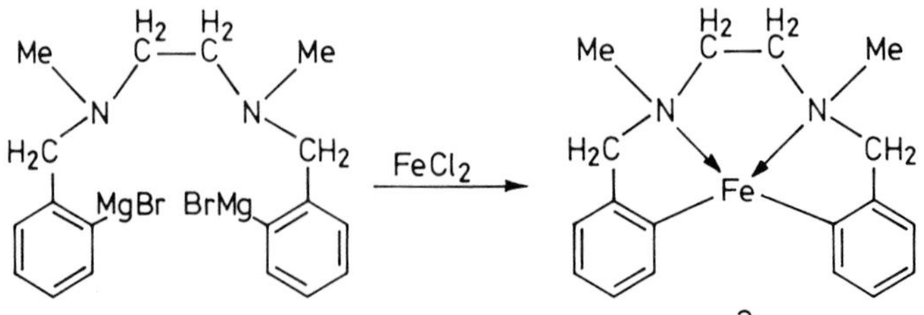

[10]

2

Several iron complexes have been made by both methods II.A and II.B. The former method is to be preferred as it usually gives higher yields of the alkyl complex.

The formation of reactive organoiron species is indicated by the colour changes which accompany the addition of iron halides to solutions of organometallic reagents. A deep-red colour develops when a solution of 1,2,3,4-tetrachlorobutadienyllithium in THF is mixed with a solution of $FeCl_3$ in the same solvent at -110°C. The red solutions are stable below -80°C; they exhibit characteristic coupling reactions which contrast with those of the lithium reagent, *e.g.* *(115)* eq. [11].

$$CHCl{=}CCl{-}CCl{=}CClLi \xrightarrow[(2)\ PhCHO]{(1)\ FeCl_3} CHCl{=}CCl{-}(CCl{=}CCl)_2{-}CCl{=}CHCl$$
$$CHCl{=}CCl{-}CCl{=}CClLi \xrightarrow{(1)\ PhCHO} CHCl{=}CCl{-}CCl{=}CCl{-}CH(OH){-}Ph \qquad [11]$$

The formation of $Ar_2Fe{\cdot}THF$ (Ar = Ph, Mesityl) in the reaction of $FeCl_2$ with the appropriate Grignard reagent has been claimed but the complexes were not isolated *(163)*. Cross coupling reactions exhibited by solutions containing FeI_2 and CH_3Li in the molar ratio 1:3 *(53)* are analogous to those of anionic manganese and copper alkyls *(52)* (eq. [12]) and are

$$R_3MnLi + R'X \longrightarrow R{-}R' \qquad [12]$$

believed to indicate the formation of the unstable trimethyliron(II) ion.

C. *INSERTION REACTIONS*

The formation of iron-carbon σ-bonds by an insertion reaction usually involves either acyl complex formation *via* CO insertion into an existing iron-carbon bond or insertion of some carbon-containing species into an Fe-H or Fe-C bond. Insertion reactions of Fe-C σ-bonds are discussed in Section IV. The name "insertion reaction" refers to a general class of reactions (eq. [13]). It carries no mechanistic implication and

$$L_aMX + nY \longrightarrow L_bM(Y)_nX \qquad [13]$$

merely serves to describe the outcome of the reaction. Some insertion reactions leading to Fe-C σ-bonds are shown in Figure 2. The stereochemistries of the products from CF_3-C≡CH were established by ^{1}H- and ^{19}F-NMR spectroscopy (Section III.A.2).

Bis(1,1,2,2-tetrafluoroethyl)tetracarbonyliron(II) has been prepared by a double-insertion reaction between $H_2Fe(CO)_4$ and tetrafluoroethylene. The *cis*-configuration predicted on the basis of the four carbonyl stretches in the IR spectrum of the compound was confirmed by an X-ray crystallographic study which showed it to have the structure 3 *(40)*.

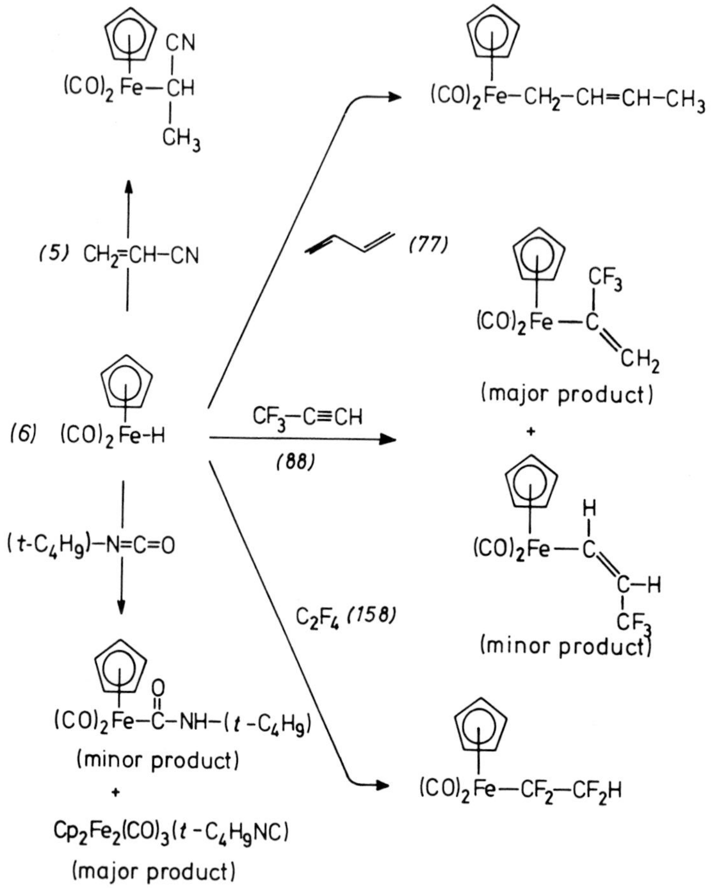

Fig. 2: Insertion reactions of $CpFe(CO)_2H$.

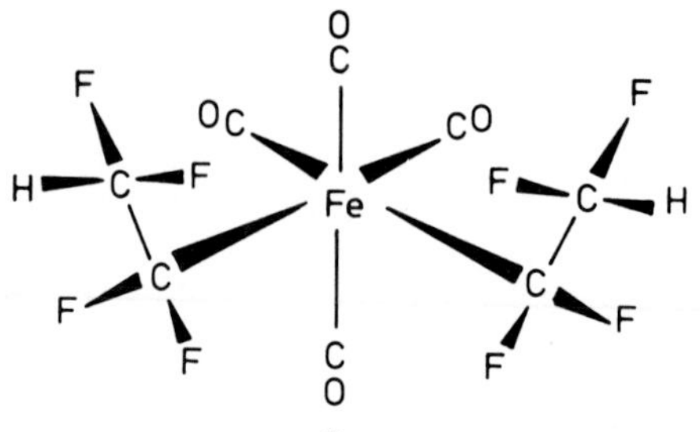

3

D.   REACTIONS AT CARBON-BONDED LIGANDS

The overall electron drift to the metal from co-ordinated molecules such as CO and olefins enhances their reactivity towards nucleophilic reagents; reactivity is even greater if additionally the complex carries a formal positive charge *(117,127)*. The utilization of nucleophilic attack at co-ordinated carbon for the formation of iron-carbon σ-bonds has been confined to cationic complexes of the type $[CpFeL_2L']^+X^-$, *e.g. (32)* eqs. [14] to [18].

$$[CpFe(CO)_3]^+ + CH_3O^- \longrightarrow \underset{(CO)_2}{CpFe} - \underset{O}{\overset{\|}{C}} - OCH_3 \qquad (32) \quad [14]$$

$$[CpFe(CO)_3]^+ + H-NRR' \longrightarrow \underset{(CO)_2}{CpFe} - \underset{O}{\overset{\|}{C}} - NRR'$$

$$R, R' = H, Me, Et, i-Pr, n-Bu, \text{+}CH_2\text{+}_4, \text{+}CH_2\text{+}_5 \qquad (32) \quad [15]$$

$$[CpFeL_2(C_2H_4)]^+ + H^- \longrightarrow C_2H_5FeL_2Cp$$
$$L = CO \ (78), \ P(OPh)_3 \ (85) \qquad\qquad [16]$$

$$\left[\bigcirc\!\!\!-Fe[P(OPh)_3]_3 \right]^+ + R^- \longrightarrow R-\bigcirc\!\!\!-Fe[P(OPh)_3]_3 \qquad (85) \quad [17]$$

$$[CpFe(CO)(CNMe)_2]^+ + NaBH_4 \longrightarrow \underset{CO}{CpFe[(CHCNMe)_2BH_2]} \qquad (162) \quad [18]$$

When the reaction of the $[CpFe(CO)_3]^+$ cation with amines was extended to include hydrazine and some alkylhydrazines *(4)* the unstable carbonyl complexes

$$(CO)_2Fe-\underset{O}{\overset{R}{\overset{\|}{C}}}-N=N\overset{R'}{\underset{R''}{\diagup}}$$

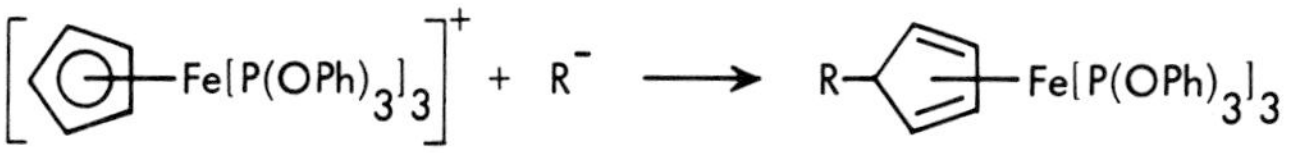

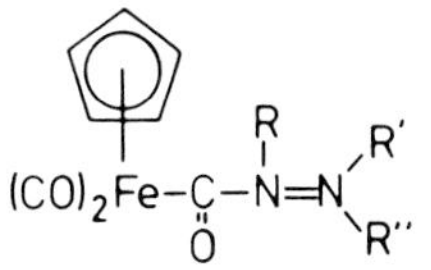

R	= H	CH_3	H	H
R'	= H	CH_3	CH_3	H
R''	= H	H	CH_3	CH_3

were obtained; those with R = H liberated ammonia to form the corresponding isocyanate complex $CpFe(CO)_2NCO$.

At least four reaction pathways are possible in the reaction between pentafluorophenyllithium and the ions $[CpFe(CO)-LL']^+$ (L = L' = CO; L = CO, L' = CNMe and L = L' = CNMe) *viz.* attack of the $C_6F_5^-$ ion, at the metal atom, at co-ordinated carbon or at the Cp ring and reduction of the cation to the dinuclear $[CpFe(CO)_2]_2$ *(160,161)*. The first three possibilities are illustrated for the $[CpFe(CO)_2CNMe]^+$ ion in Figure 3.

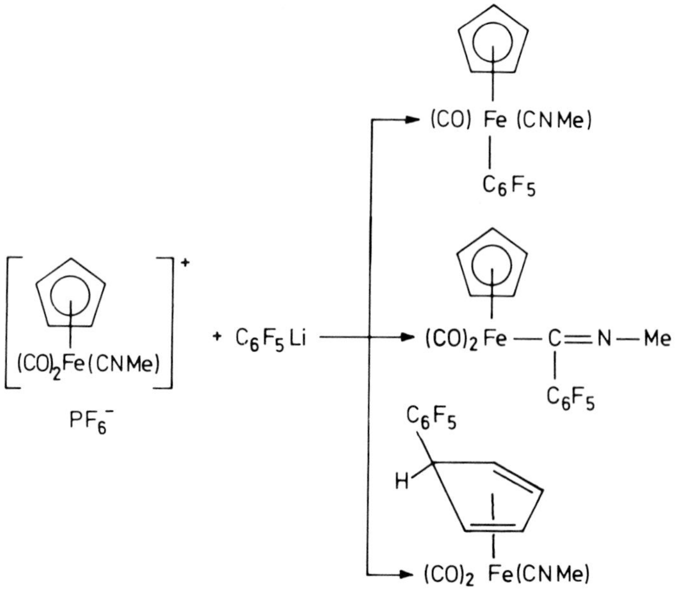

Fig. 3: Reactions of [CpFe(CO)$_2$CNMe]$^+$ with C_6F_5Li (ref. 161).

The *exo* configuration of the C_6F_5 substituent in the cyclopentadiene ligand was established by the absence of the characteristic *exo* C-H stretching frequency.

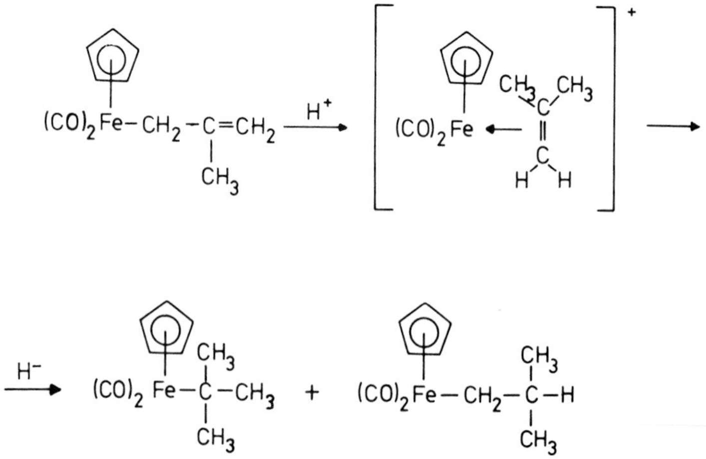

[19]

The very great difference in reactivity of the butyl complexes formed by H^- attack on the butene cation $[CpFe(CO)_2(Me_2C{=}CH_2)]^+$ has been exploited in the preparation of the *tert*-butyl complex. Treatment of the mixture of isomers with gaseous HCl selectively destroyed the *iso*-butyl complex *(71)* (eq. [19]).

E. OXIDATIVE ADDITION REACTIONS

The term "oxidative addition" has come to be used to describe a rather widespread class of reactions in which both the oxidation number and co-ordination number of the metal atom increase *(87)*. Such reactions are most common for d^7, d^8, and d^{10} metal complexes and in the case of iron will therefore be most likely for the d^8 zerovalent-metal complexes, especially the carbonyls.

Pentacarbonyliron undergoes an oxidative addition reaction with perfluoroalkyl iodides *(104,122)* (eq. [20]).

$$R_fI + Fe(CO)_5 \xrightarrow{50°C} R_fFe(CO)_4I$$

$$R_fFe(CO)_4I \xrightarrow{75°C} [R_fFe(CO)_3I]_2 \qquad [20]$$

$$R_f = C_2F_5,\ n\text{-}C_3H_7$$

The low reaction temperatures contrast with those (approx. 200°C) normally required for reactions of perfluoroalkyl iodides. Since infrared absorption bands characteristic of bridging carbonyl groups are absent from the spectra of the dinuclear complexes, dimerization most probably occurs *via* iron-iodine-iron bridges *(104)*.

The addition of HX (X = Cl, Br, I) to mono-olefin complexes of iron is a potential oxidative addition route to σ-alkyliron complexes. Mono-olefin complexes are usually destroyed by HCl affording alkane and Fe^{2+} ion *(116,144)*, but the use of an olefin containing a suitably-sited additional donor group, *e.g.* the Ph_2P group in $o\text{-}Ph_2P\text{-}C_6H_4\text{-}CH{=}CH_2$, allows the isolation of the corresponding alkyl complex 4 *(16)*.

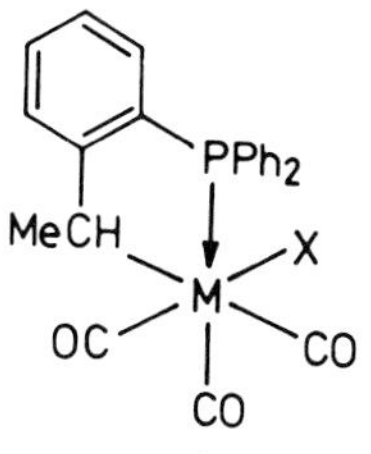

4

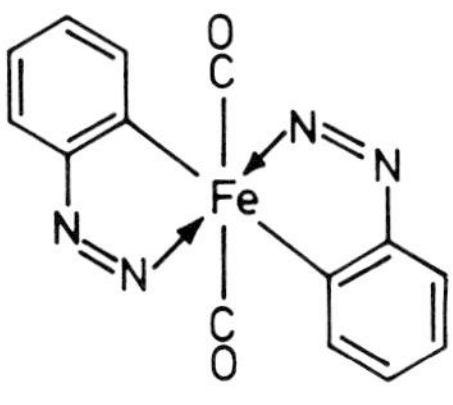

5

The co-ordination of the N=N double bond may be an initial step in the formal oxidative addition reaction between $Fe(CO)_5$ and azobenzene which yields a mixture of complexes including 5 *(13)*.

An intramolecular (oxidative addition)-(elimination) reaction is believed to occur when the η^2-ethylene complex $(\eta^2\text{-}C_2H_4)Fe(diphos)_2$ is exposed to UV irradiation. The reaction product exhibits a multiplet ^{1}H-NMR signal centered on τ 24.2 in addition to signals due to aromatic hydrogens; it has an IR absorption at 1893 cm^{-1} attributable to a metal-hydrogen stretching vibration. On the basis of this evidence structure 6 was proposed *(89)*. Although such metal-ligand hydrogen transfers are of interest in connection with studies of

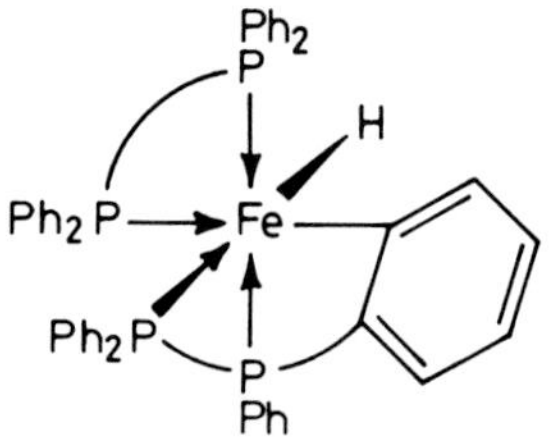

6

$Ph_2P\frown PPh_2 = Ph_2P\text{-}CH_2\text{-}CH_2\text{-}PPh_2$

metal-catalysed hydrogen-deuterium exchange reactions in aromatic compounds *(135)*, they appear to be of little synthetic value in organoiron chemistry.

Interesting but as yet little exploited zerovalent iron candidates for oxidative addition reactions are $(bipy)_2Fe(O)$ and $(diphos)_2Fe(O)$ produced according to eqs. [21] and [22].

$$(bipy)_2FeEt_2 \xrightarrow{\Delta} (bipy)_2Fe(O) \qquad (169) \qquad [21]$$

$$Fe(acac)_3 + diphos + AlMe_3 \longrightarrow (diphos)_2Fe(O) \qquad (24) \qquad [22]$$

The diphos complex reacts readily with molecules such as CO, H_2, SO_2, and acetylenes thus establishing its tendency to undergo oxidative addition reactions. Preliminary studies of its reactions with organic halides indicate it to be a fruitful source of organoiron complexes *(24)*.

F. MISCELLANEOUS REACTIONS

The solvent was established as the source of the unique hydrogen of the product from the reaction [23]. When d_8-

$$(CF_3)_2CN_2 + [CpFe(CO)_2]_2 \longrightarrow (CF_3)_2CH\text{-}Fe(CO)_2Cp \qquad [23]$$

tetrahydrofuran was used as solvent the product was $(CF_3)_2$-CD-Fe(CO)$_2$Cp *(50)*.

A transfer of aryl groups from mercury to iron is report-ed to occur in the reaction [24]. The products are deeply coloured crystalline solids with melting points above 100°C *(123)*.

$$Ar_2Hg + Fe \longrightarrow Ar_2Fe + Hg$$

$$Ar = PhCH_2, \; p\text{-}MeC_6H_4, \; p\text{-}MeOC_6H_4, \; p\text{-}NH_2C_6H_4 \qquad [24]$$

An attempt to prepare a π-bonded cyclopropenyl complex from the reaction between triphenylcyclopropenyl bromide and the [Fe(CO)$_3$NO]$^-$ ion led instead *(45)* to the maroon acyl com-plex

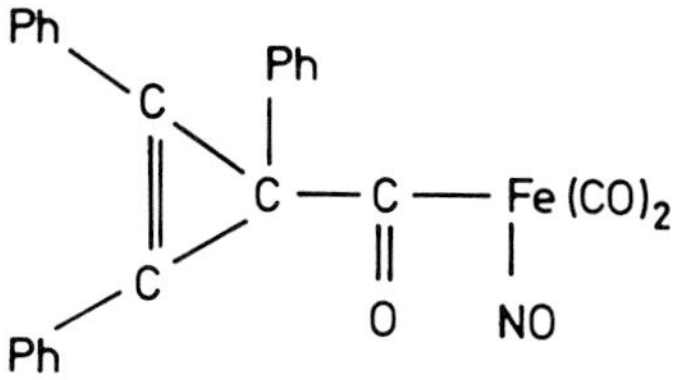

The compound ClCH$_2$-Fe(CO)$_2$Cp has proved to be a fruitful source of Fe-C complexes of the type XCH$_2$-Fe(CO)$_2$Cp. The CH$_2$Cl group reacts with thiols and alcohols to give for exam-ple EtOCH$_2$-Fe(CO)$_2$Cp, CH$_2$=CH-CH$_2$-O-CH$_2$-Fe(CO)$_2$Cp, and EtSCH$_2$-Fe(CO)$_2$Cp *(82)*.

Transition metal ions promote the formation of free radi-cals from a variety of organic compounds (see Section IV.E). Usually they undergo typical radical reactions but it has proved possible to trap the methyl, ethyl, and phenyl radical oxidation products of organic hydrazines using an iron(III) macrocyclic ligand combination (eq. [25]). Complex $\underline{7}$ (R = Me)

$$R\text{-}NH\text{-}NH_2 \xrightarrow{O_2} \left[R\text{-}N=NH\right] \xrightarrow{O_2} R^{\bullet} + N_2 \uparrow$$

$$\left[Fe(C_{22}H_{22}N_4)\,NCS\right] \text{ in MeCN} \qquad [25]$$

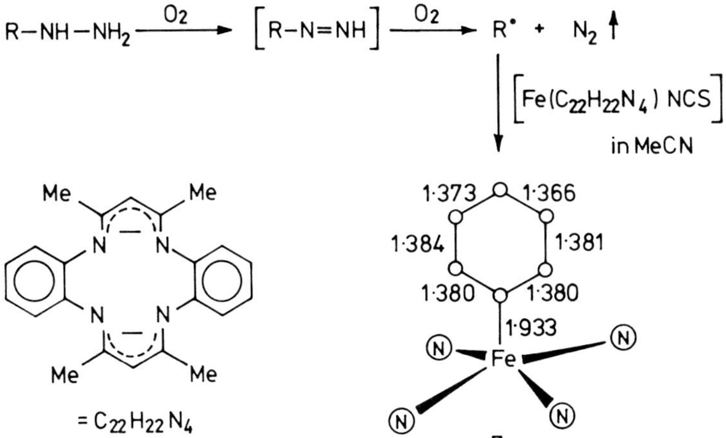

has μ_{eff} = 2.21 corresponding to low-spin Fe(III), this contrasts with the high spin nature of R-Fe(III)(salen) *(68)*. An X-ray crystal structure determination on 7 (R = Ph) established it as a five co-ordinate square pyramidal iron complex. The small displacement of the iron atom (0.23 Å) above the mean plane defined by the four nitrogens of the macrocyclic ligand is consistent with both the oxidation state and spin state of the iron. A very strong Fe-C σ-bond is indicated by the Fe-C distance of 1.933 Å, (see also Table 1) *(174)*.

A novel route to Fe-C σ-bonds is provided by the reaction between (η^4-cyclobutadiene)tricarbonyliron complexes and fluoroolefins *(22)*. Cyclobutenyl complexes are formed with the ring linked to the metal atom by a fluorocarbon residue (eq. [26]).

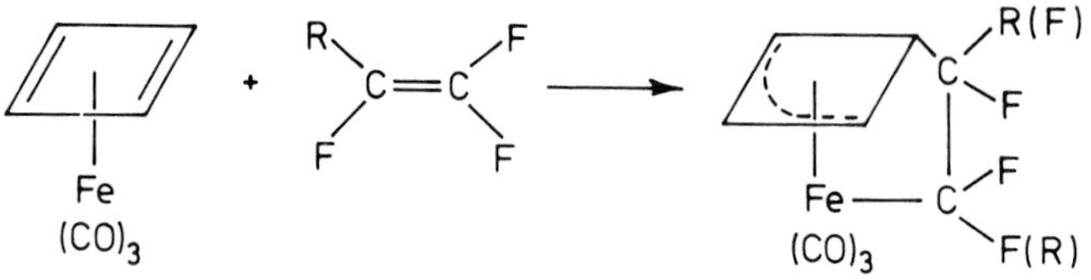

[26]

A simple 1,2-addition may also occur in the reaction with hexafluorobuta-1,3-diene but the final product is a trifluoromethyl derivative (8) which may be formed *via* a 1,3 fluorine shift.

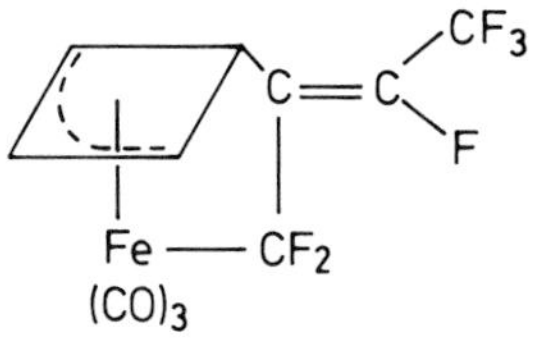

8

Opening of the cyclopropane ring in the reaction of vinylcyclopropanes with iron carbonyls makes two additional electrons available for bonding to the iron atom(s). These electrons can combine with the two olefinic π electrons to produce a 1,3-diene complex *via* a hydrogen shift or, if this is blocked by the incorporation of C(4) into a bridgehead position, one electron enters into σ-bond formation with the metal, and the other along with olefinic π electrons forms an η^3-allyl system. Representative examples of the formation of iron-carbon σ-bonds from vinylcyclopropanes and iron carbonyls are given below *(10,11,63,64)* (eqs. [27] to [29]).

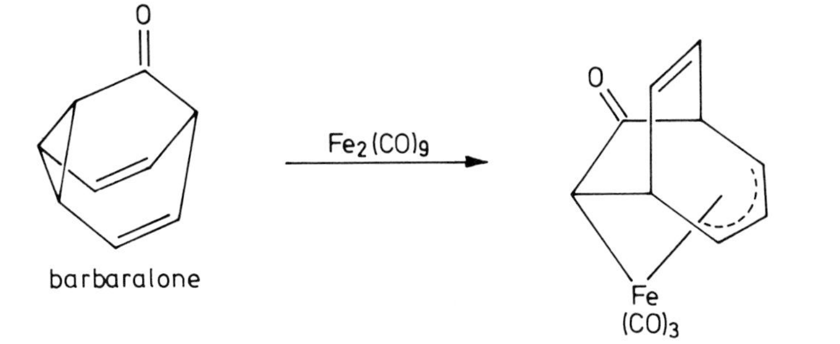

[27]

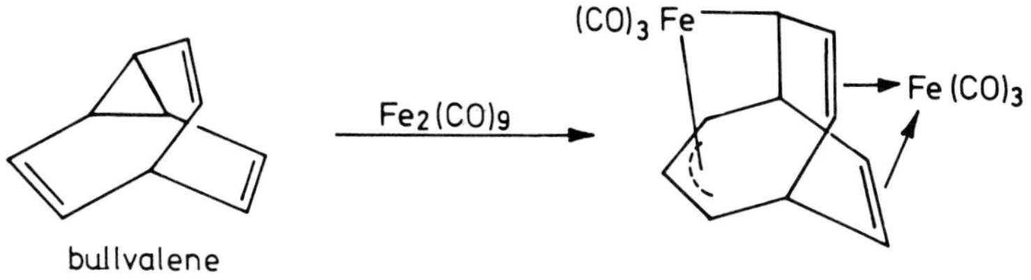

[28]

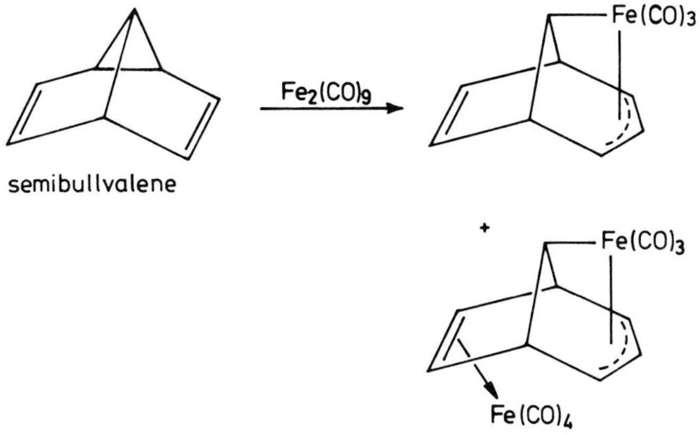

[29]

III. PHYSICAL PROPERTIES

A. SPECTROSCOPIC PROPERTIES

1. Infrared Spectra

Absorptions in the 600-5000 cm^{-1} region arise principally from vibrations of the organic ligands. C-H stretching and bending modes in organoiron complexes are unexceptional, definitive assignments are sometimes difficult to make due to interferences from absorptions of ancillary ligands such as Ph_3P and Cp. The σ-bonded cyclopentadienyl ring in $(\eta^1-C_5H_5)-Fe(CO)_2Cp$ was distinguished from the centrosymmetrically bound

Cp ligand by its more complex IR spectrum. The spectrum of the complex exhibited multiple C-H stretches and a free double bond absorption near 1600 cm^{-1} *(136)*.

Iron acyl complexes are characterized by a ketonic stretching frequency in the range 1580-1700 cm^{-1} (ν(C=O) in ketones $\sim$ 1725 cm^{-1}) the lower ν(C=O) in the metal complexes is attributed to a flow of electrons from the metal into a vacant orbital of the carbonyl group but the precise nature of this metal-carbonyl group interaction remains obscure. The appearance or disappearance of the C=O band can be used to determine whether a carbonylation or decarbonylation reaction has occurred (Section IV.B.2). An increase in the terminal metal-carbonyl ν(CO) can be used to diagnose the increase in formal oxidation number of iron which accompanies the formation of complex cations such as $[CpFe(CO)_2(C_2H_4)]^+$ *(78)*.

The high polarity of the C-F bond ensures that infrared-active C-F absorptions are more intense than their C-H counterparts. An extensive series of assignments have been reported for organofluorine ligands *(137,139)*, they have been used in conjunction with ^{19}F-NMR data (Section III.A.2) to show that the product of the reaction between perfluoroallyl chloride and the $[CpFe(CO)_2]^-$ anion is a perfluoropropenyl complex and not the anticipated σ-allyl complex *(103,104)*.

The observation of at least three carbonyl absorptions in the spectra of the monomeric complexes $R_fFe(CO)_4I$, (R_f = C_2F_5,*n*-C_3F_7) indicate a *cis* structure (C_{2v} local symmetry, four CO absorptions expected) rather than the more symmetrical *trans* structure (D_{4h} local symmetry, one CO absorption expected *(137)*. Spectroscopic data on the region below 600 cm^{-1} where Fe-C stretching and bending vibrations are likely to occur are scarce. Bands at 522 and 527 cm^{-1} in the spectrum of CH_3-FeI-$(CO)_2(PMe_3)_2$ have been assigned to ν(Fe-C) and one at 1174 cm^{-1} to the symmetric bending mode of the σ-CH_3 group *(59,132)*.

2. Nuclear Magnetic Resonance Spectra

The chemical shifts of alkyl ligand protons lie in the range τ 7.5-9.0 for both main-group and transition metal derivatives *(121)*. Comparable chemical shift differences and proton-proton coupling constants for hydrocarbon complexes give rise to complex second order spectra in many cases.

The spectrum of $(\eta^1$-$C_5H_5)Fe(CO)_2(\eta^5$-$C_5H_5)$ (9) consists of two singlets at τ 5.6 (η^5-C_5H_5) and τ 4.3 (η^1-C_5H_5) at room temperature, as the temperature is lowered the (η^1-C_5H_5) signal broadens and eventually collapses completely at about -25°C. As the temperature is lowered further, new

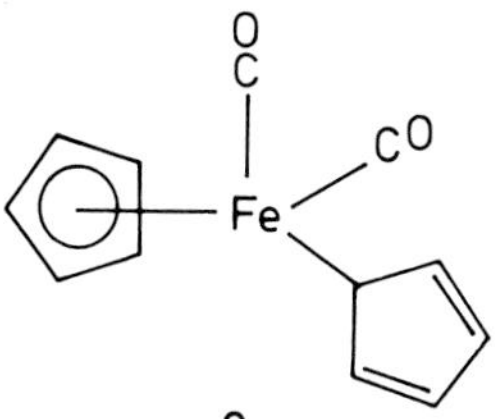

9

bands appear. Two bands of relative intensity 2, centered on ∿ τ 3.7, and a broad band of relative intensity 1, centered on ∿ τ 6.5. The structure of the spectrum is that expected for an A_2B_2H system of nuclei *(15,80)*. At room temperature the metal is exchanging $(\eta^1-C_5H_5)$ ligand sites at a rate which is fast on the NMR time scale probably *via* a sequence of 1,2-shifts; $(\eta^1-C_5H_5)Fe(CO)_2Cp$ is thus an example of a fluxional organometallic molecule (see also the NMR chapter). The structure of the compound 9 has been confirmed by X-ray diffraction *(15)*.

The methylene protons of $Ph-CH_2-Fe(CO)_2Cp$ give rise to a singlet proton resonance indicating that rotation about the metal-carbon σ-bond is rapid on the NMR time scale at ambient temperatures *(156)*. Both $Ph-CH_2-Fe(CO)(PPh_3)Cp$ *(65)* and Me_3-$Si-CH_2-Fe(CO)(PPh_3)Cp$ (10) *(133)* have an asymmetric iron centre and therefore even with rapid rotation the methylene protons would be inquivalent. The methylene resonances appear as the AB part of an ABX spectrum (X = ^{31}P) with vicinal P-H coupling constants of 2 Hz and 13 Hz for the trimethylsilyl derivative, indicating that it exists exclusively in the conformation with the maximum separation of the Cp and $SiMe_3$ groups.

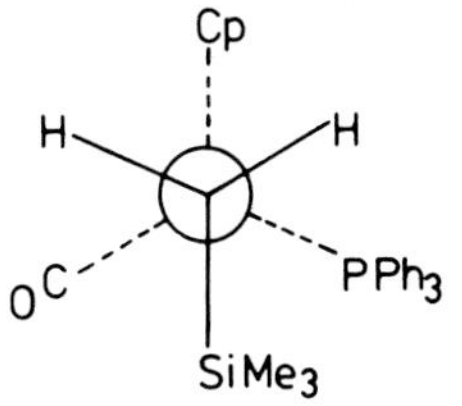

10

Chemical, X-ray crystallographic, and spectroscopic evidence has been presented to support the contention that electron delocalization from metal *d*-orbitals into the π* antibonding orbitals of the aromatic ring is significant in aryl-

[a] σ_I^0 and σ_R^0 are Taft inductive and resonance parameters, respectively *(153)*. Values of $\sigma_I^0 = -0.13$ and $\sigma_R^0 = -0.30$ *(21)* and $\sigma_I^0 = -0.24$ and $\sigma_R^0 = -0.29$ have also been reported.

transition metal complexes *(131,150)*. NMR data on *m*- and *p*-FC_6H_4-Fe(CO)$_2$Cp do show that the metal substituent is both a σ- and a π-electron donor (σ_I^0 = -0.25, σ_R^0 = -0.29)[a] but the relative invariance of σ_R^0 (-0.22 to -0.29) compared with σ_I^0 (+0.12 to +0.60) over a range of 27 neutral compounds and the σ_R^0 value of -0.09 for C_6H_5 *(173)* was taken to suggest that the π bonding contribution is unlikely to be a significant stabilizing effect in metal aryl complexes especially those with strong π acceptor ancillary ligands, but that it may become important where such ligands are absent.

The α-CF_2 fluorines of σ-fluorocarbon complexes of the transition metals experience a marked deshielding effect due to the metal; they absorb some 50 ppm to low field of the corresponding signal for main-group metal compounds *(158)*. A similar but less marked effect is observed for the *ortho*-fluorines in complexes of polyfluoroaromatic ligands. It has been suggested that the fluorocarbon derivatives possess low-lying excited electronic states which can be "mixed" with the ground state of the M-C bond by the magnetic field. This mixing produces a substantial contribution to the paramagnetic screening constant σ_p resulting in large downfield shifts for the α-CF_2 fluorine nuclei *(138)*. Presumably a similar effect operates for fluoroaromatic complexes. Diagnostically, the α-CF_2 shift has been used for example to distinguish between possible structures for the product from perfluoroallyl chloride and NaFe(CO)$_2$Cp. The CF_2=CF-CF_2- ligand should give rise to three signals in the intensity ratio 2:1:2 with one of the more intense signals appearing at low field. The observed spectrum has three signals in the intensity ratio 3:1:1 centered on 66, 86, and 166 ppm relative to CCl_3F at 0 ppm *(103, 137)*. The lower field lower intensity signal is assigned to a CF group in α-position to a transition metal. The strongest absorption is a 1:1:1:1 quartet indicative of two doublet splittings, while the weaker signals are doublets of 1:3:3:1 quartets indicating the presence of a CF_3 group. Thus, the ligand is the perfluoropropenyl ligand CF_3-CF=CF-. A high value for J_{FF} = 131 Hz establishes the *trans* geometry of the complex. Detailed ^{19}F-NMR data on fluorocarbon complexes are recorded elsewhere *(30,66,102)*.

3. Mass Spectra

Accurate molecular weights can be obtained from mass spectra providing it can be assumed that the peak at highest m/e value is due to the *parent* or molecular ion. This assumption is not always justified for organoiron complexes some of which decompose prior to ionization in the mass spectrometer; due caution must be exercised in the interpretation of the spectra of these compounds. For example, the close similarity between the spectra of Ph-CO-Fe(CO)$_2$Cp and Ph-Fe(CO)$_2$Cp indicate that decarbonylation of the benzoyl complex has occurred and that the spectra are of the same compound *(111)*. This is surprising since Ph-CO-Fe(CO)$_2$Cp does not undergo *thermal* decarbonylation. The mass spectrometry of σ-organoiron compounds has in the main been confined to complexes of the type R-Fe(CO)$_2$Cp *(27,111)*; representative fragmentation schemes for some of these are illustrated in Figure 4. The major features of these fragmentations are thermal decomposition to ferrocene

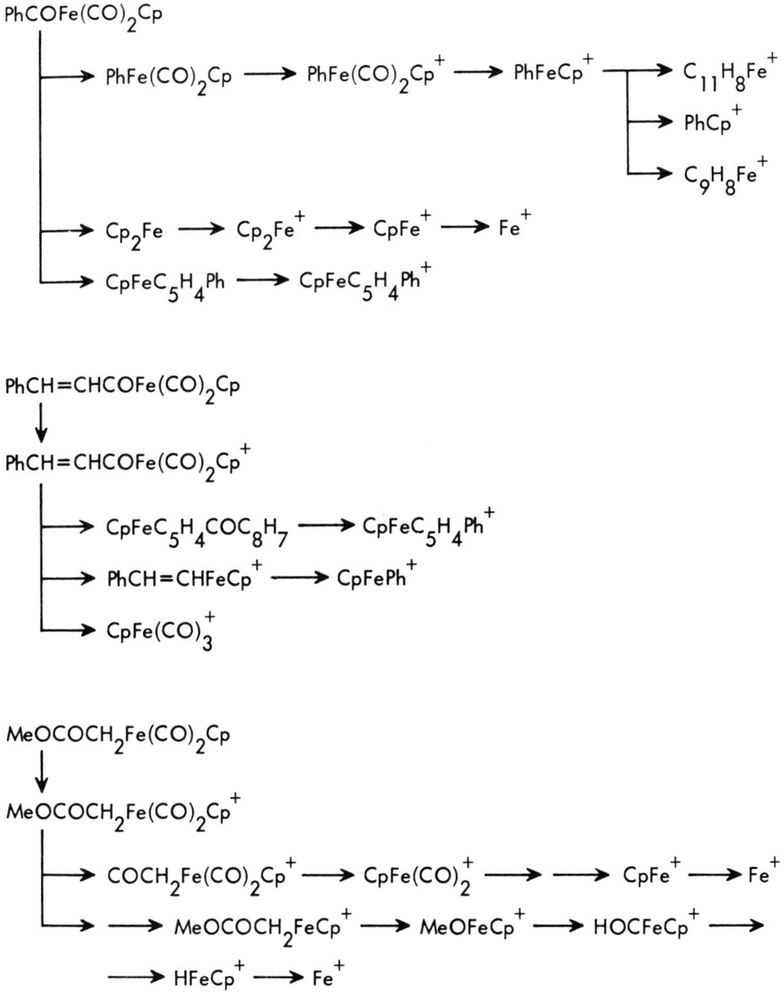

Fig. 4: Representative schematic fragmentation pathways in the mass spectra of R-Fe(CO)$_2$Cp complexes (ref. 111).

and substituted ferrocenes, the latter arising from metal-Cp ring organic ligand transfers, transfer of ligand substituents to the metal also occurs. Expulsion of C_2H_2 occurs from $CpFe^+$ ions and from ligands which possess a -CH=CH- group. Other fragmentation processes include loss of CO, Cp, Fe, and H. Fragmentation processes which characterize fluorocarbon complexes of iron include loss of the neutral molecules HF and FeF_2 *(110)*.

The ion $COCH_2Fe(CO)_2Cp^+$ from the ester derivative $MeOCOCH_2Fe(CO)_2Cp$ may have a π-bonded structure analogous to those proposed for the cationic complexes produced by protonation of complexes of the type XCH_2-$Fe(CO)_2Cp$ (X = CN, OMe) (Section IV.C.1).

The base peaks in the mass spectra of Ph-CH_2-$Fe(CO)_2Cp$ and MeC_6H_4-CH_2-$Fe(CO)_2Cp$ occur at m/e 212 and 226, respectively, corresponding to the ions $C_7H_8FeCp^+$ and $C_8H_9FeCp^+$. It has been suggested that the C_7H_8 ligand is bound to the iron atom as the π-tropylium cation *(26)* although a π-benzyl cation structure is favoured for the $C_8H_9FeCp^+$ ion *(90)*.

Tabular surveys of mass spectroscopic data on compounds which contain iron-carbon σ-bonds are available *(27,34)*.

B. *MISCELLANEOUS PHYSICAL PROPERTIES*

1. Magnetic Moments

Since ligands which are effective in stabilizing iron-carbon σ-bonds are also those which exert a high combined ligand-field effect, *e.g.* CO, Cp, and Ph_3P, it is not surprising that most stable σ-organoiron complexes are diamagnetic. Exceptions include Ph-CH_2-Fe(salen) (1) *(68)*, μ_{eff} = 5.87 B.M. corresponding to high spin Fe(III) (Section II.A); Fe(*o*-C_6H_4-CH_2-N(Me)-CH_2)$_2$ (2) *(118)*, μ_{eff} = 4.45 B.M. corresponding to tetrahedral, high-spin Fe(II) (Section II.B); and $(PEt_2Ph)_2$-$Fe(C_6Cl_5)_2$ *(39)*, μ_{eff} = 3.6 B.M. corresponding to square-planar Fe(II) (Section II.B).

2. Crystal Structures

The paucity of structural data on compounds with Fe-C σ-bonds and the known large variations in the σ-bond radius of octahedral Fe(II) prevent any conclusions being drawn regarding the detailed nature of the iron-carbon bond. Some representative data are summarized in Table 1.

An X-ray diffraction study of HOOC-CH_2-$Fe(CO)_2Cp$ revealed some novel structural features *(8,81)*. The C-O distances are the same within experimental error (1.32 Å). A difference of 0.05-0.10 Å is usually observed in carboxylic

Table 1: Iron-carbon σ-bond lengths.

Compound	Fe-C distance [Å]	*Ref.*
$(CO)_2Fe(\eta^5\text{-}C_5H_4)\text{-}CH_2\text{-}Fe(CO)_4$	2.12	*(124)*
$(Ph_2C{=}C)[Fe(CO)_4]_2$	1.98	*(125)*
$(\eta^1\text{-}C_5H_5)Fe(CO)_2Cp$	2.11	*(15)*
$(\eta^1\text{-}C_5H_5)Fe(CO)(Ph_3P)Cp$	2.12	*(12)*
$CpFe(CO)_2\text{-}CH{=}CH\text{-}CH{=}CH\text{-}Fe(CO)_2Cp$	1.98	*(41,42,57)*
cis-$(HCF_2\text{-}CF_2\text{-})_2Fe(CO)_4$	2.07	*(40)*
$CF_2\text{-}CF{=}CF\text{-}CF_2\text{-}Fe(CO)_4$	2.07	*(91)*
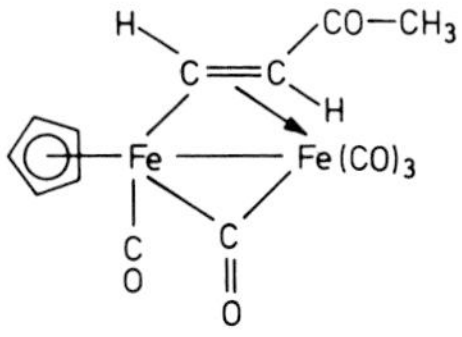	2.09	*(3)*
$PhFe(C_{22}H_{22}N_4)$ (<u>7</u>)	1.933	*(174)*

acids. The intermolecular O-H-O distance of 2.48 ± 0.02 Å is one of the shortest known. These features and the high pK_a (6.75 ± 0.08, in water) of the acid have been interpreted in terms of a direct donation of electrons by the metal atom to the carboxyl group. This is a metal-β-carbon interaction of the type proposed to account for the ready thermal decomposition and loss of hydride ion from ethyl and higher-alkyl metal complexes (Sections IV.A.2 and IV.C.2) *(84)*.

IV. REACTIONS OF IRON-CARBON σ-BONDS

A. *CLEAVAGE REACTIONS*

1. Reactions with Acids and Protic Solvents

Cleavage by a proton source is the most general reaction of σ-bonded organometallic compounds. It is usually represented by

$$R\text{-}M + H^+ \longrightarrow R\text{-}H + M^+ \qquad [30]$$

but the fate of the organic ligand depends on its structure. Thus, the methyl group of $CH_3\text{-}Fe(CO)_2Cp$ is lost as methane and the $[CpFe(CO)_2H_2O]^+$ cation is formed *(136)*. Organic groups with unsaturated β-carbon atoms undergo anomalous protonation reactions (Section IV.C.1).

Protic solvents such as water and alcohol cleave iron-carbon σ-bonds. Solvolysis of $Et_2Fe(bipy)_2$ is reported to be faster in alcohol than in water *(169)*. This is more likely to be due to the greater solubility of the complex in alcohol rather than to an intrinsically greater reactivity of the alcohol solvent since alcohols usually solvolyze metal-carbon bonds less rapidly than water in homogeneous solution. Mass spectrometry established C_2H_5D as the sole gaseous product from the deuteriolysis of $Et_2Fe(bipy)_2$, this contrasts with thermal decomposition which yields a 1:1 mixture of ethane and ethylene, and with acid cleavage which yields a 2:1 mixture of ethane and butane *(169)*. Complex 2 liberates N,N'-dimethyl-N,N'-dibenzylethylenediamine on methanolysis *(118)*.

2. Thermal cleavage

A great many of the known σ-organometallic compounds contain tertiary phosphine, η^5-cyclopentadienyl or carbonyl ancillary ligands; indeed, simple transition-metal alkyls or aryls without stabilizing ligands are highly labile, often so much so that they cannot be isolated. This led to the wide acceptance of the view that the transition metal-to-carbon bond is inherently unstable. The effects of stabilizing ligands have been attributed variously to their ability to enhance the acceptor capacity of the transition metal *via* L ← M π bonding, thus strengthening the metal-carbon interaction, to the increased separation between filled and unfilled metal-centered orbitals that such π bonding would cause bringing with it an increased resistance to homolytic dissociation, and to the involvement of normally non-bonding *d*-orbitals in π bonding interactions, thus making these orbitals less accessible to attacking reagents. However, the range of kinetically stable σ-organo-transition metal complexes is wider than expected on the basis of these considerations; also, stabilizing ligands are not always necessary. Furthermore, alternatives to the homolytic dissociation pathway for the destruction of the metal carbon bond have been identified *(62,147, 148,149,166)*; indeed one of these, a concerted bond cleavage involving fragmentation *via* β-elimination of alkene, appears to be the most common route (eq. [31]).

$$L_nM-CH_2-CH_2-R \rightleftharpoons L_nMH + CH_2{=}CHR \qquad [31]$$

This is the reverse of the olefin insertion discussed in section II.C. It has been suggested that the activation energy of this (presumably concerted) reaction can be lowered by a metal β-carbon interaction *(25)*. Such an interaction may account for the lower thermal stabilities of ethyl and higher alkyl complexes relative to their methyl analogues. For example the thermal stabilities of the complexes $R\text{-}Fe(CO)_2Cp$ change according to the series R = Me∿Ph>>Et>*i*-Pr. The inaccessibility of the β-elimination pathway may also account for the relatively high thermal stability of methyl, phenyl, benzyl, vinyl, and alkynyl complexes. A mixture of alkane and alkene is usually obtained from the thermal decomposition of a σ-alkyliron complex, *e.g.* eq. [32] *(169)*.

$$Et_2Fe(bipy)_2 \longrightarrow C_2H_6 + C_2H_4 \qquad [32]$$

This mixture could arise from disproportionation of alkyl radicals produced in the homolytic dissociation of the iron-carbon bond. Alternatively, the alkene may be formed *via* a β-elimination process, and the alkane by metal-hydride attack on the original alkyl complex in a concerted bimolecular elimination process. This sequence of reactions has been established *(166)* for the thermal decomposition of (*n*-butyl-1,1-d_2)-(tri-*n*-butylphosphine)copper(I) and (*n*-butyl-2,2-d_2)-(tri-*n*-butylphosphine)copper(I). Decomposition of the former yields 1-butene-d_2 and decomposition of the latter 1-butene-d_1. These observations are consistent only with β-hydride elimination.

Concerted bond cleavage processes require an increase in the co-ordination number of the metal atom in the transition state. The involvement of metal orbitals in bonding to ligands such as η^5-cyclopentadienyl and carbonyl as in $CpFe(CO)_2R$ will render the orbitals less accessible for this process.

Fluoroolefins are the chief volatile products from the thermal decomposition of perfluoroalkyliron complexes, *e.g.* eq. [33] *(104)*. When the decomposition of σ-(trifluoro-

$$\begin{array}{l} (C_2F_5)Fe(CO)_4 \longrightarrow CF_3\text{-}CF{=}CF\text{-}CF_3 + CF_3\text{-}CF_2\text{-}CF{=}CF_2 \\ C_2F_5Fe(CO)_4I \longrightarrow CF_2{=}CF_2 \end{array} \qquad [33]$$

methyl)(iodo)tetracarbonyliron is carried out in the presence of $CH_2{=}CH_2$ or $CF_2{=}CF_2$ the cyclopropanes $\overline{CH_2\text{-}CF_2\text{-}CH_2}$ or $\overline{CF_2\text{-}CF_2\text{-}CF_2}$ are formed indicating the elimination of difluorocarbene from the fluorocarbon complex.

3. Halogens and Alkyl Halides

The organic ligand is usually liberated as the appropriate halide when an iron-carbon σ-bond is cleaved with halogen, *e.g.* eq. [34] *(136)*.

$$RFe(CO)_2Cp + I_2 \longrightarrow RI + IFe(CO)_2Cp \qquad [34]$$

$$R = Me,\ Et,\ Ph$$

However, butane is the organic product from I_2 cleavage of $Et_2Fe(bipy)_2$ *(169)*, the same product was obtained from MeI cleavage. The stability of the iron-fluorocarbon bond in perfluoroalkyl complexes is illustrated by their resistance to attack by I_2 at room temperature, cleavage occurs at higher temperatures.

4. Metal Halides

Cleavage of organometallic compounds with mercuric halide and characterization of the resulting organomercuric halide is a classical method for the detection and characterization of transition metal-carbon bonds. Examples of its application to iron-carbon σ-bonds are shown in eqs. [35] *(128)* and [36].

$$PhFe(CO)_2Cp + HgCl_2 \longrightarrow PhHgCl + ClFe(CO)_2Cp \qquad [35]$$

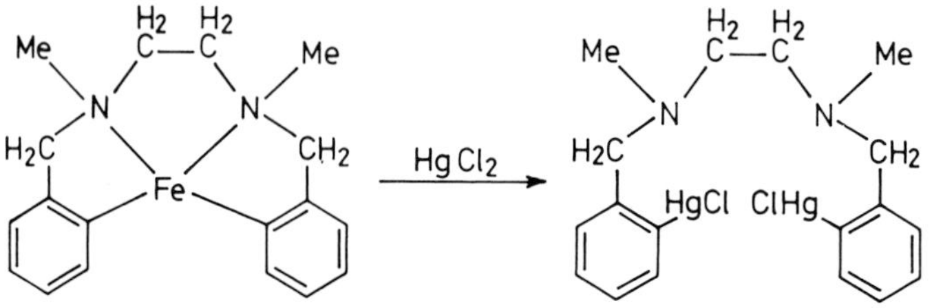

[36]

The mercuric halide cleavage of *threo*-Me_3C-CHD-CHD-$Fe(CO)_2Cp$ occurs with retention of the configuration at the carbon atom *(167)*.

5. Trifluoroacetonitrile

The iron-carbon σ-bond in $MeFe(CO)_2Cp$ is cleaved by CF_3-CN; the methyl group is replaced by a trifluoroacetiminato ligand, and substitution of one of the carbonyl ligands by CF_3CN occurs to give <u>11</u> *(112)*. Mass spectrometry established

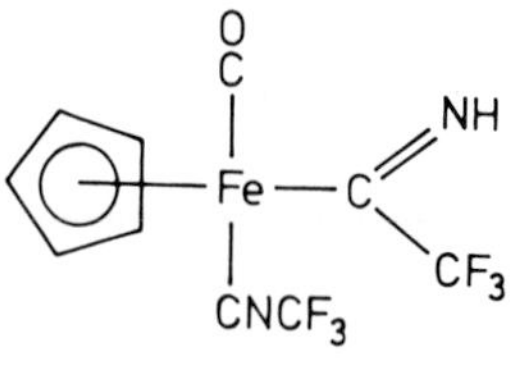

<u>11</u>

the presence of the unique hydrogen atom, further confirmation coming from the ^{1}H-NMR spectrum of 11 which showed a nitrogen-quadrupole broadened signal at τ -1.3 in addition to the Cp signal.

6. $Se(SeCN)_2$

Cleavage of $Ph-CH_2-Fe(CO)_2Cp$ with $Se(SeCN)_2$ gave NCSe-$Fe(CO)_2Cp$ and NC-$Fe(CO)_2Cp$ *(95)*.

B. INSERTION REACTIONS

Electrophilic reagents, *e.g.* SO_2, CO, tetracyanoethylene, R-NCO, R-NSO, react with iron-alkyls or iron-aryls to give the products of formal insertion of the electrophile into the iron-carbon bond. σ-Allyl complexes offer two sites for electrophilic attack; in addition, there is the possibility of rearrangement of the primary insertion product.

1. Sulphur Dioxide

The products of SO_2 insertion into the iron-carbon σ-bonds of R-$Fe(CO)_2Cp$ were established as metal sulphinate complexes by the sequence of reactions *(17,19)* shown in eq.

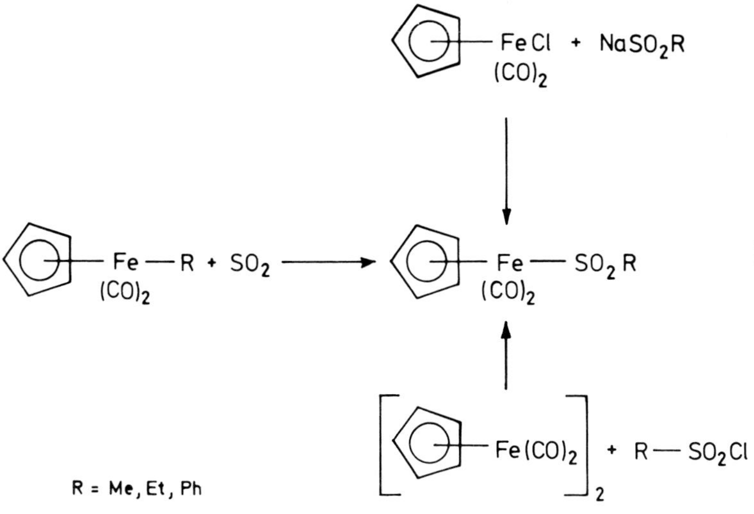

[37]

[37]. SO_2 insertion occurred in liquid SO_2 as solvent or with gaseous SO_2 in pentane solutions. Predictions of an S-sulphinate structure based on empirical correlations of ν(SO_2,sym) and ν(SO_2,asym) of the complexes with those of sulphones {-S(=O)$_2$-} rather than with those of sulphinate

esters {-OS(=O)-} were confirmed by an X-ray crystallographic study of Ph-CH=CH-CH_2-S(O)$_2$-Fe(CO)$_2$(η^5-C_5Me_5) (12) *(44)*. The rate of SO_2 insertion into the Fe-C bond of CpFeCO(L)R

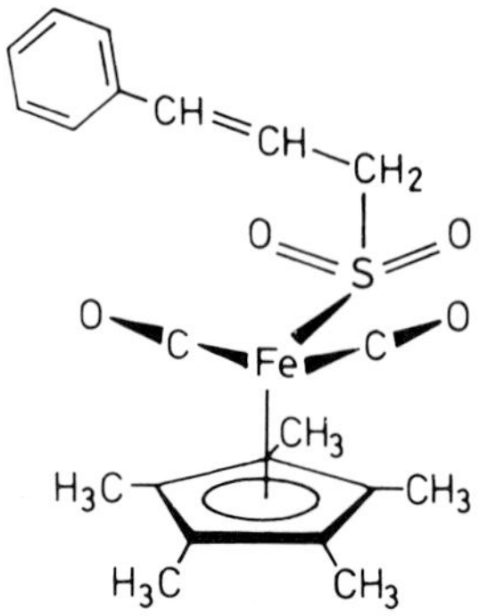

12

increases with increasing ease of electron release from R and with increasing basicity of the ancillary ligand L *(74, 92)*. Thus, electrophilic attack at the metal-carbon bond is indicated. Low temperature NMR studies of the reaction between SO_2 and CpFe(CO)$_2$-CH_2-Ph have revealed the intermediacy of an O-sulphinate which isomerized subsequently to the S-sulphinate at higher temperatures *(93,94)*. The mechanism of SO_2 insertion is thought to be as shown in eq. [38]. The

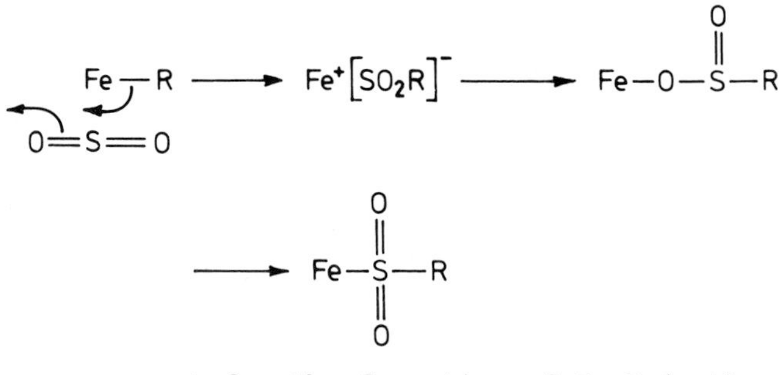

[38]

ion-pair can account for the formation of Fe-I in the presence of KI and for the stereochemistry of the insertion reaction, inversion at carbon and retention at iron *(67)*.

There appears to be only one example of SO_2 insertion involving complexes other than R-Fe(CO)$_2$Cp. The blue bis-(ethyl) complex Et_2Fe(bipy)$_2$ undergoes rapid reaction with SO_2 to give the known (S-sulphinate)$_2$Fe(bipy)$_2$ *(24,120)*.

The possibility of insertion accompanied by rearrangement is an added complication in the reactions of σ-allyliron complexes with SO_2, *e.g.* eq. [39] *(168)* (see also Figure 5). Which sulphinate isomer is predominant in the product mixture

$$R'\text{-}CH{=}CH\text{-}CHR\text{-}Fe(CO)_2Cp + SO_2 \longrightarrow$$

$$R'\text{-}CH{=}CH\text{-}CHR\text{-}S(O)_2\text{-}FeCp(CO)_2 + R\text{-}CH{=}CH\text{-}CHR'\text{-}S(O)_2\text{-}FeCp(CO)_2 \quad [39]$$

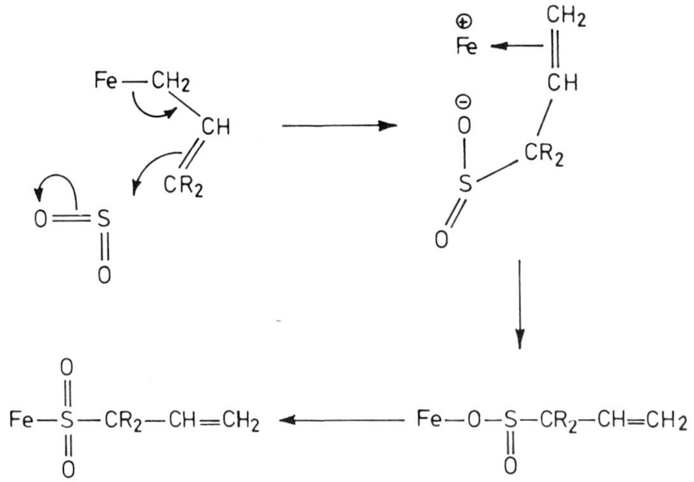

Fig. 5: Insertion of SO_2 with rearrangement in allyl complexes.

depends on the nature of R and R' and also on the substituents, if any, on the Cp ring. Bulky substituents on C(3) of the allyl ligand and electron releasing substituents such as methyl groups on the Cp ring favour direct insertion without rearrangement. Rearrangement occurs at low temperature or in inert solvents. Isomers cannot be interconverted under the conditions of SO_2 insertion, thus they must be produced *via* independent reaction pathways, namely attack at the two electron rich sites. However, where only one product is formed, isomerization may still have occurred.

The 1:1 adduct formed by reaction between $Me\text{-}C{\equiv}C\text{-}CH_2\text{-}Fe(CO)_2Cp$ and SO_2 *(142,143,155)* has been variously formulated as the allenyl(oxy)sulphinyl complex $CH_2{=}C{=}C(Me)OS(O)\text{-}Fe(CO)_2Cp$ and as the allenyl-O-sulphinate complex $CH_2{=}CH{=}C(Me)S(O)OFe(CO)_2Cp$, however, it was shown to have the structure 13 by X-ray crystallography *(43)*. An allenyl-O-sulphinate intermediate may be involved by analogy with the $Ph_3Sn\text{-}CH_2\text{-}C{\equiv}CH/SO_2$ system from which $CH_2{=}C{=}CHS(O)OSnPh_3$ has been isolated *(69)*.

The 1:1 adduct formed between $Me\text{-}C{\equiv}C\text{-}CH_2\text{-}Fe(CO)_2Cp$ and N-thionylaniline may have a structure analogous to 13 *(141)*.

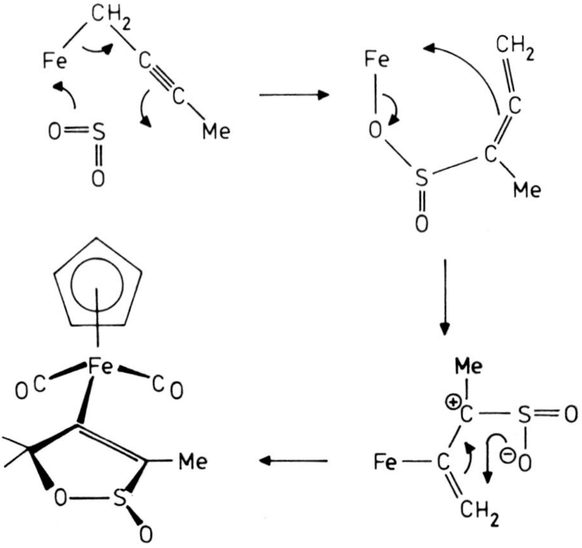

13

[40]

2. Carbon Monoxide

The conversion of alkylmetal intermediates into acylmetal compounds plays a key role in hydroformylation and related carbonylation reactions. In organoiron chemistry the formal insertion of CO into an iron-carbon bond may be induced thermally, photochemically, or by reaction with a ligand such as Ph_3P, *e.g.* eqs. [41] to [45].

The *cis* alkyl migration mechanism for "insertion" reactions of CO is now generally accepted. Assuming the principle of microscopic reversibility to hold, *cis* migration is established for iron-alkyls by the formation of *cis*-$IFe(CH_3)$-

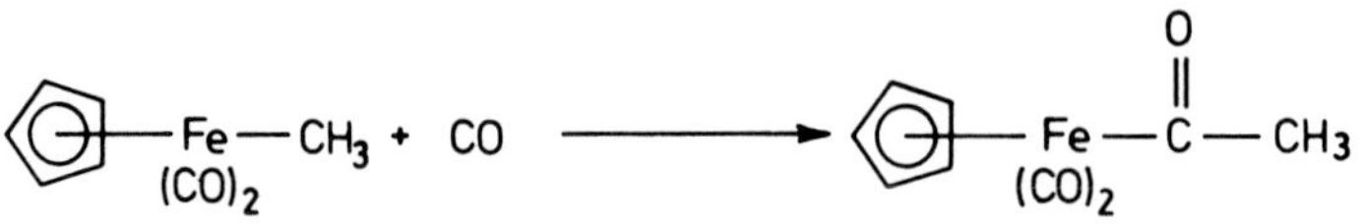

(46) [41]

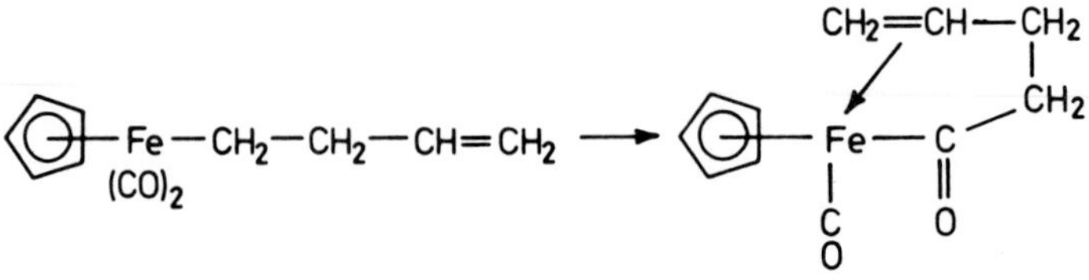

(86) [42]

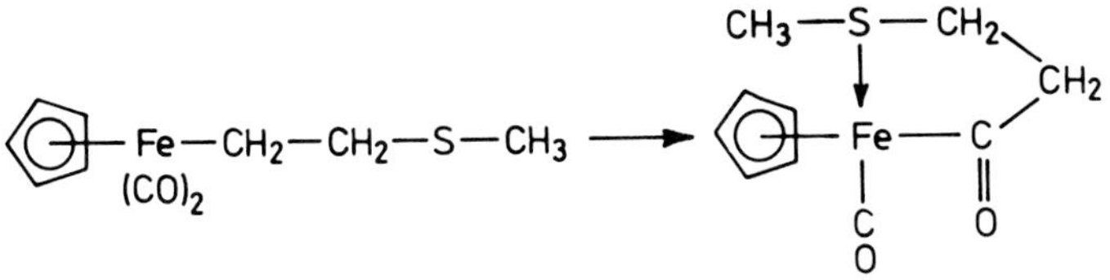

(113) [43]

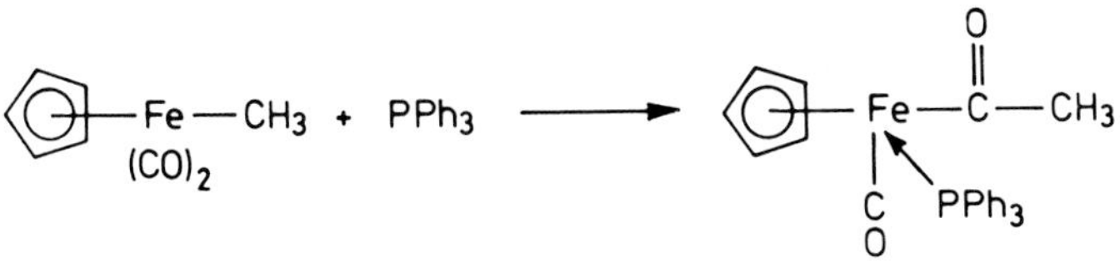

(18,33) [44]

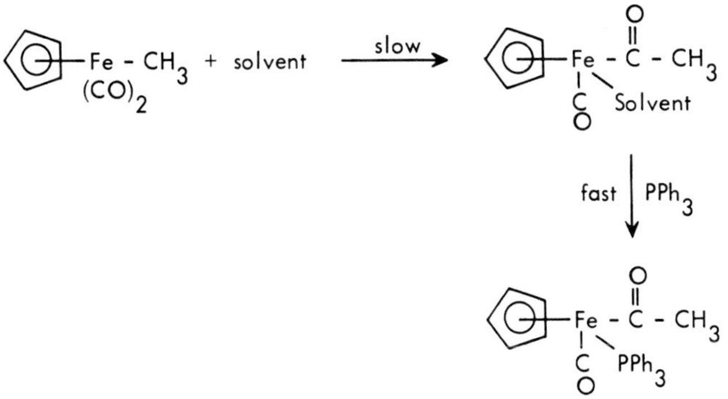

[45]

$(CO)_2(PR_3)_2$ in the decarbonylation of *trans*-$IFe(COCH_3)(CO)_2$-$(PR_3)_2$ (R = Me) *(132)*.

A kinetic study of the phosphine induced insertion indicated solvent participation in the rate determining step followed by rapid displacement of the solvent by the ligand (eq. [45]) *(33)*. An alternative proposal *(159)* that substitution of CO by Ph_3P occurs first seems less likely in view of the close similarity between the rates of disappearance of the starting material and appearance of the product *(33)*.

Carbonylation reactions have been used to produce acyl-metal complexes which have the iron atom as the centre of asymmetry, *e.g.* eq. [46] *(31)*.

Carbon monoxide insertion does not occur in reactions between fluorocarbon-iron complexes of the type $R_fFe(CO)_4I$ and $(R_f)_2Fe(CO)_4$ and ligands such as Ph_3P, diphos, and bipy, *e.g.* eq. [47] *(140)*. This presumably is a reflection of the greater strength of fluorocarbon-transition metal bonds relative to their hydrocarbon analogues.

Decarbonylation, the reverse of the carbon monoxide in-

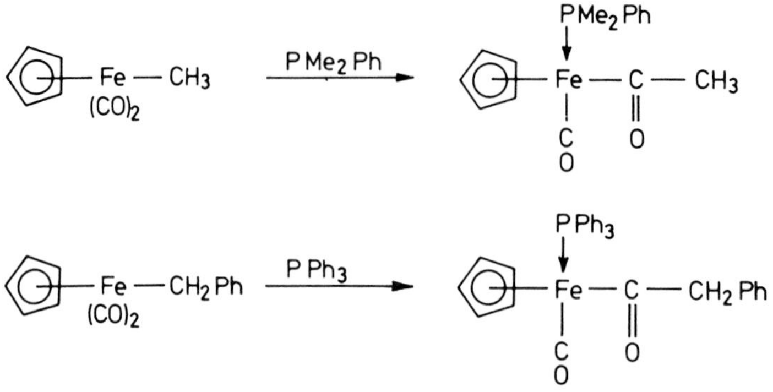

[46]

$$n\text{-}C_3F_7Fe(CO)_4I + Ph_3P \longrightarrow n\text{-}C_3F_7Fe(PPh_3)(CO)_3I + CO$$

$$(n\text{-}C_3F_7)_2Fe(CO)_4 + bipy \longrightarrow (n\text{-}C_3F_7)_2Fe(bipy)(CO)_2 + 2CO \qquad [47]$$

sertion, may be accomplished photochemically (Section II.A), in a limited number of cases thermally, or with $(Ph_3P)_3RhCl$, *e.g.* eq. [48] *(1,2)*.

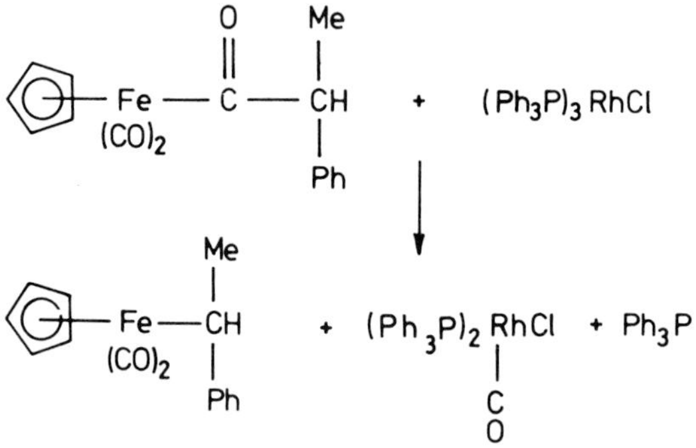

[48]

Although the perfluoroacyl compounds $(R_fCO)_2Fe(CO)_4$ undergo ready decarbonylation to the corresponding perfluoroalkyl compounds, the presence of ligands other than CO on the metal substantially increases the resistance of the acyl complex to loss of CO, thus C_2F_5-CO-$Fe(CO)_2Cp$ has not been decarbonylated *(101)*. The Cp ligand is a poorer acceptor of electrons than CO, and it may be that the increased electron density on the metal atom results in a strengthening of the metal-acyl group bond.

3. Tetracyanoethylene

The presence of four strongly electron withdrawing cyano

groups in the molecule makes tetracyanoethylene a powerful electrophile and a strong π-acid. It undergoes several different types of reaction with compounds containing iron-carbon σ-bonds including 1,2-, 1,3-, and 1,4-addition, π-complex formation, and both alkyl and metal migration reactions.

The products of the reaction between R-Fe(CO)(L)Cp and $C_2(CN)_4$ depend on the nature of R and L. When R = Me and L = Ph_3P, CO insertion occurs to give an acyl ligand and $C_2(CN)_4$ occupies the vacant co-ordination site to give the dark green complex 14 *(152)*. Analogous products are formed when

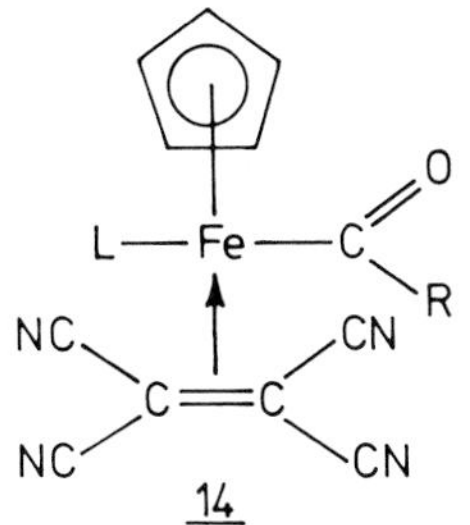

14

R = Et, *n*-Pr, L = Ph_3P; and R = Me, L = $(PhO)_3P$. Isomeric insertion products are obtained when R = CH_2-Ph, L = CO *(151)*. The yellow isomer (12 %) has C≡O and C≡N absorptions in its IR spectrum, its NMR spectrum exhibits signals due to Cp, methylene, and phenyl protons. These data support structure 15

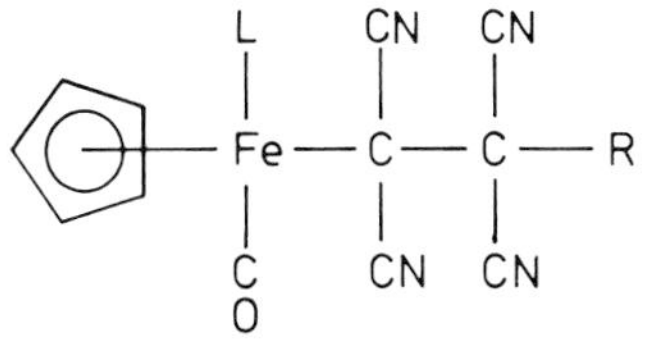

15

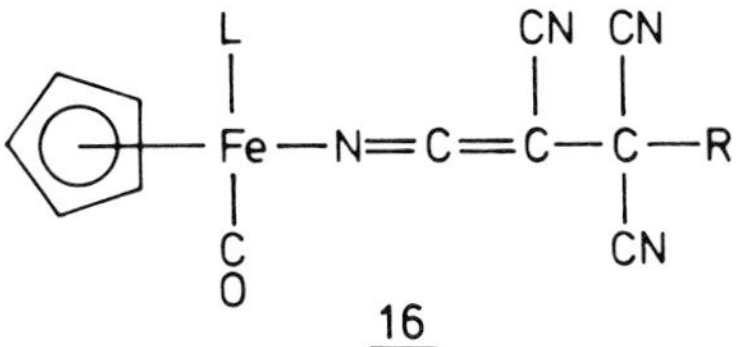

16

derived from 1,2-insertion of $(CN)_2C{=}C(CN)_2$ into the Fe-CH_2-C_6H_5 bond. The red isomer (42 %) has C≡O and C≡N absorptions and in addition two strong bands at 2151 and 1296 cm^{-1} attributable to the asymmetric and symmetric N=C=C stretches of a keteniminato structure, 16, *(14)*. Analogues of 16 are obtained for R = CH_2-Ph, L = Ph_3P or *n*-Bu_3P. Complex 14 (R = Et) undergoes thermal rearrangement to 16 (R = Et) at 65-70°C. It is not known whether this apparent transfer of C_2H_5 from CO to co-ordinated $C_2(CN)_4$ proceeds inter- or intra-molecularly *(152)*. When the organic group R has an unsaturated β-carbon atom as in $CH_2{=}CMe$-CH_2-Fe$(CO)_2$Cp and Me-C≡C-CH_2-Fe$(CO)_2$Cp the ligand isomerizes and undergoes 1,3-addition of tetracyanoethylene to afford complexes for which structures 17 and 18 have been proposed *(152)*.

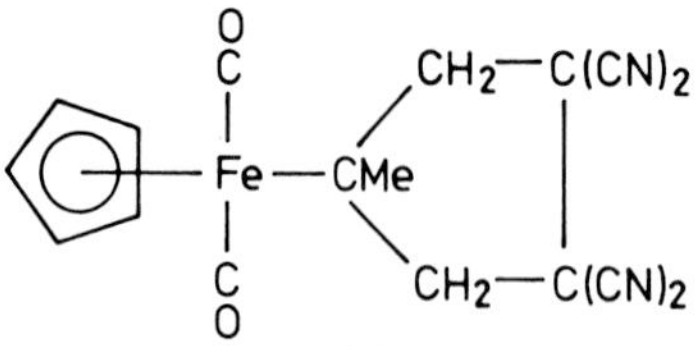

17

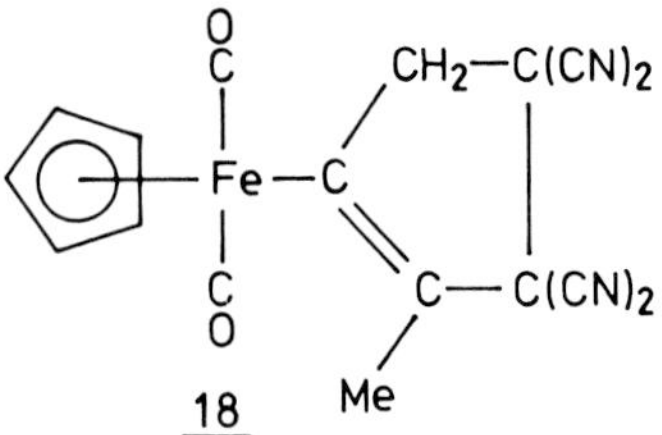

18

4. Cyclohexyl- and *tert*-Butylisocyanides

Differences in reactivity between methyl and benzyl complexes are found for the reaction between R-Fe(CO)$_2$Cp (R = Me or CH$_2$-Ph) and the alkylisocyanides $C_6H_{11}NC$ and *t*-BuNC *(172)*. The methyl complex undergoes ligand promoted CO insertion (Section IV.B.2) with *t*-BuNC to give Me-CO-Fe(*t*-BuNC)(CO)Cp. Carbon monoxide elimination occurs in the reaction of the benzyl complex with $C_6H_{11}NC$. Analytical and molecular weight data indicated the presence of three molecules of isocyanide per atom of iron in the monomeric product CpFe(CO)(CNC$_6$H$_{11}$)$_3$. The infrared spectrum shows five absorptions at 1892, 1880, 1606, 1560, and 1503 cm^{-1} but none in the region of 2100 cm^{-1}. The bands at 1892 and 1880 cm^{-1} were assigned to the terminal carbonyl group but no reason was suggested for the appearance of two bands. The other three bands were assigned to ν(C=N) of C_6H_{11}-N=C units. Three different environments for the cyclohexyl groups were indicated by three NMR signals at τ 7.4, 6.6, and 5.7, attributable to the α-protons of the C_6H_{11} groups. Structures 19, 20, and 21 were suggested *(172)*.

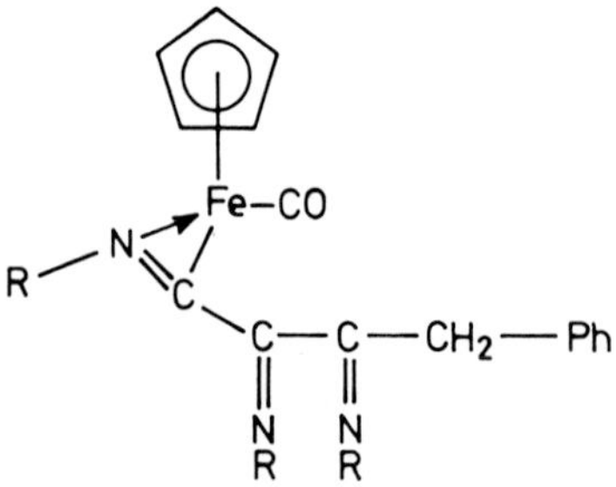

19

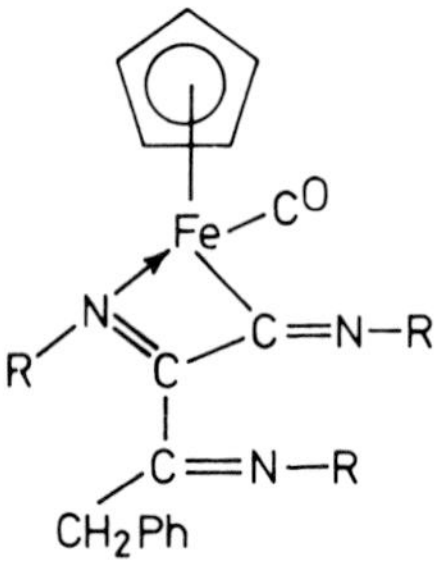

20

5. R-NCO

Alkyl isocyanates RNCO are insufficiently electrophilic

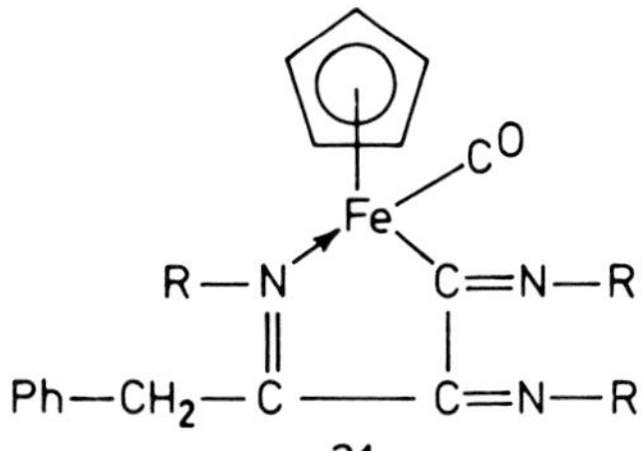

21

to react with alkyl iron complexes. When R = 2,5-dichloro-
phenyl, SO_2Cl, SO_2-C_6H_4-CH_3, or SO_2OMe reaction does occur
but only with σ-allyl- and -propargyl complexes *(56,72)*,
eqs. [49] and [50].

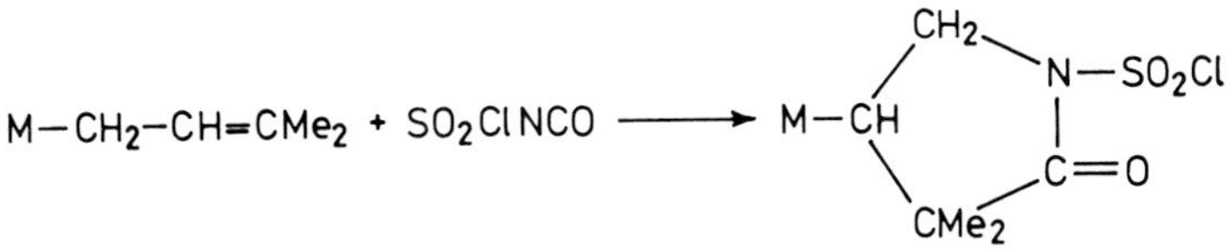

$$M-CH_2-CH=CMe_2 + SO_2ClNCO \longrightarrow M-CH \begin{array}{c} CH_2 \\ \diagdown \\ CMe_2 \end{array} \begin{array}{c} N-SO_2Cl \\ | \\ C=O \end{array} \qquad [49]$$

$$M-CH_2-C\equiv C-Me + CH_3C_6H_4SO_2NCO \longrightarrow M-C \begin{array}{c} CMe \\ \diagup \\ CH_2 \end{array} \begin{array}{c} C=O \\ | \\ N-SO_2C_6H_4-CH_3 \end{array} \qquad [50]$$

C.   σ,π REARRANGEMENTS

1.   Protonation Reactions

Acid treatment of the complexes R-Fe(CO)$_2$Cp (R = CH_2-CH=
CH_2, -CH_2-CN, -CH_2-CO-R) results in protonation of the ligand. The protonated ligand acts as a π-electron donor to
the metal which increases its formal oxidation number by one
(eqs. [51] to [54]. Bands in the region 3100-3400 and 2600-
2400 cm^{-1} in the IR spectra of the cation 22 and its deuterio
analogue have been assigned to N-H and N-D stretches, respectively. Similarly, bands at about 1660 and 1470 cm^{-1} were
assigned to N-H and N-D bending frequencies *(5)*.

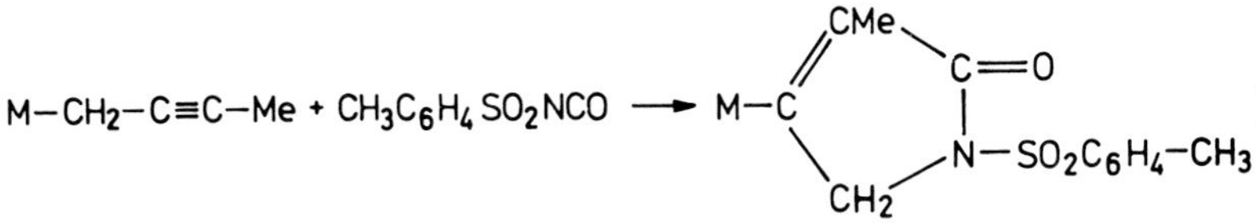

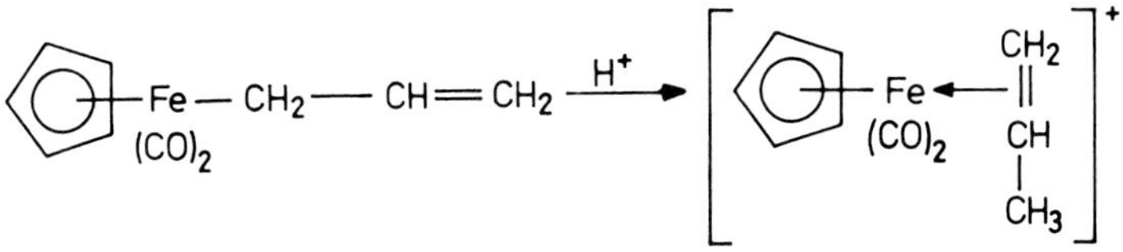

(77)   [51]

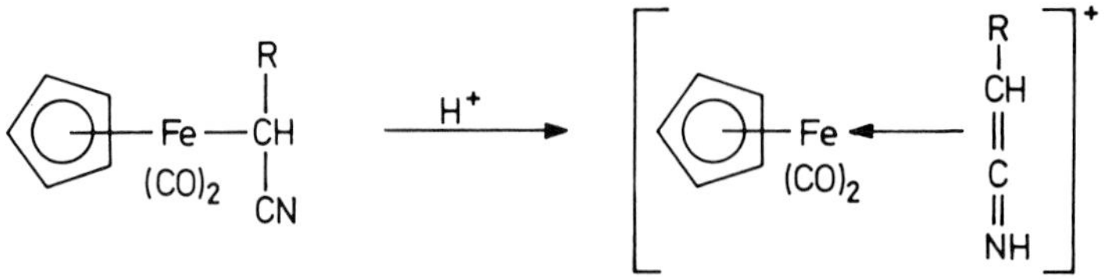

22

(5) [52]

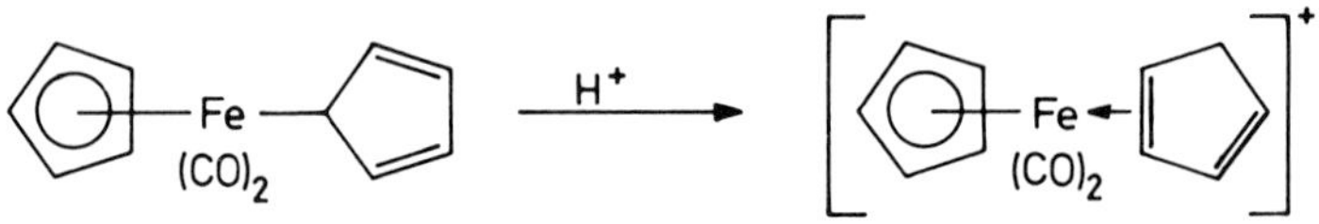

(79) [53]

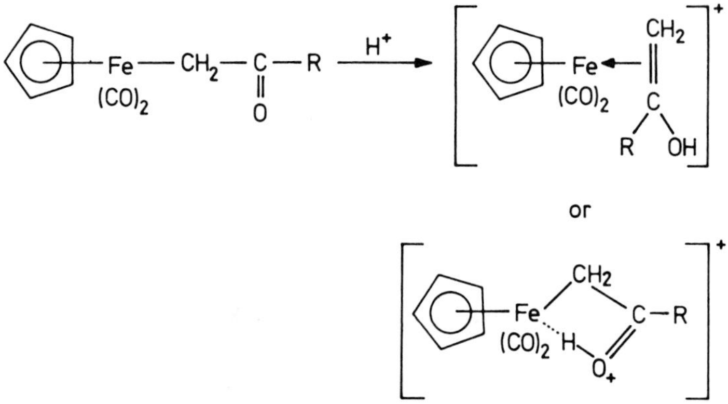

(7) [54]

Protonation of the σ-propargyl complex $HC{\equiv}C{-}CH_2{-}Fe(CO)_2{-}$Cp with dry HCl yielded an unstable cationic species which was isolated as the impure salt $[(CH_2{=}C{=}CH_2)Fe(CO)_2)Cp]{-}(SbCl_6)$ *(6)*. The aqueous media used in the isolation of the salt caused the cation to undergo nucleophilic attack by water at the allene ligand liberating acetone. Chromatography of the propargyl complex on acid-washed alumina caused it to isomerize to the corresponding alkynyl complex $CH_3{-}C{\equiv}C{-}Fe(CO)_2Cp$. It was proposed that the isomerization proceeded by way of an intermediate allene cation *(100)* (eq. [55]).

$$HC{\equiv}C{-}CH_2{-}Fe(CO)_2Cp \xrightarrow{+H^+} [(CH_2{=}C{=}CH_2)Fe(CO)_2Cp]^+ \xrightarrow{-H^+} CH_3{-}C{\equiv}C{-}Fe(CO)_2Cp \qquad [55]$$

When the propargyl complex was treated with ethanolic HCl it added the elements of H_2O to form the known complex $Me{-}CO{-}CH_2{-}Fe(CO)_2Cp$. Similar treatment of the alkynyl complex gave Et-CO-Fe$(CO)_2$Cp *(100)*.

The electron drift to the acyl carbonyl group in metal acyl complexes (Section III.A.1) should enhance its reactivity towards electrophilic reagents. In keeping with this, the complexes Me-CO-Fe(CO)(L)Cp (L = CO or Ph_3P) undergo reversible protonation with HCl forming cations which may have the coordinated carbene structure 23 *(83)*. The cationic metal

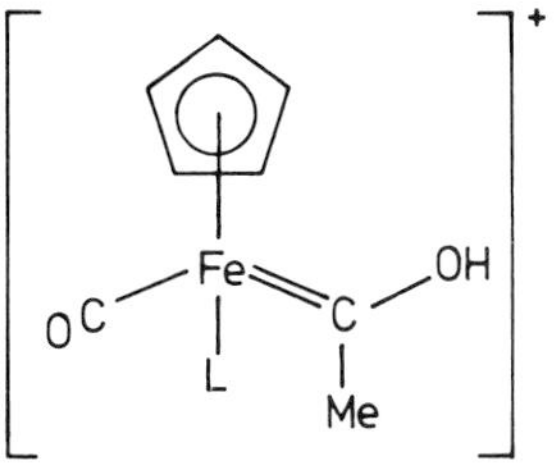

23

carbene complex $[CH_2Fe(CO)_2Cp]^+BF_4^-$ has been postulated as an intermediate in the reaction [56]. Norcarane or *cis*-1,2-

$$MeOCH_2Fe(CO)_2Cp + HBF_4 \longrightarrow MeFe(CO)_2Cp \qquad [56]$$

dimethylcyclopropane are formed if the reaction is performed in the presence of cyclohexene or *cis*-but-2-ene, respectively, but the carbene complex could not be isolated *(99)*.

2. Hydride Abstraction Reactions

The ease with which hydride ion is abstracted from alkyl (but not methyl) complexes of iron provides additional evidence for the metal-β-carbon interaction proposed to account for the relatively low thermal stability of ethyl- and higher-alkyl-iron complexes (Section IV.A.2) *(76)*. Triphenylmethyl tetrafluoroborate has generally been employed as the hydride abstracting reagent. The lack of isomerization in the reversible hydride abstraction/hydride addition reaction of *i*-Pr-Fe$(CO)_2$Cp indicates that addition of a metal hydride species to the (η^2-olefin)metal cation is unlikely to be involved in the hydride addition reaction (eq. [57]). Moreover, the formation of a deuterated cation from DMe_2C-Fe$(CO)_2$Cp shows that the abstracted hydrogen atom does come from a β-carbon atom *(78)* (eq. [58]).

The cationic complex $[(C_4H_6)Fe(CO)_2Cp]^+PF_6^-$ obtained *via* hydride ion abstraction from CH_2=CH-CH_2-CH_2-Fe$(CO)_2$Cp was assigned the η^2-diene complex structure 24 on the basis of its IR spectrum which has bands at 1621 and 1515 cm^{-1} attributable to an unco-ordinated C=C stretch and a co-ordinated C=C stretch, respectively. Treatment of the complex 24 with

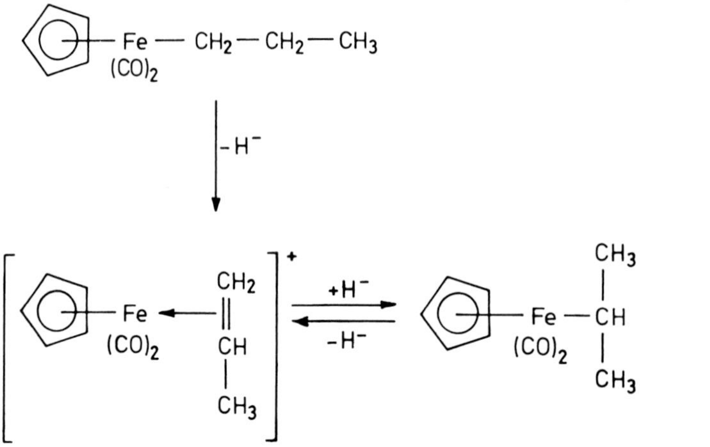

[57]

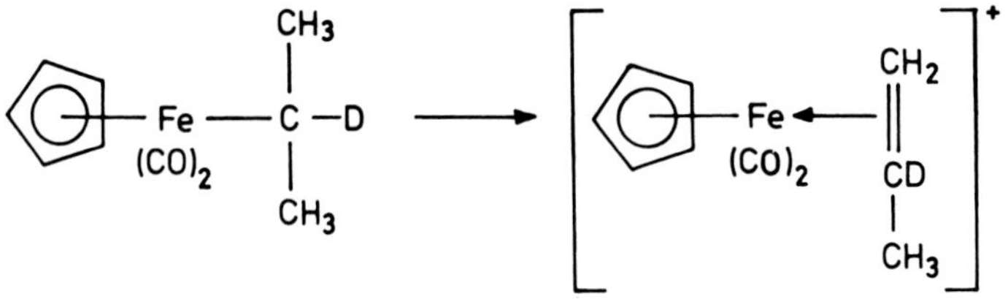

[58]

triphenylphosphine at 90°C yielded butadiene *(86)*.

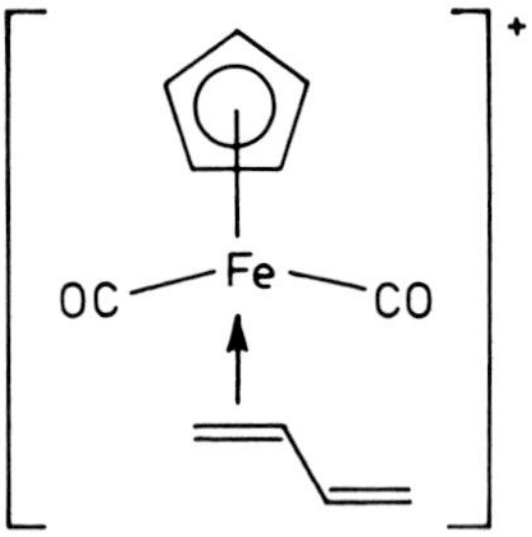

24

3. Photochemical Reactions

Thermal decomposition of $(\eta^1\text{-}C_3H_5)Fe(CO)_2Cp$ either in the solid state or in solution in an inert solvent gives $[CpFe(CO)_2]_2$ as the major iron-containing product. However, ultraviolet irradiation of the pure compound results in the loss of one molecule of CO; the olefinic double bond of the allyl group becomes co-ordinated to the metal. The ^1H-NMR spectrum of the product has three signals centered on τ 5.84 (1 H), 7.33 (2 H), and 9.32 (2 H) in addition to the Cp signal at 5.94 (5 H) *(1)*. This indicates that the formal σ- and π-bonds of the structure 25 have become delocalized and that the C_3H_5 group functions as an (η^3-allyl) ligand as in 26

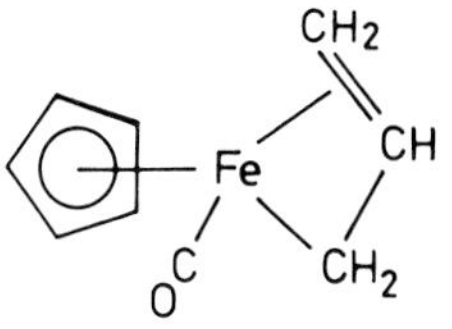

25

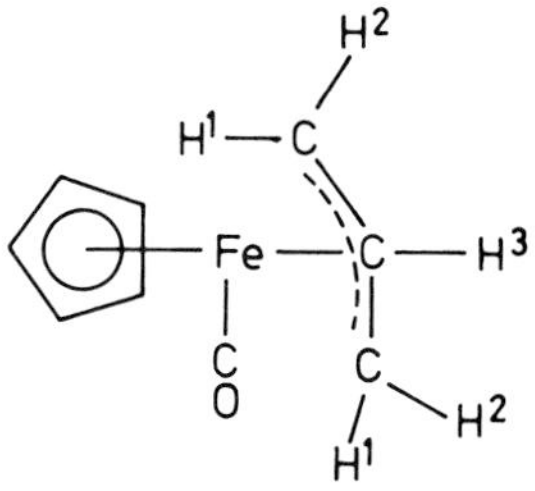

26

(75). The high-field absorption is due to the *anti*-protons H^1. Irradiation of $CH_2{=}CH{-}CH_2{-}CH_2{-}Fe(CO)_2Cp$ yields the 2-methylallyl complex (η^3-2-Me-C_3H_4)$Fe(CO)_2Cp$ *(86)*. Similar σ- to π-allyl conversions have been observed for the complex (η^1-C_3H_5)$Fe(CO)_3$(η^3-C_3H_5) derived from allyl bromide and the $[(\eta^3\text{-allyl})Fe(CO)_3]^-$ anion *(129)*, and also for Cr, Mo, W, Mn, and Co complexes.

It has been proposed that an (η^3-benzyl)iron cation $[C_8H_9FeCp]^+$ is responsible for the base peak (m/e 226) in the mass spectrum of Me-C_6H_4-$Fe(CO)_2Cp$ (Section III.A.3) *(90)*. Photochemical decarbonylation of complexes of the type R-CO-$Fe(CO)_2Cp$ (R = $CH{=}CH_2$, Ph) provides a convenient route to the corresponding R-$Fe(CO)_2Cp$ complexes *(108)*.

Photolysis of η^1-benzyl complexes of Mo and W affords the corresponding η^3-benzyl complexes, but with (η^1-benzyl)-(η^5-cyclopentadienyl)dicarbonyliron the major photolysis product is the dimer $[CpFe(CO)_2]_2$ along with ca. 1 % of ferrocenylbenzylketone *(130)*. The mechanism of this unusual ring acylation reaction is obscure.

D. *STEREOCHEMISTRY OF REACTIONS*

Data on the stereochemical changes occuring at transition metal carbon bonds are usually lacking. Experimental problems of thermal lability of the metal-carbon bond, the lack of suitable enantiomerically pure substrate molecules and of reference reactions of proven stereochemistry, and the difficulty of assigning absolute stereochemistries to molecules for which there are no X-ray crystallographic data have all contributed to this state of affairs.

Recently, NMR and CD spectroscopy have been applied with some success to the study of reactions at iron-carbon bonds. One approach takes advantage of small differences between the vicinal coupling constants J(H-H') of *erythro*- (>8 Hz) and *threo*- (<7 Hz) diastereoisomers of compounds of the type $(CH_3)_3C$-CHD-CHDX (X = Cl, Br, $Fe(CO)_2Cp$, etc.) *(20)*. The couplings are the weighted average of couplings in each of the conformers contributing to the diastereoisomeric pair (X = $Fe(CO)_2Cp$) 27.

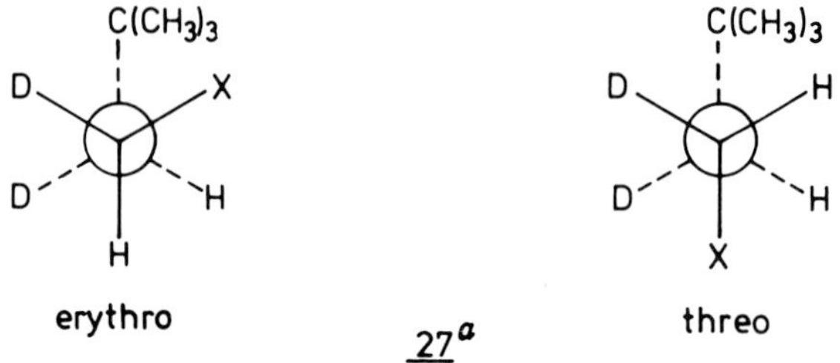

27[a]

The stereochemistry at carbon of reactions at the Fe-C bond of $(CH_3)_3C$-CHD-CHD-$Fe(CO)_2Cp$ are summarized in Figure 6. Three types of reaction were identified, *viz.* migration of the alkyl group to co-ordinated carbon monoxide induced by ligand nucleophilic attack or by oxidation and occurring with retention, electrophilic attack with retention, and electrophilic attack with inversion. Oxidatively induced migration with retention was also observed for the cyclohexene derivatives 28 and 29. The mechanism of these oxidative migrations is unknown. A possibility is that electron transfer to the oxidizing agent makes the iron atom strongly electron-withdrawing, thus weakening the metal-carbonyl π bonding and rendering the carbon of the carbonyl group susceptible to intramolecular nucleophilic attack by the alkyl group.

Insertion of SO_2 into the Fe-C bond occurs with inversion at carbon in reaction [60] which is > 95 % stereoselective. In order to throw more light on the reaction mechanism, the stereochemistry at iron was investigated. Only a single pair of diastereotopic methyl groups is evident in the NMR spectrum of the sulphinates 31, this showed the SO_2 insertion reaction to be highly stereoselective but did not establish whether the stereoselectivity was associated with inversion or retention of configuration at the iron atom *(9)*. Some epimerization (ca. 20 %) of the chiral iron occurred when liquid SO_2 was the reaction medium. Retention of configuration at the iron atom in the insertion of SO_2 into the Fe-C bond of 32 was adduced from the similarity between its circular dichroism spectrum and that of its sulphinate 33. The absolute stereochemistries were assigned on the basis of com-

---

[a] *trans* conformers are shown, these are the most favoured energetically.

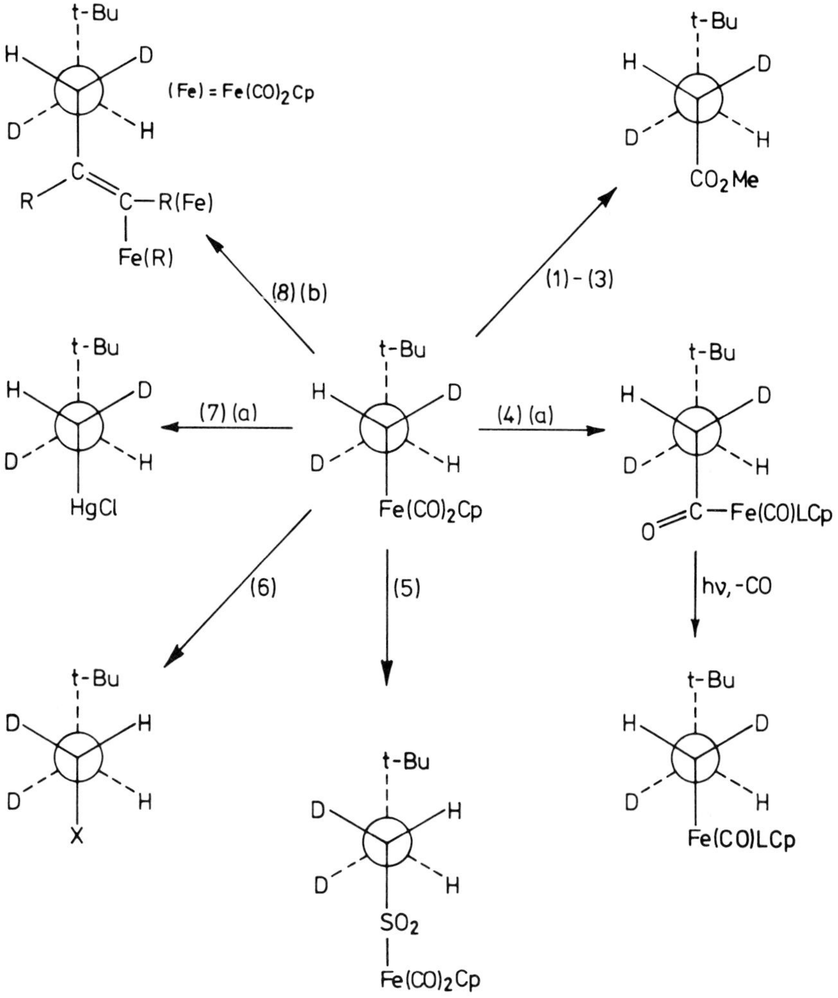

(1) Br_2, MeOH; (2) Ce(IV), MeOH; (3) O_2, MeOH;
(4) L = PPh_3, RNC; (5) SO_2; (6) Br_2(-20°C);
(7) $HgCl_2$; (8) R-C≡C-R, R = CO_2-Me
(a) Reaction carried out on the *threo* diastereoisomer.
(b) Stereochemistry at the C=C double bond uncertain.

Fig. 6: Stereochemistry of reactions of $(CH_3)_3C$-CHD-CHD-Fe-$(CO)_2$Cp (from refs. 20, 167).

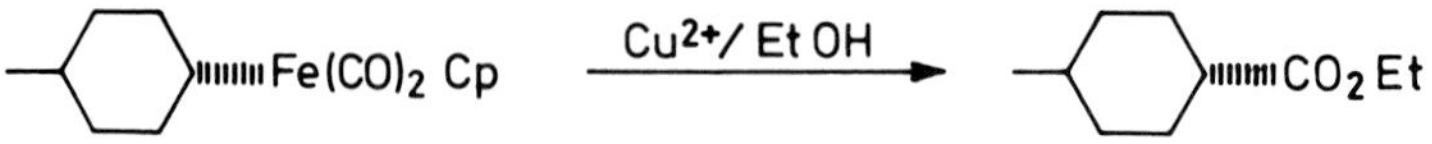

28

[59]

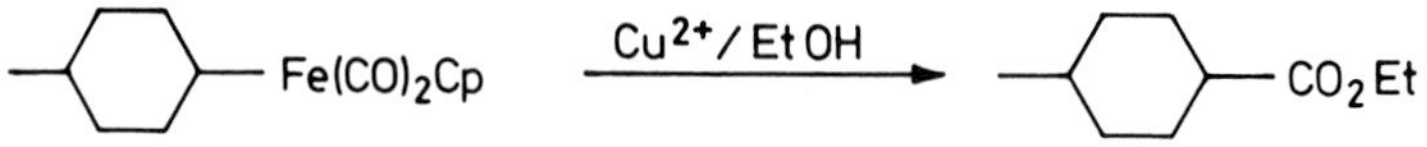

29

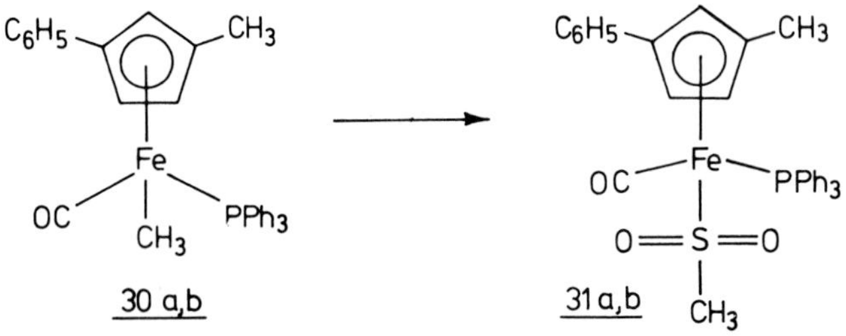

30 a,b

31 a,b

[60]

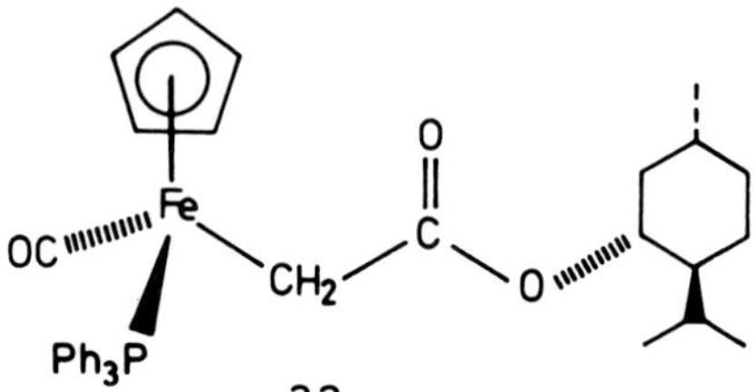

32

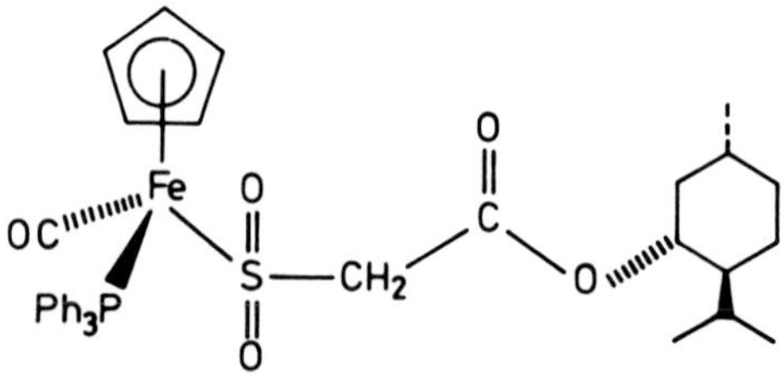

33

parison between the circular dichroism spectra and that of $[CpFe(Ph_3P)(CO)\text{-}CMe{=}NH\text{-}CH(CH_3)Ph]^+[BF_4]^-$ *(58)* for which the stereochemistry has been established by X-ray diffraction.

Inversion at carbon in SO_2 insertion into the Fe-C bond can be accomodated by backside electrophilic attack of SO_2 without the necessity of prior co-ordination of SO_2. The retention at iron is less easy to explain. It has been suggested that it may reflect the configurational stability of the iron centre in an ion-pair intermediate.

The stereochemical outcome of reactions between Fe-C bonds and reagents of the type EX (where E = HgX, X, H; X = halogen) depends on the reaction conditions and the reagent. Retention, inversion, and epimerization have all been observed. Clearly, this class of reaction is not suitable as a probe of the stereochemistry of reactions at metal-carbon bonds. In a series of reactions of complex 30 with HI, I_2, and HgI_2, the stereoselectivity of the formation of the cleavage product followed the order $I_2 > HI \gtrsim HgI_2$ *(9)*. In each case partial epimerization had occurred; this was highest with HI. The formation of a stereochemical non-rigid intermediate such as 34 formed *via* reversible oxidative addition of E^+I^- to 30 was suggested in order to account for the partial epimerization. Stereochemical non-rigidity has been observed in the related molecules $CpMo(CO)_2LX$ and $CpMo(CO)L_2X$.

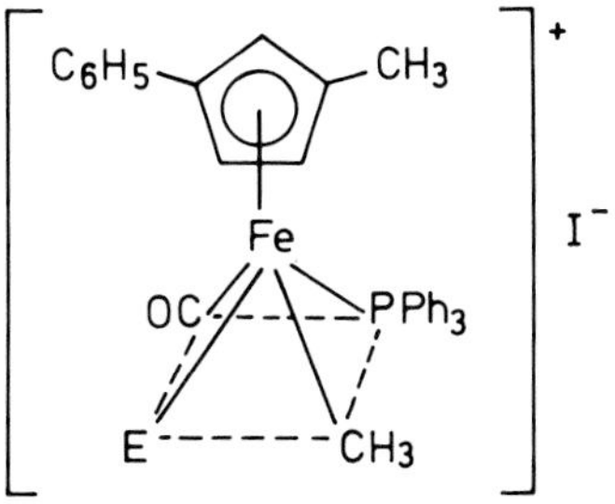

34

The collapse of 34 to the halogenated product could occur *via* reductive elimination of ECH_3.

E. CATALYSIS

Iron alkyls have been invoked as intermediates in a variety of reactions such as the oligomerization and polymerization of olefins and acetylenes catalysed by mixtures of iron salts and organoaluminium compounds. There are no reports of the characterization of these presumably highly labile species.

The hydrogenation of olefins *(119,146,147)* in the presence of iron alkyls probably occurs *via* iron hydrides formed by β-elimination from the alkyl group (scheme [61]). Concentrations of $FeCl_n$ (n = 2 or 3) as low as 10^{-5} mol l^{-1} are sufficient to catalyse the reaction [62] *(154)*. The major reaction is disproportionation, only when there are no β-hydrogens are coupling products formed (eq. [63]).

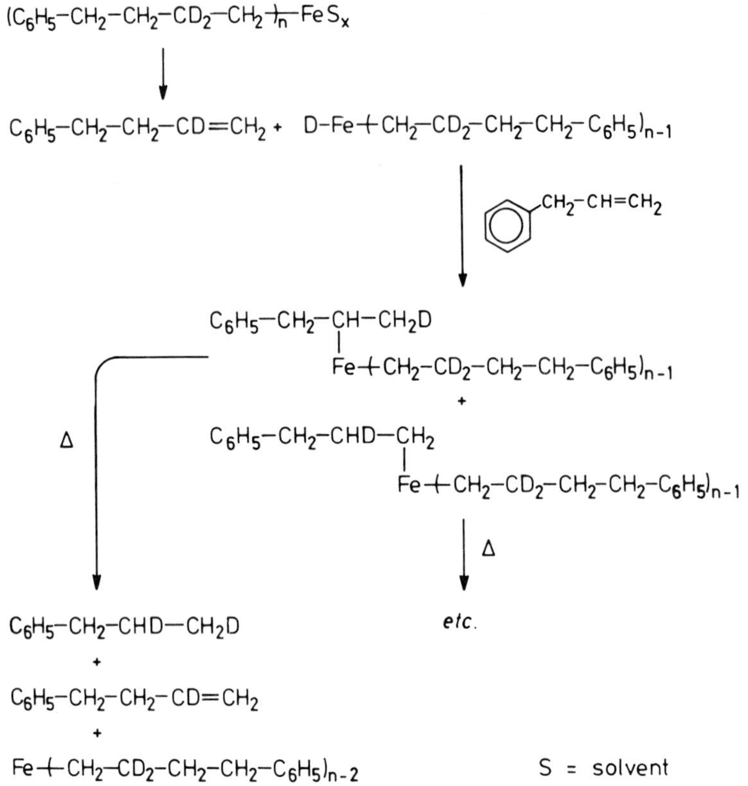

$$(C_6H_5-CH_2-CH_2-CD_2-CH_2 \overset{}{\underset{n}{\rightarrow}} Fe\,S_x$$

$$\downarrow$$

$$C_6H_5-CH_2-CH_2-CD=CH_2 + D-Fe(CH_2-CD_2-CH_2-CH_2-C_6H_5)_{n-1}$$

$$\downarrow \quad \text{(phenyl)} \, CH_2-CH=CH_2$$

$$C_6H_5-CH_2-\underset{\underset{Fe(CH_2-CD_2-CH_2-CH_2-C_6H_5)_{n-1}}{|}}{CH}-CH_2D$$

$$+$$

$$C_6H_5-CH_2-\underset{\underset{Fe(CH_2-CD_2-CH_2-CH_2-C_6H_5)_{n-1}}{|}}{CHD}-CH_2 \qquad \qquad [61]$$

$$\Delta \qquad \qquad \downarrow \Delta$$

$$C_6H_5-CH_2-CHD-CH_2D \qquad \qquad etc.$$

$$+$$

$$C_6H_5-CH_2-CH_2-CD=CH_2$$

$$+$$

$$Fe(CH_2-CD_2-CH_2-CH_2-C_6H_5)_{n-2} \qquad \qquad S = solvent$$

$$C_2H_5MgBr + C_2H_5Br \xrightarrow{FeCl_n} C_2H_6 + C_2H_4 + MgBr_2 \qquad [62]$$

$$C_2H_5MgBr + (CH_3)_3CBr \overset{\nearrow\ C_2H_4 + (CH_3)_3CH + MgBr_2}{\underset{\searrow\ C_2H_6 + (CH_3)_2CH=CH_2 + MgBr_2}{}} \qquad [63]$$

$Et_2Fe(bipy)_2$ has been found to catalyse the oligomerization and polymerization of butadiene. A detailed investigation of the reaction showed that the catalytic species is $Fe(bipy)_2$ formed *via* elimination of ethylene and ethane from the dialkyl complex *(170,171)*.

REFERENCES

1. Alexander, J.J., and Wojcicki, A., *J. Organometal. Chem.*, *15*, P 23 (1968).
2. Alexander, J.J., and Wojcicki, A., *Inorg. Chim. Acta.*, *5*, 655 (1971).
3. Andrianov, V.G., and Struchkov, Yu.T., *Chem Commun.*, *1968*, 1590.
4. Angelici, R.J., and Busetto, L., *J. Amer. Chem. Soc.*, *91*, 3197 (1969).
5. Ariyaratne, J.K.P., and Green, M.L.H., *J. Chem. Soc.*, *1963*, 2976.
6. Ariyaratne, J.K.P., and Green, M.L.H., *J. Organometal. Chem.*, *1*, 90 (1963).
7. Ariyaratne, J.K.P., and Green, M.L.H., *J. Chem. Soc.*, *1964*, 1.
8. Ariyaratne, J.K.P., Bjerrum, A.M., Green, M.L.H., Ishaq, M., Prout, C.K., and Swanwick, M.G., *J. Chem. Soc. A*, *1969*, 1309.
9. Attig, T.G., and Wojcicki, A., *J. Amer. Chem. Soc.*, *96*, 262 (1974).
10. Aumann, R., *Angew. Chem.*, *83*, 175 (1971); *Angew. Chem. Int. Ed. Engl.*, *10*, 188 (1971).
11. Aumann, R., *Angew. Chem.*, *83*, 176 (1971); *Angew. Chem. Int. Ed. Engl.*, *10*, 189 (1971).
12. Avoyan, R.L., Chapovskii, Yu.A., and Struchkov, Yu.T., *Zh. Strukt. Khim.*, *7*, 900 (1966); *J. Struct. Chem.*, *7*, 838 (1966).
13. Bagga, M., Flannigan, W.T., Knox, G.R., and Pauson, P.L., *J. Chem. Soc. C*, *1969*, 1534.
14. Beck, W., Hieber, W., and Neumair, G., *Z. anorg. allgem. Chem.*, *344*, 285 (1966).
15. Bennett, M.J., Jr, Cotton, F.A., Davison, A., Faller, J.W., Lippard, S.J., and Morehouse, S.M., *J. Amer. Chem. Soc.*, *88*, 4371 (1966).
16. Bennett, M.A., Robertson, G.B., Tomkins, I.B., and Whimp, P.O., *J. Organometal. Chem.*, *32*, C 19 (1971).
17. Bibler, J.P., and Wojcicki, A., *J. Amer. Chem. Soc.*, *86*, 5051 (1964).
18. Bibler, J.P., and Wojcicki, A., *Inorg. Chem.*, *5*, 889 (1966).
19. Bibler, J.P., and Wojcicki, A., *J. Amer. Chem. Soc.*, *88*, 4862 (1966).
20. Bock, P.L., Boschetto, D.J., Rasmussen, J.R., Demers, J.P., and Whitesides, G.M., *J. Amer. Chem. Soc.*, *96*, 2814 (1974).
21. Bolton, E.S., Knox, G.R., and Robertson, C.G., *J. Chem. Soc. D, Chem. Commun.*, *1969*, 664.

22. Bond, A., and Green, M., *J. Chem. Soc. Dalton Trans.*, *1972*, 763.
23. Bowden, F.L., and Lever, A.B.P., *Organometal. Chem. Rev.*, *3*, 227 (1968).
24. Bowden, F.L., and Johnson, D.K., unpublished results.
25. Braterman, P.S., and Cross, R.J., *J. Chem. Soc. Dalton Trans.*, *1972*, 657; and refs. therein.
26. Bruce, M.I., *J. Organometal. Chem.*, *10*, 495 (1967).
27. Bruce, M.I., *Advan. Organometal. Chem.*, *6*, 273 (1968).
28. Bruce, M.I., and Stone, F.G.A., *Angew. Chem.*, *80*, 835 (1968); *Angew. Chem. Int. Ed. Engl.*, *7*, 747 (1968).
29. Bruce, M.I., Harbourne, D.A., Waugh, F., and Stone, F.G.A., *J. Chem. Soc. A*, *1968*, 895.
30. Bruce, M.I., *J. Chem. Soc. A*, *1968*, 1459.
31. Brunner, H., and Schmidt, E., *Angew. Chem.*, *81*, 570 (1969); *Angew. Chem. Int. Ed. Engl.*, *8*, 616 (1969).
32. Busetto, L., and Angelici, R.J., *Inorg. Chim. Acta*, *2*, 391 (1968).
33. Butler, I.S., Basolo, F., and Pearson, R.G., *Inorg. Chem.*, *6*, 2074 (1967).
34. Cais, M., and Lupin, M.S., *Advan. Organometal. Chem.*, *8*, 211 (1970).
35. Chambers, R.D., and Chivers, T., *Organometal. Chem. Rev.*, *1*, 279 (1966).
36. Chapovskii, Yu.A., Lokshin, B.V., Makarova, L.G., Nesmeyanov, A.N., and Polovyanyuk, I.V., *Dokl. Akad. Nauk. SSSR*, *166*, 1125 (1966); *Dokl. Chem.*, *166*, 213 (1966).
37. Chatt, J., and Shaw, B.L., *J. Chem. Soc.*, *1959*, 705.
38. Chatt, J., and Shaw, B.L., *J. Chem. Soc.*, *1960*, 1718.
39. Chatt, J., and Shaw, B.L., *J. Chem. Soc.*, *1961*, 285.
40. Churchill, M.R., *Inorg. Chem.*, *6*, 185 (1967).
41. Churchill, M.R., Wormald, J., Giering, W.P., and Emerson, G.F., *Chem. Commun.*, *1968*, 1217.
42. Churchill, M.R., and Wormald, J., *Inorg. Chem.*, *8*, 1936 (1969).
43. Churchill, M.R., Wormald, J., Ross, D.A., Thomasson, J.E., and Wojcicki, A., *J. Amer. Chem. Soc.*, *92*, 1795 (1970).
44. Churchill, M.R., and Wormald, J., *Inorg. Chem.*, *10*, 572 (1971).
45. Coffey, C.E., *J. Amer. Chem. Soc.*, *84*, 118 (1962).
46. Coffield, T.H., Kozikowski, J., and Closson, R.D., *Int. Conf. Coord. Chem.*, *London 1959*, *Chem. Soc. Spec. Publ. 13*, p. 126.
47. Collman, J.P., *Accounts Chem. Res.*, *1*, 136 (1968).
48. Collman, J.P., Winter, S.R., and Komoto, R.G., *J. Amer. Chem. Soc.*, *95*, 249 (1973).

49. Collman, J.P., and Winter, S.R., *J. Amer. Chem. Soc.*, *95*, 4089 (1973).
50. Cooke, J., Cullen, W.R., Green, M., and Stone, F.G.A., *J. Chem. Soc. A*, *1969*, 1872.
51. Cooke, M.P., Jr., *J. Amer. Chem. Soc.*, *92*, 6080 (1970).
52. Corey, E.J., and Posner, G.H., *J. Amer. Chem. Soc.*, *90*, 5615 (1968).
53. Corey, E.J., and Posner, G.H., *Tetrahedron Lett.*, *1970*, 315.
54. Cotton, F.A., *Chem. Rev.*, *55*, 551 (1955).
55. Cross, R.J., *Organometal. Chem. Rev.*, *2*, 97 (1967).
56. Cutler, A., Fish, R.W., Giering, W.P., and Rosenblum, M., *J. Amer. Chem. Soc.*, *94*, 4354 (1972).
57. Davis, R.E., *Chem. Commun.*, *1968*, 1218.
58. Davison, A., Krusell, W.C., and Michaelson, R.C., *J. Organometal. Chem.*, *72*, C 7 (1974).
59. Dempster, A.B., Powell, D.B., and Sheppard, N., *J. Chem. Soc. A*, *1970*, 1129.
60. Dessy, R.E., Pohl, R.L., and King, R.B., *J. Amer. Chem. Soc.*, *88*, 5121 (1966).
61. Dub, M., *Organometallic Compounds*, *Vol. I*, p. 193, Springer, Berlin 1966.
62. Dvorak, J., O'Brien, R.J., and Santo, W., *J. Chem. Soc. D, Chem. Commun.*, *1970*, 411.
63. Ehntholt, D., Rosan, A., and Rosenblum, M., *J. Organometal. Chem.*, *56*, 315 (1973).
64. Eisenstadt, A., *Tetrahedron Lett.*, *1972*, 2005.
65. Faller, J.W., and Anderson, A.S., *J. Amer. Chem. Soc.*, *91*, 1550 (1969).
66. Fields, R., in E.F. Mooney (Ed.), *Annual Reports NMR Spectroscopy*, *Vol. Va*, p. 99, Academic Press, New York 1972.
67. Flood, T.C., and Miles, D.L., *J. Amer. Chem. Soc.*, *95*, 6460 (1973).
68. Floriani, C., and Calderazzo, F., *J. Chem. Soc. A*, *1971*, 3665.
69. Fong, C.W., and Kitching, W., *J. Organometal. Chem.*, *22*, 107 (1970).
70. Gerloch, M., Lewis, J., Mabbs, F.E., and Richards, A., *J. Chem. Soc. A*, *1968*, 112.
71. Giering, W.P., and Rosenblum, M., *J. Organometal. Chem.*, *25*, C 71 (1970).
72. Giering, W.P., Raghu, S., Rosenblum, M., Cutler, A., Ehntholt, D., and Fish, R.W., *J. Amer. Chem. Soc.*, *94*, 8251 (1972).
73. Goodfellow, R.J., Green, M., Mayne, N., Rest, A.J., and Stone, F.G.A., *J. Chem. Soc. A*, *1968*, 177.
74. Graziani, M., and Wojcicki, A., *Inorg. Chim. Acta.*,

4, 347 (1970).
75. Green, M.L.H., and Nagy, P.L.I., *Advan. Organometal. Chem.*, *2*, 325 (1962).
76. Green, M.L.H., and Nagy, P.L.I., *J. Amer. Chem. Soc.*, *84*, 1310 (1962).
77. Green, M.L.H., and Nagy, P.L.I., *J. Chem. Soc.*, *1963*, 189.
78. Green, M.L.H., and Nagy, P.L.I., *J. Organometal. Chem.*, *1*, 58 (1963).
79. Green, M.L.H., and Nagy, P.L.I., *Z. Naturforsch.*, *B 18*, 162 (1963).
80. Green, M.L.H., Ishaq, M., and Mole, T., *2nd Int. Conf. Organometal. Chem., Madison 1965, Abstr. Proc.*, p. 91.
81. Green, M.L.H., Ariyaratne, J.K.P., Bjerrum, A.M., Ishaq, M., and Prout, C.K., *Chem. Commun.*, *1967*, 430.
82. Green, M.L.H., Ishaq, M., and Whiteley, R.N., *J. Chem. Soc. A*, *1967*, 1508.
83. Green, M.L.H., and Hurley, C.R., *J. Organometal. Chem.*, *10*, 188 (1967).
84. Green, M.L.H., in G.E. Coates, M.L.H. Green, and K. Wade (Eds.), *Organometallic Compounds, Vol. II*, p. 220, Methuen, London 1968.
85. Green, M.L.H., and Whiteley, R.N., *J. Chem. Soc. A*, *1971*, 1943.
86. Green, M.L.H., and Smith, M.J., *J. Chem. Soc. A*, *1971*, 3220.
87. Halpern, J., *Accounts Chem. Res.*, *3*, 386 (1970).
88. Harbourne, D.A., and Stone, F.G.A., *J. Chem. Soc. A*, *1968*, 1765.
89. Hata, G., Kondo, H., and Miyake, A., *J. Amer. Chem. Soc.*, *90*, 2278 (1968).
90. Hawthorne, J.D., Mays, M.J., and Simpson, R.N.F., *J. Organometal. Chem.*, *12*, 407 (1968).
91. Hitchcock, P.B., and Mason, R., *Chem. Commun.*, *1967*, 242.
92. Jacobson, S.E., and Wojcicki, A., *J. Amer. Chem. Soc.*, *93*, 2535 (1971).
93. Jacobson, S.E., Reich-Rohrwig, P., and Wojcicki, A., *J. Chem. Soc. D, Chem. Commun.*, *1971*, 1526.
94. Jacobson, S.E., Reich-Rohrwig, P., and Wojcicki, A., *Inorg. Chem.*, *12*, 717 (1973).
95. Jennings, M.A., and Wojcicki, A., *J. Organometal. Chem.*, *14*, 231 (1968).
96. Jetz, W., and Angelici, R.J., *J. Organometal. Chem.*, *35*, C 37 (1972).
97. Johnson, M.D., and Mayle, C., *J. Chem. Soc. D, Chem. Commun.*, *1969*, 192.
98. Jolly, P.W., Bruce, M.I., and Stone, F.G.A., *J. Chem.*

Soc., 1965, 5830.
99. Jolly, P.W., and Pettit, R., *J. Amer. Chem. Soc., 88*, 5044 (1966).
100. Jolly, P.W., and Pettit, R., *J. Organometal. Chem., 12*, 491 (1968).
101. Jolly, P.W., and Stone, F.G.A., unpublished results.
102. Jones, K., and Mooney, E.F., in E.F. Mooney (Ed.), *Annual Reports NMR Spectroscopy, Vol. III*, p. 340, Academic Press, New York 1970.
103. Kaesz, H.D., King, R.B., and Stone, F.G.A., *Z. Naturforsch., B 15*, 763 (1960).
104. King, R.B., Stafford, S.L., Treichel, P.M., and Stone, F.G.A., *J. Amer. Chem. Soc., 83*, 3604 (1961).
105. King, R.B., Treichel, P.M., and Stone, F.G.A., *Proc. Chem. Soc., 1961*, 69.
106. King, R.B., *J. Amer. Chem. Soc., 85*, 1918 (1963).
107. King, R.B., *Advan. Organometal. Chem., 2*, 157 (1964).
108. King, R.B., and Bisnette, M.B., *J. Organometal. Chem., 2*, 15 (1964).
109. King, R.B., and Bisnette, M.B., *J. Organometal. Chem., 2*, 38 (1964).
110. King, R.B., *J. Amer. Chem. Soc., 89*, 6368 (1967).
111. King, R.B., *J. Amer. Chem. Soc., 90*, 1417 (1968).
112. King, R.B., and Panell, K.H., *J. Amer. Chem. Soc., 90*, 3984 (1968).
113. King, R.B., *Accounts Chem. Res., 3*, 417 (1970).
114. King, R.B., and Saran, M.S., *J. Amer. Chem. Soc., 95*, 1811 (1973).
115. Köbrich, G., and Büttner, H., *J. Organometal. Chem., 18*, 117 (1969).
116. Koerner von Gustorf, E., Jun, M.J., and Schenck, G.O., *Z. Naturforsch., B 18*, 503 (1963).
117. Kruck, T., and Noak, M., *Chem. Ber. 97*, 1693 (1964).
118. Küpper, F.-W., *J. Organometal. Chem., 13*, 219 (1968).
119. Light, J.R.C., and Zeiss, H.H., *J. Organometal. Chem., 21*, 517 (1970).
120. Lindner, E., Lorenz, I.-P., and Vitzthum, G., *Angew. Chem., 83*, 213 (1971); *Angew. Chem. Int. Ed. Engl., 10*, 193 (1971).
121. Maddox, M.L., Stafford, S.L., and Kaesz, H.D., *Advan. Organometal. Chem., 3*, 1 (1965).
122. Manuel, T.A., Stafford, S.L., and Stone, F.G.A., *J. Amer. Chem. Soc., 83*, 249 (1961).
123. Martynova, V.F., *Zh. Obshch. Khim., 32*, 2702 (1962); *J. Gen. Chem. USSR 32*, 2660 (1962).
124. Meunier-Piret, J., Piret, P., and van Meersche, M., *Acta Crystallogr., 19*, 85 (1965).
125. Mills, O.S., and Redhouse, A.D., *J. Chem. Soc. A*,

1968, 1282

126. Mowat, W., Shortland, A., Hill, N.J., Yagupsky, G., and Wilkinson, G., *J. Chem. Soc. Dalton Trans., 1972,* 533.

127. Muetterties, E.L., *Inorg. Chem., 4,* 1841 (1965).

128. Nesmeyanov, A.N., Chapovsky, Yu.A., Polovyanyuk, I.V., and Makarova, L.G., *J. Organometal. Chem., 7,* 329 (1967).

129. Nesmeyanov, A.N., and Kritskaya, I.I., *J. Organometal. Chem., 14,* 387 (1968).

130. Nesmeyanov, A.N., Chenskaya, T.B., Babakhina, G.M., and Kritskaya, I.I., *Izv. Akad. Nauk SSSR, Ser. Khim., 1970,* 1187; *Bull. Acad. Sci. USSR, Div. Chem. Ser. 1970,* 1129.

131. Nesmeyanov, A.N., Leshcheva, I.F., Polovyanyuk, I.V., Ustynyuk, Yu.A., and Makorova, L.G., *J. Organometal. Chem., 37,* 159 (1972); and refs. therein.

132. Pankowski, M., and Bigorgne, M., *J. Organometal. Chem., 30,* 227 (1971).

133. Pannell, K.H., *J. Chem. Soc. D, Chem. Commun., 1969,* 1346.

134. Parshall, G.W., and Mrowca, J.J., *Advan. Organometal. Chem., 7,* 157 (1968).

135. Parshall, G.W., *Accounts Chem. Res., 3,* 139 (1970).

136. Piper, T.S., and Wilkinson, G., *J. Inorg. Nucl. Chem., 3,* 104 (1956).

137. Pitcher, E., and Stone, F.G.A., *Spectrochim. Acta., 17,* 1244 (1961).

138. Pitcher, E., Buckingham, A.D., and Stone, F.G.A., *J. Chem. Phys., 36,* 124 (1962).

139. Pitcher, E., and Stone, F.G.A., *Spectrochim. Acta., 18,* 585 (1962).

140. Plowman, R.A., and Stone, F.G.A., *Inorg. Chem., 1,* 518 (1962).

141. Robinson, P.W., and Wojcicki, A., *J. Chem. Soc. D, Chem. Commun., 1970,* 951.

142. Ross, D.A., Ph. D. Thesis, Ohio State University, 1970; *Diss. Abstr., B 31,* 3905 (1970/71).

143. Roustan, J.-L., and Charrier, C., *C.R. Acad. Sci., Ser. C., 268,* 2113 (1969).

144. Schenck, G.O., Koerner von Gustorf, E., and Jun, M.J., *Tetrahedron Lett., 1962,* 1059.

145. Siegl, W.O., and Collman, J.P., *J. Amer. Chem. Soc., 94,* 2516 (1972).

146. Sneeden, R.P.A., and Zeiss, H.H., *J. Organometal. Chem., 19,* 93 (1969).

147. Sneeden, R.P.A., and Zeiss, H.H., *J. Organometal. Chem., 22,* 713 (1970).

148. Sneeden, R.P.A., and Zeiss, H.H., *J. Organometal. Chem., 26,* 101 (1971).
149. Sneeden, R.P.A., and Zeiss, H.H., *J. Organometal. Chem., 27,* 89 (1971).
150. Stewart, R.P., and Treichel, P.M., *J. Amer. Chem. Soc., 92,* 2710 (1970); and refs. therein.
151. Su, S.R., Hanna, J.A., and Wojcicki, A., *J. Organometal. Chem., 21,* P 21 (1970).
152. Su, S.R., and Wojcicki, A., *J. Organometal. Chem., 31,* C 34 (1971).
153. Taft, R.W., Price, E., Fox, I.R., Lewis, I.C., Andersen, K.K., and Davis, G.T., *J. Amer. Chem. Soc., 85,* 709 (1963).
154. Tamura, M., and Kochi, J., *J. Organometal. Chem., 31,* 289 (1971).
155. Thomasson, J.E., Robinson, P.W., Ross, D.A., and Wojcicki, A., *Inorg. Chem., 10,* 2130 (1971).
156. Thomson, J., Keeney, W., and Baird, M.C., *J. Organometal. Chem., 40,* 205 (1972).
157. Treichel, P.M., Chaudhari, M.A., and Stone, F.G.A., *J. Organometal. Chem., 1,* 98 (1963).
158. Treichel, P.M., and Stone, F.G.A., *Advan. Organometal. Chem., 1,* 143 (1964).
159. Treichel, P.M., Shubkin, R.L., Barnett, K.W., and Reichard, D., *Inorg. Chem., 5,* 1177 (1966).
160. Treichel, P.M., and Shubkin, R.L., *Inorg. Chem., 6,* 1328 (1967).
161. Treichel, P.M., and Stenson, J.P., *Inorg. Chem., 8,* 2563 (1969).
162. Treichel, P.M., Stenson, J.P., and Benedict, J.J., *Inorg. Chem. 10,* 1183 (1971).
163. Tsutsui, M., *Ann. N.Y. Acad. Sci., 93,* 135 (1961).
164. Tsutsui, M., Hancock, M., Ariyoshi, J., and Levy, M.N., *J. Amer. Chem. Soc., 91,* 5233 (1969).
165. Whitesides, G.M., and Boschetto, D.J., *J. Amer. Chem. Soc., 91,* 4313 (1969).
166. Whitesides, G.M., Stedronsky, E.R., Casey, C.P., and San Filippo, J., Jr., *J. Amer. Chem. Soc., 92,* 1426 (1970).
167. Whitesides, G.M., and Boschetto, D.J., *J. Amer. Chem. Soc., 93,* 1529 (1971).
168. Wojcicki, A., *Accounts Chem. Res., 4,* 344 (1971).
169. Yamamoto, A., Morifuji, K., Ikeda, S., Saito, T., Uchida, Y., and Misono, A., *J. Amer. Chem. Soc., 90,* 1878 (1968).
170. Yamamoto, T., Yamamoto, A., and Ikeda, I., *Bull. Chem. Soc. Jap., 45,* 1104 (1972).
171. Yamamoto, T., Yamamoto, A., and Ikeda, I., *Bull. Chem.*

Soc. Jap., 45, 1111 (1972).

172. Yamamoto, Y., and Yamazaki, H., *Inorg. Chem., 11,* 211 (1972).

173. Zhdanov, Yu.A., and Minkin, V.I., *Korrelyateinonyi analiz i organicheskoi khimii, Izd. Rostovskogo,* Universiteta, 1966, p. 384 - 388 (quoted as ref. 14 in ref. 131).

174. Goedken, V.L., Peng, S.-M., and Park, Y.-a., *J. Amer. Chem. Soc., 96,* 284 (1974).

MONOOLEFIN IRON COMPLEXES

By R.B. KING

Department of Chemistry, University of Georgia
Athens, Georgia 30602, U.S.A.

TABLE OF CONTENTS

ISBN 0-12-417101-X (v.1)

I. INTRODUCTION

This chapter discusses iron complexes in which carbon-carbon double bonds are individually bonded to iron atoms. Important groups of complexes of this type include compounds of the type (olefin)$Fe(CO)_4$, which may be regarded as substitution products of pentacarbonyliron, and cationic carbonylcyclopentadienyliron derivatives of the general type $[(\eta^5\text{-}C_5H_5)Fe(CO)_2(\eta^2\text{-olefin})]^+$. This chapter will also include some iron complexes of dienes, such as butadiene, and polyenes, such as fulvenes, in which the carbon-carbon double bonds are individually bonded to different iron atoms. Iron complexes of dienes, trienes, and polyenes, in which two or more double bonds are bonded together to the same iron atom(s) will be discussed in other chapters of this book.

II. OLEFIN-TETRACARBONYLIRON DERIVATIVES

Among the most important olefin-iron derivatives are the olefin-tetracarbonyliron derivatives. These compounds have the general formula $(\eta^2\text{-olefin})Fe(CO)_4$. They may be regarded as substitution products of pentacarbonyliron in which one of the equatorial carbonyl groups is replaced by the *dihapto*-coordinated olefin. Complexes of the type $(\eta^2\text{-olefin})Fe(CO)_4$ are known mainly for unsubstituted ethylene and for substituted ethylenes containing substituents more electronegative than hydrogen. In addition several dienes and trienes can form complexes of the type $L[Fe(CO)_4]_n$ (n = 1 or 2) in which the carbon-carbon double bonds are individually coordinated to $Fe(CO)_4$ units.

A. *PREPARATION OF OLEFIN-TETRACARBONYLIRON DERIVATIVES*

The first known olefin-tetracarbonyliron complex to be prepared was (acrylonitrile)tetracarbonyliron, $(CH_2{=}CH\text{-}CN)\text{-}Fe(CO)_4$, which was reported in 1960 by Kettle and Orgel *(120)*. They obtained this complex by treatment of $Fe_2(CO)_9$ with excess boiling acrylonitrile or by exposing a mixture of $Fe(CO)_5$ and excess acrylonitrile to sunlight. Subsequent crystallographic studies by Luxmore and Truter *(155,156)* indicated structure 1 (X = CN), in which the carbon-carbon double bond of the acrylonitrile is coordinated to a five-coordinate trigonal bipyramidal iron atom in one of the equatorial positions.

The yield in the original preparation of (acrylonitrile)-tetracarbonyliron was rather poor (2-3 %). This apparently discouraged the immediate extension of such preparations to

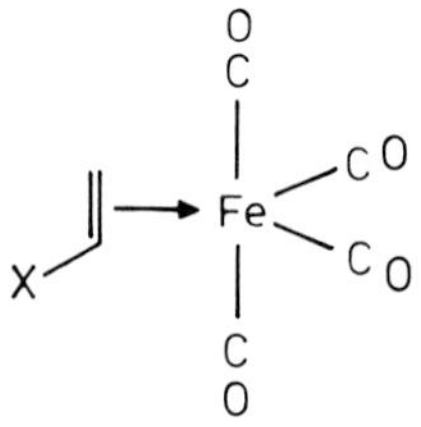

1

other systems. However, in 1963 *(228)* a group of workers at the Cyanamid European Research Institute developed general conditions for reactions of electronegatively substituted olefins with $Fe_2(CO)_9$ in benzene at 40-45°C to give the corresponding (olefin)$Fe(CO)_4$ derivatives in rather good yields (60-93 %) according to equation [1].

$$Fe_2(CO)_9 + \text{olefin} \longrightarrow (\text{olefin})Fe(CO)_4 + Fe(CO)_5 \qquad [1]$$

Examples of electronegatively substituted olefins included in this study *(228)* are maleic anhydride, maleimide, fumaric acid, maleic acid, methyl esters of maleic and fumaric acids, acrylic acid and its methyl and ethyl esters, acrylamide, and acrolein. The products were reasonably stable yellow solids. Some of them could be purified by vacuum sublimation. Similar thermal reactions have been used subsequently to prepare tetracarbonyliron complexes of ethylene *(164)*, tetramethoxyethylene *(109)*, and various *cis*- and *trans*-dihaloethylenes *(97)*.

The second successful method used in the original preparation of (acrylonitrile)tetracarbonyliron was the ultraviolet irradiation of pentacarbonyliron with acrylonitrile. This method was subsequently extended to the preparation of (olefin)$Fe(CO)_4$ derivatives of maleic anhydride *(208)*, dimethyl maleate *(208)*, dimethyl fumarate *(208)*, methyl methacrylate *(134)*, vinyl acetate *(134)*, vinyl chloride *(137)*, styrene *(137)*, propylene *(137)*, and vinyl ethyl ether *(137)*. In relatively recent work excess $Fe(CO)_5$ has been recommended as the solvent for the photochemical preparation of the acrylic acid complex (CH_2=CH-COOH)$Fe(CO)_4$ *(50)*.

The effectiveness of ultraviolet irradiation in the preparation of various olefin-tetracarbonyliron derivatives from the olefin and pentacarbonyliron undoubtedly arises from the ability for the ultraviolet irradiation to dissociate one of the carbonyl groups from $Fe(CO)_5$ without heating the system to a temperature above the decomposition temperature of the resulting (olefin)$Fe(CO)_4$ derivative. γ-Radiation has been found to be similarly effective *(117,133,136)*. Thus, γ-irradiation of $Fe(CO)_5$ with olefins in benzene solution has been used for preparing (olefin)$Fe(CO)_4$ complexes of maleic anhydride, dimethyl fumarate, vinyl acetate, and methyl methacrylate *(132,135)*. During the course of this study $Fe(CO)_5$ was found to inhibit strongly the γ-radiation induced polymerization of vinyl acetate and methyl methacrylate.

An unstable tetracarbonyliron complex of vinyl alcohol has been prepared. However, because of the instability of vinyl alcohol with respect to tautomerization to acetaldehyde, indirect methods are necessary *(220)*. Reaction of $Fe_2(CO)_9$ with trimethylsilyl vinyl ether gives the corresponding tetracarbonyliron derivative, [$(CH_3)_3SiO$-CH=CH_2]-$Fe(CO)_4$, a solid stable only below 0°C. Hydrolysis of this complex with trifluoroacetic acid in acetone solution at -90°C gives (vinyl alcohol)tetracarbonyliron, (CH_2=CH-OH)-$Fe(CO)_4$, (1, X = OH). This complex was identified by its ^{1}H-NMR spectrum and its reactions with methyl isocyanate and triphenyl phosphite to give (O-vinyl-N-methylcarbamate)tetracarbonyliron, [CH_3-NH-CO-CH=CH_2]$Fe(CO)_4$, and acetaldehyde, respectively. However, (vinyl alcohol)tetracarbonyliron decomposed above -70°C.

Some miscellaneous olefin-tetracarbonyliron derivatives have been obtained by other methods. Thus, treatment of the

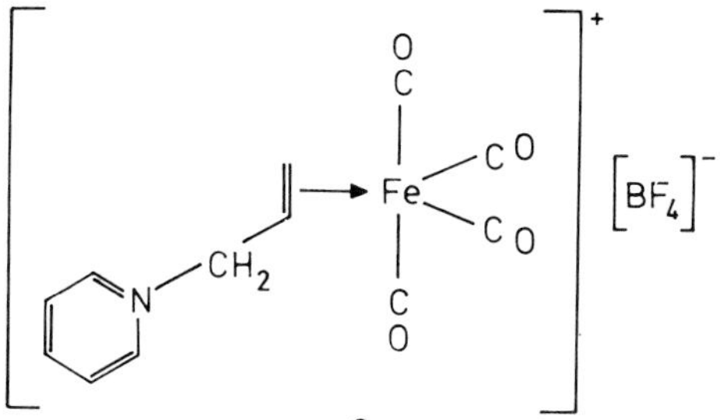

2

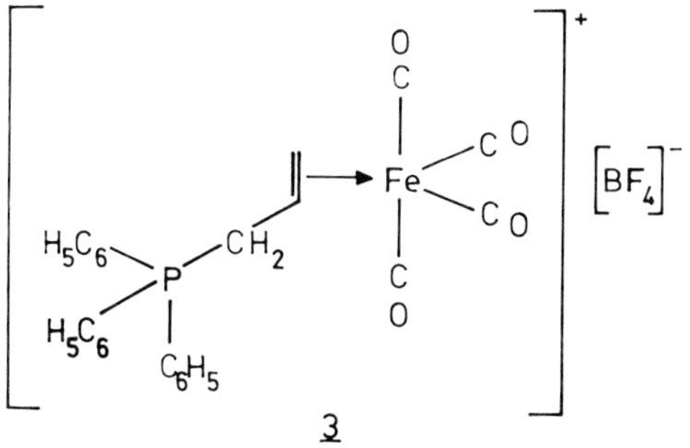

3

(η^3-allyl)-tetracarbonyliron cation, $[(\eta^3\text{-}C_3H_5)Fe(CO)_4]^+$ (as its tetrafluoroborate salt), with a Lewis base such as pyridine or triphenylphosphine gives tetracarbonyliron complexes of the allylpyridinium (2) and allyltriphenylphosphonium (3) cations *(229)*. Unstable olefin-tetracarbonyliron derivatives $(R\text{-}CH_2\text{-}CH{=}CH\text{-}CH_3)Fe(CO)_4$ (R = phenyl, benzyl, and cyclohexyl) are similarly obtained by treatment of the allyl-tetracarbonyliron cation $[(\eta^3\text{-}CH_3\text{-}CH{\cdots}CH{\cdots}CH_2)Fe(CO)_4]^+$ with the corresponding organocadmium compound, R_2Cd, in tetrahydrofuran at 0°C *(183)*.

Olefin-tetracarbonyliron complexes have also been postulated as intermediates in the reaction of benzyl chloride with $Fe_3(CO)_{12}$ in the presence of appropriate coordinating olefins such as acrylonitrile *(189)*. The ultimate products from such reactions are benzyl derivatives of the olefin introduced.

B. *POLYOLEFIN COMPLEXES CONTAINING TETRACARBONYLIRON UNITS*

Reactions of polyolefins (dienes, trienes, etc.) with carbonyliron complexes under sufficiently mild conditions may give products in which tetracarbonyliron units are individually bonded to one or more of the double bonds of the polyolefin. Such reactions must be carried out under mild conditions in order to prevent loss of carbonyl groups from the $Fe(CO)_4$ units to give metal complexes containing diene-tricarbonyliron units, discussed elsewhere in this book.

Butadiene has long been known to react with carbonyliron reagents under relatively vigorous conditions to give (butadiene)tricarbonyliron *(105,188)*. However, if the reaction between butadiene and iron carbonyls is carried out under relatively mild conditions (below 40°C) using $Fe_2(CO)_9$ as the most reactive of the three iron carbonyls, then the two tetracarbonyliron complexes $(C_4H_6)Fe(CO)_4$ (4) and $(C_4H_6)[Fe(CO)_4]_2$ (5) can be obtained in yields of 60 % and 24 %, respectively *(163)*. The mononuclear complex 4 is a thermally unstable liquid, which nevertheless can be distilled at 28.2°C/1.5 Torr. The binuclear complex 5 is an orange-yellow

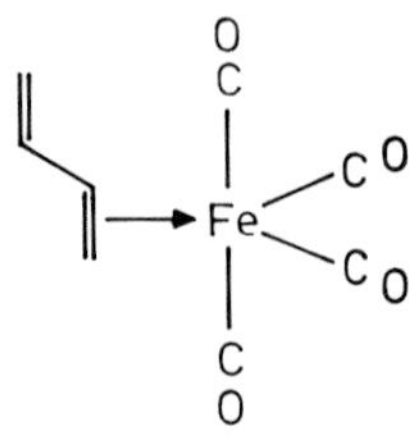

4

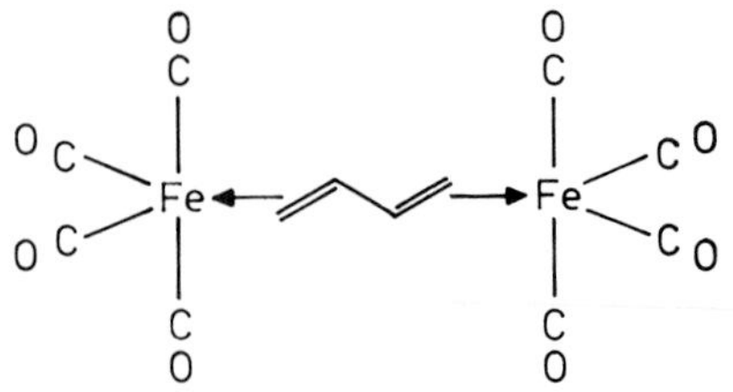

5

solid, which is fairly stable in the solid state. The schematically indicated structure 5 for μ-(1,2-η^2:3,4-η^2-C_4H_6)-$[Fe(CO)_4]_2$ with a planar *s-trans*-butadiene carbon system bonded to two $Fe(CO)_4$ groups in the equatorial positions has been confirmed by X-ray crystallography *(130)*. Tetracarbonyliron complexes of 2-methylbuta-1,3-diene and *trans*- and *cis*-1,3-pentadiene have been obtained by reactions of the 1,3-dienes with $Fe_2(CO)_9$ for relatively short times using infrared spectroscopy to follow the reaction *(79)*. Protonation of these 1,3-diene-tetracarbonyliron complexes gives good yields of the corresponding (η^3-allyl)-tetracarbonyliron cations, isolated as stable tetrafluoroborate salts.

The carbon-carbon double bonds of 1,5-cyclooctadiene can be individually bonded to tetracarbonyliron units. Ultraviolet irradiation of $Fe(CO)_5$ with 1,5-cyclooctadiene for relatively short periods of time gives the relatively unstable liquid $(C_8H_{12})Fe(CO)_4$ which can be purified by crystallization from pentane at -120°C *(139)*. The presence of coordinated 1,5-cyclooctadiene is supported by degradation with cerium(IV). Decomposition of $(C_8H_{12})Fe(CO)_4$ on standing leads to the more stable yellow crystalline binuclear complex $(C_8H_{12})[Fe(CO)_4]_2$ *(139)*. An X-ray crystallographic study of this binuclear complex indicates that the 1,5-cyclooctadiene molecule is bonded to two tetracarbonyliron units through its double bonds in the chair-conformation with each of the coordinated carbon-carbon double bonds exactly in an equatorial position of the trigonal bipyramidal iron *(147)*.

Complexes with a fulvene carbon-carbon double bond bonded to a tetracarbonyliron unit have been among the several products isolated from reactions of fulvenes with carbonyliron reagents. Thus, complexes of the type (fulvene)$[Fe(CO)_4]_2$ (6, R = C_6H_5 or *p*-ClC_6H_4) are among the several products ob-

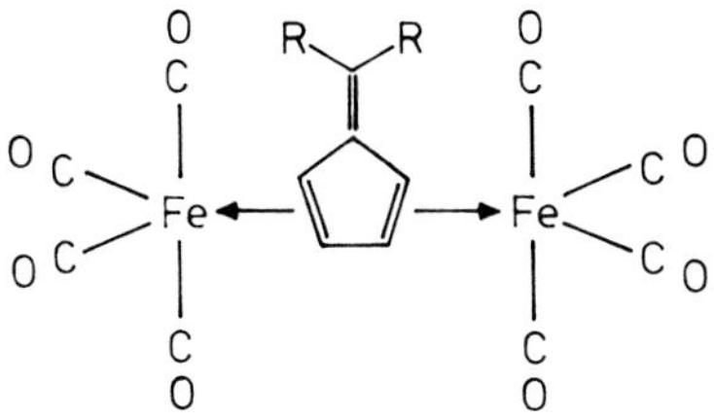

6

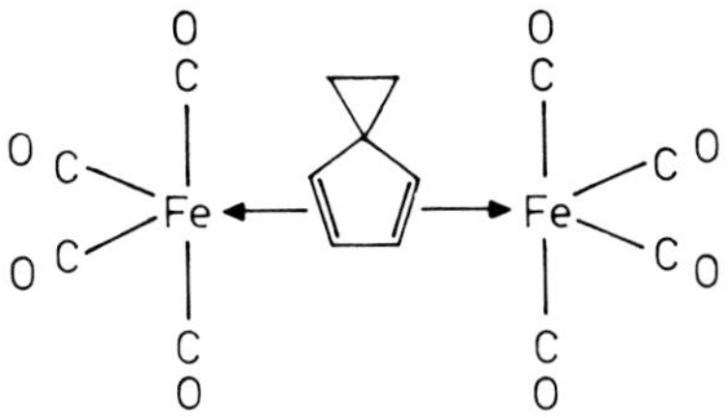

7

tained from these fulvenes and $Fe_2(CO)_9$ *(226,227)*. The two $Fe(CO)_4$ units have been shown by X-ray crystallography *(19)* to be in *trans*-positions in these complexes. A similar bis(tetracarbonyliron) complex 7 is formed as an unstable intermediate in the reaction of spiro[2.4]hepta-4,6-diene with $Fe_2(CO)_9$ to give ultimately (6-methylfulvene)hexacarbonyldiiron *(64)*.

1,3,5-Hexatriene reacts with $Fe_2(CO)_9$ at 40°C to form the mono- and bis(tetracarbonyliron) derivatives $(C_6H_8)Fe(CO)_4$ and $(C_6H_8)[Fe(CO)_4]_2$, respectively, in addition to the tricarbonyliron complex $(C_6H_8)Fe(CO)_3$ *(164)*.

The reaction of acenaphthylene with carbonyliron reagents has long been known to give a binuclear complex $(C_{12}H_8)Fe_2(CO)_5$ *(46,121)*. However, if the reaction between acenaphthylene and iron carbonyls is carried out under relatively mild conditions using $Fe_2(CO)_9$ in tetrahydrofuran, which readily generates $Fe(CO)_4$ units *(53,54)*, a product of the stoichiometry $(C_{12}H_8)Fe(CO)_4$ is obtained as yellow-orange crystals *(55)*. X-ray crystallography of this product indicates it to have structure 8 in which the non-benzenoid carbon-carbon

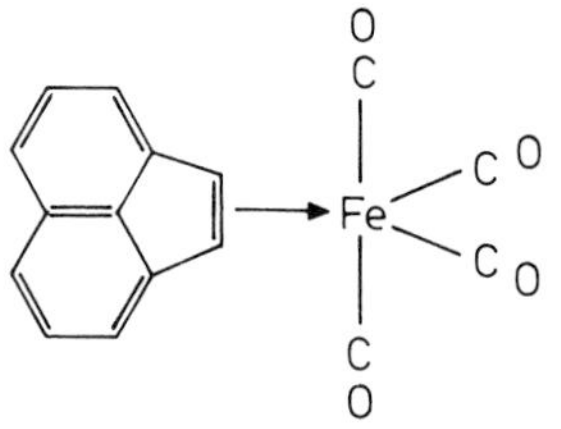

8

double bond of the acenaphthylene unit is bonded to an equatorial position of the carbonyliron trigonal bipyramid as in other (olefin)$Fe(CO)_4$ derivatives.

Olefin-tetracarbonyliron derivatives have also been obtained from reactions of styrene derivatives with iron carbonyls. Ultraviolet irradiation of various styrene derivatives with $Fe(CO)_5$ gives the corresponding styrene-tetracarbonyliron complexes 9 as air-sensitive yellow complexes in addition to

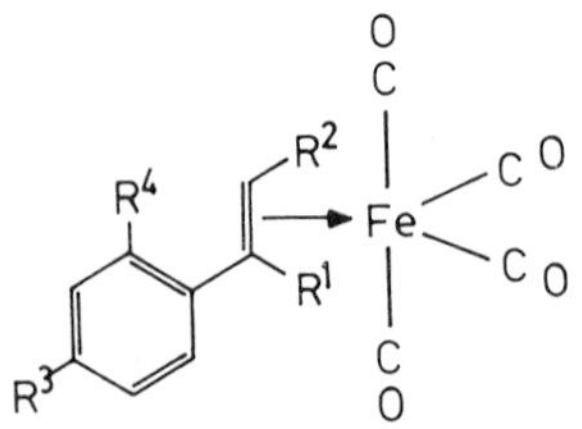

9

R_1	R_2	R_3	R_4
H	H	H	H
H	CH_3	H	H
CH_3	H	H	H
C_6H_5	H	H	H
H	*p*-$CH_3OC_6H_4$	H	H
H	C_6H_5	H	H
H	CH_3	CH_3	H
H	CH_3	Cl	H
H	CH_3	OCH_3	H
H	cyclopropyl	OCH_3	H
H	H	H	Br

compounds containing one or two tricarbonyliron units bonded to the styrene system *(222,223)*. In the case of *o*-bromostyrene a tricarbonylferraindene-tricarbonyliron derivative is also formed with bromine elimination. The compounds with $Fe(CO)_3$ units are discussed in more detail elsewhere in this book.

Certain allenes and cumulenes form tetracarbonyliron complexes involving only one carbon-carbon double bond of the cumulene system. Reaction of $Fe_2(CO)_9$ with tetramethylallene gives a mixture of (2,4-dimethylpenta-1,3-diene)tricarbonyliron (10), formed by a hydrogen shift reaction, and (tetramethylallene)tetracarbonyliron (11). The ^{1}H-NMR spectrum of

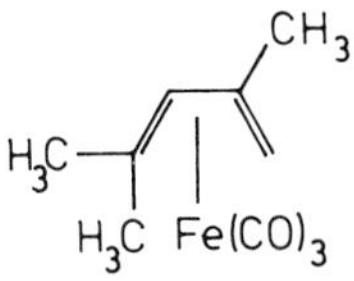

10

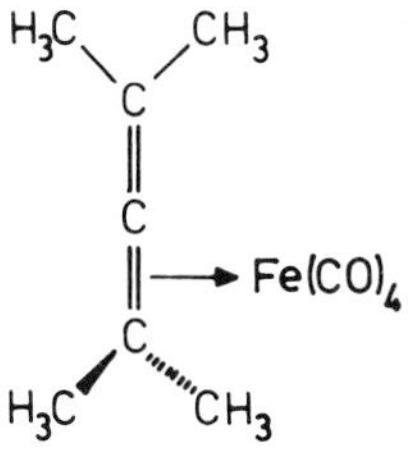

11

the latter complex shows only a single methyl resonance at room temperature, but three methyl resonances in a 1:1:2 ratio at -60°C. This indicates that at room temperature the tetracarbonyliron unit is rapidly moving from one π molecular orbital of the tetramethylallene to the second orthogonal one. The activation energy for this process is 9.0 ± 2.0 kcal *(24)*.

Tetraphenylbutatriene reacts with $Fe_2(CO)_9$ at room temperature to give both the tetracarbonyliron complex $[(C_6H_5)_2C{=}C{=}C{=}C(C_6H_5)_2]Fe(CO)_4$ and the hexacarbonyldiiron complex $[(C_6H_5)_2C{=}C{=}C{=}C(C_6H_5)_2]Fe_2(CO)_6$ *(116)*. X-ray crystallography of the former complex indicates that the center carbon-carbon double bond of the triene system is bonded to the tetracarbonyliron unit as in 12 *(29,30)*. The terminal carbon atoms of

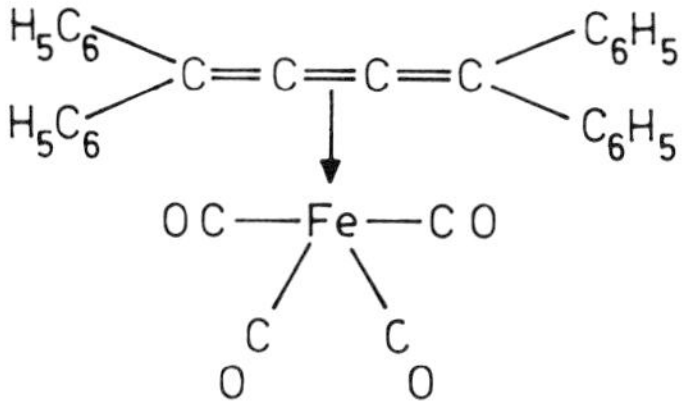

12

the butatriene chain are bent away from the iron atom so that the angles subtended at the central carbon atoms are 151°. The largest cumulene which has been reacted with iron carbonyls is tetra(*tert*-butyl)hexapentaene, which reacts with $Fe_3(CO)_{12}$ to form both a yellow tetracarbonyliron derivative and a red hexacarbonyldiiron derivative *(128)*. The ^{13}C-NMR spectrum of the tetracarbonyliron complex indicates structure 13, in which

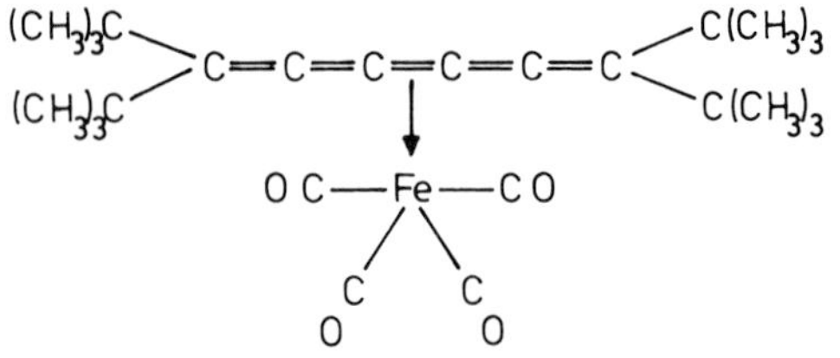

13

again the center carbon-carbon double bond is bonded to the $Fe(CO)_4$ unit. Despite the presence of five carbon-carbon double bonds, four of which are *not* complexed to a metal atom, this cumulene-tetracarbonyliron complex is remarkably stable. It thus sublimes in vacuum without decomposition. The protective effect of the four large *tert*-butyl groups in the chemistry of tetra(*tert*-butyl)hexapentaene is very significant.

C. REACTIONS OF OLEFIN-TETRACARBONYLIRON DERIVATIVES

Olefin-tetracarbonyliron derivatives react with Lewis base ligands with displacement of the coordinated olefin according to the general equation [2] where α is the fraction

$$\begin{aligned}&(\text{olefin})Fe(CO)_4 + (2-\alpha)L \\ &\longrightarrow \alpha\ Fe(CO)_4L + (1-\alpha)Fe(CO)_3L_2 + \text{olefin} + (1-\alpha)CO\end{aligned} \qquad [2]$$

of monosubstituted complex obtained during the reaction. Cardaci has studied the details of reactions of this type using mainly (styrene)tetracarbonyliron and ligands such as carbon monoxide *(37,40)*, triphenylphosphine *(39,40,41)*, triphenylarsine *(39)*, triphenylstibine *(38,39)*, and pyridine *(39)*. With triphenylstibine and pyridine only the monosubstituted $LFe(CO)_4$ was observed (*i.e.* $\alpha = 1$ in equation [2]). With triphenylarsine, the amount of disubstituted derivative was relatively small. With triphenylphosphine, however, appreciable amounts of the disubstituted derivative $[(C_6H_5)_3P]_2Fe(CO)_3$ were obtained. The amount of disubstituted derivative formed in the reaction of triphenylphosphine with (styrene)tetracarbonyliron was found to decrease as the carbon monoxide pressure was increased. On the basis of Cardaci's kinetic studies, a reaction mechanism was proposed involving first dissociation of the olefin from (olefin)$Fe(CO)_4$ to give a reactive $Fe(CO)_4$ fragment followed by further dissociation of carbon monoxide from the $Fe(CO)_4$ fragment to give an $Fe(CO)_3$ fragment *(40)*. In the cases of different complexes of the type $(CH_2{=}CHX)Fe(CO)_4$ the rate of reaction according to equation [2] with carbon monoxide as the entering ligand (α is not readily determined in this case) was found to increase with decreasing electron withdrawing power of the X substituent *(37)*.

Reactions can also be performed on functional groups attached to the olefins coordinated to tetracarbonyliron units without rupture of the olefin-iron bond. Thus, treatment of (acrylic acid)tetracarbonyliron, $(CH_2{=}CH{-}COOH)Fe(CO)_4$, with phosphorus trichloride in hexane gives a 90-95 % yield of the corresponding acid chloride complex, $(CH_2{=}CH{-}COCl)Fe(CO)_4$ *(60)*. Reaction of this acid chloride with ammonia in hexane gives an excellent yield of the acrylamide complex, $(CH_2{=}CH{-}CO{-}NH_2)Fe(CO)_4$, previously obtained *(228)* from acrylamide and $Fe_2(CO)_9$. A similar amide, $(CH_2{=}CH{-}CO{-}NC_5H_{10})Fe(CO)_4$, is analogously obtained using piperidine. The pK_a of acrylic acid in its complex is 5.16 as compared with 4.10 for the uncomplexed acid *(60)*, indicating a decrease in the acidity of acrylic acid when it is complexed with an $Fe(CO)_4$ unit. Despite this lowering of acidity of acrylic acid when bonded to iron in the complex $(CH_2{=}CH{-}COOH)Fe(CO)_4$, the complex protonates strong hard Lewis bases such as ammonia, piperidine, and pyrrolidine to form the corresponding salts of the $(CH_2{=}CH{-}COO^-)Fe(CO)_4$ anion. Soft Lewis bases react with (acrylic acid)tetracarbonyliron to form 1:1 adducts $(CH_2{=}CH{-}COOH)Fe(CO)_4L$ (L = $(CH_3O)_3P$, $C_2H_5(C_6H_5)_2P$, $(CH_3)_2(C_6H_5)P$, $(C_4H_9)_3P$, $(C_6H_{11})_3P$, and C_5H_5N) without carbon monoxide evolution. Hydrogen bonding of the Lewis base to the acidic proton of acrylic acid is suggested to account for the existance of these complexes *(60)*.

D. PHYSICAL AND SPECTROSCOPIC STUDIES ON OLEFIN-TETRACARBONYLIRON DERIVATIVES

The general interest in metal-olefin bonding *(56,104, 106)* has led to a variety of physical and spectroscopic studies on olefin-tetracarbonyliron complexes with the ultimate objective of elucidating their structure and bonding. Techniques which have been used to study these compounds include vibrational (IR and Raman) spectroscopy, mass spectroscopy, electron diffraction, nuclear magnetic resonance, electron spin resonance of radical anions produced by electrochemical reduction, Mössbauer spectroscopy, and ion cyclotron resonance.

The vibrational spectrum of (ethylene)tetracarbonyliron was investigated *(2)*. However, the relative instability of this compound made Raman spectra only possible at low temperatures. In this complex the (strongly coupled) CH_2 in-plane symmetric deformation and C=C stretching frequencies were 1510 and 1193 cm^{-1}, with the Fe-(C_2H_4) stretch and tilts at 361, 305, and 401 cm^{-1}.

The tetracarbonyliron complexes of maleic anhydride *(3,152)* and N-methylmaleimide *(153)* are considerably more stable than (ethylene)tetracarbonyliron. Therefore, extensive Raman data in addition to infrared data could be obtained on the former complexes. The carbon-carbon double bond stretching frequencies in maleic anhydride and N-methylmaleimide each decrease by more than 200 cm^{-1} upon complexing with the $Fe(CO)_4$ fragment. This suggests a strong interaction with the metal. The decrease in the infrared intensities of the out-of-plane CH modes after coordination to the metal is explained by the lowering of the effective positive charge on the olefinic protons due to the electron back-donation from metal to ligand.

The charge distribution in (olefin)$Fe(CO)_4$ derivatives has also been determined by dipole moment measurements *(214)*. Values around 3 D were found for complexes of a variety of olefins with electronegative substituents such as ketones, aldehydes, and nitriles.

In another study *(138)* the thermal stabilities of (olefin)$Fe(CO)_4$ derivatives have been related both to the position of the highest infrared ν(CO) frequency and to the ionization potential of the free olefin. The fragmentation patterns of tetracarbonyliron complexes of maleic anhydride, the dimethyl esters of maleic and fumaric acids, propylene, styrene, acrylonitrile, vinyl ethyl ether, vinyl halides, and various dihalo- and trihaloethylenes have been described *(138)*. In most cases loss of the olefin ligand from the molecular ion is competitive with the usual stepwise loss of carbonyl groups.

The olefin-tetracarbonyliron derivatives have been included as representatives of equatorially substituted $LFe(CO)_4$ derivatives in an extensive recent infrared study *(61)* of substituted carbonyliron derivatives. In the equatorially substituted (olefin)$Fe(CO)_4$ complexes the two carbonyl groups opposite to the olefin ligand exhibit somewhat larger force constants than the two carbonyl ligands *trans* to each other. The absolute intensities of the infrared ν(CO) frequencies were also measured, but the interpretation of the MCO group dipole moments calculated from these data is not clear.

Gas phase electron diffraction data have been obtained on (ethylene)tetracarbonyliron *(63)*. The resulting radial distribution curves are consistent with the accepted trigonal bipyramidal model in which the ethylene occupies an equatorial site.

A characteristic feature of $Fe(CO)_5$ is the extremely low activation barrier towards the intramolecular rearrangement of its five carbonyl groups. This is indicated by a single sharp line in the ^{13}C-NMR spectrum of $Fe(CO)_5$ down to -170°C. Recent studies on the temperature dependence of the ^{13}C-NMR spectra have shown that substitution of an equatorial CO group in $Fe(CO)_5$ by an olefin to form an (olefin)$Fe(CO)_4$ complex can raise appreciably its activation barrier towards intramolecular rearrangement *(146,232)*. Thus, the ^{13}C-NMR spectra of the tetracarbonyliron complexes of diethyl fumarate, diethyl maleate, and ethyl acrylate at -30°C, -30°C, and -80°C, respectively, exhibit the limiting 2:2, 2:1:1, and 1:1:1:1 patterns, respectively, for the metal carbonyl resonances expected for a rigid structure with no intramolecular rearrangements *(146)*. In a study of the ^{13}C-NMR spectra of a still greater variety of olefin-tetracarbonyliron complexes, the activation barrier to carbonyl site exchange was found to increase with increasing π-acceptor ability of the olefin *(232)*. In all cases the complexity of the metal carbonyl region of the ^{13}C-NMR spectrum is consistent with a trigonal bipyramidal structure in which the two olefinic carbons of the ligand occupy the equatorial plane.

Tetracarbonyliron complexes of dimethyl fumarate, acrylamide, N-phenylmaleimide, methyl cinnamate, and dimethyl maleate have been electrochemically reduced to the corresponding radical anions *(65)*. The reduction processes are electrochemically reversible and occur in the voltage region from -1.80 to -2.20 V relative to the $Ag|Ag^+$ (10^{-3}*M*) reference couple. The electron spin resonance spectra of these radical anions show hyperfine coupling of around 5 G for each of the protons remaining on the complexed olefinic carbon-carbon double bond. The infrared spectra in the metal carbonyl ν(CO) region of all of the radical anions exhibit decreases to longer wavelengths by about 100 to 150 cm^{-1} relative to the neutral compound. This suggests that some of the additional electron density of the radical anions is transmitted into the antibonding orbitals of the carbonyl groups. Conversion of an olefin-tetracarbonyliron complex into its radical anion changes its Mössbauer spectrum by increasing the isomer shift, but by decreasing, sometimes by a factor of 2, the quadrupole splitting. Other workers *(49,144)* during the course of obtaining Mössbauer data on most of the readily available neutral (olefin)$Fe(CO)_4$ derivatives have observed a linear relationship between the isomer shifts and quadrupole splittings with a negative slope.

The gas phase ion chemistry of pentacarbonyliron with various substances including deuterated ethylene has been investigated using ion cyclotron resonance spectroscopy *(78)*.

The ion $[Fe(CO)_5]^+$ was inert to C_2D_4. However, ions containing fewer carbonyl groups underwent the reactions [3]. Ions con-

$$
\begin{aligned}
Fe(CO)_n^+ + C_2D_4 &\longrightarrow Fe(C_2D_4)(CO)_{n-1}^+ + CO \\
Fe(C_2D_4)(CO)_{n-1}^+ + C_2D_4 &\longrightarrow Fe(C_2D_4)_2(CO)_{n-2}^+ + CO
\end{aligned}
\qquad [3]
$$

taining more than two C_2D_4 units were not found.

E. CHIRAL OLEFIN-TETRACARBONYLIRON DERIVATIVES

In general, olefins, even those with four different substituents of the type WXC=CYZ, are achiral because of the plane of symmetry through the two double bond carbons and the bonding atoms of the four substituents. However, if any type of olefin other than the symmetrical types $X_2C{=}CX_2$, *cis*-XYC=CXY, and $X_2C{=}CY_2$ is coordinated to a metal, this plane of symmetry will disappear thereby leading to chiral systems. Tetracarbonyliron complexes of fumaric acid *(181, 182)* and acrylic acid *(182)* have been resolved into the enantiomers by fractional crystallization of their brucine salts, followed by removal of the brucine with hydrogen chloride in acetone. A similar procedure, when applied to (maleic acid)tetracarbonyliron, failed to give an optically active product in accord with the presence of a plane of symmetry in this compound *(182)*. The crystal structures of both the racemic *(51,184)* and the optically active (-)-forms *(52)* of (fumaric acid)tetracarbonyliron have been investigated. The four carbon atoms of the fumaric acid ligand in these complexes deviate significantly from coplanarity. The absolute configuration of (—)-(fumaric acid)$Fe(CO)_4$ was determined using anomalous diffraction of Cu-K_α radiation by the iron atoms. All three crystallographically distinct molecules have been the same *(R,R)* configuration induced by the metal coordination at the olefinic carbon atoms. The circular dichroism spectra of (—)-(fumaric acid)$Fe(CO)_4$ and (+)-(acrylic acid)$Fe(CO)_4$ *(165,166)* have the same pattern but the opposite sign. The magnitude of the circular dichroism of the acrylic acid complex is about half that of the fumaric acid complex consistent with the fact that the acrylic acid complex has only one asymmetric carbon atom (that bearing the carboxyl group) but the fumaric acid complex has two asymmetric carbon atoms (each of the two olefinic carbon atoms). The following additional conclusions are readily derived from these CD data: (1) the olefins coordinated to the $Fe(CO)_4$ unit in (—)-(fumaric acid)tetracarbonyliron and (+)-(acrylic acid)tetracarbonyliron have opposite configurations; (2) each asymmetric carbon atom contributes equally to the total optical activity.

Both the vinyl and ester methyl protons in racemic (dimethyl fumarate)tetracarbonyliron are split into two resonances each in the presence of the chiral shift reagents europium(III)- and praseodymium(III)-tris-3-trifluoroacetyl-1*R*-campherates *(212)*. This provides a simple and quick method for the detection of the chirality of π-complexed prochiral olefins in cases where functional groups are present which interact with the lanthanide chiral shift reagent.

F. APPLICATIONS OF OLEFIN-TETRACARBONYLIRON DERIVATIVES

Several applications for olefin-tetracarbonyliron derivatives have been found. The tetracarbonyliron complexes of maleic anhydride, dimethyl fumarate, and ethyl cinnamate (as well as several diene-tricarbonyliron complexes) are active catalysts for the hydrogenation of methyl sorbate at 160°C/40 atm. The mechanism for this process appears to involve exchange of the methyl sorbate with the coordinated olefin *(35)*. (Dimethyl fumarate)$Fe(CO)_4$ has been shown to be efficient for the quenching of the triplet state of fluorenone *(85)*. (Maleic anhydride)$Fe(CO)_4$ has been used for the modification of proteins such as ribonuclease to provide useful metal atom probes *(84)* in biologically significant molecules.

III. BIS(OLEFIN)-TRICARBONYLIRON DERIVATIVES

The extensive variety of olefin-tetracarbonyliron derivatives which can be obtained as discussed above naturally raises the question whether a second carbonyl group of $Fe(CO)_5$ can also be replaced by an olefin to give an (olefin)$_2Fe(CO)_3$ derivative in which the two olefins are bonded to the same iron atom. However, attempts to prepare such compounds have had limited success. The experimental evidence discussed below suggests that compounds in which two monoolefins are bonded to a single iron atom are unstable with respect to coupling of the two olefinic ligands to form a new carbon-carbon bond.

The first studies on (olefin)$_2Fe(CO)_3$ complexes, like those on the (olefin)$Fe(CO)_4$ complexes discussed above, were performed using acrylonitrile as a ligand. In 1961 Schrauzer *(210)* briefly mentioned the formation of an unstable yellow substance of stoichiometry $(C_2H_3CN)_2Fe(CO)_3$ from the reaction of $Fe_3(CO)_{12}$ with acrylonitrile. In 1966 Schubert and Sheline *(211)* described the ultraviolet irradiation of $Fe(CO)_5$ with acrylonitrile in hexane solution to give a product of the same stoichiometry. However, molecular weight determinations showed that this product was the dimer $[Fe(CO)_3(CH_2{=}CH{-}CN)]_2$. A subsequent X-ray crystallography study *(235)* indicated structure 14 in which the acrylonitrile ligands act as brid-

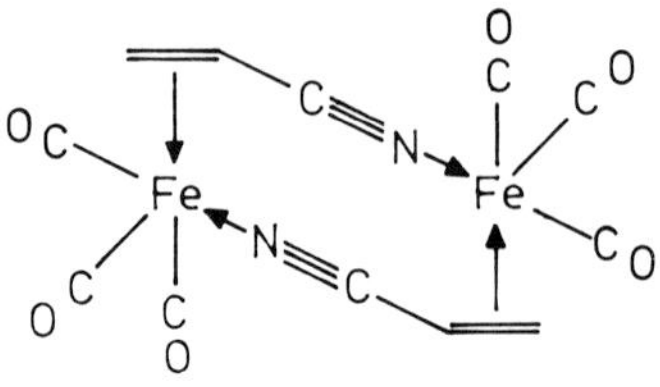

14

ges between the two iron atoms. These bridging acrylonitrile ligands bond to one iron atom using their carbon-carbon double bonds and to the other iron atom using the lone pairs of their cyano nitrogen atoms. Thus, acrylonitrile does not appear to form a true mononuclear $(olefin)_2Fe(CO)_3$ complex apparently because coordination of a second olefinic carbon-carbon double bond cannot compete with coordination of the nitrogen lone pair.

A true $(olefin)_2Fe(CO)_3$ complex has been recently obtained using methyl acrylate as the ligand *(98)*. Ultraviolet irradiation of (methyl acrylate)tetracarbonyliron with excess methyl acrylate in hexane solution at -30°C gives the bright yellow bis(methyl acrylate)tricarbonyliron, $(CH_3O{-}CO{-}CH{=}CH_2)_2{-}Fe(CO)_3$. On the basis of the infrared spectrum of this complex the two methyl acrylate ligands are assigned to equatorial positions in the trigonal bipyramid as depicted in structure 15. The ^{1}H-NMR spectrum indicates the presence of the two stereoisomers 15a and 15b.

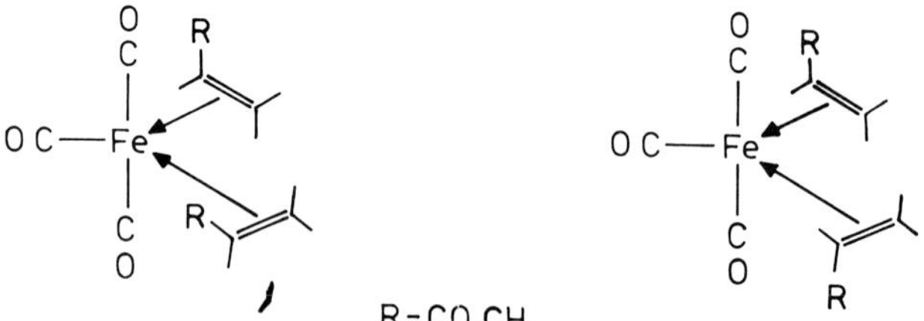

15a R = CO_2CH_3 15b

Bis(methyl acrylate)tricarbonyliron (15) is a highly reactive molecule. Upon warming to -5°C it dissociates by losing one of the olefinic ligands to form (methyl acrylate)tricarbonyliron with an apparent structure 16 involving 1-oxabutadiene coordination of the acrylic ester, but this compound also is rather unstable. At room temperature two stereoisomers of the metallacyclopentane complex (17a and 17b) are

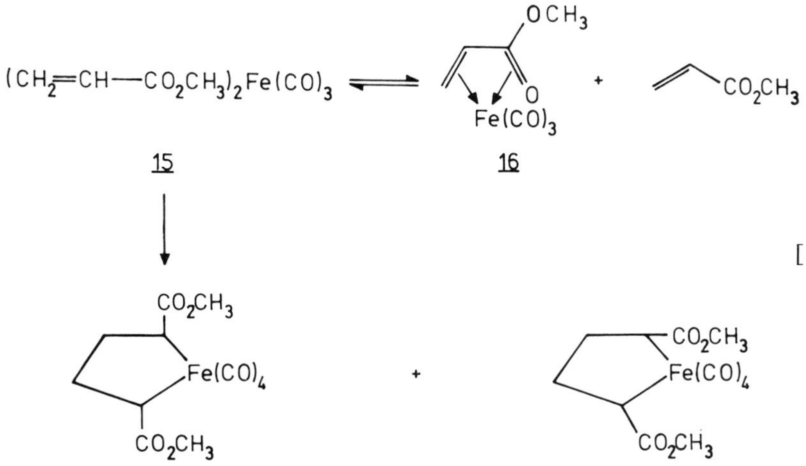

[4]

formed from 15, apparently because the proximity of the two equatorial methyl acrylate ligands leads readily to such coupling reactions (scheme [4]). Both stereoisomers 17a and 17b are readily hydrogenated with Raney nickel to dimethyl adipate. This indicates that in both stereoisomers the ester substituents are located on the carbon atoms directly bonded to the iron.

Bis(methyl acrylate)tricarbonyliron undergoes a variety of interesting reactions with potential ligands. Reaction with carbon monoxide at 20°C regenerates the tetracarbonyliron complex, $(CH_3O\text{-}CO\text{-}CH{=}CH_2)Fe(CO)_4$, with liberation of one equivalent of methyl acrylate *(98)*. Reaction with triphenylphosphine also displaces one equivalent of methyl acrylate to give $(CH_3O\text{-}CO\text{-}CH{=}CH_2)Fe(CO)_3P(C_6H_5)_3$ *(237)*, shown by X-ray crystallography to have a trigonal bipyramidal structure in which the methyl acrylate is in an equatorial position and the triphenylphosphine is in an axial position. Reaction of bis(methylacrylate)tricarbonyliron with a 1,3-diene (butadiene, isoprene, and 2,3-dimethylbutadiene) *(101,102)* leads to displacement of one of the two methyl acrylate ligands with formation of complex 18 containing an acrylate-diene adduct 1,2,3,6-*tetrahapto*-coordinated to the $Fe(CO)_3$ moiety (R = R' = H or

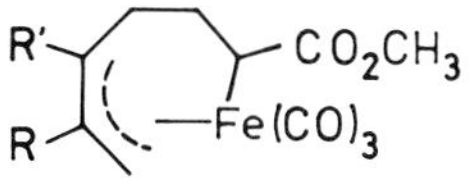

18

CH_3; R = H, R' = CH_3). This complex is also obtained by irradiation of the corresponding (1,3-diene)-tricarbonyliron with methyl acrylate. Reaction of bis(methyl acrylate)$Fe(CO)_3$ with alkynes at room temperature results in displacement of both methyl acrylate ligands to give complexes similar to those obtained from reactions of various iron carbonyls with alkynes, however, under much more vigorous conditions *(101)*.

IV. REACTIONS OF CARBONYLIRON COMPLEXES WITH FLUOROOLEFINS

The strongly electron-withdrawing effect of fluorine atoms in polyfluorinated olefins can make the chemistry of such olefins quite different from that of olefins containing mainly hydrogen or hydrocarbon substituents. For this reason, the reactions of carbonyliron complexes with such fluoroolefins is discussed in this separate section of this chapter.

The first reaction between carbonyliron complexes and a perfluoroolefin was described by Watterson and Wilkinson in 1959 *(224)*. They reported that tetrafluoroethylene reacts with $Fe_3(CO)_{12}$ to give a white volatile solid, to which they gave the formula $(C_2F_4)_2Fe(CO)_3$. This originally proposed formula was soon found to be incorrect *(112,225)*. This product was shown instead to be (octafluorotetramethylene)tetracarbonyliron (19). Thus, two tetrafluoroethylene units can couple in the presence of iron carbonyls to form a ferracyclopentane derivative completely analogous to the coupling of two methyl acrylate units to form the ferracyclopentane derivative 17 discussed above. Subsequently an improved preparation of $(C_4F_8)Fe(CO)_4$ (19) was developed *(73)* using the ultraviolet irradiation of tetrafluoroethylene with pentacarbonyliron in the presence of catalytic quantities of $Fe_3(CO)_{12}$. Reactions of $(C_4F_8)Fe(CO)_4$ with a variety of donor ligands give complexes of the types $(C_4F_8)Fe(CO)_3L$ (L = E-$(C_6H_5)_3$, E = P, As, or Sb; L = $P(OR)_3$, R = C_6H_5 or C_2H_5) and $(C_4F_8)Fe(CO)_2L_2$ (L = $P(OC_2H_5)_3$ or pyridine; L_2 = 2,2'-bipyridyl, *o*-phenanthroline, or $(C_6H_5)_2P-CH_2-CH_2-P(C_6H_5)_2$) *(73)*.

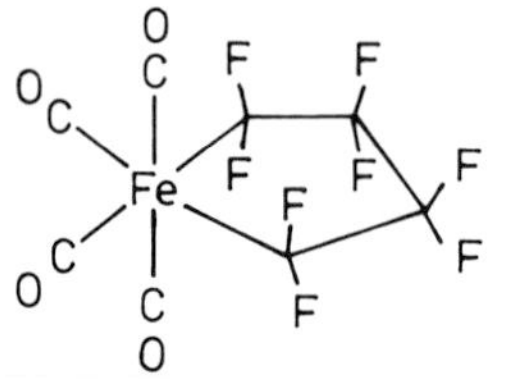

19

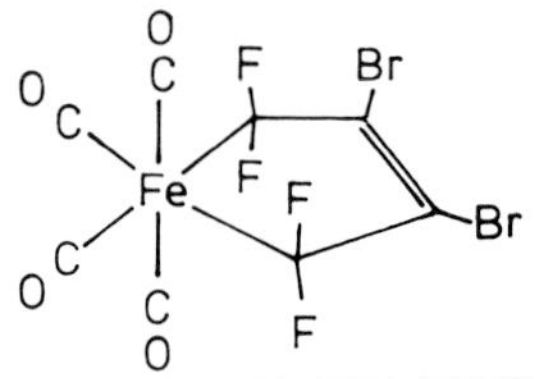

20

The formation of the octafluorotetramethylene complex 19 rather than a true tetrafluoroethylene complex from reactions between tetrafluoroethylene and iron carbonyls left uncertain for some time whether true carbonyliron η^2-complexes of polyfluoroolefins could be prepared. However, by careful use of the reaction conditions suitable for preparing (olefin)$Fe(CO)_4$ complexes from non-fluorinated olefins as discussed above (*i.e.* ultraviolet irradiation of $Fe(CO)_5$ with excess fluoroolefin or reaction with $Fe_2(CO)_9$ at room temperature), similar complexes of the type (η^2-olefin)$Fe(CO)_4$ could be prepared from a variety of fluoroolefins *(72,74,75)* including tetrafluoroethylene, the partially fluorinated olefins $CF_2{=}CFX$ (X = Cl, Br, or H), $CF_2{=}CHX$ (X = Cl, Br, and H), the trifluoromethyl substituted ethylenes $CH_2{=}CH{-}CF_3$, *trans*-$CF_3{-}CX{=}CX{-}CF_3$ (X = H and F), and $(CF_3)_2C{=}C(CF_3)_2$, and the cyclic perfluoroolefins $\overline{CF{=}CF(CF_2)_n}$ (n = 2, 3, and 4).

Most of these fluoroolefin-tetracarbonyliron derivatives are relatively unstable. Thus, perfluoropropene is removed from the complex $(CF_3{-}CF{=}CF_2)Fe(CO)_4$ upon treatment with a variety of reagents including iodine in benzene at 80°C, phosphorus pentachloride at 70°C, bromine, hydrogen chloride at 70°C, aqueous sodium hydroxide at 75°C, or 70 % aqueous sulfuric acid at 70°C. Triphenylphosphine reacts with $(CF_3{-}CF{=}CF_2)Fe(CO)_4$ to give a mediocre yield (17 %) of the substitution product $(CF_3{-}CF{=}CF_2)Fe(CO)_3P(C_6H_5)_3$. However, most (67 %) of the complexed hexafluoropropene is lost in this reaction. The tetrafluoroethylene complex $(CF_2{=}CF_2)Fe(CO)_4$ reacts with excess tetrafluoroethylene to give the ferracyclopentane complex 19 discussed above. Similar ferracyclopentane complexes could not be obtained for any other of the fluoroolefins. However, in another study *(213)* $CF_2{=}CBr_2$ was found to give not only the olefin complex $(CF_2{=}CBr_2)Fe(CO)_4$ but also the ferracyclopentane complex 20 upon irradiation with $Fe(CO)_5$.

The strong electron-withdrawing characteristics of the four fluorine atoms in tetrafluoroethylene suggests that for $(CF_2{=}CF_2)Fe(CO)_4$ the resonance structure 21b containing a metallacyclopropane ring will predominate relative to the true metal olefin complex structure 21a . The gas-phase electron diffraction of $(CF_2{=}CF_2)Fe(CO)_4$ has been interpreted

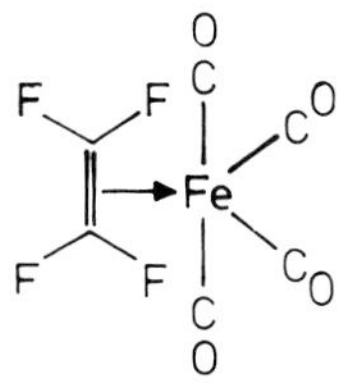

21a

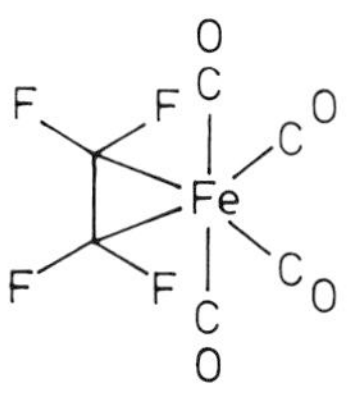

21b

(18) on the basis of the metallacyclopropane structure 21b in which both tetrafluoroethylene carbon atoms are in the equatorial plane.

Some zerovalent carbonyliron phosphite complexes also give perfluoroolefin (or more accurately, see above, metallacyclopropane) complexes upon treatment with polyfluoroolefins *(33)*. Reaction of *trans*-$[(RO)_3P]_2Fe(CO)_3$ (R = CH_3 and C_2H_5) with the fluoroolefins $CF_2{=}CFX$ (X = F, CF_3, and Cl) results in the elimination of one equivalent of carbon monoxide to give *cis*-$[(RO)_3P]_2Fe(CO)_2(CF_2{=}CFX)$ (22). However, *trans*-$[(C_2H_5O)_3P]_2Fe(CO)_3$ reacts with trifluoroethylene to give the ferracyclopentane complex 23, suggesting further the delicate balance between three- and five-membered ring formation in reactions of this type *(33)*.

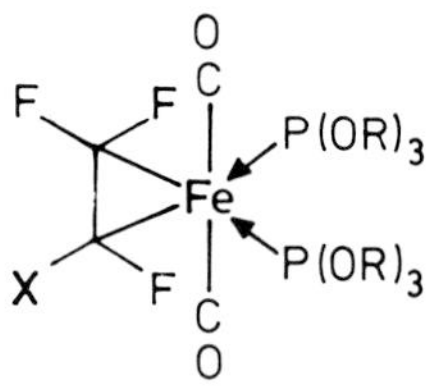

22

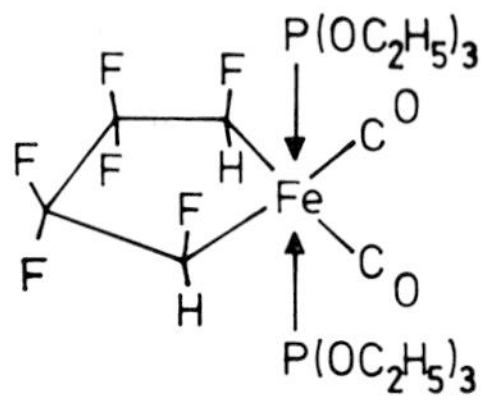

23

Reactions of carbonyliron complexes with perfluorinated 1,3-dienes may involve the two carbon-carbon double bonds either separately or simultaneously. Thus, ultraviolet irradiation of $Fe(CO)_5$ with hexafluorocyclopentadiene gives an 8 % yield of the yellow crystalline $(C_5F_6)[Fe(CO)_4]_2$ (24) involving separate reaction of the two carbon-carbon double bonds *(17)*. However, hexafluorobutadiene reacts with iron carbonyls to form the perfluoroferracyclopentene complex $(C_4F_6)Fe(CO)_4$ (25) involving simultaneous interaction of the two conjugated carbon-carbon double bonds with the metal carbonyl system *(111,114)*.

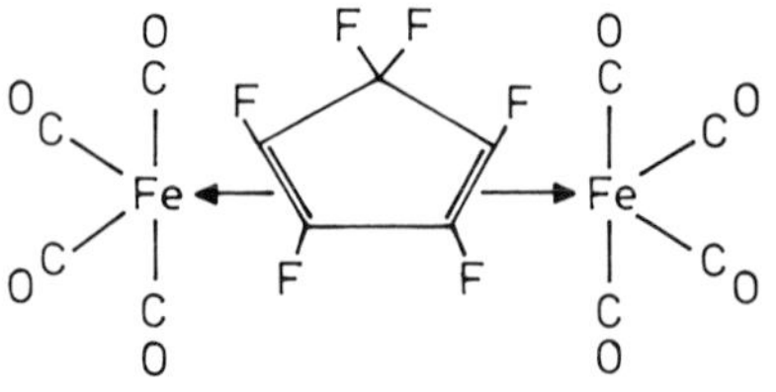

24

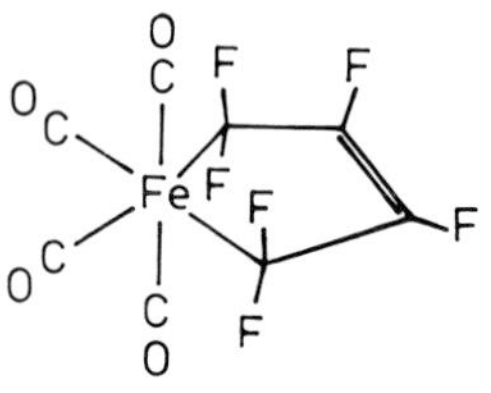

25

V. REACTIONS OF CARBONYLIRON COMPLEXES WITH UNSATURATED CYCLOPROPANE DERIVATIVES

The presence of a strained, and hence reactive, cyclopropane ring in an olefin can have a major effect on its reactivity towards carbonyliron complexes. In many, although not all, cases opening of the cyclopropane ring can occur upon reaction with iron carbonyls.

The first reported *(47)* reaction of a cyclopropane derivative with a carbonyliron complex was the reaction of triphenylcyclopropenyl bromide with the anion $[Fe(CO)_3NO]^-$ to give maroon crystals of the stoichiometry $(C_6H_5)_3C_3(CO)Fe(CO)_2NO$. Recent spectroscopic *(124)* and crystallographic *(185)* results on the apparently analogous compounds $R_3C_3COCo(CO)_3$ (R = CH_3 and C_6H_5) from $[Co(CO)_4]^-$ and the corresponding cyclopropenyl cations, $R_3C_3^+$, suggest the ketocyclobutenyl structure 26 for this iron derivative. The reaction of

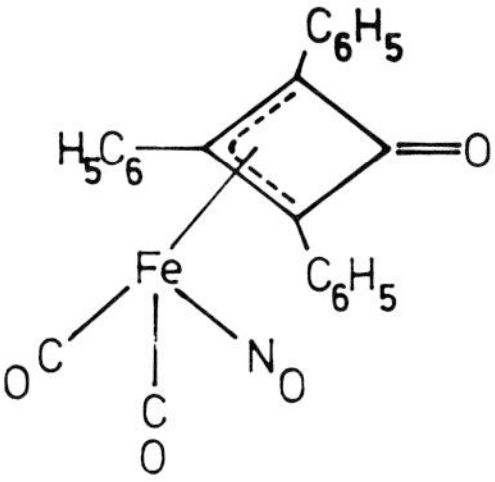

26

$[Fe(CO)_3NO]^-$ with triphenylcyclopropenyl bromide thus appears to result in carbonyl insertion into the three-membered ring to give a four-membered ring.

Another example of ring opening upon reaction of cyclopropene derivatives with carbonyliron complexes is the formation of the vinylketene derivative 27a or 27b from 1,3,3-

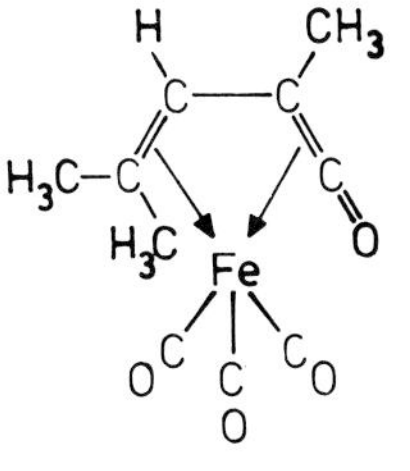

27a

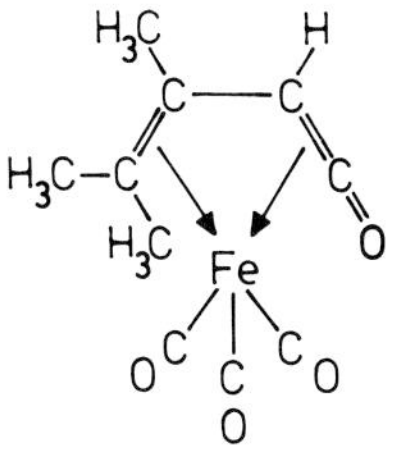

27b

trimethylcyclopropene and $Fe_3(CO)_{12}$ *(123)*. The available spectroscopic data on the resulting yellow crystalline

$(C_7H_{10}O)Fe(CO)_3$ do not differentiate between isomers 27a and 27b.

In contrast to the above reactions, carbonyliron complexes may react with methylenecyclopropanes derived from Feist's acid under sufficiently mild conditions to form products in which the cyclopropane ring is retained. Thus, reactions of *trans*- or *cis*-(dimethyl methylenecyclopropane-2,3-dicarboxylate) (28a or 28b, R = CH_3) with $Fe_2(CO)_9$ at room temperature lead stereospecifically to the formation of the corresponding olefin-tetracarbonyliron complexes 29a (R = CH_3) or 29b (R = CH_3), respectively, with retention of

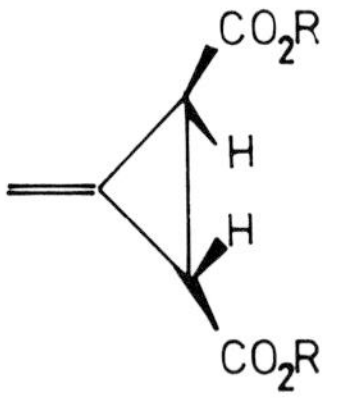

28a

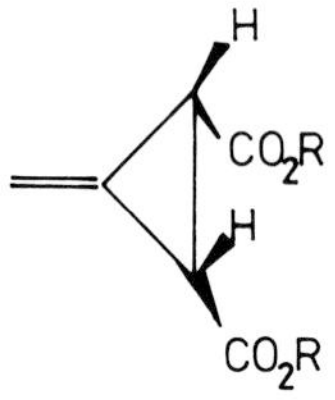

28b

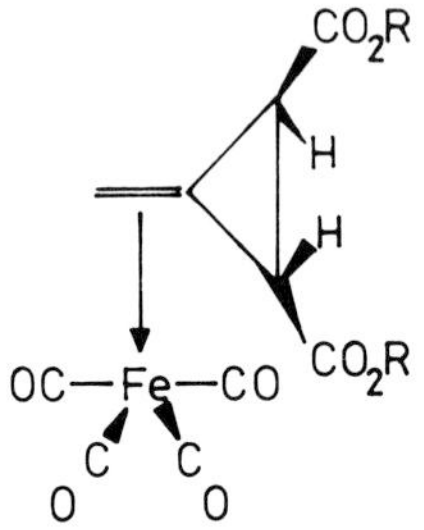

29a

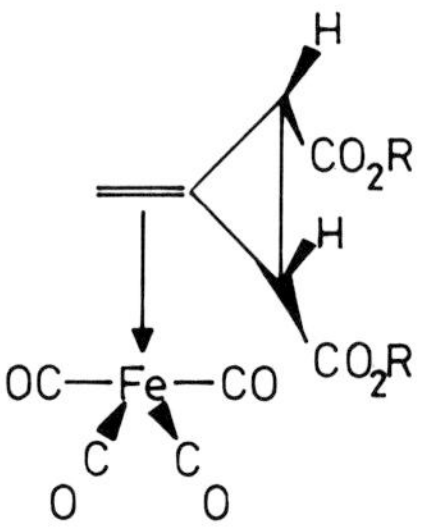

29b

the cyclopropane ring *(231)*. The *endo* stereochemistry of the product 29b (R = CH_3) from the *cis*-isomer 28b (R = CH_3) and $Fe_2(CO)_9$ was subsequently confirmed by X-ray crystallography *(230)*. In the case of the corresponding diethyl esters treatment of the *endo*-$Fe(CO)_4$ complex 29b (R = C_2H_5) with ethanolic sodium ethoxide leads to epimerization giving the corresponding *trans*-complex 29a (R = C_2H_5) *(118)*.

Several interesting rearrangements have been observed upon heating the methylenecyclopropane-$Fe(CO)_4$ complexes 29a (R = CH_3) and 29b (R = CH_3) *(231)*. The major products formed from 29a (R = CH_3) and 29b (R = CH_3) upon heating them alone in boiling toluene or with $Fe_2(CO)_9$ under milder conditions are the tricarbonyliron complexes of dimethyl *cis*- and *trans*-buta-1,3-diene-1,2-dicarboxylate. These reactions again are stereospecific, since the *trans*-isomer 29a ($R = CH_3$) of the methylenecyclopropane complex only forms the *cis*-isomer 30a and the *endo*-isomer 29b ($R = CH_3$) only forms the *trans*-isomer 30b upon such treatment. The *trans*-complex 29a also gives

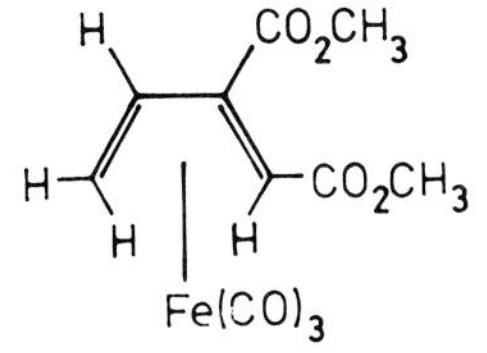

30a

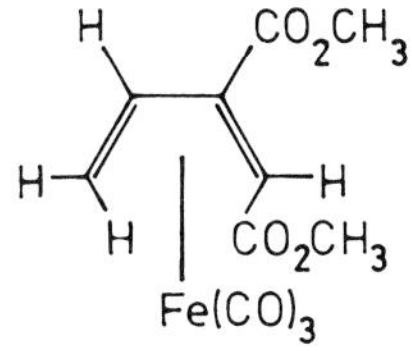

30b

the binuclear η^3-allylic complex 31 upon reaction with excess $Fe_2(CO)_9$ and the η^3-allylic-σ-acyl complex 32 upon ultraviolet irradiation *(231)*.

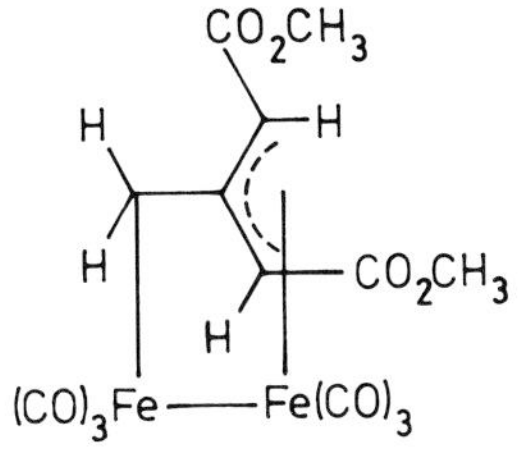

31

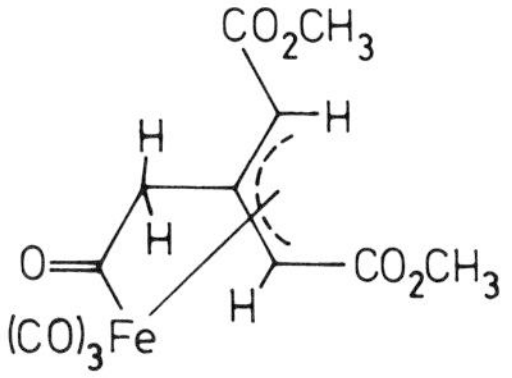

32

The *trans*-substituted diester 28a ($R = C_2H_5$) has no plane of symmetry and therefore can be resolved into the enantiomers. The corresponding free acid 28a ($R = H$) was resolved and then esterified to the optically active diethyl ester with ethereal diazoethane. Treatment of this ester with $Fe_2(CO)_9$ gave an optically active tetracarbonyliron complex 29a ($R = C_2H_5$). Degradation of this $Fe(CO)_4$ complex with cupric bromide in benzene was found to regenerate the diethyl ester 28a ($R = C_2H_5$) with no loss of its optical activity *(118)*.

Similar reactions of the 2,3-bis(hydroxymethyl)methyl-

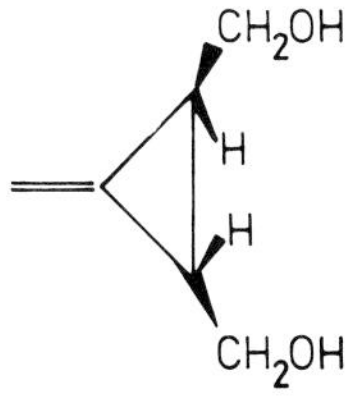

33a

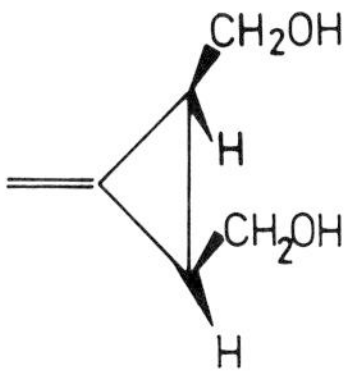

33b

enecyclopropanes 33a and 33b with carbonyliron complexes lead to opening of the cyclopropane ring *(88)*. Thus, treatment of either isomer 33a or 33b with $Fe_2(CO)_9$ in diethyl ether at room temperature gives the orange 1,4-pentadiene-lactone-$Fe(CO)_3$ complex 34 as the major product ($\sim$50 %) and the related tetracarbonyliron derivative 35 as a minor product ($\sim$16 %).

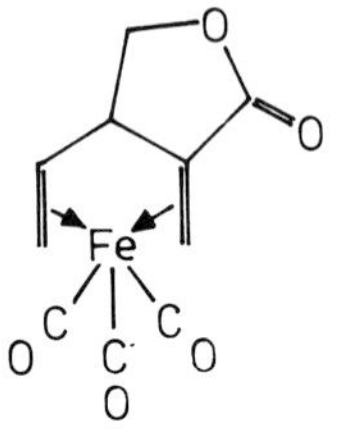

34

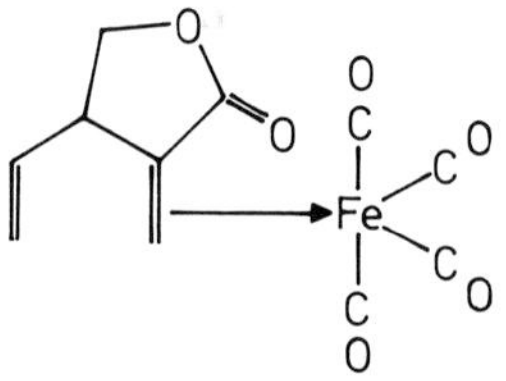

35

The yield of 35 relative to that of 34 can be improved by carrying out the reaction under a carbon monoxide atmosphere. The structure 34 has been confirmed by X-ray crystallography *(88)*.

Reactions of carbonyliron complexes with vinylcyclopropane derivatives give numerous products in which the cyclopropane ring has opened. Ultraviolet irradiation of vinylcyclopropane with pentacarbonyliron at -50°C gives a 10:1 mixture of yellow liquid (vinylcyclopropane)tetracarbonyliron (36) and yellow crystalline [1,2,3,6-η-(hex-1-en-6-one-3,6-diyl)]tricarbonyliron (37) *(13)*. The vinylcyclopropane com-

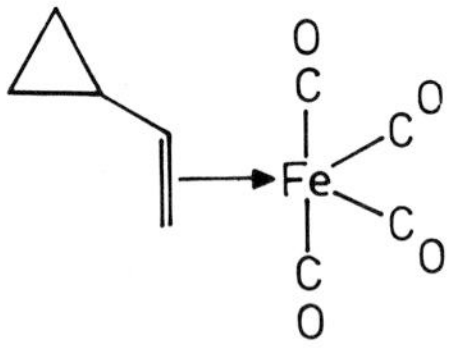

36

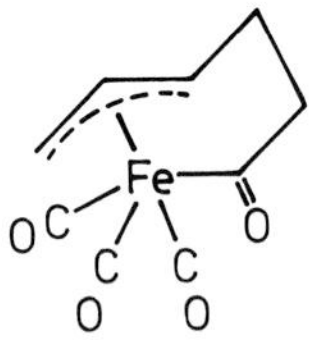

37

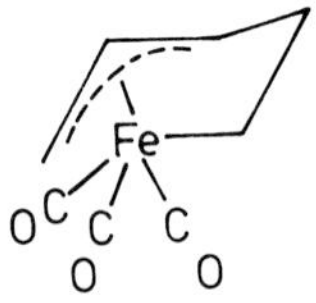

38

plex 36 decomposes above 0°C regenerating vinylcyclopropane. The complex 37 undergoes reversible decarbonylation at 25°C to give yellow liquid [1,2,3,5-η-(pent-1-ene-3,5-diyl)]tricarbonyliron (38). Under more vigorous conditions 37 gives 2-cyclohexenone, and 36 gives a 3:1 mixture of the two isomeric (1,3-pentadiene)tricarbonyliron complexes *(13)*.

A variety of products were obtained by reactions of methylene-spiranes with iron carbonyls *(206)*. Reaction of 1,1-dicyclopropylethylene with $Fe(CO)_5$ gives a 1:1 mixture of the diene-$Fe(CO)_3$ complexes 39 and 40 upon heating and

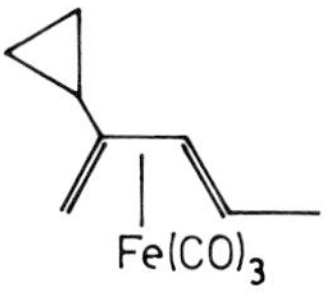

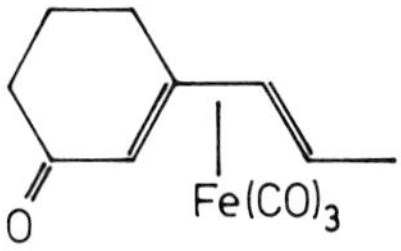

39 40

predominantly 40 upon ultraviolet irradiation. Reaction of the dispiroolefin 41 with $Fe(CO)_5$ gives the yellow tricarbonyliron σ,π-complex 42 as the major product and 1,1-ethano-7-keto-$\Delta^{8,9}$-octalin 43 as a minor product. An analogous car-

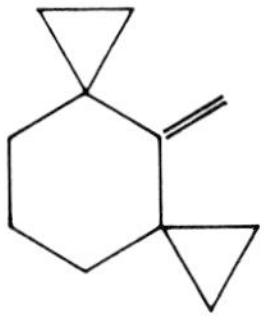

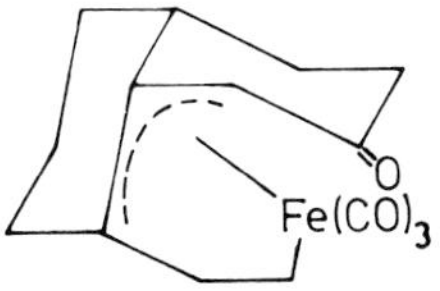

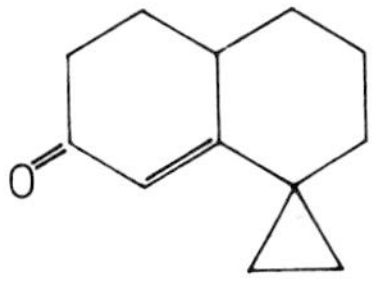

41 42 43

bonyl insertion occurs upon irradiation of the spirane 44 with $Fe(CO)_5$ giving 45 and its $Fe(CO)_4$ complex 46 as the major products, and 47 and 48 as minor products. Irradiation

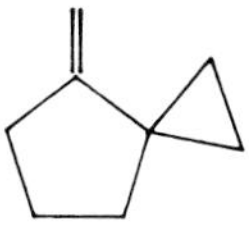

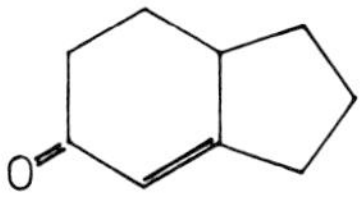

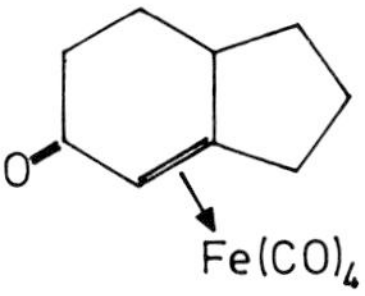

44 45 46

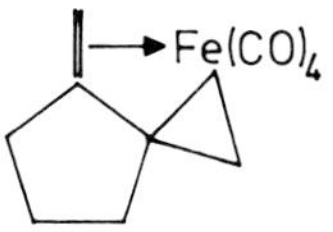

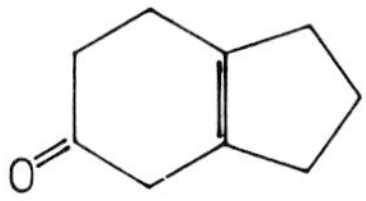

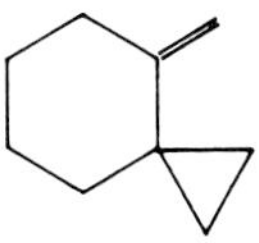

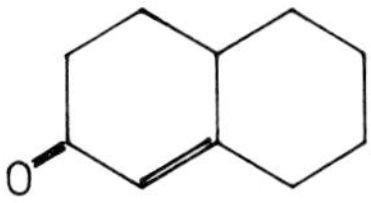

47 48 49 50

of 4-methylene-spiro[2.5]octane (49) with $Fe(CO)_5$ gives a high yield of the conjugated enone 50 without formation of an isolable carbonyliron complex.

Some reactions of bicyclic vinylcyclopropane derivatives with iron carbonyls have been investigated *(12)*. Reaction of bicyclo[3.1.0]hex-2-ene (51) with $Fe_2(CO)_9$ in diethyl ether (scheme [5]) at 30°C gives a mixture of the yellow air-sensitive liquid olefin-tetracarbonyliron complex 52 and the yel-

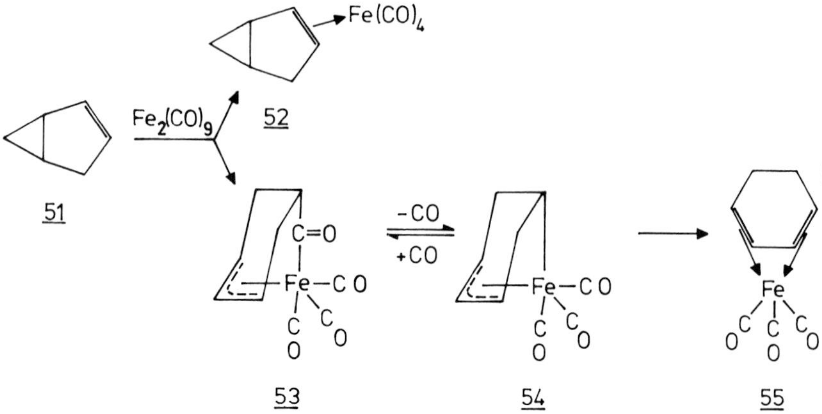

[5]

low crystalline η^3-allyl-σ-acyl complex 53. Heating 53 in decane solution at 130°C for several hours results in decarbonylation and proton migration to give (1,3-cyclohexadiene)-tricarbonyliron (55) presumably through an intermediate 54.

Reaction of bicyclo[4.1.0]hept-2-ene (56) with excess $Fe_2(CO)_9$ in diethyl ether at 30°C gives the yellow volatile η^3-allyl-σ-alkyl complex 57 in 75 % yield. Carbonylation of 57 at 20°C/100 atm for 6 days results in carbonyl insertion to give 58, which is stable at -15°C, but which reverts to 57

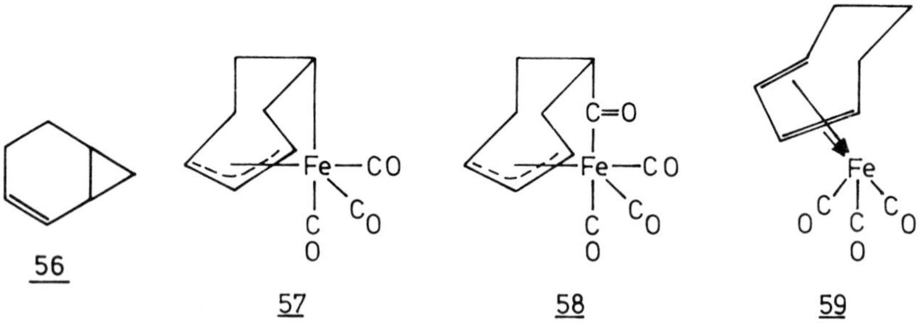

at room temperature in the absence of excess carbon monoxide. Pyrolysis of 57 at 120°C gives (1,3-cycloheptadiene)tricarbonyliron (59). Aqueous sodium borohydride reduction of the (cycloheptadienyl)tricarbonyliron cation at 0°C gives a 2:1 mixture of 57 and 59, thereby providing a more convenient preparation of 57.

Reactions of carbonyliron complexes with several polycyclic hydrocarbons containing vinylcyclopropane units have been investigated. For example, photolysis of semibullvalene (60) with $Fe(CO)_5$ at -50°C in diethyl ether gives the $Fe(CO)_4$ complex 61 and the η^3-allyl-σ-alkyl complex 62 *(14)*. Upon mild heating (40°C) 61 undergoes disproportionation liberating semibullvalene (60) to give the stable yellow crystalline bis(tetracarbonyliron) complex 63, which can also be obtained

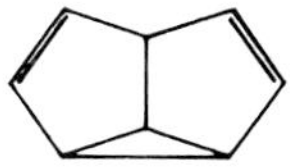

60

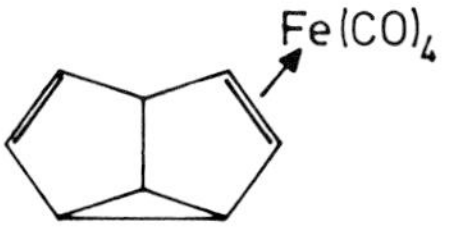

61

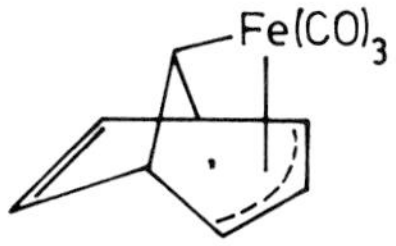

62

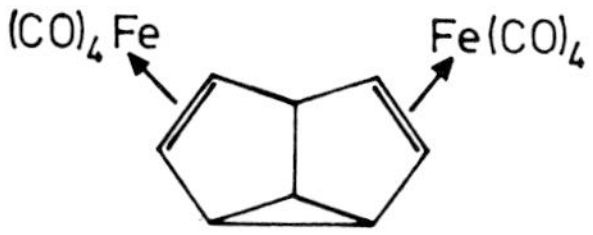

63

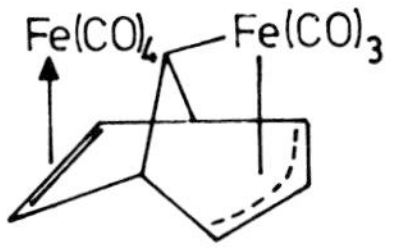

64

from semibullvalene and $Fe_2(CO)_9$ *(67)*. Reaction of semibullvalene either photochemically with $Fe(CO)_5$ in diethyl ether at 25°C *(14)* or thermally with $Fe_2(CO)_9$ at room temperature *(67)* also gives the heptacarbonyldiiron derivative 64. Similar products have also been obtained by analogous carbonyliron reactions with other more complex polycyclic hydrocarbons containing vinylcyclopropane units such as barbaralone *(71)*, bullvalene *(10,11,16)*, and homosemibullvalene *(15)*.

VI. REACTIONS OF CARBONYLIRON COMPLEXES WITH OTHER STRAINED RING OLEFINS

Strained cyclic olefins such as norbornenes and cyclobutenes exhibit a pronounced tendency to react with carbonyliron complexes under formation of cyclopentanones *(236)* according to equation [6]. This type of reaction has been

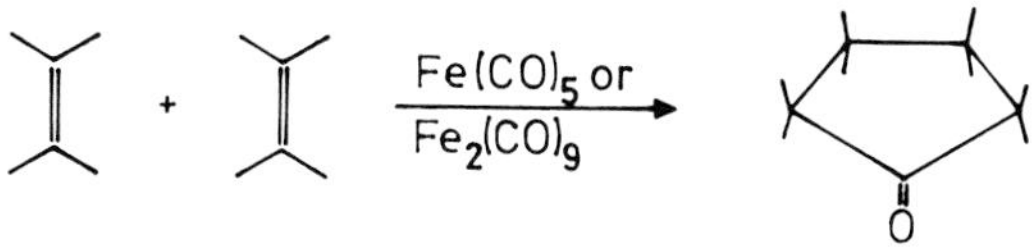

[6]

studied for a variety of norbornene derivatives *(86,157,158,*

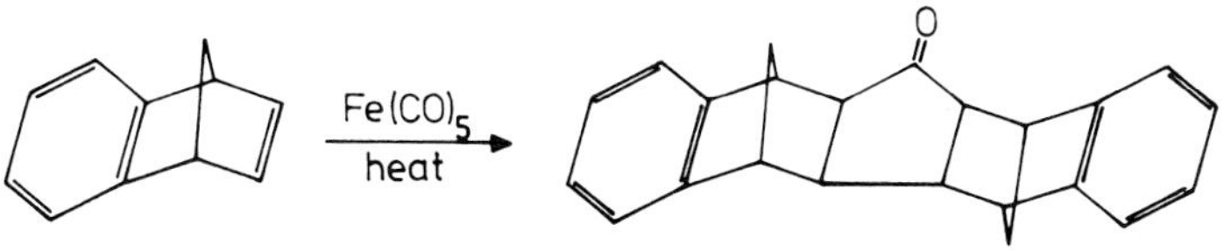

[7]

159,160,215), as exemplified by reaction [7], and can be remarkably stereospecific. In one case only one out of 36 possible stereoisomers was obtained *(86)*. $Fe(CO)_2(NO)_2$ is also effective for certain reactions of this type *(154)*. The oxygen-bridged 1,4-dihydro-1,4-epoxynaphthalene does not give a

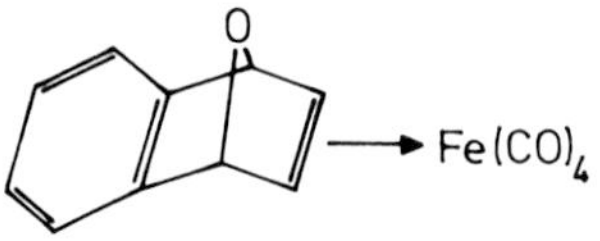

65

polycyclic cyclopentanone derivative on reaction with carbonyliron complexes but instead naphthalene is formed through the relatively stable (olefin)$Fe(CO)_4$ complex 65 which decomposes only at ca. 60°C.

Reaction of *cis*-cyclobutene-3,4-dicarboxylic anhydride with $Fe_2(CO)_9$ yields a stable (olefin)$Fe(CO)_4$ complex. The corresponding dimethyl ester 66, however, gives the tricyclic ferracyclopentane complex 67 with the carbonyl group of one of the ester functions occupying a coordination site at the metal (scheme [8]). The photoreaction of 66 with $Fe(CO)_5$ leads to a similar product 68 exhibiting a different orientation of the ester groups as the result of a 1,3-hydrogen shift. Analogously to the ferracyclopentane complex 17 derived from methyl acrylate, carbonylation with carbon monoxide at 70°C/55 atm converts 67 and 68 into the corresponding cyclopentanones 69 and 70, and treatment with hydrogen/Raney-nickel gives the isomeric bicyclobutyls 71 and 72, respectively. Ferracyclopentanes analogous to 17, 67, and 68, and bis(olefin)-tricarbonyliron complexes like 15 have also been proposed to be intermediates in the synthesis of cyclopentanones from norbornene derivatives and carbonyliron complexes, but in these cases they are obviously too unstable to be isolated.

Reactions of carbonyliron complexes with several other cyclobutene derivatives have been described. Treatment of *cis*-bicyclo[4.2.0]oct-7-ene (73) with $Fe_2(CO)_9$ in boiling hexane (eq. [9]) results in rearrangement to give *cis*-bicyclo-

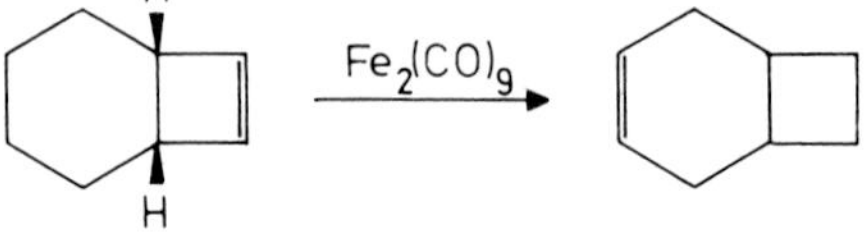

[9]

73 74

[4.2.0]oct-2-ene (74) *(36)*. This rearrangement has been postulated to involve hydrogen transfer from the six-membered ring to the cyclobutene through the iron center without open-

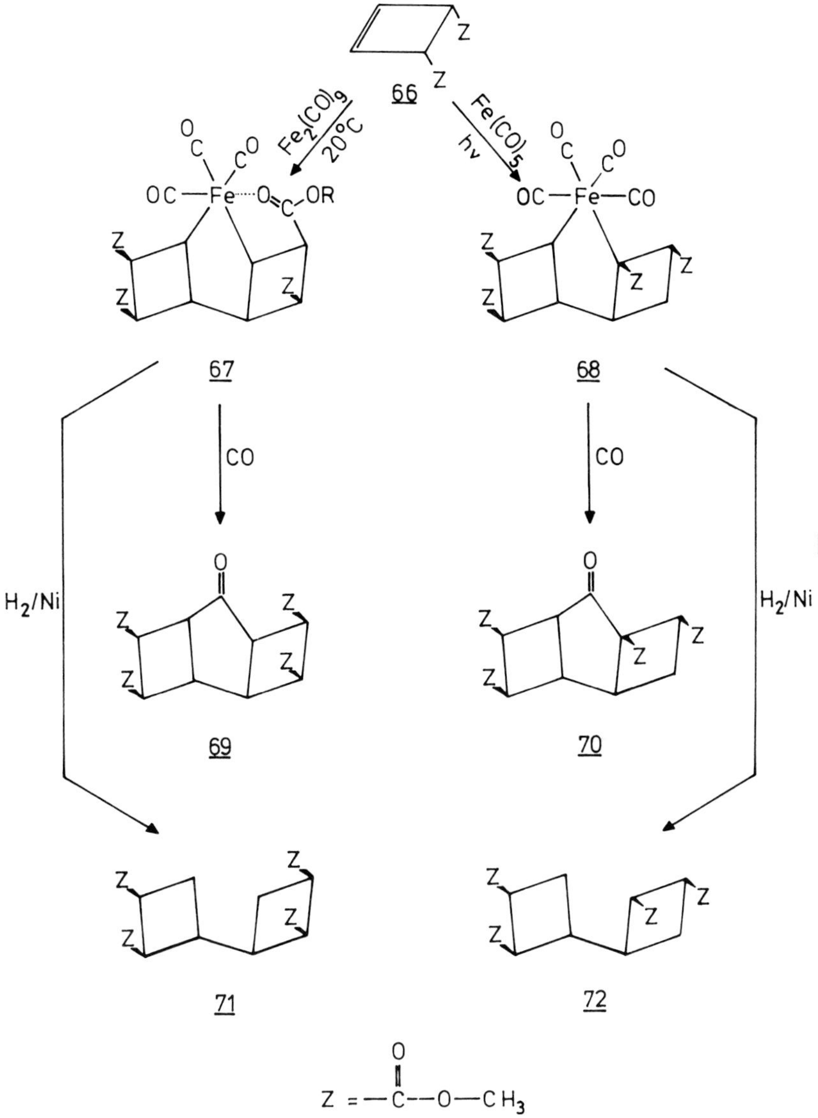

[8]

ing the four-membered ring *(36)*.

Reaction of 3,4-dimethylenecyclobutene (75, X = H) with $Fe_2(CO)_9$ in tetrahydrofuran at room temperature *(129)* gives the symmetrical complex 76 (X = H) in which only one of the three double bonds of the triene is bonded to the metal (scheme [10]). Mild warming of 76 (X = H) results in rear-

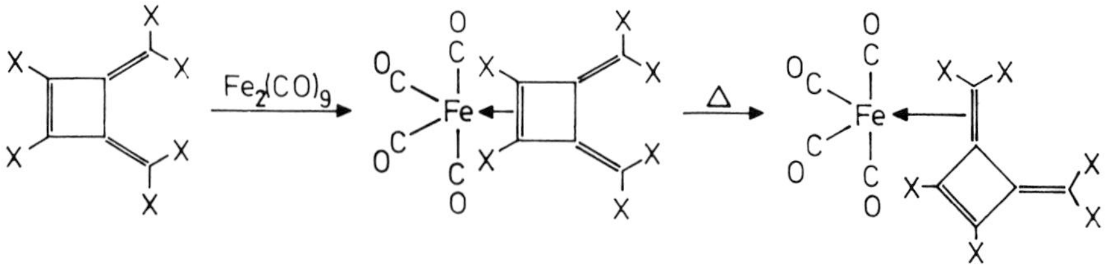

75 76 77 [10]

rangement to the unsymmetrical isomer 77 (X = H) as shown by NMR spectra, whereas more vigorous heating of 76 or 77 (X = H) leads to complete decomposition.

A similar reaction of perchloro-3,4-dimethylenecyclobutene (75, X = Cl) with $Fe_2(CO)_9$ gives a much more stable yellow crystalline tetracarbonyliron complex. The positions of the six lines in its chlorine nuclear quadrupole resonance (NQR) spectrum indicate this complex to be the unsymmetrical derivative 77 (X = Cl) *(161)*. However, the ^{13}C-NMR spectrum of 77 (X = Cl) exhibits only three lines from the C_6Cl_6 ligand. This suggests rapid (on the NMR time scale) interchange of the $Fe(CO)_4$ unit between both of the exocyclic carbon-carbon double bonds of the perchloro-3,4-dimethylenecyclobutene ligand. However, this interesting interpretation of the ^{13}C-NMR spectrum depends critically upon the correct interpretation of the NQR spectrum. Refluxing the complex 77 (X = Cl) in hexane results partially in decomposition to the free chlorocarbon 75 (X = Cl) and partially in decarbonylation to form a yellow diene-tricarbonyliron derivative $(C_6Cl_6)Fe(CO)_3$. A perchloroolefin-tetracarbonyliron complex similar to 77 (X = Cl) is obtained by the reaction of perchlorofulvene with $Fe_2(CO)_9$ in hexane at 70°C to give a $(C_6Cl_6)Fe(CO)_4$ derivative. The ^{13}C-NMR and chlorine NQR spectra of this compound indicate that the tetracarbonyliron group is bonded to one of the double bonds in the five-membered ring.

Carbonyliron complexes have been used for the ring expansion of α-pinene (78) *(217)*. Thus, treatment of either

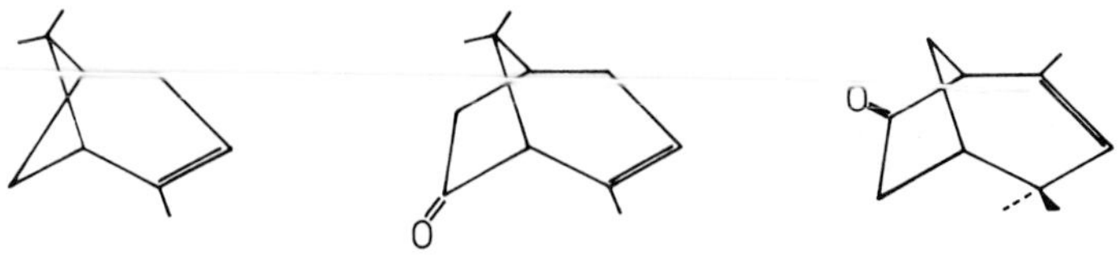

78 79 80

α-pinene or β-pinene with an equimolar quantity of $Fe(CO)_5$ results in the stereospecific formation of the ketones 79 and 80.

VII. REACTIONS OF CARBONYLIRON COMPLEXES WITH OLEFINS CONTAINING FUNCTIONAL GROUPS

Reactions of carbonyliron complexes with olefins containing functional groups are of particular interest when the functional group can also participate in the bonding to the iron atom. For example, olefinic ketones sometimes form carbonyliron complexes in which both the carbon-carbon and carbon-oxygen double bonds are coordinated to the iron atom. The most important classes of such compounds forming carbonyliron complexes are unsaturated ketones, and olefinic phosphines and arsines.

A. REACTIONS OF CARBONYLIRON COMPLEXES WITH UNSATURATED KETONES

Carbonyliron complexes may react with α,β-unsaturated ketones containing the C=C-C=O structural unit under mild conditions by stepwise replacement of two carbonyl groups according to the reaction sequence [11]. Suitably mild con-

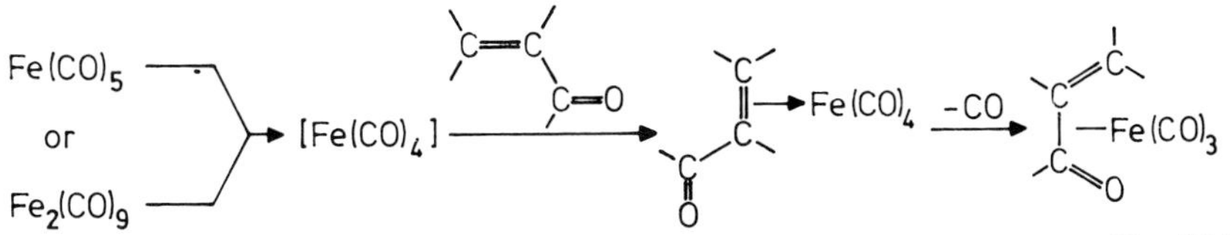

Fe(CO)5
or
Fe2(CO)9
[Fe(CO)4]
Fe(CO)4
-CO
Fe(CO)3

81 [11]

ditions for effecting such reactions include $Fe_2(CO)_9$ at ∿40°C in an inert solvent such as benzene *(31,113,142,167,170, 198)* and ultraviolet irradiation with $Fe(CO)_5$ at room temperature *(32,141)*.

The first (enone)$Fe(CO)_3$ derivative to be prepared in which both the carbon-carbon and carbon-oxygen double bonds are attached to the iron atom as depicted in 81 was the cinnamaldehyde derivative *(216)*. This complex was reported in 1964 by Stark, Lancaster, Murdoch, and Weiss *(216)* as a product obtained by heating the corresponding $Fe(CO)_4$ complex to 60°C for 15 h. Subsequently a group of Russian workers *(170)* obtained not only the tetracarbonyliron derivatives but also the tricarbonyliron complexes of both the *cis*- and *trans*-isomers of β-benzoylacrylic acid methyl ester, C_6H_5-CO-CH=CH-CO_2CH_3. However, alkyl and aryl β-chlorovinyl ketones of the type R-CO-CH=CHCl (R = CH_3, C_6H_5, *p*-CH_3-C_6H_4, *p*-CH_3O-C_6H_4, and *p*-BrC_6H_4) were found to give only the corresponding tetracarbonyliron complexes upon reaction with $Fe_2(CO)_9$ in benzene at 40°C *(167)*. More recently enone-tricarbonyliron complexes of the type 81 have been obtained from

CH_2=CH-CO-C_6H_5 *(198)*, C_6H_5-CH=CH-CO-CH_3 *(31,113,141)*, C_6H_5-CH=CH-CO-C_6H_5 *(31,113)*, 2,6-dibenzylidenecyclohexanone *(31, 113)*, $(C_6H_5)(CH_3)$C=CH-CO-C_6H_5 *(113)*, isobutylidenemeldrumic acid *(142)*, pinocarvone *(141)*, and pulegone *(141)*. This chemistry has also been extended to the preparation of enone-tricarbonyliron complexes (81) of the ferrocenylvinylketones Fc-CH=CH-CO-C_6H_5 and Fc-CO-CH=CH-C_6H_5 *(173,177)* [Fc = $(\eta^5$-$C_5H_5)Fe(\eta^5$-$C_5H_4)$].

Recently Cardaci *(43)* has studied the kinetics of the formation of (enone)$Fe(CO)_3$ derivatives from the corresponding (enone)$Fe(CO)_4$ complexes using the enone ligands C_6H_5-CH=CH-CO-R (R = H, CH_3, and C_6H_5). The results were interpreted in terms of the equation [12] where α corresponds to

$$(C_6H_5{-}CH{=}CH{-}CO{-}R)Fe(CO)_4 \longrightarrow \alpha\,(C_6H_5{-}CH{=}CH{-}CO{-}R)Fe(CO)_3 + (1-\alpha)\,Fe(CO)_5 + (1-\alpha)\,C_6H_5{-}CH{=}CH{-}CO{-}R + (2\alpha-1)\,CO \qquad [12]$$

the fraction of the $Fe(CO)_3$ complex formed per mole of the $Fe(CO)_4$ derivative. The reaction was found to be first order with respect to the tetracarbonyliron complex. The first order rate constants decrease with an increase in concentration of the liberated C_6H_5-CH=CH-CO-R and were found to be a function of CO concentration from experiments done in the presence of added carbon monoxide. The ratio $(1-\alpha)/\alpha$ changes linearly with 1/[C_6H_5-CH=CH-CO-R] and also changes with CO pressure according to a square law. These results were interpreted to exclude an intramolecular chelation mechanism. A mechanism involving an $Fe(CO)_3$ intermediate, which is responsible for the formation of the tricarbonyliron complexes, was suggested as an alternative.

Reactions of (enone)$Fe(CO)_3$ derivatives with various Lewis bases have been investigated *(42,172,221)*. With ligands such as triphenylphosphine the first step of such reactions consists of ligand addition to form the corresponding (enone)$Fe(CO)_3L$ derivative with displacement of the coordinated carbon-oxygen double bond. The (enone)$Fe(CO)_3L$ derivative may either react with excess ligand to displace completely the enone ligand to give $L_2Fe(CO)_3$ or may undergo loss of one carbonyl group to give the dicarbonyl complex, (enone)$Fe(CO)_2L$. The net result is that ligands such as triphenylphosphine can displace either the enone ligand or one carbonyl group in the (enone)$Fe(CO)_3$ complexes.

The relative lability of the enone ligand in the (enone)$Fe(CO)_3$ complexes make these complexes good sources under mild conditions of reactive $Fe(CO)_3$ groups for the preparation of various tricarbonyliron complexes. For example, $(C_6H_5$-CH=CH-CO-$CH_3)Fe(CO)_3$ was used to prepare tricarbonyliron complexes of heptafulvene derivatives, which are too unstable to heat and ultraviolet irradiation to react with any of the normal carbonyliron complexes under conditions where their carbonyl groups could be replaced *(113)*. Also, the reactions of $(C_6H_5\text{-}CH{=}CH\text{-}CO\text{-}CH_3)Fe(CO)_3$ with various alkynes including $R\text{-}C{\equiv}C\text{-}R$ ($R = CH_3$, C_2H_5, and C_6H_5) and macrocyclic alkadiynes were found to give exclusively tricarbonylferrole-tricarbonyliron derivatives unaccompanied by the variety of other complex products usually obtained from alkynes and iron carbonyls *(126)*.

The coordinated ketone group in (enone)$Fe(CO)_3$ derivatives is susceptible to reaction with the electrophile $[CH_3CO]^+[BF_4]^-$ *(174)*. Thus, treatment of the tricarbonyliron complexes $(R'\text{-}CH{=}CH\text{-}CO\text{-}R)Fe(CO)_3$ ($R = CH_3$, $R' = H$; $R = H$, $R' = C_6H_5$; $R = R' = C_6H_5$) with $[CH_3CO]^+[BF_4]^-$ in nitromethane gives cations formulated as 82 containing a coordinated carbonyl

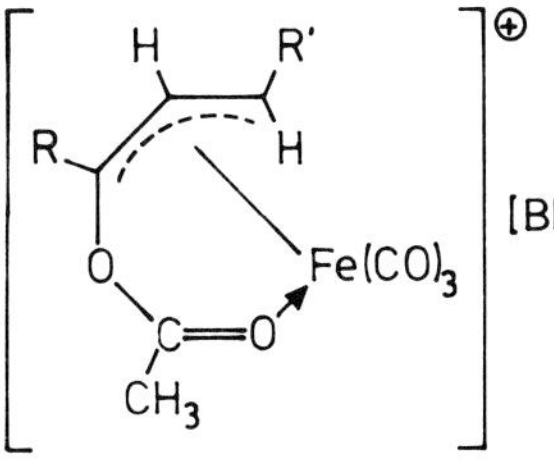

82

from the acetyl group. Reaction of these complexes with me-

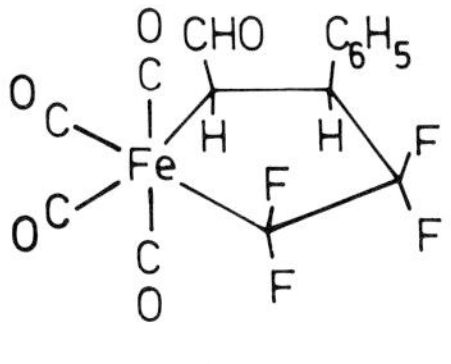

83a

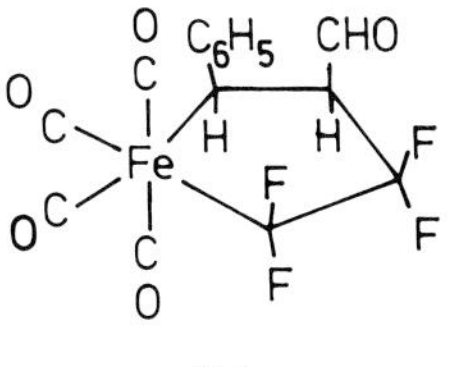

83b

thanol regenerates the starting (enone)$Fe(CO)_3$ derivative.

Ultraviolet irradiation of both the $Fe(CO)_4$ and the $Fe(CO)_3$ complexes of *trans*-cinnamaldehyde with excess tetrafluoroethylene gives the pale yellow crystalline metallocycle shown by ^1H- and ^{19}F-NMR spectroscopy to be a 4:5 mixture of the isomers 83a and 83b, respectively *(26,27)*.

B. REACTIONS OF CARBONYLIRON COMPLEXES WITH OLEFINS CONTAINING TRIVALENT NITROGEN ATOMS

Olefins containing adjacent trivalent nitrogen atoms may

function as chelate ligands in carbonyliron chemistry using both their carbon-carbon double bond and their trivalent nitrogen atom for bonding to the iron atom.

Reactions of 1,2-dimethyl- and 1-methyl-2-phenyl-1,2-dihydropyridazine-3,6-diones (84, R = CH_3 or C_6H_5) with $Fe_2(CO)_9$ at or below room temperature give a thermally unstable tetracarbonyliron derivative 85 in which only the py-

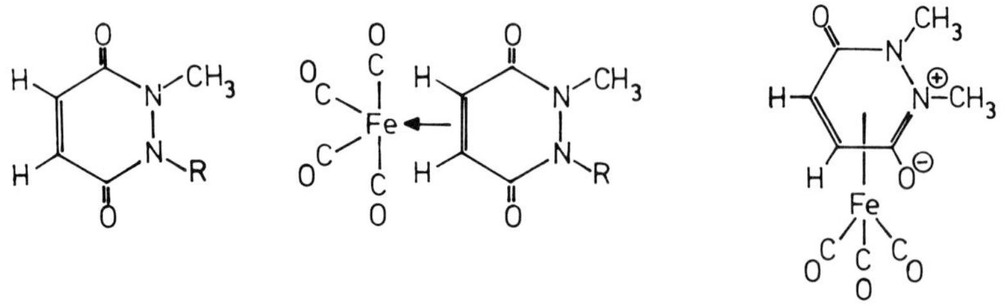

84 85 86

ridazine carbon-carbon double bond is coordinated to the iron atom. Upon heating or ultraviolet irradiation this $Fe(CO)_4$ complex (85, R = CH_3) readily loses one carbonyl group to give an $Fe(CO)_3$ complex shown by X-ray crystallography to have the zwitterionic structure 86 with *tetrahapto*-coordination of the dihydropyridazine ligand through three adjacent carbons and one nitrogen *(175)*.

In another type of reaction, the treatment of N-allylpyrazole with $Fe_2(CO)_9$ in hexane at room temperature gives yellow $(C_3H_5N_2C_3H_3)Fe(CO)_3$ (87). This compound was also one

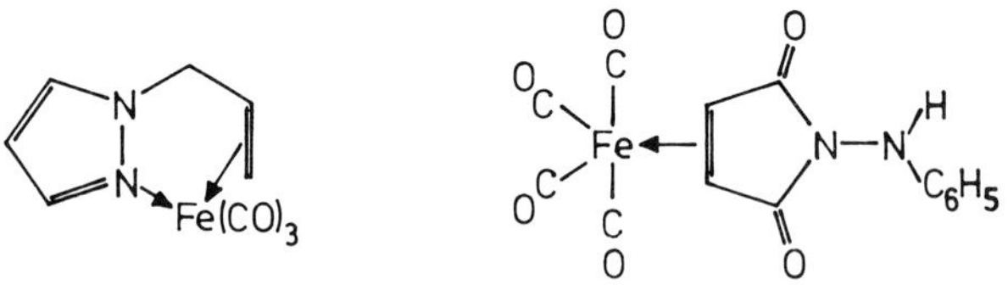

87 88

of several products obtained from (allyl)tricarbonyliron iodide and potassium tris(pyrazolyl)borate *(127)*.

The maleic acid monohydrazide complexes (R-NH-NH-CO-CH=CH-CO_2H)$Fe(CO)_4$ can be prepared either by reaction of the ligands with $Fe_2(CO)_9$ in acetic acid at 35-40°C or by reactions of the hydrazines with (maleic anhydride)tetracarbonyliron *(176)*. Cyclization of the phenylhydrazide complex (R = C_6H_5) with acetic anhydride at room temperature gives (N-anilinomaleimide)tetracarbonyliron (88) as the major carbonyliron product.

C. REACTIONS OF CARBONYLIRON COMPLEXES WITH OLEFINS ALSO CONTAINING SULFUR ATOMS

Reaction of thiete sulfone with $Fe_2(CO)_9$ in boiling ether or with $Fe(CO)_5$ in benzene assisted by ultraviolet irradiation gives the corresponding tetracarbonyliron complex 89. Heating this complex with excess thiete sulfone in

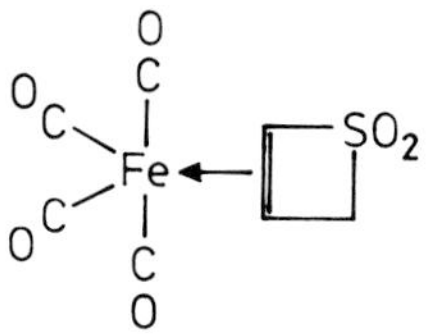

89

boiling hexane results in reduction of some of the sulfone groups to give eventually $Fe_3(CO)_9S_2$ *(162)*.

Reaction of 2,5-dihydrothiophene-1-oxide with $Fe_2(CO)_9$ in boiling diethyl ether for 2.5 h results in the formation of a tricarbonyliron complex. X-ray crystallography indicates structure 91 for this complex. In this structure both the

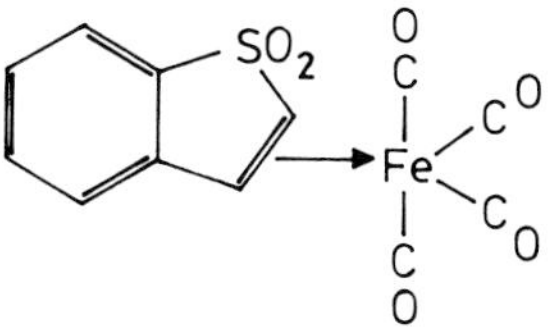

90

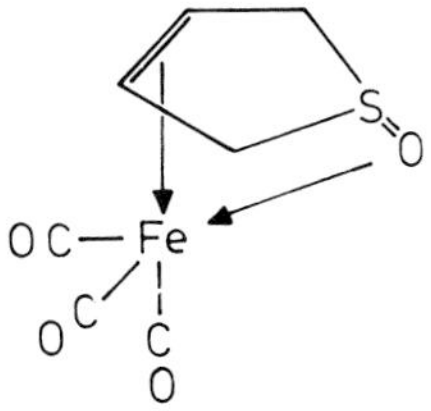

91

carbon-carbon double bond and the sulfoxide oxygen of the dihydrothiophene oxide are coordinated to the iron atom. Spectroscopic evidence is presented for an intermediate tetracarbonyliron derivative in this reaction. However, this $Fe(CO)_4$ complex appears to have the sulfoxide oxygen rather than the carbon-carbon double bond of the dihydrothiophene oxide bonded to the tetracarbonyliron unit *(66)*.

Reaction of benzo[b]thiophene-1,1-dioxide with carbonyliron complexes leads to a mixture of the corresponding $Fe(CO)_4$ and $Fe(CO)_3$ derivatives *(103)*. X-ray crystallography shows the tetracarbonyliron complex to have structure 90 in which the double bond of the five-membered ring of the benzothiophene dioxide is coordinated to an $Fe(CO)_4$ group in the equatorial position.

Reaction of the β-oxovinyl sulfones *cis*- and *trans*-R-CO-CH=CH-SO_2R' (R = CH_3, R' = C_6H_5; R = R' = C_6H_5; R = C_6H_5,

R' = *p*-CH_3-C_6H_4) with $Fe_2(CO)_9$ leads to the corresponding tetracarbonyliron complexes in which the β-oxovinyl sulfone carbon-carbon double bond is bonded to the iron atom *(169)*. The two strongly electronegative substituents on the carbon-carbon double bond make these complexes unusually stable relative to most other (olefin)$Fe(CO)_4$ derivatives.

D. REACTIONS OF OLEFINIC PHOSPHORUS, ARSENIC, AND BORON COMPOUNDS WITH CARBONYLIRON COMPLEXES INVOLVING THE CARBON-CARBON DOUBLE BOND

The phosphorus substituted olefin (2-vinylphenyl)(diphenyl)phosphine, (*o*-CH_2=CH-C_6H_4)P(C_6H_5)$_2$, reacts with $Fe_3(CO)_{12}$ to form both the tricarbonyl complex [(*o*-CH_2=CH-C_6H_4)P-(C_6H_5)$_2$]$Fe(CO)_3$ and the dicarbonyl complex [(*o*-CH_2=CH-C_6H_4)P-(C_6H_5)$_2$]$_2$$Fe(CO)_2$ *(22)*. The $Fe(CO)_3$ complex has structure 92

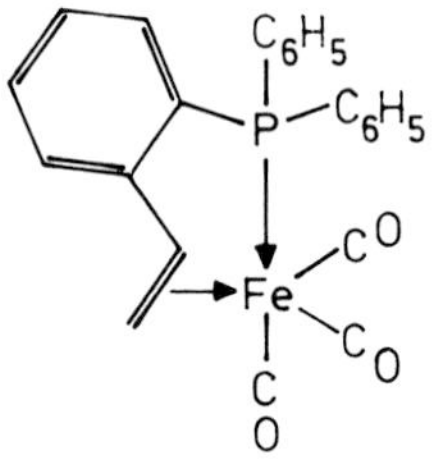

92

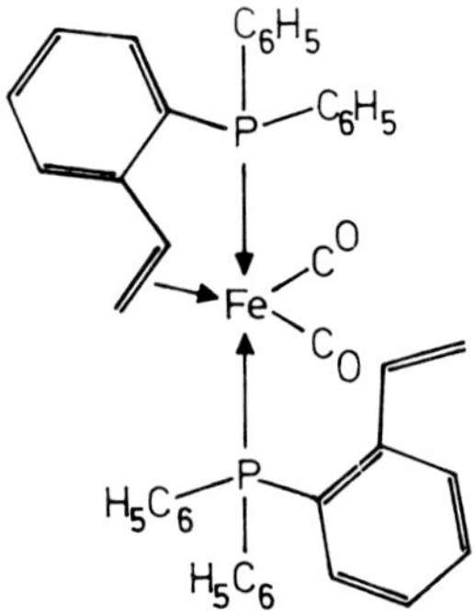

93

in which both the trivalent phosphorus atom and the carbon-carbon double bond of the olefinic phosphine are bonded to the iron atom. The structure of the $Fe(CO)_2$ complex was shown by X-ray crystallography *(22,190)* to be 93 in which one olefinic phosphine is bonded to the pentacoordinate iron atom

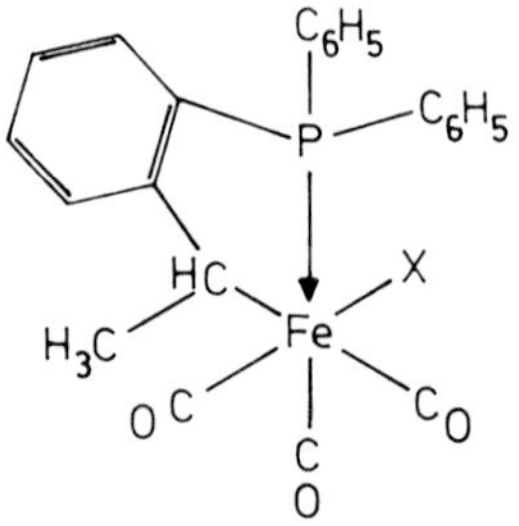

94

both through the phosphorus atom in an axial position and a carbon-carbon double bond in an equatorial position, and the second olefinic phosphine is bonded to the iron atom only through its phosphorus atom, which is coordinated to the second axial position. Reactions of the tricarbonyliron complex 92 with the hydrogen halides HX (X = Cl and Br) in hexane result in addition of hydrogen to the coordinated carbon-carbon double bond to give the octahedral iron σ-alkyl halide complexes 94 *(23)*.

Several reactions of 92 with unsaturated fluorocarbons have been investigated. Ultraviolet irradiation of 92 with the fluoroolefins tetrafluoroethylene *(27)*, hexafluoropropene *(89)*, trifluoroethylene *(89)*, and chlorotrifluoroethylene *(89)* gives the crystalline 1:1 adducts shown to be the metallocycles 95 (X = F, CF_3, H, and Cl, respectively). An analogous

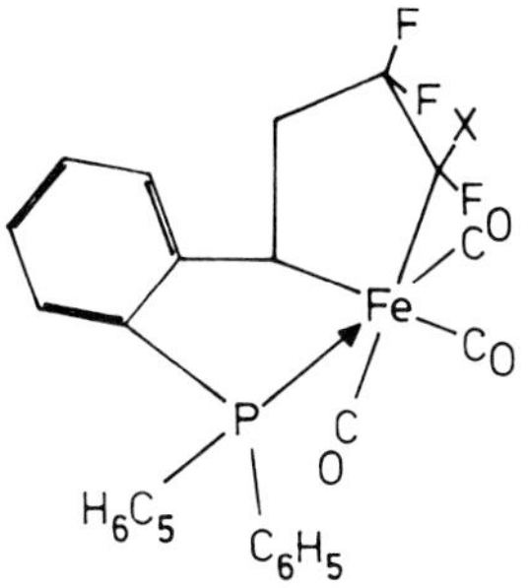

95

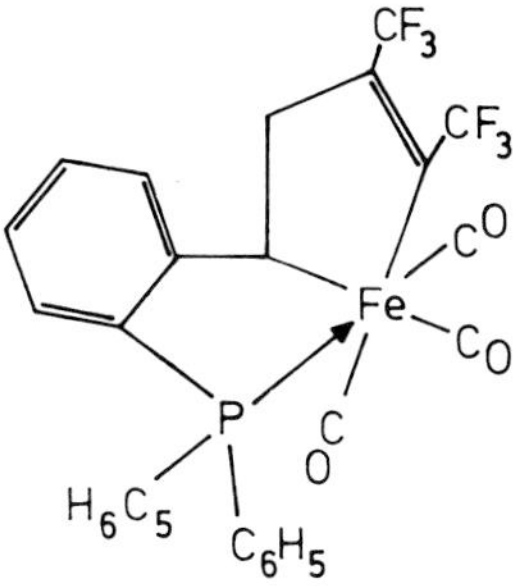

96

metallocycle 96 *(28)* is obtained from a similar ultraviolet irradiation of 92 with hexafluorobut-2-yne.

Olefinic ditertiary phosphines and arsines with the structural unit *cis*-(RR'E)CR"=CR"(ERR') (E = P or As; often 2 R" are the ends of a perfluoropolymethylene bridge) react with carbonyliron reagents to give hexacarbonyldiiron complexes with the general structure 97 apparently containing an iron-iron dative bond. In this structure both group V atoms and the carbon-carbon double bond are coordinated to the iron atoms *(45,57,68,69)*. In other iron complexes isolated from reactions of this type only the group V atoms are coordinated to the metal atom. Often in order to achieve the necessary *cis*-configuration of the donor group V atoms about the carbon-carbon double bond, ligands prepared by reactions of perfluorocycloolefins such as perfluorocyclobutene and perfluorocyclopentene *(45,68,69)* with appropriate phosphorus and/or arsenic reagents are used. The structure of the compound 98 prepared from a perfluorocyclobutene-derived ditertiary arsine has been confirmed by X-ray crystallography *(68,69)*. Compounds

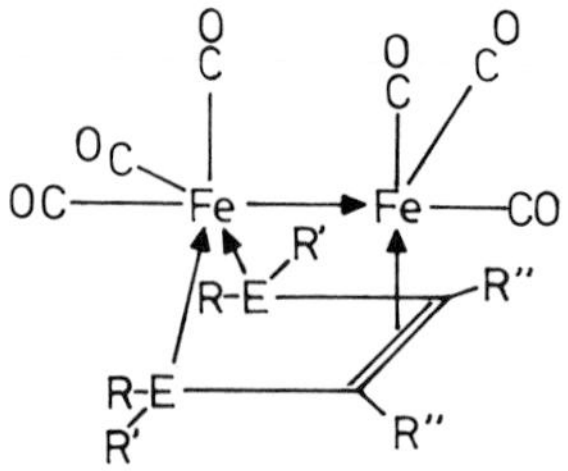

97

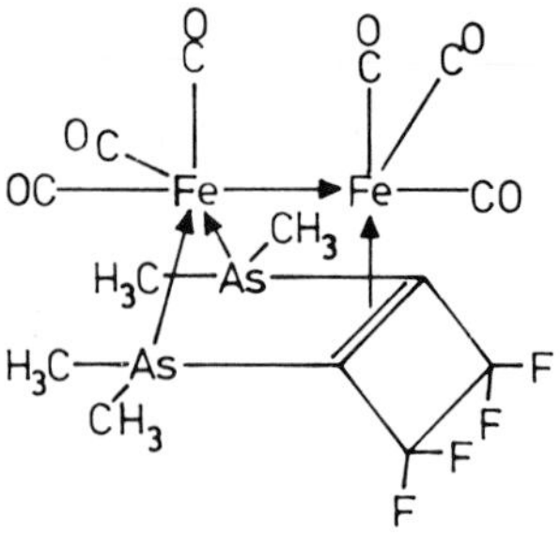

98

of the type 97 including 98 react further with trivalent phosphorus and arsenic ligands with substitution of additional carbonyl groups. The compound 99 is an example of a further

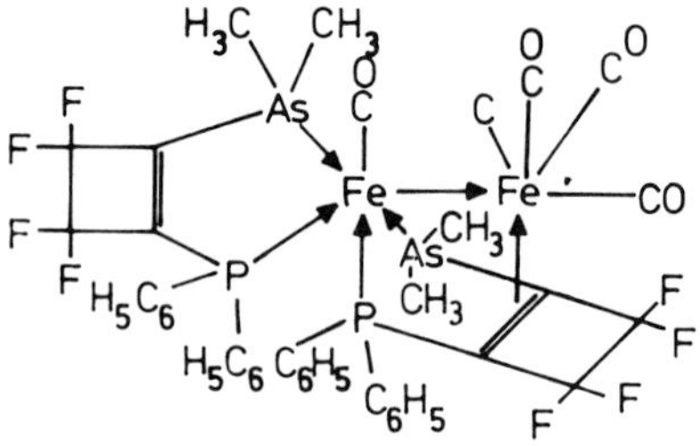

99

substitution product in which the carbon-carbon double bond of one of the two olefinic bidentate ligands is still bonded to an iron atom. The structure of 99 has been confirmed by X-ray crystallography *(70)*.

The Mössbauer spectra of 98 and some related compounds of the type 97 were analysed using perturbations in magnetic fields up to 50 kG to assist with the assignments of the four observed lines and to deduce the signs of the quadrupole coupling constants at both iron sites *(199)*.

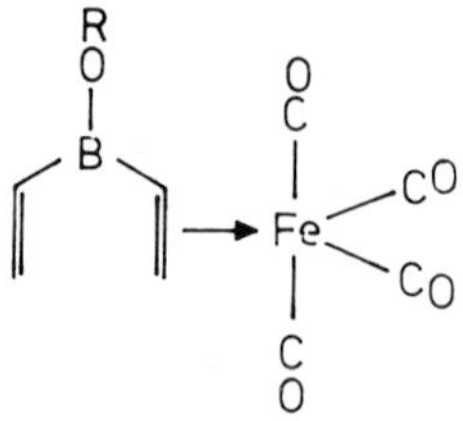

100

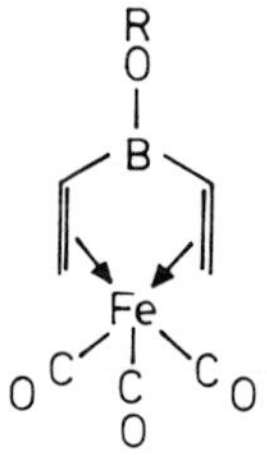

101

The scope of the reported reactions of carbonyliron reagents with vinylboron derivatives is rather limited. Reactions of the (alkoxy)divinylboranes $RO\text{-}B(CH{=}CH_2)_2$ ($R = n\text{-}C_4H_9$ and $C_6H_5\text{-}CH_2$) with $Fe_2(CO)_9$ give labile tetracarbonyliron complexes with structure 100 in which only one double bond is coordinated to the iron atom. Ultraviolet irradiation of the compounds 100 ($R = n\text{-}C_4H_9$ and $C_6H_5\text{-}CH_2$) in diethyl ether at 10°C results in decarbonylation to give the likewise unstable tricarbonyliron complexes 101 *(110)*.

VIII. CYCLOPENTADIENYL-OLEFIN-IRON COMPLEXES

The cationic olefin complexes $[(\eta^5\text{-}C_5H_5)Fe(CO)_2(\eta^2\text{-olefin})]^+$ are very favourable species in cyclopentadienyliron chemistry. These complexes are stable, formed by a considerable variety of olefins in numerous types of reactions, and are frequently isolated easily as salts of large anions, particularly hexafluorophosphate.

A. PREPARATION OF DICARBONYL-CYCLOPENTADIENYL-OLEFIN-IRON CATIONS AND RELATED COMPOUNDS

The first method used for the preparation of $[(\eta^5\text{-}C_5H_5)\text{-}Fe(CO)_2(\eta^2\text{-olefin})]^+$ cations was the reaction between dicarbonyl-cyclopentadienyliron halides and the olefin in the presence of a strong Lewis acid catalyst such as aluminium halides according to scheme [13] *(76,77)*. After hydrolysis the

$$(C_5H_5)Fe(CO)_2X + \text{olefin} + AlX_3 \longrightarrow [(C_5H_5)Fe(CO)_2(\text{olefin})]^+[AlX_4]^- \qquad [13]$$

stable hexafluorophosphate salts can be isolated by addition of ammonium hexafluorophosphate. Olefins which can be used in this reaction include ethylene *(76)*, propylene, *cis*-butene-2, octadecene-1, cyclohexene, and cyclooctene *(77)*. Other acidic metal halides such as the anhydrous chlorides of titanium, indium, zinc, or iron(III) can be substituted for the anhydrous aluminium halides indicated above *(77)*. A similar type of reaction with the 1,3-dienes butadiene and 1,3-cyclohexadiene using the milder catalyst $ZnCl_2$ to minimize polymerization of the diene gives complexes of the type $[(\eta^5\text{-}C_5H_5)Fe(CO)_2(1,2\text{-}\eta^2\text{-diene})]^+$, in which only one carbon-carbon double bond of the diene is coordinated to the iron atom *(77)*. In another variation of this reaction type, the treatment of $(C_5H_5)Fe[P(OC_6H_5)_3]_2I$ with ethylene in the presence of silver tetrafluoroborate as the halogen acceptor was found to give the tetrafluoroborate salt of the cation $\{(C_5H_5)Fe[P(OC_6H_5)_3]_2(C_2H_4)\}^+$ *(95)*.

A second method which was subsequently developed for the preparation of $[(\eta^5\text{-}C_5H_5)Fe(CO)_2(\eta^2\text{-olefin})]^+$ cations is the abstraction of hydride from the β-carbon atom of the alkyls $R'\text{-}CH_2\text{-}CH(R)\text{-}Fe(CO)_2(\eta^5\text{-}C_5H_5)$ with triphenylmethyl salts such as the perchlorate or the safer tetrafluoroborate *(92,94)*. The ethylene and propene complexes have both been prepared in this manner. The success of this method is a consequence of the relative ease of hydride removal from *sp*^{3} carbon atoms directly bonded to carbon atoms, which themselves are bonded to transition metals. The necessary iron alkyls can be prepared by reactions of the halides $R'\text{-}CH_2\text{-}CH(R)X$ with $NaFe(CO)_2\text{-}(C_5H_5)$.

A third method for preparing $[(\eta^5\text{-}C_5H_5)Fe(CO)_2(\eta^2\text{-olefin})]^+$ derivatives is the protonation of η^1-allyl derivatives, which can be prepared from $NaFe(CO)_2(C_5H_5)$ and the corresponding allyl halides or less frequently from $HFe(CO)_2\text{-}(C_5H_5)$ and 1,3-dienes *(80,90,93)*. Thus, treatment of the η^1-allyl derivatives $R\text{-}CH{=}CH\text{-}CH_2\text{-}Fe(CO)_2(C_5H_5)$ with strong proton donors such as hydrogen chloride in hydrocarbon solvents or tetrafluoroboric acid in acetic or propionic anhydride gives the corresponding olefin complexes $[(\eta^5\text{-}C_5H_5)Fe(CO)_2(\eta^2\text{-}CH_2{=}CH\text{-}CH_2R)]^+$ (R = H, CH_3). Electrophiles other than protons have recently been shown also to react readily with (η^1-allyl)dicarbonyl(cyclopentadienyl)iron derivatives to give various substituted $[(\eta^5\text{-}C_5H_5)Fe(CO)_2(\eta^2\text{-olefin})]^+$ cations *(59)*. Reactions of the unsubstituted allyl derivative $CH_2{=}CH\text{-}CH_2\text{-}Fe(CO)_2(C_5H_5)$ with various electrophiles, for example, proceed as follows:

(1) Treatment with trimethyloxonium tetrafluoroborate in liquid SO_2 gives $[(\eta^5\text{-}C_5H_5)Fe(CO)_2(\eta^2\text{-}CH_2{=}CH\text{-}CH_2\text{-}SO_2CH_3)]^+\text{-}[BF_4]^-$. In this connection spectroscopic and conductivity data *(44)* have been used to infer the presence of closely related unstable zwitterionic olefin complexes such as $[(C_5H_5)\text{-}Fe^+(CO)_2(H_2C{=}CH\text{-}CH_2SO_2^-)]$ in solutions of $CH_2{=}CH\text{-}CH_2\text{-}Fe(CO)_2\text{-}(C_5H_5)$ in liquid sulfur dioxide. Such a zwitterionic intermediate should be readily alkylated, possibly at the sulfur atom, by trimethyloxonium tetrafluoroborate.

(2) Treatment with trimethyloxonium tetrafluoroborate in dichloromethane solution proceeds differently from the corresponding reaction in liquid sulfur dioxide to give the 1-butene derivative $[(\eta^5\text{-}C_5H_5)Fe(CO)_2(\eta^2\text{-}CH_2{=}CH\text{-}CH_2\text{-}CH_3)]^+\text{-}[BF_4]^-$.

(3) Treatment with tropylium tetrafluoroborate in dichloromethane gives $[(\eta^5\text{-}C_5H_5)Fe(CO)_2(\eta^2\text{-}CH_2{=}CH\text{-}CH_2\text{-}C_7H_7)]^+[BF_4]^-$

(4) Treatment with trichlorocyclopropenium hexachloroantimonate in dichloromethane gives $[(\eta^5\text{-}C_5H_5)Fe(CO)_2(\eta^2\text{-}CH_2{=}CH\text{-}$

CH_2-C_3Cl_3)]$^+$[$SbCl_6$]$^-$.
(5) Treatment with dialkoxycarbonium ions such as [(CH_3O)$_2$-CH]$^+$[PF_6]$^-$ in dichloromethane gives [(η^5-C_5H_5)Fe(CO)$_2$(η^2-CH_2=CH-CH_2-CH(OCH_3)$_2$)]$^+$[PF_6]$^-$.
(6) Treatment with bromine in dichloromethane at -78°C followed by treatment with hexafluorophosphoric acid etherate at the low temperature does not cleave the iron-carbon bond but instead gives [(η^5-C_5H_5)Fe(CO)$_2$(η^2-CH_2=CH-CH_2Br)]$^+$[PF_6]$^-$.

Protonation of other R-Fe(CO)$_2$(C_5H_5) derivatives sometimes can give [(η^5-C_5H_5)Fe(CO)$_2$(η^2-olefin)]$^+$ derivatives containing rather unusual olefinic ligands. Thus, the reaction of NaFe(CO)$_2$(C_5H_5) with ClCH_2-CN gives the cyanomethyl derivative NC-CH_2-Fe(CO)$_2$(C_5H_5). Protonation of this cyanomethyl derivative using hydrogen chloride in hydrocarbon solvents gives the cation [(η^5-C_5H_5)Fe(CO)$_2$(CH_2=C=NH)]$^+$ containing a complexed ketenimine *(80)*. Similarly, the reaction of HFe(CO)$_2$(C_5H_5) with CH_2=CH-CN gives the adduct (NC)(CH_3)-CH-Fe(CO)$_2$(C_5H_5), which on protonation gives the substituted ketenimine complex [(η^5-C_5H_5)Fe(CO)$_2$(CH_3-CH=C=NH)]$^+$ *(8)*. Protonation of (C_5H_5)Fe(CO)$_2$-CH_2-CHO with hydrogen chloride in an inert solvent with trifluoromethanesulfonic acid in dichloromethane, or with hexafluorophosphoric acid diethyl etherate gives the vinyl alcohol complex [(η^5-C_5H_5)-Fe(CO)$_2$(η^2-CH_2=CH-OH)]$^+$ *(9,58)*. Protonation of the propargyl complexes R-C≡C-CH_2-Fe(CO)$_2$(C_5H_5) (R = CH_3, C_6H_5) similarly gives the coordinated allene complexes [(η^5-C_5H_5)Fe(CO)$_2$-(η^2-CH_2=C=CHR)]$^+$ in which the unsubstituted carbon-carbon double bond is coordinated to the iron *(20,151)*.

Some complexes of the type [(η^5-C_5H_5)Fe(CO)$_2$(η^2-olefin)]$^+$ can be prepared by olefin exchange from another more readily prepared derivative such as the isobutene complex *(80)* [(η^5-C_5H_5)Fe(CO)$_2$(η^2-CH_2=C(CH_3)$_2$]$^+$[BF_4]$^-$ which has been recommended for this purpose *(59,81,178)*. Thus, heating the isobutene complex briefly (~10 min) with other olefins in 1,2-dichloroethane solution has been used to prepare [(η^5-C_5H_5)Fe(CO)$_2$-(η^2-olefin)]$^+$[BF_4]$^-$ complexes containing ethylene, cyclohexene, cycloheptene, cyclooctene, 1,3-cyclohexadiene, 1,4-cyclohexadiene, norbornadiene, and acenaphthylene. A similar reaction of this isobutene complex with 1,5-cyclooctadiene gives almost equal amounts of the 1:1 and 2:1 metal-olefin complexes. This reaction has been used to protect one double bond of norbornadiene from bromination, hydrogenation, and mercuration reactions. The free olefin can be removed from the iron atom by treatment with sodium iodide in acetone after completion of the desired reaction *(178)*.

This method of preparing [(η^5-C_5H_5)Fe(CO)$_2$(η^2-olefin)]$^+$ complexes by olefin exchange with the isobutene complex [(η^5-C_5H_5)Fe(CO)$_2$(η^2-CH_2=C(CH_3)$_2$]$^+$[BF_4]$^-$ is necessarily limited to the preparation of those olefin complexes which are thermally stable in solution under conditions used to effect exchange with the isobutene complex *(195)*. An alternative method without this limitation which is useful for the preparations of olefin complexes having a greater variety of functional groups is based on the reaction of $NaFe(CO)_2(C_5H_5)$ with epoxides *(82,196)* according to scheme [14]. If the intermediate

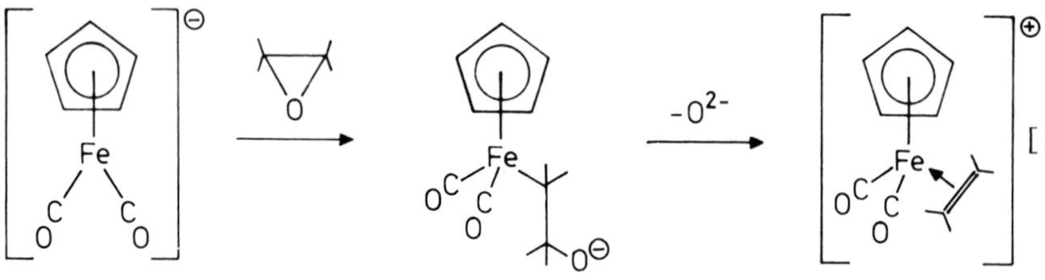

[14]

alkoxide is converted to the olefin complex by treatment with fluoroboric acid, then the stereochemistry of the olefin is retained *(82)*. However, if the intermediate alkoxide is converted to the corresponding olefin complex by pyrolysis, then the stereochemistry of the olefin is inverted *(196)*. Examples of $[(\eta^5\text{-}C_5H_5)Fe(CO)_2(\eta^2\text{-olefin})]^+$ cations which have been prepared by this method include derivatives of ethylene, propylene, 1-butene, *cis*- and *trans*-2-butene, *cis*- and *trans*-2-pentene, cyclohexene, butadiene (1,2-η^2-coordinated), acrolein, ethyl crotonate, 4-vinyl-cyclohexene *(82,196)*, and methyl vinyl ketone *(193)*.

The ethylene complex $[(\eta^5\text{-}C_5H_5)Fe(CO)_2(\eta^2\text{-}C_2H_4)]^+$ has also been obtained by treatment of $NaFe(CO)_2(C_5H_5)$ with $ClCH_2OCH_3$ followed by protonation with tetrafluoroboric acid. This reaction is believed to proceed through the unstable carbene complex $[(C_5H_5)Fe(CO)_2(CH_2)]^+$ *(115)*. However, this reaction does not appear to have any preparative value since the reactions discussed above provide several much better routes to the ethylene complex.

Some $[(\eta^5\text{-}C_5H_5)Fe(CO)_2(\eta^2\text{-olefin})]^+$ complexes derived from olefins unstable in the free state have been prepared by using indirect methods. Thus, rather unstable complexes of cyclobutadiene and benzocyclobutadiene appear to be accessible. Reaction of $NaFe(CO)_2(C_5H_5)$ with *cis*-3,4-dichlorocyclobutene at -78°C in a 1:1 molar ratio gives the thermally unstable complex <u>102</u> (X = Cl). The cyclobutadiene complex <u>103</u> appears to be generated by chloride abstraction from <u>102</u> (X = Cl) with silver hexafluorophosphate in dichloromethane *(202,203)*, but it is too unstable for isolation in the pure state. Its presence in the reaction mixture can be demonstrated by isolating stable products from its Diels-Alder reaction with cyclopentadiene and 1,3-diphenylisobenzofuran,

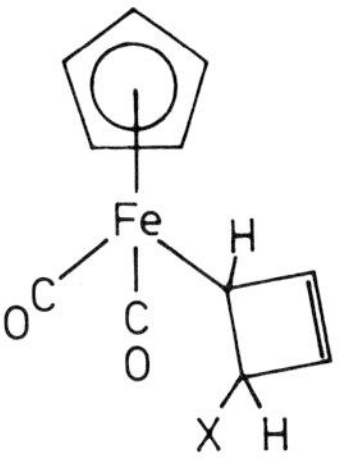

102

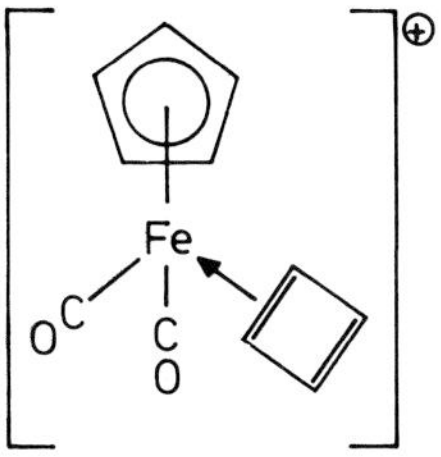

103

and from its dimerization to form the binuclear *syn*-tricyclooctadiene complex 104. The Diels-Alder adducts and complex

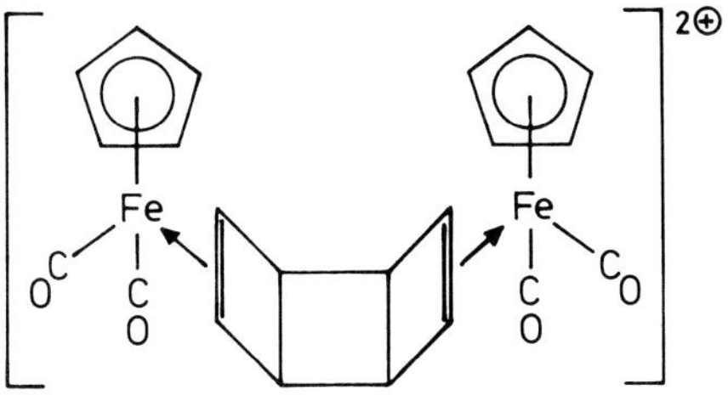

104

104 were characterized by reductive demetallation with NaFe-$(CO)_2(C_5H_5)$ to give known olefins and $[(C_5H_5)Fe(CO)_2]_2$. This demetallation method may prove to be useful for efficiently liberating the olefin from other $[(\eta^5-C_5H_5)Fe(CO)_2(\eta^2-olefin)]^+$ complexes. The much more stable binuclear cyclobutadiene complex $\{[(\eta^5-C_5H_5)Fe(CO)_2]_2-\mu-(1,2-\eta^2:3,4-\eta^2-C_4H_4)\}^{2+}$ has also been prepared *(203)*.

Some related chemistry on cyclobutenyliron derivatives also leads to derivatives of the type $[(\eta^5-C_5H_5)Fe(CO)_2(\eta^2-olefin)]^+$ *(204)*. For example, protonation of the binuclear complex 102 (X = $Fe(CO)_2(C_5H_5)$) with tetrafluoroboric acid in acetic anhydride gives a salt of stoichiometry $\{[(C_5H_5)Fe(CO)_2]_2(C_4H_5)\}^+[BF_4]^-$. The infrared spectrum indicates two non-equivalent $Fe(CO)_2(C_5H_5)$ groups suggesting structure 105.

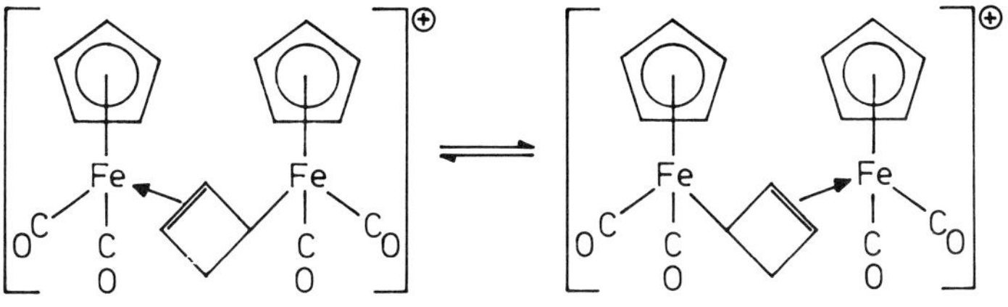

105a 105b

However, the ^{1}H-NMR spectrum indicates that 105 is a fluxional system with rapid interchange between the equivalent structures 105a and 105b. Protonation of the binuclear complex $(C_5H_5)Fe(CO)_2$-CH=CH-CH=CH-$Fe(CO)_2(C_5H_5)$ with hydrogen chloride in dichloromethane gives the binuclear butadiene dication $\{[(\eta^5\text{-}C_5H_5)Fe(CO)_2]_2\text{-}\mu\text{-}(1,2\text{-}\eta^2\text{:}3,4\text{-}\eta^2\text{-}C_4H_6)\}^{2+}$ in which the two butadiene double bonds are individually coordinated to different iron atoms *(204)*.

The benzocyclobutadiene complex 106 might be expected to be obtained by hydride abstraction from the benzocyclobutenyl complex 107 (R = H). However, hydride is abstracted

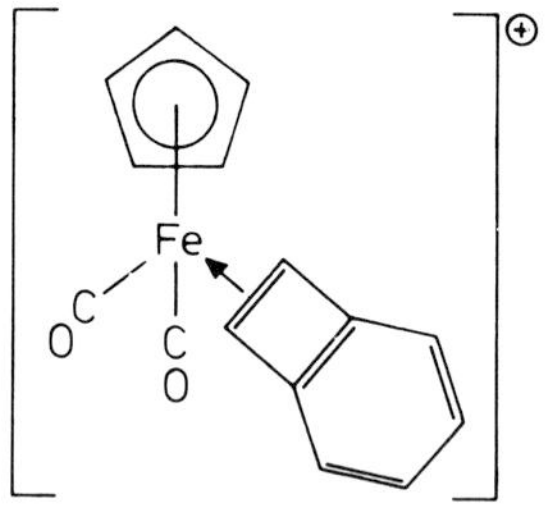

106

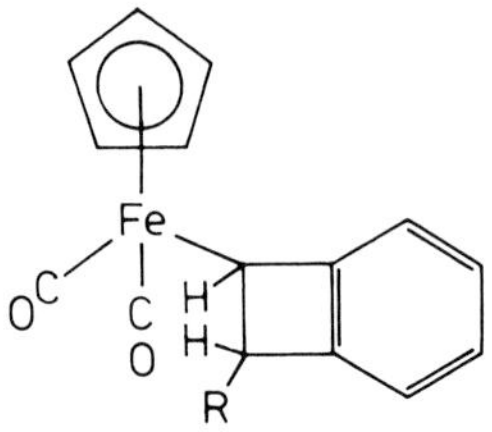

107

anomalously from the α- rather than the β-carbon atom of this benzocyclobutenyl complex 107 (R = H) to give the metallocarbenium ion *(48,200,201,205)*. However, removal of $[(C_5H_5)\text{-}Fe(CO)_2]^-$ from the binuclear complex 107 (R = $Fe(CO)_2(C_5H_5)$) *(125)* gives the unstable benzocyclobutadiene complex 106 which can be identified by its reaction with the nucleophilic reagents $LiBH_4$ in tetrahydrofuran, sodium bicarbonate in methanol, and triphenylphosphine to give the *trans*-benzocyclobutenyl derivatives 107 (R = H, OCH_3, and $(C_6H_5)_3\overset{+}{P}$, respectively) *(201,205)*.

Heptafulvene is another unstable hydrocarbon that has been stabilized by formation of a $[(C_5H_5)Fe(CO)_2(\text{olefin})]^+$ derivative *(119)*. Treatment of 7-cycloheptatrienylmethyl

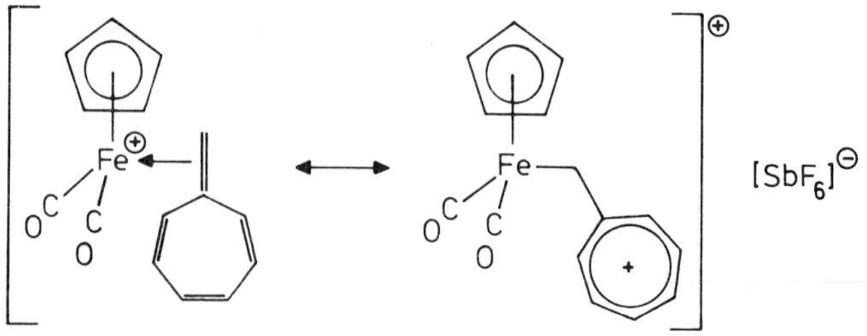

108a 108b

p-toluenesulfonate with $NaFe(CO)_2(C_5H_5)$ gives the 7-cycloheptatrienylmethyliron derivative $(C_7H_7CH_2)Fe(CO)_2(C_5H_5)$. Abstraction of hydride from this complex with triphenylmethyl hexafluoroantimonate in dichloromethane gives red-black crystals of the hexafluoroantimonate of the heptafulvene complex $[(C_5H_5)Fe(CO)_2(C_7H_6{=}CH_2)]^+[SbF_6]^-$. Spectroscopic data and an X-ray structure analysis require that 108 be represented as a hybrid structure (108a $\leftrightarrow$ 108b) rather than as an η^2-alkene complex (108a).

Some cyclopentadienyl-olefin-iron complexes of types other than $[(\eta^5\text{-}C_5H_5)Fe(CO)_2(\eta^2\text{-olefin})]^+$ have also been prepared. Reaction of $NaFe(CO)_2(C_5H_5)$ with *p*-CH_3-C_6H_4-SO_2-OCH_2-$C(CH_3)_2$-$CH{=}CH_2$ gives a rather unstable σ-alkyliron complex $CH_2{=}CH$-$C(CH_3)_2$-CH_2-$Fe(CO)_2(C_5H_5)$ (109). Ultraviolet irradiation of this complex results in decarbonylation to give the cyclic olefin complex 110. Upon standing complex 109 decomposes to give the orange crystalline cyclic acyl derivative 111 *(96)*. The methyl groups on the carbon adjacent to the car-

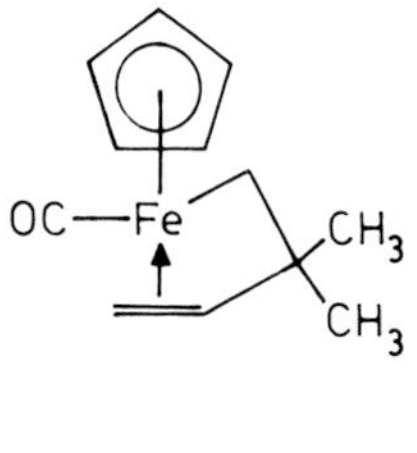

110

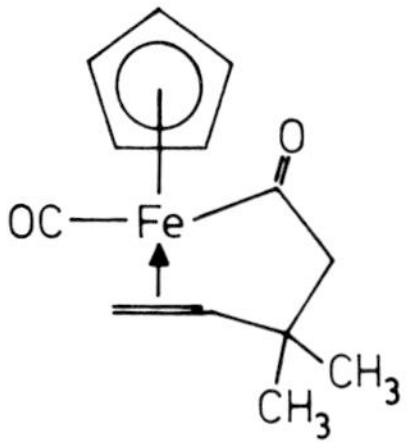

111

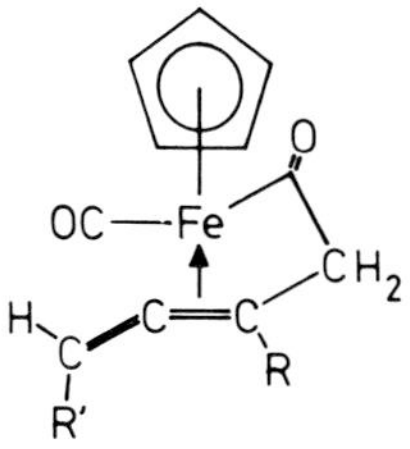

112

bon-carbon double bond in these systems are essential to avoid hydrogen shift reactions to give η^3-allyl derivatives. Reaction of $NaFe(CO)_2(C_5H_5)$ with the halides R'-$CH{=}C{=}C(R)$-CH_2-CH_2Br in tetrahydrofuran at room temperature gives the liquid σ-alkyl complexes R'-$CH{=}C{=}C(R)$-CH_2-CH_2-$Fe(CO)_2(C_5H_5)$ ($R = R' = H$; $R = H$, $R' = CH_3$; $R = CH_3$, $R' = H$). These compounds in tetrahydrofuran at 30°C rearrange to the crystalline isomeric derivatives 112 in which the allenic carbon chain is bonded to the iron atom through both an acyl-metal σ-bond and an η^2-olefin-metal bond *(21,197)*.

Ultraviolet irradiation of the triphenyltin derivative $(C_6H_5)_3SnFe(CO)_2(C_5H_5)$ with ethylene in benzene solution is reported *(143)* to give the monosubstituted complex $(C_6H_5)_3Sn$-$Fe(CO)(\eta^2\text{-}C_2H_4)(\eta^5\text{-}C_5H_5)$. Ultraviolet irradiation of the η^1-cyclopentadienyl derivative 113 with hexafluorobut-2-yne in hexane solution results in a Diels-Alder addition of the alkyne to the η^1-cyclopentadienyl ring followed by rearrangement of the metal coordination to give an 80-90 % yield of crystalline 114 containing a chelating norbornadiene ligand

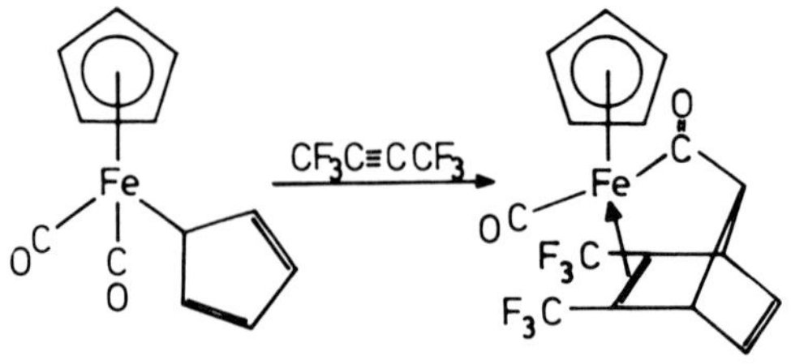

113 114

which bonds to the iron both through one carbon-carbon double bond and through an acyl carbonyl group from its 7-carbon atom *(62)*.

B. REACTIONS OF DICARBONYL-CYCLOPENTADIENYL-OLEFIN-IRON CATIONS AND RELATED COMPOUNDS WITH NUCLEOPHILES

Compounds of the type $[(\eta^5\text{-}C_5H_5)Fe(CO)_2(\eta^2\text{-olefin})]^+$ react with nucleophiles in three different ways as exemplified by the equations [15], [16], and [17] given below for the propene complex:

(1) Addition to the olefin to form a β-substituted alkyl derivative (eq. [15]).

$$[(\eta^5\text{-}C_5H_5)Fe(CO)_2(\eta^2\text{-}CH_2{=}CH\text{-}CH_3)]^+ + X^- \longrightarrow CH_3\text{—}CHX\text{—}CH_2\text{—}Fe(CO)_2(\eta^5\text{-}C_5H_5) \quad [15]$$

(2) Deprotonation of the coordinated olefin to form an η^1-allyl derivative (eq. [16]).

$$[(\eta^5\text{-}C_5H_5)Fe(CO)_2(\eta^2\text{-}CH_2{=}CH\text{-}CH_3)]^+ + X^- \longrightarrow CH_2{=}CH\text{—}CH_2\text{—}Fe(CO)_2(\eta^5\text{-}C_5H_5) + HX \quad [16]$$

(3) Displacement of the olefin to form a dicarbonyl-cyclo-

$$[(\eta^5\text{-}C_5H_5)Fe(CO)_2(\eta^2\text{-}CH_2{=}CH\text{—}CH_3)]^+ + X^- \longrightarrow X\text{—}Fe(CO)_2(\eta^5\text{-}C_5H_5) + CH_3\text{—}CH{=}CH_2 \quad [17]$$

pentadienyliron derivative (eq. [17]).

Use of an olefin in reaction [17] corresponds to the preparation of $[(\eta^5\text{-}C_5H_5)Fe(CO)_2(\eta^2\text{-olefin})]^+$ derivatives by the olefin displacement reaction as discussed above.

One of the methods for preparing $[(\eta^5\text{-}C_5H_5)Fe(CO)_2\text{-}(\eta^2\text{-olefin})]^+$ derivatives discussed above is the hydride abstraction from the β-carbon atoms in $R\text{-}Fe(CO)_2(C_5H_5)$ alkyls. This type of reaction is reversible. For example, reduction of the cations $[(\eta^5\text{-}C_5H_5)Fe(CO)_2(\eta^2\text{-}CH_2{=}CH\text{-}R)]^+$ (R = H and

CH_3) with sodium borohydride in tetrahydrofuran gives the corresponding alkyls CH_3-CHR-Fe$(CO)_2(C_5H_5)$ (R = H and CH_3) *(91)*. Since the propene complex $[(\eta^5-C_5H_5)Fe(CO)_2(\eta^2-CH_2=CH-CH_3)]^+$ can be prepared by hydride abstraction from the *n*-propyl complex $CH_3-CH_2-CH_2-Fe(CO)_2(C_5H_5)$, but is converted to the *i*-propyl complex $(CH_3)_2CH-Fe(CO)_2(C_5H_5)$ upon hydride reduction; the two step sequence of hydride abstraction followed by hydride addition can be useful for converting straight chain iron alkyls to the isomeric branched chain iron alkyls. The use of this method for preparing branched chain iron alkyls is further illustrated by the sodium borohydride reduction of the isobutene complex $[(\eta^5-C_5H_5)Fe(CO)_2-(\eta^2-CH_2=C(CH_3)_2)]^+$ to give the *tert*-butyl complex $(CH_3)_3C-Fe(CO)_2(C_5H_5)$ *(80)*, a compound not directly accessible from $NaFe(CO)_2(C_5H_5)$ and *tert*-butyl chloride *(94)*.

Reactions of $[(\eta^5-C_5H_5)Fe(CO)_2(\eta^2\text{-olefin})]^+$ complexes with other nucleophiles can give R-$Fe(CO)_2(C_5H_5)$ derivatives not directly accessible from $NaFe(CO)_2(C_5H_5)$ reactions *(150)*. Reaction of $[(\eta^5-C_5H_5)Fe(CO)_2(\eta^2-CH_2=CH_2)]^+$ with ammonia gives the dialkylated product $[(C_5H_5)Fe(CO)_2-CH_2-CH_2-]_2NH_2^+$ which regenerates the ethylene cation upon treatment with hydrochloric acid *(131)*. Reactions of $[(\eta^5-C_5H_5)Fe(CO)_2(\eta^2-CH_2=CH_2)]^+$ with sodium carbonate in methanol, with methylamine, and with *tert*-butyl mercaptan in the presence of potassium carbonate give the β-substituted iron alkyls $R-CH_2-CH_2-Fe(CO)_2(C_5H_5)$ (R = OCH_3, $NHCH_3$, and $SC(CH_3)_3$, respectively), which likewise regenerate the ethylene cation upon treatment with hydrochloric acid *(34,150)*. Reaction of $[(\eta^5-C_5H_5)-Fe(CO)_2(\eta^2-CH_2=CH_2)]^+[BF_4]^-$ with triphenylphosphine in nitromethane solution gives the air-stable phosphonium salt $[(C_5H_5)Fe(CO)_2-CH_2-CH_2-P(C_6H_5)_3]^+[BF_4]^-$ *(150)*.

The enolates lithium diethyl malonate, lithium diethyl methylmalonate, and lithium ethyl acetoacetate all react with $[(\eta^5-C_5H_5)Fe(CO)_2(\eta^2-CH_2=CH_2)]^+$ in tetrahydrofuran to form the corresponding complexes $(C_5H_5)Fe(CO)_2-CH_2-CH_2-R$ (R = $CH(CO_2C_2H_5)_2$, $C(CH_3)(CO_2C_2H_5)_2$, and $CH(COCH_3)(CO_2C_2H_5)$, respectively) *(191)*. The rather exotic nucleophile $[(C_5H_5)Mo(CO)_3-CH_2-CH_2O]^-$ reacts with $[(\eta^5-C_5H_5)Fe(CO)_2(\eta^2-CH_2=CH_2)]^+$ to give the bimetallic derivative $(C_5H_5)Mo(CO)_3-CH_2-CH_2-O-CH_2-CH_2-Fe(CO)_2(C_5H_5)$, which decomposes slowly over a period of weeks with ethylene evolution *(131)*.

The nature of the product obtained by reaction of $[(\eta^5-C_5H_5)Fe(CO)_2(\eta^2-CH_2=CH_2)]^+$ with cyanide seems to depend upon the reaction conditions. Reaction of $[(\eta^5-C_5H_5)Fe(CO)_2(\eta^2-CH_2=CH_2)]^+$ with potassium cyanide in acetone appears to give exclusively $(C_5H_5)Fe(CO)_2CN$ with ethylene elimination *(34)*. However, the corresponding reaction with tetraethylammonium cyanide in acetonitrile also gives a modest yield (21 %) of

$(C_5H_5)Fe(CO)_2$-CH_2-CH_2-CN *(131)*. The yields of 2-cyanoethyl complexes are higher from reactions of cyanide ion with (cyclopentadienyl)(ethylene)-iron cations in which one or both of the carbonyliron groups are substituted by trivalent phosphorus ligands, apparently because the coordinated ethylene is more firmly bound when carbonyl groups are replaced by weaker π-acceptors. Thus, the triphenylphosphite complex $\{(\eta^5$-$C_5H_5)Fe[P(OC_6H_5)_3]_2(\eta^2$-$CH_2{=}CH_2)\}^+$ gave a 69 % yield of the corresponding 2-cyanoethyl complex $(C_5H_5)Fe[P(OC_6H_5)_3]_2$-$CH_2$-$CH_2$-CN upon treatment with tetraethylammonium cyanide in acetonitrile *(131)*. Reaction of $[(\eta^5$-$C_5H_5)Fe(CO)P(C_6H_5)_3$-$(\eta^2$-$CH_2{=}CH_2)]^+[BF_4]^-$ with potassium cyanide in ethanol gave a 78 % yield of the corresponding 2-cyanoethyl derivative $(C_5H_5)Fe(CO)_2P(C_6H_5)_3$-$CH_2$-$CH_2$-CN *(187)*.

Some reactions of $[(\eta^5$-$C_5H_5)Fe(CO)_2(\eta^2$-$CH_2{=}CH_2)]^+$ with other pseudohalides have been investigated. Reaction of this cation with sodium azide was first believed *(34)* to give the cyanato complex $[(\eta^5$-$C_5H_5)Fe(CO)(\eta^2$-$C_2H_4)NCO]$. However, this reaction was subsequently shown to give the azido complex $(C_5H_5)Fe(CO)_2N_3$ with ethylene elimination *(192)*. The kinetics of the reactions of the cations $[(C_5H_5)Fe(CO)_2L]^+$ (L = CO, η^2-C_2H_4, and $P(C_6H_5)_3$) with azide were found to be first order with respect to both the iron cation and the azide anion *(87)*. The rate constants for the sequence L = CO, $P(C_6H_5)_3$, and C_2H_4 are in the ratio 1:67:300 *(87)*. Reaction of $[(C_5H_5)Fe(CO)_2$-$(CH_2{=}CH_2)]^+$ with potassium cyanate results in ethylene elimination to give $(C_5H_5)Fe(CO)_2NCO$ *(34)*.

Several reactions of the acenaphthylene cation 115 with

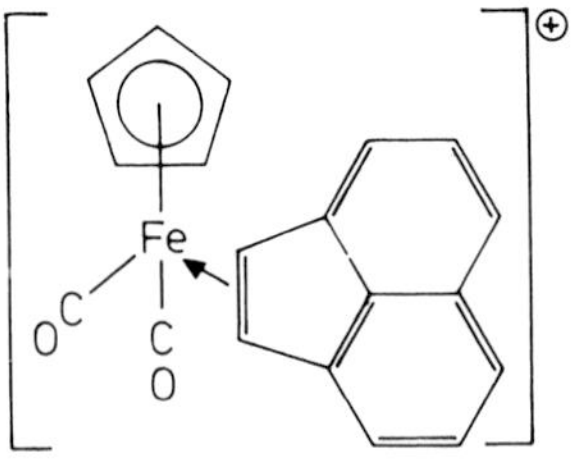

115

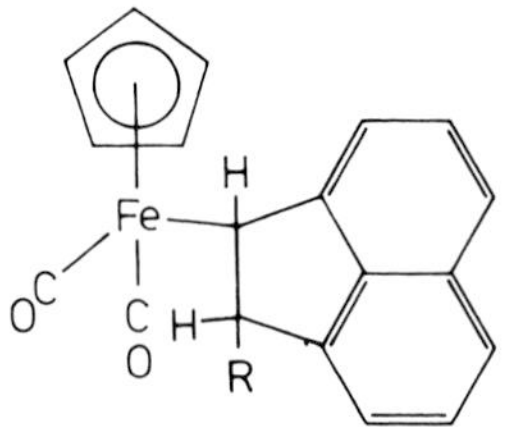

116

nucleophiles have been investigated *(179)*. Reactions of 115 with the nucleophiles $(CH_3)_3CSH$ in the presence of sodium bicarbonate and isobutyraldehyde pyrrolidine enamine result in stereospecific *trans*-addition to the coordinated acenaphthylene to give the alkyls 116 (R = $(CH_3)_3CS$ and H-CO-$C(CH_3)_2$ [after hydrolysis], respectively). However, reactions of 115 with the nucleophiles CH_3OH/Na_2CO_3, $(C_6H_5)_3P$, and sodium iodide result in liberation of acenaphthylene to give $[(C_5H_5)$-$Fe(CO)_2]_2$ (with methanol), $[(C_5H_5)Fe(CO)_2P(C_6H_5)_3]^+$ (with triphenylphosphine), and $(C_5H_5)Fe(CO)_2I$ (with sodium iodide).

Deprotonation of olefin-iron complexes of the type $[(\eta^5-C_5H_5)Fe(CO)_2(\eta^2-R-CH=CH-CH_2-R')]^+$ in dichloromethane solution with a tertiary amine such as triethylamine or diisopropylethylamine gives the corresponding (η^1-allyl)iron derivatives $R'-CH=CH-CHR-Fe(CO)_2(C_5H_5)$ *(59,83)*. This deprotonation occurs preferentially *exo* (*trans*) to the metal-ligand bond in contrast to the deprotonation of cationic (η^3-allyl)-iron complexes in which *endo* protons are preferentially removed *(83)*. This type of deprotonation occurs quantitatively with the cyclopentene and cyclohexene cations but fails with the corresponding cycloheptene cation. In the preferred conformation of the cycloheptene complex in which the pendant $(C_5H_5)Fe(CO)_2$ group lies *exo* to the ring, no allylic protons *trans* to the iron-olefin bond are available in sharp contrast to the cyclopentene and cyclohexene complexes.

The ability of $[(\eta^5-C_5H_5)Fe(CO)_2(\eta^2\text{-olefin})]^+$ derivatives to react readily with certain nucleophiles with liberation of the olefin can be useful for the preparation of certain $(C_5H_5)Fe(CO)_2R$ derivatives by nucleophilic substitution reactions. Reaction of $[(\eta^5-C_5H_5)Fe(CO)_2(\eta^2-CH_2=CH-CH_3)]^+$ $[PF_6]^-$ with halide-free methyllithium in diethyl ether gives the σ-methyl derivative $CH_3-Fe(CO)_2(C_5H_5)$ as the exclusive organoiron product *(7)*. However, reaction of the propene cation with phenyllithium in diethyl ether results both in deprotonation to give $CH_2=CH-CH_2-Fe(CO)_2(C_5H_5)$ (38 % yield) and in propene displacement to give $C_6H_5-Fe(CO)_2(C_5H_5)$ (45 % yield). These two products could be separated readily by chromatography. Formation of the η^1-allyl by-product in this reaction presumably could have been avoided by using the ethylene-iron complex rather than the propylene-iron complex as the starting material.

The cyclohexene complex $[(\eta^5-C_5H_5)Fe(CO)_2(\eta^2-C_6H_{10})]^+$ is a good starting material for preparing unusual compounds containing iron-boron bonds by reactions with appropriate nucleophiles *(207,234)*. Thus, reactions of this cyclohexene complex with the boron nucleophiles $[7,8-B_9H_{10}CHP]^-$, $[7,8-B_9H_{10}As_2]^-$, $[B_{10}H_{12}P]^-$, $[B_{10}H_{12}As]^-$, $[B_{10}H_{13}]^-$, and $[7,8-B_9C_2H_{12}]^-$ result in the formation of the corresponding (C_5H_5)-$Fe(CO)_2R$ derivative (R = group derived from the boron cage anion used) in at least 40 % yields. Many of these iron-boron derivatives could not be prepared from the corresponding boron nucleophiles and $(C_5H_5)Fe(CO)_2I$ *(234)*. This suggests some real preparative value for the cyclohexene complex in certain reactions with nucleophiles.

Some reactions of the allene complexes $[(\eta^5-C_5H_5)Fe$-$(CO)_2(\eta^2-CH_2=C=CHR)]^+$ ($R = CH_3$, C_6H_5) with nucleophiles have been investigated *(151)*. Sodium borohydride reacts with these cations in tetrahydrofuran by attacking the 1-carbon atom to give the substituted σ-vinyl derivatives R-CH=C(CH_3)-Fe$(CO)_2$-(C_5H_5) (117, R = CH_3, C_6H_5; X = H). Reactions of the allene cations with diethylamine in pentane proceeds similarly to give the aminovinyl derivatives R-CH=C[-CH_2-N$(C_2H_5)_2$]-Fe-$(CO)_2$(C_5H_5) (117, R = CH_3, C_6H_5; X = N$(C_2H_5)_2$). These reactions are not stereospecific since mixtures of *cis*- and *trans*-isomers are produced (117a and 117b). Reactions of the allene

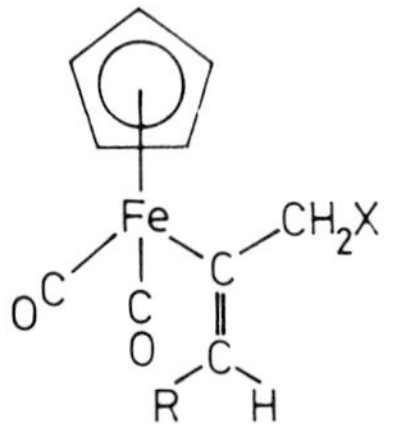

117a

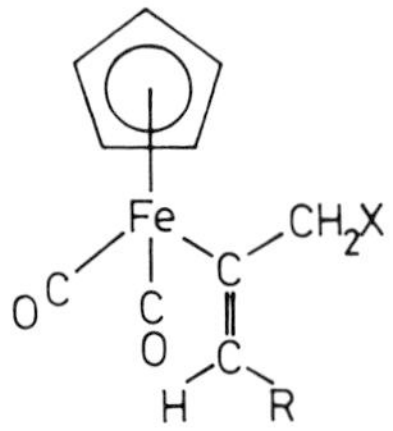

117b

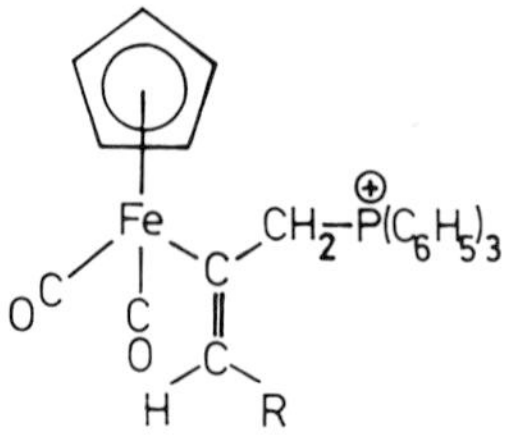

118

cations with triphenylphosphine also result in attack at the 1-carbon atom to give the corresponding phosphonium cations 118 (R = CH_3, C_6H_5). Reactions of the allene cations with sodium methoxide result in attack at the 2-carbon atom of the complexed allene to give the ketones R-CH_2-CO-CH_2-Fe$(CO)_2$-(C_5H_5) (119, R = CH_3, C_6H_5) as the major products. Reaction

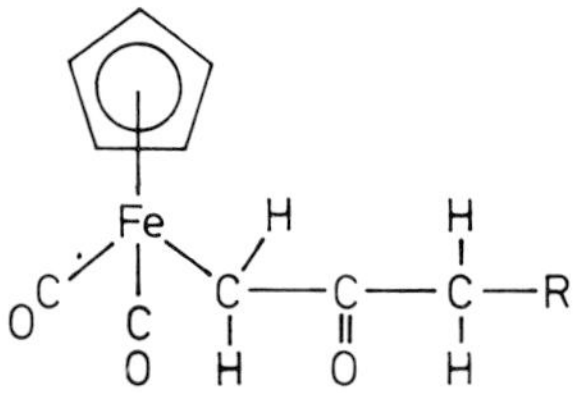

119

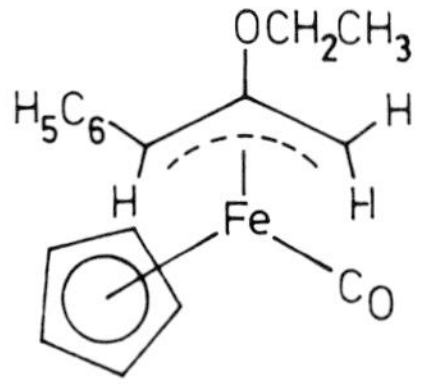

120

of $[(\eta^5\text{-}C_5H_5)Fe(CO)_2(\eta^2\text{-}CH_2{=}C{=}CH\text{-}C_6H_5)]^+$ with sodium ethoxide gives the orange crystalline η^3-allyl derivative 120.

The reactions of the methylvinylketone complex 121 with some nucleophiles follow a different pattern *(193)*. Thus, the reaction of an acetonitrile solution of the tetrafluoroborate of 121 with cyclohexanone lithium enolate at -78°C gives a 45 % yield of the adduct 122 (R = H) in a Michael-type reaction. Refluxing 122 in boiling methylene chloride in the presence of basic alumina results in the elimination of the dicarbonyl(cyclopentadienyl)iron unit to give the octalone 123 (R = H). The corresponding methyl derivatives 122 and 123 (R = CH_3) can be prepared by an analogous method using the regiospecifically generated enolate formed by addition of li-

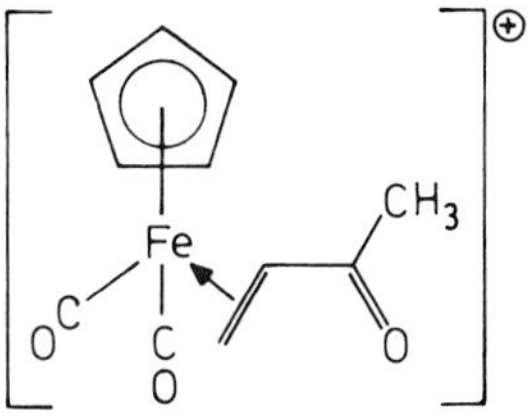

121

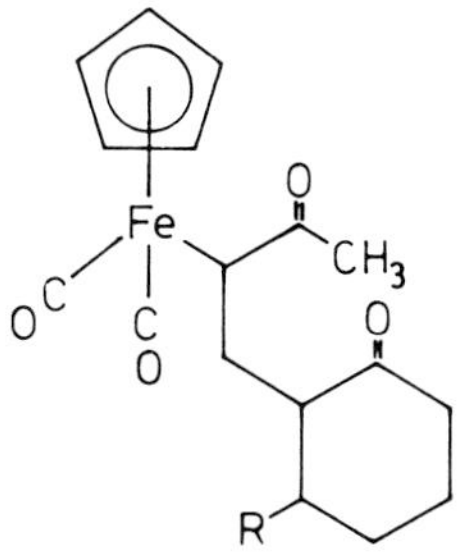

122

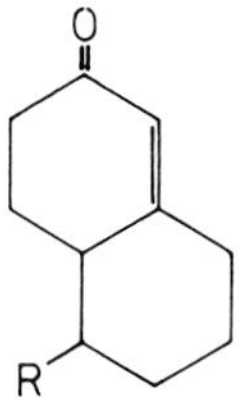

123

thium dimethylcuprate to cyclohexenone. Reactions of 121 with cyclohexanone enamines at 0°C also give 122 (R = H) *(193)*.

The ethylene complex $[(\eta^5\text{-}C_5H_5)Fe(CO)_2(\eta^2\text{-}C_2H_4)]^+$ failed to show any exchange with ^{14}C-labelled ethylene after 6 days at 30°C *(180)*. The ^{13}C-NMR spectrum of the propene complex $[(\eta^5\text{-}C_5H_5)Fe(CO)_2(\eta^2\text{-}CH_2{=}CH\text{-}CH_3)]^+$ has been reported *(6)*.

IX. CYCLOBUTADIENE-OLEFIN-IRON COMPLEXES

Ultraviolet irradiation of (cyclobutadiene)tricarbonyliron, $(\eta^4\text{-}C_4H_4)Fe(CO)_3$, with the olefins dimethyl maleate and dimethyl fumarate gives the corresponding cyclobutadiene-olefin-iron complexes $(\eta^4\text{-}C_4H_4)Fe(CO)_2(\eta^2\text{-}L)$ (L = dimethyl maleate and dimethyl fumarate) *(186)*. Oxidation of these complexes with cerium(IV) leads to formation of the bicyclo[2.2.0]hexene derivatives identical to the known adducts of the olefins with the cyclobutadiene generated by oxidation of $(\eta^4\text{-}C_4H_4)Fe(CO)_3$ *(209)*.

X. CARBONYLIRON DERIVATIVES WITH BRIDGING VINYL GROUPS

Several types of compounds are known where a vinyl group bridges two metal atoms, at least one of which is iron. In these cases the vinyl group coordinates to one of the metals through its carbon-carbon double bond and to the second metal by forming a metal-carbon σ-bond.

The first compounds of this type were obtained in 1961 *(122)* by reactions of carbonyliron complexes with vinyl sulfides. Thus, reactions of $Fe_3(CO)_{12}$ with the vinyl sulfides $CH_2{=}CH\text{-}SR$ (R = CH_3, C_2H_5, $CH{=}CH_2$, and $CH(CH_3)_2$) give the red liquid hexacarbonyldiiron complexes $(RS\text{-}CH{=}CH_2)Fe_2(CO)_6$. Their ^1H-NMR spectra indicate the structures 124 (R = CH_3, C_2H_5, $CH{=}CH_2$, and $CH(CH_3)_2$) in which a vinyl group bridges two iron atoms.

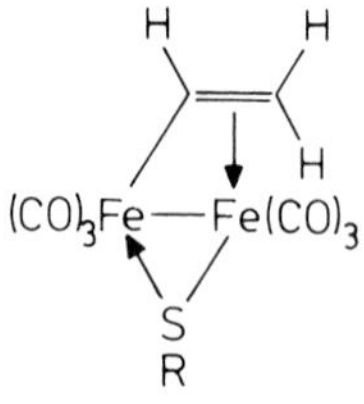

124

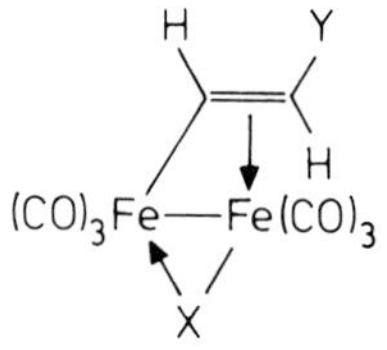

125

A series of compounds was subsequently prepared with structures similar to 124 but with a halogen bridge instead of the alkylthio bridge. Thus, ultraviolet irradiation of the *cis*- or *trans*-1,2-dihaloethylene-tetracarbonyliron complexes $(XCH{=}CHY)Fe(CO)_4$ (X = Y = Cl or Br; X = Br, Y = F) leads to the products 125 (X = Y = Cl or Br; X = Br, Y = F) containing both a bridging *trans*-2-halovinyl group and a bridging halogen atom *(100,140)*. The X-ray crystal structure of the compound 125 (X = Y = Br) has been determined *(148)*. This represents the first X-ray confirmation of structures of this type.

Nesmeyanov and co-workers have prepared compounds in which a vinyl group acts as a bridge between a carbonyliron group and a cyclopentadienyl-metal unit containing iron or tungsten. For example, reactions of the 2-chlorovinyl ketones R-CO-CH=CHCl with $NaFe(CO)_2(C_5H_5)$ give the derivatives R-CO-CH=CH-$Fe(CO)_2(C_5H_5)$ (R = CH_3 and C_6H_5) containing iron-carbon σ-bonds. These cyclopentadienyliron derivatives then react

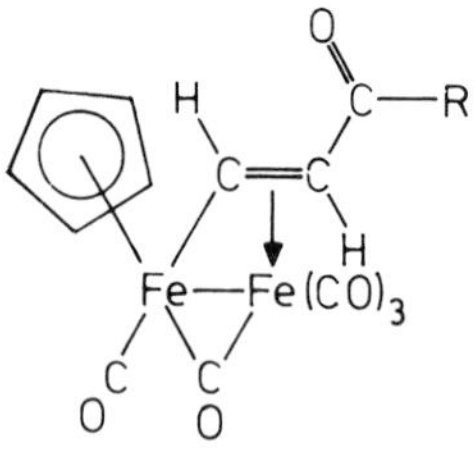

126

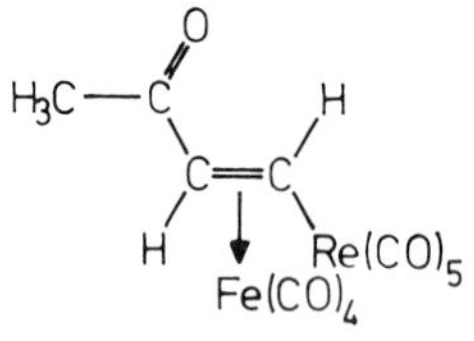

127

with $Fe_2(CO)_9$ in benzene at 35-40°C to give the green binuclear derivatives $(\eta^5\text{-}C_5H_5)(CO)Fe(\mu\text{-}CO)(\mu\text{-}\eta^1{:}\eta^2\text{-}CH{=}CH\text{-}CO\text{-}R)\text{-}Fe(CO)_3$ *(168)*. X-ray crystallography of the methyl derivative *(4,5)* indicates structure 126 (R = CH_3) containing a bridging $\eta^1{:}\eta^2$-vinyl group, a bridging carbonyl group, and a metal-metal bond.

Some related compounds can be prepared in which a substituted vinyl group bridges an iron atom with another metal *(171)*. In some cases, however, bridging vinyl derivatives are formed in which the metal-metal bond and the bridging carbonyl group are both absent. Thus, the reaction of *trans*-CH_3-CO-CH=CH-Re$(CO)_5$ with $Fe_2(CO)_9$ gives 127 which is a compound of this type. Reaction of the cyclopentadienyltungsten derivative C_6H_5-CO-CH=CH-W$(CO)_3(C_5H_5)$ with $Fe_2(CO)_9$ gives products of both types, 128 and 129. In addition the bimetallic

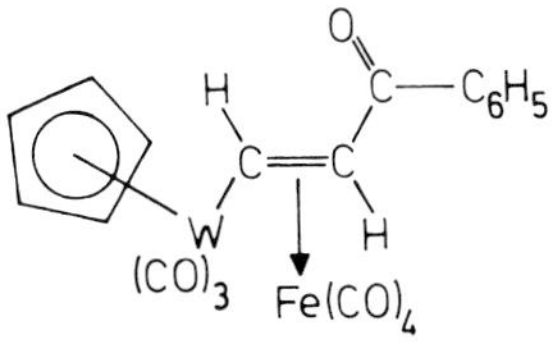

128

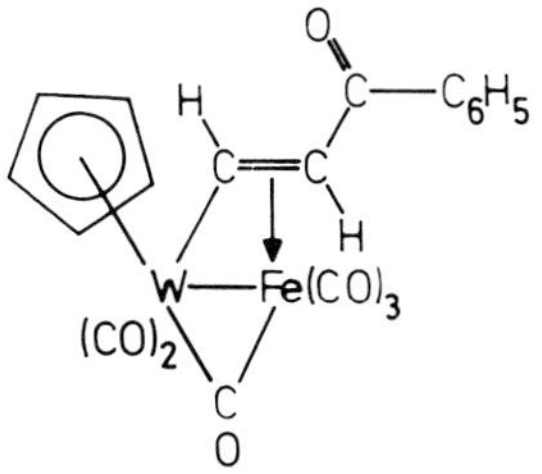

129

(enone)Fe$(CO)_3$ complex 130 is obtained.

The σ-acyliron derivatives R-CH=CH-CO-Fe$(CO)_2(C_5H_5)$ are also useful starting materials for preparing derivatives containing bridging vinyl groups *(171)*. Reactions of R-CH=CH-CO-Fe$(CO)_2(C_5H_5)$ (R = H and C_6H_5) with $Fe_2(CO)_9$ in benzene at 40°C give the tetracarbonyliron complexes 131 (R = H, C_6H_5).

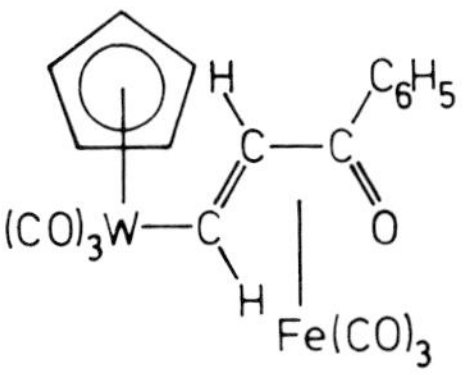

130

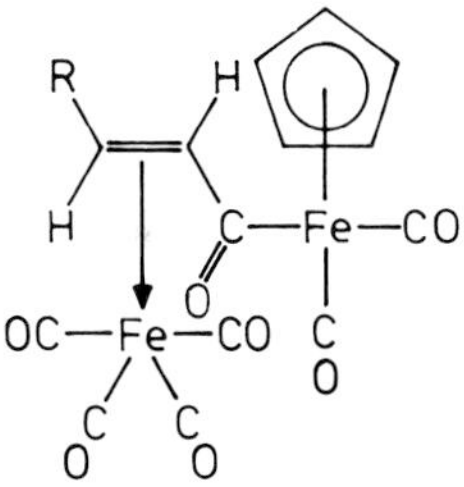

131

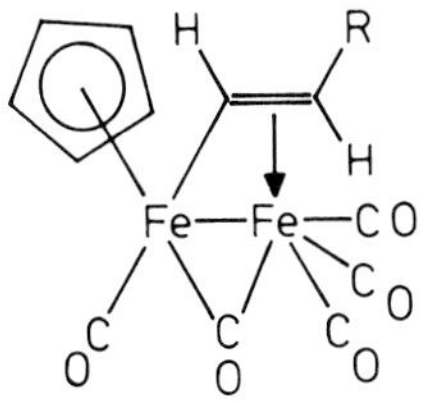

132

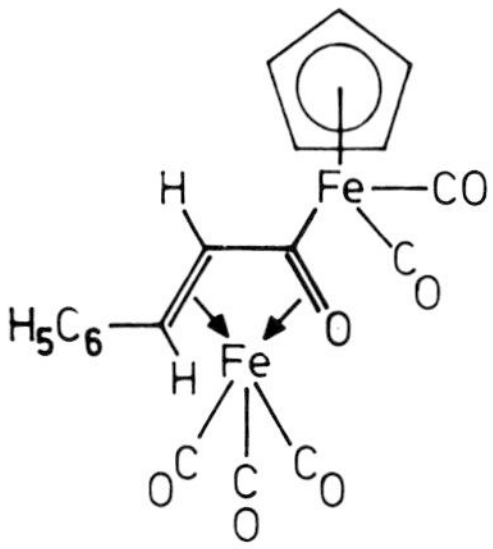

133

Heating the cinnamoyl derivative 131 (R = C_6H_5) in benzene at 60°C results in the loss of two carbonyl groups to give the dark green styryl derivative 132 (R = C_6H_5), which can also be prepared by heating C_6H_5-CH=CH-Fe(CO)$_2$(C_5H_5) with $Fe_2(CO)_9$ in benzene. Ultraviolet irradiation of the cinnamoyl derivative 131 proceeds entirely differently to give the orange (enone)Fe(CO)$_3$ complex η^4-[C_6H_5-CH=CH-CO-Fe(CO)$_2$(C_5H_5)]-Fe(CO)$_3$ (133) in which an acyl carbonyl is σ-bonded to one iron atom and η^2-bonded to another one. The green unsubstituted bridging vinyl derivative 132 (R = H) can be prepared by heating CH_2=CH-Fe(CO)$_2$(C_5H_5) with $Fe_2(CO)_9$.

XI. REACTIONS OF IRON COMPLEXES WITH TETRACYANOETHYLENE

The chemistry of tetracyanoethylene is completely different from that of ethylene or even tetrafluoroethylene because of the combined electron-withdrawing and mesomeric effects of the four cyano substituents. Therefore, its iron chemistry is best discussed separately.

Ferrocene forms a charge-transfer complex with tetracyanoethylene *(194)*. The X-ray crystal structure of this 1:1 complex *(1)* indicates an interaction between one of the cyclopentadienyl rings of ferrocene and the tetracyanoethylene carbon-carbon double bond in this complex with no new true chemical bond formation. Tetracyanoethylene is well-known to form similar charge-transfer complexes with various aromatic hydrocarbons.

Transition metal complexes of the type R-Fe(CO)$_2$(C_5H_5) react readily with tetracyanoethylene, but in most cases insertion products rather than olefin complexes are obtained *(218,219)*. Thus, the alkyls R-Fe(CO)$_2$(C_5H_5) (R = CH_3, C_2H_5, n-C_3H_7, CH_2-C_6H_5, and CH(CH_3)(C_6H_5)) react with tetracyanoethylene in dichloromethane at 25°C to give mixtures of the cyanoalkyls R-C(CN)$_2$-C(CN)$_2$-Fe(CO)$_2$(C_5H_5) and the isomeric keteniminates R-C(CN)$_2$-C(CN)=C=N-Fe(CO)$_2$(C_5H_5). However, the substituted alkyls R-Fe(CO)(PR'_3)(C_5H_5) (R' = C_6H_5, R = CH_3, C_2H_5, and n-C_3H_7; R' = n-C_4H_9 and OC_6H_5, R = CH_3) react with

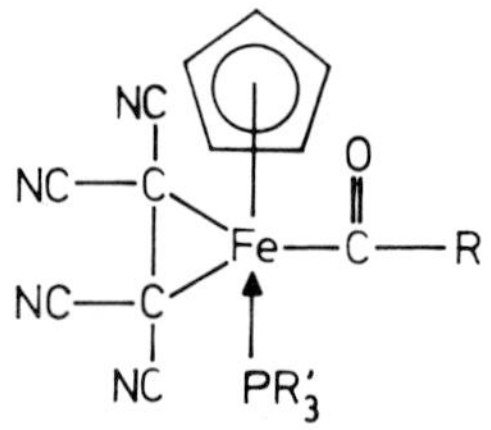

134

tetracyanoethylene in benzene at 5-10°C to form green 1:1 adducts. These adducts correspond to tetracyanoethylene olefin complexes of acyliron derivatives, but are best represented as the metallacyclopropane derivatives 119.

XII. OLEFIN-IRON COMPLEXES CONTAINING NEITHER CARBONYL NOR CYCLOPENTADIENYL LIGANDS

The analogy between the chemistry of $Fe(CO)_5$ and the corresponding trifluorophosphine complex $Fe(PF_3)_5$ suggested a study of the reactions of the latter with various olefins *(145)*. Ultraviolet irradiation of $Fe(PF_3)_5$ in diethyl ether solution with several olefins containing electronegative substituents gives (η^2-olefin)$Fe(PF_3)_4$ derivatives (olefin = acrylonitrile, crotononitrile, styrene, and methyl acrylate) completely analogous to the (η^2-olefin)$Fe(CO)_4$ derivatives discussed earlier in this chapter. All of these (η^2-olefin)-$Fe(PF_3)_4$ complexes are intense yellow low-melting solids. They are remarkably air-stable and can be sublimed or distilled unchanged. The stability of the (η^2-olefin)$Fe(PF_3)_4$ derivatives may arise from the impossibility of their decomposition by olefin elimination to give the unknown $Fe_3(PF_3)_{12}$. This contrasts with the (η^2-olefin)$Fe(CO)_4$ derivatives, which have a relatively easy decomposition pathway to give $Fe_3(CO)_{12}$. Similar ultraviolet irradiation of $Fe(PF_3)_5$ with several α,β-unsaturated carbonyl compounds containing the structural unit -C=C-C=O leads to (η^4-enone)$Fe(PF_3)_3$ derivatives (enone = crotonaldehyde, methyl vinyl ketone, and even methyl methacrylate) which are more sensitive to air, moisture, and heat than the (η^2-olefin)$Fe(PF_3)_4$ derivatives.

Some complexes of the type $(bid)_2Fe(\eta^2$-olefin) are known in which bid is a bidentate ligand, particularly the ditertiary phosphine $(C_6H_5)_2P-CH_2-CH_2-P(C_6H_5)_2$ (abbreviated here as diphos). Reaction of iron(III) acetylacetonate with ethoxydiethylaluminium in the presence of diphos in diethyl ether at 0°C gives an unstable red complex. Repeated crystallization of this complex from benzene/ether gives the violet ethylene complex $(diphos)_2Fe(\eta^2-C_2H_4)$ *(108)*. Ultraviolet irradiation of $(diphos)_2Fe(\eta^2-C_2H_4)$ results in elimination of ethylene to give a product of stoichiometry $(diphos)_2Fe$. However, NMR indicates this product to contain an iron-hydrogen bond. It therefore must be formulated as $HFe[-C_6H_4-P(C_6H_5)-CH_2-CH_2-P(C_6H_5)_2](diphos)$. Hydrogen transfer from a ligand phenyl group to an iron atom thus appears to occur in the formation of this complex. Upon treatment with I_2 and H_2 the ethylene complex $(diphos)_2Fe(\eta^2-C_2H_4)$ undergoes oxidative addition reactions with ethylene elimination to give (diphos)$_2$FeI$_2$ and (diphos)$_2$FeH$_2$, respectively *(108)*. Reaction of (diphos)$_2$Fe(η^2-C_2H_4) with $HSiCl_3$ can give either (diphos)Fe-(H)($SiCl_3$), (diphos)$_2$Fe(H)($SiCl_3$), or (diphos)$_2$Fe($SiCl_3$)$_2$, in all cases with ethylene elimination, depending upon the reaction conditions *(149)*. In accord with its high reactivity, the ethylene complex (diphos)$_2$Fe(η^2-C_2H_4) is an active catalyst for the addition of ethylene to butadiene to form 1:1 adducts consisting of hexa-1,*cis*-4-diene, hexa-1,3-diene, hexa-2,4-diene, and hexa-1,5-diene, and 2:1 adducts *(107)*.

Several iron alkyl and hydride reactions with various olefins have also been reported. Reaction of the complex $H_2Fe(N_2)L_3$ (L = ethyldiphenylphosphine) with ethylene gives an unstable complex formulated as (H)C_2H_5-FeL_2L' (L' = solvent or L) which decomposes rapidly at 30°C. An ethylene complex $H_2Fe(C_2H_4)L_2L'$ is proposed as an intermediate *(25)*. The iron alkyl (dipy)$_2$Fe(C_2H_5)$_2$ appears to form olefin complexes with tetracyanoethylene, maleic anhydride, and acrylamide, but their stoichiometries and structures are obscure *(233)*.

Acknowledgement

A fellowship of the Max-Planck-Gesellschaft during the time this article was written at the Institut für Strahlenchemie im Max-Planck-Institut für Kohlenforschung (Mülheim a.d. Ruhr, Germany) is gratefully acknowledged.

REFERENCES

1. Adman, E., Rosenblum, M., Sullivan, S., and Margulis, T.N., *J. Amer. Chem. Soc.*, *89*, 4540 (1967).
2. Andrews, D.C., and Davidson, G., *J. Organometal. Chem.*, *35*, 161 (1972).
3. Andrews, D.C., and Davidson, G., *J. Organometal. Chem.*, *74*, 441 (1974).
4. Andrianov, V.G., and Struchkov, Yu.T., *Chem. Commun.*, *1968*, 1590.
5. Andrianov, V.G., and Struchkov, Yu.T., *Zh. Strukt. Khim.*, *9*, 845 (1968); *J. Struct. Chem.*, *9*, 737 (1968).
6. Aris, K.R., Aris, V., and Brown, J.M., *J. Organometal. Chem.*, *42*, C 67 (1972).
7. Aris, K.R., Brown, J.M., and Taylor, K.A., *J. Chem. Soc. Dalton Trans.*, *1973*, 2222.
8. Ariyaratne, J.K.P., and Green, M.L.H., *J. Chem. Soc.*, *1963*, 2976.
9. Ariyaratne, J.K.P., and Green, M.L.H., *J. Chem. Soc.*, *1964*, 1.
10. Aumann, R., *Angew. Chem.*, *83*, 175, (1971); *Angew. Chem. Int. Ed. Engl.*, *10*, 188 (1971).
11. Aumann, R., *Angew. Chem.*, *83*, 176, (1971); *Angew. Chem. Int. Ed. Engl.*, *10*, 189 (1971).
12. Aumann, R., *J. Organometal. Chem.*, *47*, C 29 (1973).
13. Aumann, R., *J. Amer. Chem. Soc.*, *96*, 2631 (1974).
14. Aumann, R., *J. Organometal. Chem.*, *66*, C 6 (1974).
15. Aumann, R., *J. Organometal. Chem.*, *77*, C 33 (1974).
16. Aumann, R., *Chem. Ber.*, *108*, 1974 (1975).
17. Banks, R.E., Harrison, T., Haszeldine, R.N., Lever, A.B.P., Smith, T.F., and Walton, J.B., *Chem. Commun.*, *1965*, 30.
18. Beagley, B., Schmidling, D.G., and Cruickshank, D.W.J., *Acta. Crystallogr.*, *B 29*, 1499 (1973).
19. Behrens, U., *J. Organometal. Chem.*, *107*, 103 (1976).
20. Benaim, J., Mêrour, J.Y., and Roustan, J.L., *C.R. Acad. Sci., Ser. C*, *272*, 789 (1971).
21. Benaim, J., Mêrour, J.Y., and Roustan, J.L., *Tetrahedron Lett.*, *1971*, 983.
22. Bennett, M.A., Robertson, G.B., Tomkins, I.B., and Whimp, P.O., *J. Chem. Soc. D, Chem. Commun.*, *1971*, 341.
23. Bennett, M.A., Robertson, G.B., Tomkins, I.B., and Whimp, P.O., *J. Organometal. Chem.*, *32*, C 19 (1971).
24. Ben-Shoshan, R., and Pettit, R., *J. Amer. Chem. Soc.*, *89*, 2231 (1967).
25. Bianco, V.D., Doronzo, S., and Aresta, M., *J. Organometal. Chem.*, *42*, C 63 (1972).
26. Bond, A., Green, M., Lewis, B., and Lowrie, S.F.W.,

J. Chem. Soc. D, Chem. Commun., 1971, 1230.
27. Bond, A., Lewis, B., and Green, M., *J. Chem. Soc. Dalton Trans., 1975,* 1009.
28. Bottrill, M., Goddard, R., Green, M., Hughes, R.P., Lloyd, M.K., Lewis, B., and Woodward, P., *J. Chem. Soc. Chem. Commun., 1975,* 253.
29. Bright, D., and Mills, O.S., *Chem. Commun., 1966,* 211.
30. Bright, D., and Mills, O.S., *J. Chem. Soc. A, 1971,* 1979.
31. Brodie, A.M., Johnson, B.F.G., Josty, P.L., and Lewis, J., *J. Chem. Soc. Dalton Trans., 1972,* 2031.
32. Brookhart, M., *American Chemical Society, Southeastern Regional Meeting,* Charleston (South Carolina) November 1973, Abstracts.
33. Burt, R., Cooke, M., and Green, M., *J. Chem. Soc. A, 1970,* 2975.
34. Busetto, L., Palazzi, A., Ros, R., and Belluco, U., *J. Organometal. Chem., 25,* 207 (1970).
35. Cais, M., and Maoz, N., *J. Chem. Soc. A, 1971,* 1811.
36. Cann, K., and Barborak, J.C., *J. Chem. Soc. Chem. Commun., 1975,* 190.
37. Cardaci, G., and Narciso, V., *J. Chem. Soc. Dalton Trans., 1972,* 2289.
38. Cardaci, G., *Int. J. Chem. Kinetics, 5,* 805 (1973).
39. Cardaci, G., *Inorg. Chem., 13,* 368 (1974).
40. Cardaci, G., *Inorg. Chem., 13,* 2974 (1974).
41. Cardaci, G., *J. Organometal. Chem., 76,* 385 (1974).
42. Cardaci, G., and Concetti, G., *J. Organometal. Chem., 90,* 49 (1974).
43. Cardaci, G., *J. Amer. Chem. Soc., 97,* 1412 (1975).
44. Chen, L.S., Su, S.R., and Wojicki, A., *J. Amer. Chem. Soc., 96,* 5655 (1974).
45. Chia, L.S., Cullen, W.R., Sams, J.R., and Scott, J.C., *Can. J. Chem., 53,* 2232 (1975).
46. Churchill, M.R., and Wormald, J., *Chem. Commun., 1968,* 1597.
47. Coffey, C.E., *J. Amer. Chem. Soc., 84,* 118 (1962).
48. Cohen, L., Giering, W.P., Kenedy, D., Magatti, C.V., and Sanders, A., *J. Organometal. Chem., 65,* C 57 (1974).
49. Collins, R.L., and Pettit, R., *J. Chem. Phys., 39,* 3433 (1963).
50. Conder, H.L., and Darensbourg, M.Y., *J. Organometal. Chem., 67,* 93 (1974).
51. Corradini, P., Pedone, C., and Sirigu, A., *Chem. Commun., 1966,* 341.
52. Corradini, P., Pedone, C., and Sirigu, A., *Chem. Commun., 1968,* 275.
53. Cotton, F.A., and Troup, J.M., *J. Amer. Chem. Soc., 96,* 3438, (1974).

54. Cotton, F.A., and Troup, J.M., *J. Amer. Chem. Soc., 96,* 4422 (1974).
55. Cotton, F.A., and Lahuerta, P., *Inorg. Chem., 14,* 116 (1975).
56. Cramer, R., *J. Amer. Chem. Soc., 86,* 217 (1964).
57. Cullen, W.R., and Mihichuk, L., *Can. J. Chem., 51,* 936 (1973).
58. Cutler, A., Raghu, S., and Rosenblum, M., *J. Organometal. Chem., 77,* 381 (1974).
59. Cutler, A., Ehntholt, D., Lennon, P., Nicholas, K., Marten, D.F., Madhavarao, M., Raghu, S., Rosan, A., and Rosenblum, M., *J. Amer. Chem. Soc., 97,* 3149 (1975).
60. Darensbourg, D.J., Tappan, J.E., and Marwedel, B.J., *J. Organometal. Chem., 54,* C 39 (1973).
61. Darensbourg, D.J., Nelson, H.H., III, and Hyde, C.L., *Inorg. Chem., 13,* 2135 (1974).
62. Davidson, J.L., Green, M., Stone, F.G.A., and Welch, A.J., *J. Chem. Soc. Chem. Commun., 1975,* 286.
63. Davis, M.I., and Speed, C.S., *J. Organometal. Chem., 21,* 401 (1970).
64. DePuy, C.H., Kobal, V.M., and Gibson, D.H., *J. Organometal. Chem., 13,* 266 (1968).
65. Dessy, R.E., Charkoudian, J.C., Abeles, T.P., and Rheingold, A.L., *J. Amer. Chem. Soc., 92,* 3947 (1970).
66. Eekhof, J.H., Hogeveen, H., Kellogg, R.M., and Schudde, E.P., *J. Organometal. Chem., 105,* C 35 (1976).
67. Ehntholt, D., Rosan, A., and Rosenblum, M., *J. Organometal. Chem., 56,* 315 (1973).
68. Einstein, F.W.B., Cullen, W.R., and Trotter, J., *J. Amer. Chem. Soc., 88,* 5670 (1966).
69. Einstein, F.W.B., and Trotter, J., *J. Chem. Soc. A, 1967,* 824.
70. Einstein, F.W.B., and Jones, R.D.G., *Inorg. Chem., 12,* 255 (1973).
71. Eisenstadt, A., *Tetrahedron Lett., 1972,* 2005.
72. Fields, R., Germain, M.M., Haszeldine, R.N., and Wiggans, P.W., *Chem. Commun., 1967,* 243.
73. Fields, R., Germain, M.M., Haszeldine, R.N., and Wiggans, P.W., *J. Chem. Soc. A, 1970,* 1964.
74. Fields, R., Germain, M.M., Haszeldine, R.N., and Wiggans, P.W., *J. Chem. Soc. A, 1970,* 1969.
75. Fields, R., Godwin, G.L., and Haszeldine, R.N., *J. Organometal. Chem., 26,* C 70 (1971).
76. Fischer, E.O., and Fichtel, K., *Chem. Ber., 94,* 1200 (1961).
77. Fischer, E.O., and Fichtel, K., *Chem. Ber., 95,* 2063 (1962).
78. Foster, M.S., and Beauchamp, J.L., *J. Amer. Chem. Soc.,*

97, 4808 (1975).

79. Gibson, D.H., and Vonnahme, R.L., *J. Organometal. Chem.*, *70*, C 33 (1974).

80. Giering, W.P., and Rosenblum, M., *J. Organometal. Chem.*, *25*, C 71 (1970).

81. Giering, W.P., and Rosenblum, M., *J. Chem. Soc. D, Chem. Commun.*, *1971*, 441.

82. Giering, W.P., Rosenblum, M., and Tancrede, J., *J. Amer. Chem. Soc.*, *94*, 7170 (1972).

83. Giering, W.P., Raghu, S., Rosenblum, M., Cutler, A., Ehntholt, D., and Fish, R.W., *J. Amer. Chem. Soc.*, *94*, 8251 (1972).

84. Giese, R.W., and Vallee, B.L., *J. Amer. Chem. Soc.*, *94*, 6199 (1972).

85. Gilbert, A., Kelly, J.M., and Koerner von Gustorf, E., *Mol. Photochem.*, *6*, 225 (1974).

86. Grandjean, J., Laszlo, P., and Stockis, A., *J. Amer. Chem. Soc.*, *96*, 1622 (1974).

87. Graziani, M., Busetto, L., and Palazzi, A., *J. Organometal. Chem.*, *26*, 261 (1971).

88. Green, M., Hughes, R.P., and Welch, A.J., *J. Chem. Soc. Chem. Commun.*, *1975*, 487.

89. Green, M., Lewis, B., Daly, J.J., and Sanz, F., *J. Chem. Soc. Dalton Trans.*, *1975*, 1118.

90. Green, M.L.H., and Nagy, P., *Proc. Chem. Soc.*, *1961*, 378.

91. Green, M.L.H., and Nagy, P.L.I., *J. Amer. Chem. Soc.*, *84*, 1310 (1962).

92. Green, M.L.H., and Nagy, P.L.I., *Proc. Chem. Soc.*, *1962*, 74.

93. Green, M.L.H., and Nagy, P.L.I., *J. Chem. Soc.*, *1963*, 189.

94. Green, M.L.H., and Nagy, P.L.I., *J. Organometal. Chem.*, *1*, 58 (1963).

95. Green, M.L.H., and Whiteley, R.N., *J. Chem. Soc. A*, *1971*, 1943.

96. Green, M.L.H., and Smith, M.J., *J. Chem. Soc. A*, *1971*, 3220.

97. Grevels, F.-W., and Koerner von Gustorf, E., *Justus Liebigs Ann. Chem.*, *1973*, 1821.

98. Grevels, F.-W., Schulz, D., and Koerner von Gustorf, E., *Angew. Chem.*, *86*, 558 (1974); *Angew. Chem. Int. Ed. Engl.*, *13*, 534 (1974).

99. Grevels, F.-W., Foulger, B.E., Leitich, J., Schulz, D., and Koerner von Gustorf, E.A., XVIth Int. Conf. Coord. Chem., Dublin, Ireland, 1974, Abstracts, 4.22.

100. Grevels, F.-W., and Koerner von Gustorf, E., *Justus Liebigs Ann. Chem.*, *1975*, 547.

101. Grevels, F.-W., Feldhoff, U., and Schneider, K., VIIth

Int. Conf. Organometal. Chem., Venice, Italy, 1975, Abstracts, 192.
102. Grevels, F.-W., unpublished results, 1975-1976.
103. Guilard, R., and Dusausoy, Y., *J. Organometal. Chem., 77*, 393 (1974).
104. Gusev, A.I., and Struchkov, Yu.T., *Zh. Strukt. Khim., 11*, 368 (1970); *J. Struct. Chem., 11*, 340 (1970).
105. Hallam, B.F., and Pauson, P.L., *J. Chem. Soc., 1958*, 642.
106. Hartley, F.R., *Angew. Chem., 84*, 657 (1972); *Angew. Chem. Int. Ed. Engl., 11*, 596 (1972).
107. Hata, G., and Miyake, A., *Bull. Chem. Soc. Jap., 41*, 2762 (1968).
108. Hata, G., Kondo, H., and Miyake, A., *J. Amer. Chem. Soc., 90*, 2278 (1968).
109. Herberhold, M., and Brabetz, H., *Z. Naturforsch., B 26*, 656 (1971).
110. Herberich, G.E., and Müller, H., *Angew. Chem., 83*, 1020 (1971); *Angew. Chem. Int. Ed. Engl., 10*, 937 (1971).
111. Hitchcock, P.B., and Mason, R., *Chem. Commun., 1967*, 242.
112. Hoehn, H.H., Pratt, L., Watterson, K.F., and Wilkinson, G., *J. Chem. Soc., 1961*, 2738.
113. Howell, J.A.S., Johnson, B.F.G., Josty, P.L., and Lewis, J., *J. Organometal. Chem., 39*, 329 (1972).
114. Hunt, R.L., Roundhill, D.M., and Wilkinson, G., *J. Chem. Soc. A, 1967*, 982.
115. Jolly, P.W., and Pettit, R., *J. Amer. Chem. Soc., 88*, 5044 (1966).
116. Joshi, K.K., *J. Chem. Soc. A, 1966*, 598.
117. Jun, M.J., Dissertation, Technische Hochschule Aachen, 1966.
118. Kagan, J., Lin, W.-L., Cohen, S.M., and Schwartz, R.N., *J. Organometal. Chem., 90*, 67 (1975).
119. Kerber, R.C., and Ehntholt, D.J., *J. Amer. Chem. Soc., 95*, 2927 (1973).
120. Kettle, S.F.A., and Orgel, L.E., *Chem. Ind. (London), 1960*, 49.
121. King, R.B., and Stone, F.G.A., *J. Amer. Chem. Soc., 82*, 4557 (1960).
122. King, R.B., Treichel, P.M., and Stone, F.G.A., *J. Amer. Chem. Soc., 83*, 3600 (1961).
123. King, R.B., *Inorg. Chem., 2*, 642 (1963).
124. King, R.B., and Efraty, A., *J. Organometal. Chem., 24*, 241 (1970).
125. King, R.B., Efraty, A., and Zipperer, W.C., *J. Organometal. Chem., 38*, 121 (1972).
126. King, R.B., and Ackermann, M.N., *J. Organometal. Chem., 60*, C 57 (1973).
127. King, R.B., and Bond, A., *J. Amer. Chem. Soc., 96*, 1343

(1974).

128. King, R.B., and Harmon, C.A., *J. Organometal. Chem., 88,* 93 (1975).

129. King, R.B., and Harmon, C.A., *J. Amer. Chem. Soc., 98,* 2409 (1976).

130. Klanderman, K.A., Dissertation, University of Wisconsin, 1965.

131. Knoth, W.H., *Inorg. Chem., 14,* 1566 (1975).

132. Koerner von Gustorf, E., Jun, M.J., Huhn, H., and Schenck, G.O., *Angew. Chem., 75,* 1120 (1963).

133. Koerner von Gustorf, E., Jun, M.J., Köller, H., and Schenck, G.O., *Industrial Uses of Large Radiation Sources,* International Atomic Energy Agency, Vienna, 1963, Vol. II, p. 73.

134. Koerner von Gustorf, E., Jun, M.J., and Schenck, G.O., *Z. Naturforsch., B 18,* 503 (1963).

135. Koerner von Gustorf, E., Jun, M.J., and Schenck, G.O., *Z. Naturforsch., B 18,* 767 (1963).

136. Koerner von Gustorf, E., Köller, H., Jun, M.J., and Schenck, G.O., *Atomstrahlung in Medizin und Technik,* Thiemig, München 1964, p. 300.

137. Koerner von Gustorf, E., Henry, M.C., and Di Pietro, C., *Z. Naturforsch., B 21,* 42 (1966).

138. Koerner von Gustorf, E., Henry, M.C., and McAdoo, D.J., *Justus Liebigs Ann. Chem., 707,* 190 (1967).

139. Koerner von Gustorf, E., and Hogan, J.C., *Tetrahedron Lett., 1968,* 3191.

140. Koerner von Gustorf, E., Grevels, F.-W., and Hogan, J.C., *Angew. Chem., 81,* 918 (1969); *Angew. Chem. Int. Ed. Engl., 8,* 899 (1969).

141. Koerner von Gustorf, E., Grevels, F.-W., Krüger, C., Olbrich, G., Mark, F., Schulz, D., and Wagner, R., *Z. Naturforsch., B 27,* 392 (1972).

142. Koerner von Gustorf, E., Jaenicke, O., and Polansky, O.E., *Z. Naturforsch., B 27,* 575 (1972).

143. Kolobova, N.E., Skripkin, V.V., and Anisimov, K.N., *Izv. Akad. Nauk SSSR, Ser. Khim., 1970,* 2225; *Bull. Acad. Sci. USSR, Div. Chem. Ser., 1970,* 2095.

144. Korecz, L., and Burger, K., *Acta. Chim. Acad. Sci. Hung., 58,* 253 (1968).

145. Kruck, T., and Knoll, L., *Chem. Ber., 106,* 3578 (1973).

146. Kruczynski, L., LiShingMan, L.K.K., and Takats, J., *J. Amer. Chem. Soc., 96,* 4006 (1974).

147. Krüger, C., *J. Organometal. Chem., 22,* 697 (1970).

148. Krüger, C., Tsay, Y.-H., Grevels, F.-W., and Koerner von Gustorf, E., *Israel J. Chem., 10,* 201 (1972).

149. Lappert, M.F., and Speier, G., *J. Organometal. Chem., 80,* 329 (1974).

150. Lennon, P., Madhavarao, M., Rosan, A., and Rosenblum, M., *J. Organometal. Chem.*, *108*, 93 (1976).
151. Lichtenberg, D.W., and Wojcicki, A., *J. Organometal. Chem.*, *94*, 311 (1975).
152. Lokshin, B.V., Aleksanyan, V.T., and Klemenkova, Z.S., *J. Organometal. Chem.*, *70*, 437 (1974).
153. Lokshin, B.V., Aleksanyan, V.T., Klemenkova, Z.S., Rybin, L.V., and Gubenko, N.T., *J. Organometal. Chem.*, *74*, 97 (1974).
154. Lombardo, L., Wege, D., and Wilkinson, S.P., *Aust. J. Chem.*, *27*, 143 (1974).
155. Luxmoore, A.R., and Truter, M.R., *Proc. Chem. Soc.*, *1961*, 466.
156. Luxmoore, A.R., and Truter, M.R., *Acta. Crystallogr.*, *15*, 1117 (1962).
157. Mantzaris, J., and Weissberger, E., *Tetrahedron Lett.*, *1972*, 2815.
158. Mantzaris, J., and Weissberger, E., *J. Amer. Chem. Soc.*, *96*, 1873 (1974).
159. Mantzaris, J., and Weissberger, E., *J. Amer. Chem. Soc.*, *96*, 1880 (1974).
160. Mantzaris, J., and Weissberger, E., *J. Org. Chem.*, *39*, 726 (1974).
161. Matsukura, T., Mano, K., and Fujino, A., *Bull. Chem. Soc. Jap.*, *48*, 2464 (1975).
162. McCaskie, J.E., Chang, P.L., Nelsen, T.R., and Dittmer, D.C., *J. Org. Chem.*, *38*, 3963 (1973).
163. Murdoch, H., and Weiss, E., *Helv. Chim. Acta.*, *45*, 1156 (1962).
164. Murdoch, H.D., and Weiss, E., *Helv. Chim. Acta.*, *46*, 1588 (1963).
165. Musco, A., Paiaro, G., and Palumbo, R., *Ric. Sci.*, *39*, 417 (1969).
166. Musco, A., Palumbo, R., and Paiaro, G., *Inorg. Chim. Acta.*, *5*, 157 (1971).
167. Nesmeyanov, A.N., Akhmed, K., Rybin, L.V., Rybinskaya, M.I., and Ustynyuk, Yu.A., *Dokl. Akad. Nauk SSSR*, *175*, 1070 (1967); *Dokl. Chem.*, *175*, 718 (1967).
168. Nesmeyanov, A.N., Rybin, L.V., Rybinskaya, M.I., Kaganovich, V.S., Ustynyuk, Yu.A., and Leshcheva, I.F., *Zh. Obshch. Khim.*, *38*, 1471 (1968); *J. Gen. Chem. USSR*, *38*, 1424 (1968).
169. Nesmeyanov, A.N., Rybin, L.V., Rybinskaya, M.I., Gubenko, N.T., Leshcheva, I.F., and Ustynyuk, Yu.A., *Zh. Obshch. Khim.*, *38*, 1476 (1968); *J. Gen. Chem. USSR*, *38*, 1428 (1968).
170. Nesmeyanov, A.N., Rybin, L.V., Rybinskaya, M.I., Gubenko, N.T., Leshcheva, I.F., and Ustynyuk, Yu.A.,

Zh. Obshch. Khim., *39*, 2091 (1969); *J. Gen. Chem. USSR*, *39*, 2045 (1969).

171. Nesmeyanov, A.N., Rybinskaya, M.I., Rybin, L.V., Kaganovich, V.S., and Petrovskii, P.V., *J. Organometal. Chem.*, *31*, 257 (1971).

172. Nesmeyanov, A.N., Rybin, L.V., Gubenko, N.T., Petrovskii, P.V., and Rybinskaya, M.I., *Zh. Obshch. Khim.*, *42*, 2473 (1972); *J. Gen. Chem. USSR*, *42*, 2465 (1972).

173. Nesmeyanov, A.N., Shul'pin, G.B., Fedorov, L.A., Petrovsky, P.V., and Rybinskaya, M.I., *J. Organometal. Chem.*, *69*, 429 (1974).

174. Nesmeyanov, A.N., Rybin, L.V., Gubenko, N.T., Rybinskaya, M.I., Petrovskii, P.V., *J. Organometal. Chem.*, *71*, 271 (1974).

175. Nesmeyanov, A.N., Rybinskaya, M.I., Rybin, L.V., Arutyunyan, A.V., Kuz'mina, L.G., and Struchkov, Yu.T., *J. Organometal. Chem.*, *73*, 365 (1974).

176. Nesmeyanov, A.N., Rybinskaya, M.I., Rybin, L.V., and Arutyunyan, A.V., *Zh. Obshch. Khim.*, *44*, 604 (1974); *J. Gen. Chem. USSR*, *44*, 578 (1974).

177. Nesmeyanov, A.N., Shul'pin, G.B., Rybin, L.V., Gubenko, N.T., Rybinskaya, M.I., Petrovskii, P.V., and Robas, V.I., *Zh. Obshch. Khim.*, *44*, 2032 (1974); *J. Gen. Chem. USSR*, *44*, 1994 (1974).

178. Nicholas, K.M., *J. Amer. Chem. Soc.*, *97*, 3254 (1975).

179. Nicholas, K.M., and Rosan, A.M., *J. Organometal. Chem.*, *84*, 351 (1975).

180. Paiaro, G., and Panunzi, A., *Ric. Sci.*, *Parte II*, *Sez. A*, *4*, 601 (1964).

181. Paiaro, G., Palumbo, R., Musco, A., and Panunzi, A., *Tetrahedron Lett.*, *1965*, 1067.

182. Paiaro, G., and Palumbo, R., *Gazz. Chim. Ital.*, *97*, 265 (1967).

183. Pearson, A.J., *Tetrahedron Lett.*, *1975*, 3617.

184. Pedone, C., and Sirigu, A., *Acta Crystallogr.*, *23*, 759 (1967).

185. Potenza, J., Johnson, R., Mastropaolo, D., and Efraty, A., *J. Organometal. Chem.*, *64*, C 13 (1974).

186. Reeves, P., Henery, J., and Pettit, R., *J. Amer. Chem. Soc.*, *91*, 5888 (1969).

187. Reger, D.L., *Inorg. Chem.*, *14*, 660 (1975).

188. Reihlen, O., Gruhl, A., Hessling, G., and Pfrengle, O., *Justus Liebigs Ann. Chem.*, *482*, 161 (1930).

189. Rhee, I., Ryang, M., and Tsutsumi, S., *J. Organometal. Chem.*, *9*, 361 (1967).

190. Robertson, G.B., and Whimp, P.O., *J. Chem. Soc. Dalton Trans.*, *1973*, 2454.

191. Rosan, A., Rosenblum, M., and Tancrede, J., *J. Amer.*

Chem. Soc., *95*, 3062 (1973).
192. Rosan, A., and Rosenblum, M., *J. Organometal. Chem.*, *80*, 103 (1974).
193. Rosan, A., and Rosenblum, M., *J. Org. Chem.*, *40*, 3621 (1975).
194. Rosenblum, M., Fish, R.W., and Bennett, C., *J. Amer. Chem. Soc.*, *86*, 5166 (1964).
195. Rosenblum, M., *Accounts Chem. Res.*, *7*, 122 (1974).
196. Rosenblum, M., Saidi, M.R., and Madhavarao, M., *Tetrahedron Lett.*, *1975*, 4009.
197. Roustan, J.L., Benaim, J., Charrier, C., and Mérour, J.Y., *Tetrahedron Lett.*, *1972*, 1953.
198. Rybinskaya, M.I., Rybin, L.V., Gubenko, N.T., and Nesmeyanov, A.N., *Zh. Obshch. Khim.*, *41*, 2020 (1971); *J. Gen. Chem. USSR*, *41*, 2041 (1971).
199. Sams, J.R., and Scott, J.C., *J. Chem. Soc. Dalton Trans.*, *1974*, 2265.
200. Sanders, A., Cohen, L., Giering, W.P., Kenedy, D., and Magatti, C.V., *J. Amer. Chem. Soc.*, *95*, 5430 (1973).
201. Sanders, A., Magatti, C.V., and Giering, W.P., *J. Amer. Chem. Soc.*, *96*, 1610 (1974).
202. Sanders, A., and Giering, W.P., *J. Amer. Chem. Soc.*, *97*, 919 (1975).
203. Sanders, A., and Giering, W.P., *J. Organometal. Chem.*, *104*, 49 (1976).
204. Sanders, A., and Giering, W.P., *J. Organometal. Chem.*, *104*, 67 (1976).
205. Sanders, A., Bauch, T., Magatti, C.V., Lorenc, C., and Giering, W.P., *J. Organometal. Chem.*, *107*, 359 (1976).
206. Sarel, S., Felzenstein, A., Victor, R., and Yovell, J., *J. Chem. Soc. Chem. Commun.*, *1974*, 1025.
207. Sato, F., Yamamoto, T., Wilkinson, J.R., and Todd, L.J., *J. Organometal. Chem.*, *86*, 243 (1975).
208. Schenck, G.O., Koerner von Gustorf, E., and Jun, M.J., *Tetrahedron Lett.*, *1962*, 1059.
209. Schmidt, E.K.G., *Chem. Ber.*, *108*, 1609 (1975).
210. Schrauzer, G.N., *Chem. Ber.*, *94*, 642 (1961).
211. Schubert, E.H., and Sheline, R.K., *Inorg. Chem.*, *5*, 1071 (1966).
212. Schurig, V., *Tetrahedron Lett.*, *1976*, 1269.
213. Seel, F., and Röschenthaler, G.V., *Z. anorg. allgem. Chem.*, *386*, 297 (1971).
214. Sorriso, S., and Cardaci, G., *J. Chem. Soc. Dalton Trans.*, *1975*, 1041.
215. Speert, A., Gelan, J., Anteunis, M., Marchand, A.P., and Laszlo, P., *Tetrahedron Lett.*, *1973*, 2271.
216. Stark, K., Lancaster, J.E., Murdoch, H.D., and Weiss, E., *Z. Naturforsch.*, *B 19*, 284 (1964).

217. Stockis, A., and Weissberger, E., *J. Amer. Chem. Soc.*, *97*, 4288 (1975).
218. Su, S.R., and Wojcicki, A., *J. Organometal. Chem.*, *31*, C 34 (1971).
219. Su, S.R., and Wojcicki, A., *Inorg. Chem.*, *14*, 89 (1975).
220. Thyret, H., *Angew. Chem.*, *84*, 581 (1972); *Angew. Chem. Int. Ed. Engl.*, *11*, 520 (1972).
221. Vessieres, A., and Dixneuf, P., *Tetrahedron Lett.*, *1974*, 1499.
222. Victor, R., Ben-Shoshan, R., and Sarel, S., *Tetrahedron Lett.*, *1970*, 4257.
223. Victor, R., Ben-Shoshan, R., and Sarel, S., *J. Chem. Soc. D, Chem. Commun.*, *1971*, 1241.
224. Watterson, K.F., and Wilkinson, G., *Chem. Ind. §London/*, *1959*, 991.
225. Watterson, K.F., and Wilkinson, G., *Chem. Ind. §London/*, *1960*, 1358.
226. Weiss, E., and Hübel, W., *Angew. Chem.*, *73*, 298 (1961).
227. Weiss, E., and Hübel, W., *Chem. Ber.*, *95*, 1186 (1962).
228. Weiss, E., Stark, K., Lancaster, J.E., and Murdoch, H.D., *Helv. Chim. Acta*, *46*, 288 (1963).
229. Whitesides, T.H., Arhart, R.W., and Slaven, R.W., *J. Amer. Chem. Soc.*, *95*, 5792 (1973).
230. Whitesides, T.H., Slaven, R.W., and Calabrese, J.C., *Inorg. Chem.*, *13*, 1895 (1974).
231. Whitesides, T.H., and Slaven, R.W., *J. Organometal. Chem.*, *67*, 99 (1974).
232. Wilson, S.T., Coville, N.J., Shapely, J.R., and Osborn, J.A., *J. Amer. Chem. Soc.*, *96*, 4038 (1974).
233. Yamamoto, T., Yamamoto, A., and Ikeda, S., *Bull. Chem. Soc. Jap.*, *45*, 1104 (1972).
234. Yamamoto, T., Todd, L.J., *J. Organometal. Chem.*, *67*, 75 (1974).
235. Ziegler, M.L., *Angew. Chem.*, *80*, 239 (1968); *Angew. Chem. Int. Ed. Engl.*, *7*, 222 (1968).
236. Weissberger, E., and Laszlo, P., *Accounts Chem. Res.*, *9*, 209 (1976).
237. Krüger, C., and Tsay, Y.-H., *Cryst. Struct. Commun.*, *5*, 219 (1976).
238. Krüger, C., and Tsay, Y.-H., *Cryst. Struct. Commun.*, *5*, 215 (1976).
239. Grevels, F.-W., Feldhoff, U., Leitich, J., and Krüger, C., *J. Organometal. Chem.*, *118*, 79 (1976).

ALLYL IRON COMPLEXES

By R.B. KING

Department of Chemistry, University of Georgia
Athens, Gerorgia 30602, U.S.A.

TABLE OF CONTENTS

ISBN 0-12-417101-X (v.1)

I. INTRODUCTION

This chapter discusses carbonyliron complexes containing the 1,2,3-*trihapto*-allyl (η^3-allyl) unit (1) in which exactly three adjacent carbon atoms of a hydrocarbon unit are coordinated to an iron atom. Important types of carbonyliron derivatives containing one such η^3-allyl unit include derivatives of the types (η^3-allylic)$Fe(CO)_3X$ (X = halide or other one-electron donor ligand), $[(\eta^3\text{-allylic})Fe(CO)_4]^+$, and ($\eta^3$-allylic)$Fe(CO)_2NO$ in which the iron atom has the favoured 18-electron noble gas configuration, as well as electron-deficient $[(\eta^3\text{-allylic})Fe(CO)_3]^+$ derivatives in which the iron atom has only a 16-electron configuration. Compounds of the type $(\eta^3\text{-allylic})_2Fe(CO)_2$ are known containing two η^3-allylic ligands coordinated to a single iron atom. An unstable compound $(C_3H_5)_3Fe$ containing three allyl ligands bonded to a single iron atom is also known.

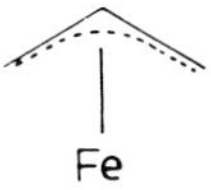

1

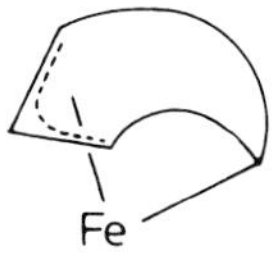

2a

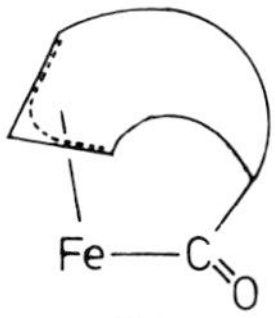

2b

Organoiron derivatives containing a 1,2,3-*trihapto*-allylic unit as part of a more complex ligand system have also been prepared. Numerous compounds of the type 2a are known containing a bidentate four-electron donor chelating ligand which is coordinated to the iron atom by both an η^3-allyl group and a metal-carbon σ-bond. Many complexes of the type 2a readily undergo carbon monoxide insertion reactions to give the corresponding chelating η^3-allyl-σ-acyl derivatives of type 2b. Complexes of the types 2a and 2b generally arise either from reactions of cycloolefin-carbonyliron complexes or from reactions of carbonyliron reagents with vinylcyclopropane derivatives, including polycyclic hydrocarbons such as semibullvalene and bullvalene. A hexacarbonyldiiron unit can bond to a hydrocarbon network with two sets of η^3-allylic bonds in different ways to give the following types of complexes:

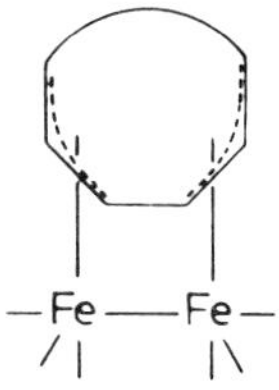

3a

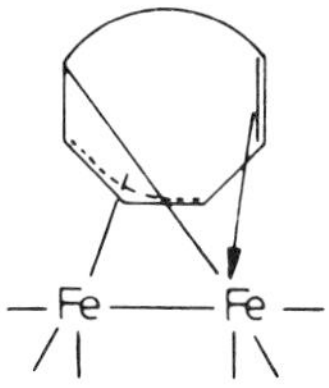

3b

(1) The (cyclic triene)-hexacarbonyldiiron derivatives 3a in which six carbon atoms of two η^3-allylic units are part of the same ring system. In addition to these 1-3:4-6-bis(η^3-allylic)iron complexes the less symmetrical 1,5,6-η^3:2-4-η^3-derivatives 3b are formed. Such complexes generally arise from reactions of appropriate cyclic trienes with carbonyliron reagents, *e.g.* $Fe_2(CO)_9$, under mild conditions. They will be discussed only briefly in this chapter and in more detail in the chapter on triene and tetraene complexes. (2) Tetramethyleneethane-hexacarbonyldiiron complexes of the type 4

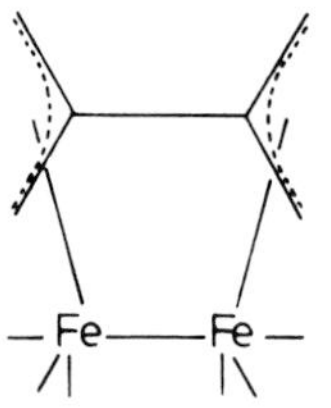

4

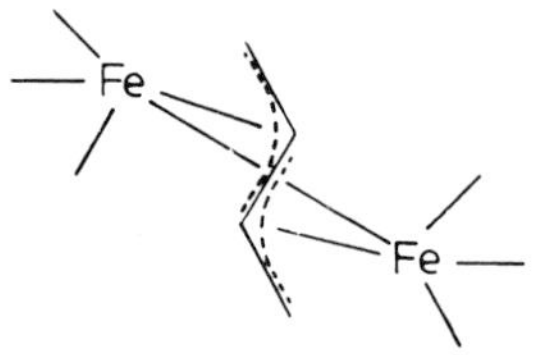

5

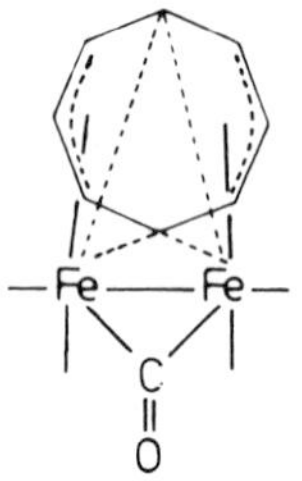

6

in which the center carbons of the two η^3-allylic units are linked together (rather than the outer carbons as in type 3a). Such complexes generally are obtained from reactions of allenes with carbonyliron reagents. (3) Butatriene-hexacarbonyldiiron derivatives of the type 5 in which the two η^3-allylic units are orthogonal and share two atoms of the four-carbon chain. Such complexes generally are prepared from reactions of butatriene derivatives with carbonyliron reagents. (4) Still another type of bis(η^3-allylic)carbonyliron derivative is type 6 which may arise from reactions of cyclic tetraenes, such as cyclooctatetraene, with carbonyliron complexes. Compounds of the type 6 contain two η^3-allylic units separated by single sp^2 carbon atoms, which contribute to the bridging between the two iron atoms. All of the binuclear bis(η^3-allylic) derivatives of the types 3a, 3b, 4, 5, and 6, contain an iron-iron bond.

Several general review articles are available on the chemistry of η^3-allyl-metal compounds *(50,91,106)*. In the most recent one *(50)* the stereochemistry of the metal-allyl bond of the type 1 is discussed in some detail. The C-C-C angle of the η^3-allyl group in 1 is usually close to 120°. The plane defined by the three-carbon skeleton of the η^3-allyl group is usually not perpendicular to the plane defined by the metal atom and the terminal carbon atoms of the η^3-allyl group. Values for this angle may be as large as 110° *(50)*. The substituted η^3-allyl ligand is generally non-planar. There are several examples of conformational isomerism based on two different orientations of the allyl group relative to other ligands in the molecule. For example, such isomerism is found in the iron compounds $(\eta^3\text{-}C_3H_5)Fe(CO)_3X$. Isomerism can also result from the non-equivalence of the *syn* and *anti* sites in an η^3-allylic group of the complex type 1.

II. η^3-ALLYLIC CARBONYLIRON CATIONS AND HALIDES

This section discusses the preparations as well as the physical, spectroscopic, and chemical properties of complexes of the types $(\eta^3\text{-allylic})Fe(CO)_3X$, $[(\eta^3\text{-allylic})Fe(CO)_3]^+$, and $[(\eta^3\text{-allylic})Fe(CO)_4]^+$ containing simple acyclic 1,2,3-*trihapto*-allylic ligands.

A. *PREPARATIVE METHODS*

Compounds of the type $(\eta^3\text{-allylic})Fe(CO)_3X$ are most frequently prepared by oxidative addition reactions of zerovalent carbonyliron complexes with the corresponding allyl halides. In addition some $(\eta^3\text{-allylic})Fe(CO)_3X$ derivatives can be prepared by addition of hydrogen halides to diene-tricarbonyliron derivatives. Related protonation reactions of $(diene)Fe(CO)_3$ derivatives in the absence of halide first give the $[(\eta^3\text{-allylic})Fe(CO)_3]^+$ cations, which may then undergo disproportionation or carbon monoxide addition to give the more stable $[(\eta^3\text{-allylic})Fe(CO)_4]^+$ cations. Other addition reactions of unsaturated fluoroolefins to $(diene)Fe(CO)_3$ derivatives may result in formation of η^3-allyl-σ-alkyl complexes of type 2a. η^3-Allylic-carbonyliron derivatives also arise from some reactions of $(diene)Fe(CO)_3$ derivatives with other electrophiles such as boron trifluoride or acyl halides.

1. Reactions of Allylic Halides with Carbonyliron Complexes and Other Zerovalent Iron Derivatives

The following conditions are generally useful for reactions of carbonyliron reagents with allylic halides to give the corresponding $(\eta^3\text{-allylic})Fe(CO)_3X$ derivatives: (1) reaction of the allylic halide with pentacarbonyliron at a closely controlled temperature *(161,165,183)*; (2) ultraviolet irradiation of the allylic halide with pentacarbonyliron *(110,111)*; (3) reaction of enneacarbonyldiiron with allylic halides around 40°C *(155)*. The reaction of pentacarbonyliron with allyl iodide at 40-45°C is probably the most effective way to get $(\eta^3\text{-}C_3H_5)Fe(CO)_3I$ *(165,183)*. However, it is necessary to control the temperature of this reaction very carefully to avoid decomposition to iron(II) iodide. The photochemical method *(110,111)* avoids this difficulty and has been used to prepare a wider range of (η^3-allylic)$Fe(CO)_3X$ derivatives including (η^3-R-$CHCHCH_2$)$Fe(CO)_3X$ (R = H, X = Cl, Br, I; R = CH_3, X = Cl, Br; R = CH_3-CO_2, X = Br) and [η^3-$CH_2C(CH_3)CH_2$]$Fe(CO)_3Cl$. The use of enneacarbonyldiiron without ultraviolet irradiation may be desirable in certain cases *(155)*. The carboranyl derivative 7 has been

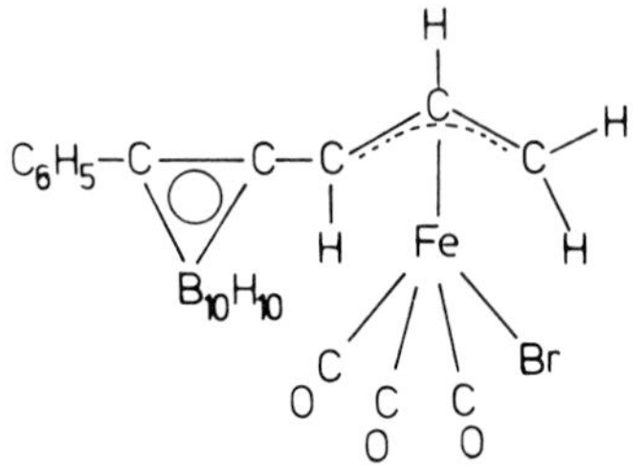

7

prepared by reaction of *trans*-1-(phenyl-*o*-carboranyl)-3-bromoprop-1-ene with enneacarbonyldiiron at 40°C *(212)*.

Some related reactions of carbonyliron complexes with allylic halides proceed anomalously to give products other than the (η^3-allylic)$Fe(CO)_3X$ derivatives. For example, reactions of enneacarbonyldiiron with the 2-methoxyallyl halides give not only the halide complexes [η^3-$CH_2C(OCH_3)CH_2$]$Fe(CO)_3X$ but also result in hydrogen halide elimination to give the vinylketene complex 8 *(114)*. The yield of 8 is maximized by

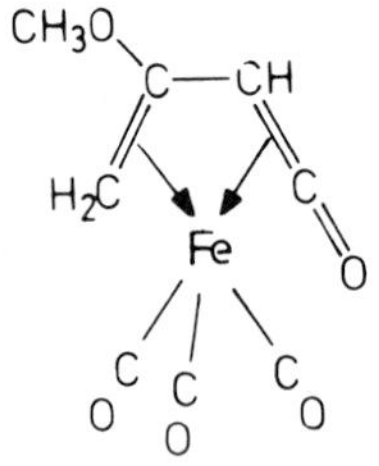

8

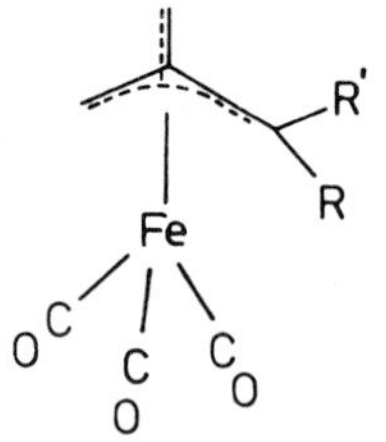

9

using 2-methoxyallyl chloride rather than the corresponding bromide or iodide. Reactions of the allylic dihalides $RR'C{=}C(CH_2Cl)_2$ with enneacarbonyldiiron give the corresponding trimethylenemethane-tricarbonyliron derivatives [(RR'C)C-$(CH_2)_2$]$Fe(CO)_3$ (9, R = R' = H, CH_3, C_2H_5-CO_2; R = C_6H_5 and CH_3-CO_2, R' = H) *(79,89)*. Thermal decomposition of 2-methylallylic derivatives of the type [η^3-R-$CHC(CH_3)CH_2$]$Fe(CO)_3Cl$ (R = H, C_6H_5) also produces the corresponding trimethylenemethane-tricarbonyliron derivatives (9, R = H, R' = H and C_6H_5, respectively) *(78)*. Reactions of 3,4-dichlorocyclobutenes with enneacarbonyldiiron give the corresponding cyclobutadiene-tricarbonyliron derivatives *(88,182)*, which are discussed elsewhere in this book.

Similar oxidative addition reactions of $Fe_2(CO)_9$ with allylic alcohols result in the formation of products of type 2b containing η^3-allyl-σ-acyl ligands. Thus, reactions of but-2-ene-1,4-diol or *cis*-4-chlorobut-2-en-1-ol with $Fe_2(CO)_9$ give 10 (R = R' = H) with elimination of water and hydrogen

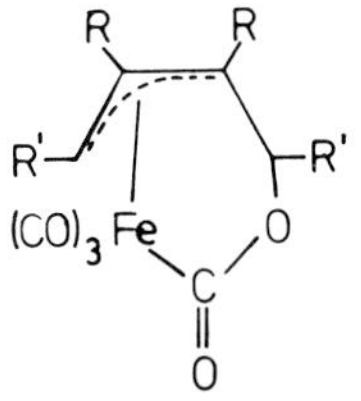

10

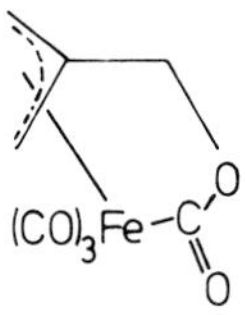

11

chloride, respectively *(110,156)*. Similar reactions of Fe_2-$(CO)_9$ with 2,3-dimethylbut-2-ene-1,4-diol and with hex-3-ene-2,4-diol give the substituted derivatives 10 (R = CH_3, R' = H and R = H, R' = CH_3, respectively) *(156)*. The related compound 11 can be obtained by the reaction of 2-hydroxymethyl-prop-2-en-3-ol with enneacarbonyldiiron.

Analogous reactions of enneacarbonyldiiron with benzylic halides have been investigated *(164,170)*. However, the carbon-carbon double bonds in benzene and naphthalene rings are not reactive enough to participate in the formation of η^3-benzyl and η^3-naphthylmethyl ligands in these carbonyl-iron reactions *(164,170)*. Instead, reactions of the benzylic bromides R-CH_2Br with enneacarbonyldiiron result in coupling of the R-CH_2 unit with carbon monoxide insertion to give the corresponding ketones $(R\text{-}CH_2)_2CO$ (R = phenyl, 3,5-dimethyl-phenyl, 4-nitrophenyl, 1- and 2-naphthyl, and 4-bromo-2-naphthyl). Halides of the type R_2CHBr (R = C_6H_5 and C_6F_5; also 9-bromofluorene) couple to form the corresponding R_2CH-CHR_2 derivatives upon similar reactions with $Fe_2(CO)_9$. Reactions of the perfluoroarylmethyl halides R_f-CH_2X (R_f = pentafluorophenyl, X = Br; R_f = heptafluoro-2-naphthyl, X = Cl) with enneacarbonyldiiron give the corresponding $(R_f\text{-}CH_2)_2$-$Fe(CO)_4$ derivatives as stable yellow crystalline solids *(170)*. However, an η^3-allylic derivative is obtained by the treatment of (2-bromomethyl)naphthalene with enneacarbonyldiiron to give a stable yellow crystalline solid tentatively formulated as 12 *(164)*.

Some related syntheses of (η^3-allylic)$Fe(CO)_3X$ derivatives involve the transfer of an allyl group from another metal to iron. Thus, reactions of the readily available η^3-allylic palladium halides $[(\eta^3\text{-}R\text{-}CHCHCH_2)PdX]_2$ with ennea-

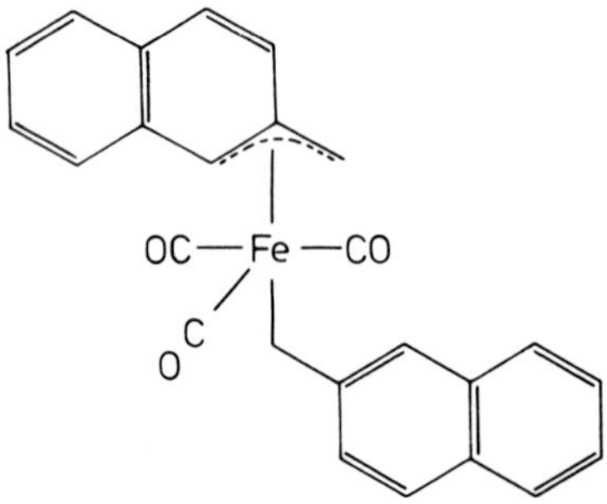

12

carbonyldiiron at room temperature give the corresponding (η^3-R-CHCHCH$_2$)Fe(CO)$_3$X derivatives (R = H, X = Cl, Br, I; R = CH$_3$, X = Cl; R = C$_6$H$_5$, X = Cl) *(165,167)*. A similar reaction of [(η^3-C$_3$H$_5$)NiBr]$_2$ with Fe$_2$(CO)$_9$ in diethyl ether at room temperature gives (η^3-C$_3$H$_5$)Fe(CO)$_3$Br. The iodide (η^3-C$_3$H$_5$)Fe(CO)$_3$I can also be obtained by the reaction of enneacarbonyldiiron with the bis(η^3-allyl) derivatives of nickel, palladium, and platinum followed by treatment with iodine *(167)*, or by the reaction of Fe(CO)$_4$I$_2$ with allyltrimethyltin *(2)*.

Similar oxidative addition reactions between allylic halides and other iron(0) derivatives have been investigated. Ultraviolet irradiations of Fe(PF$_3$)$_5$ with allyl bromide and with allyl iodide give the corresponding (η^3-C$_3$H$_5$)Fe(PF$_3$)$_3$X (X = Br, I) derivatives *(139)*. However, a similar ultraviolet irradiation of Fe(PF$_3$)$_5$ with allyl chloride results in coupling of the allyl groups with a hydrogen shift to give (1,3-hexadiene)tris(trifluorophosphine)iron *(139)*. Reactions of the trimethyl phosphite complex Fe[P(OCH$_3$)$_3$]$_5$ with allylic halides give cations of the type $\{(\eta^3\text{-allyl})Fe[P(OCH_3)_3]_4\}^+$ *(153)*. However, a similar reaction of Fe[P(OCH$_3$)$_3$]$_5$ with benzyl iodide does not form a stable benzyliron derivative but instead results in coupling to give dibenzyl and [(CH$_3$O)$_3$P]$_2$FeI$_2$ *(153)*.

2. Conversion of Olefin-carbonyliron Complexes to η^3-Allylic Derivatives

Most of these reactions involve addition to various diene-tricarbonyliron derivatives with concurrent rupture of one of the four iron-carbon bonds to the diene to give the 1,2,3-*trihapto*-allylic system. Similar reactions have also been found with trimethylenemethane derivatives.

Reactions of the diene-tricarbonyliron complexes (η^4-R-

$CH{=}CH{-}CR'{=}CH_2)Fe(CO)_3$ (R = R' = H; R = CH_3, R' = H; R = C_6H_5, R' = CH_3) with hydrogen chloride in an inert solvent result in the formation of the corresponding (η^3-allylic)$Fe(CO)_3Cl$ derivatives 13 *(116)*. The stereochemistry of this addition is not clear *(86,205)*, although formulation as the indicated *anti*-isomer 13 appears more probable.

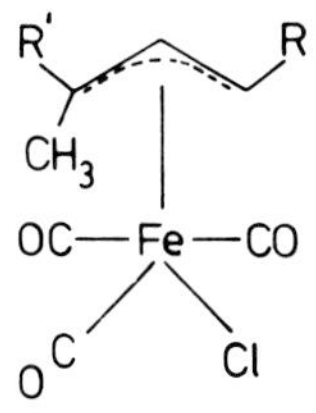

13

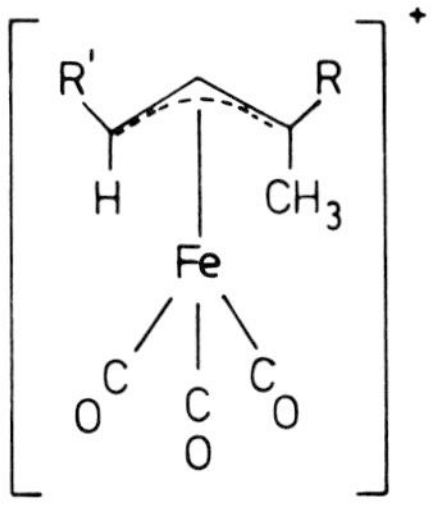

14

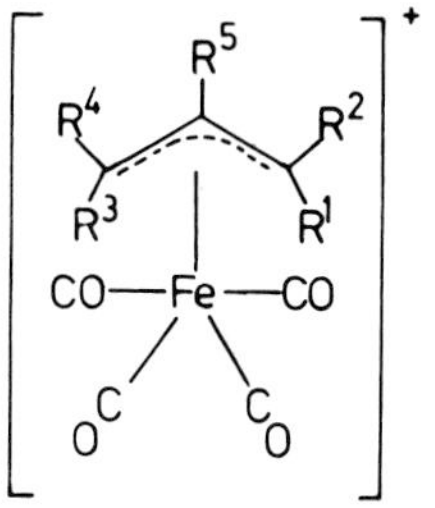

15

A similar protonation of diene-tricarbonyliron complexes in the absence of a coordinating anion such as halide leads first to the corresponding η^3-allylic-tricarbonyliron cations. Salts of the type 14 (R = R' = H; R = CH_3, R' = H; R = H, R' = CH_3; R = H, R' = C_6H_5) have been obtained by protonation of the tricarbonyliron complexes of butadiene, isoprene, *trans*-piperylene, and 1-phenylbutadiene, respectively, using tetrafluoroboric, perchloric, or hexachloroantimonic acids in nitromethane solution *(85)*. Alternatively, the η^3-allylic-tricarbonyliron cations 14 (R = R' = H; R = H, R' = CH_3) have also been prepared by removal of chloride from the corresponding (η^3-allylic)$Fe(CO)_3Cl$ derivative using silver perchlorate or silver tetrafluoroborate in an inert solvent *(86)*. These η^3-allylic-tricarbonyliron cations have a deficient 16-electron configuration but nevertheless are isolable as pale yellow crystalline tetrafluoroborate salts. The salts derived from the tricarbonyliron complexes of butadiene (14, R = R' = H) and *trans*-piperylene (14, R = H, R' = CH_3) react with water to give 2-butanone and 2-pentanone, respectively, through rearrangement of intermediate enols *(85)*. However, a similar hydrolysis of the salt 14 (R = CH_3, R' = H) gives dimethylvinylcarbinol, $CH_2{=}CH{-}C(OH)(CH_3)_2$, which cannot rearrange analogously to a ketone because of the absence of hydrogen on the carbon atom bearing the hydroxyl group *(85)*.

Although the electron-deficient $[(\eta^3\text{-allylic})Fe(CO)_3]^+$ cations can be isolated as solid salts, they are unstable in strong acid solution with respect to ligand reorganization to give the corresponding $[(\eta^3\text{-allylic})Fe(CO)_4]^+$ salts *(94)*. Thus, the infrared ν(CO) and ^{1}H-NMR spectra of a freshly prepared solution of (butadiene)tricarbonyliron in trifluoroacetic acid indicates the presence of the η^3-allylic-tricarbonyliron cation $[(\eta^3\text{-}C_4H_7)Fe(CO)_3]^+$ (14, R = R' = H) discussed above. However, upon standing in the strong acid the infrared ν(CO) and ^{1}H-NMR spectra change to those of the corresponding η^3-allylic-tetracarbonyliron cation $[(\eta^3\text{-}C_4H_7)\text{-}Fe(CO)_4]^+$ (15, $R^1 = CH_3$, $R^2 = R^3 = R^4 = R^5 = H$). This cation can be isolated as its stable tetrafluoroborate salt by stirring (butadiene)tricarbonyliron at room temperature with a six-fold excess of tetrafluoroboric acid in acetic anhydride for 2.5 h followed by precipitation with cold diethyl ether. A similar protonation of (isoprene)tricarbonyliron gives the η^3-allylic-tetracarbonyliron cation 15 ($R^1 = R^2 = CH_3$, $R^3 = R^4 = R^5 = H$) *(94)*. Dissolving this cation in trifluoroacetic acid results in the production of its tautomer 15 ($R^1 = R^5 = CH_3$, $R^2 = R^3 = R^4 = H$) to the extent of 10 % as indicated by the ^{1}H-NMR spectrum of the solution *(95)*. A similar protonation of (2,4-dimethylpenta-1,3-diene)tricarbonyliron gives the η^3-allylic-tetracarbonyliron cation 15 ($R^1 = CH(CH_3)_2$, $R^2 = R^3 = R^4 = H$, $R^5 = CH_3$) *(94)*. The isomeric cation 15 ($R^1 = R^2 = R^3 = R^4 = CH_3$, $R^5 = H$) is obtained by protonation of (tetramethylallene)tetracarbonyliron *(93)*. Dissolving either of these isomeric $[(\eta^3\text{-}C_7H_{13})Fe(CO)_4]^+$ cations in trifluoroacetic acid results in the production of an equilibrium mixture of the two cations *(95)*. The similar protonation of (2,3-dimethylbutadiene)tricarbonyliron to give the corresponding η^3-allylic-tetracarbonyliron cation 15 ($R^1 = R^2 = R^5 = CH_3$, $R^3 = R^4 = H$) proceeds at a much slower rate than the above protonations to give only a 5 % yield after 3.5 h *(94)*.

These preparations of $[(\eta^3\text{-allylic})Fe(CO)_4]^+$ cations by the protonation of diene-tricarbonyliron complexes all have the inherent disadvantage that 25 % of the starting material is consumed to provide the fourth carbonyl group. Two methods have been devised to circumvent this inefficiency. In the first such method, the 1,3-diene is allowed to react with $Fe_2(CO)_9$ only until the tetracarbonyliron complex is produced *(96)*. Protonation of this complex then gives good yields of the corresponding η^3-allylic-tetracarbonyliron cation 15. In the second such method the more readily available and stable diene-tricarbonyliron complex is used, but the protonation is carried out in a carbon monoxide atmosphere in order to introduce the fourth carbonyl group *(206)*. In both of these reactions an acid with a weakly coordinating anion such as tetrafluoroboric acid is used. If an acid with a coordinating anion is used, the initially formed η^3-allylic-tetracarbonyliron derivative is unstable with respect to carbonyl displacement by the anion. Thus, treatment of (butadiene)tetracarbonyliron with hydrogen chloride gives the yellow unstable ionic chloride $[(\eta^3\text{-}C_4H_7)Fe(CO)_4]^+Cl^-$ which rapidly decomposes at room temperature with carbon monoxide evolution to give the non-ionic tricarbonyliron derivative (η^3-CH_3-$CHCHCH_2$)$Fe(CO)_3Cl$ (13, R = R' = H) *(154)*.

Under conditions similar to those used in these preparations of η^3-allylic-tetracarbonyliron cations, some isomerization reactions can occur *(97)*. For example, treatment of (*anti*-1-methylallyl)tetracarbonyliron tetrafluoroborate (15, $R^1 = CH_3$, $R^2 = R^3 = R^4 = R^5 = H$) with trifluoroacetic acid or sulfur dioxide at 60°C for 36 h results in rearrangement to the corresponding *syn*-isomer (15, $R^2 = CH_3$, $R^1 = R^3 = R^4 = R^5 = H$). The *anti,syn*-1,3-dimethyl cation 15 ($R^1 = R^4 = CH_3$, $R^2 = R^3 = R^5 = H$) undergoes a similar isomerization to the corresponding *syn,syn*-isomer 15 ($R^1 = R^3 = R^5 = H$, $R^2 = R^4 = CH_3$). In the case of the *anti*-1-isopropyl-2-methyl cation 15 ($R^1 = CH(CH_3)_2$, $R^2 = R^3 = R^4 = H$, $R^5 = CH_3$) the corresponding isomerization is more difficult and requires about 6 days at 70°C to produce 15 ($R^1 = R^3 = R^4 = H$, $R^2 = CH(CH_3)_2$, $R^5 = CH_3$).

The intermediates in the protonation of diene-tricarbonyliron complexes have been investigated by NMR spectroscopy *(208,211)*. Covalent η^3-allylic-tricarbonyliron trifluoroacetates appear to be intermediates in protonations in trifluoroacetic acid solution *(208)*.

Other reactions of diene-tricarbonyliron derivatives with electrophilic reagents also lead to η^3-allylic derivatives. Treatment of (butadiene)tricarbonyliron with boron trifluoride in sulfur dioxide solution gives the yellow crystalline η^3-allylic derivative 16 of which the structure has been deter-

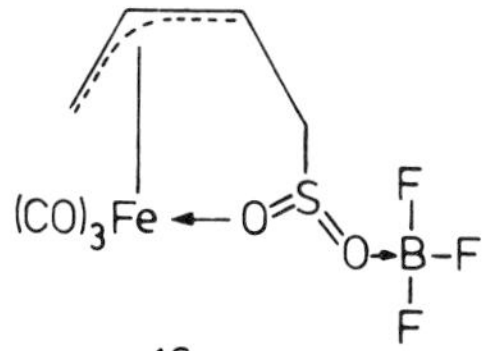

16

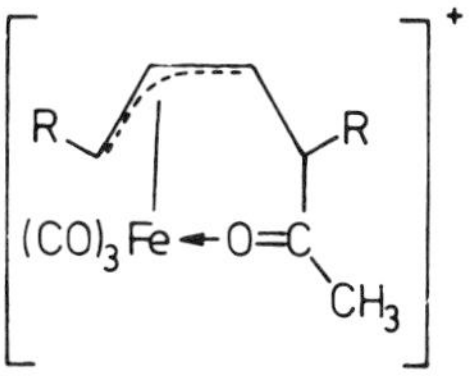

17

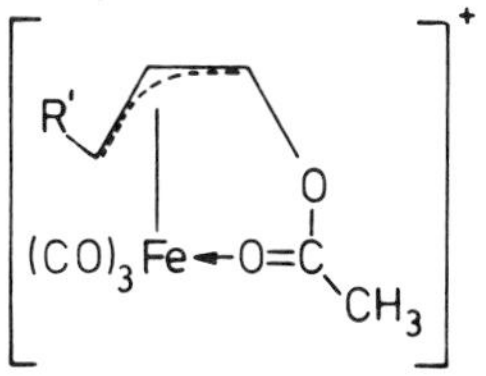

18

mined by X-ray crystallography *(43,45)*. An intermediate in the Friedel-Crafts acetylation of (butadiene)tricarbonyliron *(98)* is the η^3-allylic cation shown by X-ray crystallography *(109)* to have structure 17 (R = H). The analogous intermediate from the Friedel-Crafts acetylation of (*trans,trans*-2,4-hexadiene)tricarbonyliron has been shown by X-ray crystallography to have structure 17 ($R = CH_3$) indicating stereospecific *endo* attack *(99)*. Related η^3-allylic derivatives 18 ($R = CH_3$, R' = H; R = H, $R' = C_6H_5$; $R = R' = C_6H_5$) can be obtained by the acetylation of the tricarbonyliron or tetracarbonyliron complexes of the enones R'-CH=CH-CO-R with acetyl tetrafluoroborate in nitromethane *(171)*.

Some carbonyliron derivatives containing η^3-allyl-σ-alkyl ligands have been obtained by reactions of diene-tricarbonyliron complexes with various fluorocarbons. Thus, the η^3-allyl-σ-alkyl derivatives 19 (X = F, CF_3; R = R' = H; R =

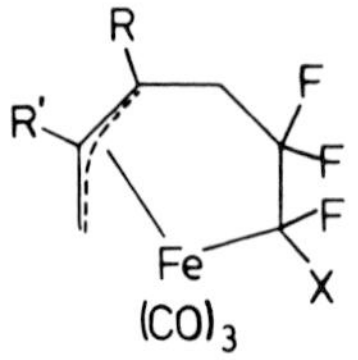

19

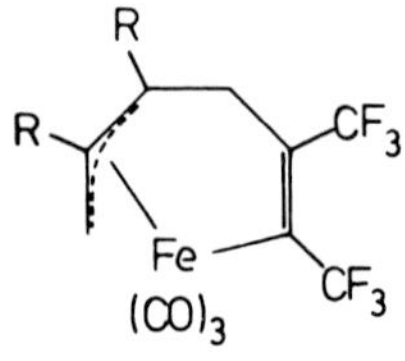

20

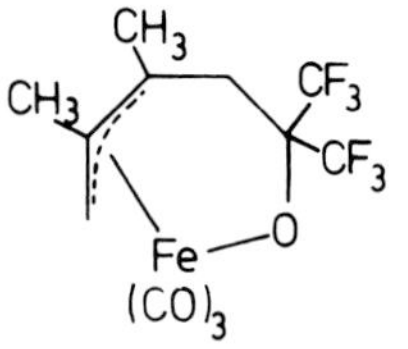

21

H, R' = CH_3; R = R' = CH_3) are obtained by the ultraviolet irradiations of tetrafluoroethylene *(23,26)* and hexafluoropropene *(104)*, respectively, with the tricarbonyliron complexes of butadiene, isoprene, and 2,3-dimethylbutadiene, respectively. A similar ultraviolet irradiation of chlorotrifluoroethylene with (2,3-dimethylbutadiene)tricarbonyliron gives the related η^3-allyl-σ-alkyl derivative 19 (X = Cl, R = R' = CH_3) *(104)*. Similarly, the ultraviolet irradiation of hexafluorobutyne with the tricarbonyliron complexes of butadiene and 2,3-dimethylbutadiene gives the η^3-allyl-σ-alkenyl derivatives 20 (R = H, CH_3) *(28)*. Ultraviolet irradiation of hexafluoroacetone with (2,3-dimethylbutadiene)tricarbonyliron gives the 1:1 adduct with structure 21 containing a chelating ligand which bonds to the iron both through an η^3-allylic group and an alkoxide oxygen atom *(101)*. A similar ultraviolet irradiation of hexafluoroacetone with (isoprene)tricarbonyliron gives a yellow 2:1 adduct 22 which rearranges to 23 upon heating in hexane to 80°C *(104)*.

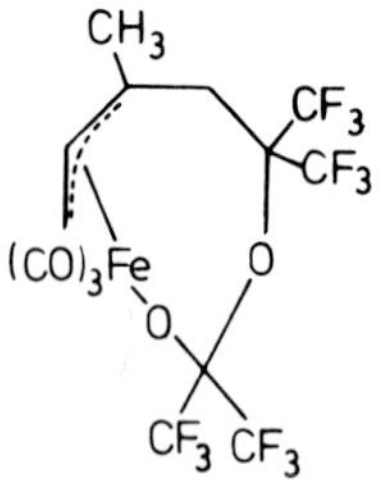

22

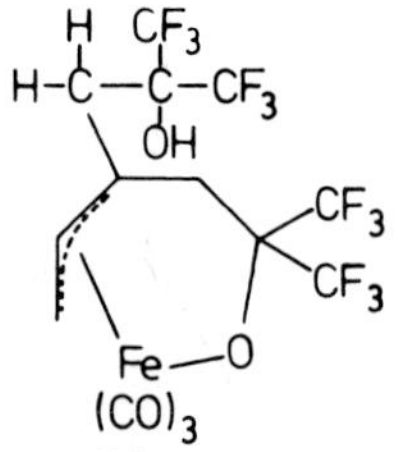

23

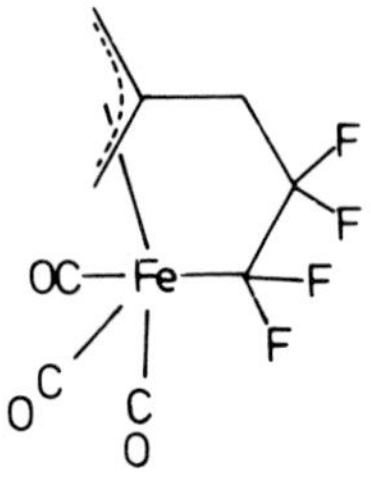

24

Some η^3-allylic-carbonyliron derivatives have been prepared from trimethylenemethane-tricarbonyliron derivatives. Thus (trimethylenemethane)tricarbonyliron, $[(CH_2)_3C]Fe(CO)_3$, reacts with one equivalent of bromine to give the (2-bromomethyl)allyl derivative $[\eta^3\text{-}BrCH_2\text{-}C(CH_2)_2]Fe(CO)_3Br$ *(78,79)*, with hydrogen chloride to give the methallyl derivative $[\eta^3\text{-}CH_3\text{-}C(CH_2)_2]Fe(CO)_3Cl$ *(79)*, and with tetrafluoroethylene under ultraviolet irradiation to give the crystalline 1:1 adduct 24 *(23)*.

B. *PHYSICAL AND SPECTROSCOPIC STUDIES*

One of the first research groups to prepare (η^3-allyl)-tricarbonyliron iodide *(183)* observed that the ^{1}H-NMR spectrum requires the presence of two different types of allyl groups present in unequal quantities. They therefore suggested an equilibrium between the monomer (η^3-C_3H_5)Fe(CO)$_3$I and the dimer [(η^3-C_3H_5)Fe(CO)$_3$I]$_2$. However, subsequent studies *(62, 165,166)* have shown that this explanation cannot be correct. In this connection a critical observation is that reliable molecular weight determinations in either benzene or cyclohexane *(62)* always give values corresponding to the monomer. For this reason an alternative explanation based on the presence of the two stereoisomers 25a and 25b (L = CO, X = I) for (η^3-C_3H_5)Fe(CO)$_3$I is indicated *(62,165,166)*. The stereo-

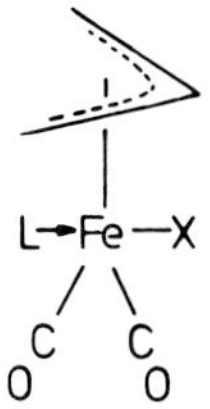

25a

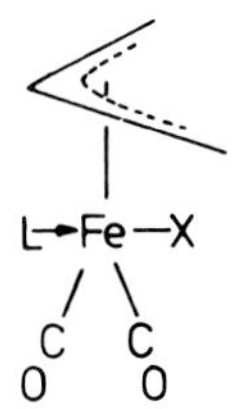

25b

isomers 25a and 25b differ only in the orientation of the η^3-allyl group relative to the X group. Existence of 25a and 25b as separately detectable species is only possible because of restricted rotation around the iron-allyl bond. The ^{13}C-NMR spectrum *(185)* of (η^3-C_3H_5)Fe(CO)$_3$I also shows resonances arising from the non-equivalent allyl and carbonyl carbons in both stereoisomers 25a and 25b (L = CO, X = I). The ^{1}H-NMR spectra of the other (η^3-allyl)tricarbonyliron halides (η^3-C_3H_5)Fe(CO)$_3$X (X = Br, Cl) also indicate the presence of stereoisomers 25a and 25b (L = CO; X = Br, Cl) *(166)*. However, the ^{1}H-NMR spectra of the corresponding nitrate, (η^3-C_3H_5)Fe(CO)$_3NO_3$, only indicate the presence of one stereoisomer *(166)*. The ^{1}H-NMR spectra *(166)* of the substituted η^3-allyl derivatives [η^3-CH_3-C(CH_2)$_2$]Fe(CO)$_3$Cl and [η^3-BrC(CH_2)$_2$]Fe(CO)$_3$Br indicate the presence of only a single isomer. The ^{1}H-NMR spectra of the phosphite complexes (η^3-C_3H_5)Fe(CO)$_2$[P(OR)$_3$]I (R = CH_3, C_2H_5) also indicate the presence of two stereoisomers *(62)*, possibly 25a and 25b (L = P(OCH_3)$_3$, P(OC_2H_5)$_3$).

X-ray crystallography has been used to determine the structures of some η^3-allyl-carbonyliron derivatives. The structure of isomer 25a (L = CO, X = I) has been determined *(146,147)*. The symmetry of the η^3-allyl group in 25a is confirmed. The structures of the two polymorphic modifications of the triphenylphosphine derivative $(\eta^3-C_3H_5)Fe(CO)_2-[P(C_6H_5)_3]I$ have also been determined *(148)*. Both modifications have essentially the same molecular structure 25a (X = I, L = $P(C_6H_5)_3$) in which the phosphine ligand is in the position *trans* to the iodine atom. Replacement of a carbonyl ligand with triphenylphosphine in $(\eta^3-C_3H_5)Fe(CO)_3I$ to give $(\eta^3-C_3H_5)Fe(CO)_2[P(C_6H_5)_3]I$ results in a shortening of the iron-iodine distance from 2.75 Å to 2.65 Å in accord with the lower π-acceptor strength of triphenylphosphine relative to carbon monoxide.

Some other physical and spectroscopic studies on η^3-allyl-carbonyliron derivatives have been reported. The infrared specific intensities of the carbonyl stretching frequencies of the η^3-allyl-tricarbonyliron halides $(\eta^3-R-C_3H_4)-Fe(CO)_3X$ (R = H, X = Cl, Br; R = CH_3, X = Cl, Br; R = CH_3-CO_2, X = Br) have been determined *(172)*. The Mössbauer spectra of various η^3-allyl-tricarbonyliron derivatives have also been investigated *(27,138)*. In going from $(\eta^3-C_3H_5)Fe(CO)_3Cl$ to $(\eta^3-C_3H_5)Fe(CO)_3I$, the values of both the isomer shifts and quadrupole splittings change significantly.

The mass spectra of a variety of η^3-allyl-tricarbonyliron derivatives of the type $(\eta^3-R-C_3H_4)Fe(CO)_3X$ (X = NO_3, R = H, 1-CH_3, 2-CH_3; X = Cl, R = H, 1-CH_3, 2-CH_3, 1-C_6H_5; X = Br, R = H, 1-CH_3, 2-CH_3, 1-C_6H_5, 2-C_6H_5, 2-Br; X = I, R = H, 2-CH_3) have been investigated *(131,169)*. The molecular ions are only intense for the iodides. The molecular ions decay along two principal routes involving Fe-X and Fe-CO bond rupture giving the families of ions $[RC_3H_4Fe(CO)_nX]^+$ (n = 2,1, 0) and $[RC_3H_4Fe(CO)_n]^+$ (n = 3,2,1,0). In the case of $(\eta^3-C_3H_5)Fe(CO)_3I$ the families of ions $[C_2H_2Fe(CO)_n]^+$ (n = 3,2,1, 0) resulting from methyl iodide elimination, and $[Fe(CO)_n]^+$ resulting from loss of both the η^3-allyl group and the iodine atom are also found *(131)*. The ratio of the ion intensities from Fe-CO and Fe-X bond rupture depends upon the nature of the substituent R and the ligand X. As the electronegativity of X is increased in the series I<Br<Cl<NO_3, rupture of the Fe-X bond over rupture of the Fe-CO bond is favoured.

C. *CHEMICAL REACTIONS*

Chemical reactions of the η^3-allyl-tricarbonyliron halides can involve either carbonyl displacement, halide displacement, or both of these processes. In some of the reactions of η^3-allyl-tricarbonyliron halides loss of the allyl group also occurs. Reactions of the η^3-allyl-carbonyliron cations, on the other hand, generally involve nucleophilic attack on the η^3-allyl ligand to give substituted η^2-olefin derivatives. In view of the generally different types of reactions of η^3-allyl-tricarbonyliron halides and of η^3-allylcarbonyliron cations, the chemistry of these two classes of compounds is discussed separately.

1. Reactions of η^3-Allyl-tricarbonyliron Halides

One carbonyl group in the halides $(\eta^3\text{-}C_3H_5)Fe(CO)_3X$ is readily replaced with trivalent phosphorus ligands by reactions at room temperature *(104,110,165)*. In this way the compounds $(\eta^3\text{-}C_3H_5)Fe(CO)_2(PR_3)X$ ($R = C_6H_5$, X = Br, I; $R = OCH_3$, OC_2H_5, X = I) have been prepared. However, reaction of $(\eta^3\text{-}C_3H_5)Fe(CO)_3I$ with tris(dimethylamino)phosphine at room temperature results in loss of both allyl and iodide to give *trans*-$[(Me_2N)_3P]_2Fe(CO)_3$ *(127)*. Reactions of $(\eta^3\text{-}C_3H_5)\text{-}Fe(CO)_3X$ (X = Br, I) with dimethylsulfoxide at room temperature results in the formation of dimethylsulfoxide complexes of iron(II) and iron(III) halides containing neither allyl nor carbonyl groups *(165)*.

Some examples of halide exchange reactions of η^3-allyl-tricarbonyliron halides are known. Thus, the bromide $(\eta^3\text{-}C_3H_5)Fe(CO)_3Br$ can be obtained by treatment of the corresponding iodide with tetraethylammonium bromide in chloroform *(161)*. Tetramethylammonium halides in methanol are useful for converting $[\eta^3\text{-}CH_3\text{-}C(CH_2)_2]Fe(CO)_3Cl$ into the corresponding bromide *(169)* and for converting $(\eta^3\text{-}C_3H_5)Fe(CO)_3I$ into the corresponding chloride *(161)*.

In other cases reactions of η^3-allyl-tricarbonyliron halides with silver salts are useful for preparing other $(\eta^3\text{-}C_3H_5)Fe(CO)_3X$ derivatives. Reaction of $(\eta^3\text{-}C_3H_5)Fe(CO)_3I$ with silver nitrate in nitromethane gives the yellow nitrate $(\eta^3\text{-}C_3H_5)Fe(CO)_3NO_3$ *(161)*. The corresponding (η^3-2-methylallyl)tricarbonyliron nitrate can be prepared analogously in methanol solution *(169)*. However, the unsubstituted $(\eta^3\text{-}C_3H_5)Fe(CO)_3NO_3$ easily decomposes in methanol to give diallyl ketone and a second incompletely characterized ketone. Both ketones were isolated as their 2,4-dinitrophenylhydrazones *(163)*. Reactions of $(\eta^3\text{-}C_3H_5)Fe(CO)_3I$ with the silver perfluorocarboxylates $R_f\text{-}CO_2Ag$ ($R_f = CF_3$, C_2F_5) in dichloromethane at room temperature form the corresponding perfluorocarboxylates $(\eta^3\text{-}C_3H_5)Fe(CO)_3CO_2R_f$ as yellow crystalline solids *(129)*. The reactions of the trifluoroacetate $(\eta^3\text{-}C_3H_5)Fe(CO)_3CO_2CF_3$ with several trivalent phosphorus ligands have been investigated *(130)*. However, the only pure product isolated was the monocarbonyl $CF_3\text{-}CO_2\text{-}Fe(CO)[$*cis*-$(C_6H_5)_2P\text{-}CH{=}CH\text{-}P(C_6H_5)_2](\eta^3\text{-}C_3H_5)$.

Some reactions of (η^3-allyl)tricarbonyliron iodide give novel β-diketone complexes *(73,187)*. Thus, the treatment of (η^3-C_3H_5)$Fe(CO)_3I$ with the β-diketones R-CO-CH_2-CO-R' (R = R' = CH_3, C_6H_5; R = CF_3, R' = thienyl) either in the presence of diethylamine in ethanol solution or with the sodium enolates of the β-diketones at 60°C results in loss of all three carbonyl groups to give the red crystalline complexes (η^3-C_3H_5)Fe(RCOCHCOR')$_2$ (26, R = R' = CH_3, C_6H_5; R = CF_3, R' =

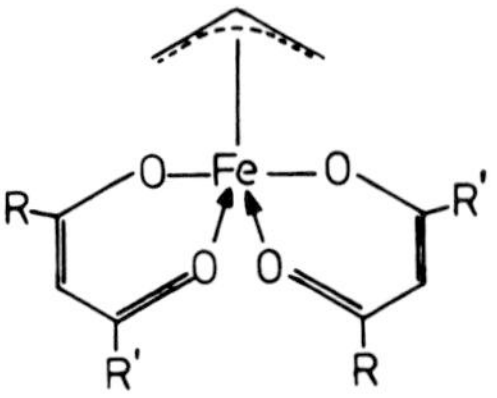

26

thienyl). These complexes may be viewed as isoelectronic to iron(III) β-diketonates in which an η^3-allyl group replaces one of the three β-diketonate ligands.

Some unsual products have been obtained by reactions of (η^3-C_3H_5)$Fe(CO)_3I$ with potassium polypyrazolylborates *(133, 134)*. Reaction of (η^3-C_3H_5)$Fe(CO)_3I$ with potassium bispyrazolylborate in 1:1 diethyl ether/tetrahydrofuran at room temperature results in displacement of both the allyl and the iodide to give yellow-orange $[H_2B(C_3H_3N_2)_2]_2Fe(CO)_2$ (27).

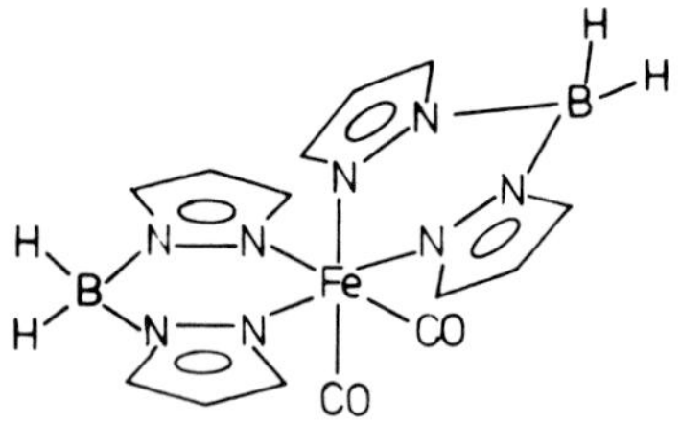

27

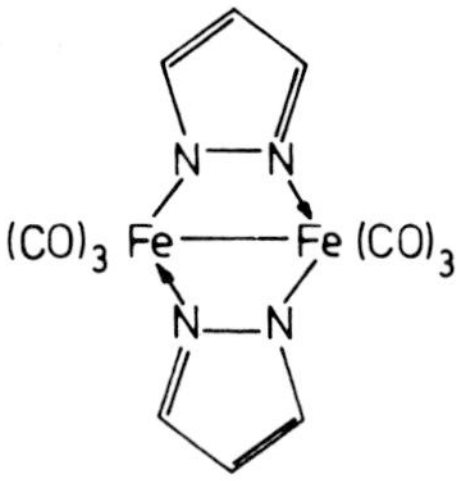

28

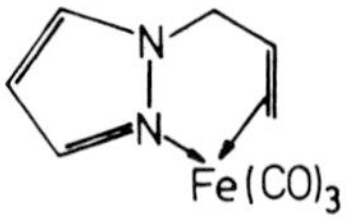

29

Reaction of (η^3-C_3H_5)$Fe(CO)_3I$ with potassium trispyrazolylborate under similar conditions gives the following products: (1) violet $[HB(C_3H_3N_2)_3]_2Fe$, (2) orange $[C_3H_3N_2Fe(CO)_3]_2$ (28) containing two bridging pyrazole ligands, (3) yellow (C_3H_5-$N_2C_3H_3$)$Fe(CO)_3$ (29), (4) yellow *trans*-CH_3-CH=CH-$Fe(CO)_2[(C_3H_3N_2)_3BH]$ (30) in which the η^3-allyl group has rearranged to a σ-propenyl group, (5) yellow *cis*-CH_3-CH=CH-CO-$Fe(CO)_2[(C_3H_3N_2)_3BH]$ (31). Heating 31 in boiling hexane results in its quantitative decarbonylation to give 30.

The reductions of the halides (η^3-C_3H_5)$Fe(CO)_3X$ have been studied both polarographically and chemically. Mössbauer

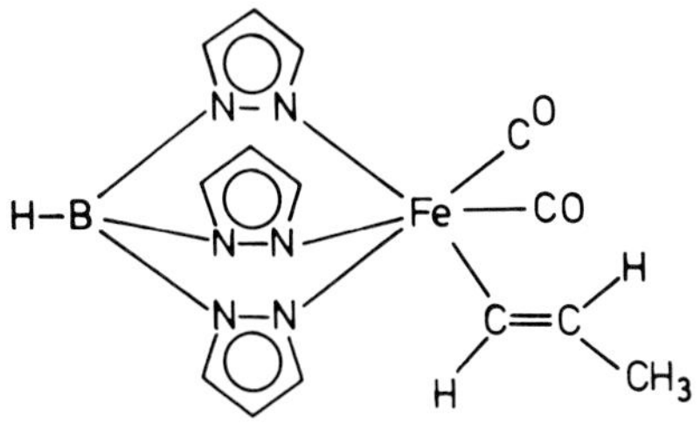

30

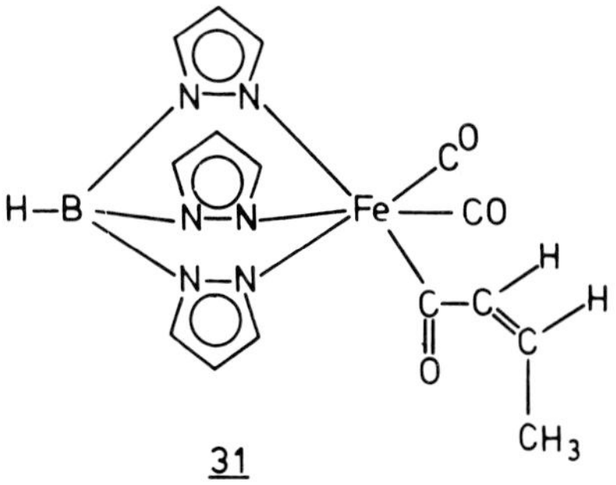

31

spectra of the reduction products have also been investigated *(27)*. Electrochemical studies on $(\eta^3\text{-}C_3H_5)Fe(CO)_3X$ (X = Cl, Br, I) in acetonitrile solution or in a freshly prepared dimethylformamide solution indicate two successive one-electron reductions to give first the radical $[(\eta^3\text{-}C_3H_5)Fe(CO)_3]^{\bullet}$ and then the anion $[(\eta^3\text{-}C_3H_5)Fe(CO)_3]^-$ *(108)*. The radical $[(\eta^3\text{-}C_3H_5)Fe(CO)_3]^{\bullet}$ can be isolated as a red extremely air-sensitive volatile solid either by the reduction of $(\eta^3\text{-}C_3H_5)$-$Fe(CO)_3Br$ with $NaMn(CO)_5$ or $NaFe(CO)_2(\eta^5\text{-}C_5H_5)$, by chromatography of $(\eta^3\text{-}C_3H_5)Fe(CO)_3X$ on deactivated alumina, or as a by-product from the reaction of allyl chloride with enneacarbonyldiiron *(157)*. Concentrated solutions of $(\eta^3\text{-}C_3H_5)$-$Fe(CO)_3$ are red whereas dilute solutions are green indicating an equilibrium between a reactive paramagnetic monomeric radical (green) and a diamagnetic dimer (red). The $[(\eta^3\text{-}C_3H_5)$-$Fe(CO)_3]^{\bullet}$ radical is also generated in tetrahydrofuran solution *(64)* by reduction of $(\eta^3\text{-}C_3H_5)Fe(CO)_3I$ with ytterbium, samarium, yttrium, or manganese. Similar reductions of $(\eta^3\text{-}1\text{-}R\text{-}C_3H_4)Fe(CO)_3X$ (R = $CH_3\text{-}CO_2$, CH_3) give the corresponding radicals $[(\eta^3\text{-}1\text{-}R\text{-}C_3H_4)Fe(CO)_3]^{\bullet}$ but these are much less stable than the unsubstituted radicals *(157)*. Similar reduction processes have also been used to prepare the phosphine substituted radicals $[(\eta^3\text{-}C_3H_5)Fe(CO)_2PR_3]^{\bullet}$ (R = C_6H_5, C_4H_9) and $\{(\eta^3\text{-}C_3H_5)Fe(CO)[P(n\text{-}C_4H_9)_3]_2\}^{\bullet}$ which appear as pure monomers. The electron spin resonance spectra of these radicals have been observed *(157)*. In the phosphine substituted η^3-allyl-carbonyliron radicals the phosphorus hyperfine splitting is in the range of 17 to 14 Gauss.

The second reduction product of the halides is the anion $[(C_3H_5)Fe(CO)_3]^-$ *(108)*. This anion is formed upon reduction of the halides $(\eta^3\text{-}C_3H_5)Fe(CO)_3X$ with sodium amalgam. It has been characterized by its infrared spectrum in the ν(CO) region, by its Mössbauer spectrum, and by its reaction with allyl chloride to give bis(η^3-allyl)dicarbonyliron (see below). Reaction of $[(\eta^3\text{-}C_3H_5)Fe(CO)_3]^-$ with triphenyltin chloride was found to give hexaphenylditin rather than a product with an iron-tin bond.

(η^3-Allyl)tricarbonyliron bromide is one of several allyl transition metal derivatives which have been studied as a catalyst for the coupling of allyl bromide or crotyl chloride with organomagnesium compounds *(173)*.

2. Reactions of η^3-Allyl-tetracarbonyliron Cations

Some reactions of η^3-allyl-tetracarbonyliron cations with nucleophiles have been investigated. Reactions of η^3-allyl-tetracarbonyliron cations with triphenylphosphine give *cis*-allyltriphenylphosphonium salts *(206)*. In related reactions η^3-allyl-tetracarbonyliron cations have been used to allylate diethylamine, the anion of acetylacetone, and 1-phenylethylamine *(206)*. η^3-Allyl-tetracarbonyliron cations also react with pyridine to give allylpyridinium derivatives *(206)*. These reactions proceed rapidly in tetrahydrofuran at room temperature using the tetrafluoroborate salts of the η^3-allyl-tetracarbonyliron cations *(206)*. Reaction of the cation $[(\eta^3\text{-}CH_3\text{-}CHCHCH_2)Fe(CO)_4]^+$ obtained by the protonation of butadiene with dialkylcadmium derivatives in diethyl ether or tetrahydrofuran gives the corresponding unstable yellow liquid η^2-olefin-tetracarbonyliron derivatives $[\eta^2\text{-}(R\text{-}CH_2\text{-}CH{=}CH\text{-}CH_3)]Fe(CO)_4$, which are readily oxidized in air to the corresponding free olefins *(181)*.

An interesting transformation of (η^2-tetramethylallene)-tetracarbonyliron to substituted η^4-butadiene-tricarbonyl-iron derivatives through η^3-allyl-tetracarbonyliron cations has been reported *(93)*. Friedel-Crafts acylation of (η^2-tetramethylallene)tetracarbonyliron with the acid chlorides R-COCl (R = CH_3, C_6H_5) in dichloromethane solution in the presence of aluminium chloride gives the cations 32 (R = CH_3, C_6H_5) isolated as their tetrachloroaluminate salts. Heating

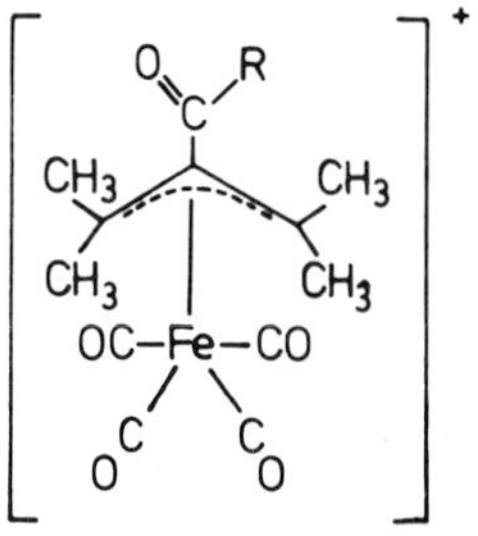

32

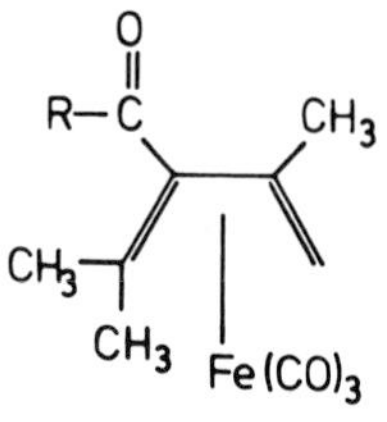

33

these η^3-allyl-tetracarbonyliron derivatives in acetone solution results in both deprotonation and decarbonylation to give the substituted diene-tricarbonyliron derivatives 33 (R = CH_3, C_6H_5).

III. BIS(η^3-ALLYLIC) CARBONYLIRON DERIVATIVES

The only known compound of this type with acyclic *tri-hapto*-allyl groups is the unsubstituted $(\eta^3\text{-}C_3H_5)_2Fe(CO)_2$ (34). This complex can be prepared as a yellow-orange crystalline solid volatile at 20-25°C/5-10 Torr by the reduction of $(\eta^3\text{-}C_3H_5)Fe(CO)_3I$ with sodium amalgam in tetrahydrofuran followed by reaction of the resulting $Na[(\eta^3\text{-}C_3H_5)Fe(CO)_3]$ with allyl bromide at room temperature *(162,165)*. This complex 34 is unstable at room temperature even in an inert

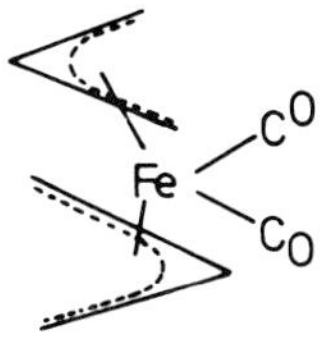

34

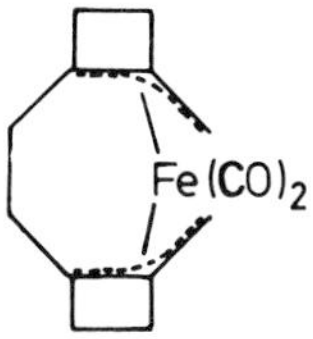

35

atmosphere and can decompose explosively to 1,5-hexadiene (biallyl) upon exposure to air. However, $(\eta^3\text{-}C_3H_5)_2Fe(CO)_2$ (34) is unreactive towards Lewis base ligands such as triphenylphosphine, triphenyl phosphite, and pyridine. The ^{1}H-NMR spectrum of $(\eta^3\text{-}C_3H_5)_2Fe(CO)_2$ is temperature dependent *(26,165)*. At -70°C the allyl groups are non-equivalent in the ^{1}H-NMR spectrum whereas at +20°C both allyl groups become equivalent in the ^{1}H-NMR spectrum. The rotation barrier in 34 around the iron-allyl bonds can be estimated at 5 kcal/mole on the basis of these ^{1}H-NMR data *(26,165)*.

The instability of $(\eta^3\text{-}C_3H_5)_2Fe(CO)_2$ (34) is also indicated by the observation that it is not isolated as a product from the reaction of allyl halides with any of the potassium iron carbonylates $K_2Fe(CO)_4$, $KHFe(CO)_4$, or $K_2Fe_2(CO)_8$ in ethanol *(200)*. Such reactions give a good yield of propene as well as significant amounts of 1,5-hexadiene as a by-product.

Compounds of the type $(\eta^3\text{-allyl})_2Fe(CO)_2$ appear to be more stable if the two (η^3-allyl) units are part of a chelating ligand. Thus, a major product from the reaction of 1,2-dimethylenecyclobutane with dodecacarbonyltriiron in boiling benzene is an air-stable yellow volatile solid of the stoichiometry $(C_{12}H_{16})Fe(CO)_2$ *(137)*. A careful analysis of the ^{13}C-NMR spectrum of this complex suggests structure 35 in which the $C_{12}H_{16}$ unit functions as a bis(η^3-allyl) ligand.

IV. η^3-ALLYLIC-CARBONYLIRON COMPLEXES AS REACTION INTERMEDIATES

η^3-Allyl-tricarbonyliron hydride derivatives of the general type 36 have been postulated as intermediates in the

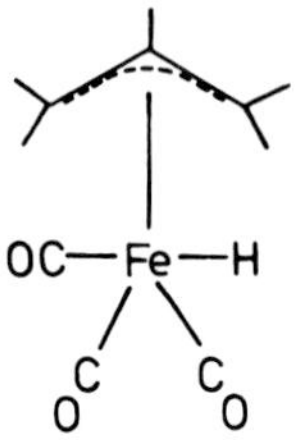

36

carbonyliron catalysed isomerizations of terminal to internal olefins *(38,143)*, of allyl alcohol to propionaldehyde *(63,85, 113)*, and of 1,4-dienes to 1,3-dienes *(3)*. Experimental evidence in favour of this postulate includes the following: (1) The observation of intramolecular 1,3 hydrogen shifts in the isomerization of 3-ethylpent-1-ene-(*3-d*$_1$) to 3-ethylpent-2-ene catalysed by dodecacarbonyltriiron *(38)*; (2) the formation of CH_2D-CH_2-CDO upon isomerization of CH_2=CH-CD_2OH with pentacarbonyliron *(113)*; (3) the rearrangement of *endo*-α-1-hydroxy-5,6-dihydro(dicyclopentadiene) to tetrahydro(dicyclopentadien)-1-one with pentacarbonyliron *(63)*; (4) the distribution of the deuterium in the (1,3-cyclohexadiene)tricarbonyliron formed by reaction of pentacarbonyliron with 1,4-cyclohexadiene-*(3,3,6,6-d*$_4$*)* *(3)*.

V. η^3-ALLYL NITROSYLIRON DERIVATIVES

Previous sections of this chapter have discussed η^3-allyl-carbonyliron derivatives of the types (η^3-allyl)-$Fe(CO)_3X$, $[(\eta^3\text{-allyl})Fe(CO)_4]^+$, and $(\eta^3\text{-allyl})_2Fe(CO)_2$ all of which have the favoured 18-electron noble gas configuration. Additional η^3-allyl-carbonyliron derivatives with the 18-electron configuration include the nitrosyls of the type (η^3-allyl)$Fe(CO)_2NO$.

Two general methods are available for conversion of carbonyliron complexes into (η^3-allyl)$Fe(CO)_2NO$ derivatives. In the most convenient method pentacarbonyliron is first allowed to react with sodium nitrite to form the salt Na-$[Fe(CO)_3NO]$. Reaction of this salt with allylic halides in diethyl ether gives the corresponding (η^3-R-C_3H_4)$Fe(CO)_2NO$ derivatives (R = H, 1-CH_3, 2-CH_3, 1-Cl, 2-Cl, 2-Br, 1-C_6H_5) *(30,36,39)*. In a variant of this method suitable for certain substituted η^3-allylic derivatives the salt $Na[Fe(CO)_3NO]$ is allowed to react with acetic acid at -78°C. The resulting unstable $HFe(CO)_3NO$ is then treated at -78°C with a 1,3-diene, and the mixture is allowed to warm to room temperature. Such reactions using butadiene and isoprene give the allylic derivatives $(\eta^3\text{-RR'CCHCH}_2)Fe(CO)_2NO$ (R = H, R' = CH_3; R = R' = CH_3) *(39)*.

The second general method for the conversion of carbonyliron complexes to $(\eta^3\text{-allyl})Fe(CO)_2NO$ derivatives uses first the reaction of $Fe_2(CO)_9$ with the allylic halide to give the corresponding $(\eta^3\text{-allylic})Fe(CO)_3X$ derivatives as discussed above. Reactions of these halides with nitric oxide result in the replacement of the halide and one of the carbonyl groups to give the corresponding $(\eta^3\text{-allylic})Fe(CO)_2NO$ derivatives *(30,158)*. This method has been used to prepare the derivatives $(\eta^3\text{-R-CHCHCH}_2)Fe(CO)_2NO$ (R = H, CH_3, $CH_3\text{-}CO_2$). The reaction of $(\eta^3\text{-}C_3H_5)Fe(CO)_2[P(C_6H_5)_3]I$ with nitric oxide results in the displacement of the triphenylphosphine rather than a carbonyl group to give $(\eta^3\text{-}C_3H_5)Fe(CO)_2NO$ *(30,158)*.

The compounds of the type $(\eta^3\text{-allylic})Fe(CO)_2NO$ are dark red air-sensitive liquids which are stable under nitrogen at 0°C and which can be readily distilled in vacuum at room temperature. The $(\eta^3\text{-allyl})Fe(CO)_2NO$ derivatives are often more readily characterized as their red crystalline triphenylphosphine derivatives $(\eta^3\text{-allyl})Fe(CO)(NO)[P(C_6H_5)_3]$, which are obtained by warming with triphenylphosphine in toluene *(36)* or tetrahydrofuran *(39)* solutions.

A more detailed study of the reactions of $(\eta^3\text{-allylic})$-$Fe(CO)_2NO$ derivatives with phosphines indicates that these reactions proceed through an intermediate *monohapto*-allyl derivative of the type $(\eta^1\text{-allylic})Fe(CO)_2(NO)L$ *(35,36,37)*. For example, triphenylphosphine reacts rapidly with the 2-chloroallyl derivative $[\eta^3\text{-ClC}(CH_2)_2]Fe(CO)_2NO$ in benzene at room temperature without gas evolution to give a red solid indicated by its infrared ν(CO) frequencies and ^{1}H-NMR spectra to be the η^1-allyl derivative $H_2C{=}CCl\text{-}CH_2\text{-}Fe(CO)_2(NO)$-$[P(C_6H_5)_3]$. However, in toluene at 50°C this complex loses one equivalent of carbon monoxide to form the η^3-allylic derivative $[\eta^3\text{-ClC}(CH_2)_2]Fe(CO)(NO)[P(C_6H_5)_3]$. In similar reactions of $(\eta^3\text{-R-}C_3H_4)Fe(CO)_2NO$ (R = H, 1-CH_3, 2-CH_3, 1-C_6H_5) the loss of carbon monoxide upon reaction with trivalent phosphorus ligands was found to be too rapid for detection of an η^1-allylic intermediate, even by spectroscopic methods. However, in the reactions of $(\eta^3\text{-R-}C_3H_4)Fe(CO)_2NO$ (R = 1-Cl, 1-CN, 2-Cl, 2-Br) with the trivalent phosphorus ligands $(C_6H_5)_3P$, $(n\text{-}C_4H_9)_3P$, $(C_6H_5)_2PC_2H_5$, $P(OC_2H_5)_3$, and $P(OCH_2)_3\text{-}CC_2H_5$, the intermediate η^1-allyl derivative can be detected by infrared spectroscopy in the ν(CO) and ν(NO) regions immediately after mixing the reactants in toluene solution but before carbon monoxide evolution begins. Reactions between $(\eta^3\text{-RC}_3H_4)Fe(CO)_2NO$ (R = 1-Cl, 2-Cl, 2-Br) with the phosphites $(RO)_3P$ (R = CH_3, C_2H_5) also result in insertion of the liberated carbon monoxide to give the corresponding 2-butenoyl complexes 37. The relative rates of the reactions of $(\eta^3\text{-R-C}_3H_4)Fe(CO)_2NO$ with these ligands occur in the

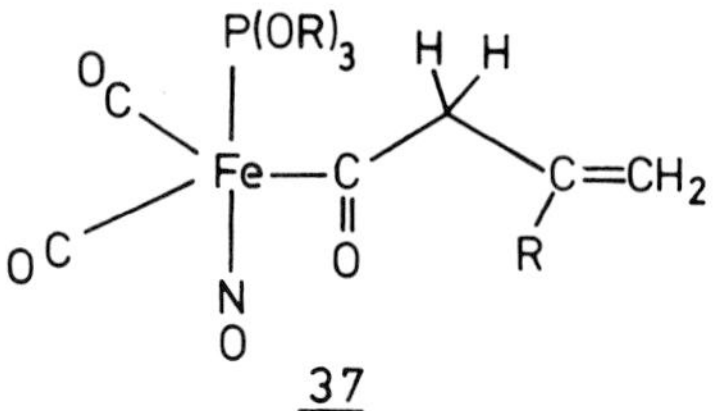

37

sequence R = 1-CH_3, 2-CH_3 < H < 1-Cl, 2-Cl. This is in accord with expectations based on the inductive effects of the substituents but is in contrast to the corresponding $(\eta^3$-allylic$)Co(CO)_3$ derivatives where some anomalies are observed *(51)*.

Some physical and spectroscopic studies have been carried out on η^3-allyl-carbonyl-nitrosyliron derivatives. The vibrational spectra of $(\eta^3\text{-C}_3H_5)Fe(CO)_2NO$ and $(\eta^3\text{-C}_3D_5)$-$Fe(CO)_2NO$ have been studied in the liquid and solid states *(177)*. The spectra can be interpreted on the basis of C_s molecular symmetry. The assignment of some of the bands in the allylic fragment is based on ν_H/ν_D isotopic shifts. The Mössbauer and infrared spectra of $(\eta^3\text{-C}_3H_4R)Fe(CO)_2NO$ (R = H, 1-CH_3, 2-CH_3, 1-Cl) indicate that the carbonyl and nitrosyl groups absorb the inductive effects of allyl substituents, leaving the metal *s* electron density relatively unaffected *(49)*. From dipole moment measurements *(197)* on the complexes $[\eta^3\text{-RC}(CH_2)_2]Fe(CO)_2L$ (R = H, L = CO; R = H, L = $P(C_6H_5)_3$; R = CH_3, L = CO; R = Cl, L = CO; R = Br, L = CO) the following conclusions have been made: (1) The value of the group moment of $[(\eta^3\text{-C}_3H_5)Fe]$ is ca 1.6 D if μ(Fe-CO) = 0.5 D, μ(Fe-NO) = 1.3 D, and the CO-Fe-CO(NO) angle is 98°, with the positive charge on the iron in all cases; (2) a *trans* conformation (38) is most likely for $[\eta^3\text{-RC}(CH_2)_2]Fe(CO)_2NO$ (R =

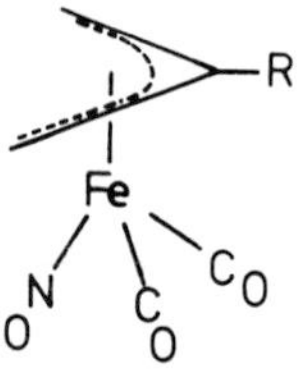

38

Cl, Br, CH_3). Polarographic reduction of $(\eta^3\text{-C}_3H_4R)Fe(CO)_2NO$

(R = H, 1-CH_3, 2-CH_3, 1-Cl, 2-Cl, 2-Br) indicates two reduction waves, but only the first wave was found to be amenable to detailed study *(176)*. Even the first wave arises from an irreversible process which appears to consist of reductive removal of the allyl group to give the unstable $[Fe(CO)_2NO]^-$ which forms the stable $[Fe(CO)_3NO]^-$ through either disproportionation or CO abstraction from unreduced (η^3-allylic)-$Fe(CO)_2NO$.

The η^3-allylic derivative (η^3-C_3H_5)$Fe(CO)_2NO$ is one of several metal carbonyl nitrosyls that have been evaluated for the catalytic dimerization of butadiene and isoprene under relatively mild conditions *(34)*. Dimerization of butadiene with 1 % by weight of (η^3-C_3H_5)$Fe(CO)_2NO$ at 100°C for 5 h gives nearly a quantitative conversion to 4-vinylcyclohex-1-ene. A similar dimerization of isoprene with (η^3-C_3H_5)-$Fe(CO)_2NO$ gives a mixture of 4-isopropenyl-1-methylcyclohex-1-ene, 1,5-dimethyl-5-vinylcyclohex-1-ene, and 1,4-dimethyl-4-vinylcyclohex-1-ene. The iodide (η^3-C_3H_5)$Fe(CO)_3I$ was found to be inactive as a catalyst for the formation of 4-vinylcyclohex-1-ene from butadiene.

A carbonyl-free η^3-allyl-nitrosyliron complex has also been prepared. Reaction of $[Fe(NO)_2Cl]_2$ with tetraallyltin in a mixture of benzene and tetrahydrofuran gives orange crystalline $[(\eta^3\text{-}C_3H_5)Fe(NO)_2]_2SnCl_2$ (39) *(144)*. This com-

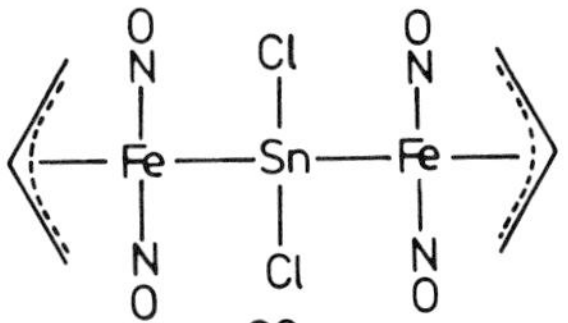

39

pound is relatively air-stable but slowly hydrolyzes in water. Apparently no other η^3-allyl-dinitrosyliron derivatives of the type $RFe(NO)_2(\eta^3\text{-}C_3H_5)$ are known.

VI. η^3-ALLYLIC IRON COMPOUNDS FROM ALLENES AND CUMULENES

Reactions of allene with carbonyliron complexes under various conditions give complexes containing η^3-allyl-iron bonds. Treatment of allene with enneacarbonyldiiron in an autoclave at 50°C gives a mixture of orange liquid (C_3H_4)-$Fe_2(CO)_7$ and red crystalline (C_6H_8)$Fe_2(CO)_6$ *(23)*. Reaction of the liquid (C_3H_4)$Fe_2(CO)_7$ with triphenylphosphine gives the two crystalline derivatives (C_3H_4)$Fe_2(CO)_6P(C_6H_5)_3$ and (C_3H_4)-$Fe_2(CO)_5[P(C_6H_5)_3]_2$. The molecular structure of (C_3H_4)Fe_2-$(CO)_6P(C_6H_5)_3$ has been found by X-ray crystallography to be 40 (L = $P(C_6H_5)_3$) *(65)*. In this structure the C_3H_4 unit is

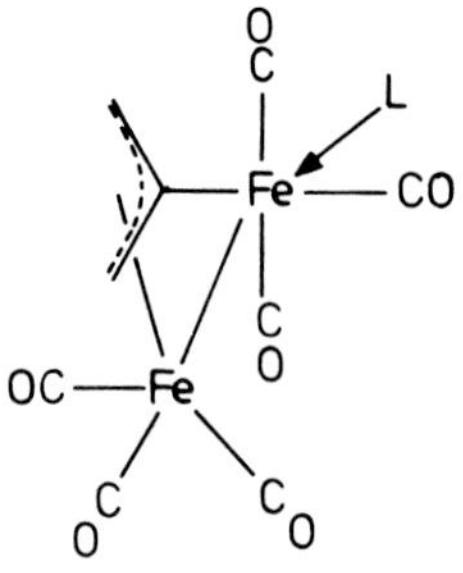

40

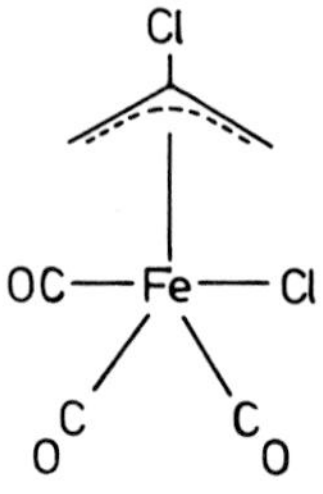

41

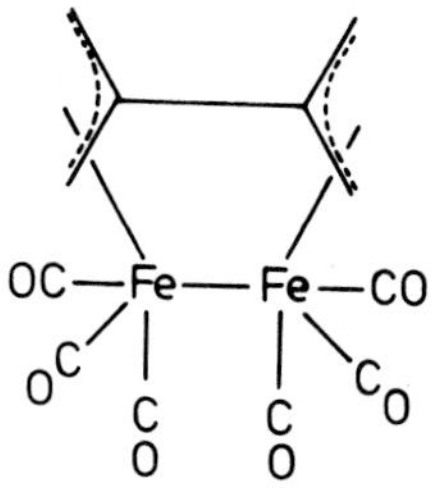

42

bonded to the $Fe(CO)_3$ group as a *trihapto* ligand. In addition the center carbon of the C_3H_4 unit forms a σ-bond with the iron in the $Fe(CO)_4$ group. Thus the C_3H_4 ligand in 40 is an example of an η^3-allyl-σ-alkyl ligand in which the two modes of bonding are orthogonal and thus must involve different metal atoms. The compound $(C_3H_4)Fe_2(CO)_7$ (40, L = CO) can also be obtained by reaction of $H_2C{=}CCl{-}CH_2Cl$ with $Fe_2(CO)_9$ to give $[\eta^3\text{-}ClC(CH_2)_2]Fe(CO)_3Cl$ (41) followed by dehalogenation with excess $Fe_2(CO)_9$ *(23)*.

The second product from the reaction of allene with enneacarbonyldiiron, red crystalline $(C_6H_8)Fe_2(CO)_6$, is also obtained by reaction of allene with $Fe_3(CO)_{12}$ in hexane solution at 85-90°C *(159,160,175)*. This complex is formulated as the tetramethyleneethane complex 42 in which the tetramethyleneethane ligand functions as a bis(η^3-allylic) ligand as in structure 4. The ^1H-NMR spectrum of 42 is temperature dependent *(160)*. A similar reaction between 1-phenylallene and enneacarbonyldiiron *(175)* gives three $[(C_6H_5)_2C_6H_6]Fe_2(CO)_6$ products which appear to be isomers of the diphenyl derivative of 42. Since six different disubstituted derivatives of 42 of the type $(R_2C_6H_6)Fe_2(CO)_6$ are possible, this system is obviously a very complicated one.

Reaction of excess allene with dodecacarbonyltriiron at 85-90°C gives not only 42 but also a yellow complex (isomer A) of stoichiometry $(C_9H_{12})Fe_2(CO)_6$ *(175)*. At 120°C this reaction also gives a second $(C_9H_{12})Fe_2(CO)_6$ isomer (isomer B) which also can be obtained by the pyrolysis of the (C_9H_{12})-$Fe_2(CO)_6$ isomer A in toluene for several minutes. However, pyrolysis of either isomer A or B of $(C_9H_{12})Fe_2(CO)_6$ in boiling toluene for several hours gives a third isomer C. Isomer A has also been obtained by treatment of $(C_6H_8)Fe_2(CO)_6$ with excess allene at 90-100°C. X-ray crystallography *(210)* indicates structure 44 for isomer B and structure 45 for isomer C. On the basis of spectroscopic and chemical evidence structure 43 is postulated for isomer A. Isomer A (43) can arise by allene insertion into the iron-iron bond and one of

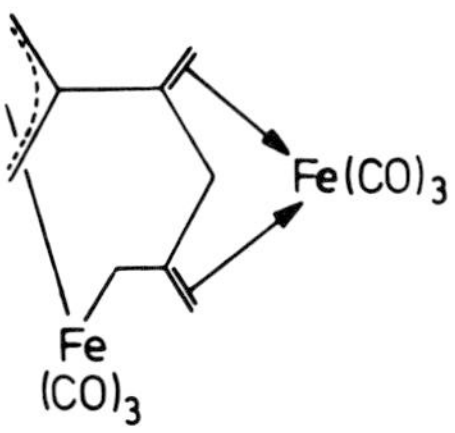

43

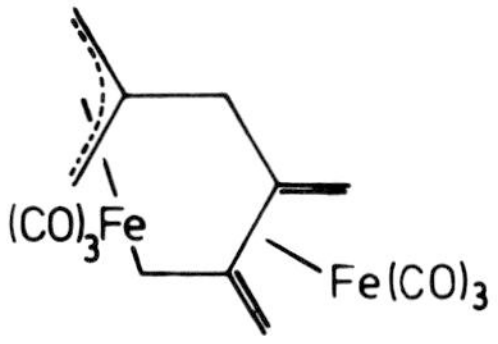

44

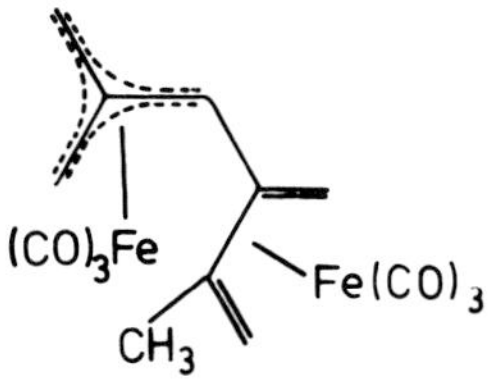

45

the iron-carbon bonds of 42. Upon conversion of isomer A (43) to isomer B (44) the C_9H_{12} skeleton is maintained but differently coordinated by eight of its carbon atoms to the two iron atoms. However, conversion of isomer B (44) to isomer C (45) involves rearrangement of the C_9H_{12} ligand to an isomeric η^8-C_9H_{12} ligand by a hydrogen shift *(210)*.

Some reactions of cyclic allenes with carbonyliron complexes have also been found to give products containing η^3-allyl-iron bonds. Reaction of 1,2-cycloundecadiene with enneacarbonyldiiron gives a crystalline heptacarbonyldiiron derivative $(C_{11}H_{18})Fe_2(CO)_7$. X-ray crystallography *(141)* indicates this complex to have structure 46 which is closely

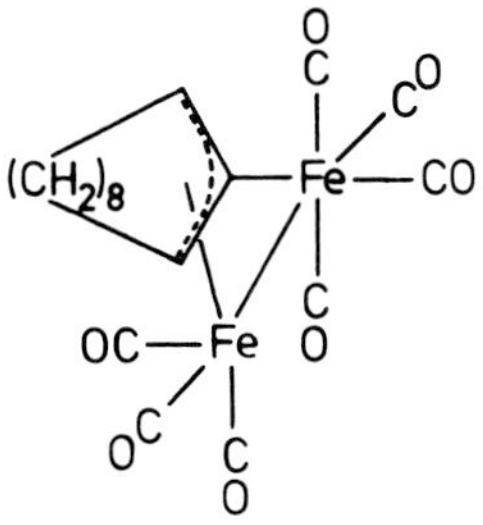

46

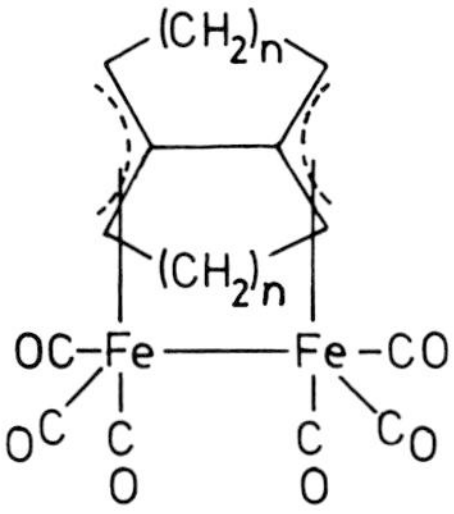

47

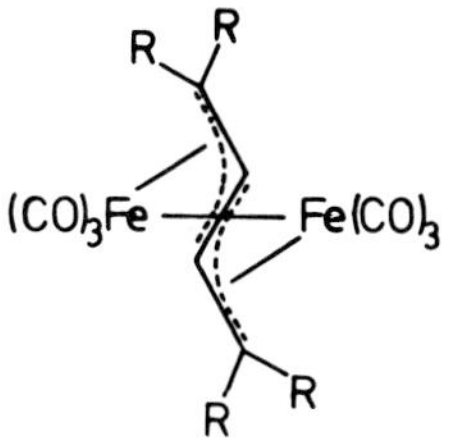

48

related to structure 40. Reactions of the macrocyclic diallenes 1,2,6,7-cyclodecatetraene and 1,2,9,10-cyclohexadecatetraene result in intramolecular transannular cyclization to give the red-orange bicyclic tetramethyleneethane-hexacarbonyldiiron derivatives $(C_{10}H_{12})Fe_2(CO)_6$ (47, n = 2) and $(C_{16}H_{24})Fe_2(CO)_6$ (47, n = 5), respectively *(135)*. The structures of 47 are closely related to the acyclic 42.

Compounds of the type 48 containing two orthogonal 1,2,3-*trihapto*-allyl units sharing two carbon atoms (structure 5) are obtained from reactions of carbonyliron reagents with butatrienes. Thus, the reaction of dodecacarbonyltriiron with tetraphenylbutatriene gives the tetraphenyl derivative 48 (R = C_6H_5) *(128,159)*. Since unsubstituted butatriene is unstable and difficult to handle, the unsubstituted complex $(C_4H_4)Fe_2(CO)_6$ (48, R = H) is best obtained by treatment of

1,4-dibromobutane with zinc in the presence of dodecacarbonyltriiron *(29,128,159)*. Reaction of tetra-*tert*-butylhexapentaene with dodecacarbonyltriiron gives a more complex derivative (48, 2R = $=C[C(CH_3)_3]_2$) containing the same type of butatriene-hexacarbonyldiiron system *(136)*. In this last reaction the terminal carbon-carbon double bonds, which are highly protected sterically by the *tert*-butyl substituents, are not involved in the complex formation.

VII. η^3-ALLYLIC CARBONYLIRON COMPOUNDS FROM REACTIONS OF UNSATURATED CARBOCYCLIC SYSTEMS WITH CARBONYLIRON COMPLEXES

Reactions of unsaturated cyclic hydrocarbons with carbonyliron complexes and further transformations of the resulting products give numerous η^3-allylic carbonyliron derivatives with diverse structures. Some of the simpler reactions of this type provide η^3-allylic carbonyliron complexes where the 1,2,3-*trihapto*-allyl unit is part of a carbocyclic ring. Other reactions give products containing a bidentate four-electron donor chelating ligand in which one arm bonds to the iron through an η^3-allylic bond and the other arm to the iron through a simple iron-carbon σ-alkyl or σ-acyl bond (see structures 2a and 2b, respectively). Binuclear bis(η^3-allyl)-hexacarbonyldiiron complexes (type 3a), the unsymmetrical derivatives 3b, and bis(η^3-allyl)-pentacarbonyldiiron complexes (type 6) also arise from reactions of cyclopolyolefins with carbonyliron reagents.

In this section, this rather extensive chemistry will first be classified as to whether the reacting carbocyclic compound is monocyclic or polycyclic, and then as to whether the resulting η^3-allylic carbonyliron complex is mononuclear or binuclear.

A. *η^3-ALLYLIC CARBONYLIRON COMPOUNDS FROM REACTIONS OF MONOCYCLIC OLEFINS WITH CARBONYLIRON COMPLEXES*

Reactions of cyclic dienes with carbonyliron complexes followed by further transformations of the resulting diene-tricarbonyliron derivatives can lead either to simple η^3-allylic-carbonyliron derivatives or to chelating η^3-allyl-σ-alkyl or η^3-allyl-σ-acyl derivatives. Reactions of cyclic trienes and tetraenes with carbonyliron complexes frequently give binuclear carbonyliron derivatives containing two η^3-allyl-iron bonds.

1. Mononuclear η^3-Allylic-carbonyliron Compounds from Reactions of Monocyclic Olefins with Carbonyliron Complexes

Addition reactions to cyclobutadiene-carbonyliron complexes provide sources of cyclobutenyl-carbonyliron derivatives. Reaction of the dicarbonyl(cyclobutadiene)nitrosyliron salt $[(\eta^4\text{-}C_4H_4)Fe(CO)_2NO][PF_6]$ with the tertiary phosphines $R_2R'P$ ($R = R' = CH_3$, C_2H_5, $n\text{-}C_3H_7$, $n\text{-}C_4H_9$, C_6H_5; $R = CH_3$, $R' = C_6H_5$; $R = C_6H_5$, $R' = CH_3$) in acetone at 20°C forms the orange-red crystalline adducts indicated to be the cyclobutenyl derivatives 49 on the basis of their infrared, ^1H-NMR,

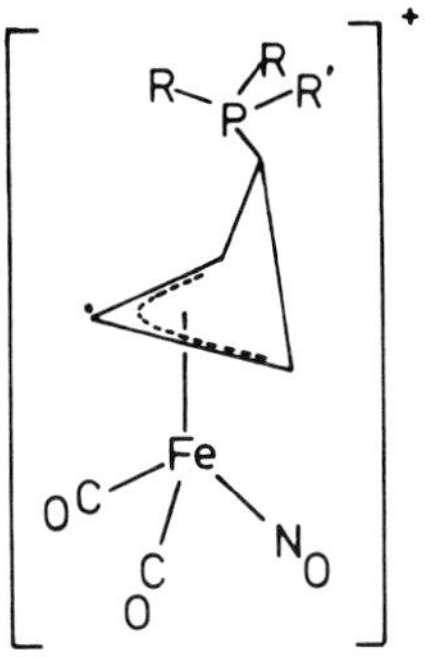

49

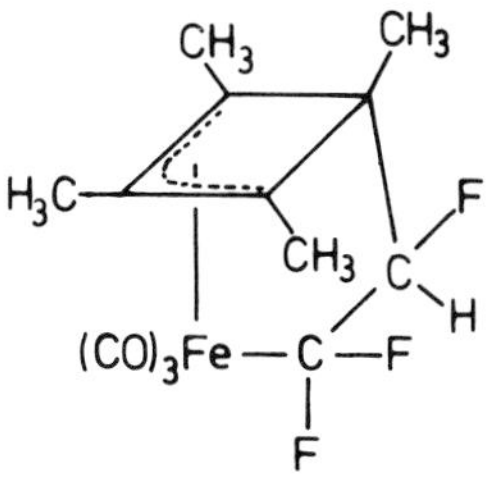

50

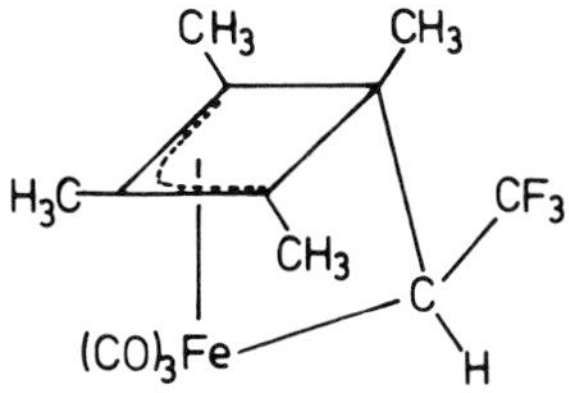

51

and Mössbauer spectra as well as an X-ray crystallography study on the trimethylphosphine derivative *(75)*. Ultraviolet irradiation of (tetramethylcyclobutadiene)tricarbonyliron with trifluoroethylene gives a mixture of two isomeric 1:1 adducts *(25)*. One of these products is the cyclobutenyl complex 50, a normal trifluoroethylene insertion product similar to compounds formed from many photochemical reactions of diene-tricarbonyliron complexes with fluoroolefins. The second 1:1 adduct is indicated by its ^1H- and ^{19}F-NMR spectra to be the isomeric cyclobutenyl complex 51 in which the trifluoroethylene has rearranged to 2,2,2-trifluoroethylidene before inserting into the cyclobutadiene-iron bond *(25)*. Pyrolysis of 50 in hexane solution at 100°C gives a mixture of the red-black (trifluoromethyltetramethylcyclopentadienyl)-dicarbonyliron dimer $\{[\eta^5\text{-}CF_3(CH_3)_4C_5]Fe(CO)_2\}_2$ and the keto-cyclopentenyl derivative 52. The latter complex is prepared in better yield (ca. 80 %) by insertion of CO into 51 by treatment with carbon monoxide at 80°C/100 atm *(25)*.

Some $(\eta^3\text{-allylic})Fe(CO)_2NO$ derivatives have been prepared containing cyclic η^3-allylic ligands *(39)*. Reactions of $NaFe(CO)_3NO$ with 3-chlorocyclopentene or 3-bromocyclohexene give the corresponding cycloalkenyl complexes 53 (L = CO, n = 2,3) as red distillable liquids. Reactions of these red

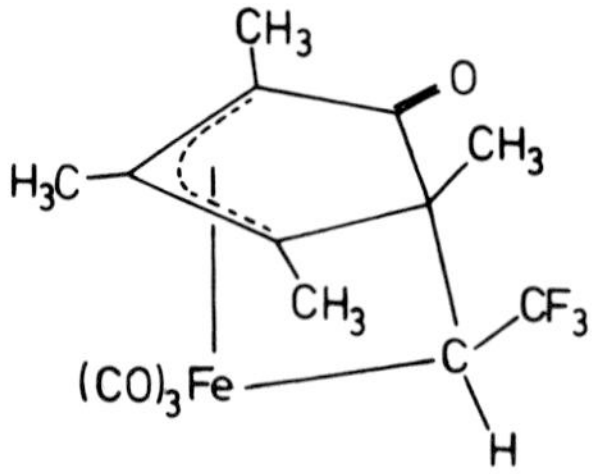

52

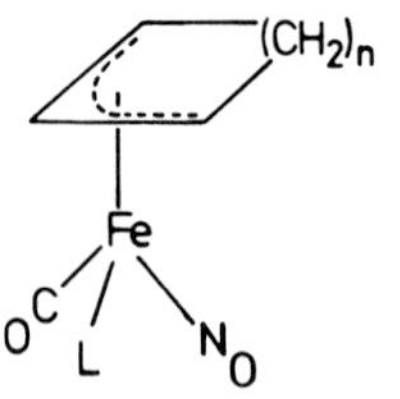

53

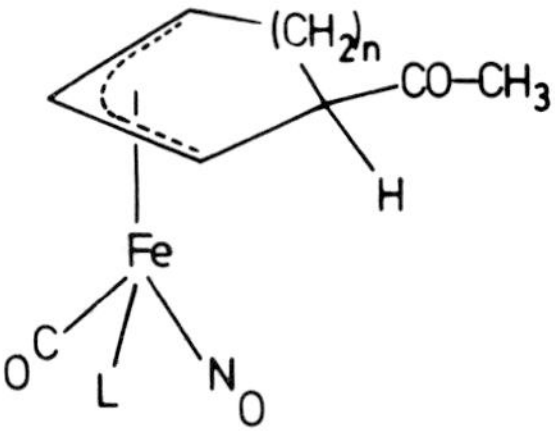

54

liquids with triphenylphosphine in boiling benzene give the more stable red crystalline substitution products 53 (L = $P(C_6H_5)_3$, n = 2,3). Reaction of a mixture of $NaFe(CO)_3NO$ and methyl iodide (which generates CH_3-$Fe(CO)_3NO$) with cyclopentadiene or cyclohexadiene in diethyl ether gives the corresponding red liquid (1-3-η^3-4-acetylcycloalkenyl) derivatives 54 (L = CO, n = 1,2) *(39)*. Reaction of triphenylcyclopropenyl bromide with $NaFe(CO)_3NO$ gives maroon [η^3-$(C_6H_5)_3$-C_3CO]$Fe(CO)_3NO$ *(52)* indicated to be the ketocyclobutenyl derivative 55 by analogy with the studies on the closely related ketocyclobutenyl-tricarbonylcobalt derivatives (η^3-R_3C_3CO)$Co(CO)_3$ (R = CH_3, C_6H_5) *(132,184)*.

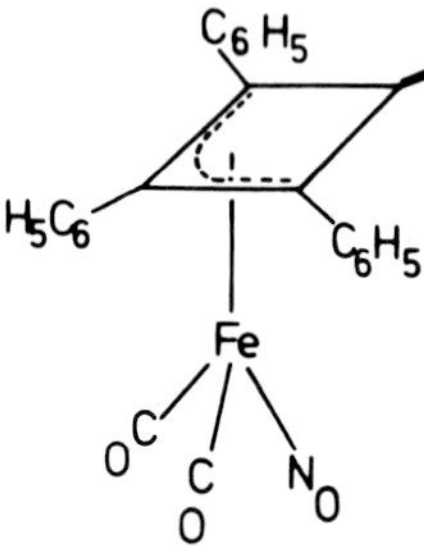

55

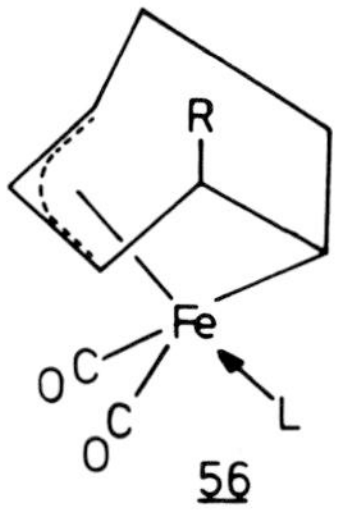

56

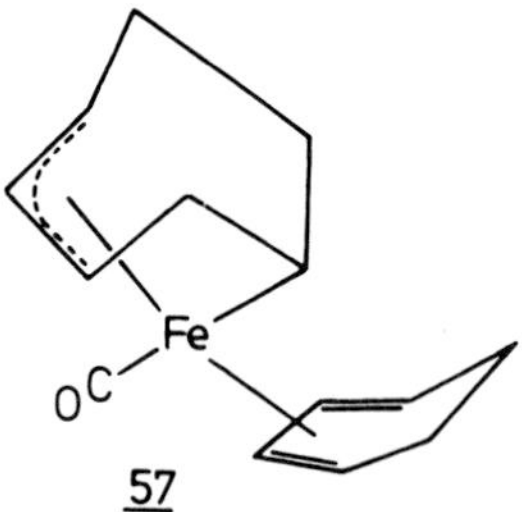

57

Some cyclic η^3-allylic carbonyliron derivatives have been obtained by reactions of cycloheptadienyl-carbonyliron compounds *(12,74)*. Sodium borohydride reduction of the (cycloheptadienyl)tricarbonyliron cation gives a mixture of (1,3-cycloheptadiene)tricarbonyliron and the chelating η^3-allyl-σ-acyl derivative 56 (R = H, L = CO). The yield of 56 (R = H, L = CO) relative to the isomeric cycloheptadiene complex can be maximized if the reaction with sodium borohydride is carried out in aqueous solution rather than in a mixture of dichloromethane and ethanol. Analogous sodium borohydride reductions of the substituted cycloheptadienyl complexes [(η^5-C_7H_9)$Fe(CO)_2E(C_6H_5)_3$][PF_6] give the corresponding η^3-allyl-σ-acyl derivatives 56 (R = H, L = $(C_6H_5)_3E$, E = As or P) free from the isomeric 1,3-cycloheptadiene derivatives *(74)*.

Similar cyano derivatives (56, R = CN, L = CO or $(C_6H_5)_3E$, E = P or As) have been obtained by using potassium cyanide rather than sodium borohydride as the nucleophile. A related monocarbonyliron derivative $(1,2,3,5\text{-}\eta^4\text{-}C_7H_{10})Fe(CO)(\eta^4\text{-}C_6H_8)$ (57) has been obtained by the following reaction sequence: (a) ultraviolet irradiation of (cycloheptadienyl)tricarbonyliron tetrafluoroborate with 1,3-cyclohexadiene in dichloromethane to give $[(\eta^5\text{-}C_7H_9)Fe(CO)(\eta^4\text{-}C_6H_8)][BF_4]$, (b) reduction of this salt with sodium borohydride to give (η^4-1,4-cycloheptadiene)(η^4-1,3-cyclohexadiene)carbonyliron, (c) thermal rearrangement of this complex to give 57. Heating 57 in solution to 90°C gives an equilibrium mixture containing only 10 % of 57 mixed with 90 % of (η^4-1,3-cyclohexadiene)-(η^4-1,3-cycloheptadiene)carbonyliron (see the chapter on diene-iron complexes for further details) *(119,123)*.

The anion $[(\eta^3\text{-}C_3H_5)Fe(CO)_3]^-$ obtained by the sodium amalgam reduction of the halides $(\eta^3\text{-}C_3H_5)Fe(CO)_3X$ was mentioned earlier in this chapter. The related (cycloheptatrienyl)tricarbonyliron anion (58) *(71,72,142,192)* is conveniently prepared as a deep red solution by treatment of (cycloheptatriene)tricarbonyliron with a stoichiometric amount of potassium *tert*-butoxide in tetrahydrofuran at room temperature *(72)*. Reaction of 58 with allyl halides results in coupling to give yellow crystalline ditropylbis(tricarbonyliron), $(OC)_3Fe(C_7H_7\text{-}C_7H_7)Fe(CO)_3$, rather than an allyliron derivative *(72)*. Reactions of 58 with the tropylium cations $[(C_7H_7)\text{-}M(CO)_3]^+$ (M = Cr, Mo, W) give the corresponding mixed metal ditropyl derivatives $(OC)_3Fe(C_7H_7\text{-}C_7H_7)M(CO)_3$ (M = Cr, Mo, W) *(72)*.

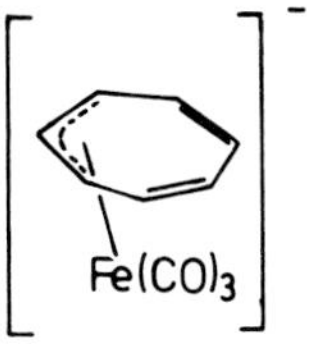

58

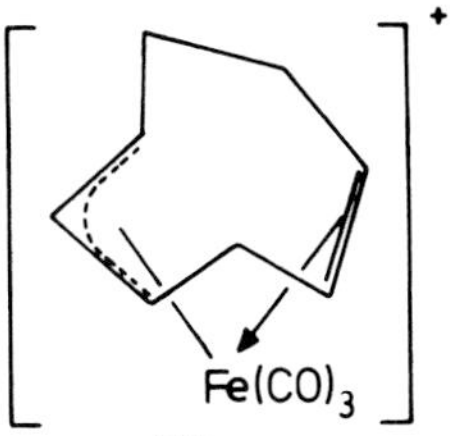

59

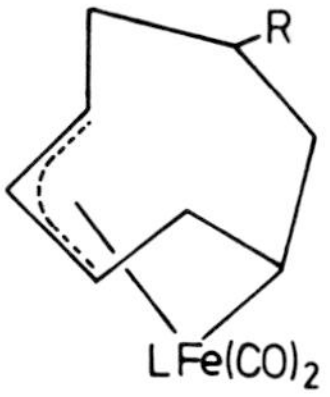

60

A series of cyclic 1-3-η^3-allylic-carbonyliron derivatives is available from (1,5-cyclooctadiene)tricarbonyliron. Hydride abstraction from this complex with triphenylmethyl tetrafluoroborate in dichloromethane solution gives the stable yellow η^3-allyl-η^2-olefin cation 59 *(56,69)*. Reduction of 59 with sodium borohydride in tetrahydrofuran gives a mixture of 10 % of (1,5-cyclooctadiene)tricarbonyliron and 90 % of the η^3-allyl-σ-alkyl complex 60 (L = CO, R = H). Similar reactions of 59 with other nucleophiles such as potassium cyanide *(69)*, sodium acetylacetonate *(69)*, sodium diethyl malonate

(69), sodium diethyl phenylmalonate *(69)*, sodium azide *(194)*, and sodium methoxide *(194)* give mixtures of substituted 1,5-cyclooctadiene-tricarbonyliron derivatives and substituted η^3-allyl-σ-alkyl complexes 60 (L = CO; R = CN, $CH(COCH_3)_2$, $CH(CO_2C_2H_5)_2$, $C(C_6H_5)(CO_2C_2H_5)_2$, N_3, and OCH_3, respectively). However, reaction of 59 with potassium iodide results in carbonyl displacement to give the iodide $(C_8H_{11})Fe(CO)_2I$ (61)

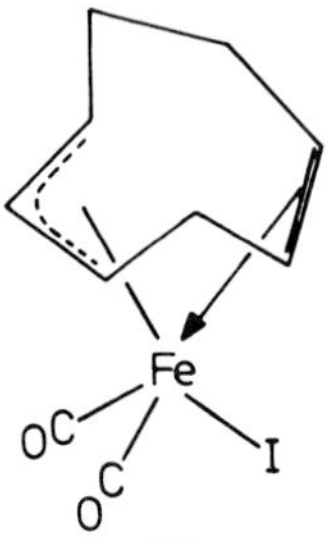

61

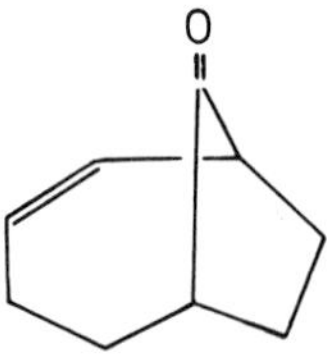

62

(194). Reaction of the η^3-allyl-σ-alkyl complex 60 (L = CO, R = H) with triphenylphosphine in boiling cyclohexane for 20 min results in the replacement of one carbonyl group to form the triphenylphosphine derivative 60 (L = $(C_6H_5)_3P$, R = H) in contrast to (1,5-cyclooctadiene)tricarbonyliron which reacts with triphenylphosphine to give *trans*-$[(C_6H_5)_3P]_2Fe(CO)_3$ with liberation of 1,5-cyclooctadiene *(56,69)*. Kinetic studies *(118,120)* have indicated that the reaction of the η^3-allyl-σ-alkyl complex 60 (L = CO, R = H) with triphenylphosphine or triphenyl phosphite proceeds by a CO-dissociative mechanism. Observed pseudo-first order rate constants for reactions of 60 (L = CO, R = H) with several more nucleophilic phosphorus ligands such as $P(OR)_3$ (R = CH_3, C_2H_5, $CH(CH_3)_2$), $CH_3P(C_6H_5)_2$, $(CH_3)_2PC_6H_5$, $(C_2H_5)_2PC_6H_5$, and $P[C(CH_3)_3]_3$ are given by the relation $k_{obs} = k_1 + k_2[L]$. The ligand-independent term corresponds to formation of the carbonyl substituted derivative 60. The ligand-dependent term corresponds the formation of the complexes $L_3Fe(CO)_2$ and the transannular ketone 62 *(118, 120)*. The transannular ketone 62 is also obtained by treatment of 60 (L = CO, R = H) with carbon monoxide at 25°C/85 atm *(56,69)*. Reactions of cyclodiene- or cyclopolyene-$Fe(CO)_3$ derivatives with olefins which contain electronegative substituents, such as tetrafluoro- or tetracyanoethylene, can give η^3-allyl-σ-alkyl carbonyliron derivatives. These types of reactions follow a different course depending upon whether the starting $Fe(CO)_3$ complex contains any uncoordinated carbon-carbon double bonds. Using cyclodiene-$Fe(CO)_3$ complexes as the starting material the η^3-allyl-iron bond in the product comes from the cyclic diene but the σ-alkyl-iron bond comes from the reacting olefin. However, in the products

obtained from the $Fe(CO)_3$ complexes of polyenes such as cycloheptatriene and cyclooctatetraene, both the η^3-allyl-iron and σ-alkyl-iron bonds are formed using carbon atoms of the cyclopolyene.

Ultraviolet irradiations of (1,3-cyclohexadiene)tricarbonyliron with the fluoroolefins tetrafluoroethylene *(26)* and trifluoroethylene *(104)* give the 1:1 η^3-allyl-σ-alkyl adducts 63 (X = F, H) (see the chapter on diene-iron complexes for

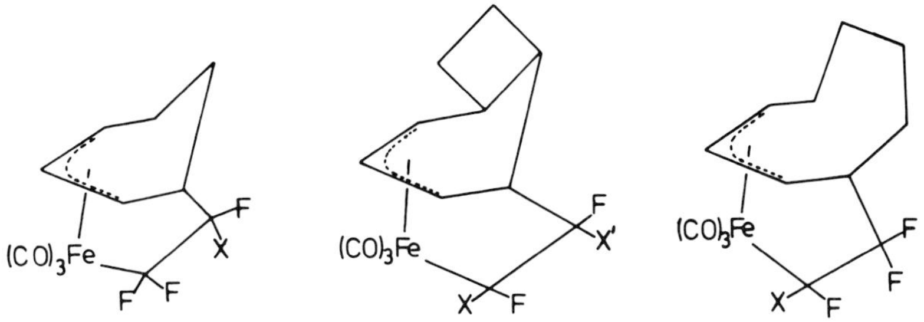

63 64 65

further details). Similarly, the bicyclic 1,3-cyclohexadiene derivative (bicyclo[4.2.0]octadiene)tricarbonyliron forms the related η^3-allyl-σ-alkyl derivatives 64 (X = X' = F; X = CF_3, X' = F; X = F, X' = H) upon ultraviolet irradiation with tetrafluoroethylene *(26)*, hexafluoropropene *(104)*, and trifluoroethylene *(104)*, respectively. Ultraviolet irradiation of (1,3-cyclooctadiene)tricarbonyliron with tetrafluoroethylene *(26)* and with hexafluoropropene *(104)* forms the η^3-allyl-σ-alkyl derivatives 65 (X = F and CF_3, respectively). However, (1,3-cyclooctadiene)tricarbonyliron does not react with chlorotrifluoroethylene upon ultraviolet irradiation *(104)*.

Cycloheptatriene and cyclooctatetraene contain additional carbon-carbon double bonds outside the 1,3-diene system. (Cycloheptatriene)tricarbonyliron reacts rapidly with tetracyanoethylene, 1,1-dicyano-2,2-bis(trifluoromethyl)ethylene, and *trans*-1,2-dicyano-1,2-bis(trifluoromethyl)ethylene to give the crystalline 1:1 adducts 66 ($R^1 = R^2 = R^3 = R^4$ = CN; $R^1 = R^2$ = CN, $R^3 = R^4 = CF_3$; $R^1 = R^4$ = CN, $R^2 = R^3 = CF_3$). In these reactions *exo* 1,3-addition of the olefin to the (cycloheptatriene)tricarbonyliron unit to form a 1,2,3,5-*tetrahapto* system has occurred *(102)*. The molecular structure 66 of the 1:1 adduct of tetracyanoethylene and (cycloheptatriene)tricarbonyliron has been confirmed by X-ray crystallography *(204)*. A similar 1:1 adduct 67 was obtained from (cycloheptatriene)tricarbonyliron and hexafluoroacetone *(102)*.

(Cyclooctatetraene)tricarbonyliron was first reported to form a 1:1 adduct with tetracyanoethylene. The structure of

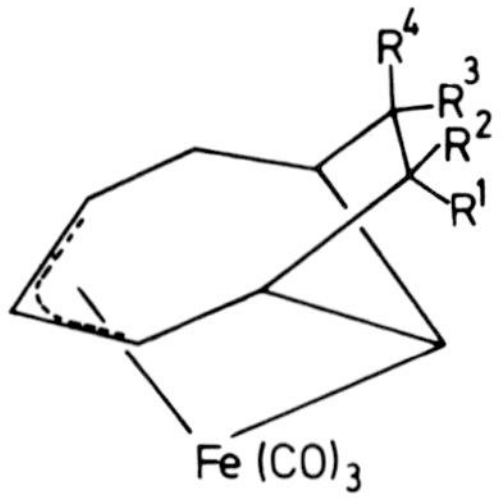

66

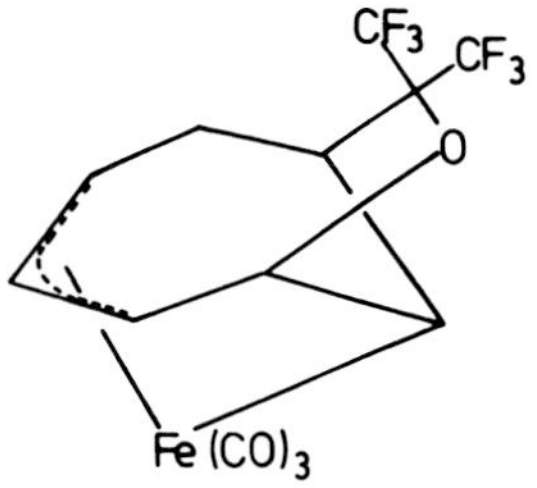

67

this adduct was first erroneously *(66,67,186,195)* believed to be a Diels-Alder 1,4-adduct and later *(100)* a 1,2-adduct. However, X-ray crystallography *(178)* confirms a subsequently proposed *(76)* formulation of this complex as the 1,3-adduct 68 ($R^1 = R^2 = R^3 = R^4 = H$). Tetracyanoethylene also reacts

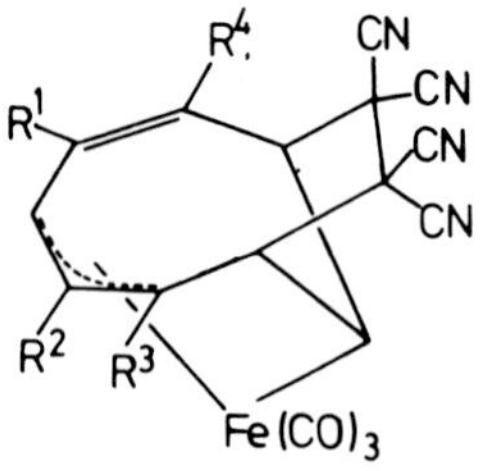

68

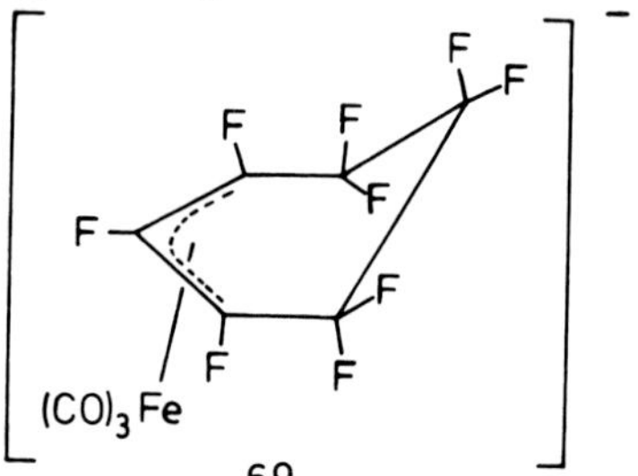

69

similarly with the tricarbonyliron complexes of methyl- *(102)*, bromo- *(102)*, phenyl- *(102)*, and methoxycarbonylcyclooctatetraene *(179)*. The reaction of tetracyanoethylene with (methylcyclooctatetraene)tricarbonyliron gives a mixture of 71 % of isomer 68 ($R^1 = R^2 = R^4 = H$, $R^3 = CH_3$) and 22 % of isomer 68 ($R^1 = R^3 = R^4 = H$, $R^2 = CH_3$). The reaction of tetracyanoethylene with (phenylcyclooctatetraene)tricarbonyliron gives a mixture of 39 % of isomer 68 ($R^1 = C_6H_5$, $R^2 = R^3 = R^4 = H$), 16 % of isomer 68 ($R^1 = R^2 = R^4 = H$, $R^3 = C_6H_5$), and 23 % of isomer 68 ($R^1 = R^3 = R^4 = H$, $R^2 = C_6H_5$). The reaction of tetracyanoethylene with (methoxycarbonylcyclooctatetraene)tricarbonyliron gives a mixture of 23 % of isomer 68 ($R^1 = CO_2$-CH_3, $R^2 = R^3 = R^4 = H$) and 64 % of isomer 68 ($R^1 = R^2 = R^3 = H$, $R^4 = CO_2$-CH_3). Oxidation of these 1:1 adducts of tetracyanoethylene with (cyclooctatetraene)tricarbonyliron derivatives using cerium(IV) gives high yields of dihydrotetracyanotriquinacenes *(179)*.

A fluorinated η^3-allylic-tricarbonyliron anion has been prepared from a perfluoro-1,3-diene complex *(180)*. Reaction of (octafluorocyclohexa-1,3-diene)tricarbonyliron with cesium fluoride in tetrahydrofuran followed by metathesis with aqueous tetramethylammonium chloride gives the yellow tetramethylammonium salt of the (η^3-nonafluorocyclohexenyl)tricarbonyliron anion (69).

2. Binuclear η^3-Allylic-carbonyliron Compounds from Reactions of Monocyclic Polyolefins with Carbonyliron Complexes

Two major types of binuclear η^3-allylic-carbonyliron compounds are formed from reactions of monocyclic polyolefins with carbonyliron complexes. Cyclic trienes react with $Fe_2(CO)_9$ or $Fe_3(CO)_{12}$ to form hexacarbonyldiiron complexes of the types 3a and 3b using all six sp^2 carbons of the triene system. Cyclooctatetraene reacts with $Fe_2(CO)_9$ to give a bis(η^3-allyl)-pentacarbonyldiiron derivative (structure 6) in which six of the eight sp^2 carbons are used for the two 1-3-η^3-allylic systems and the two remaining sp^2 carbon atoms bridge the two iron atoms. Further details on the reactions of cyclic trienes and tetraenes with carbonyliron complexes are given in the chapter on triene and tetraene complexes.

a. *(Triene)$Fe_2(CO)_6$ Derivatives*

Cyclic trienes which form (triene)$Fe_2(CO)_6$ derivatives upon reaction with enneacarbonyldiiron at room temperature include cycloheptatriene *(87)*, 7-methoxycyclohepta-1,3,5-triene *(87)*, 1,3,5-cyclooctatriene *(87)*, cyclooctatetraene *(87)*, bis(trimethylsilyl)cyclooctatriene *(68)*, and *cis*-bicyclo[6.1.0]nonatriene *(70,188)*. The (triene)$Fe_2(CO)_6$ derivatives of cyclooctatriene and cyclooctatrienone have also been obtained by heating the triene with dodecacarbonyltriiron *(126)*.

X-ray crystallographic studies indicate some subtle but significant differences in the bis(η^3-allylic) bonding of these cyclic trienes to the hexacarbonyldiiron unit. Thus, the $(C_7H_8)Fe_2(CO)_6$ obtained from cycloheptatriene and enneacarbonyldiiron has been found to be the symmetrical 1,2,3-*trihapto*-4,5,6-*trihapto* derivative 70 *(57)*. The ^{13}C-NMR-

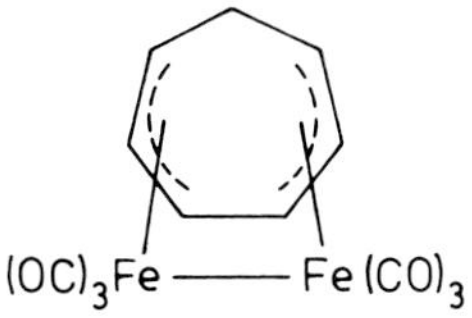

70

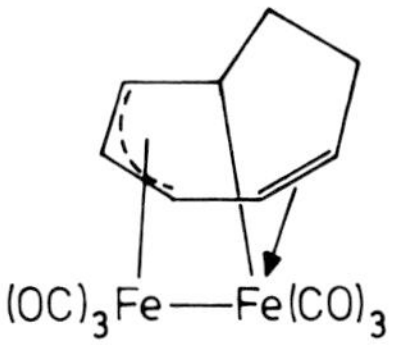

71

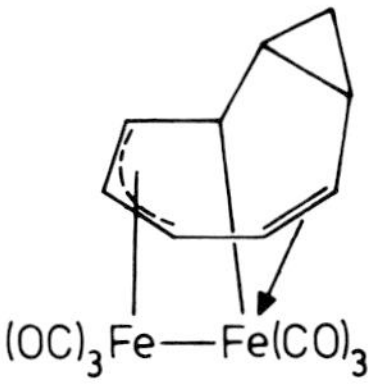

72

spectrum of $(C_7H_8)Fe_2(CO)_6$ (70) exhibits at -60°C three carbonyl resonances of 1:1:1 relative intensities. These broaden uniformly and coalesce to a singlet at +40°C *(60)*.

Localized scrambling on an $Fe(CO)_3$ unit has been used to account for these observations. The $(C_8H_{10})Fe_2(CO)_6$ and the $(C_9H_{10})Fe_2(CO)_6$ obtained by reactions of enneacarbonyldiiron with 1,3,5-cyclooctatriene *(55)* and bicyclo[6.1.0]nonatriene *(199)*, respectively, have been found to have the unsymmetrical 1,2,6-*trihapto*-3,4,5-*trihapto* structures 71 and 72, respectively. The ^1H-NMR spectra of these unsymmetrical complexes 71 and 72 at ambient temperature indicate a plane of symmetry which is not consistent with these solid state structures *(61,70,199)*. Fluxional properties are thus indicated. However, cooling the solutions of 71 and 72 below ∿ -100°C results in a low temperature limiting NMR spectrum not consistent with the unsymmetrical solid state structures. A detailed study of the fluxional processes in the cyclooctatriene complex 71 has been made from a temperature dependence study of its ^{13}C-NMR spectrum *(61)*. Three different fluxional processes have been identified. At the lowest temperatures (below -65°C) the ^{13}C-NMR spectrum is in complete agreement with the skew structure 71 observed in the crystal. There are eight separate ring carbon signals and six separate carbonyl resonances. The first set of changes, which are observed from -65°C to about +10°C, results from a so-called "twitching" process in which the two enantiomorphic forms of the structure interconvert by the minimal movement of the hexacarbonyldiiron group relative to the cyclooctatriene ring without interchanging the ends of the hexacarbonyldiiron moiety. Simultaneously, the three carbonyl groups on one of the iron atoms are scrambled among themselves. Above +10°C the high temperature ^{13}C-NMR spectrum also indicates the scrambling of the three carbonyl groups on the other iron atom among themselves.

The $(triene)Fe_2(CO)_6$ complexes from cycloheptatriene (70) and 1,3,5-cyclooctatriene (71) undergo hydride abstraction with triphenylmethyl tetrafluoroborate at room temperature to give the cations $[(C_7H_7)Fe_2(CO)_6]^+$ (73) and $[(C_8H_9)Fe_2(CO)_6]^+$

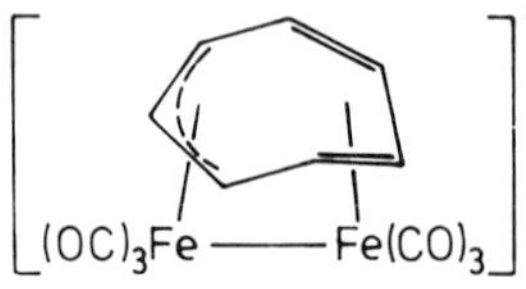

73

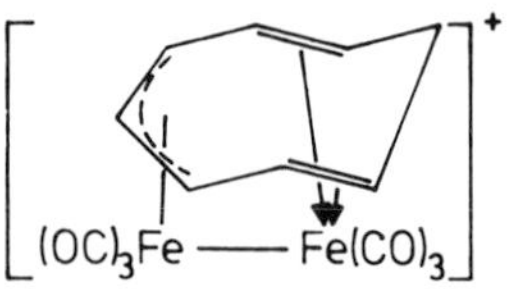

74

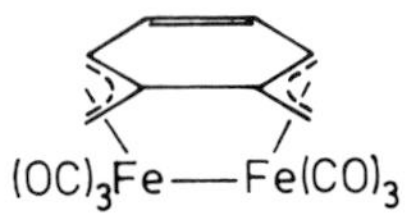

75

(74), respectively *(87,126)*. The cation 73 is also obtained by the treatment of the 7-methoxycycloheptatriene derivative $(C_7H_7OCH_3)Fe_2(CO)_6$ with tetrafluoroboric acid *(87)*. Sodium borohydride reduction of 73 regenerates 70. Despite the unsymmetrical structure of 73 its ^1H-NMR spectrum at room temperature exhibits only a single sharp ring ^1H-NMR resonance indicative of a fluxional system *(87)*. The $[(C_8H_9)Fe_2(CO)_6]^+$ cation reacts with nucleophiles to give neutral substituted (1,3,5-cyclooctatriene)-hexacarbonyldiiron derivatives $(C_8H_9X)Fe_2(CO)_6$ (X = H, N_3, CN) as well as the cations $[(C_8H_9R)Fe_2(CO)_6][BF_4]$ (R = pyridine, triphenylphosphine) *(5)*. These substituted cyclooctatriene-hexacarbonyldiiron derivatives have a similar asymmetric structure as the unsubstituted 71. Interconversion of the two enantiomers of these substituted derivatives can be studied by the temperature dependence of their ^{1}H-NMR spectra *(5)*.

The ultraviolet irradiation of (*o*-quinodimethane)tricarbonyliron with excess pentacarbonyliron gives three isomeric $(C_8H_8)Fe_2(CO)_6$ complexes in comparable yields. One of these products appears to be the bis(η^3-allylic) complex 75 containing a tetramethyleneethane unit (structure 4) similar to products such as 42 obtained from reactions of allenes with carbonyliron complexes *(202)*.

b. *(Tetraene)$Fe_2(CO)_5$ Derivatives*

In some cases reactions of cyclooctatetraene with carbonyliron reagents give (tetraene)$Fe_2(CO)_5$ derivatives containing two η^3-allylic-iron bonds. Reaction of cyclooctatetraene with enneacarbonyldiiron gives three isomeric $(C_8H_8)Fe_2(CO)_6$ derivatives *(124)*. One of these products is the bis(η^3-allylic) derivative 76 mentioned above. A second

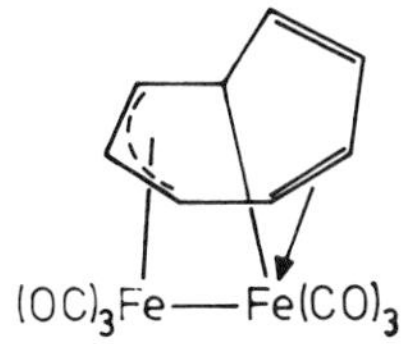

76

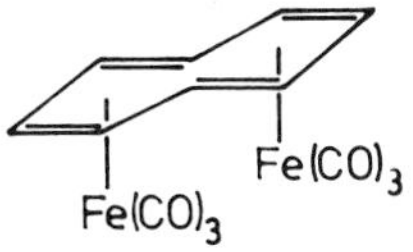

77

$(C_8H_8)Fe_2(CO)_6$ isomer is the *cis*-bis(diene-tricarbonyliron) complex 77. Heating either 76 or 77 in carbon tetrachloride or benzene solution results in rapid decarbonylation to give black $(C_8H_8)Fe_2(CO)_5$ *(124)*. An X-ray crystallographic study on $(C_8H_8)Fe_2(CO)_5$ indicates structure 78 (R = H, M = Fe) in which the eight sp^2 carbon atoms of the cyclooctatetraene ligand form two 1-3-η^3-allylic units leaving two sp^2 carbons to bridge between the two iron atoms *(92)*. These sp^2 carbons of the C_8H_8 ring along with the single bridging carbonyl group lead to three bridging groups in $(C_8H_8)Fe_2(CO)_5$ (78) similar to the three bridging carbonyl groups in $Fe_2(CO)_9$. The ^{1}H-NMR spectrum of $(C_8H_8)Fe_2(CO)_5$ (78, R = H, M = Fe) exhibits a single sharp resonance *(124)* indicating fluxional properties. A similar substituted derivative, $[(CH_3)_4C_8H_4]Fe_2(CO)_5$ (78,

M = Fe, R = CH_3), is obtained as black crystals from the

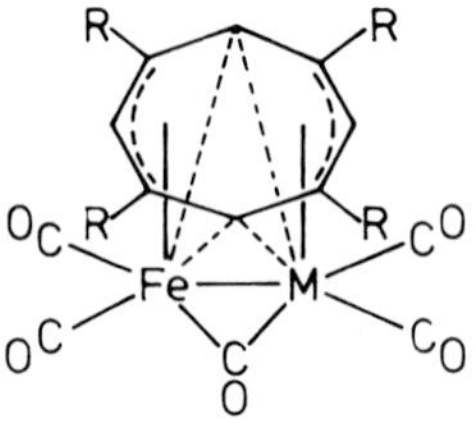

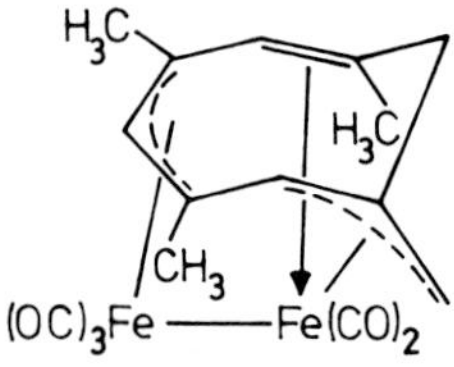

78 79

reaction of 1,3,5,7-tetramethylcyclooctatetraene with enneacarbonyldiiron *(53)*. The reaction of 1,3,5,7-tetramethylcyclooctatetraene with $Fe_3(CO)_{12}$ gives not only a low yield of $[(CH_3)_4C_8H_4]Fe_2(CO)_5$ (78, M = Fe, R = CH_3) but also a very low yield of an isomeric red crystalline $(C_{12}H_{16})Fe_2(CO)_5$ derivative shown by X-ray crystallography *(54)* to be the 1,3,5-trimethyl-7-methylenecycloocta-1,3,5-triene complex 79 containing two 1-3-η^3-allylic systems and an additional complexed carbon-carbon double bond.

A cyclooctatetraene dimetal pentacarbonyl derivative containing both iron and ruthenium has also been prepared *(1)*. Reaction of (cyclooctatetraene)tricarbonyliron with dodecacarbonyltriruthenium in boiling xylene for 12 h gives the mixed metal derivative $(C_8H_8)FeRu(CO)_5$ (78, M = Ru, R = H). This is one of the rare examples of a bridging carbonyl group between a first row and a second row transition metal. The ^1H-NMR spectrum of $(C_8H_8)FeRu(CO)_5$ exhibits only a single sharp resonance indicating a fluxional molecule similar to $(C_8H_8)Fe_2(CO)_5$ *(1)*.

B. η^3-ALLYLIC CARBONYLIRON COMPOUNDS FROM REACTIONS OF POLYCYCLIC OLEFINS WITH CARBONYLIRON COMPLEXES

The types of cyclic polyenes that have been shown to form η^3-allylic-carbonyliron compounds upon reactions with carbonyliron derivatives may be classified as follows: (1) Planar polycyclic systems such as isoindene, acenaphthylene, azulene, and naphthacene; (2) bridged polycyclic systems such as bicyclo[3.2.2] nonadiene.

1. η^3-Allylic-carbonyliron Compounds from Reactions of Planar Polycyclic Systems with Carbonyliron Complexes

Reaction of the isoindene complex 80 *(190)* with aluminium chloride gives the yellow crystalline η^3-allyl-σ-alkyl derivative 81 resulting from insertion of carbon monoxide into the five-membered ring of the isoindene system *(121)*. The complex 81 undergoes further carbon monoxide insertion at

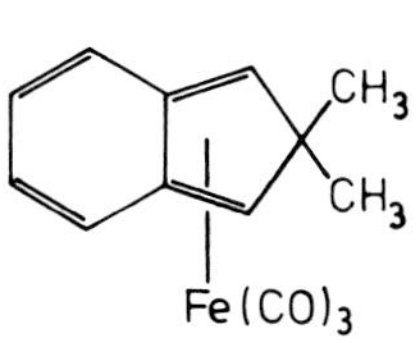

80

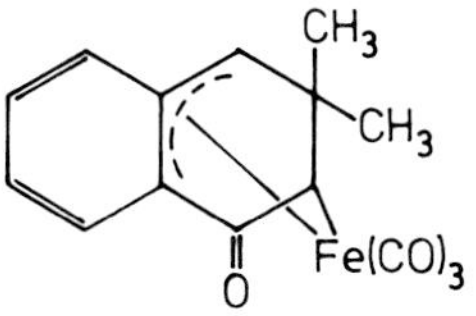

81

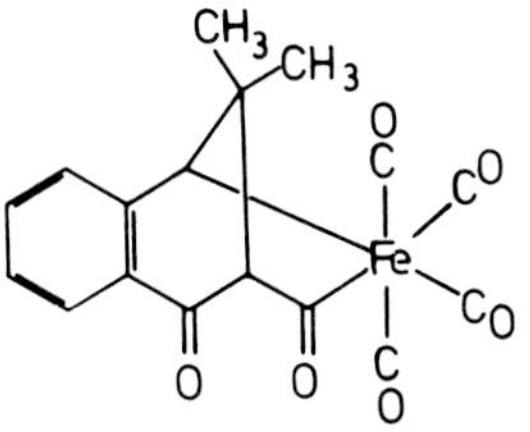

82

high pressures (30°C/80 atm) to give the tetracarbonyliron derivative 82. However, at atmospheric pressure 82 readily loses carbon monoxide to revert to 81.

Reaction of acenaphthylene with dodecacarbonyltriiron gives a dark red-purple solid, first *(125)* believed to be $(C_{12}H_8)Fe_2(CO)_6$ but later suggested *(128)* on the basis of its mass spectrum to be $(C_{12}H_8)Fe_2(CO)_5$. An X-ray diffraction study *(42,44)* on this complex indicates structure 83. In this

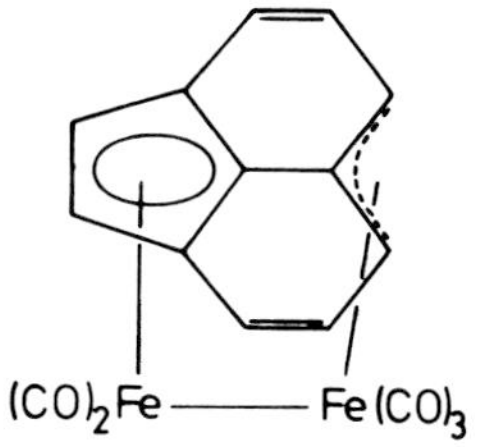

83

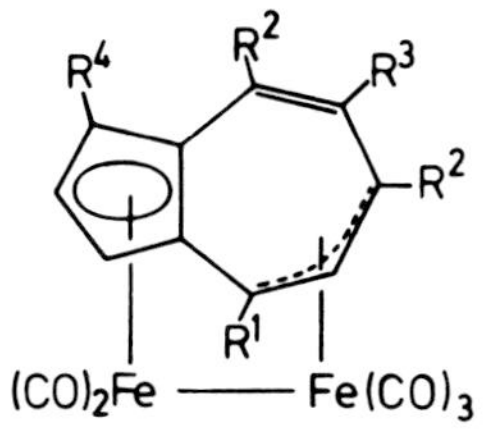

84

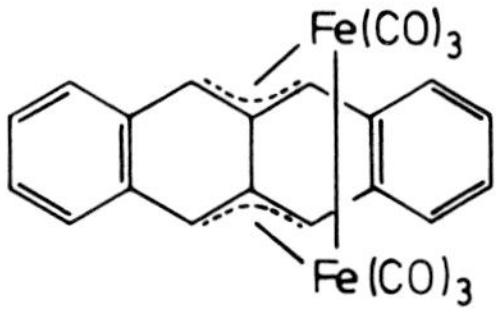

85

structure the acenaphthylene functions as an *octahapto* ligand bonding to one iron atom through an η^3-allyl bond and to the other iron atom through an η^5-cyclopentadienyl bond. A study of the temperature-dependence of the ^{13}C-NMR spectrum of 83 indicates that the carbonyl groups of the $Fe(CO)_3$ unit scramble among themselves but do not exchange with those on the $Fe(CO)_2$ unit *(60)*. Another series of complexes containing an *octahapto* planar hydrocarbon bonded to one iron atom through an η^3-allyl system and to the other iron atom through an η^5-cyclopentadienyl system includes the (azulene)$Fe_2(CO)_5$ derivatives 84 ($R^1 = R^2 = R^3 = R^4 = H$; $R^1 = R^2 = CH_3$, $R^3 = R^4 = H$; $R^1 = R^4 = CH_3$, $R^2 = H$, $R^3 = CH(CH_3)_2$) which are obtained by heating the corresponding azulene with excess pentacarbonyliron *(31,32,46)*. The structure of the unsubstituted derivative 84 ($R^1 = R^2 = R^3 = R^4 = H$) has been confirmed by X-ray crystallography *(41)*.

The tetracyclic benzenoid hydrocarbon naphthacene reacts with dodecacarbonyltriiron in boiling benzene to give a hexacarbonyldiiron derivative shown from its NMR spectrum to be the bis(η^3-allylic) derivative 85 *(19)*. The relative orientation of the two η^3-allylic units is the same as that found in the compound 42 obtained from allene and $Fe_2(CO)_9$ or $Fe_3(CO)_{12}$ (see above).

2. η^3-Allylic-carbonyliron Compounds from Reactions of Bridged Polycyclic Systems with Carbonyliron Complexes

A series of cations of the general type 86 (X = $-CH_2-$ *(83)*, -CH=CH- *(80)*, $-CH_2-CO-$ *(80)*, and *o*-C_6H_4 *(81)*) may be prepared as their stable yellow tetrafluoroborate salts by protonation of the corresponding (bridged bicyclic diene)-

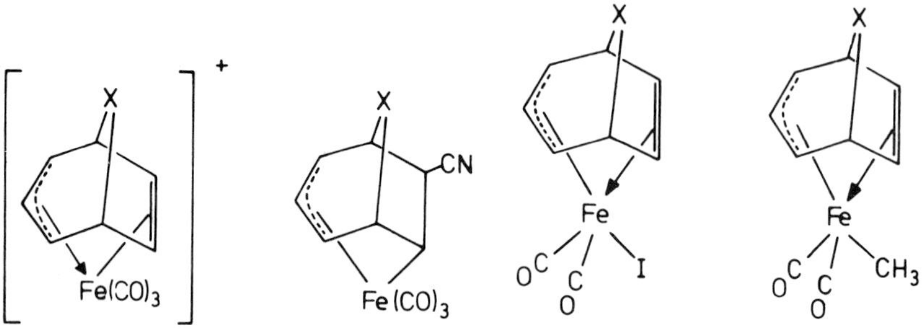

86 87 88 89

tricarbonyliron derivatives containing hydroxy or methoxy substituents with tetrafluoroboric acid in acetic anhydride (see chapter on diene-iron complexes for further details). In all of these cations one of the bridges is coordinated to the tricarbonyliron group through an η^3-allylic bond and another of the bridges is coordinated to the same tricarbonyliron group through an η^2-olefin-iron bond. These cations react with aqueous potassium cyanide with rupture of half of the olefiniron bond to give the corresponding η^3-allyl-σ-alkyl derivatives 87 (X = $-CH_2-$, -CH=CH-, and *o*-C_6H_4) as stable yellow crystalline solids *(83)*. However, these cations react with potassium iodide in acetone with evolution of carbon monoxide to give the corresponding maroon crystalline iodides 88 (X = $-CH_2-$, -CH=CH-, and *o*-C_6H_4). Reaction of the iodide 88 (X = $-CH_2-$) with methyllithium gives the corresponding σ-methyl derivative 89 (X = $-CH_2-$) in 90 % yield as a stable yellow crystalline solid *(83)*.

Reactions of α- or β-pinene with pentacarbonyliron at 160°C results in the stereospecific insertion of a carbonyl group into the cyclobutane ring *(198)*. Successive η^3-allylic-iron intermediates of the types 90 and 91 are proposed for this reaction. The bis(η^3-allylic)-hexacarbonyldiiron complex 93 has been proposed as an intermediate in the rearrangement of 92 (eq. [1]) to the furan derivative 94 upon heating with pentacarbonyliron in boiling di-*n*-butyl ether *(4)*.

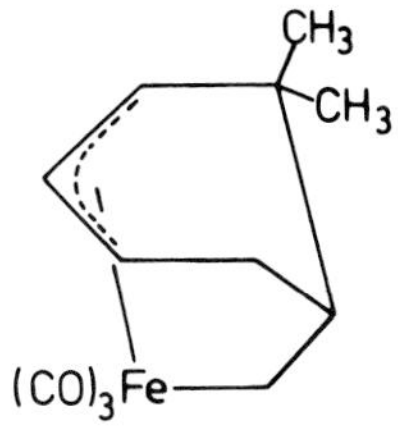

90

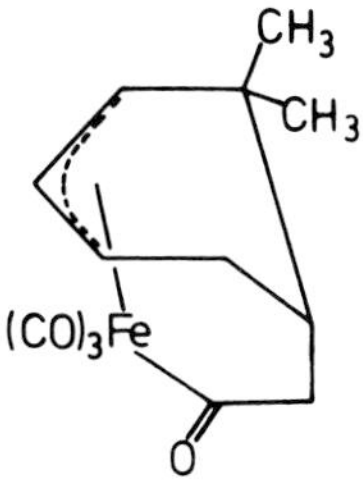

91

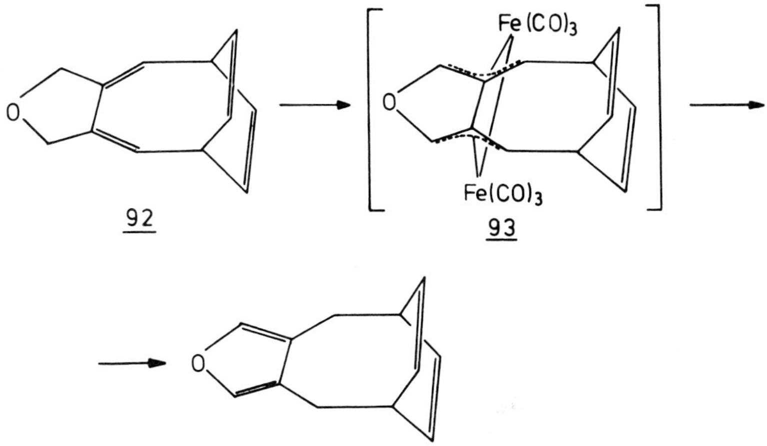

92 93 94

[1]

VIII. η^3-ALLYLIC CARBONYLIRON COMPOUNDS FROM REACTIONS OF CYCLOPROPANE DERIVATIVES WITH CARBONYLIRON COMPLEXES

Some reactions of carbonyliron complexes with cyclopropanes containing adjacent unsaturation lead to products in which an η^3-allylic unit is bonded to iron. Although such products have been obtained from methylenecyclopropanes, a much larger variety of η^3-allylic-carbonyliron compounds has been obtained from reactions of vinylcyclopropanes with carbonyliron reagents.

A. η^3-ALLYLIC CARBONYLIRON COMPOUNDS FROM REACTIONS OF METHYLENECYCLOPROPANES WITH CARBONYLIRON COMPLEXES

Reactions of dimethyl *trans*-methylenecyclopropane-2,3-dicarboxylate (95) with carbonyliron complexes can lead to η^3-allylic-carbonyliron compounds in addition to olefin-

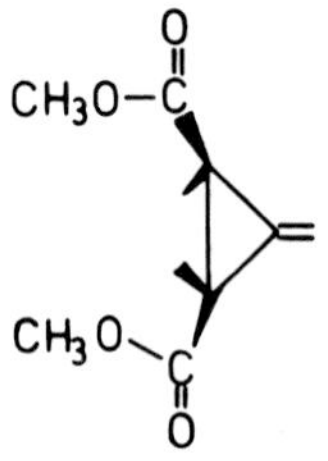

95

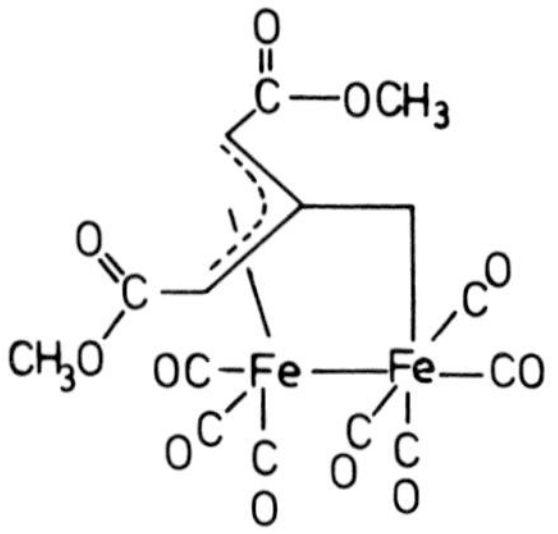

96

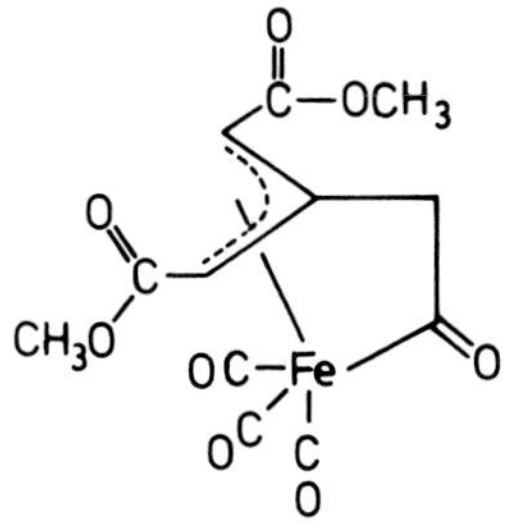

97

and diene-carbonyliron complexes discussed elsewhere in this book *(207)*. The olefin-tetracarbonyliron complex prepared from 95 and enneacarbonyldiiron reacts with excess $Fe_2(CO)_9$ to give a low yield of the deep red crystalline binuclear η^3-allylic derivative 96 and can be photolyzed in hexane solution to give the unstable yellow liquid mononuclear η^3-allylic derivative 97 as one of the products. The complex 96 is also obtained in very low yields by reaction of the tetracarbonyliron complex of the *cis*-isomer of 95 with excess enneacarbonyldiiron.

B. *η^3-ALLYLIC CARBONYLIRON COMPOUNDS FROM REACTIONS OF VINYLCYCLOPROPANES WITH CARBONYLIRON COMPLEXES*

Reactions of vinylcyclopropanes with carbonyliron complexes give a wide variety of η^3-allyl-σ-alkyl derivatives (structure 2a) which in some cases can be carbonylated to the corresponding η^3-allyl-σ-acyl derivatives (structure 2b). Such carbonyliron complexes have also been obtained from a wide variety of polycyclic compounds containing vinylcyclopropane structural units such as semibullvalene, barbaralone, and bullvalene.

1. η^3-Allylic-carbonyliron Compounds from Reactions of Monocyclic Vinylcyclopropanes with Carbonyliron Complexes

Ultraviolet irradiation of vinylcyclopropane with two equivalents of pentacarbonyliron in dilute diethyl ether solution at -50°C gives a 9 % yield of the yellow crystalline η^3-allyl-σ-acyl derivative 98 in addition to an 80 % yield of (η^2-vinylcyclopropane)tetracarbonyliron *(14)*. The η^3-allyl-σ-acyl derivative 98 is unstable at room temperature in the absence of excess carbon monoxide with respect to decarbonylation to the corresponding yellow liquid η^3-allyl-σ-alkyl derivative 99. Formation of 98 from vinylcyclopropane and carbonyliron complexes involves insertion of a carbonyliron

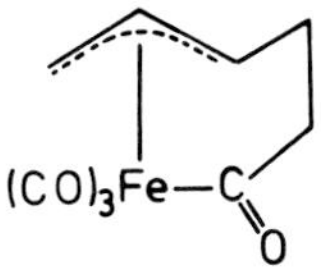

98

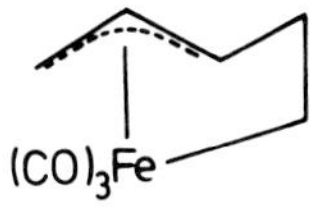

99

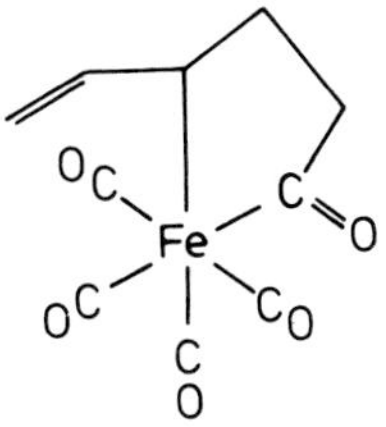

100

unit into a cyclopropane carbon-carbon bond adjacent to the vinyl carbon-carbon double bond, possibly through an intermediate such as 100. The η^3-allyl ligand is then readily generated through displacement of a carbonyl group by the vinyl double bond. Similar reaction pathways are frequently found in the reactions of carbonyliron reagents with the more complex vinylcyclopropane derivatives discussed below.

2. η^3-Allylic Carbonyliron Compounds from Reactions of Polycyclic Vinylcyclopropanes with Carbonyliron Complexes

Among the simplest polycyclic vinylcyclopropanes are the dihydrosemibullvalenes containing an additional fused saturated ring (101, n = 2,3) since their vinylcyclopropane system

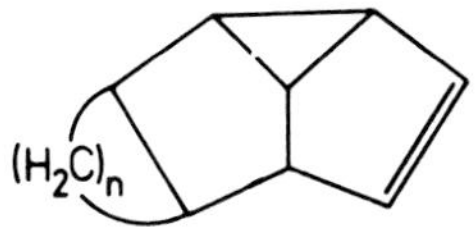

101

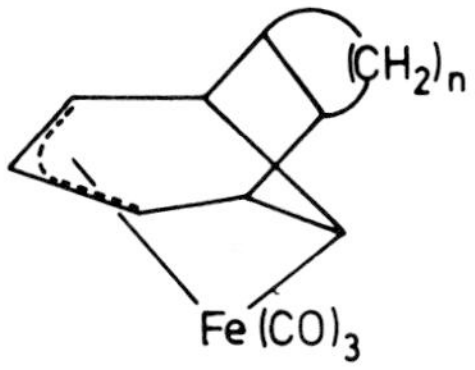

102

is the only potentially reactive site *(151)*. These hydrocarbons react with enneacarbonyldiiron in boiling benzene to give the corresponding η^3-allyl-σ-alkyl complexes 102 (n = 2,3) in at least 75 % yield. Formation of 102 from 101 and carbonyliron complexes corresponds exactly to the formation of 99 from unsubstituted vinylcyclopropane and a carbonyliron complex.

Semibullvalene (103) has a structure similar to 101 except for the absence of the additional fused cycloalkane ring and the presence of a second carbon-carbon double bond adjacent to the cyclopropane ring. The major product obtained from enneacarbonyldiiron and semibullvalene (103) in boiling benzene is the yellow liquid η^3-allyl-σ-alkyl derivative 104 *(149)*. The formation of this bicyclo[3.2.1]octane derivative 104 from 103 is a consequence of the tendency for carbonyliron groups to bond to chelating η^3-allyl-σ-alkyl ligands, since palladium(II), which only forms η^3-allyl deriv-

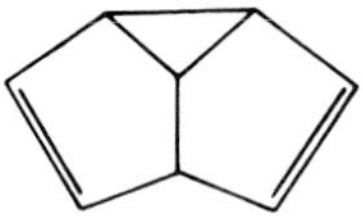

103

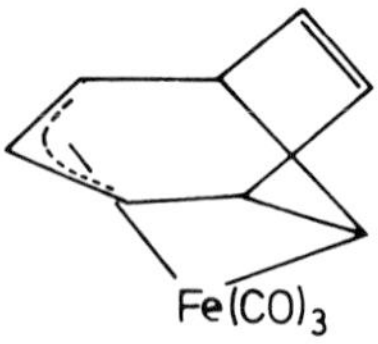

104

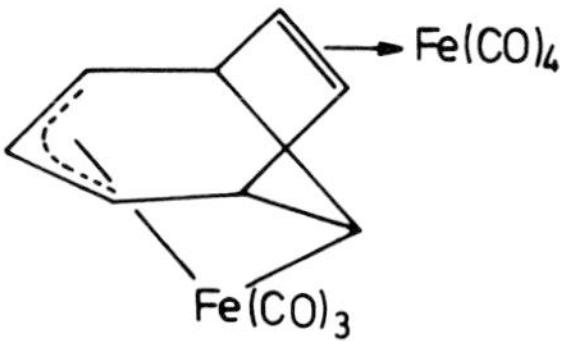

105

atives without forming σ-alkyl derivatives, reacts with semibullvalene (103) to form a bicyclo[3.3.0]octadiene derivative rather than a bicyclo[3.2.1]octadiene derivative *(152)*. The reaction of semibullvalene (103) with enneacarbonyldiiron at room temperature gives several additional products *(77)*. One of these is 105, the tetracarbonyliron complex of 104. Reaction of 104 with enneacarbonyldiiron at room temperature also gives 105. The reaction of benzosemibullvalene (106) with enneacarbonyldiiron in boiling benzene proceeds exactly like

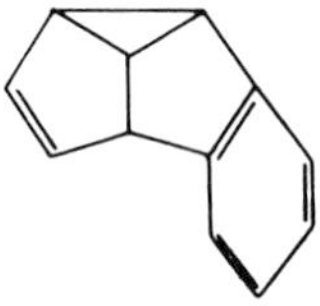

106

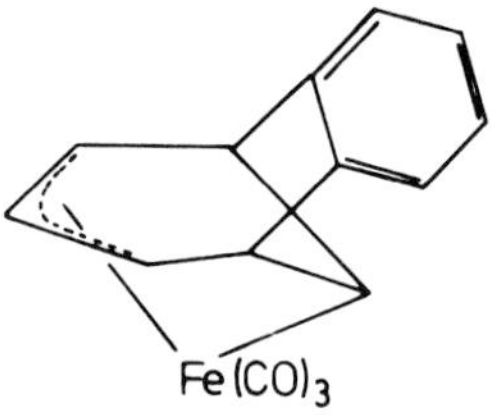

107

the corresponding reactions of semibullvalene (103) and even vinylcyclopropane to give a similar η^3-allyl-σ-alkyl complex 107 *(151)*.

The reaction of dibenzosemibullvalene (108) with carbonyliron reagents clarifies some of the above reactions of vinylcyclopropane derivatives *(150,201)*. In 108 the only carbon-carbon double bonds present are parts of benzenoid

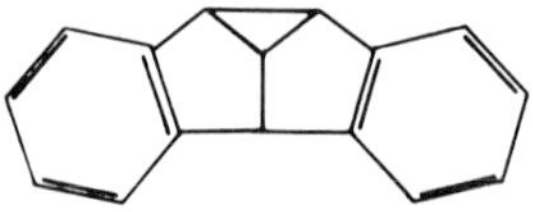

108

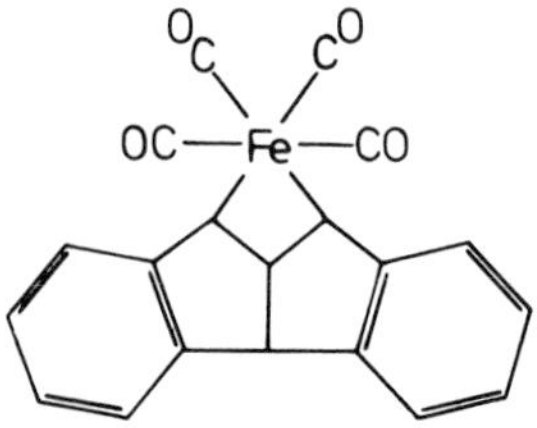

109

rings. Such carbon-carbon double bonds are considerably less reactive than isolated carbon-carbon double bonds. When 108 reacts with enneacarbonyldiiron in benzene at 60°C, the cyclopropane ring is opened to give the ferretane derivative 109 as a very stable yellow crystalline solid which survives concentrated hydrochloric acid at 80°C for 3 days. Decarbonylation of 109 to form the corresponding η^3-allyl-σ-alkyl derivative analogous to 104 and 107 is unfavourable because the available carbon-carbon double bonds to form the η^3-allylic unit (actually an η^3-benzylic unit) are all parts of benzenoid systems.

The polycyclic divinylcyclopropane ketone barbaralone (110) also reacts analogously with enneacarbonyldiiron to give a good yield of the corresponding η^3-allyl-σ-alkyl derivative 111 *(82)*. The structure of 111 has been confirmed by independ-

110

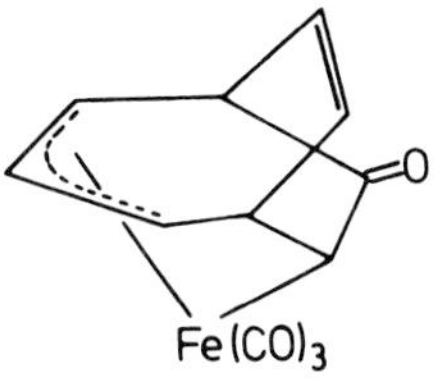

111

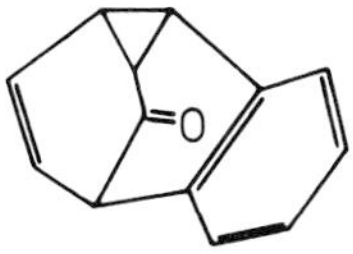

112

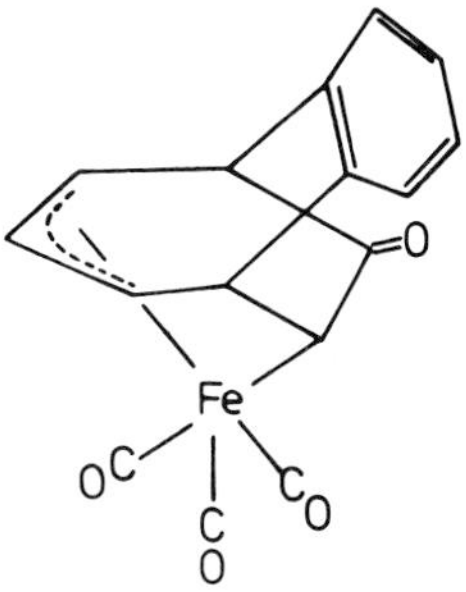

113

ent X-ray crystallography studies on both triclinic *(59)* and monoclinic *(203)* crystalline modifications. The complex 111 derived from barbaralone must have a relatively high tendency of formation since it is also obtained by treatment of (cyclooctatetraene)tricarbonyliron with aluminium chloride in benzene *(112,122)*. This reaction provides a useful synthesis of barbaralone (110) from cyclooctatetraene since degradation of 111 with carbon monoxide at 120°C/100 atm regenerates barbaralone nearly quantitatively. The complex 111 is also found in low yield among several other products in the mixture obtained from bicyclo[6.2.0]deca-2,4,6-triene and enneacarbonyldiiron *(58)*. Reaction of benzobarbaralone (112) with enneacarbonyldiiron also proceeds in the usual manner for vinylcyclopropane systems to give the η^3-allyl-σ-alkyl derivative 113 *(84)*.

The reactions of bullvalene (114) with carbonyliron reagents are extremely complex, but form many products containing 1-3-η^3-allylic ligands *(15)*. Reaction of bullvalene with enneacarbonyldiiron in diethyl ether at 30°C for 12 h in an evacuated flask gives the following products listed in order of their elution from a chromatography column: (a) The yellow 119 in 20 % yield in which the $C_{10}H_{10}$ ligand is bonded to each iron atom as an η^3-allyl-σ-alkyl ligand *(7)*, (b) the red diferratricyclododecatriene derivative 122 in 3 % yield *(9)*, (c) yellow (tricarbonylferrole)tricarbonyliron (123) in 1 % yield, (d) the yellow 115 in 39 % yield in which the

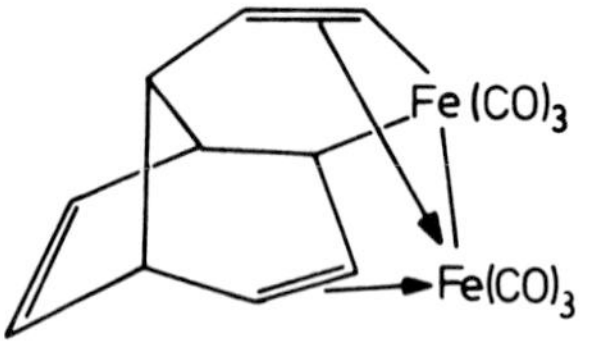

122

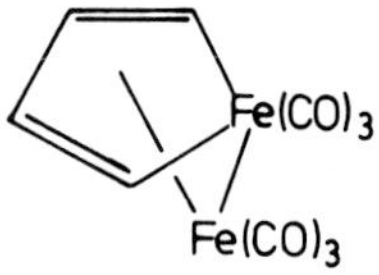

123

$C_{10}H_{10}$ hydrocarbon is bonded to one iron as a diene and to the other as an η^3-allyl-σ-alkyl ligand *(6)*, (e) the ochre

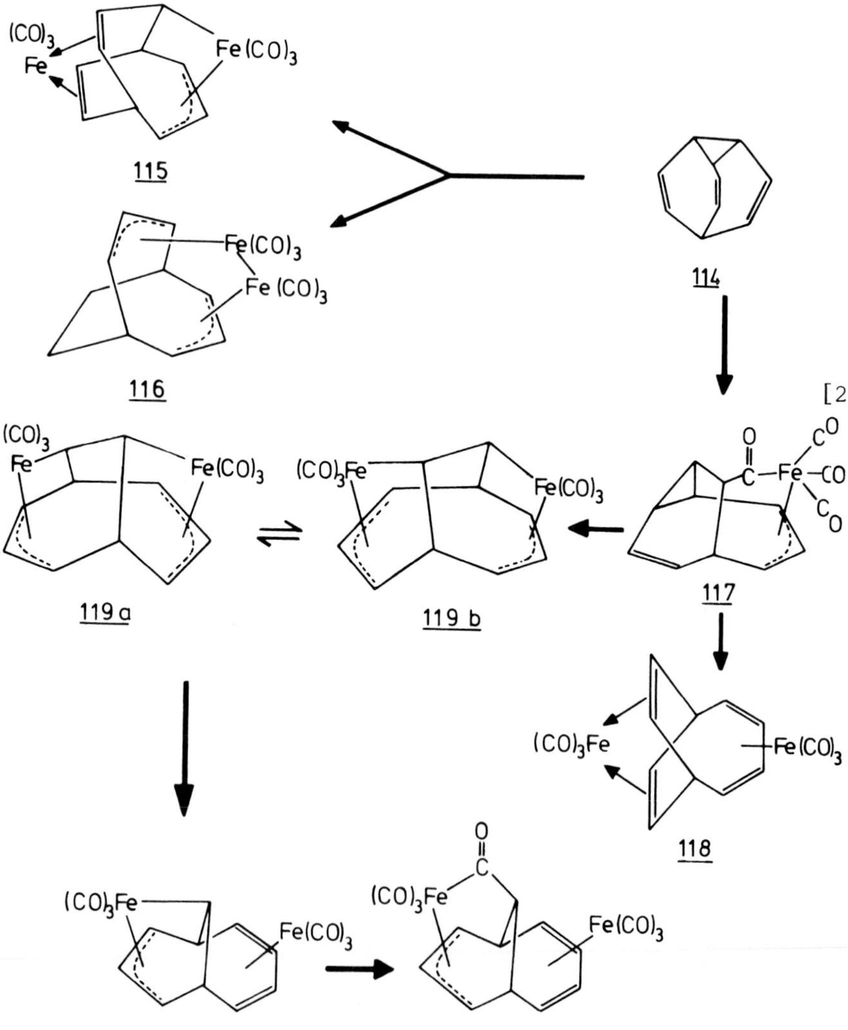

[2]

118 in 26 % yield in which the $C_{10}H_{10}$ hydrocarbon is bonded to both iron atoms as dienes, (f) the red bis(η^3-allyl)-hexacarbonyldiiron derivative 116 in 3 % yield. If bullvalene is added to the solution obtained by the ultraviolet irradiation of pentacarbonyliron in cyclohexene at -78°C, then the reaction proceeds exclusively along the pathway leading to 115 and 116 with the yield of 116 being increased to 13 % (scheme [2]). Ultraviolet irradiation of bullvalene with pentacarbonyliron in moist benzene gives as the major products a 59 % yield of the acyl derivative 117 and a 29 % yield of the hydroxy derivative 124 arising from the water in the benzene. The compound 117 still contains a vinylcyclopropane unit and is the precursor to the bis(η^3-allyl-σ-alkyl) derivative 119 by reaction of this vinylcyclopropane unit with carbonyliron reagents in the usual manner *(15)*. Reaction of the hydroxy derivative 124 with hexafluorophosphoric acid in diethyl ether gives the hexafluorophosphate salt of the stable cation 125 containing a chelating η^3-allyl-η^2-olefin ligand

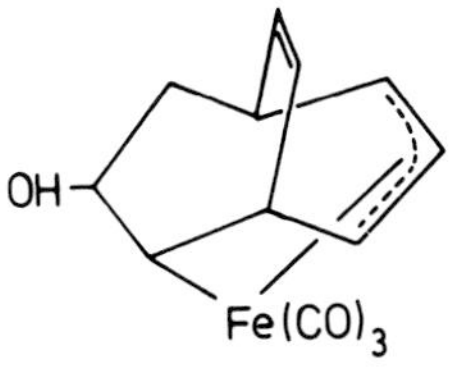

124

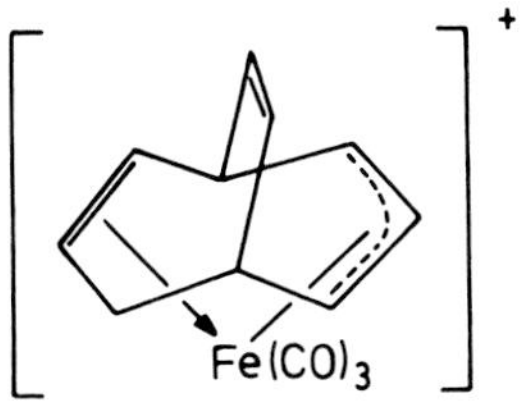

125

similar to the cations discussed above *(15)*. The cation 125 can also be obtained by oxidizing a mixture of 115 and hexafluorophosphoric acid in diethyl ether with ferric chloride. Pyrolysis of the bis(η^3-allyl-σ-alkyl) derivative 119 in octane at 120°C results in rearrangement to give yellow 120 in nearly quantitative yield *(8,15)* Treatment of 120 with carbon monoxide at ∿ 25°C/100 atm results in insertion of a carbonyl group into the σ-alkyl-iron bond to give the corresponding σ-acyl derivative 121. The structures of 115 *(115)*, 118 *(196)*, 119 *(203)*, and 121 *(203)* have been confirmed by X-ray crystallography. ^{1}H-NMR spectroscopy indicates that the asymmetric complex 115 is stable to racemization *(6,15)*. However, the two enantiomers of the bis(η^3-allyl-σ-alkyl) derivative, 119a and 119b, are shown by ^{1}H-NMR spectroscopy to interconvert rapidly above room temperature, although a limiting spectrum corresponding to a frozen enantiomer is seen at 0°C *(7,15)*. The ^{1}H-NMR spectrum of 122 is temperature dependent in the range from 6°C to 80°C indicating a rapid degenerate valence isomerization at higher temperatures *(9, 15)*.

The reaction of the azabullvalene 126 with enneacarbonyldiiron has also been reported *(20)*. Treatment of 126 with

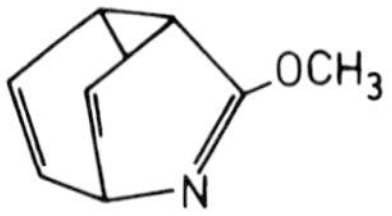

126

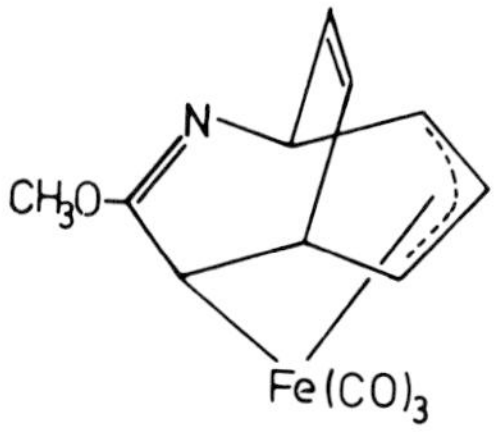

127

$Fe_2(CO)_9$ in benzene at 40°C for 1 h gives the yellow-brown liquid η^3-allyl-σ-alkyl derivative 127 in addition to (methoxy-cyclooctatetraene)tricarbonyliron.

The relative reactivities of different η^3-allyl-σ-alkyl derivatives towards carbon monoxide insertion have been examined *(10)*. Thus, the complexes 120 and 128 (from enneacarbonyldiiron and homosemibullvalene *(11)*) react rapidly with carbon monoxide in hexane even at room temperature to form the corresponding acyls 121 and 129, respectively. Similar insertion reactions do not occur with 104 and 105. Reaction of 120 and 128 with triphenylphosphine for a few seconds at room temperature also results in carbonyl insertion into the σ-alkyl-iron bond to give the triphenylphosphine derivatives 130 and 131, respectively.

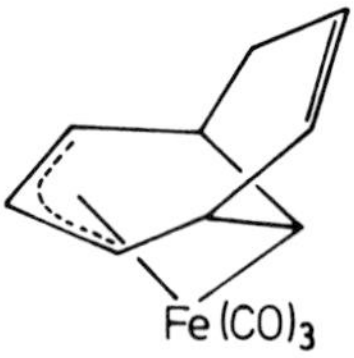

128

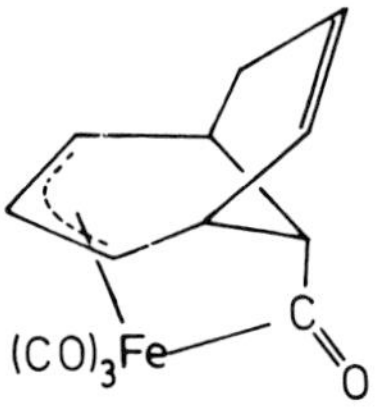

129

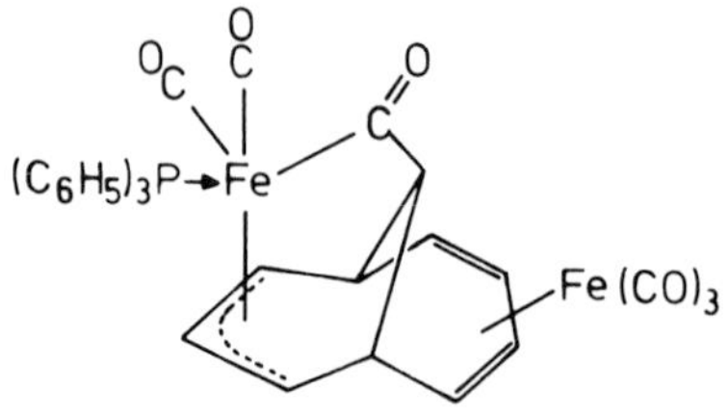

130

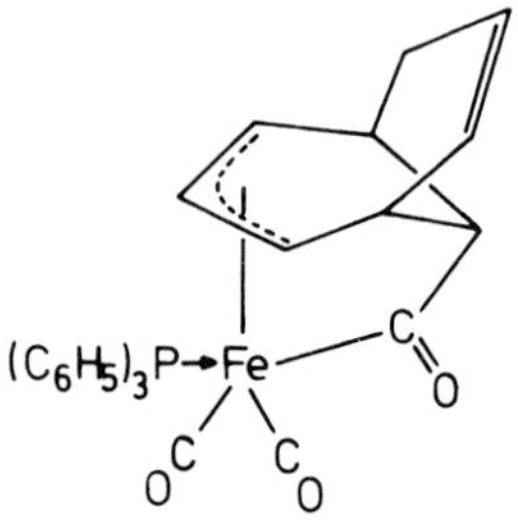

131

Some reactions of carbonyliron complexes with polycyclic vinylcyclopropanes containing two cyclopropane rings have also been investigated. Reaction of 132 with enneacarbonyldiiron

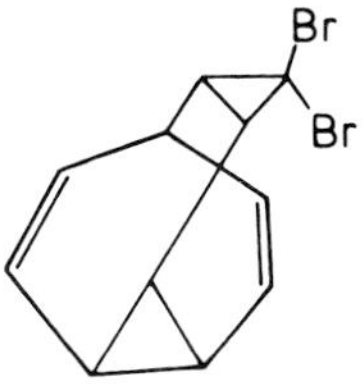

132

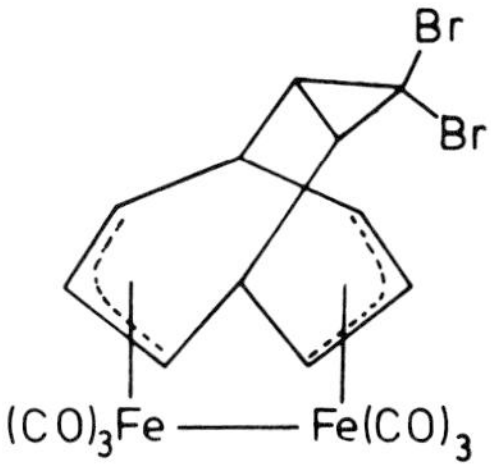

133

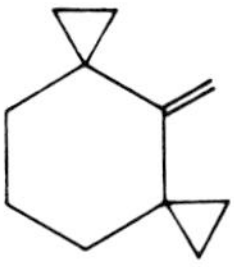

134

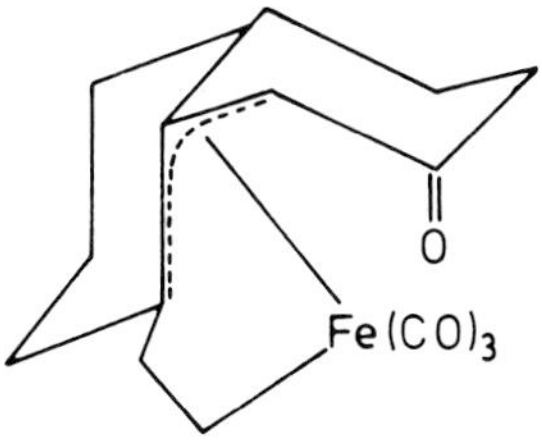

135

in diethyl ether results only in the opening of the non-brominated cyclopropane ring to give the red bis(η^3-allylic) derivative 133 *(11)*. The major product from the ultraviolet irradiation of the dispirane 134 with pentacarbonyliron is the pale yellow crystalline η^3-allyl-σ-alkyl derivative 135 *(193)*.

IX. η^3-ALLYLIC CARBONYLIRON COMPOUNDS FROM REACTIONS OF HETEROCYCLIC DERIVATIVES WITH CARBONYLIRON COMPLEXES

Vinyloxiranes and vinylaziridines react with carbonyliron complexes with ring opening to form η^3-allylic derivatives *(13)*. Thus, ultraviolet irradiation of the vinyloxiranes 136 (R = H, CH_3; R' = H) with pentacarbonyliron in ben-

136

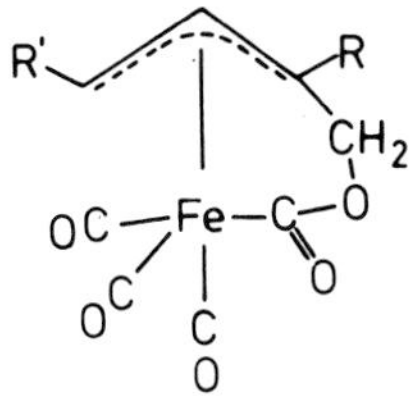

137

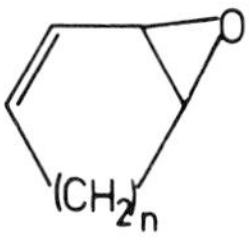

138

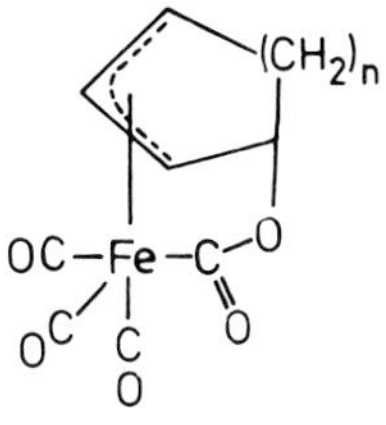

139

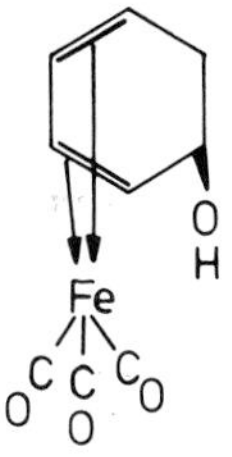

140

zene gives the corresponding lactones 137 (R = H, CH_3; R' = H) as bright yellow solids. Similarly, the ultraviolet irradiations of the bicyclic vinyloxiranes 138 (n = 1,2) with pentacarbonyliron give the corresponding lactones 139 (n = 1, 2). The compounds 137 (R = H, CH_3; R' = H) are stable in boiling benzene. However, 139 (n = 2) undergoes decarbonylation with hydrogen migration to give *endo*-(hydroxycyclohexadiene)tricarbonyliron (140). Studies on the photochemical reactions of pentacarbonyliron with the four stereoisomers of 136 (R = R' = CH_3; *cis* and *trans* around the carbon-carbon double bond, and *cis* and *trans* on the epoxide carbons) to form the corresponding lactones 137 indicate that this reaction is stereospecific *(40)*. Ultraviolet irradiation of cyclooctatetraene epoxide (9-oxabicyclo[4.2.1]nona-2,4,7-triene) with pentacarbonyliron in diethyl ether *(17)* results in the precipitation of the lactone 141 as colourless crystals. Heating

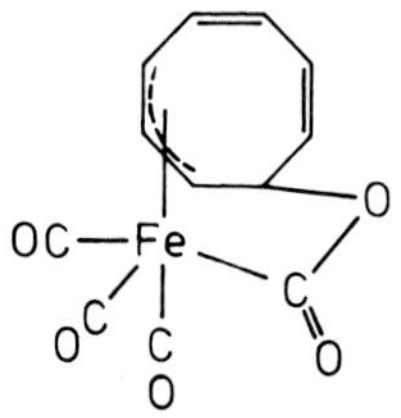

141

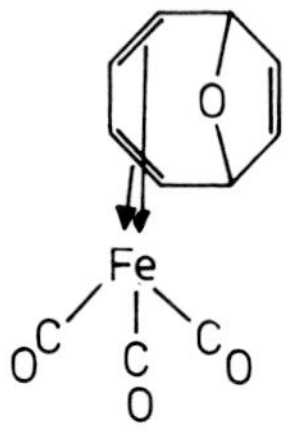

142

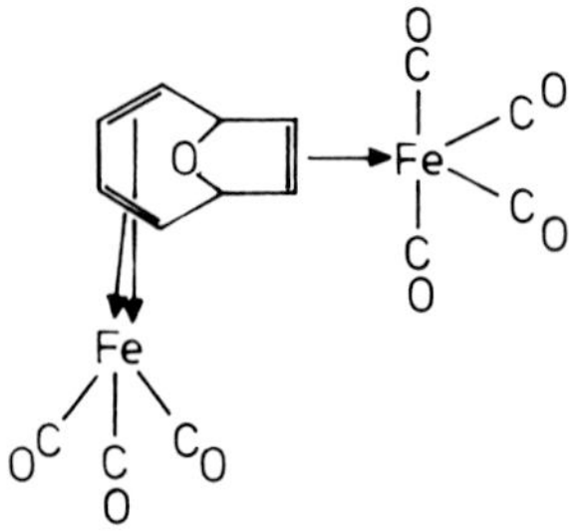

143

141 in chloroform at 40°C for 50 min results in rearrangement to the yellow crystalline diene-tricarbonyliron derivative 142. Ultraviolet irradiation of cyclooctatetraene epoxide with pentacarbonyliron at room temperature gives the binuclear olefin complex 143 *(17)*.

Ultraviolet irradiation of vinylaziridines with pentacarbonyliron in benzene solution also results in opening of the three-membered ring *(13)*. Thus, the aziridine 144 upon

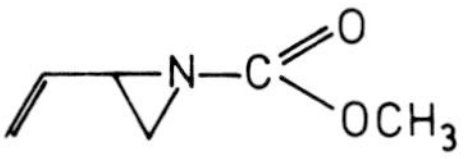

144

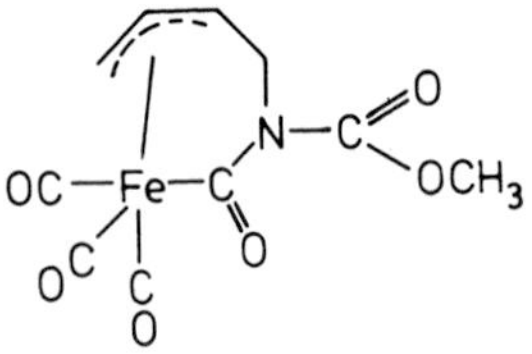

145

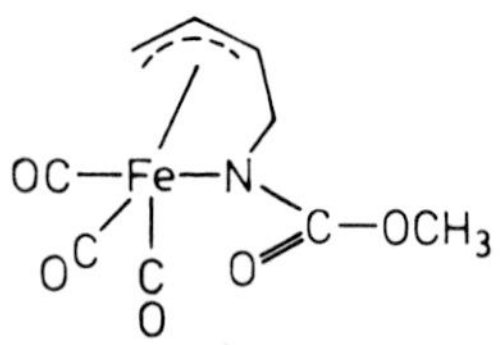

146a

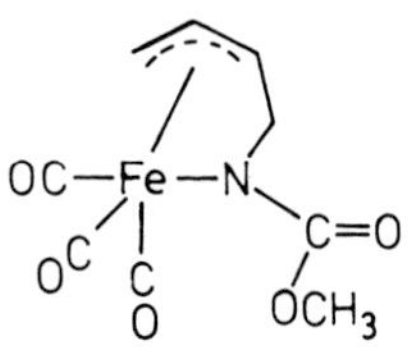

146b

ultraviolet irradiation with pentacarbonyliron gives a 71 % yield of the pale yellow lactam 145. Heating 145 at 60°C for 20 min results in quantitative decarbonylation to give an inseparable 2:1 mixture of the stereoisomers 146a and 146b.

Ultraviolet irradiation of 2,7-dimethyloxepin with pentacarbonyliron in diethyl ether at -60°C gives *o*-xylene, 2,6-dimethylphenol, a 5 % yield of (2,7-dimethyloxepin)tricarbonyliron, and a 1 % yield of a red hexacarbonyldiiron complex

(16). X-ray crystallography of this last complex indicates structure 147 in which the oxepin ring has opened.

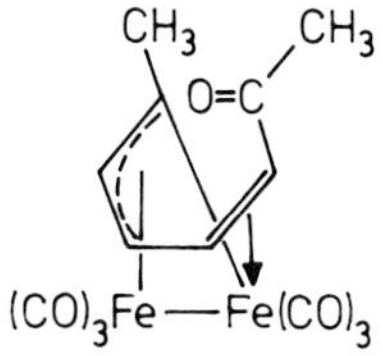

147

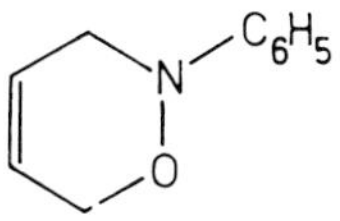

148

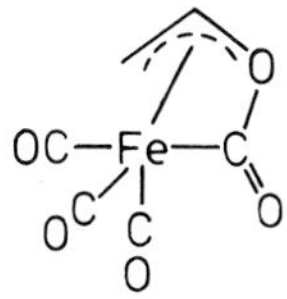

149

Reaction of the heterocycle 148 with enneacarbonyldiiron in benzene solution at 42°C for 30 min in the presence of a stoichiometric amount of water results in aniline elimination to give 149, identical to a product, discussed above, obtained from $Fe_2(CO)_9$ and but-2-ene-1,4-diol *(21)*. Reaction of 149 with primary amines RNH_2 ($R = C_6H_5$, CH_3) in the presence of alumina results in elimination of water to give the corresponding lactam 150 ($R = C_6H_5$, CH_3). Heating 150 ($R = C_6H_5$) in boiling methanol results in rearrangement to 151 by an intramolecular 1,4-hydrogen shift *(21)*.

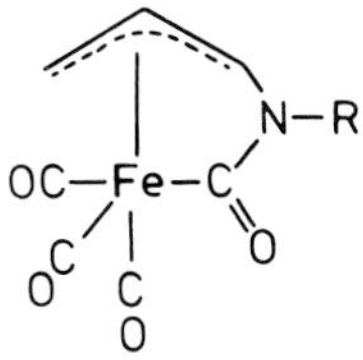

150

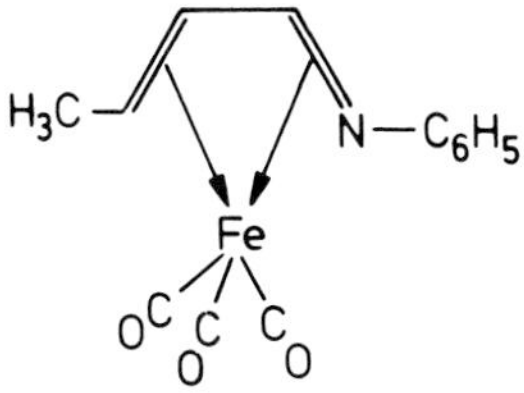

151

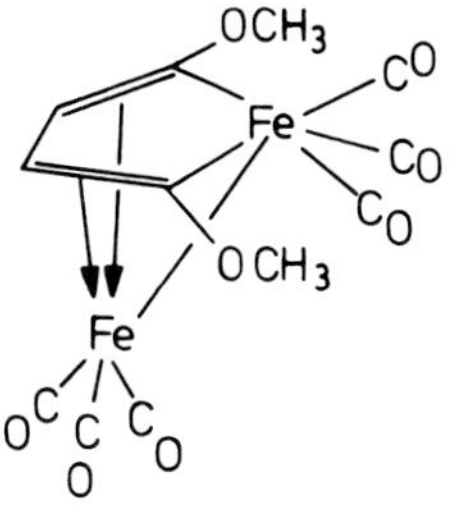

152

Several reactions of the tricarbonylferrole-tricarbonyliron derivative 152 give products with ligands which are 1-3-η^3-allylic coordinated to iron. Thus, reaction of 152 with

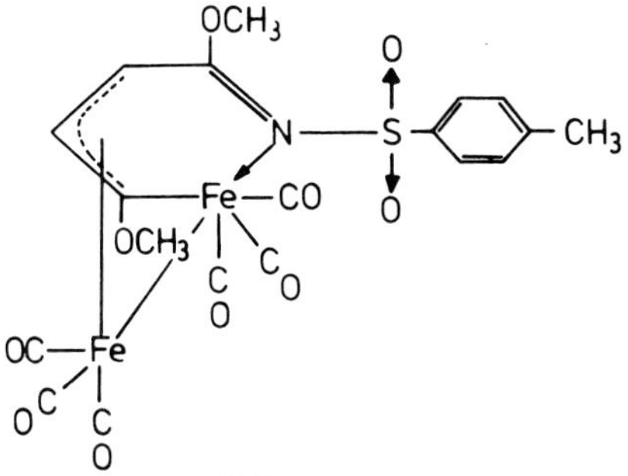

153

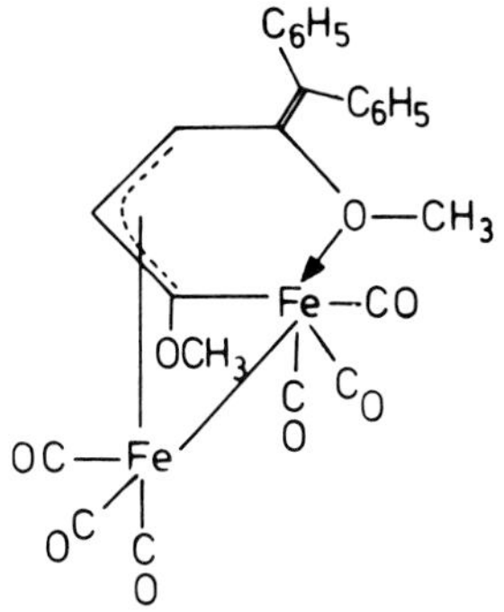

154

dichloramine-T (CH_3-C_6H_4-SO_2NCl_2) gives the η^3-allylic derivative 153 *(189)*. Similarly, ultraviolet irradiation of 152 with diphenyldiazomethane gives a complex mixture of products *(18,117)*. One of these products, a dark brown solid, has been shown by X-ray crystallography to be 154.

X. η^3-ALLYLIC IRON COMPOUNDS ALSO CONTAINING η^5-CYCLOPENTADIENYL RINGS

Several compounds of the type (η^3-allylic)Fe(CO)(η^5-C_5H_5) can be prepared by the decarbonylation of the corresponding σ-allylic derivatives. Thus, reaction of the sodium salt $NaFe(CO)_2(C_5H_5)$ with allyl chloride gives the η^1-allyl derivative CH_2=CH-CH_2-$Fe(CO)_2$(η^5-C_5H_5) (155, R = H). The reaction of H-$Fe(CO)_2$(η^5-C_5H_5) with butadiene gives the related η^1-allylic derivative CH_3-CH=CH-CH_2-$Fe(CO)_2$(η^5-C_5H_5)

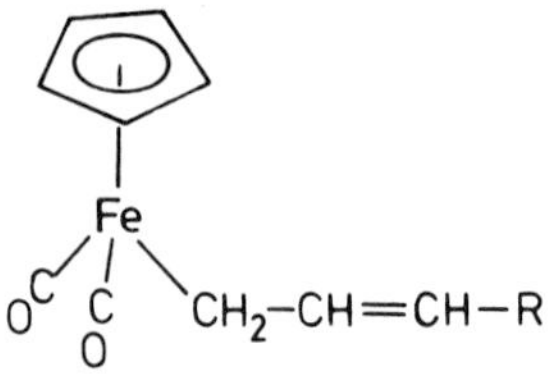

155

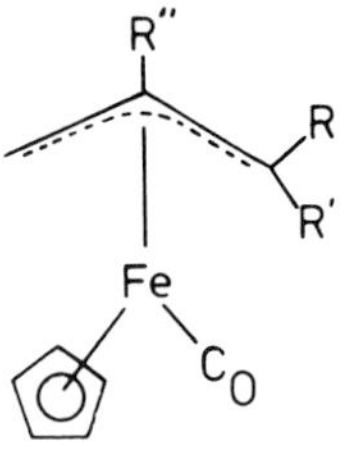

156

(155, R = CH_3). Both of these η^1-allyl derivatives are yellow liquids which oxidize readily in air and decompose above 60°C without forming an η^3-allyl derivative. However, ultraviolet irradiation of the η^1-allyl derivative 155 (R = H) either alone or in cyclohexane solution gives the corresponding η^3-allyl complex 156 (R = R' = R" = H) *(105)*. ^{1}H-NMR and infrared spectral data indicate this complex to be a single isomer with the indicated orientation of the allyl ligand *(90,105)*. Ultraviolet irradiation of the substituted derivative 155 (R = CH_3) gives the corresponding η^3-allylic complex suggested by its complex ^{1}H-NMR spectrum to be a mixture of the *syn*- (156, R = CH_3, R' = R" = H) and *anti*- (156, R' = CH_3, R = R" = H) isomers *(105,145)*. The corresponding 2-methylallyl derivative 156 (R = R' = H, R" = CH_3) has been prepared by completely analogous methods *(90)* and has been shown by its ^{1}H-NMR and infrared spectra to be a single isomer.

Some homo-π-allylic derivatives have also been investigated. Reaction of $NaFe(CO)_2(C_5H_5)$ with 1-chlorobut-3-ene gives the corresponding σ-butenyl complex CH_2=CH-CH_2-CH_2-$Fe(CO)_2$(η^5-C_5H_5). Ultraviolet irradiation of this compound in cyclohexane solution results not only in decarbonylation but also in a hydrogen shift to give the η^3-allylic derivative 156 (R' = CH_3, R = R" = H; mixed with R = CH_3, R' = R" = H) *(107,145)*. However, if this hydrogen shift is blocked by methyl substitution then the reaction proceeds differently. Thus, ultraviolet irradiation of CH_2=CH-C(CH_3)$_2$-CH_2-Fe(CO)$_2$-(η^5-C_5H_5) in cyclohexane solution gives the yellow liquid homo-π-allylic derivative 157 *(107)*.

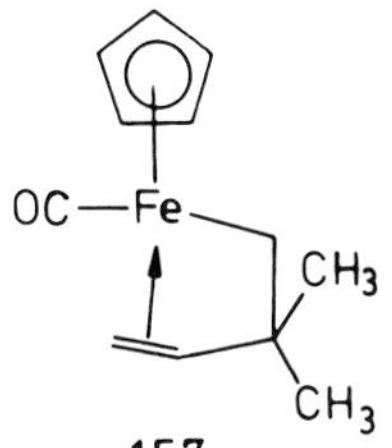

157

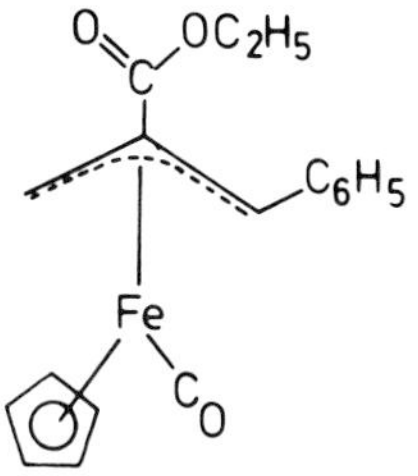

158

An unusual preparation of an (η^3-allylic)Fe(CO)(η^5-C_5H_5) complex is the reaction of the phenylallene cation [(η^5-C_5H_5)-Fe(CO)$_2$(η^2-CH_2=C=CH-C_6H_5)][BF_4] with sodium ethoxide in ethanol to give {(η^5-C_5H_5)Fe(CO)[η^3-C_6H_5-CHC(CO_2-C_2H_5)CH_2]} (158) in 67 % yield *(140)*.

The preparation of some cyclopentadienyl-carbonyliron complexes containing cyclic 1-3-η^3-allylic ligands has been investigated. Reaction of NaFe(CO)$_2$(C_5H_5) with tropylium tetrafluoroborate in tetrahydrofuran at -80°C gives a low yield of red crystalline (η^5-C_5H_5)Fe(CO)(η^3-C_7H_7) *(47,48)* indicated to have structure 159 containing an η^3-cycloheptatrienyl ring. The ^{1}H-NMR spectrum of the seven-membered ring of 159 is characteristic of fluxional molecules. Thus, at room temperature in toluene-d_8 the C_7H_7 ring of 159 exhibits a single resonance which splits to give four resonances of relative areas 2:2:2:1 at -50°C. The proton averaging is believed to occur by a sequence of rapid 1,2-shifts *(48)*. An attempt to prepare an η^3-benzyl complex of iron containing an 1-3-η^3-allylic unit by the ultraviolet irradiation of the σ-benzyl complex C_6H_5-CH_2-Fe(CO)$_2$(η^5-C_5H_5) was unsuccessful *(168)*. Prolonged ultraviolet irradiation of this complex was found to lead to [(η^5-C_5H_5)Fe(CO)$_2$]$_2$ and (η^5-C_5H_5)Fe(η^5-C_5H_4-CO-CH_2-C_6H_5) as the only identifiable cyclopentadienyliron products.

Some cyclopentadienyliron derivatives containing an 1-3-η^3-allylic cyclopentenone ring have been prepared from haloalkylallenes *(22,191)*. Reaction of NaFe(CO)$_2$(C_5H_5) with the allenic bromides R-CH=C=CR-(CH_2)$_2$-Br followed by pyrolysis in tetrahydrofuran at 50°C for 48 h results in cyclization to give the corresponding η^3-allylic cyclopentenone complexes

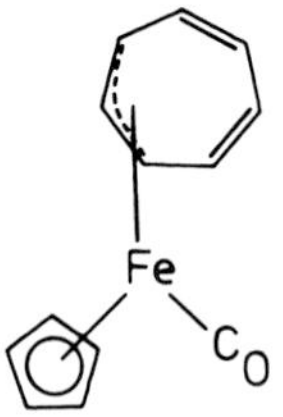
159

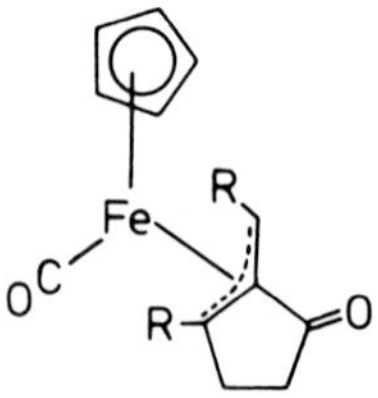
160

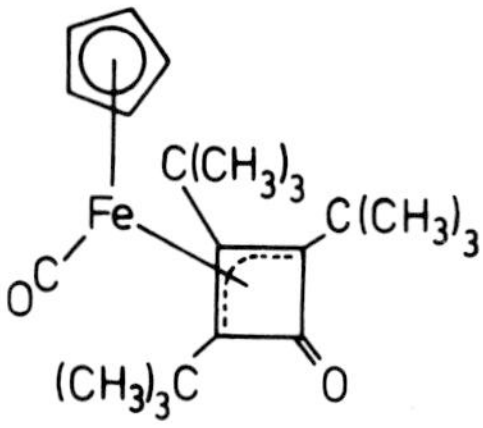
161

160 (R = H, CH_3) as stable crystalline solids.

Reaction of $NaFe(CO)_2(C_5H_5)$ with tri-*tert*-butylcyclopropenyl tetrafluoroborate in tetrahydrofuran at room temperature gives a 70 % yield of the yellow-orange crystalline ketocyclobutenyl complex 161 analogous to 55 discussed earlier in this chapter *(103)*.

XI. TRIS(ALLYL)IRON

Tris(allyl)iron is obtained as very thermally unstable golden yellow to dark brown plates by the reaction of iron(III) chloride with allylmagnesium chloride in diethyl ether at -78°C. Pure crystalline tris(allyl)iron begins to decompose above -60°C and decomposes violently above -10°C. Reaction of tris(allyl)iron with carbon monoxide gives a volatile bright yellow product of stoichiometry $(C_3H_5)_2Fe(CO)_3$ believed to be the σ-propenyl complex 162 *(209)*. A mixture of

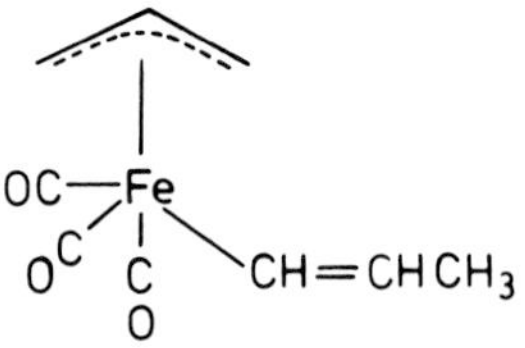
162

tris(allyl)iron and triethylphosphine in diethyl ether at -78°C reacts with molecular nitrogen to form a dinitrogen-iron complex of unknown nature. An infrared spectrum of the resulting mixture exhibited a band at 2038 cm^{-1} in the region expected for dinitrogen coordinated to iron *(33)*.

Acknowledgement

A fellowship of the Max-Planck-Gesellschaft during the time this article was written at the Institut für Strahlenchemie im Max-Planck-Institut für Kohlenforschung (Mülheim a.d. Ruhr, Germany) is gratefully acknowledged.

References

1. Abel, E.W., and Moorhouse, S., *Inorg. Nucl. Chem. Lett.*, *6*, 621 (1970).
2. Abel, E.W., and Moorhouse, S., *J. Chem. Soc. Dalton Trans.*, *1973*, 1706.
3. Alper, H., LePort, P.C., and Wolfe, S., *J. Amer. Chem. Soc.*, *91*, 7553 (1969).
4. Altman, J., and Ginsburg, D., *Tetrahedron*, *27*, 93 (1971).
5. Aumann, R., and Winstein, S., *Angew. Chem.*, *82*, 667 (1970); *Angew. Chem. Int. Ed. Engl.*, *9*, 638 (1970).
6. Aumann, R., *Angew. Chem.*, *83*, 175 (1971); *Angew. Chem. Int. Ed. Engl.*, *10*, 188 (1971).
7. Aumann, R., *Angew. Chem.*, *83*, 176 (1971); *Angew. Chem. Int. Ed. Engl.*, *10*, 189 (1971).
8. Aumann, R., *Angew. Chem.*, *83*, 177 (1971); *Angew. Chem. Int. Ed. Engl.*, *10*, 190 (1971).
9. Aumann, R., *Angew. Chem.*, *83*, 583 (1971); *Angew. Chem. Int. Ed. Engl.*, *10*, 560 (1971).
10. Aumann, R., *Angew. Chem.*, *84*, 583 (1972); *Angew. Chem. Int. Ed. Engl.*, *11*, 522 (1972).
11. Aumann, R., and Lohmann, B., *J. Organometal. Chem.*, *44*, C 51 (1972).
12. Aumann, R., *J. Organometal. Chem.*, *47*, C 29 (1973).
13. Aumann, R., Fröhlich, K., and Ring, H., *Angew. Chem.*, *86*, 309 (1974); *Angew. Chem. Int. Ed. Engl.*, *13*, 275 (1974).
14. Aumann, R., *J. Amer. Chem. Soc.*, *96*, 2631 (1974).
15. Aumann, R., *Chem. Ber.*, *108*, 1974 (1975).
16. Aumann, R., Averbeck, H., and Krüger, C., *Chem. Ber.*, *108*, 3336 (1975).
17. Aumann, R., and Averbeck, H., *J. Organometal. Chem.*, *85*, C 4 (1975).
18. Bagga, M.M., Ferguson, G., Jeffreys, J.A.D., Mansell, C.M., Pauson, P.L., Robertson, I.C., and Sime, J.G., *J. Chem. Soc. D, Chem. Commun.*, *1970*, 672.
19. Bauer, R.A., Fischer, E.O., and Kreiter, C.G., *J. Organometal. Chem.*, *24*, 737 (1970).
20. Becker, Y., Eisenstadt, A., and Shvo, Y., *J. Chem. Soc. Chem. Commun.*, *1972*, 1156.
21. Becker, Y., Eisenstadt, A., and Shvo, Y., *Tetrahedron*, *30*, 839 (1974).
22. Benaïm, J., Mérour, J.Y., and Roustan, J.L., *Tetrahedron Lett.*, *1971*, 983.
23. Ben-Shoshan, R., and Pettit, R., *Chem. Commun.*, *1968*, 247.
24. Bond, A., Green, M., Lewis, B., and Lowrie, S.F.W., *J.

Chem. Soc. D, Chem. Commun., 1971, 1230.

25. Bond, A., Green, M., and Taylor, S.H., *J. Chem. Soc. Chem. Commun., 1973,* 112.

26. Bond, A., Lewis, B., and Green, M., *J. Chem. Soc. Dalton Trans., 1975,* 1109.

27. Borshagovskii, B.V., Gol'danskii, V.I., Gubin, S.P., Denisovich, L.I., and Stukan, R.A., *Teor. Eksp. Khim., 5,* 372 (1969); *Theor. Exp. Chem., 5,* 240 (1969).

28. Bottrill, M., Goddard, R., Green, M., Hughes, R.P., Lloyd, M.K., Lewis, B., and Woodward, P., *J. Chem. Soc. Chem. Commun., 1975,* 253.

29. Bright, D., and Mills, O.S., *J. Chem. Soc. Dalton Trans., 1972,* 2465.

30. Bruce, R., Chaudhary, F.M., Knox, G.R., and Pauson, P.L., *Z. Naturforsch., B 20,* 73 (1965).

31. Burton, R., Green, M.L.H., Abel, E.W., and Wilkinson, G., *Chem. Ind. (London), 1958,* 1592.

32. Burton, R., Pratt, L., and Wilkinson, G., *J. Chem. Soc., 1960,* 4290.

33. Campbell, C.H., Dias, A.R., Green, M.L.H., Saito, T., and Swanwick, M.G., *J. Organometal. Chem., 14,* 349 (1968).

34. Candlin, J.P., and Janes, W.H., *J. Chem. Soc. C, 1968,* 1856.

35. Cardaci, G., Murgia, S.M., and Foffani, A., *J. Organometal. Chem., 37,* C 11 (1972).

36. Cardaci, G., and Foffani, A., *J. Chem. Soc. Dalton Trans., 1974,* 1808.

37. Cardaci, G., *J. Chem. Soc. Dalton Trans., 1974,* 2452.

38. Casey, C.P., and Cyr, C.R., *J. Amer. Chem. Soc., 95,* 2248 (1973).

39. Chaudhari, F.M., Knox, G.R., and Pauson, P.L., *J. Chem. Soc. C,* 1967, 2255.

40. Chen, K.-N., Moriarty, R.M., DeBoer, B.G., Churchill, M.R., and Yeh, H.J.C., *J. Amer. Chem. Soc., 97,* 5602 (1975).

41. Churchill, M.R., *Inorg. Chem., 6,* 190 (1967).

42. Churchill, M.R., and Wormald, J., *Chem. Commun., 1968,* 1597.

43. Churchill, M.R., Wormald, J., Young, D.A.T., and Kaesz, H.D., *J. Amer. Chem. Soc., 91,* 7201 (1969).

44. Churchill, M.R., and Wormald, J., *Inorg. Chem., 9,* 2239 (1970).

45. Churchill, M.R., and Wormald, J., *Inorg. Chem., 9,* 2430 (1970).

46. Churchill, M.R., *Progr. Inorg. Chem., 11,* 53 (1970).

47. Ciappenelli, D., and Rosenblum, M., *J. Amer. Chem. Soc., 91,* 3673 (1969).

48. Ciappenelli, D., and Rosenblum, M., *J. Amer. Chem. Soc., 91,* 6876 (1969).
49. Clarke, H.L., and Fitzpatrick, N.J., *J. Organometal. Chem., 66,* 119 (1974).
50. Clarke, H.L., *J. Organometal. Chem., 80,* 155 (1974).
51. Clarke, H.L., *J. Organometal. Chem., 80,* 369 (1974).
52. Coffey, C.E., *J. Amer. Chem. Soc., 84,* 118 (1962).
53. Cotton, F.A., and Musco, A., *J. Amer. Chem. Soc., 90,* 1444 (1968).
54. Cotton, F.A., and Takats, J., *J. Amer. Chem. Soc., 90,* 2031 (1968).
55. Cotton, F.A., and Edwards, W.T., *J. Amer. Chem. Soc., 91,* 843 (1969).
56. Cotton, F.A., Deeming, A.J., Josty, P.L., Ullah, S.S., Domingos, A.J.P., Johnson, B.F.G., and Lewis, J., *J. Amer. Chem. Soc., 93,* 4624 (1971).
57. Cotton, F.A., DeBoer, B.G., and Marks, T.J., *J. Amer. Chem. Soc., 93,* 5069 (1971).
58. Cotton, F.A., and Troup, J.M., *J. Amer. Chem. Soc., 95,* 3798 (1973).
59. Cotton, F.A., and Troup, J.M., *J. Organometal. Chem., 76,* 81 (1974).
60. Cotton, F.A., Hunter, D.L., and Lahuerta, P., *Inorg. Chem., 14,* 511 (1975).
61. Cotton, F.A., Hunter, D.L., and Lahuerta, P., *J. Amer. Chem. Soc., 97,* 1046 (1975).
62. Cotton, J.D., Doddrell, D., Heazlewood, R.L., and Kitching, W., *Aust. J. Chem., 22,* 1785 (1969).
63. Cowherd, F.G., and von Rosenberg, J.L., *J. Amer. Chem. Soc., 91,* 2157 (1969).
64. Crease, A.E., and Legzdins, P., *J. Chem. Soc. Chem. Commun., 1973,* 775.
65. Davis, R.E., *Chem. Commun., 1968,* 248.
66. Davison, A., McFarlane, W., and Wilkinson, G., *Chem. Ind. (London), 1962,* 820.
67. Davison, A., McFarlane, W., Pratt, L., and Wilkinson, G., *J. Chem. Soc., 1962,* 4821.
68. Davison, J.B., and Bellama, J.M., *Inorg. Chim. Acta, 14,* 263 (1975).
69. Deeming, A.J., Ullah, S.S., Domingos, A.J.P., Johnson, B.F.G., and Lewis, J., *J. Chem. Soc. Dalton Trans., 1974,* 2093.
70. Deganello, G., Maltz, H., and Kozarich, J., *J. Organometal. Chem., 60,* 323 (1973).
71. Deganello, G., Uguagliati, P., Calligaro, L., Sandrini, P.L., and Zingales, F., *Inorg. Chim. Acta, 13,* 247 (1975).
72. Deganello, G., Boschi, T., and Toniolo, L., *J.*

Organometal. Chem., *97*, C 46 (1975).
73. Domrachev, G.A., Sorokin, Yu.A., Razuvaev, G.A., and Suvorova, O.N., *Dokl. Akad. Nauk SSSR*, *183*, 1085 (1968); *Dokl. Chem.*, *183*, 1066 (1968).
74. Edwards, R., Howell, J.A.S., Johnson, B.F.G., and Lewis, J., *J. Chem. Soc. Dalton Trans.*, *1974*, 2105.
75. Efraty, A., Potenza, J., Sandhu, S.S., Johnson, R., Mastropaolo, M., Bystrek, R., Denney, D.Z., and Herber, R.H., *J. Organometal. Chem.*, *70*, C 24 (1974).
76. Ehntholt, D.J., and Kerber, R.C., *J. Organometal. Chem.*, *38*, 139 (1972).
77. Ehntholt, D., Rosan, A., and Rosenblum, M., *J. Organometal. Chem.*, *56*, 315 (1973).
78. Ehrlich, K., and Emerson, G.F., *J. Chem. Soc. D, Chem. Commun.*, *1969*, 59.
79. Ehrlich, K., and Emerson, G.F., *J. Amer. Chem. Soc.*, *94*, 2464 (1972).
80. Eisenstadt, A., and Winstein, S., *Tetrahedron Lett.*, *1970*, 4603.
81. Eisenstadt, A., *J. Organometal. Chem.*, *38*, C 32 (1972).
82. Eisenstadt, A., *Tetrahedron Lett.*, *1972*, 2005.
83. Eisenstadt, A., *J. Organometal. Chem.*, *60*, 335 (1973).
84. Eisenstadt, A., *Tetrahedron*, *30*, 2353 (1974).
85. Emerson, G.F., and Pettit, R., *J. Amer. Chem. Soc.*, *84*, 4591 (1962).
86. Emerson, G.F., Mahler, J.E., and Pettit, R., *Chem. Ind. (London)*, *1964*, 836.
87. Emerson, G.F., Mahler, J.E., Pettit, R., and Collins, R., *J. Amer. Chem. Soc.*, *86*, 3590 (1964).
88. Emerson, G.F., Watts, L., and Pettit, R., *J. Amer. Chem. Soc.*, *87*, 131 (1965).
89. Emerson, G.F., Ehrlich, K., Giering, W.P., and Lauterbur, P.C., *J. Amer. Chem. Soc.*, *88*, 3172 (1966).
90. Faller, J.W., Johnson, B.V., and Dryja, T.P., *J. Organometal. Chem.*, *65*, 395 (1974).
91. Fischer, E.O., and Werner, H., *Z. Chem.*, *2*, 174 (1962).
92. Fleischer, E.B., Stone, A.L., Dewar, R.B.K., Wright, J.D., Keller, C.E., and Pettit, R., *J. Amer. Chem. Soc.*, *88*, 3158 (1966).
93. Gibson, D.H., Vonnahme, R.L., and McKiernan, J.E., *J. Chem. Soc. D, Chem. Commun.*, *1971*, 720.
94. Gibson, D.H., and Vonnahme, R.L., *J. Amer. Chem. Soc.*, *94*, 5090 (1972).
95. Gibson, D.H., and Vonnahme, R.L., *J. Chem. Soc. Chem. Commun.*, *1972*, 1021.
96. Gibson, D.H., and Vonnahme, R.L., *J. Organometal. Chem.*, *70*, C 33 (1974).
97. Gibson, D.H., and Erwin, D.K., *J. Organometal. Chem.*,

86, C 31 (1975).

98. Greaves, E.O., Knox, G.R., and Pauson, P.L., *J. Chem. Soc. D, Chem. Commun.*, *1969*, 1124.

99. Greaves, E.O., Knox, G.R., Pauson, P.L., Toma, S., Sim, G.A., and Woodhouse, D.I., *J. Chem. Soc. Chem. Commun.*, *1974*, 257.

100. Green, M., and Wood, D.C., *J. Chem. Soc. A*, *1969*, 1172.

101. Green, M., and Lewis, B., *J. Chem. Soc. Chem. Commun.*, *1973*, 114.

102. Green, M., Heathcock, S., and Wood, D.C., *J. Chem. Soc. Dalton Trans.*, *1973*, 1564.

103. Green, M., and Hughes, R.P., *J. Chem. Soc. Chem. Commun.*, *1975*, 862.

104. Green, M., Lewis, B., Daly, J.J., and Sanz, F., *J. Chem. Soc. Dalton Trans.*, *1975*, 1118.

105. Green, M.L.H., and Nagy, P.L.I., *J. Chem. Soc.*, *1963*, 189.

106. Green, M.L.H., and Nagy, P.L.I., *Advan. Organometal. Chem.*, *2*, 325 (1964).

107. Green, M.L.H., and Smith, M.J., *J. Chem. Soc. A*, *1971*, 3220.

108. Gubin, S.P., and Denisovich, L.I., *J. Organometal. Chem.*, *15*, 471 (1968).

109. Hardy, A.D.U., and Sim, G.A., *J. Chem. Soc. Dalton Trans.*, *1972*, 2305.

110. Heck, R.F., and Boss, C.R., *J. Amer. Chem. Soc.*, *86*, 2580 (1964).

111. Heck, R.F., U.S. Patent 3,338,936 (1967).

112. Heil, V., Johnson, B.F.G., Lewis, J., and Thompson, D.J., *J. Chem. Soc. Chem. Commun.*, *1974*, 270.

113. Hendrix, W.T., Cowherd, F.G., and von Rosenberg, J.L., *Chem. Commun.*, *1968*, 97.

114. Hill, A.E., and Hoffmann, H.M., *J. Chem. Soc. Chem. Commun.*, *1972*, 574.

115. Huttner, G., and Regler, D., *Chem. Ber.*, *105*, 3936 (1972).

116. Impastato, F.J., and Ihrman, K.G., *J. Amer. Chem. Soc.*, *83*, 3726 (1961).

117. Jeffreys, J.A.D., Willis, C.M., Robertson, I.C., Ferguson, G., and Sime, J.G., *J. Chem. Soc. Dalton Trans.*, *1973*, 749.

118. Johnson, B.F.G., Lewis, J., and Twigg, M.V., *J. Organometal. Chem.*, *52*, C 31 (1973).

119. Johnson, B.F.G., Lewis, J., Matheson, T.W., Ryder, I.E., and Twigg, M.V., *J. Chem. Soc. Chem. Commun.*, *1974*, 269.

120. Johnson, B.F.G., Lewis, J., and Twigg, M.V., *J. Chem. Soc. Dalton Trans.*, *1974*, 241.

121. Johnson, B.F.G., Lewis, J., and Thompson, D.J., *Tetrahedron Lett.*, *1974*, 3789.
122. Johnson, B.F.G., Lewis, J., Thompson, D.J., and Heil, B., *J. Chem. Soc. Dalton Trans.*, *1975*, 567.
123. Johnson, B.F.G., Lewis, J., Ryder, I.E., and Twigg, M.V., *J. Chem. Soc. Dalton Trans.*, *1976*, 421.
124. Keller, C.E., Emerson, G.F., and Pettit, R., *J. Amer. Chem. Soc.*, *87*, 1388 (1965).
125. King, R.B., and Stone, F.G.A., *J. Amer. Chem. Soc.*, *82*, 4557 (1960).
126. King, R.B., *Inorg. Chem.*, *2*, 807 (1963).
127. King, R.B., *Inorg. Chem.*, *2*, 936 (1963).
128. King, R.B., *J. Amer. Chem. Soc.*, *88*, 2075 (1966).
129. King, R.B., and Kapoor, R.N., *J. Organometal. Chem.*, *15*, 457 (1968).
130. King, R.B., and Kapoor, R.N., *J. Inorg. Nucl. Chem.*, *31*, 2169 (1969).
131. King, R.B., *Org. Mass Spectr.*, *2*, 401 (1969).
132. King, R.B., and Efraty, A., *J. Organometal. Chem.*, *24*, 241 (1970).
133. King, R.B., and Bond, A., *J. Organometal. Chem.*, *46*, C 53 (1972).
134. King, R.B., and Bond, A., *J. Amer. Chem. Soc.*, *96*, 1343 (1974).
135. King, R.B., and Harmon, C.A., *J. Organometal. Chem.*, *86*, 239 (1975).
136. King, R.B., and Harmon, C.A., *J. Organometal. Chem.*, *88*, 93 (1975).
137. King, R.B., and Harmon, C.A., *J. Amer. Chem. Soc.*, *98*, 2409 (1976).
138. Korecz, L., and Burger, K., *Acta Chim. Acad. Sci. Hung.*, *58*, 253 (1968).
139. Kruck, T., and Knoll, L., *Z. Naturforsch.*, *B 28*, 34 (1973).
140. Lichtenberg, D.W., and Wojcicki, A., *J. Organometal. Chem.*, *94*, 311 (1975).
141. Lindley, P.F., and Mills, O.S., *J. Chem. Soc. A*, *1970*, 38.
142. Maltz, H., and Kelly, B.A., *J. Chem. Soc. D, Chem. Commun.*, *1971*, 1390.
143. Manuel, T.A., *J. Org. Chem.*, *27*, 3941 (1962).
144. Maxfield, P.L., *Inorg. Nucl. Chem. Lett.*, *6*, 707 (1970).
145. Merour, J.-Y., Charrier, C., Roustan, J.-L., and Benaim, J., *C. R. Acad. Sci., Ser. C*, *273*, 285 (1971).
146. Minasyants, M.Kh., Struchkov, Yu.T., Kritskaya, I.I., and Avoyan, R.L., *Zh. Strukt. Khim.*, *7*, 903 (1966); *J. Struct. Chem.*, *7*, 840 (1966).

147. Minasyants, M.Kh., and Struchkov, Yu.T., *Zh. Strukt. Khim.*, *9*, 665 (1968); *J. Struct. Chem.*, *9*, 577 (1968).
148. Minasyants, M.Kh., Andrianov, V.G., and Struchkov, Yu.T., *Zh. Strukt. Khim.*, *9*, 1055 (1968); *J. Struct. Chem.*, *9*, 939 (1968).
149. Moriarty, R.M., Yeh, C.-L., and Ramey, K.C., *J. Amer. Chem. Soc.*, *93*, 6709 (1971).
150. Moriarty, R.M., Chen, K.-N., Yeh, C.-L., Flippen, J.L., and Karle, J., *J. Amer. Chem. Soc.*, *94*, 8944 (1972).
151. Moriarty, R.M., Yeh, C.-L., Chen, K.-N., and Srinivasan, R., *Tetrahedron Lett.*, *1972*, 5325.
152. Moriarty, R.M., Yeh, C.-L., Chen, K.-N., Yeh, E.L., Ramey, K.C., and Jefford, C.W., *J. Amer. Chem. Soc.*, *95*, 4756 (1973).
153. Muetterties, E.L., and Rathke, J.W., *J. Chem. Soc. Chem. Commun.*, *1974*, 850.
154. Murdoch, H.D., and Weiss, E., *Helv. Chim. Acta*, *45*, 1156 (1962).
155. Murdoch, H.D., and Weiss, E., *Helv. Chim. Acta*, *45*, 1927 (1962).
156. Murdoch, H.D., *Helv. Chim. Acta*, *47*, 936 (1964).
157. Murdoch, H.D., and Lucken, E.A.C., *Helv. Chim. Acta*, *47*, 1517 (1964).
158. Murdoch, H.D., *Z. Naturforsch.*, *B 20*, 179 (1965).
159. Nakamura, A., Kim, P.-J., and Hagihara, N., *J. Organometal. Chem.*, *3*, 7 (1965).
160. Nakamura, A., *Bull. Chem. Soc. Jap.*, *39*, 543 (1966).
161. Nesmeyanov, A.N., Kritskaya, I.I., and Fedin, E.I., *Dokl. Akad. Nauk SSSR*, *164*, 1058 (1965); *Dokl. Chem.*, *164*, 973 (1965).
162. Nesmeyanov, A.N., Kritskaya, I.I., Ustynyuk, Yu.A., and Fedin, E.I., *Dokl. Akad. Nauk SSSR*, *176*, 341 (1967); *Dokl. Chem.*, *176*, 808 (1967).
163. Nesmeyanov, A.N., Kritskaya, I.I., Kudryavtsev, R.V., and Lyakhovetskii, Yu.I., *Izv. Akad. Nauk SSSR, Ser. Khim.*, *1967*, 418; *Bull. Acad. Sci. USSR, Div. Chem. Ser.*, *1967*, 396.
164. Nesmeyanov, A.N., Kritskaya, I.I., Zol'nikova, G.P., Ustynyuk, Yu.A., Babakhina, G.M., and Vainbert, A.M., *Dokl. Akad. Nauk SSSR*, *182*, 1091 (1968); *Dokl. Chem.*, *182*, 903 (1968).
165. Nesmeyanov, A.N., and Kritskaya, I.I., *J. Organometal. Chem.*, *14*, 387 (1968).
166. Nesmeyanov, A.N., Ustynyuk, Yu.A., Kritskaya, I.I., and Shchembelov, G.A., *J. Organometal. Chem.*, *14*, 395 (1968).
167. Nesmeyanov, A.N., Gubin, S.P., and Rubezhov, A.Z., *J. Organometal. Chem.*, *16*, 163 (1969).

168. Nesmeyanov, A.N., Chenskaya, T.B., Babakhina, G.M., and Kritskaya, I.I., *Izv. Akad. Nauk SSSR, Ser. Khim., 1970,* 1187; *Bull. Acad. Sci. USSR, Div. Chem. Ser., 1970,* 1129.

169. Nesmeyanov, A.N., Nekrasov, Yu.S., Avakyan, N.P., and Kritskaya, I.I., *J. Organometal. Chem., 33,* 375 (1971).

170. Nesmeyanov, A.N., Zol'nikova, G.P., Babakhina, G.M., Kritskaya, I.I., and Yakobson, G.G., *Zh. Obshch. Khim., 43,* 2007 (1973); *J. Gen. Chem. USSR, 43,* 1993 (1973).

171. Nesmeyanov, A.N., Rybin, L.V., Gubenko, N.T., Rybinskaya, M.I., and Petrovskii, P.V., *J. Organometal. Chem., 71,* 271 (1974).

172. Noack, K., *Helv. Chim. Acta, 45,* 1847 (1962).

173. Ohbe, Y., Takagi, M., and Matsuda, T., *Tetrahedron, 30,* 2669 (1974).

174. Otsuka, S., Nakamura, A., and Tani, K., *J. Chem. Soc. A, 1968,* 2248.

175. Otsuka, S., Nakamura, A., and Tani, K., *J. Chem. Soc. A, 1971,* 154.

176. Paliani, G., Murgia, S.M., and Cardaci, G., *J. Organometal. Chem., 30,* 221 (1971).

177. Paliani, A., Poletti, A., Cardaci, G., Murgia, S.M., and Cataliotti, R., *J. Organometal. Chem., 60,* 157 (1973).

178. Paquette, L.A., Ley, S.V., Broadhurst, M.J., Truesdell, D., Fayos, J., and Clardy, J., *Tetrahedron Lett., 1973,* 2943.

179. Paquette, L.A., Ley, S.V., Maiorana, S., Schneider, D.F., Broadhurst, M.J., and Boggs, R.A., *J. Amer. Chem. Soc., 97,* 4658 (1975).

180. Parshall, G.W., and Wilkinson, G., *J. Chem. Soc., 1962,* 1132.

181. Pearson, A.J., *Tetrahedron Lett., 1975,* 3617.

182. Pettit, R., and Henery, J., *Org. Syn., 50,* 21 (1970).

183. Plowman, R.A., and Stone, F.G.A., *Z. Naturforsch., B 17* 575 (1962).

184. Potenza, J., Johnson, R., Mastropaolo, D., and Efraty, A., *J. Organometal. Chem., 64,* C 13 (1974).

185. Randall, E.W., Rosenberg, E., and Milone, L., *J. Chem. Soc. Dalton Trans., 1973,* 1672.

186. Rausch, M.D., and Schrauzer, G.N., *Chem. Ind. (London), 1959,* 957.

187. Razuvaev, G.A., Domrachev, G.A., Suvorova, O.N., and Abakumova, L.G., *J. Organometal. Chem., 32,* 113 (1971).

188. Reardon, E.J., Jr., and Brookhardt, M., *J. Amer. Chem. Soc., 95,* 4311 (1973).

189. Rodrique, L., van Meerssche, M., and Piret, P., *Acta*

Crystallogr., *B 25*, 519 (1969).
190. Roth, W.R., and Meier, J.D., *Tetrahedron Lett.*, *1967*, 2053.
191. Roustan, J.L., Benaîm, J., Charrier, C., and Mérour, J.Y., *Tetrahedron Lett.*, *1972*, 1953.
192. Sunder, S., and Bernstein, H.J., *Inorg. Chem.* *13*, 2274 (1974).
193. Sarel, S., Felzenstein, A., Victor, R., and Yovell, J., *J. Chem. Soc. Chem. Commun.*, *1974*, 1025.
194. Schiavon, G., Paradisi, C., and Boanini, C., *Inorg. Chim. Acta*, *14*, L 5 (1975).
195. Schrauzer, G.N., and Eichler, S., *Angew. Chem.*, *74*, 585 (1962); *Angew. Chem. Int. Ed. Engl.*, *1*, 454 (1962).
196. Schrauzer, G.N., Glockner, P., Reid, K.I.G., and Paul, I.C., *J. Amer. Chem. Soc.*, *92*, 4479 (1970).
197. Sorriso, S., Cardaci, G., and Murgia, S.M., *Z. Naturforsch.* *B 27*, 1316 (1972).
198. Stockis, A., and Weissberger, E., *J. Amer. Chem. Soc.*, *97*, 4288 (1975).
199. Takats, J., *J. Organometal. Chem.*, *90*, 211 (1975).
200. Takegami, Y., Watanabe, Y., Mitsudo, T., and Okajima, T., *Bull. Chem. Soc. Jap.*, *42*, 1992 (1969).
201. Tam, S.W., *Tetrahedron Lett.*, *1974*, 2385.
202. Victor, R., and Ben-Shoshan, R., *J. Organometal. Chem.*, *80*, C 1 (1974).
203. Wang, A.H.-J., Paul, I.C., and Aumann, R., *J. Organometal. Chem.*, *69*, 301 (1974).
204. Weaver, J., and Woodward, P., *J. Chem. Soc. A*, *1971*, 3521.
205. Whitesides, T.H., and Arhart, R.W., *J. Amer. Chem. Soc.*, *93*, 5296 (1971).
206. Whitesides, T.H., Arhart, R.W., and Slaven, R.W., *J. Amer. Chem. Soc.*, *95*, 5792 (1973).
207. Whitesides, T.H., and Slaven, R.W., *J. Organometal. Chem.*, *67*, 99 (1974).
208. Whitesides, T.H., and Arhart, R.W., *Inorg. Chem.*, *14*, 209 (1975).
209. Wilke, G., Bogdanović, B., Hardt, P., Heimbach, P., Keim, W., Kröner, M., Oberkirch, W., Tanaka, K., Steinrücke, E., Walter, D., and Zimmermann, H., *Angew. Chem.*, *78*, 157 (1966); *Angew. Chem. Int. Ed. Engl.*, *5*, 151 (1966).
210. Yasuda, N., Kai, Y, Yasuoka, N., Kasai, N., and Kakudo, M., *J. Chem. Soc. Chem. Commun.*, *1972*, 157.
211. Young, D.A.T., Holmes, J.R., and Kaesz, H.D., *J. Amer. Chem. Soc.*, *91*, 6968 (1969).
212. Zakharkin, L.I., Kazantsev, A.V., and Litovchenko, L.E., *Izv. Akad. Nauk SSSR, Ser. Khim.*, *1971*, 2050; *Bull. Acad. Sci. USSR, Div. Chem. Ser.*, *1971*, 1932.

THE ORGANIC CHEMISTRY OF IRON, VOLUME 1

DIENE IRON COMPLEXES

By R.B. KING

Department of Chemistry, University of Georgia
Athens, Georgia 30602, U.S.A.

TABLE OF CONTENTS

ISBN 0-12-417101-X (v.1)

I. INTRODUCTION

This chapter discusses iron complexes with dienes. Most known diene-iron complexes are 1,3-diene-tricarbonyliron complexes in which all four carbon atoms of the 1,3-diene are coordinated to the tricarbonyliron unit. The first compound of this type, (butadiene)tricarbonyliron, was prepared by Reihlen *et al.* *(342)* in 1930 although its nature was not recognized until much later. 1,3-Diene-tricarbonyliron complexes are now known for acyclic 1,3-dienes, cyclopentadienes, cyclopentadienones, 1,3-cyclohexadienes, cyclohexadienones, cycloheptadienes, vinylketenes, phospholes, thiophene dioxides, silacyclopentadienes, and perfluoro-cyclohexa-1,3-diene. In addition adjacent carbon-carbon double bonds of polyenes such as 1,3,5-trienes, fulvenes, cycloheptatriene, tropone, 1,3,5-cyclooctatriene, cyclooctatetraene, vinylbenzenes, anthracene, azepines, and oxepines have been found to bond to tricarbonyliron units in a 1,2,3,4-*tetrahapto* manner. These complexes are briefly mentioned in this chapter but are discussed in greater detail in the chapter on triene and tetraene complexes.

Most types of non-conjugated dienes including 1,4-pentadienes and 1,4-cyclohexadienes rearrange to the corresponding conjugated dienes upon reaction with carbonyliron reagents and eventually form 1,3-diene-tricarbonyliron complexes. However, norbornadiene and 1,5-cyclooctadiene form stable tricarbonyliron derivatives in which the non-conjugated arrangement of carbon-carbon double bonds is preserved. A detailed review *(334)* and two shorter articles *(333,335)* on the chemistry of diene-tricarbonyliron derivatives were published by Pettit during 1963-1965. However, since that time major developments in this field have made these reviews obsolete.

The stability of the diene-tricarbonyliron unit is sufficiently great that a variety of chemical reactions can be performed on diene-tricarbonyliron systems without rupture of the diene-tricarbonyliron bond. Diene-tricarbonyliron complexes appear to be particularly reactive towards electrophilic reagents. Thus, protonation or hydride abstraction reactions on various diene-tricarbonyliron derivatives can lead to important hydrocarbon-tricarbonyliron cations including the allyl-tricarbonyliron and cyclohexadienyl-tricarbonyliron cations. Reactions of these cations, particularly with various nucleophiles, is a field of growing importance. Some diene-tricarbonyliron derivatives even form tractable substitution products under the Friedel-Crafts or Vilsmeier reaction conditions thereby giving ketones or aldehydes still containing the diene-tricarbonyliron unit.

Other types of diene-iron complexes are of much more

limited importance. However, the field of bis(diene)-monocarbonyliron derivatives, which has only developed since 1970, promises to be of increasing significance for both synthesis and catalysis.

II. DIENE-TRICARBONYLIRON DERIVATIVES AND RELATED COMPOUNDS

This section discusses the preparation and properties of the numerous complexes of the type (diene)$Fe(CO)_3$. Complexes from acyclic dienes will be treated first followed by complexes derived from the cyclic dienes cyclopentadiene, cyclopentadienone, cyclohexadiene, cyclohexadienone, cycloheptadiene, cycloheptadienone, cyclooctadiene, bicyclo[4.2.0]octadiene, and norbornadiene. Diene-tricarbonyliron derivatives obtained from rearrangements of cyclopropane and cyclobutane derivatives will next be discussed followed by diene-tricarbonyliron complexes derived from heterocyclic systems, vinylsilicon and vinylboron derivatives, and perfluorinated dienes. Finally some general studies on carbonyl substitution and some physical and spectroscopic studies on diene-tricarbonyliron complexes will be discussed.

A. PREPARATION OF DIENE-TRICARBONYLIRON DERIVATIVES FROM ACYCLIC DIENES

The first diene-tricarbonyliron complexes were prepared by Reihlen, Gruhl, v. Hessling, and Pfrengle and reported in 1930 *(342)*. These authors found that treatment of butadiene with pentacarbonyliron in a bomb at 135°C for 24 h gave a liquid of stoichiometry $(C_4H_6)Fe(CO)_3$. This liquid was stable to air and could be distilled at atmospheric pressure at 120 - 180°C without decomposition, although distillation under pressure was preferable for purification. Similar reactions with isoprene and 2,3-dimethylbutadiene gave ill defined liquids of approximate stoichiometries $(diene)_2Fe(CO)_3$ which apparently could not be obtained pure. In 1942 carbonyliron complexes of butadiene, isoprene, and 2,3-dimethylbutadiene were patented *(377)* as antiknock agents for motor fuels.

Following this original discovery of (butadiene)tricarbonyliron no further work was done on this or other diene-tricarbonyliron complexes until after the discovery and elucidation of the nature of ferrocene more than twenty years later. In 1958 one of the co-discoverers of ferrocene, Pauson, along with his co-worker Hallam, successfully repeated the early (butadiene)tricarbonyliron preparation of Reihlen and co-workers *(199)*. On the basis of their studies including the relative chemical and thermal stability of (butadiene)-

tricarbonyliron, they postulated the now familiar *tetrahapto*-structure 1a. The chemical studies of Hallam and Pauson

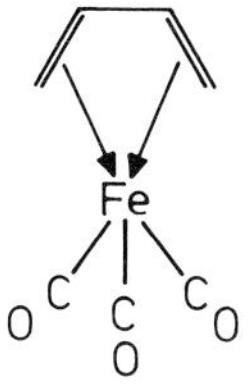

1a

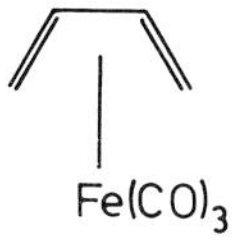

1b

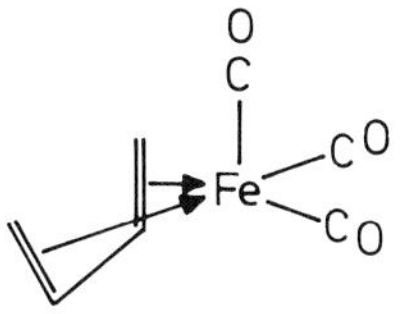

1c

(199) as well as a ^{1}H-NMR study published shortly thereafter *(187)* suggested that the butadiene ligand remains intact when complexed with a tricarbonyliron unit in (butadiene)tricarbonyliron. The ultraviolet spectrum of (butadiene)tricarbonyliron was reported shortly after this work *(280)*.

Subsequently the structure of (butadiene)tricarbonyliron was confirmed by X-ray crystallography *(299,300)*. The X-ray studies indicate that the coordination of the iron atom in (butadiene)tricarbonyliron is square pyramidal with a carbonyl group in the apical position as in structure 1c. However, for convenience structures of the types 1a or 1b will be used in this chapter to represent diene-tricarbonyliron complexes.

Since this original work numerous 1,3-dienes have been shown to react with various carbonyliron complexes to form (diene)$Fe(CO)_3$ derivatives. In this connection a variety of carbonyliron reagents have been used. Pentacarbonyliron is the least expensive of the carbonyliron reagents, but generally requires relatively high temperatures (typically around 130°C) for reactions with dienes to form tricarbonyliron derivatives. Not only are some types of diene-tricarbonyliron complexes unstable at such temperatures but also some dienes themselves form dimers through Diels Alder reactions with themselves, *etc.*, at such elevated temperatures. In some cases the volatilities of such a diene dimer and the corresponding (diene)$Fe(CO)_3$ complexes are so similar that their separation is very difficult. This is the probable explanation for the products of stoichiometries (diene)$_2Fe(CO)_3$ from isoprene and 2,3-dimethylbutadiene in the early work by Reihlen and coworkers *(342)*. These products were probably mixtures of dimers (and possibly other oligomers) of the 1,3-diene and the (diene)$Fe(CO)_3$ derivative. Subsequent investigations *(239)* have shown that the reaction of isoprene (2-methylbutadiene) with pentacarbonyliron gives not only (isoprene)tricarbonyliron but also dimers of isoprene such as dipentene. If the reaction between isoprene and pentacarbonyliron is carried out near room temperature using ultraviolet irradiation rather than by heating to 130°C as in the original work *(342)*, then the formation of isoprene dimers can be minimized *(239)*. Since this original work, photochemical reactions of 1,3-dienes with carbonyliron complexes have been used extensively for the preparation of other (diene)$Fe(CO)_3$ complexes. However, in some cases prolonged irradiation of 1,3-dienes with pentacarbonyliron can also lead to displacement of an additional two carbonyl groups from the (diene)$Fe(CO)_3$ complex to give $(diene)_2FeCO$ complexes. These complexes will be discussed in detail later in this chapter.

The need for relatively high temperatures or ultraviolet irradiation to form (diene)$Fe(CO)_3$ derivatives can be avoided by substituting a more reactive carbonyliron derivative for pentacarbonyliron. Dodecacarbonyltriiron reacts well with many 1,3-dienes around 80°C (*i.e.* in boiling benzene). Enneacarbonyldiiron reacts with butadiene even at 40°C *(305)*, but the major products are (butadiene)tetracarbonyliron and (butadiene)bis(tetracarbonyliron) in which the butadiene double bonds are coordinated individually to tetracarbonyliron units (see the chapter on olefin complexes). (Benzalacetone)tricarbonyliron, $(\eta^4\text{-}C_6H_5\text{-}CH{=}CH\text{-}CO\text{-}CH_3)Fe(CO)_3$ appears to be a good starting material for the preparation of tricarbonyliron complexes under mild conditions (*e.g.* in toluene solution at 50°C for 6 h in one case *(212)*). Dienetricarbonyliron complexes can be prepared from pentacarbonyliron and the diene at or below room temperature if trimethylamine N-oxide is added to the system as an oxidizing agent *(365)*. In cases where the diene is inconveniently unstable to use as a reagent for the reaction with carbonyliron derivatives, the reaction of the corresponding α- or β-unsaturated alcohol with enneacarbonyldiiron or dodecacarbonyltriiron in the presence of copper sulfate can be used *(318)*. This reaction involves dehydration of the alcohol.

Some reactions of simple acyclic dienes with carbonyliron complexes lead to rearrangements, in which hydrogen shifts frequently occur. Thus, *cis*-1,3-pentadiene, *trans*-1,3-pentadiene, and 1,4-pentadiene all react with pentacarbonyliron to form (*trans*-1,3-pentadiene)tricarbonyliron (2) *(148,*

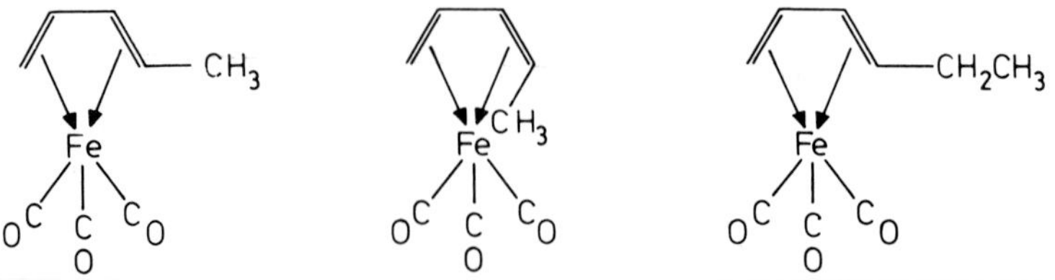

2 3 4

239). However, (*cis*-1,3-pentadiene)tricarbonyliron (3) can be prepared by treatment of *cis*-1,3-pentadiene with $Fe_2(CO)_9$ in boiling diethyl ether *(249)*. The much lower reaction temperature apparently prevents the isomerization of 3 to 2. Reactions of *cis*-1,4-hexadiene *(238)* and even 1,5-hexadiene *(148)* with pentacarbonyliron give (*trans*-1,3-hexadiene)tricarbonyliron (4) as the major product in addition to small amounts of (*trans*,*trans*-2,4-hexadiene)tricarbonyliron (5). A similar reaction of *trans*,*trans*-2,4-hexadiene with pentacarbonyliron

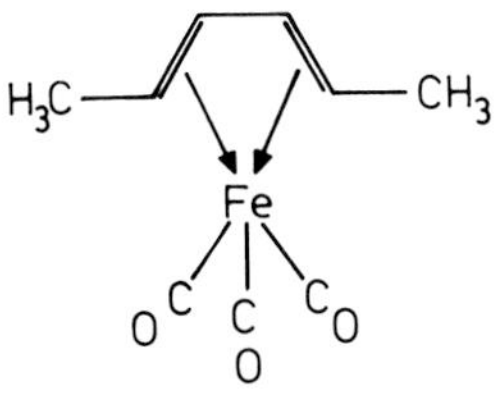

5

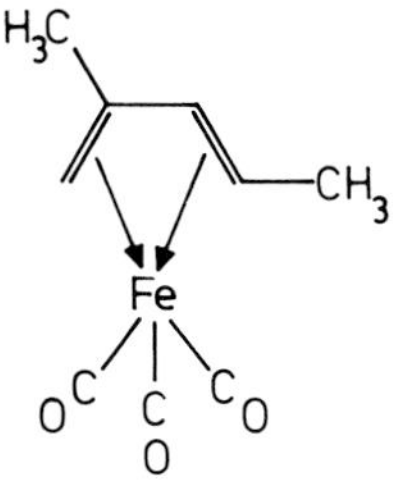

6

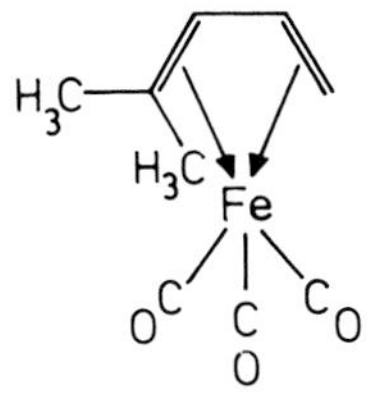

7

leads exclusively to the corresponding carbonyliron complex 5 without any rearrangements *(148)*. Reaction of 4-methylpenta-1,3-diene with pentacarbonyliron at 90 - 100°C for 140 h results exclusively in rearrangement to give (*trans*-2-methylpenta-1,3-diene)tricarbonyliron (6), whereas reaction of the same 1,3-diene with $Fe_2(CO)_9$ in boiling diethyl ether gives the corresponding diene-tricarbonyliron complex 7 without rearrangement *(249)*. Both 2,5-dimethylhexa-1,5-diene and 2,5-dimethylhexa-2,4-diene react with pentacarbonyliron to give (*trans*-2,5-dimethylhexa-1,3-diene)tricarbonyliron (8). 2-Methylhexa-1,5-diene reacts with pentacarbonyliron to

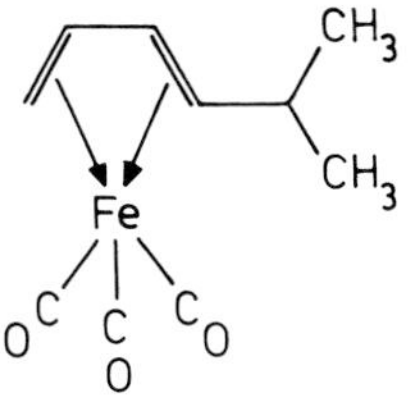

8

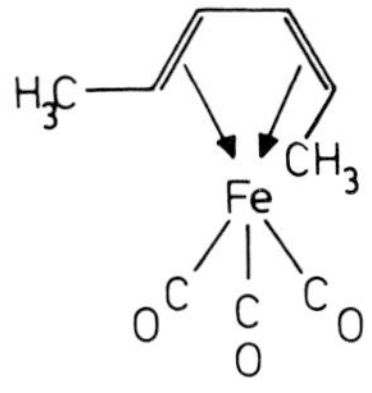

9

give (2-methylhexa-1,3-diene)tricarbonyliron. *Cis*,*trans*-2,4-hexadiene reacts with pentacarbonyliron without rearrangement to give (*cis*,*trans*-2,4-hexadiene)tricarbonyliron (9). These reactions appear to be controlled by kinetic rather than thermodynamic factors *(319)*. They demonstrate the tendency of acyclic 1,4-dienes and 1,5-dienes to rearrange to 1,3-diene-tricarbonyliron derivatives upon reaction with pentacarbonyliron. Furthermore, 1,5-hexadiene derivatives tend to form products in which the tricarbonyliron group is at the end of the C_6-chain *(148)*.

The effects of chlorine substitution on the formation of butadiene-tricarbonyliron derivatives have been investigated. Both of the isomeric monochlorobutadienes and the 1,4- and 1,2-dichlorobutadienes form the corresponding diene-tricarbonyliron complexes in reasonable yields (14 - 27 %) when allowed to react with dodecacarbonyltriiron in tetrahydrofuran solution *(64)*. Under similar conditions 2,3-dichlorobutadiene gives only a 4.5 % yield of the corresponding tricarbonyliron complex. On the other hand neither *trans,trans*- nor *cis,cis*-1,2,3,4-tetrachlorobutadiene forms tricarbonyliron complexes under similar conditions. The trichlorobutadienes *trans*-1,2,3-trichlorobutadiene and *trans,trans*-1,2,4-trichlorobutadiene form the corresponding tricarbonyliron complexes in low yield (1.5 - 3.3 %) with dodecacarbonyltriiron in tetrahydrofuran *(65)*. However, no tricarbonyliron complex can be prepared analogously from *trans*-1,1,4-trichlorobutadiene. A tricarbonyliron complex can be prepared readily from *cis, cis*-1,4-dichloro-2,3-dimethylbutadiene *(63)*. These results suggest that in order for substituted chlorobutadienes to form the corresponding tricarbonyliron complexes, no more than three chlorine atoms can be present and the butadiene cannot contain a terminal CCl_2 group. These results have been explained on the basis that excessive chlorine substitution can make it more difficult for the 1,3-diene to adopt the *s-cis*-conformation necessary for complex formation *(65)*.

Cleavage of the carbon-bromine bond can complicate the reactions of 2-bromobutadiene with carbonyliron complexes *(190)*. One of the products from the reaction of 2-bromobutadiene with either enneacarbonyldiiron or dodecacarbonyltriiron is the expected tricarbonyliron complex 10 (X = Br). In addi-

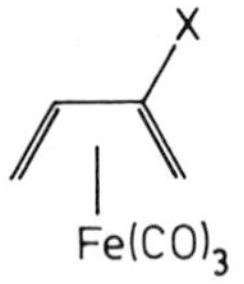

10

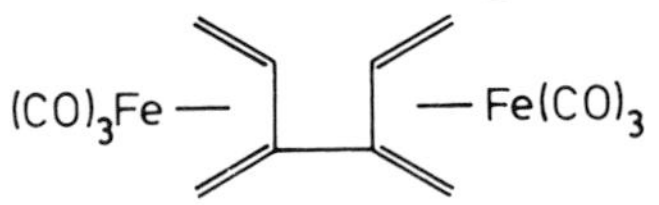

11

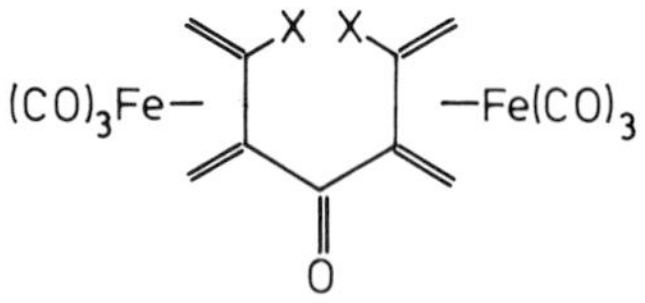

12

tion a complex mixture of (butadiene)tricarbonyliron (1), the coupling product (2,2'-bibutadienyl)bis(tricarbonyliron) (11), the coupling product with carbon monoxide insertion [bis(2-butadienyl)ketone]bis(tricarbonyliron) (12, X = H), and a more complex derivative of possible structure 13 is obtained from the reaction between 2-bromobutadiene and enneacarbonyldiiron in boiling hexane. From a similar reaction in boiling diethyl ether, the carboxylic acid derivative 14 can also be isolated. In contrast, the corresponding reaction of 1-bromobutadiene with enneacarbonyldiiron gives only the correspond-

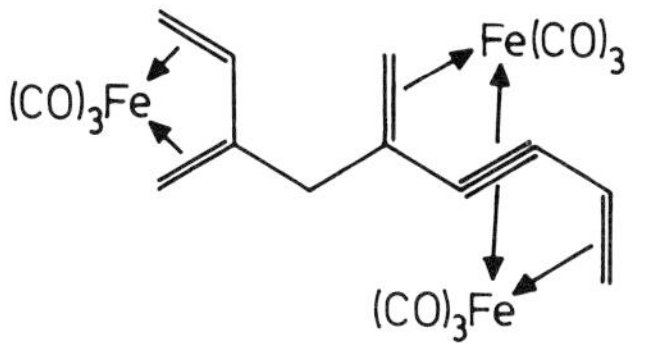

13

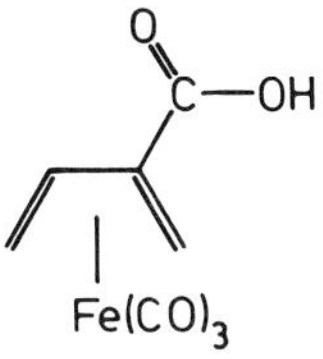

14

ing diene-tricarbonyliron complex without any carbon-bromine bond cleavage reactions.

Recent work *(320)* has shown that analogous carbon-chlorine bond cleavage can also occur in certain reactions of chlorobutadienes with carbonyliron complexes, especially if enneacarbonyldiiron rather than dodecacarbonyltriiron is used *(320)*. Thus, the reaction of 2,3-dichlorobutadiene with enneacarbonyldiiron in boiling hexane results in dechlorination to give a mixture of (2-chlorobutadiene)tricarbonyliron (10, X = Cl), (butatriene)bis(tricarbonyliron) (15), and the binu-

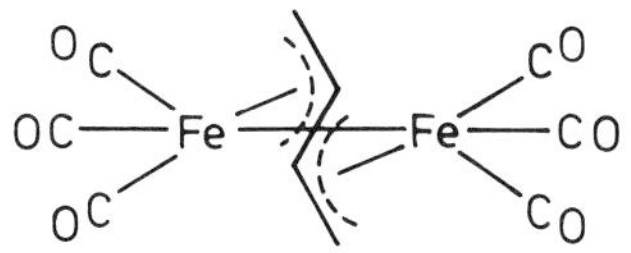

15

clear complex 12 (X = Cl). An intermediate with only one of the carbon-carbon double bonds of the chlorobutadiene bonded to a tetracarbonyliron unit is suggested for this dechlorination reaction. The observations made during this study *(320)* suggest that a chlorine substituent on a central atom of an η^4-coordinated 1,3-diene is unreactive but a chlorine substituent on a coordinated double bond of an η^2-coordinated 1,3-diene is reactive towards insertion of a carbonyliron species into the carbon-chlorine bond.

In most cases adjacent carbonyl or cyano groups do not prevent a 1,3-diene from forming the corresponding tricarbonyliron complex. Thus, the diene-tricarbonyliron complexes (η^4-R-CH=CH-CH=CH-R')$Fe(CO)_3$ (R = CH_3, R' = CO_2CH_3, $CO_2C_2H_5$, CHO, or CN; R = R' = CO_2CH_3) can be prepared by reaction of the corresponding 1,3-diene either with dodecacarbonyltriiron in boiling benzene *(239)* or with pentacarbonyliron in a high boiling solvent such as di-*n*-butyl ether *(79,80,136)*. The tricarbonyliron derivatives of sorbic and muconic acids, (R-CH=CH-CH=CH-CO_2H)$Fe(CO)_3$ (R = CH_3, CO_2H), can be obtained by alkaline hydrolysis of the corresponding methyl or ethyl ester tricarbonyliron complexes *(80,136)*.

(Sorbic acid)tricarbonyliron, (CH_3-CH=CH-CH=CH-CO_2H)-$Fe(CO)_3$, lacks a plane of symmetry. It has been separated into the expected optical antipodes by fractional crystallization of its salt with (*S*)-α-phenethylamine followed by regeneration of the now optically active free acid complex with 6 *N* aqueous hydrochloric acid *(307,308)*. The X-ray crystal structure of racemic (sorbic acid)tricarbonyliron has been reported *(144)*.

Conditions have been found for the separation by high speed liquid chromatography of tricarbonyliron complexes of various hexadienone and heptadienone derivatives *(179)*. A 50 cm × 2 mm DuPont Permaphase column was used with a 20 % aqueous methanol mobile phase at 25 - 30°C. Efficient separations of several pairs of *cis*- and *trans*-isomers could be achieved with retention times between 2 and 10 minutes.

Hydroxybutadiene-tricarbonyliron complexes have also been prepared. However, indirect methods are necessary, since the free hydroxybutadienes are not stable. Thus, reaction of 2-acetoxybutadiene with enneacarbonyldiiron in benzene gives the corresponding yellow air-stable tricarbonyliron complex. Treatment of this complex with methyllithium in diethyl ether followed by acidification gives the air sensitive and relatively unstable (2-hydroxybutadiene)tricarbonyliron (16)

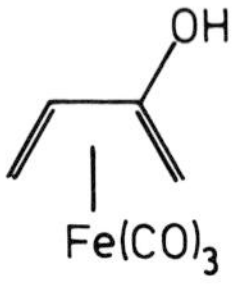

16

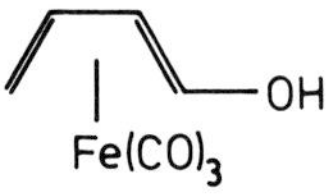

17

(125). A similar sequence of reactions can also be used to prepare (*syn*-1-hydroxybutadiene)tricarbonyliron (17). However, an attempt to prepare (*anti*-1-hydroxybutadiene)tricarbonyliron by an analogous method led to a rearrangement to the corresponding *syn*-isomer 17. These hydroxybutadiene-tricarbonyliron complexes have been characterized by their reactions with benzoyl bromide in the presence of a base to give the corresponding benzoyl derivatives *(125)*.

Reactions starting from α-pyrone provide another source of acyclic diene-tricarbonyliron derivatives containing oxygen functional groups *(127)*. Thus, treatment of the α-pyrone

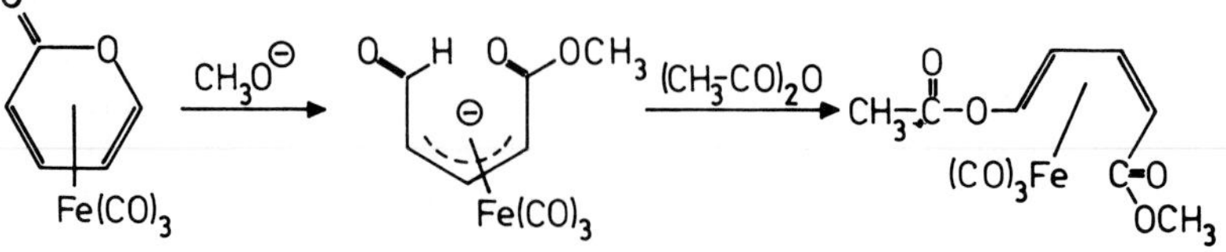

18 19 20 [1]

complex 18 with methoxide ion (eq. [1]) gives the anion 19 which is acetylated with acetic anhydride to give 20 containing both methoxycarbonyl and acetate groups. Similarly the ketones 21 (R = CH_3, C_6H_5) can be obtained by ring opening of

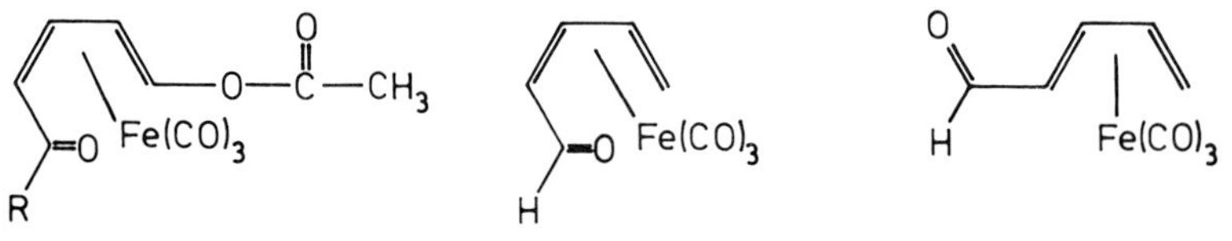

21 22a 22b

the α-pyrone complex 18 with the corresponding alkyllithium compound RLi followed by acetylation with acetic anhydride. Reduction of 18 with $LiAlH_4$ followed by hydrolysis gives the aldehyde 22a which readily isomerizes to 22b *(127)*.

Hydroxyalkyldiene-tricarbonyliron derivatives are known and are important for the preparation of acyclic pentadienyltricarbonyliron cations. In some cases such complexes can be prepared by the direct reaction of the corresponding dienol with pentacarbonyliron. Thus, reaction of CH_2=CH-CH=CH-CH_2OH with pentacarbonyliron gives the corresponding dienol-tricarbonyliron complex *(285)*. In other cases reaction of a dienone-tricarbonyliron complex with sodium borohydride or an alkylmagnesium halide can be used to convert the ketone into an alcohol without rupture of the diene-iron bond. For example, the aldehyde CH_3-CH=CH-CH=CH-CHO can be converted into the dienol-tricarbonyliron complex (CH_3-CH=CH-CH=CH-CH(CH_3)OH)-$Fe(CO)_3$ either first by reacting with methylmagnesium iodide to form the dienol CH_3-CH=CH-CH=CH(CH_3)OH followed by reaction of this dienol with pentacarbonyliron or by first reacting the dienal with pentacarbonyliron to form (CH_3-CH=CH-CH=CH-CHO)$Fe(CO)_3$ followed by treatment of this complex with methylmagnesium iodide *(285)*.

Some diene-tricarbonyliron derivatives containing aminoalkyl substituents have also been prepared *(369)*. Thus, conversion of the alcohol 23 to the corresponding methanesul-

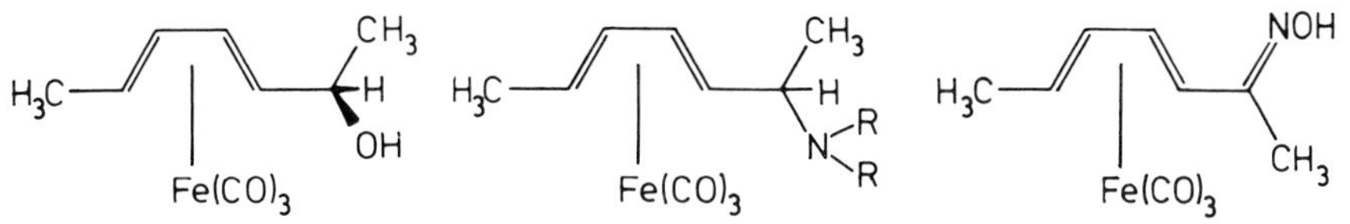

23 24 25

fonate by treatment with a mixture of methanesulfonyl chloride and trimethylamine followed by treatment of this methanesulfonate with a secondary amine, R_2NH, gives the corresponding (6-aminohepta-2,4-diene)tricarbonyliron derivatives 24 (R = H, CH_3) as a mixture of *endo*- and *exo*-isomers.

The products can also be obtained by reaction of the free 6-aminohepta-2,4-diene with pentacarbonyliron at 120°C or (24, R = H) by $LiAlH_4$ reduction of the corresponding oxime 25.

The presence of aryl substituents does not interfere with the ability of 1,3-dienes to form tricarbonyliron complexes. Exhaustive heating of 1,4-diphenylbutadiene with dodecacarbonyltriiron gives the orange crystalline tricarbonyliron complex $(C_6H_5\text{-}CH{=}CH\text{-}CH{=}CH\text{-}C_6H_5)Fe(CO)_3$ (26) through an interme-

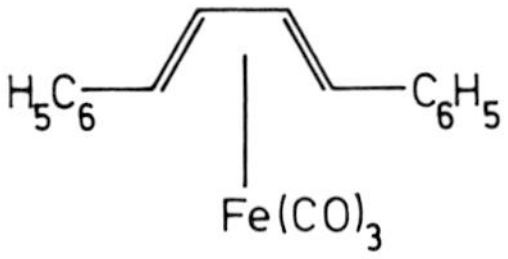

26

diate adduct containing an additional ½ mole of 1,4-diphenylbutadiene *(291)*. The crystal structure of this adduct *(122)* provides an interesting opportunity to compare the geometries of free and complexed 1,4-diphenylbutadiene in the same crystal. The uncomplexed 1,4-diphenylbutadiene has the *s-trans*-configuration whereas the complexed 1,4-diphenylbutadiene has the *s-cis*-configuration required for η^4-coordination to a tricarbonyliron group. The crystal structures of the *syn*-complex $(m\text{-}NO_2\text{-}C_6H_4\text{-}NH\text{-}CH(CH_3)\text{-}CH{=}CH\text{-}CH{=}CH\text{-}CH_3)Fe(CO)_3$ (27) and the

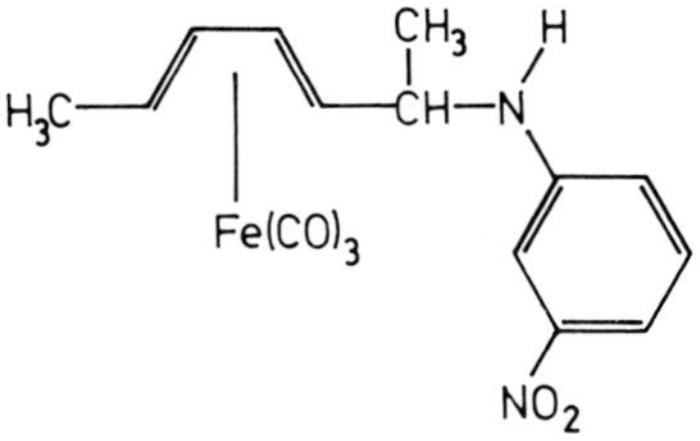

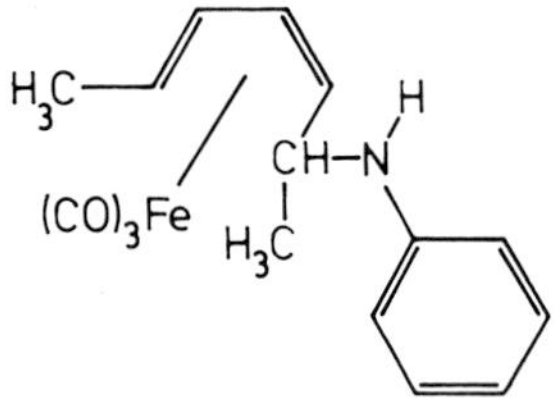

27 28

anti-complex $(C_6H_5\text{-}NH\text{-}CH(CH_3)\text{-}CH{=}CH\text{-}CH{=}CH\text{-}CH_3)Fe(CO)_3$ (28) have also been determined *(220)* to an unusually high degree of accuracy (R-factors = 2.5±0.1 %). The hydrogen atoms therefore could be located in these structures. The *anti*-hydrogen atoms deviate by 30° from the diene plane bent away from the metal whereas the *syn*-hydrogen atoms deviate by 20° bent towards the metal *(220)*.

Basicity measurements on substituted anilines *(272)* indicate that the butadiene-tricarbonyliron group as the substituent has electron withdrawing properties similar to the un-coordinated diene group, the $Fe(CO)_3$ moiety slightly reducing the electron withdrawal of the diene.

Trifluoroacetic acid was found to catalyse the isomerization of (1,5-diphenylpenta-1,3-diene)tricarbonyliron from

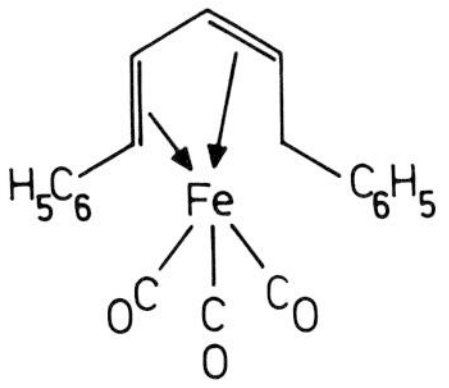

29a

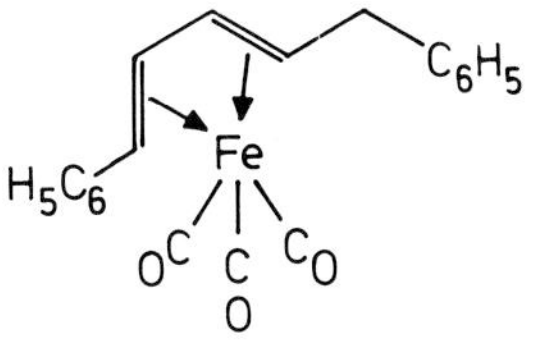

29b

the *cis*-isomer 29a to the corresponding *trans*-isomer 29b *(398)*. Labelling studies *(400)* indicated that heating 29a in benzene resulted in both isomerization to 29b and metal epimerization leading to racemization and *exo-endo* scrambling.

Reaction of 1-(*p*-nitrophenyl)butadiene with dodecacarbonyltriiron in methanol results in reduction of the nitro group to an amino group before formation of a diene-tricarbonyliron complex *(273)*.

Reaction of 1-(methyl-*o*-carboranyl)buta-1,3-diene with pentacarbonyliron in boiling di-*n*-butyl ether gives the corresponding diene-tricarbonyliron derivative 30 as an air-

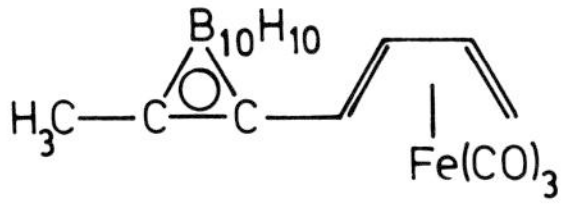

30

stable orange crystalline solid *(409)*.

Some diene-tricarbonyliron derivatives have been prepared also containing other organometallic systems. Reaction of 1,4-diphenylbutadiene with hexacarbonylchromium gives a mixture of the mono- and bis(tricarbonylchromium) derivatives. These derivatives react with dodecacarbonyltriiron or pentacarbonyliron to form the mixed metal complexes 31 and 32, respectively *(78,291)*.

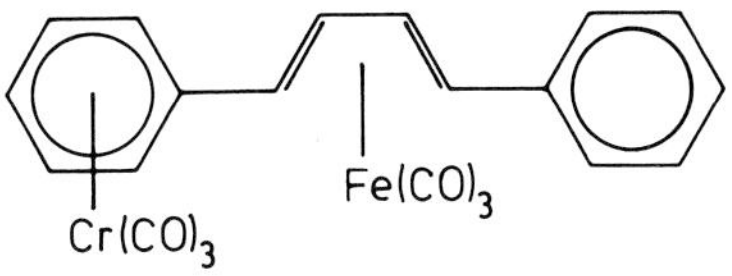

31

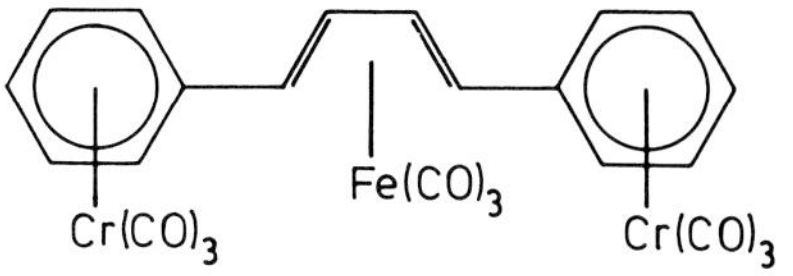

32

Ferrocenylbutenols of the type R'-CH_2-CR(OH)-CH=CH-[(η^5-C_5H_4)Fe(η^5-C_5H_5)] (R = CH_3, R' = CN; R = R' = H) react with enneacarbonyldiiron or dodecacarbonyltriiron in boiling benzene in the presence of copper sulfate to give the corresponding 1-ferrocenylbutadiene-tricarbonyliron derivatives 33 (R = CH_3, R' = CN; R = R' = H) *(314,315)*. (2-Ferrocenylbutadiene)tricarbonyliron (34) has been prepared by an analogous

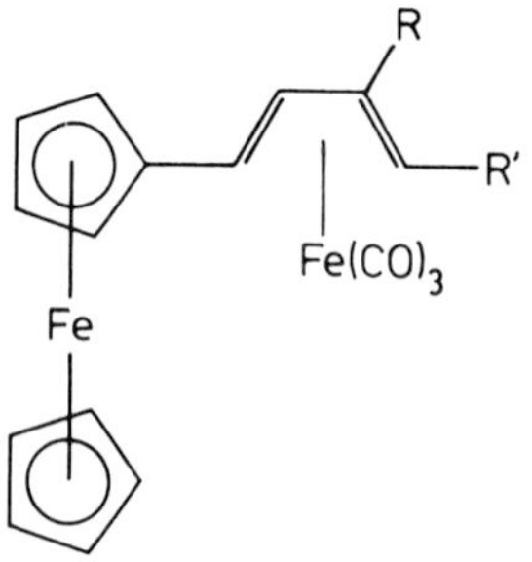

33

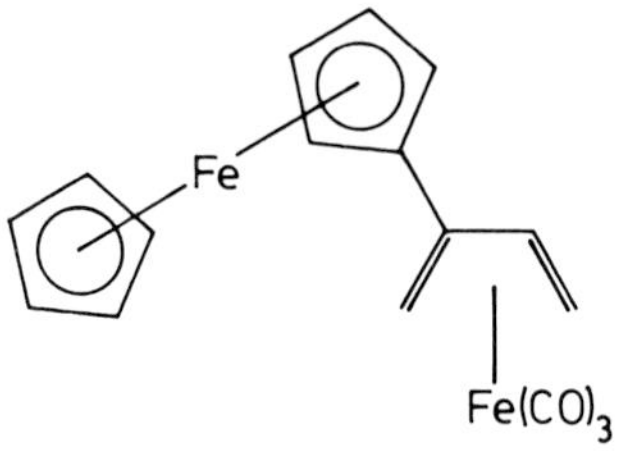

34

method *(317,318)*.

Acyclic trienes such as 1,3,5-hexatriene *(306)* and *allo*-ocimene (2,6-dimethylocta-2,4,6-triene) *(27,239)* can form tricarbonyliron complexes using only two of the three conjugated double bonds of the 1,3,5-triene. Some interesting "shift isomers" consisting of tricarbonyliron groups coordinated to different pairs of the carbon-carbon double bonds in 1,3,5-trienes of the type R-CH=CH-CH=CH-CH=CH-R' have been studied *(402,403)*. These shift isomers are discussed in more detail in the chapter on triene complexes.

Natural products containing chains with several carbon-carbon double bonds which form tricarbonyliron complexes include β-ionone *(80)*, retro-ionylidene acetate *(80)*, β-ionylidene acetate *(80)*, myrcene *(27)*, *cis*-ocimene *(27)*, and α-phellandrene *(27)*. Diene-tricarbonyliron complexes have been shown to be intermediates in the hydrogenation of methyl linolenate, a triene carboxylic ester in unsaturated fats, using pentacarbonyliron as a homogeneous catalyst *(170,171,172)*. Treatment of *cis*-1,4-polybutadiene with dodecacarbonyltriiron in boiling benzene containing some 1,2-dimethoxyethane results in the introduction of up to 20 % by weight of $Fe(CO)_3$ units into the polymer *(35)*. In many of these reactions of complex natural products and polymers with iron carbonyls to form tricarbonyliron complexes shifts of carbon-carbon double bonds along a carbon chain must occur analogous to the formation of 1,3-diene-tricarbonyliron complexes from reactions of 1,4-pentadiene and 1,5-hexadiene derivatives with iron carbonyls as discussed above.

Some reactions of allene with carbonyliron reagents give diene-tricarbonyliron derivatives. Tetramethylallene reacts with enneacarbonyldiiron *(32)* to give a mixture of (tetramethylallene)tetracarbonyliron and (2,4-dimethylpenta-1,3-diene)tricarbonyliron (35). Tetraphenylallene reacts with pentacarbonyliron in boiling isooctane to give a tricarbonyliron complex, $[(C_6H_5)_4C_3]Fe(CO)_3$, of unknown structure *(198,311, 312)*. However, the fact that this complex regenerates tetra-

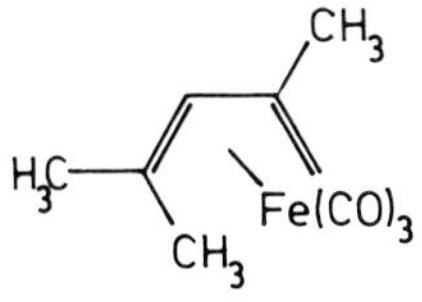

35

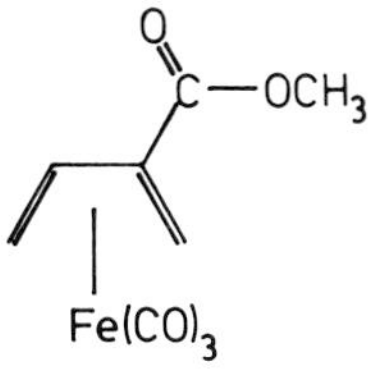

36

phenylallene upon reaction with triphenylphosphine suggests that the tetraphenylallene unit has remained intact. Reaction of unsubstituted allene $H_2C{=}C{=}CH_2$ with $Fe_2(CO)_9$ at 50°C gives two carbonyliron derivatives: the tetramethyleneethane derivative $(C_6H_8)[Fe(CO)_3]_2$ and the π-allylic derivative (C_3H_4)-$Fe_2(CO)_7$ *(33,198,312,313)*. These complexes are discussed in more detail elsewhere in this book. Tetraalkyl- and tetraarylbutatrienes normally react with iron carbonyls to form either tetracarbonyliron complexes using only the center carbon-carbon double bond of the cumulene or orthogonal bis-(η^3-allylic) derivatives using all four carbon atoms of the butatriene system *(232)*. However, in an attempt to prepare an unsubstituted butatriene complex, the reaction between $K_2Fe_2(CO)_8$ and 1,4-dichlorobut-2-yne in methanol was found instead to give a mixture of yellow crystalline (2-methoxycarbonylbutadiene)tricarbonyliron (36), and a red-brown solid formulated as (3,5-dimethylenehepta-1,6-dien-4-one)bis(tricarbonyliron) (12, X = H) in rather poor yields *(232)*.

B. REACTIONS OF (ACYCLIC DIENE)-TRICARBONYLIRON DERIVATIVES

Diene-tricarbonyliron derivatives react with several types of electrophilic reagents. They thus undergo protonation. In some cases reactions of diene-tricarbonyliron complexes with acyl halides under Friedel-Crafts conditions lead to tractable products. Reactions with fluoroolefins, fluoroalkynes, and fluoroketones may give unusual η^3-allylic derivatives. The free diene can be liberated from the corresponding diene-tricarbonyliron complexes by treatment with appropriate oxidizing agents.

Protonation of (butadiene)tricarbonyliron can lead to η^3-allylic tricarbonyliron derivatives. In the first reported reaction of this type, treatment of (butadiene)tricarbonyliron with anhydrous hydrogen chloride was found to give (η^3-1-methylallyl)tricarbonyliron chloride *(221)*. This reaction was first believed *(147)* to occur with geometric inversion to give the *syn*-isomer 37a (R = R' = H). However, subsequent work *(393)* with (1-phenyl-3-methylbutadiene)tricarbonyliron suggested an alternative mechanism giving the corresponding

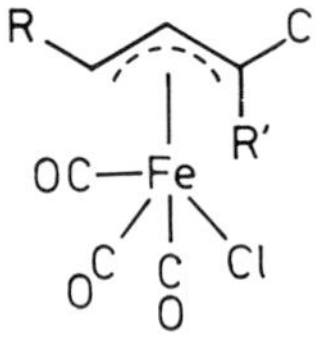

37a

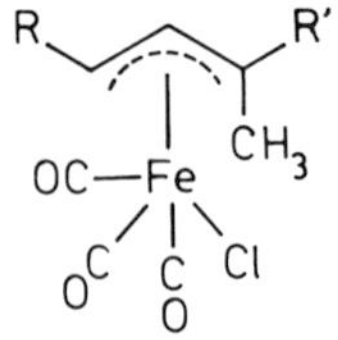

37b

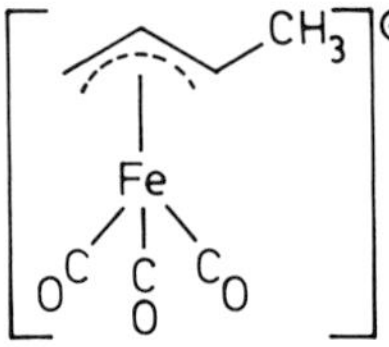

38

anti-isomer 37b ($R = C_6H_5$, $R' = H$).

A similar protonation of (butadiene)tricarbonyliron but in the absence of a coordinating anion such as chloride *(146)* leads to salts of the (η^3-1-methylallyl)tricarbonyliron cation (38). Suitable acids for effecting such protonations include HBF_4, $HClO_4$, and $HSbCl_6$. Analogous protonations of the tricarbonyliron complexes of isoprene, *trans*-piperylene, and 1-phenylbutadiene have also been effected *(146)*. As a consequence of their coordinative unsaturation these allylic tricarbonyliron cations are readily decomposed by water to give vinyl alcohols which in most cases rearrange to the tautomeric ketones *(146)*. On the basis of the ^{1}H-NMR spectra of a solution of (butadiene)tricarbonyliron in a mixture of fluorosulfonic acid and liquid sulfur dioxide, the cationic hydride 39 has been postulated as an intermediate in the

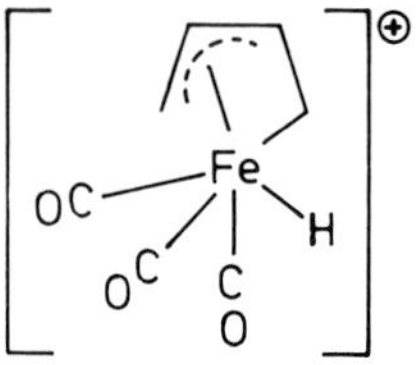

39

protonation of (butadiene)tricarbonyliron *(60,396,407)*. Rapid intramolecular proton exchange, however, has to be invoked to explain the observed ^{1}H-NMR spectrum.

The coordinatively unsaturated η^3-allyl-tricarbonyliron cations can also add a fourth carbonyl group to form coordinatively saturated η^3-allyl-tetracarbonyliron cations. In the presence of excess acid, the allylic tricarbonyliron cations can decompose through "disproportionation" with carbonyl scrambling to form the corresponding allylic tetracarbonyliron cations *(175,176)*. Thus, treatment of (butadiene)tricarbonyliron with a sixfold excess of tetrafluoroboric acid in acetic anhydride solution followed by precipitation with cold diethyl ether gives an 86 % yield of $[(\eta^3\text{-}CH_3\text{-}C_3H_4)Fe(CO)_3]^+[BF_4]^-$ *(174)*. However, other methods for preparing the allylic tetracarbonyliron cations by protonation of neutral tetracarbonyliron derivatives such as (tetramethylallene)tetracarbonyliron

(175) and diene-*tetra*carbonyliron complexes (see the chapter on olefin complexes) *(176)* do not involve the automatic loss of 25 % of the iron needed to furnish the fourth carbonyl group. Alternatively, the diene-tricarbonyliron complexes can be protonated (*e.g.* with HBF_4) in the presence of carbon monoxide to provide an efficient synthesis of the η^3-allylic tetracarbonyliron cations *(286)*.

Protonation of hydroxyalkylbutadiene-tricarbonyliron complexes (40) gives the *cis*-pentadienyl-tricarbonyliron cations (41) according to scheme [2] (R = R' = H or CH_3; R =

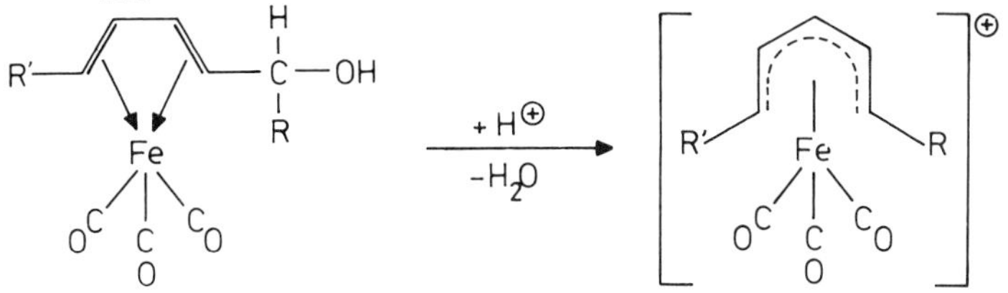

[2]

40 41

H, R' = CH_3) *(285,286)*. A requirement for this reaction is an appropriate configuration of the hydroxyalkyl group to give the indicated *cis*-stereochemistry. This protonation reaction is reversible, since treatment of the stable *cis*-pentadienyl-tricarbonyliron cation salts with water results in regeneration of the starting alcohol *(286)*. Similar reactions of the cations 41 with alcohols result in solvolysis to give the butadienyl ether complexes 42 (R = CH_3, C_2H_5). Reactions of the (*syn,syn*-1,5-dimethylpentadienyl)tricarbonyliron cation 41 (R = R' = CH_3) with strongly basic amines (isopropylamine,

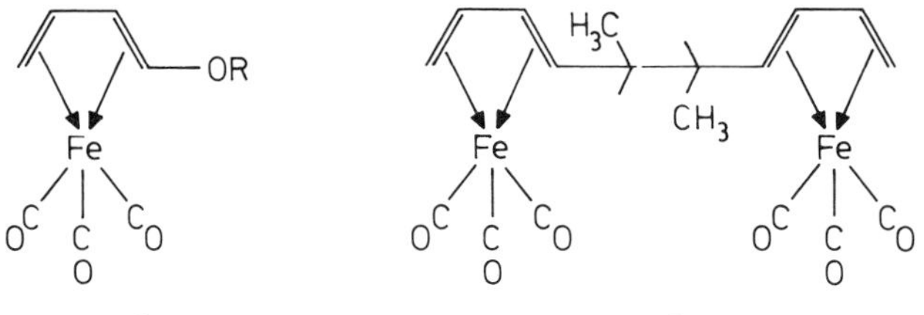

42 43

ethylamine, and methylbenzylamine) to give *cis,trans*-dienylamine-tricarbonyliron complexes, and with weakly basic amines such as aniline derivatives to give *trans,trans*-dienylamine-tricarbonyliron complexes have been studied *(282,283,284)*.

Reduction of the cation 41 (R = H, R' = CH_3) with zinc results in coupling to give two isomeric bis(diene-tricarbonyliron) derivatives of the type 43 *(286)*. Treatment of the (*cis*-pentadienyl)tricarbonyliron cation 41 (R = H, R' = CH_3) with basic alumina in dichloromethane solution at room temperature for 1 h *(8,296)* gives the complex 44, previously

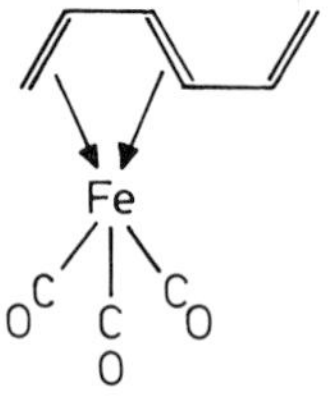

44

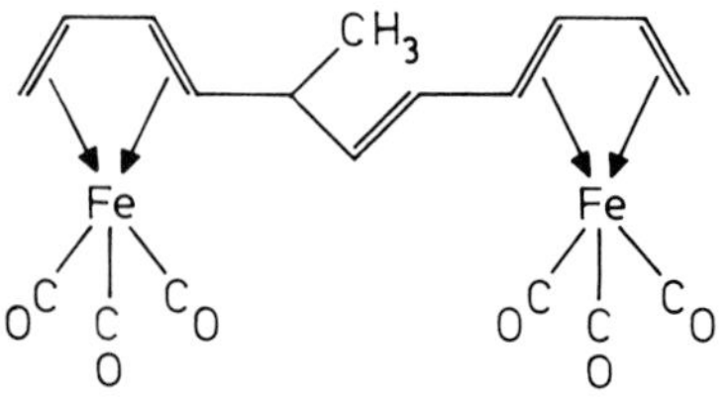

45

isolated from the reaction between $Fe_2(CO)_9$ and 1,3,5-hexatriene *(306)*. If the reaction period is extended to 64 h, then coupling occurs to give 45. *Cis*-pentadienyl-tricarbonyliron cations can also substitute an aromatic hydrogen in 1,3-dimethoxybenzene *(53)*. The stereochemistry in these hydroxyalkyldiene-tricarbonyliron systems (*e.g.* 40, R = CH_3) has been investigated in some detail *(94,168)*.

The solvolyses of 3,5-dinitrobenzoate esters of the type 46 (R = 3,5-dinitrobenzoate) in aqueous acetone have been

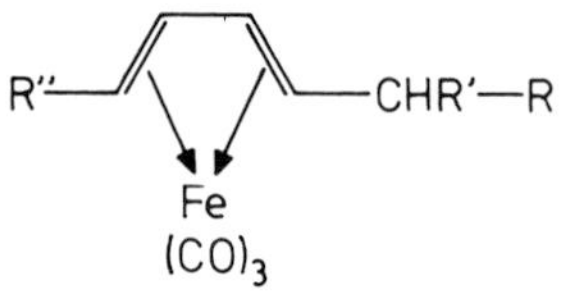

46

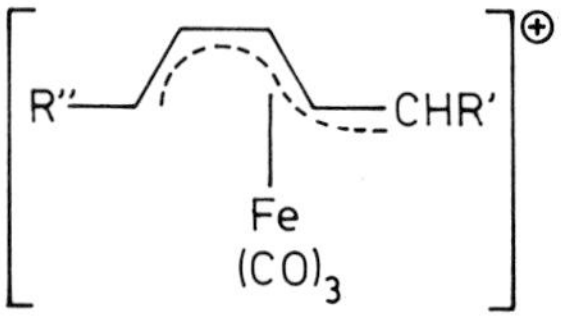

47

used to infer the existence of *trans*-pentadienyl-tricarbonyliron cations 47. However, these species, unlike the *cis*-pentadienyl-tricarbonyliron ions 41 discussed above, are too unstable for isolation *(93,95,267)*. Furthermore, recent work *(71)* indicates that these solvolyses are not completely stereospecific. Protonation of the dienal complex (CH_3-CH=CH-CH=CH-CHO)$Fe(CO)_3$ in a mixture of HSO_3F and SO_2ClF at -120°C gives a mixture of the *trans*-pentadienyl-tricarbonyliron cations 48a and 48b, identified by their NMR spectra *(58,*

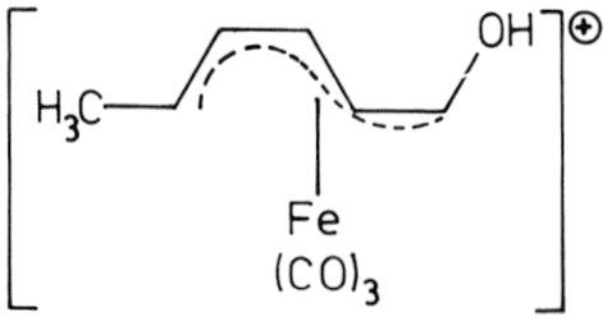

48a

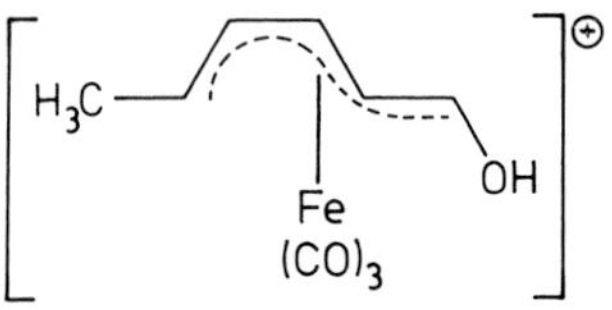

48b

277,367). When this mixture is warmed to -28°C, the isomeric *cis*-pentadienyl-tricarbonyliron cation 49 is obtained as indicated by the ^{1}H-NMR spectrum *(58,276)*. A similar low temperature NMR study *(50)* of the alcohol 50 has been used to infer the presence of the likewise unstable cross-conjugated cation

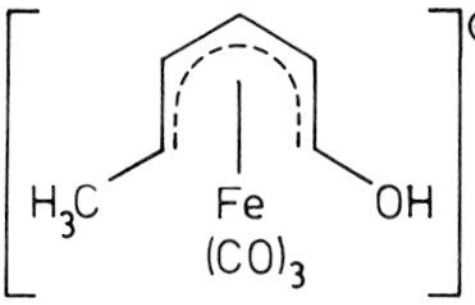

49

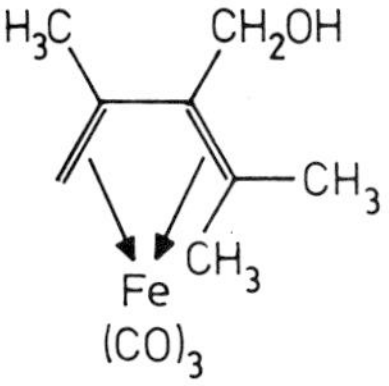

50

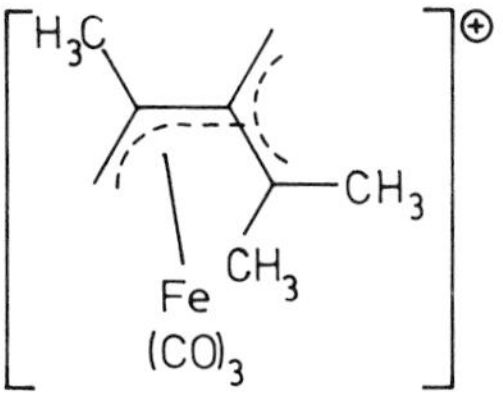

51

51. Quenching the acidic solution of 50 with water at -78°C gives the rearrangement products 52 and 53 depending upon the precise conditions *(50)*.

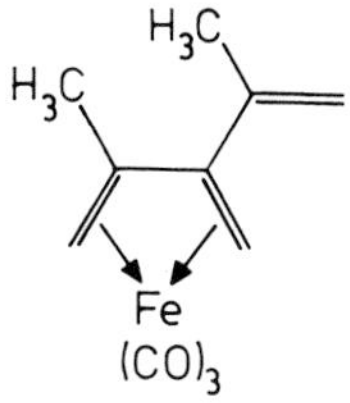

52

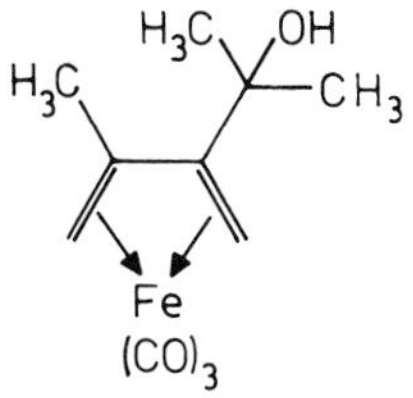

53

(Butadiene)tricarbonyliron is reactive towards Friedel-Crafts acylation. Thus, treatment of (butadiene)tricarbonyliron (1) with acetic anhydride in the presence of aluminium chloride in dichloromethane solution gives a mixture of the 1-acetyl derivative (54; R' = H, R = CH_3) and the 2-acetyl

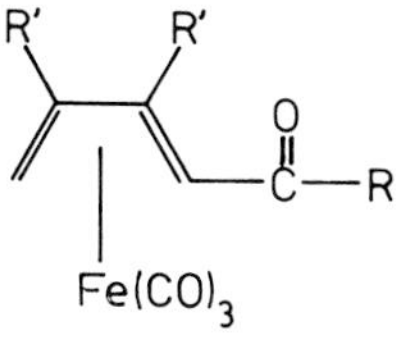

54

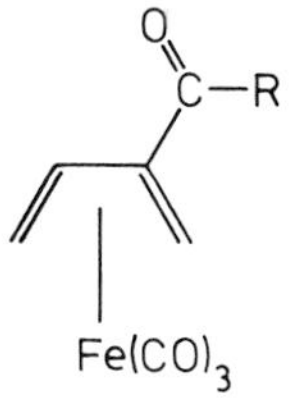

55

derivative (55, R = CH_3) *(11)*. A similar benzoylation of (butadiene)tricarbonyliron with benzoyl chloride in the presence of aluminium chloride in dichloromethane solution was found to give only the 1-benzoyl derivative (54; R' = H, R = C_6H_5). One difficulty with these reactions is the tendency for the liberated hydrogen chloride to react with the unchanged (butadiene)tricarbonyliron to give the (η^3-1-methylallyl)tricarbonyliron chloride (37, R = R' = H). Competitive acylation experiments indicate that (butadiene)tricarbonyliron has a similar reactivity towards electrophilic substitution as ferrocene and is considerably more reactive than either tricarbonyl(cyclopentadienyl)manganese or benzene

(11). (2,3-Dimethylbutadiene)tricarbonyliron has been acetylated under similar conditions to give yellow crystalline 54 (R' = CH_3) *(325)*. Condensation of (1-acetylbutadiene)tricarbonyliron (54; R = CH_3, R' = H) with benzaldehyde in the presence of sodium hydroxide gives yellow crystalline (1-cinnamoylbutadiene)tricarbonyliron *(324)*. Treatment of these acylated 1,3-diene-tricarbonyliron complexes with $LiAlH_4$ in tetrahydrofuran or diethyl ether results simultaneously in the reduction of the ketone carbonyl to an alcohol, in the removal of the diene from the iron atom, and in the hydrogenation of the two diene carbon-carbon double bonds *(324)*.

An intermediate of stoichiometry $[CH_3COC_4H_6Fe(CO)_3]$-$[AlCl_4]$ has been isolated from the Friedel-Crafts acylation of (butadiene)tricarbonyliron with acetyl chloride in the presence of aluminium chloride *(181)*. An X-ray crystallographic study on this intermediate indicates the η^3-allylic structure 56 (R = H) in which the carbonyl oxygen of the

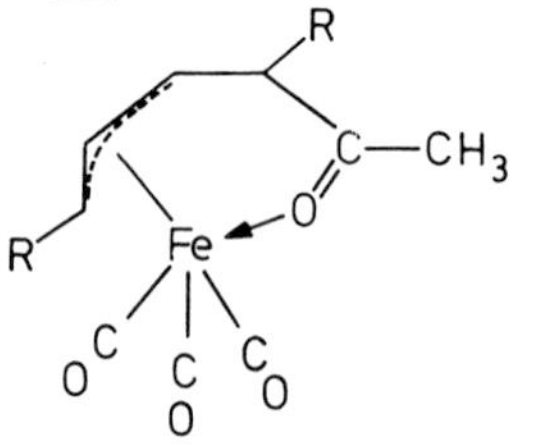

56

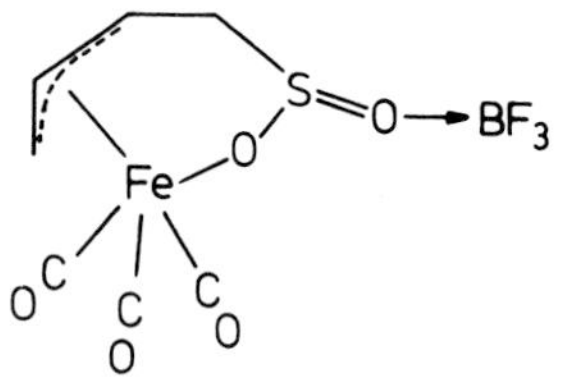

57

acetyl substituent is coordinated to the iron atom *(201)*. An X-ray study on a similar intermediate from (*trans,trans*-2,4-hexadiene)tricarbonyliron (5) indicates structure 56 (R = CH_3). This shows that (*trans,trans*-2,4-hexadiene)tricarbonyliron undergoes stereospecific *endo* attack during Friedel-Crafts substitutions *(182)*. A similar type of intermediate of stoichiometry $[C_4H_6Fe(CO)_3 \cdot SO_2 \cdot BF_3]$ has been isolated by bubbling gaseous boron trifluoride into a solution of (butadiene)tricarbonyliron in liquid sulfur dioxide. An X-ray crystallographic study on this adduct indicates the η^3-allylic sulfinato structure 57 in which one of the sulfinate oxygens is bonded to iron and the other sulfinate oxygen is bonded to boron *(89,90)*.

Fluoroolefins also add to diene-tricarbonyliron complexes to form somewhat similar types of η^3-allylic derivatives. Thus, ultraviolet irradiation of the tricarbonyliron complexes of butadiene, *trans*-1,3-pentadiene, isoprene, and 2,3-dimethylbutadiene with tetrafluoroethylene gives 1:1 adducts formulated as the η^3-allylic derivatives 58 on the basis of their ^1H- and ^{19}F-NMR spectra. The linking reaction occurs prefer-

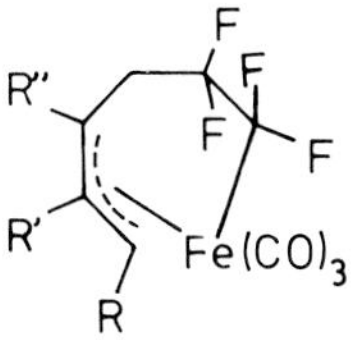

58

R	R'	R''
H	H	H
CH_3	H	H
H	CH_3	H
H	CH_3	CH_3

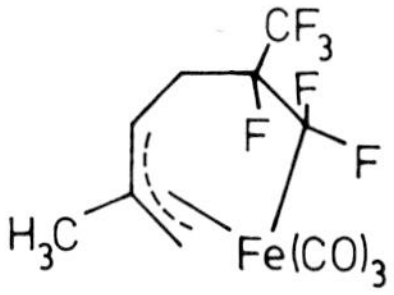

59

entially on the least substituted end of the diene *(51,52)*. A similar reaction between (isoprene)tricarbonyliron and hexafluoropropene was found to give exclusively the isomer 59 suggesting that the reaction was stereospecific *(51,185)*. This structure was confirmed by an X-ray crystallographic study *(185)*. Related oxidative linking reactions have also been obtained from hexafluoropropene and the tricarbonyliron complexes of butadiene and 2,3-dimethylbutadiene, and from chlorotrifluoroethylene and (2,3-dimethylbutadiene)tricarbonyliron *(185)*.

The fluorinated alkyne hexafluorobut-2-yne also reacts similarly with some diene-tricarbonyliron complexes. Ultraviolet irradiation of the tricarbonyliron complexes of butadiene and 2,3-dimethylbutadiene with hexafluorobut-2-yne results in the formation of the η^3-allylic derivatives 60 (R = H or CH_3) *(55)*. Heating these complexes in boiling hexane re-

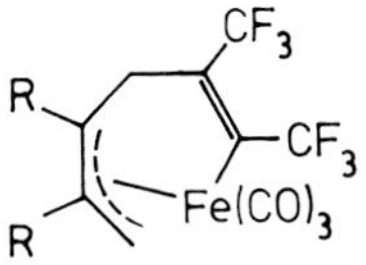

60

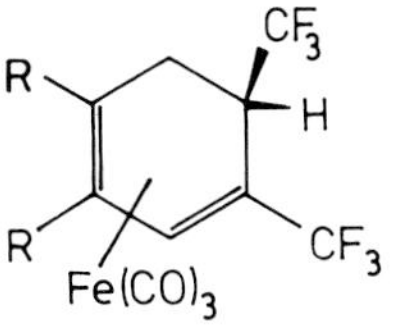

61

sults in rearrangement to give the corresponding substituted 1,3-cyclohexadiene-tricarbonyliron derivatives 61 (R = H or CH_3) *(119)*. The sequence of the reactions of a diene-tricarbonyliron derivative with hexafluorobut-2-yne to give first 60 and then 61 may be regarded as a stepwise Diels-Alder reaction of the alkyne to the 1,3-diene in which the tricarbonyliron group stabilizes an intermediate through complex formation. A similar ultraviolet irradiation of the tricarbonyliron complexes of isoprene and either *cis*- or *trans*-1,3-pentadiene with hexafluorobut-2-yne results only in displacement of the coordinated 1,3-diene to give [tetrakis(trifluoromethyl)cyclopentadienone]tricarbonyliron *(55)*.

Similar insertions to form η^3-allylic derivatives are also found in the reactions of hexafluoroacetone with various diene-tricarbonyliron complexes, but this chemistry contains additional complexities *(183,186)*. Ultraviolet irradiation of hexafluoroacetone with (2,3-dimethylbutadiene)tricarbonyliron forms an orange-yellow crystalline 1:1 adduct formulated as 62. Reaction of hexafluoroacetone with (isoprene)tricarbonyliron, however, gives a 2:1 adduct formulated on the basis of its ^{19}F-NMR spectrum as 63 in which two hexafluoroacetone molecules are linked head to tail *endo* onto the coordinated isoprene *(183,186)*. Heating the complex 63 in hexane solution to 80°C for 24 h results in migration of one of the hexafluoroacetone units to the isoprene methyl group giving the isomer

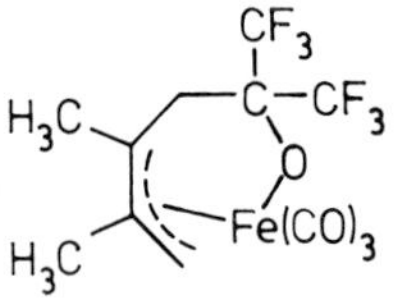

62

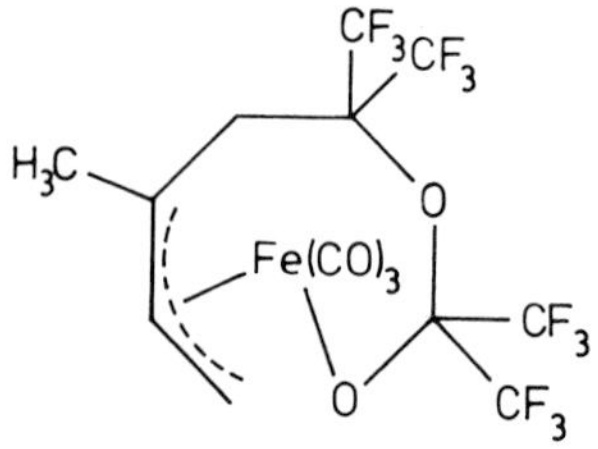

63

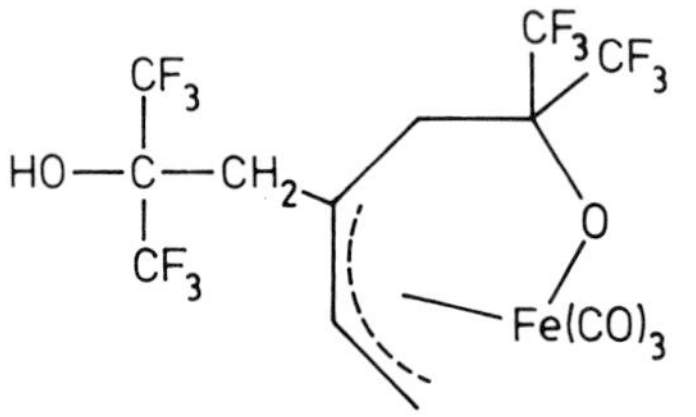

64

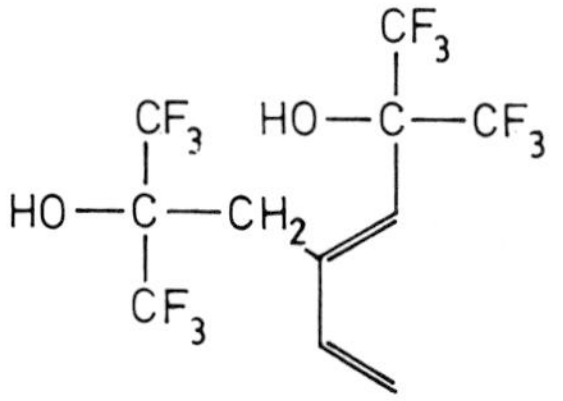

65

64. In addition the iron-free 1,3-diene 65 is isolated.

Recently some similar reactions have been observed from diene-tricarbonyliron complexes and non-fluorinated olefins. For example, ultraviolet irradiation of (butadiene)tricarbonyliron with methyl acrylate at room temperature gives the 1:1 adduct of structure 66. The same compound is also obtained

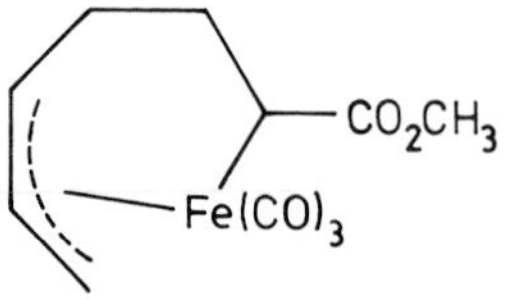

66

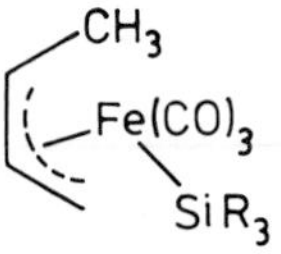

67

by a photochemical reaction of (methyl acrylate)tetracarbonyliron with butadiene or by treatment of the very reactive bis-(methyl acrylate)tricarbonyliron with butadiene below room temperature *(192,412)*.

Some very recent work has demonstrated the photochemical addition of silanes to (butadiene)tricarbonyliron to give η^3-allylic derivatives *(166)*. Thus, ultraviolet irradiation of (butadiene)tricarbonyliron with the silanes R_3SiH (R = CH_3, C_6H_5) forms the 1:1 adducts with the η^3-allylic structure 67. Mild heating of these adducts causes R_3Si transfer to give *cis*-butenyl-SiR_3, and, to some extent, dissociation back into (butadiene)tricarbonyliron and R_3SiH.

Reactions of diene-tricarbonyliron complexes containing electronegative substituents such as formyl, benzoyl, and the cyano group with deuterated alcohols in the presence of base at room temperature leads to exchange of the terminal hydrogens of the butadiene *(404)*.

The methods for liberating the complexed 1,3-dienes from their tricarbonyliron complexes are of interest both for establishing structures and for organic syntheses. In the early work *(239)* triphenylphosphine was used, but the necessary reaction conditions are too vigorous to use with many dienes. Oxidative decomposition with Fe(III) was next developed *(148)*. However, oxidation with Ce(IV) ($(NH_4)_2Ce(NO_3)_6$) in a polar solvent such as ethanol is now the most frequently used method *(149)*. However, a relatively new method using trimethylamine *N*-oxide as the oxidant in aprotic solvents such as benzene *(364)* may become the method of choice in many cases for removing the diene intact from diene-tricarbonyliron complexes because of the mildness of the conditions required and the relative unreactivity of sensitive polyenes towards excess trimethylamine *N*-oxide.

Reactions of diene-tricarbonyliron complexes with α,α'-dibromoketones also result in removal of the coordinated 1,3-diene through formation of cycloheptenone derivatives of the type 68. This reaction can be used for the synthesis of such

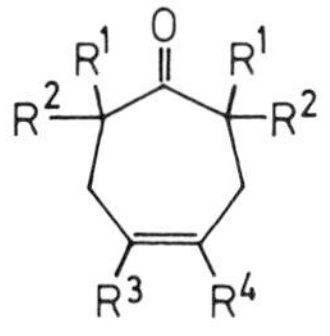

68

cycloheptenones by reacting the free 1,3-diene with the α,α'-dibromoketone in the presence of enneacarbonyldiiron without isolation of the intermediate diene-tricarbonyliron derivative *(327)*.

C. *VINYLKETENE-TRICARBONYLIRON DERIVATIVES*

Closely related to the acyclic 1,3-dienes are the vinylketenes, which are derived from the 1,3-dienes by replacement of the two substituents on one of the terminal carbon atoms of the diene system with a doubly bonded oxygen atom. Two vinylketene-tricarbonyliron derivatives have been prepared by indirect methods, but this area of chemistry does not appear to have been investigated in any detail.

The first vinylketene-tricarbonyliron complex 69 (R = H,

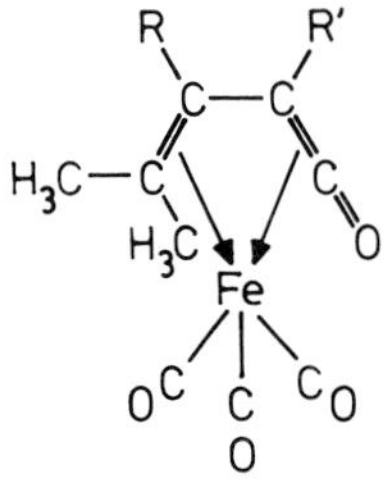

69

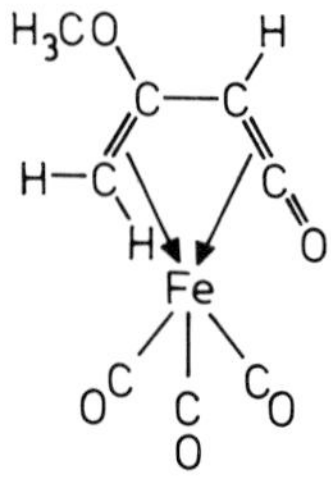

70

R' = CH_3, or R = CH_3, R' = H) was made by treatment of 1,3,3-trimethylcyclopropene with dodecacarbonyltriiron in boiling benzene *(241)*. Apparently carbonyl insertion and opening of the cyclopropene ring occur in this reaction. Nearly ten years later *(207)* a second vinylketene-tricarbonyliron complex 70 was reported from the reaction of 2-methoxyallyl chloride, $CH_2=C(OCH_3)-CH_2Cl$ with enneacarbonyldiiron in benzene at 40°C. Both of the known vinylketene-tricarbonyliron complexes 69 and 70 are reasonably stable volatile yellow crystalline solids in contrast to the very unstable free vinylketenes.

D. *(CYCLOPENTADIENE)TRICARBONYLIRON AND ITS SUBSTITUTION PRODUCTS*

The reaction of cyclopentadiene with carbonyliron reagents leads very readily to the loss of a hydrogen atom from the cyclopentadiene to form the cyclopentadienyl derivative $[(\eta^5-C_5H_5)Fe(CO)_2]_2$ *(337)*. In order to prepare (cyclopentadiene)tricarbonyliron (71, L = CO) the reaction between cyclopentadiene and carbonyliron reagents must be carried out under conditions mild enough that loss of hydrogen to form a cyclopentadienyl derivative does not occur. For this reason the discovery of (cyclopentadiene)tricarbonyliron (71, L = CO) first required the development of carbonyliron reagents capable of generating $Fe(CO)_3$ units under mild conditions. Therefore this discovery occurred long after the discovery of

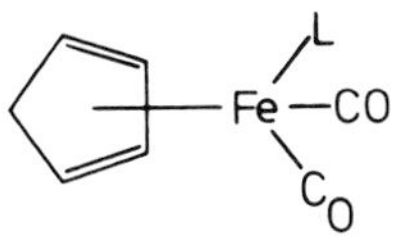

71

the reaction between cyclopentadiene and pentacarbonyliron under relatively vigorous conditions (130 - 140°C) to give the cyclopentadienyl derivative $[(\eta^5\text{-}C_5H_5)Fe(CO)_2]_2$.

The first reported synthesis of (cyclopentadiene)tricarbonyliron used the reaction in boiling benzene of pentacarbonyliron with a nickel compound then believed to be bis(cyclopentadiene)nickel, $(\eta^4\text{-}C_5H_6)_2Ni$, but subsequently shown to be (cyclopentadienyl)(cyclopentenyl)nickel, $(\eta^5\text{-}C_5H_5)Ni(\eta^3\text{-}C_5H_7)$, *(153,154)*. Thus, the original view of this reaction as an exchange of cyclopentadiene ligands from nickel to iron is not really accurate. In any case this early reaction was made obsolete by the subsequent discovery by Kochhar and Pettit *(249)* that cyclopentadiene reacts with enneacarbonyldiiron in boiling diethyl ether to give a 27 % yield of (cyclopentadiene)tricarbonyliron (71, L = CO).

(Cyclopentadiene)tricarbonyliron (71, L = CO) is a yellow liquid freezing at -6°C and purified by distillation at 30-35°C/0.2 Torr. Upon heating above 100°C it gradually loses hydrogen to form the cyclopentadienyl derivative $[(\eta^5\text{-}C_5H_5)Fe(CO)_2]_2$ identical to the product obtained from cyclopentadiene and pentacarbonyliron under more vigorous conditions. Reaction of (cyclopentadiene)tricarbonyliron (71, L = CO) with triphenylmethyl tetrafluoroborate rapidly leads to hydride abstraction to form the η^5-cyclopentadienyl derivative $[(\eta^5\text{-}C_5H_5)Fe(CO)_3]^+[BF_4]^-$ *(249)*. Reduction of $[(\eta^5\text{-}C_5H_5)Fe(CO)_3]^+[BF_4]^-$ with sodium cyanoborohydride in tetrahydrofuran regenerates 71 (L = CO) *(399)*. Using $NaBD_3CN$ instead of $NaBH_3CN$ in this reaction resulted in the isolation of deuterated 71 (L = CO) with the CH_2 groups stereospecifically deuterated in the *exo*-position *(399)*. Reduction of $[(\eta^5\text{-}C_5H_5)Fe(CO)_3]^+[BF_4]^-$ with $NaBH_4$ rather than $NaBH_3CN$ in tetrahydrofuran gives $[(\eta^5\text{-}C_5H_5)Fe(CO)_2]_2$ as the only carbonyliron derivative *(120)*.

Cyclopentadiene-tricarbonyliron derivatives in which one of the carbonyl groups is replaced by a tertiary phosphine ligand appear to be more stable than the unsubstituted (cyclopentadiene)tricarbonyliron towards hydrogen losses to form η^5-cyclopentadienyl derivatives. Reduction of the cation $[(\eta^5\text{-}C_5H_5)Fe(CO)_2P(C_6H_5)_3]^+$ with sodium borohydride in a mixture of tetrahydrofuran and diethyl ether gives the cyclopentadiene complex $[(\eta^4\text{-}C_5H_6)Fe(CO)_2P(C_6H_5)_3]$ (71, L = $P(C_6H_5)_3$) *(120)*. A low yield (1 - 2 %) of $[(\eta^4\text{-}C_5H_6)Fe(CO)_2\text{-}$

$P(C_6H_5)_3]$ has also been obtained from the sodium borohydride reduction of $\{(\eta^5\text{-}C_5H_5)Fe(CO)[P(C_6H_5)_3]Cl\}$ accompanied by much larger quantities of the hydride $(\eta^5\text{-}C_5H_5)Fe(CO)[P(C_6H_5)_3]H$ *(234)*.

Introduction of substituents into the CH_2 group of cyclopentadiene (the 5-position) also appears to stabilize the corresponding cyclopentadiene-tricarbonyliron derivatives. For example, the reaction of (acetyl pentamethyl)cyclopentadiene with enneacarbonyldiiron in pentane at room temperature gives the diene complex $[\eta^5\text{-}CH_3\text{-}CO\text{-}C_5(CH_3)_5]Fe(CO)_3$ (72) in

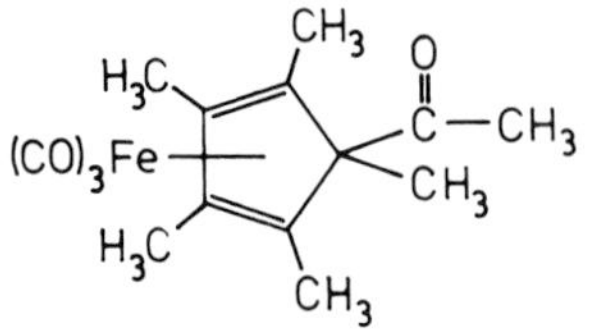

72

addition to the η^5-pentamethylcyclopentadienyl derivatives $CH_3\text{-}CO\text{-}Fe(CO)_2[\eta^5\text{-}C_5(CH_3)_5]$ and $[\eta^5\text{-}C_5(CH_3)_5Fe(CO)_2]_2$. This reaction demonstrates clearly the tendency for a substituent in the 5-position in a (cyclopentadiene)tricarbonyliron complex to migrate from carbon to iron *(245,246)*.

An opposite type of reaction occurs upon treatment of the η^5-cyclopentadienyl derivative $NaFe(CO)_2(C_5H_5)$ with *o*-carboranecarboxylic acid chlorides to give cyclopentadiene-tricarbonyliron derivatives with migration of a carboranyl group from an acyl carbonyl to a cyclopentadienyl ring *(408,410)*. Reaction of the *o*-carboranylcarboxylic acid chlorides $R\text{-}C(B_{10}H_{10})C\text{-}CO\text{-}Cl$ ($R = CH_3$, C_6H_5) with $NaFe(CO)_2$-(C_5H_5) in tetrahydrofuran at room temperature gives the cyclopentadiene complexes 73 ($R = CH_3$, C_6H_5) in 40 - 65 %

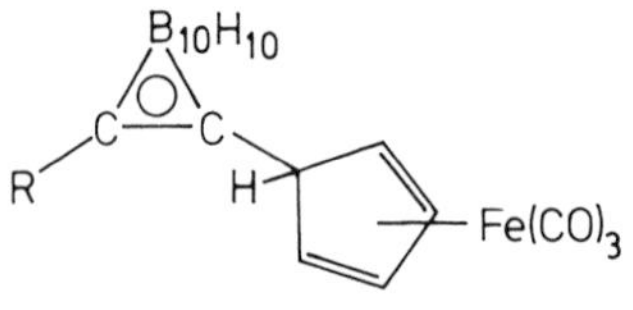

73

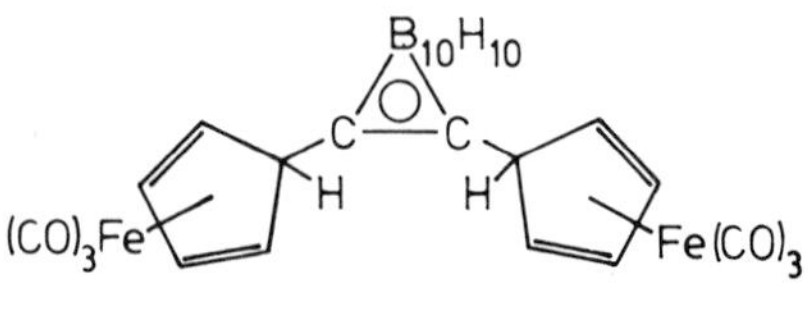

74

yields. Treatment of the unsubstituted *o*-carboranylcarboxylic acid chloride with $NaFe(CO)_2(C_5H_5)$ gives the usual σ-acyl derivative $H\text{-}C(B_{10}H_{10})C\text{-}CO\text{-}Fe(CO)_2(C_5H_5)$ which rearranges to the isomeric cyclopentadiene complex 73 (R = H) upon stirring in tetrahydrofuran solution at 20°C for 10 days. Reaction of the acid chloride of *o*-carboranedicarboxylic acid, $Cl\text{-}CO\text{-}C$-$(B_{10}H_{10})C\text{-}CO\text{-}Cl$, with $NaFe(CO)_2(C_5H_5)$ leads directly at room temperature to 74 containing two cyclopentadiene-tricarbonyliron units. Similar rearrangements to cyclopentadiene-tricarbonyliron derivatives do not occur during the corresponding reactions of $NaFe(CO)_2(C_5H_5)$ with the chlorides of the corresponding *m*-carboranecarboxylic acids.

Cyclopentadiene-tricarbonyliron derivatives are also known with fluorocarbon substituents in the 5-position. Thus, one of the products obtained in low yield from the reaction of $[(\eta^5\text{-}C_5H_5)Fe(CO)_3][BF_4]$ with pentafluorophenyllithium *(374,375)* is the substituted *exo*-pentafluorophenyl-cyclopentadiene derivative $(\eta^5\text{-}C_5H_5\text{-}C_6F_5)Fe(CO)_3$ (75, L = CO). A

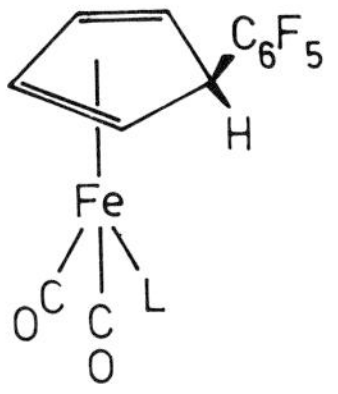

75

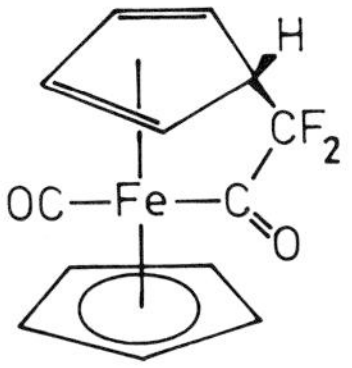

76

triphenylphosphine substitution product (75, L = $P(C_6H_5)_3$) can be obtained in 56 % yield from the corresponding reaction of pentafluorophenyllithium with $[(\eta^5\text{-}C_5H_5)Fe(CO)_2P(C_6H_5)_3]I$ *(374,375)*. Ultraviolet irradiation of the η^1-cyclopentadienyl derivative $[(\eta^5\text{-}C_5H_5)Fe(CO)_2(\eta^1\text{-}C_5H_5)]$ with tetrafluoroethylene gives an 8 % yield of a red crystalline solid shown by X-ray crystallography to be the cyclopentadiene complex 76 in which cleavage of the tetrafluoroethylene carbon-carbon bond has occurred *(116)*.

If both hydrogen atoms of the CH_2 moiety in cyclopentadiene are substituted by other groups, then the tendency for decomposition to form η^5-cyclopentadienyl derivatives can be minimized. Furthermore, if the two substituents in the 5-position on the cyclopentadiene ring are different, then the possibility for geometrical isomerism exists. For example, treatment of 5-hydroxymethyl-5-methyl-cyclopentadiene with enneacarbonyldiiron gives a mixture of the two stereoisomeric alcohols 77a and 77b in a 4:1 ratio *(302)*. These two isomers

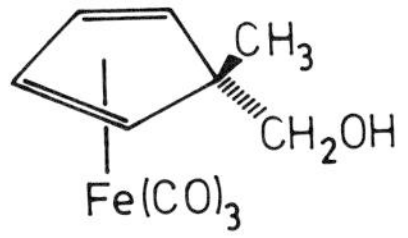

77a

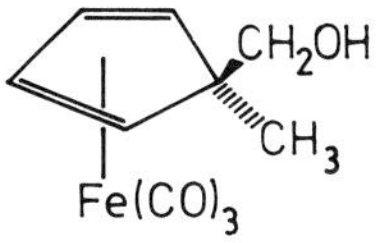

77b

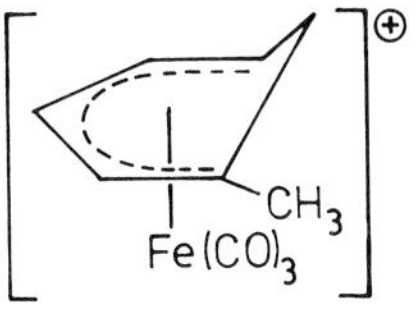

78

can be separated readily by chromatography. Heterolysis of the *p*-toluenesulfonate of the isomer 77b with tetrafluoroboric acid in acetic anhydride leads to a novel ring expansion to form the (methylcyclohexadienyl)tricarbonyliron cation 78 isolated as its hexafluorophosphate salt *(204)*. A similar treatment of the isomer 77a leads only to complete decomposition. The mass spectra of the isomer pair 77a and 77b as well as the corresponding pair of isomeric methyl ethers are distinctly different *(304)*. Stereospecific radical scission of the *exo*-substituent occurs in these mass spectra.

Both 5-positions of cyclopentadiene can be substituted by the formation of a spirane such as 79. However, even in these cases rearrangement to substituted η^5-cyclopentadienyl-carbonyliron derivatives can occur if the reaction conditions are sufficiently vigorous. For example, treatment of the spirononadiene 79 with pentacarbonyliron at ~130°C leads to rearrangement to form the tetrahydroindenyl derivative 80 *(200)*. However, treatment of 79 with enneacarbonyldiiron under much milder conditions (benzene at 80°C for 1.5 h) gives the corresponding diene-tricarbonyliron complex 81 without rearrangement *(180)*. In addition, this reaction gives a low yield (2 %) of the bridged σ-alkyl-η^5-cyclopentadienyl derivative 82 *(138)*. The reaction of enneacarbonyldi-

79

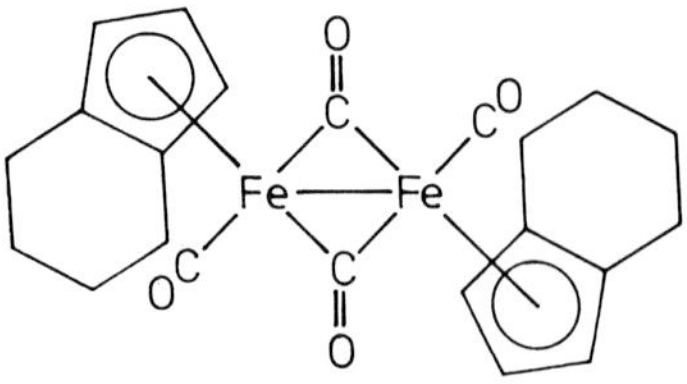

80

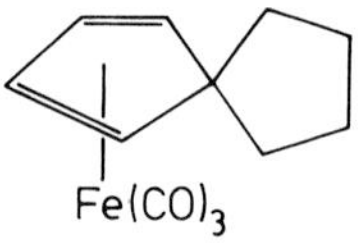

81

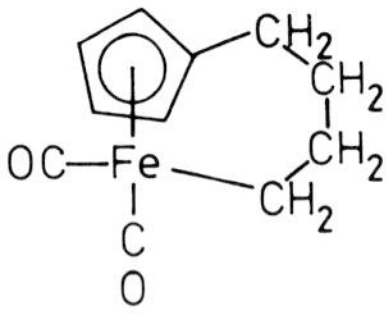

82

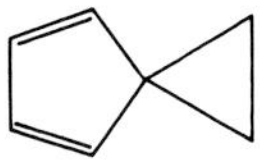

83

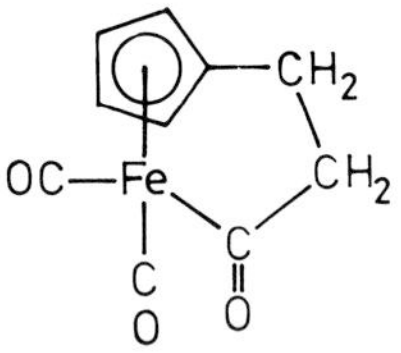

84

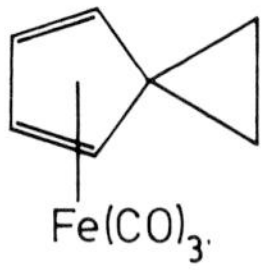

85

iron with spiro[2.4]hepta-4,6-diene (83) in boiling diethyl ether gives the yellow crystalline bridging σ-acyl-η^5-cyclopentadienyl derivative 84 *(138)*. Compound 84 was erroneously earlier identified as the diene-tricarbonyliron derivative 85 *(126)*.

1,1'-Bicyclopentenyl reacts with pentacarbonyliron in boiling ethylcyclohexane or with dodecacarbonyltriiron in boiling cyclohexane to give the corresponding tricarbonyliron derivative 86 *(292)*.

Among the many products formed from reactions of fulvenes with iron carbonyls are the fulvene-tricarbonyliron

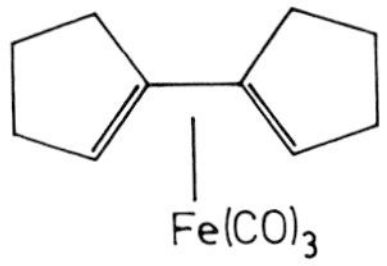

86

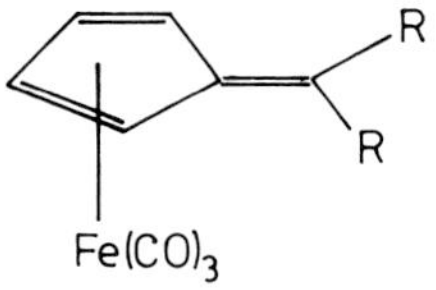

87

complexes 87 (*e.g.*, R = phenyl or other aryl groups) which may be regarded as diene-tricarbonyliron complexes *(389,392)*. This chemistry is discussed in detail elsewhere in this book.

E. *CYCLOPENTADIENONE-TRICARBONYLIRON DERIVATIVES*

Substitution of the CH_2 group in cyclopentadiene by a C=O group to give cyclopentadienone alters considerably its carbonyliron chemistry. Therefore the chemistry of cyclopentadienone-tricarbonyliron complexes is discussed in this separate section of this chapter.

The history of the chemistry of cyclopentadienone-tricarbonyliron complexes somewhat parallels that of (butadiene)-tricarbonyliron, since cyclopentadienone-tricarbonyliron derivatives were first obtained long before their nature was recognized. The first cyclopentadienone-tricarbonyliron derivatives were obtained during World War II by Reppe and co-workers from reactions of acetylenes and carbonyliron complexes *(343)*. After the discovery of ferrocene, the product of approximate stoichiometry $(C_6H_5C_2H)_2Fe(CO)_4$ from phenylacetylene and a mixture of iron and nickel carbonyls was studied in greater detail *(99,231,274)*. This product was eventually identified *(213,214,357,358,360)* as (2,5-diphenylcyclopentadienone)tricarbonyliron (88; R = R"' = C_6H_5,

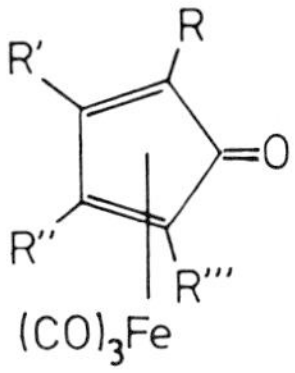

88

R' = R" = H). The corresponding product from the unsubstituted acetylene and iron carbonyls was identified as the unsubstituted (cyclopentadienone)tricarbonyliron (88 ; R = R' = R" = R"' = H) *(388,390)*.

Subsequent studies have shown that the formation of cyclopentadienone-tricarbonyliron derivatives from alkynes and carbonyliron reagents is rather general. Thus, (tetraphenylcyclopentadienone)tricarbonyliron (88, R = R' = R" = R"' =

C_6H_5) is readily obtained from diphenylacetylene and appropriate carbonyliron complexes *(213,214,358,360)*. Other diarylacetylenes give similar tetraarylcyclopentadienone-tricarbonyliron derivatives. Similarly, hexafluorobut-2-yne reacts with pentacarbonyliron in the temperature range 100-200°C *(54)* to give orange-yellow crystalline [tetrakis(trifluoromethyl)cyclopentadienone]tricarbonyliron (89, L = CO). The crystal structure of this complex *(24,86)* determined by X-ray crystallography indicates the cyclopentadienone ligand to be non-planar with the keto group bent at an angle of about 20° above the plane of the other four carbons. This suggests sig-

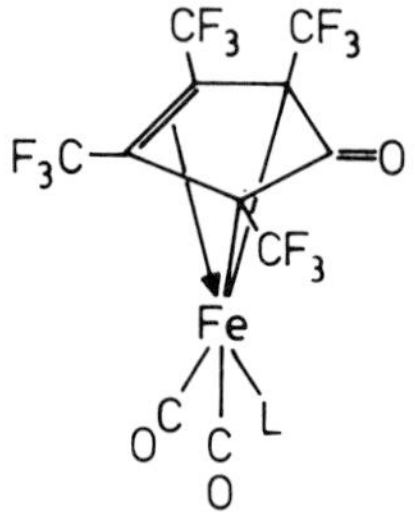

89a

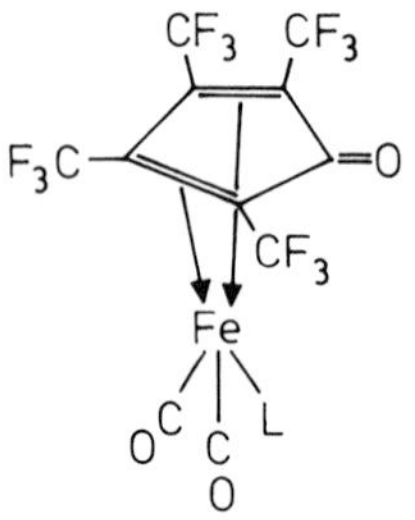

89b

nificant contribution of structure 89a (L = CO) containing two iron-carbon σ-bonds and one η^2-coordinated C=C double bond. Such structures are relatively favourable in the case of the tetrakis(trifluoromethyl)cyclopentadienone derivative because of the highly electronegative trifluoromethyl group which removes electron density from the π-system to weaken the dienone ligand as a π-donor while making the terminal carbons of the 1,3-diene unit more electronegative and hence better suited for forming strong σ-bonds with the iron.

Similar reactions of the phosphine-substituted carbonyliron complexes *trans*-$L_2Fe(CO)_3$ (L = $(C_2H_5O)_3P$ and $(CH_3)_2PC_6H_5$) using ultraviolet irradiation at room temperature in hexane solution give the substituted tetrakis(trifluoromethyl)cyclopentadienone derivatives $[(CF_3)_4C_4CO]Fe(CO)_2L$ (89, L = $(C_2H_5O)_3P$ and $(CH_3)_2PC_6H_5$) with loss of one carbonyl group and one trivalent phosphorus ligand *(72)*. Reactions of 3,3,3-trifluoropropyne *(129)* and pentafluorophenylacetylene *(130)* with pentacarbonyliron at 120°C give the corresponding 2,5-disubstituted cyclopentadienone complexes 88 (R = R''' = CF_3 or C_6F_5, respectively; R' = R" = H). Reaction of the unstable and explosive dichloroacetylene with enneacarbonyldiiron gives the yellow-orange air-stable crystalline (tetrachlorocyclopentadienone)tricarbonyliron (88, R = R' = R" = R''' = Cl) *(260)*.

A tricyclic cyclopentadienone complex 90 is one of the products isolated from the ultraviolet irradiation of penta-

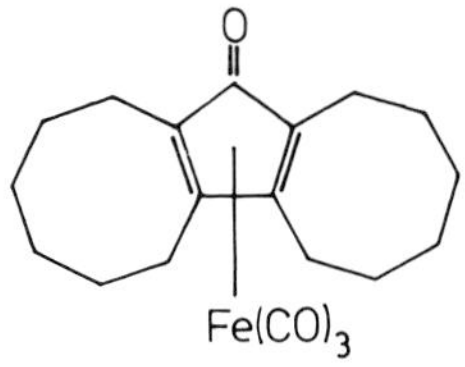

90

carbonyliron with the very reactive alkyne cyclooctyne *(256)*. A complicated trinuclear product containing one tricarbonylferrole-tricarbonyliron unit and one cyclopentadienone-tricarbonyliron unit has been isolated from the reaction of 2,4-hexadiyne with dodecacarbonyltriiron in boiling toluene *(247)*.

All of these preparations of cyclopentadienone-tricarbonyliron derivatives use alkynes as starting materials and construct the cyclopentadienone unit from two alkyne molecules and one carbonyl group. Such syntheses are attractive, since in most cases alkynes are more readily available than cyclopentadienones, many of which are unstable with respect to dimerization and other decomposition reactions even at room temperature and below. However, in cases where stable cyclopentadienones are available, they may be reacted with carbonyliron reagents to form the corresponding cyclopentadienone-tricarbonyliron complexes. For example, (tetraphenylcyclopentadienone)tricarbonyliron is available not only from diphenylacetylene and pentacarbonyliron as discussed above but also by direct reaction of tetraphenylcyclopentadienone and pentacarbonyliron *(356,387)*. Furthermore, one of the products from the reaction of cyclopentadienone diethyl ketal (*i.e.* 5,5-diethoxycyclopentadiene) with enneacarbonyldiiron in boiling pentane is the unsubstituted (cyclopentadienone)-tricarbonyliron (88, R = R' = R" = R"' = H), shown to be identical with the product originally obtained from acetylene and pentacarbonyliron *(141)*. One of the products obtained by treatment of santonin with enneacarbonyldiiron in benzene at 40°C is the tricyclic cyclopentadienone complex 91. However,

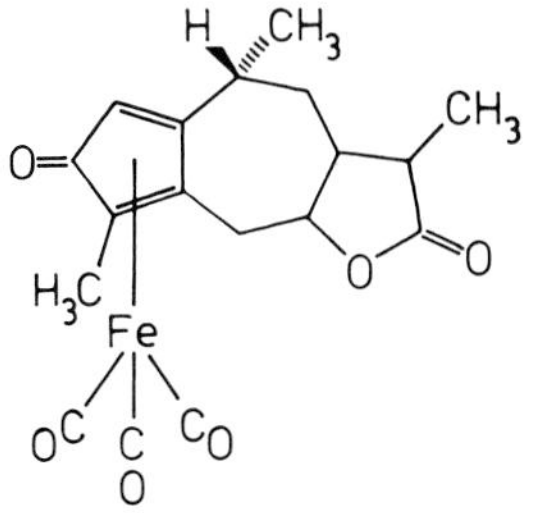

91

a more predominant product from this reaction is the cyclopentadienol-tricarbonyliron complex arising from hydrogenation of the cyclopentadienone carbonyl group in 91 *(4)*.

The cyclopentadienone complexes $(R_4C_4CO)Fe(CO)_3$ (R = C_6H_5, CF_3) and $[(CF_3)_4C_4CO]Fe(CO)_2[P(C_6H_5)_3]$ have been reduced electrochemically *(134)*, but the chemical irreversibility of even the observed one-electron reductions indicates that these reductions do not correspond to simple formation of radical anions. The hydroquinone adduct of (cyclopentadienone)tricarbonyliron has some activity as a polymerization initiator for the polymerization of methyl methacrylate in the presence of carbon tetrachloride *(25)*.

F. TRICARBONYLIRON COMPLEXES FORMED FROM DIENES CONTAINING SIX- AND SEVEN-MEMBERED RINGS

Six- and seven-membered rings containing 1,3-diene units are particularly favourable for forming diene-tricarbonyliron complexes. The shape of the ring holds the carbon-carbon double bonds in the *cisoid* configuration required for formation of diene-tricarbonyliron complexes. Furthermore, 1,3-diene complexes containing six- and seven-membered rings cannot readily decompose like cyclopentadiene complexes, which easily can lose a hydrogen to form η^5-cyclopentadienyl derivatives. Important six- and seven-membered ring compounds which form diene-tricarbonyliron complexes include 1,3-cyclohexadiene, 1,3-cycloheptadiene, cycloheptatriene, cyclohexadienone, cycloheptadienone, and tropone (cycloheptatrienone).

1. Cyclohexadiene Complexes

In their original work Hallam and Pauson *(199)* recognized that butadiene must be in the *cisoid* configuration to form (butadiene)tricarbonyliron (1). This led them to investigate the corresponding reaction of pentacarbonyliron with 1,3-cyclohexadiene, a 1,3-diene which must always be in the *cisoid* configuration, and which therefore readily should form a 1,3-diene-tricarbonyliron complex. This reaction was successful and led to the discovery of (1,3-cyclohexadiene)tricarbonyliron (92) as a stable yellow liquid freezing

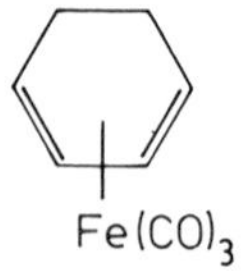

92

at 8°C. Subsequently the reaction of 1,4-cyclohexadiene with pentacarbonyliron was also shown *(12,239)* to give the same (1,3-cyclohexadiene)tricarbonyliron (92). Thus, 1,4-cyclohexadiene, like 1,4-pentadiene discussed above, rearranges to the corresponding isomeric 1,3-diene upon treatment with carbonyliron. A study of this reaction using 1,4-cyclohexadiene with both CH_2 groups deuterated gave a deuterium distribution in the (1,3-cyclohexadiene)tricarbonyliron product which suggested an η^3-allylic tricarbonyliron hydride as an intermediate *(2)*.

A variety of 1,3-cyclohexadiene-tricarbonyliron complexes are accessible by the reduction of benzenoid compounds with sodium in liquid ammonia (Birch reduction) to give the corresponding 1,4-cyclohexadienes followed by reactions of the resulting 1,4-cyclohexadienes with appropriate carbonyliron reagents to give the corresponding substituted 1,3-cyclohexadiene-tricarbonyliron complexes. Pentacarbonyliron in boiling di-*n*-butyl ether is often an effective reagent for introducing the tricarbonyliron group in the second step of the sequence. Benzenoid compounds which have been subjected to this reaction sequence include toluene *(40)*, *m*- and *p*-xylenes *(40)*, mesitylene *(239,336)*, anisole *(38,40,41)*, and the isomeric methoxytoluenes *(38,40,41)*. Similar reductions of benzoic and *o*-toluic acids followed by esterification with diazomethane gives methyl cyclohexadienecarboxylates which form the corresponding methoxycarbonyl-substituted 1,3-cyclohexadiene-tricarbonyliron complexes upon treatment with pentacarbonyliron in boiling di-*n*-butyl ether *(42)*.

The mixtures of isomeric 1,3-cyclohexadiene-carbonyliron derivatives formed in some of these reactions may be rather complex since reduction of the benzene can generally give several possible isomeric cyclohexadiene derivatives. In addition isomerization of the 1,4-cyclohexadiene to the 1,3-cyclohexadiene-tricarbonyliron can occur in several ways. Even after a 1,3-cyclohexadiene-tricarbonyliron derivative is formed, further isomerization reactions may be possible. For example, the product 93, obtained by photolysis of 1-phenyl-

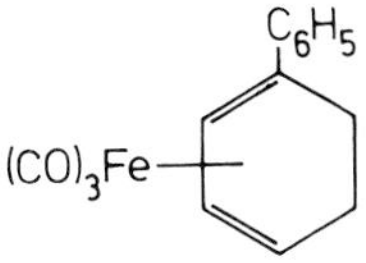

93

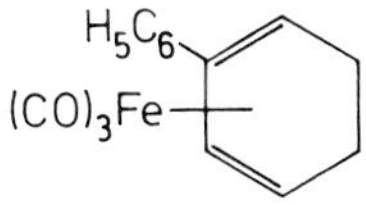

94

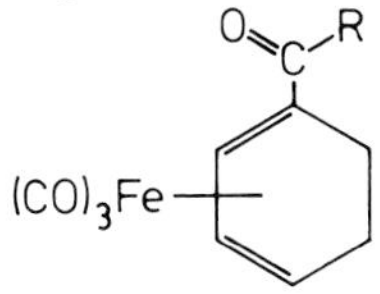

95

cyclohexa-1,3-diene with pentacarbonyliron, isomerizes to an equilibrium mixture of 93 and 94 upon heating at 145°C in xylene solution *(394,400)*. Studies of this reaction with deuterated phenylcyclohexadiene led to the suggestion of an iron hydride intermediate and involvement of one of the double bonds of the phenyl ring *(394,400)*. This isomerization is inhibited by triphenylphosphine *(400)*.

Reaction of 4-vinylcyclohexene with enneacarbonyldiiron or dodecacarbonyltriiron results in rearrangement to a mixture of isomeric (ethylcyclohexa-1,3-diene)tricarbonyliron complexes *(318)*. Reaction of a mixture of 1-(diethylamino)butadiene and a dienophile $CH_2{=}CHX$ (X = CHO, $CO{-}CH_3$) with dodecacarbonyltriiron in boiling benzene results in deamination to give the substituted 1,3-cyclohexadiene-tricarbonyliron derivatives <u>95</u> (R = H, CH_3) *(318)*.

An unusual method for preparing a 1,3-cyclohexadiene-tricarbonyliron derivative is the addition of hexafluorobutyne to the tricarbonyliron complexes of butadiene and 2,3-dimethylbutadiene followed by pyrolysis of the resulting 1:1 adduct at 80°C to give <u>61</u> (R = H and CH_3). This reaction is discussed above in greater detail *(119)*.

Steroids containing a cyclohexadiene unit which have been converted into the corresponding tricarbonyliron complexes include cholesta-2,4-diene, 3-methylcholesta-2,4-diene, 3-methoxycholesta-2,4-diene, cholesta-5,7-diene-3β-ol, ergosterol *(1)*, and acetylergosterol *(310)*. In these cases pentacarbonyliron in boiling isooctane, cyclohexane, or dibutyl ether can be used to introduce the tricarbonyliron group. The opium alkaloid thebaine, which contains a methoxycyclohexadiene ring, has also been converted into a tricarbonyliron complex *(37)*. Octahydro-*as*-indacene and decahydrophenanthrene derivatives containing a 1,4-cyclohexadiene unit have also been converted to the corresponding polycyclic 1,3-cyclohexadiene-tricarbonyliron complexes by treatment with pentacarbonyliron in boiling dibutyl ether *(109)*.

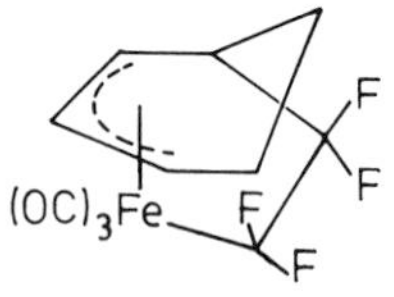

<u>96</u>

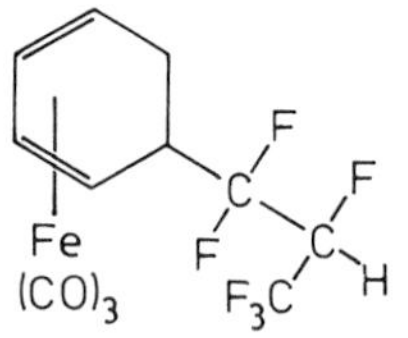

<u>97</u>

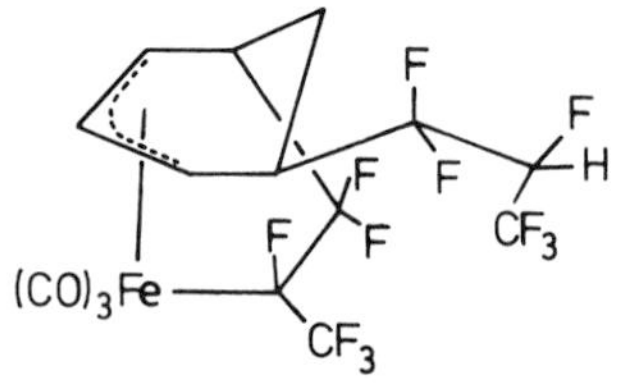

<u>98</u>

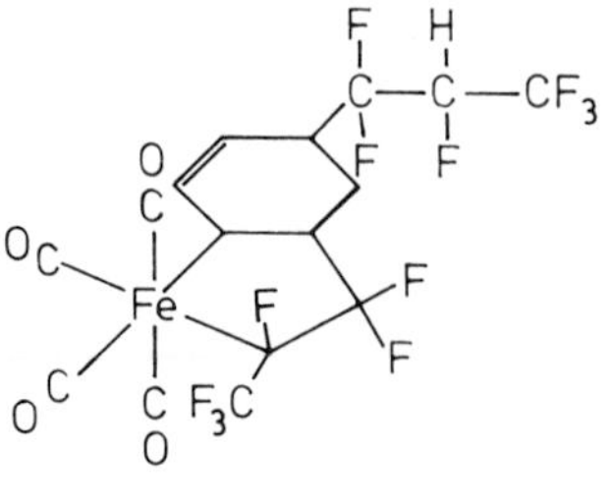

<u>99</u>

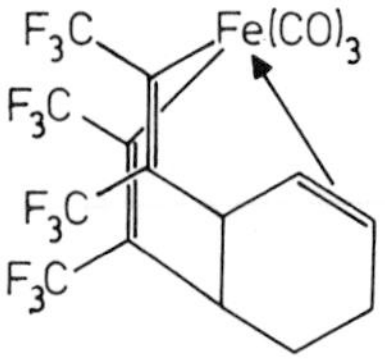

<u>100</u>

Several types of interesting products have been formed by reactions of (1,3-cyclohexadiene)tricarbonyliron (92) with various fluorocarbon derivatives. Ultraviolet irradiation of 92 with tetrafluoroethylene gave the cyclohexenyl derivative 96 *(52)*. A similar ultraviolet irradiation of 92 with hexafluoropropene gave a complex mixture from which the three products 97, 98, and 99 were isolated *(185)*. Ultraviolet irradiation of 92 with hexafluorobut-2-yne gave a 1:2 adduct formulated as the double bond alkyne insertion product 100 on the basis of the crystal structure of its ruthenium analogue *(55)*.

Oxidation of substituted 1,3-cyclohexadiene-tricarbonyliron complexes with ethanolic cupric chloride at room temperature results in liberation of the free 1,3-cyclohexadiene in good yield *(371)* in most cases. However, in some cases chlorination of the 1,3-cyclohexadiene ligand can occur.

Bicyclohexenyl reacts with pentacarbonyliron in boiling di-*n*-butyl ether to form the corresponding tricarbonyliron complex 101 *(80,292)*.

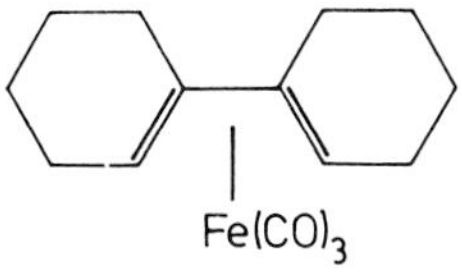

101

A characteristic feature of the mass spectra of 1,3-cyclohexadiene-tricarbonyliron complexes is the ability of H_2 loss to compete with CO loss from the molecular ion *(118, 197,406)*.

2. Diene-tricarbonyliron Derivatives from Cycloheptadienes and Cycloheptatrienes

The reaction between cycloheptatriene and pentacarbonyliron was originally *(73)* believed to give (cycloheptatriene)-dicarbonyliron. However, subsequent investigations *(74,113)* showed this to be incorrect. Instead, this reaction gives a mixture of the tricarbonyliron complexes of cycloheptatriene (102) and 1,3-cycloheptadiene (103). The amount of cyclohep-

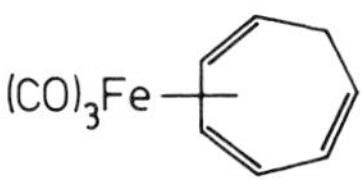

102

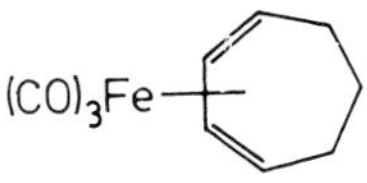

103

tadiene complex 103 relative to the cycloheptatriene complex 102 increases as the time of the reaction is increased, as might be expected. Apparently hydrogen shift reactions similar to those involved in the rearrangement of 1,4-dienes to 1,3-dienes can cause cycloheptatriene to be hydrogenated to 1,3-cycloheptadiene in the presence of carbonyliron.

Several reactions of (cycloheptatriene)tricarbonyliron (102) give products of interest. Reaction of 102 with diiodomethane in boiling diethyl ether in the presence of a zinc-copper couple cyclopropanates the uncomplexed carbon-carbon double bond to give (bicyclo[5.1.0]octadiene)tricarbonyliron (104) *(341)*. (Cycloheptatriene)tricarbonyliron (102) can be

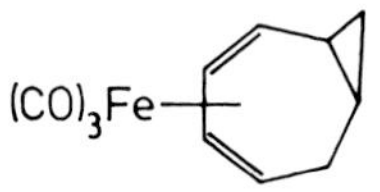

104

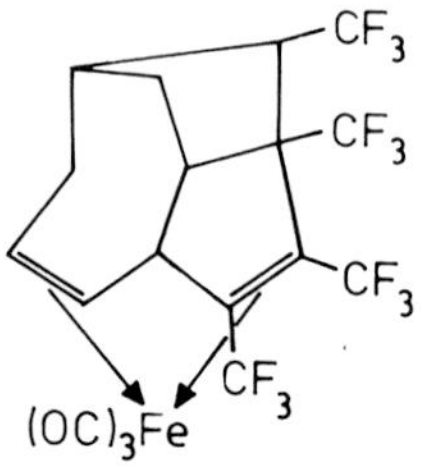

105

formylated with phosphoryl chloride in dimethylformamide at 0°C and acetylated with acetyl chloride and aluminium chloride in dichloromethane at 0°C *(226)*. Ultraviolet irradiation of (cycloheptatriene)tricarbonyliron (102) with hexafluorobut-2-yne gives a 1:2 adduct formulated as 105 on the basis of the crystal structure of the product obtained by the substitution of one carbonyl group in this adduct with the phosphite $P(OCH_2)_3CCH_3$ *(55)*.

1,1'-Bicycloheptenyl reacts with dodecacarbonyltriiron in boiling cyclohexane to give the corresponding tricarbonyliron complex *(292)*.

3. Formation and Reactions of Cyclohexadienyl-tricarbonyliron Cations

A characteristic reaction of 1,3-cyclohexadiene-tricarbonyliron complexes is hydride abstraction to give the corresponding cyclohexadienyl-tricarbonyliron cations. The first example of a reaction of this type was reported by Fischer and Fischer in 1960 *(158)*. They found that the reaction of unsubstituted (1,3-cyclohexadiene)tricarbonyliron (102) with triphenylmethyl tetrafluoroborate in dichloromethane at room temperature resulted in hydride abstraction to form the yellow crystalline air-stable tetrafluoroborate of the (cyclohexadienyl)tricarbonyliron cation (106). The stability of the

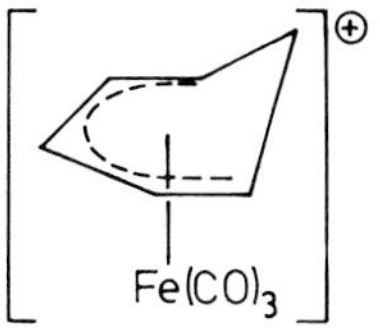

106

cation 106, in which all five carbons bonded to the iron atom are held in a mutual *cis*-conformation by the geometry of the ring, appears to be considerably greater than that of the acyclic pentadienyl-tricarbonyliron cations (*e.g.*, 41) discussed above.

The availability of numerous substituted 1,3-cyclohexadiene-tricarbonyliron complexes as discussed above has prompted the study of the corresponding substituted cyclohexadienyl-tricarbonyliron cations. The reaction of dihydromesitylene with dodecacarbonyltriiron gives a mixture of the isomeric (trimethylcyclohexadiene)tricarbonyliron complexes 107a and 107b. Triphenylmethyl tetrafluoroborate abstracts hydride from

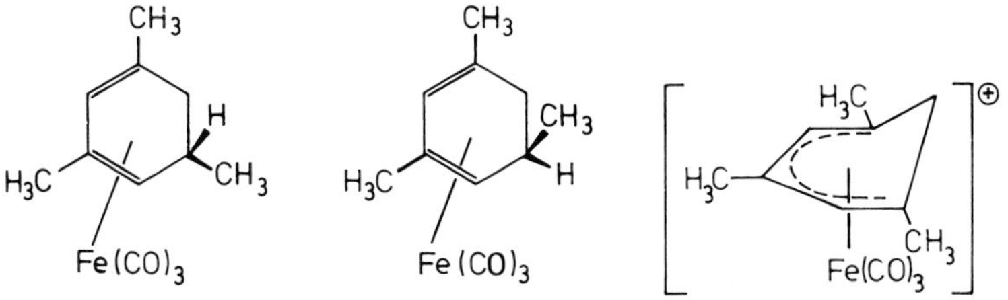

107a 107b 108

107a to give the substituted trimethylcyclohexadienyl cation 108. However, triphenylmethyl tetrafluoroborate does not react with 107b *(336)*. Reaction of the methoxycyclohexadiene complex 109 with triphenylmethyl tetrafluoroborate (eq. [3]) gives

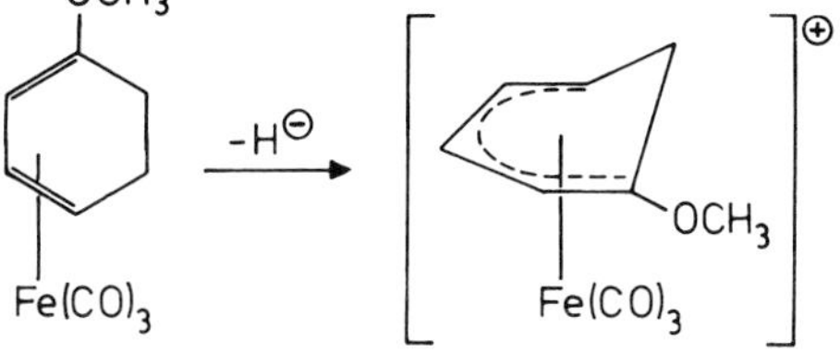

[3]

109 110

the methoxycyclohexadienyl complex 110. A similar reaction of the isomeric methoxycyclohexadiene complex 111 with triphenylmethyl tetrafluoroborate (eq. [4]) gives the methoxycyclohexadienyl complex 112. The cation 112 is stable to water

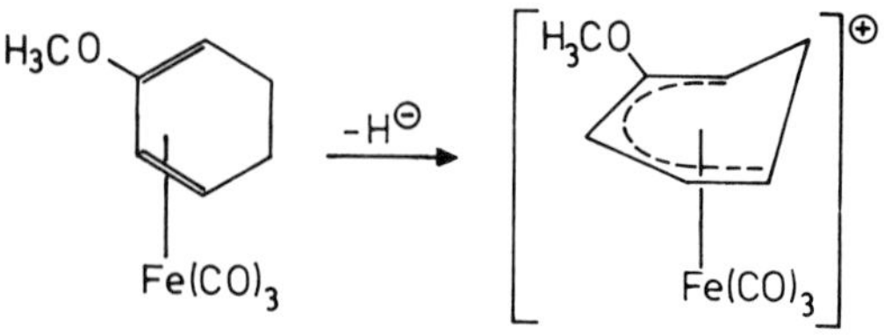

111 112 [4]

whereas the isomeric cation 110 is readily hydrolyzed by water with elimination of methanol to form (1,3-cyclohexadienone)-tricarbonyliron *(38)*. Similar studies have been made of the hydride abstraction from methoxycarbonylcyclohexadiene-tricarbonyliron derivatives *(42)*.

All of these reactions above to form cyclohexadienyl-tricarbonyliron cations involve hydride abstraction with the triphenylmethyl cation. Cyclohexadienyl-tricarbonyliron cations can also be obtained from (methoxycyclohexadiene)tricarbonyliron by methoxy abstraction with cold concentrated sulfuric acid *(39,40)*. For example, either of the methoxycyclohexadiene-tricarbonyliron derivatives 109 and 111 reacts rapidly with cold concentrated sulfuric acid to give the unsubstituted (cyclohexadienyl)tricarbonyliron cation 106. Similar reactions can be used to prepare various methylcyclohexadienyl-tricarbonyliron cations.

Preparation of the (heptamethylcyclohexadienyl)tricarbonyliron cation (113) by hydride abstraction would require a cyclohexadiene derivative not accessible by reduction of a benzenoid compound. In this exceptional case, however, the uncomplexed heptamethylcyclohexadienyl cation is available as its tetrachloroaluminate salt. Reaction of this tetrachloroaluminate salt with excess pentacarbonyliron in a sealed tube at 150°C gives some (heptamethylcyclohexadienyl)tricarbonyliron (113) tetrachloroferrate in addition to larger quantities

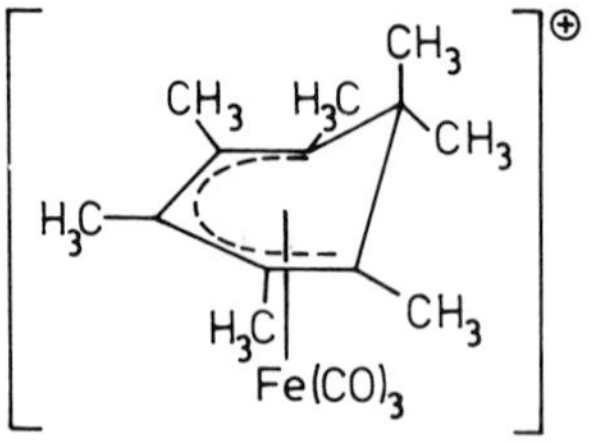

113

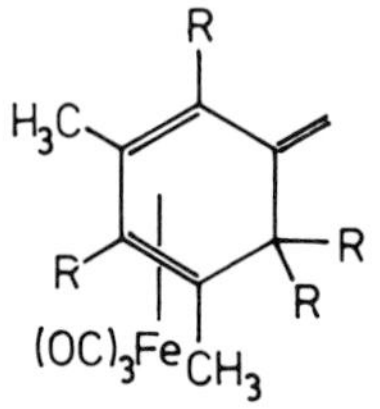

114

of the bis(hexamethylbenzene)iron(II) cation *(257)*. This tetrachloroferrate can be converted into the corresponding tetraphenylborate *(257)* and hexafluorophosphate *(363)* salts.

During the course of this work the methyl groups in positions 2 and 6 of the cations 108 and 113 were found to undergo facile deuterium exchange in D_2O *(363)*. Proton abstraction from the cations 108 and 113 using *tert*-butylamine in petroleum ether at room temperature gives the corresponding 2-methylene-cyclohexadiene-tricarbonyliron complexes 114 (R = H and CH_3, respectively) *(363)*.

Some cyclohexadienyl-tricarbonyliron cations have been prepared derived from natural products. Thus, 115 is obtained by reaction of triphenylmethyl tetrafluoroborate with the tricarbonyliron complex of either 1,3-cholestadiene or 2,4-cho-

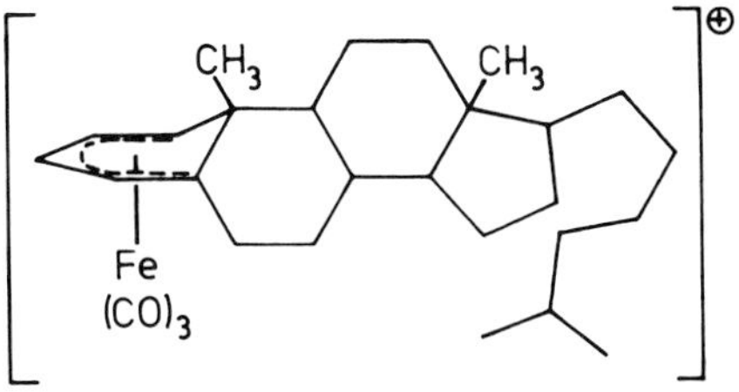

115

lestadiene *(5)*. Reaction of the tricarbonyliron derivative of thebaine with aqueous tetrafluoroboric acid gives the substituted cyclohexadienyl-tricarbonyliron salt 116 which upon

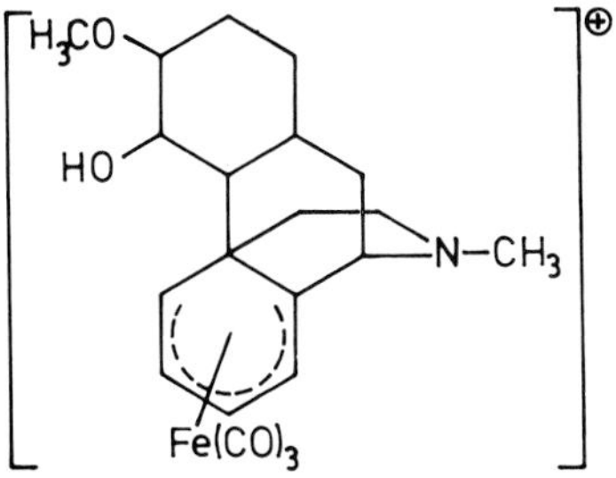

116

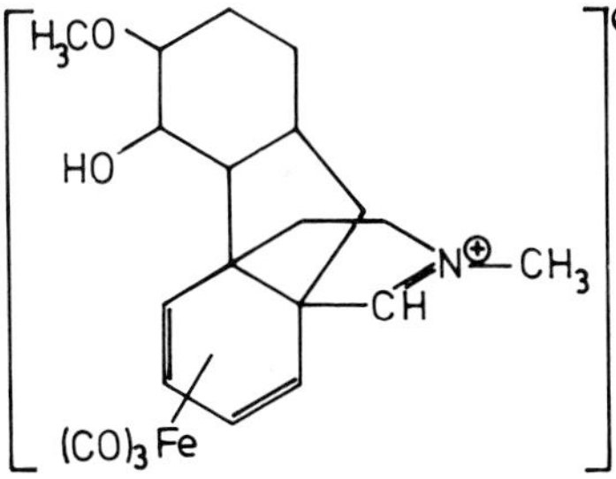

117

heating in boiling ethanol rearranges to the immonium cation 117 with ring contraction. The unusual structure of 117 has been confirmed by X-ray crystallography *(37)*.

Numerous reactions of the (cyclohexadienyl)tricarbonyliron cation (106) with nucleophiles have been investigated. Attack by the nucleophile can either occur at the metal atom with displacement of carbon monoxide to give a cyclohexadienyl-dicarbonyliron derivative or at the cyclohexadienyl ring to give a substituted cyclohexadiene-tricarbonyliron derivative.

Reaction of the (cyclohexadienyl)tricarbonyliron cation with potassium iodide in acetone proceeds through attack at

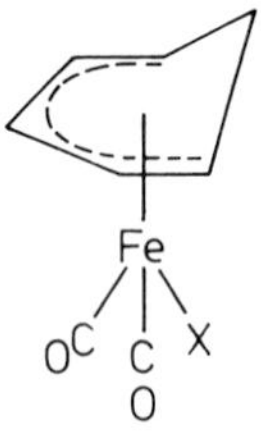

118

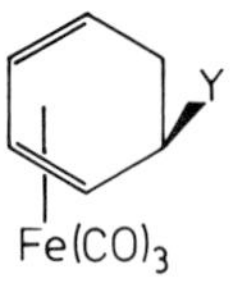

119

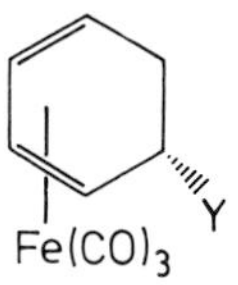

120

the metal atom to give the red-brown iodide $(\eta^5\text{-}C_6H_7)Fe(CO)_2I$ (118, X = I). This iodide reacts with potassium cyanide in methanol to give the yellow unstable cyanide $(\eta^5\text{-}C_6H_7)$-$Fe(CO)_2CN$ (118, X = CN) and with sodium amalgam in tetrahydrofuran to give the red rather unstable binuclear derivative $[(\eta^5\text{-}C_6H_7)Fe(CO)_2]_2$ *(202)*.

In most of the other reactions of the (cyclohexadienyl)-tricarbonyliron cation (106) with nucleophiles the six-membered ring is attacked to give the *exo*-substituted 1,3-cyclohexadiene-tricarbonyliron derivatives 119 (Y = group from the incoming nucleophile). Thus, the reaction of 106 with potassium cyanide in acetone gives the cyanocyclohexadiene derivative 119 (Y = CN) *(202)*. Reaction of 106 with sodium methoxide in methanol gives the *exo*-methoxycyclohexadiene derivative 119 (Y = OCH_3) *(202)* whereas prolonged heating of 106 in boiling methanol gives the isomeric *endo*-methoxycyclohexadiene derivative 120 *(208)*. These methoxycyclohexadiene-tricarbonyliron derivatives regenerate the (cyclohexadienyl)tricarbonyliron cation (106) in good yield upon treatment with triphenylmethyl tetrafluoroborate or tetrafluoroboric acid in propionic anhydride. Hydrolysis of 106 in aqueous solution in the presence of sodium hydrogen carbonate or sodium acetate gives the (hydroxycyclohexadiene)tricarbonyliron (119, Y = OH) *(38)*. However, hydrolysis with hydrated cesium fluoride in the absence of a solvent gives the binuclear ether $O[(\eta^4\text{-}C_6H_7)$-$Fe(CO)_3]_2$ (121, X = O) *(316)*. Reaction of 106 with the se-

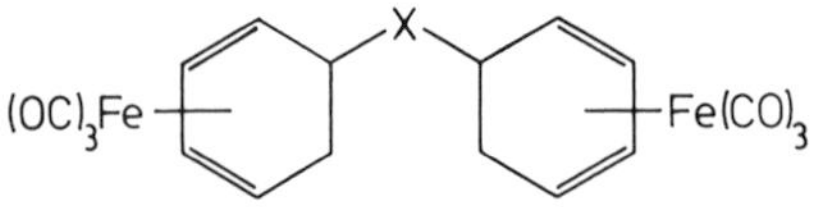

121

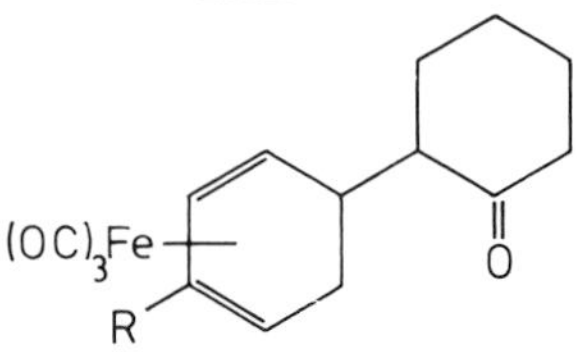

122

condary amines pyrrolidine and morpholine in aqueous solution gives the corresponding aminocyclohexadiene-tricarbonyliron derivatives 119 (Y = $N(CH_2)_4$ or $N(CH_2)_4O$, respectively) *(38)*. Compounds of the type 122 (R = H or CH_3O) can be obtained by reactions of the cyclohexadienyl-tricarbonyliron cations 106 and 112, respectively, with cyclohexanone in boiling ethanol *(41)*, or with 1-pyrrolidinylcyclohexene in either boiling acetonitrile *(41)* or aqueous acetic acid in the presence of sodium acetate *(222)*. Syntheses of analogues of 122 from cyclohexadienyl-tricarbonyliron cations and ketones or their enamines have also used acetone *(41)*, 6-methoxytetralone *(41)*, butanone *(41,222)*, methyl isopropyl ketone *(41)*, mesityl oxide *(41)*, benzaldehyde *(41)*, and cholest-4-en-3-one *(222)*. Reactions of the (cyclohexadienyl)tricarbonyliron cation 106 with diethyl malonate *(38,202)*, acetylacetone *(38)*, and dimedone *(38)* also give the corresponding cyclohexadiene-tricarbonyliron derivatives 119 (Y = $CH(CO_2-C_2H_5)_2$, $CH(CO-CH_3)_2$, and $C_8H_{11}O_2$, respectively). The kinetics of these reactions have been found *(235)* to follow the rate law $-dR/dt = k_{obs} \cdot [R] \cdot [BH]$, where R is the dienyl salt 106 and BH is acetylacetone or dimedone. Oxidation of the product obtained from 106 and acetylacetone (119, Y = $CH(CO-CH_3)_2$) with activated manganese dioxide in boiling benzene leads to a novel cyclization to form the *cis*-3a,7a-dihydrofuran-tricarbonyliron

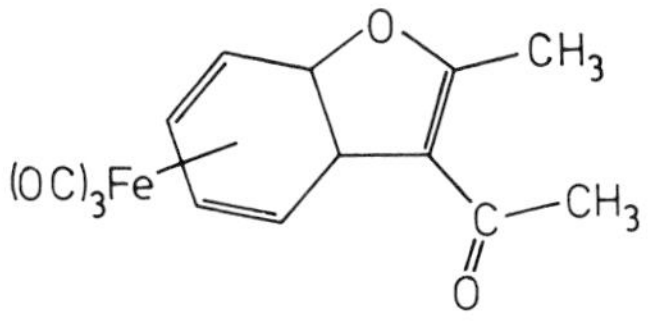

123

derivative 123. Experiments with deuterated intermediates indicate that this cyclization occurs with specific loss of the 6-*endo* proton *(43)*.

Reactions of the (cyclohexadienyl)tricarbonyliron cation (106) with various phosphorus and sulfur compounds have been investigated. Reaction of the tetrafluoroborate of 106 with triphenylphosphine results in addition to the cyclohexadienyl ring to give the triphenylphosphonium-substituted cationic cyclohexadiene-tricarbonyliron derivative 124 *(150)*. However,

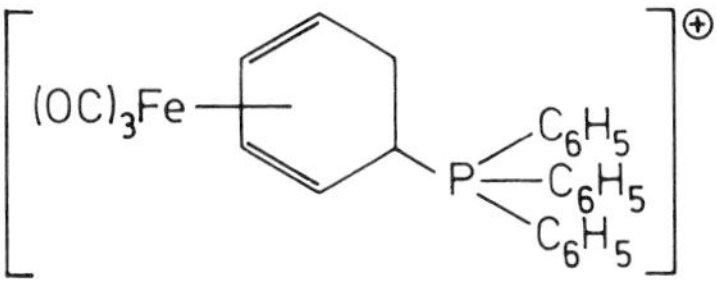

124

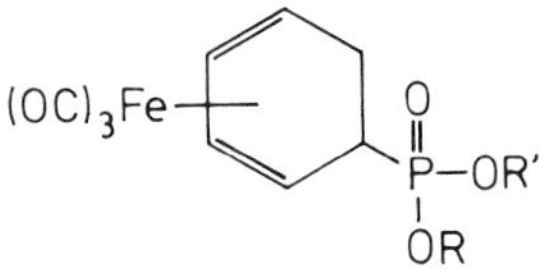

125

106 undergoes a facile Arbusov reaction with trimethyl phosphite to give the neutral phosphonate 125 (R = R' = CH_3). This phophonate is stable to mild acid treatment, unaffected by triphenylmethyl tetrafluoroborate, and converted into 125

($R = H$, $R' = CH_3$) upon treatment with cyclohexylamine followed by acid *(45)*. Reaction of 106 with aqueous hypophosphorous acid at 65°C for 2 h gives the phosphinic acid 119 ($Y = P(O)H(OH)$) which is oxidized by mercuric oxide in benzene to the corresponding phosphonic acid 125 ($R = R' = H$) *(45)*. Aqueous sodium hydrogen sulfite reacts with the (cyclohexadienyl)tricarbonyliron cation (106) to give the corresponding sulfonic acid 119 ($Y = SO_3H$), characterized as its *p*-toluidine salt. Sodium dithionite reacts with 106 to form the disulfone 126. Aqueous sodium sulfide reacts with 106 to form the binuclear sulfide $S[(\eta^4\text{-}C_6H_7)Fe(CO)_3]_2$ (121, X = S) *(45)*.

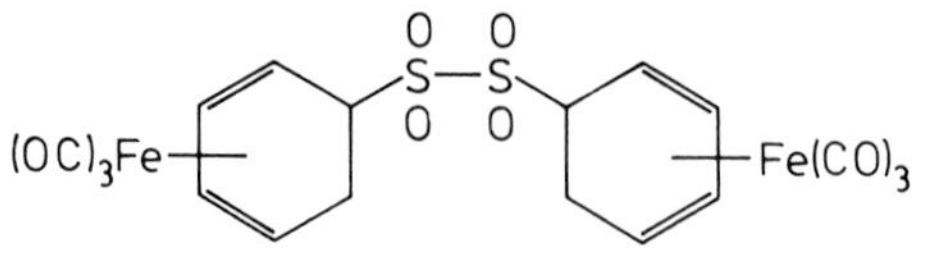

126

Several reactions of the (cyclohexadienyl)tricarbonyliron cation (106) with organometallic compounds have been investigated. Reaction of 106 with methyllithium in diethyl ether gives the 5-methylcyclohexadiene derivative 119 ($Y = CH_3$) *(38)*. However, reaction of 106 with methylmagnesium iodide results in coupling to give [bis(1,3-cyclohexadienyl)]-bis(tricarbonyliron) (127), m.p. 165 - 168°C *(38)*. An isomer

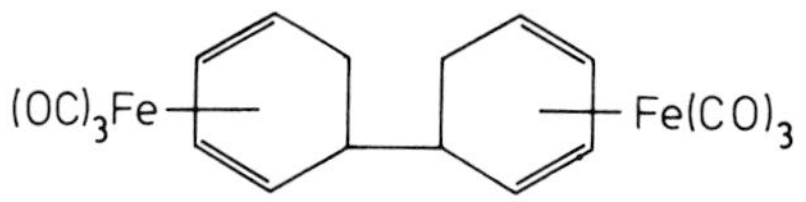

127

of 127, m.p. 120 - 122°C, can be obtained by the reductive coupling of 106 with zinc either in dioxane in the presence of sodium bromide or in tetrahydrofuran in the presence of copper *(45)*. (Cyclohexadienyl)tricarbonyliron (106) tetrafluoroborate is readily alkylated with dialkylzinc and dialkylcadmium derivatives *(46)* to give the corresponding *exo*-substituted 5-alkylcyclohexadiene derivatives $(\eta^4\text{-}R\text{-}C_6H_7)\text{-}Fe(CO)_3$ 103 (Y = allyl, isopropyl, 1-propenyl, phenyl, and benzyl) in 40 - 82 % yields. Reactions of (cyclohexadienyl)-tricarbonyliron (106) tetrafluoroborate with silicon and tin derivatives of the type $ArE(CH_3)_3$ (E = Si or Sn; Ar = 2-furyl, 2-thienyl, or XC_6H_4; X = H, *p*-CH_3O, or *p*-$N(CH_3)_2$) give the corresponding diene-substituted aromatic compounds $(\eta^4\text{-}Ar\text{-}C_6H_7)Fe(CO)_3$ *(225)*. In these reactions the aryltin compounds react more readily than their silicon analogues. In the case of the aromatic derivatives the reactivity order is $C_6H_5 < p\text{-}CH_3O\text{-}C_6H_4 < p\text{-}(CH_3)_2N\text{-}C_6H_4$ in accord with the reactivities of these systems towards other electrophilic reagents.

The (cyclohexadienyl)tricarbonyliron cation (106) has been shown to be a sufficiently reactive electrophile to substitute activated arenes including the benzenoid derivatives *(287)* 1,3,5-trimethoxybenzene and 1,3-dimethoxybenzene and the heterocycles *(236)* furan, thiophene, pyrrole, indole, *N*-methylindole, and 2-methylindole. The products are substituted cyclohexadiene-tricarbonyliron derivatives of the type 119 (Y = aryl).

Reaction of the (methoxycyclohexadienyl)tricarbonyliron cation (112) with the trialkylalkynylborates $[R_3B\text{-}C{\equiv}C\text{-}R']^-$ in tetrahydrofuran solution *(331)* gives the adducts 128 which

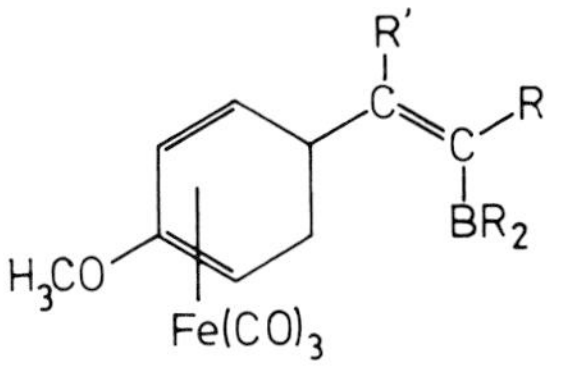

128

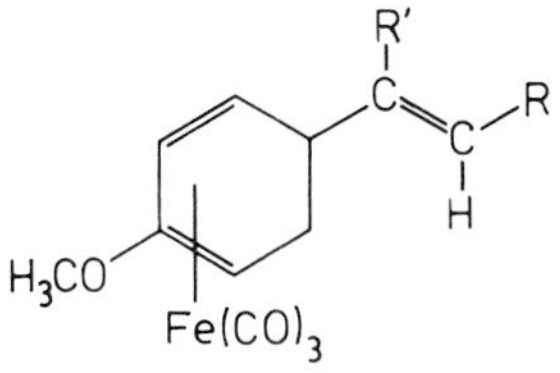

129

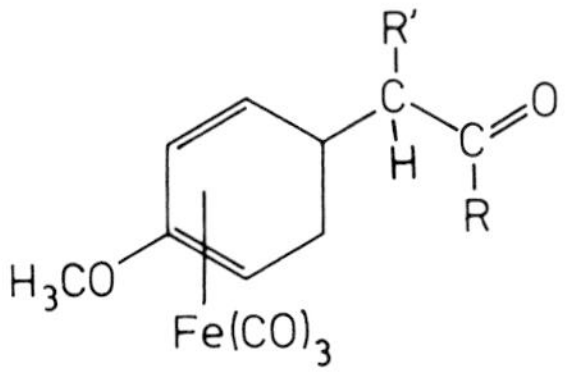

130

are hydrolyzed by isobutyric acid to 129, and which are oxidized by a limited amount of trimethylamine *N*-oxide to 130 (R = *n*-hexyl, R' = *n*-butyl; R = cyclohexyl, R' = *n*-hexyl). The addition reaction is stereo- and regiospecific but the alkyl migration step is not *(331)*.

Calculations on the (cyclohexadienyl)tricarbonyliron cation (106) using the intermediate neglect of differential overlap (INDO) scheme suggest a correlation between the bond index (or free valence) values at each dienyl carbon and the site of nucleophilic addition *(92)*.

4. Formation and Reactions of Cycloheptadienyl-Tricarbonyliron Cations

The (cycloheptadienyl)tricarbonyliron cation 131 can be easily prepared from the tricarbonyliron complexes of either 1,3-cycloheptadiene (103) or cycloheptatriene (102). Thus, hydride abstraction from (1,3-cycloheptadiene)tricarbonyliron with triphenylmethyl tetrafluoroborate in dichloromethane

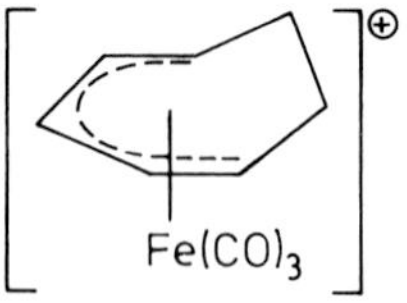

131

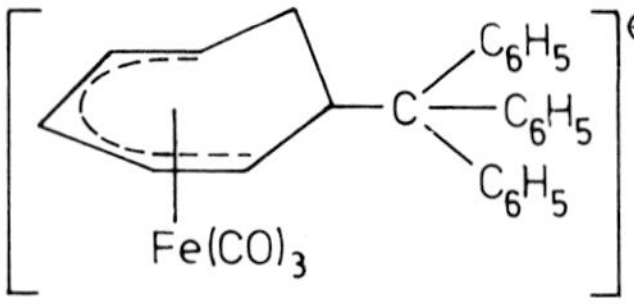

132

solution gives a nearly quantitative yield of the (cycloheptadienyl)tricarbonyliron cation (131) isolated as its tetrafluoroborate salt. The same product can also be obtained by proton addition to the uncomplexed carbon-carbon double bond in (cycloheptatriene)tricarbonyliron (102) using tetrafluoroboric acid in propionic anhydride. The latter method is generally preferred since 102 is a more readily available starting material than 103. Reaction of (cycloheptatriene)-tricarbonyliron (102) with triphenylmethyl tetrafluoroborate does not result in hydride abstraction but instead in addition of the triphenylmethyl cation to the uncomplexed carbon-carbon double bond to give the substituted cycloheptadienyl-tricarbonyliron cation 132 *(113)*.

Other electrophilic reagents can add to the uncomplexed carbon-carbon double bond in (cycloheptatriene)tricarbonyliron (102). Thus, the reaction of 102 with the acylium tetrafluoroborates [R-CO][BF_4] in dichloromethane at -78°C (R = CH_3,

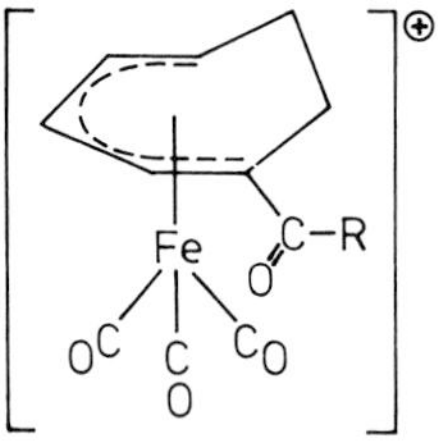

133

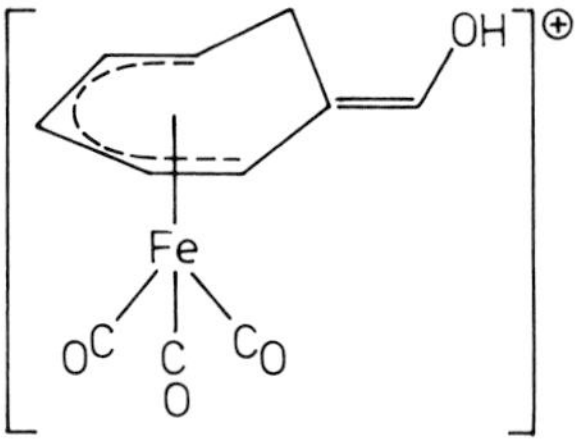

134

C_6H_5) gives the acylcycloheptadienyl cations 133 (R = CH_3, C_6H_5) *(226)*. Protonation of (formylcycloheptatriene)tricarbonyliron with hexafluorophosphoric acid gives the cation 134.

Several reactions of the (cycloheptadienyl)tricarbonyliron cation (131) with nucleophiles have been investigated. Most of this chemistry is rather similar to the corresponding chemistry of the (cyclohexadienyl)tricarbonyliron cation (106) discussed above. For example, reaction of 131 with potassium iodide results in carbon monoxide evolution to give the maroon iodide (η^5-C_7H_9)Fe(CO)$_2$I *(202)*. Reduction of this iodide with sodium amalgam in tetrahydrofuran gives the red dimer [(η^5-C_7H_9)Fe(CO)$_2$]$_2$. Reaction of (η^5-C_7H_9)Fe(CO)$_2$I with potassium cyanide gives orange crystals of the corresponding cyanide $(\eta^5\text{-}C_7H_9)Fe(CO)_2CN$ *(202)*. The cycloheptadienyl-dicarbonyliron derivatives of these types appear to be more stable than the corresponding cyclohexadienyl-dicarbonyliron derivatives.

Other reactions of the (cycloheptadienyl)tricarbonyliron cation (131) with nucleophiles involve additions to the cycloheptadienyl ring to form substituted cycloheptadiene-tricarbonyliron derivatives. Thus, reactions of the tetrafluoroborate of 131 with sodium alkoxides give products of the type 135 ($Y = OC_2H_5$, etc.), and with sodium diethylmalonate 135 ($Y = CH(CO_2C_2H_5)_2$) is obtained *(202)*. Reduction of 131 with

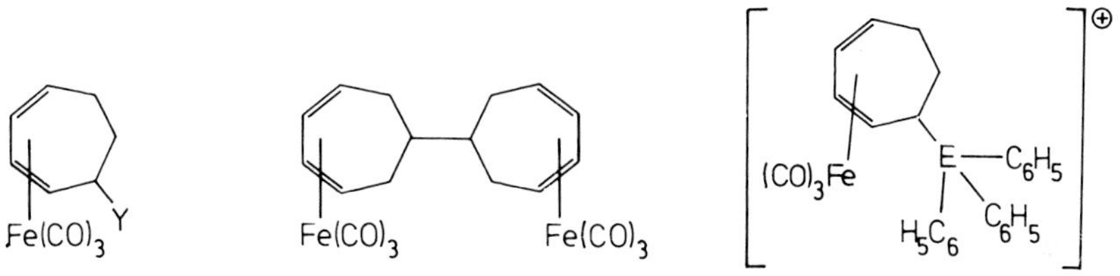

135 136 137

zinc dust in tetrahydrofuran at room temperature for 5 days results in coupling to give the yellow crystalline binuclear complex 136. Reaction of (cycloheptadienyl)tricarbonyliron tetrafluoroborate (131) with triphenylphosphine or triphenylarsine results in addition to the seven-membered ring to give the cations 137 (E = P or As) *(150)*.

Some reactions of cycloheptadienyl-tricarbonyliron cations with nucleophiles lead to products other than substituted cycloheptadiene-carbonyliron derivatives. For example, reduction of the unsubstituted $[(\eta^5\text{-}C_7H_9)Fe(CO)_3]^+$ with aqueous sodium borohydride gives a 2:1 mixture of the diene-tricarbonyliron complex 103 and the η^3-allyl-σ-alkyl complex 138 (L = CO) *(18,23)*. Carbonylation of 138 gives the

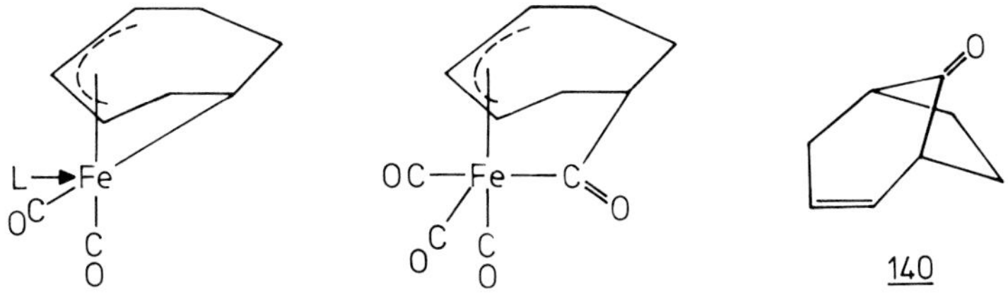

138 139 140

acyl derivative 139 at atmospheric pressure and the iron-free ketone 140 at 80°C/80 atm. Reaction of the cycloheptadienyl derivative $(\eta^5\text{-}C_7H_9)Fe(CO)_2I$ with silver hexafluorophosphate in the presence of Lewis bases gives the salts $[(\eta^5\text{-}C_7H_9)\text{-}Fe(CO)_2L][PF_6]$ ($L = P(C_6H_5)_3$, $As(C_6H_5)_3$, $Sb(C_6H_5)_3$, pyridine, $CH_3\text{-}CN$, NH_3, and $CH_2{=}CH\text{-}CN$) *(137)*. Reduction of the triphenylphosphine derivative with aqueous sodium borohydride gives the corresponding substituted η^3-allyl-σ-alkyl derivative 138 (L = $(C_6H_5)_3P$) *(137)*.

5. Tricarbonyliron Complexes of Cyclohexadienone and Cycloheptadienone Derivatives

Free cyclohexadienones are unstable with respect to tautomerism to form phenols. Nevertheless, cyclohexadienones can be stabilized as their carbonyliron complexes. For example, reaction of the (methoxycyclohexadienyl)tricarbonyliron cation, $[(\eta^5\text{-}C_6H_6\text{-}OCH_3)Fe(CO)_3]^+$ (110) with water results in the elimination of methanol to give the yellow crystalline 2,4-cyclohexadienone complex 141 *(36,38)*. The cyclohexa-

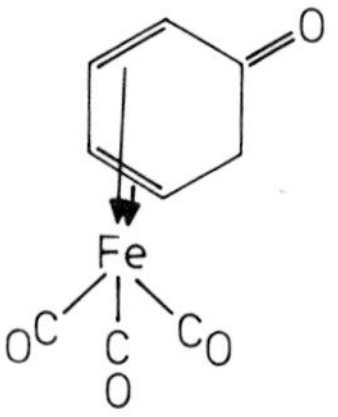

141

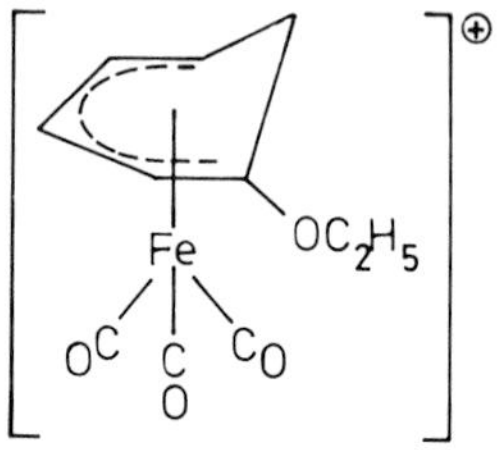

142

dienone carbonyl group exhibits a ν(CO) band at 1665 cm^{-1} in the infrared spectrum and is reduced by sodium borohydride to the corresponding alcohol. The cyclohexadienone complex 141 is a useful reagent for phenylating aromatic amines *(44)*. It thus reacts with aniline in glacial acetic acid to give diphenylamine. Alkylation of 141 with triethyloxonium tetrafluoroborate gives 142 which can be used to phenylate aliphatic amines such as cyclohexylamine *(44)*.

The 2,4-cyclohexadienone derivatives 144 (R = R" = H, R' = CH_3; R = R' = R" = CH_3) cannot tautomerize to a phenol because of the presence of two methyl groups on one carbon atom. Reactions of either cyclohexadienone 144 with penta-

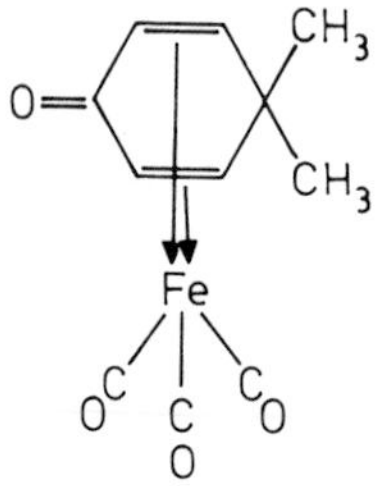

143

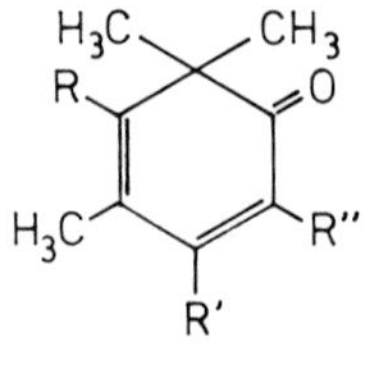

144

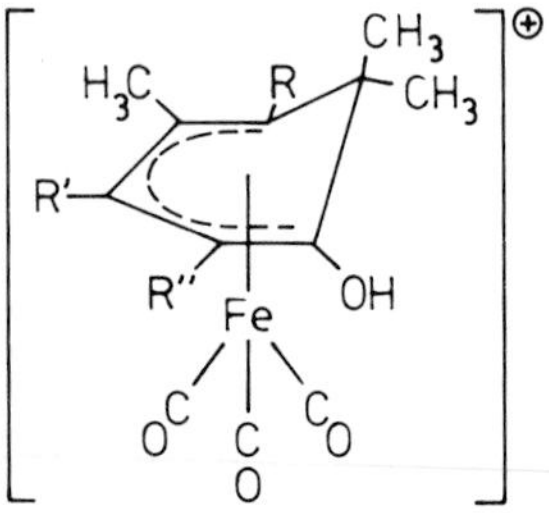

145

carbonyliron in a sealed tube at 180°C give a relatively good yield of the corresponding yellow volatile crystalline tricarbonyliron derivatives *(362)*. These compounds are protonated with strong acids to give the corresponding hydroxycyclohexadienyl-tricarbonyliron cations 145.

Tricarbonyliron complexes of cross-conjugated cyclohexadienones can also be prepared *(3)*. Thus, reaction of 4,4-dimethylcyclohexa-2,5-dienone in isooctane gives the yellow liquid tricarbonyliron complex 143. Similar tricarbonyliron complexes have been prepared analogously from cholesta-1,4-dien-3-one, androsta-1,4-dien-3,11,17-trione, santonin, and santonin oxime.

Exposure of a mixture of pentacarbonyliron and dimethylacetylene to sunlight gives large crystals of (duroquinone)-tricarbonyliron (146) *(3)*.

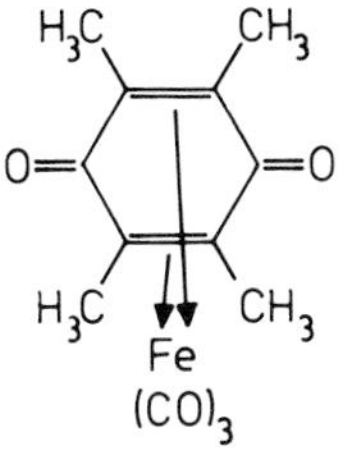

146

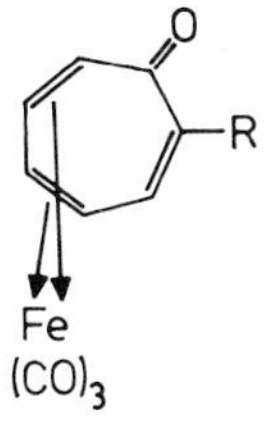

147

(Tropone)tricarbonyliron (147, R = H) can be prepared from acetylene and enneacarbonyldiiron *(391)* or from tropone and either dodecacarbonyltriiron *(242)* or enneacarbonyldiiron *(216)*. Substituted tropone-tricarbonyliron complexes (147, R = CH_3, Cl, and C_6H_5) have also been prepared from enneacarbonyldiiron and the corresponding tropone *(143)*. (2,4-Cycloheptadienone)tricarbonyliron (148) can be prepared

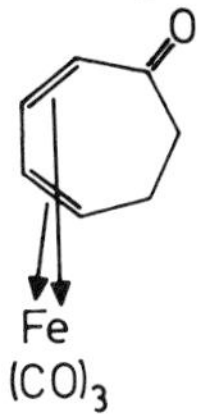

148

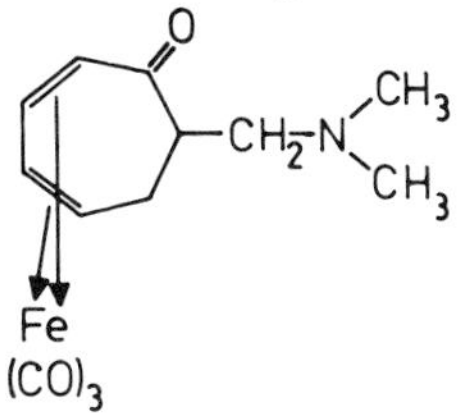

149

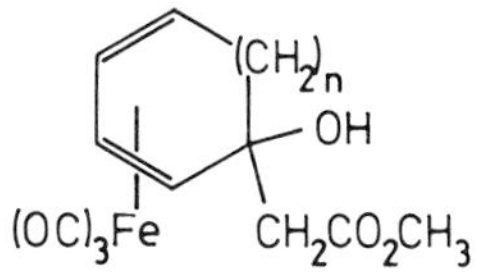

150

by hydrogenation of the uncomplexed carbon-carbon double bond in (tropone)tricarbonyliron either with molecular hydrogen in the presence of palladium over charcoal *(391)* or with triethylsilane in trifluoroacetic acid *(216)*. Reaction of the (cycloheptadienyl)tricarbonyliron cation (131) with water followed by oxidation of the resulting alcohol with chromium trioxide in pyridine also gives (2,4-cycloheptadienone)tricarbonyliron (148) *(110,275)*. A comparison of the reactions of the tricarbonyliron complexes of cyclohexadienone (141) and cycloheptadienone (148) indicates that 141 shows no enol reactivity but 148 undergoes the Mannich reaction with diethylammonium chloride and paraformaldehyde to give 149. Both 141 and 148 undergo the expected Reformatskii reaction with zinc and methyl α-bromoacetate to give the hydroxy esters 150 (n = 1 and 2, respectively) *(110,275)*.

Several other reactions of (tropone)tricarbonyliron (147, R = H) have been investigated. Treatment of 147 (R = H) with the diazoalkanes $N_2CR(R')$ (R = R' = H, CH_3) results in addition to the uncomplexed double bond to give the bicyclic cycloheptadienone-tricarbonyliron derivatives 151 (R = R' = H,

151

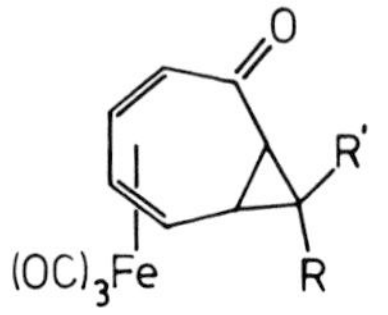

152

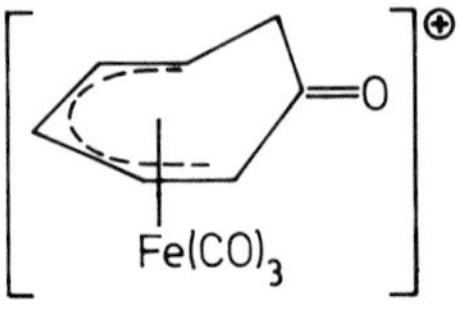

153

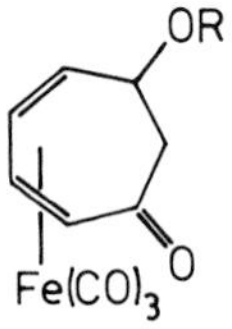

154

CH_3). Pyrolyses of the products 151 at 80 - 120°C give the corresponding homotropone complexes 152 *(169)*. The free homotropone can be liberated from these complexes by treatment with trimethylamine *N*-oxide. A similar reaction of (tropone)-tricarbonyliron (147, R = H) with the diazoalkane CH_3-CHN_2 gives a mixture of the two stereoisomers 151 (R = H, R' = CH_3 and R = CH_3, R' = H) which form the two isomeric homotropone complexes 152 (R = H, R' = CH_3 and R = CH_3, R' = H) on pyrolysis. This pyrolysis reaction is not stereospecific *(169)*. (Tropone)tricarbonyliron (147, R = H) is protonated by trifluoroacetic acid or concentrated sulfuric acid to give the ketocycloheptadienyl-tricarbonyliron cation 153 *(140,143,216)*. Hydrolysis of this cation gives the hydroxycycloheptadienone complex 154 (R = H) *(216)*. Similarly, treatment of this cation with a suspension of sodium carbonate in methanol gives the corresponding methoxycycloheptadienone complex 154 (R = CH_3) *(140)*.

Reaction of β-tropolone with dodecacarbonyltriiron in a mixture of benzene and ethanol gives yellow crystalline 155

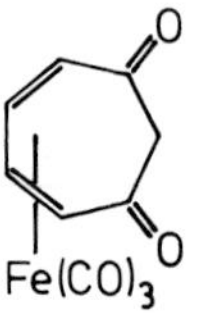

155

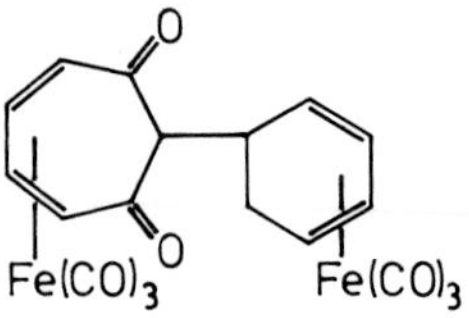

156

shown to exist entirely in its diketo form *(38)*. A novel binuclear iron complex 156, containing both cycloheptadienone-tricarbonyliron and cyclohexadiene-tricarbonyliron units, has been obtained by treatment of 155 with the tetrafluoroborate of the (cyclohexadienyl)tricarbonyliron cation (106) in aqueous alcohol *(38)*.

6. Tricarbonyliron Complexes of Benzenoid Derivatives

Tricarbonyliron complexes of benzene using two of its three double bonds have not been prepared. Furthermore, the failure of benzene to form tricarbonyliron complexes is suggested by the ability to carry out many carbonyliron reactions, even with reactive species, in benzene solution without isolating any benzene-carbonyliron derivatives. However, some arene-tricarbonyliron derivatives are known containing arenes in which one or more exocyclic double bonds are conjugated with the benzene ring. For example, styrene and various substituted styrenes form both mono- and a bis(tricarbonyliron) complexes *(378,380,381)*. Bis(tricarbonyliron) derivatives are formed by *m*- and *p*-divinylbenzenes *(291)*. Benzocyclobutadiene forms not only a mono(tricarbonyliron) complex using only the cyclobutadiene ring but also a bis-(tricarbonyliron) derivative using both its cyclobutadiene and benzene rings *(195,382)*. Some fused polycyclic hydrocarbons such as anthracene and 9-acetylanthracene form tricarbonyliron complexes *(293)*. Tricarbonyliron complexes are also formed by quinodimethane and isoindene *(229,348)*, two hydrocarbons which are very unstable in the free state. These complexes are discussed in more detail in other chapters of this book.

G. DIENE-TRICARBONYLIRON COMPLEXES OBTAINED FROM HYDROCARBONS CONTAINING EIGHT-MEMBERED RINGS

Tricarbonyliron complexes of both 1,3- and 1,5-cyclooctadiene have been prepared. However, relatively mild reaction conditions are necessary. For example, if 1,5-cyclooctadiene is heated with pentacarbonyliron at 115°C for 7 h, the 1,5-cyclooctadiene is quantitatively isomerized to 1,3-cyclooctadiene, and no tricarbonyliron complex of any cyclooctadiene is isolated *(12)*. Various early reports of the preparation of (1,5-cyclooctadiene)tricarbonyliron appear to be dubious *(219, 239,309)*. However, ultraviolet irradiation of 1,5-cyclooctadiene with pentacarbonyliron gives authenic yellow crystalline (1,5-cyclooctadiene)tricarbonyliron (157) as a yellow solid, m.p. 90°C *(250)*. The 1,5-cyclooctadiene complexes $(\eta^2\text{-}C_8H_{12})\text{-}Fe(CO)_4$ and $\mu\text{-}(1,2\text{-}\eta{:}5,6\text{-}\eta\text{-}C_8H_{12})[Fe(CO)_4]_2$ *(265)* in which one

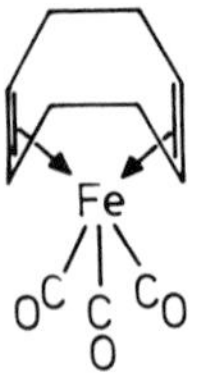

157

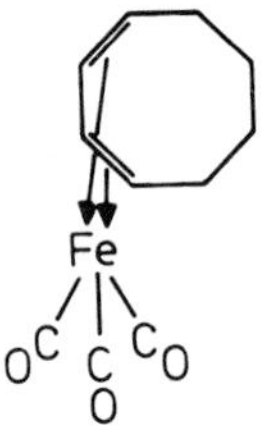

158

or both of the carbon-carbon double bonds are individually bonded to tetracarbonyliron units, also may be obtained from this reaction *(250)*. A similar ultraviolet irradiation of 1,3-cyclooctadiene with pentacarbonyliron gives (1,3-cyclooctadiene)tricarbonyliron (158) as a yellow solid, m.p. 37°C *(250)*. No intermediate 1,3-cyclooctadiene-tetracarbonyliron complexes have been obtained from this reaction. Treatment of 1,4-cyclooctadiene with pentacarbonyliron in boiling ligroin results in isomerization to 1,3-cyclooctadiene without complex formation *(370)*.

Some reactions of both of the cyclooctadiene-tricarbonyliron derivatives 157 and 158 have been investigated. Hydride abstraction from (1,5-cyclooctadiene)tricarbonyliron (157) with triphenylmethyl tetrafluoroborate gives the cation 159, isolated as its tetrafluoroborate salt *(100,124)*. Reduction of the cation 159 with sodium borohydride in water at 0°C

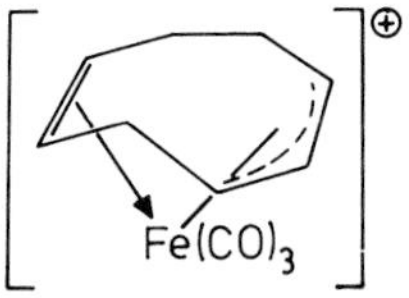

159

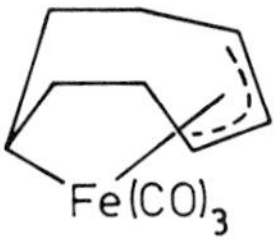

160

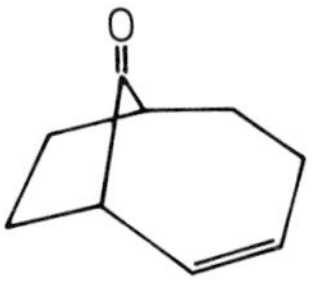

161

results in hydride addition to give a yellow crystalline $C_8H_{12}Fe(CO)_3$, m.p. 105°C, shown by its spectroscopic properties to be the η^3-allyl-σ-alkyl derivative 160 rather than a diene-tricarbonyliron derivative. Decomposition of 160 with carbon monoxide gives the bicyclic ketone 161. Reactions of the cation 159 with other nucleophiles such as aqueous sodium cyanide, the sodium derivatives of diethyl malonate, diethyl phenylmalonate, and acetylacetone, with methylmagnesium iodide in diethyl ether, allylmagnesium bromide in diethyl ether, sodium azide, and sodium methoxide give various mixtures of the substituted 1,5-cyclooctadiene-tricarbonyliron derivatives 162 and the substituted η^3-allyl-σ-alkyl derivatives 163 (Y = CN, $CH(CO_2-C_2H_5)_2$, $C(C_6H_5)(CO_2-C_2H_5)_2$, $CH(CO-CH_3)_2$, CH_3, CH_2-

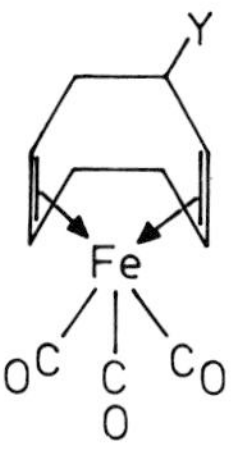

162

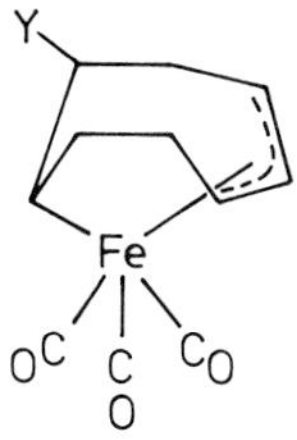

163

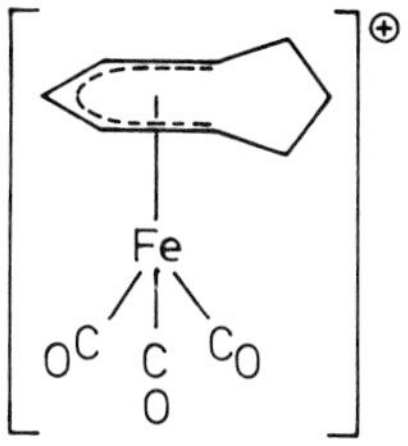

164

CH=CH_2, N_3, and OCH_3, respectively) *(100,120,354)*. The corresponding reaction of (1,3-cyclooctadiene)tricarbonyliron (158) with triphenylmethyl tetrafluoroborate in dichloromethane at room temperature gives the (1-5-η^5-cyclooctadienyl)-tricarbonyliron cation (164). Reaction of 164 with aqueous sodium borohydride leads mainly to decomposition but small yields of 158 and a yellow solid of stoichiometry $[(C_8H_{12})_2$-$CO]Fe(CO)_3$ and unknown structure can be isolated *(124)*. Reactions of either $[(C_8H_{11})Fe(CO)_3]^+$ cation 159 or 164 with iodide result in attack at a carbonyl group rather than at the eight-membered ring to give the corresponding neutral iodide $C_8H_{11}Fe(CO)_2I$ *(354)*.

Some reactions of (1,3-cyclooctadiene)tricarbonyliron (158) with fluoroolefins have been investigated *(52,185)*. Ultraviolet irradiation of 158 with tetrafluoroethylene gives a 30 % yield of the yellow crystalline tricarbonyliron derivative 165 (X = F) and a 9 % yield of colourless needles

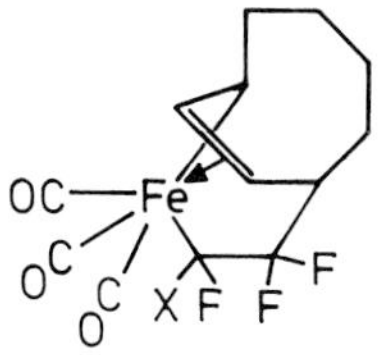

165

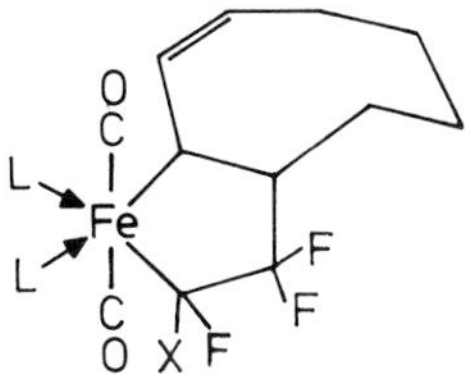

166

of the tetracarbonyliron derivative 166 (L = CO, X = F). Similar products 165 (X = CF_3) and 166 (L = CO, X = CF_3) are obtained from the reactions of (1,3-cyclooctadiene)tricarbonyliron (158) with hexafluoropropene. The studies with hexafluoropropene indicate that the initial product from this reaction is of the type 165. Mild pyrolysis of 165 in the presence of a ligand (which can be carbon monoxide from decomposition reactions) results in rearrangement to a product of the type 166 with rupture of the iron-olefin bond. Thus, heating 165 (X = CF_3) with trimethyl phosphite in boiling hexane gives a derivative of 166 (L = $P(OCH_3)_3$, X = CF_3) in which two of the carbonyl groups are replaced by trimethyl phosphite ligands *(185)*.

The reaction of 1,3,5-cyclooctatriene with carbonyliron complexes under mild conditions gives the monocyclic derivative 167 whereas the corresponding reaction of 1,3,5-cyclooctatriene with carbonyliron complexes under vigorous conditions gives the bicyclic derivative 168 *(59,289,290,297)*. A similar rearrangement of the free hydrocarbon was first reported in 1952 *(98)*. The monocyclic tricarbonyliron complex 167 is best prepared by ultraviolet irradiation of 1,3,5-cyclooctatriene with pentacarbonyliron at room temperature. The bicyclic tricarbonyliron complex 168 can be conveniently prepared by treatment of 1,3,5-cyclooctatriene with pentacar-

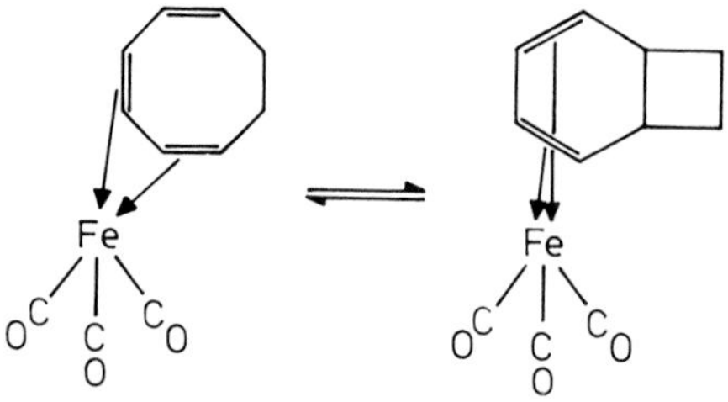

167 168 [5]

bonyliron in a high boiling solvent such as octane or ethylcyclohexane. A kinetic study of reaction [5] indicated a first order rate constant of $7 \times 10^{-5}\ s^{-1}$ corresponding to $\Delta F^{\ddagger}$ = 29.3 kcal/mole. At equilibrium less than 1 % of 167 remains indicating that K>100 and ΔG<-3.4 kcal/mole. In the corresponding equilibria in the free hydrocarbon system *(98)* the 1,3,5-cyclooctatriene is favoured over the bicyclo[4.2.0]octa-2,4-diene with K = 0.18 and ΔG = +1.1 kcal/mole. The unsubstituted bicyclic derivative 168 is also produced by the thermal isomerization of (bicyclo[5.1.0]octadiene)tricarbonyliron *(17,22,61)*. A bis(trimethylsilyl) derivative of 168 has been obtained from reactions of bis(5,8-trimethylsilyl)cycloocta-1,3,6-triene with various carbonyliron complexes *(121)*.

The effects of additional ring fusion on the equilibria in eq. [5] have been determined by studying the corresponding

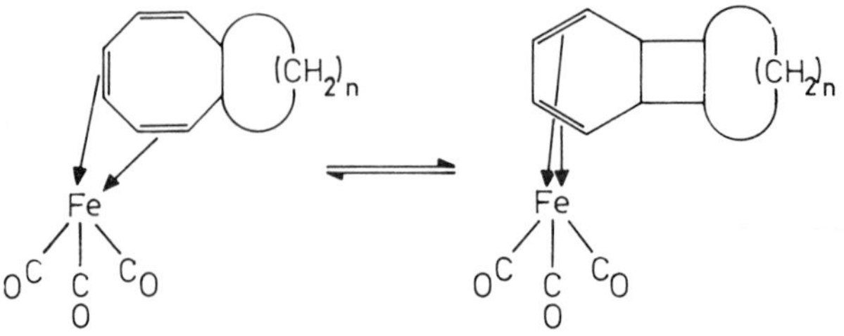

169 170 [6]

equilibria in eq. [6] for the cases n = 1, 2, 3, and 4 for the free hydrocarbons and their tricarbonyliron complexes *(101,103,355)*. For n = 1 the tricyclic tricarbonyliron derivative 170 (n = 1) can be isolated by reaction of the bicyclic hydrocarbon with (benzalacetone)tricarbonyliron in benzene at 55°C *(355)*. For n = 2 the free hydrocarbon exists exclusively in the bicyclic form. Reaction of this bicyclic hydrocarbon with enneacarbonyldiiron gives a complex mixture of exotic carbonyliron complexes *(102,104,106,107)* including the tricyclic derivative 170 (n = 2) but not the bicyclic derivative 169 (n = 2). Thus, in this case the complexed tricarbonyliron group completely changes the equilibrium of eq. [6]. The yield of 170 (n = 2) can be greatly improved (up to 82 %) by using (benzalacetone)tricarbonyliron in benzene at 65°C for the reaction with the bicyclic hydrocarbon *(355)*. For the free hydrocarbons with n = 3 the tricyclic isomer is favoured over the bicyclic isomer with K = 32 and ΔG° = -2 kcal/mole *(101, 103)*. However, treatment of these hydrocarbons (n = 3) with enneacarbonyldiiron in hexane at 50°C gives only the tricyclic isomer 170 (n = 3). The crystal structure of this iron complex indicates that both ring fusions are *cis*, and the five- and six-membered rings have an *anti*-relationship relative to the central four-membered ring *(105)*. For the free hydrocarbons with n = 4 both the bicyclic and tricyclic isomers have similar stabilities *(101, 103)*. Reaction of a mixture of the bicyclic and tricyclic hydrocarbons (n = 4) with enneacarbonyldiiron in hexane at 50°C gives the tricyclic derivative 170 (n = 4) as the only mononuclear tricarbonyliron complex *(103)*. The crystal structure of this tricarbonyliron complex indicates that the two six-membered rings are *trans* to each other with respect to the shared four-membered ring *(108)*.

Some reactions of (bicyclo[4.2.0]octadiene)tricarbonyliron (168) with various fluoroolefins have been investigated *(52,185)*. Thus, ultraviolet irradiations of 168 with tetrafluoroethylene *(52)*, trifluoroethylene *(185)*, and hexafluoropropene *(185)* give the η^3-allylic derivatives 171 (X = Y = F; X = H, Y = F; X = F, Y = CF_3).

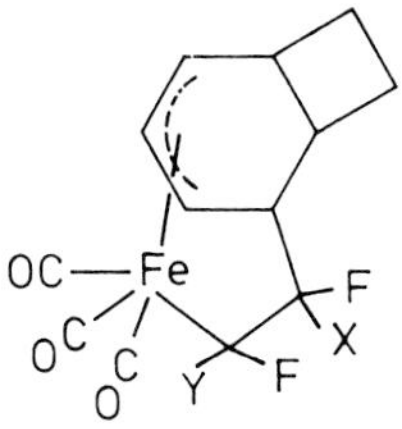

171

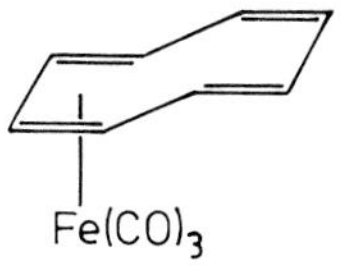

172

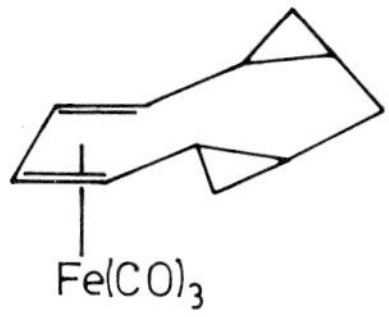

173

Cyclooctatetraene reacts with carbonyliron reagents under various conditions to give several products including both $(1\text{-}4\text{-}\eta^4\text{-}C_8H_8)Fe(CO)_3$ (172) and $\mu\text{-}(1\text{-}4\text{-}\eta^4\text{:}5\text{-}8\text{-}\eta^4\text{-}C_8H_8)\text{-}[Fe(CO)_3]_2$ with *trans* stereochemistry *(128,237,289,334,339)*. These complexes are described in detail elsewhere in this book. Reaction of (cyclooctatetraene)tricarbonyliron (172) with methylene iodide in boiling diethyl ether in the presence of a zinc-copper couple (Simmons-Smith reagent) results in addition of three CH_2 units followed by rearrangement to give a yellow crystalline adduct $(C_{11}H_{14})Fe(CO)_3$ formulated as the substituted cyclononadiene complex 173 *(341)*.

H. REACTIONS OF NORBORNADIENE AND OTHER BICYCLIC DIENES WITH CARBONYLIRON COMPLEXES

Examples have been given earlier in this chapter where non-conjugated dienes rearrange to conjugated dienes upon treatment with carbonyliron complexes. Norbornadiene is an example of a non-conjugated diene where such rearrangement is forbidden because of the instability of double bonds in bridgehead positions in bridged polycyclic systems. Furthermore, the two double bonds of the norbornadiene system are in an excellent position to coordinate to a single metal atom. Reaction of norbornadiene with pentacarbonyliron thus occurs readily to give the corresponding (norbornadiene)tricarbonyliron (174) as a volatile yellow liquid *(47,48,188,322)*. In

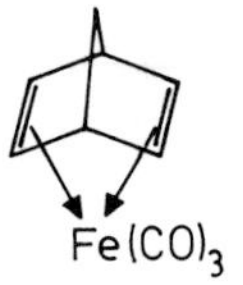

174

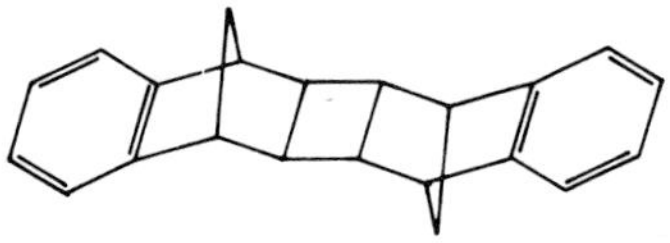

175

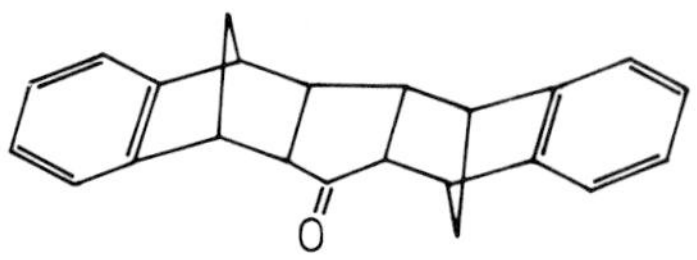

176

addition four different norbornadiene dimers and five ketones may be isolated from this reaction mixture *(47)*. The ketones arise from oligomerization of the norbornadiene with carbon monoxide insertion. A similar reaction of benzonorbornadiene with various carbonyliron reagents results in both dimerization to give 175 and dimerization with carbonyl insertion to give the polycyclic cyclopentanone 176 without the formation of a tricarbonyliron complex analogous to 174 *(278,288)*. The benzonorbornadiene double bonds within the benzenoid ring thus appear to be too unreactive to form carbonyliron complexes.

The assignments of the normal modes in the infrared and Raman spectra of (norbornadiene)tricarbonyliron (174) have been reported *(411)*.

Some 7-substituted norbornadiene-tricarbonyliron complexes have also been prepared *(215)*. Reaction of 7-benzoyl-oxynorbornadiene with pentacarbonyliron in boiling *n*-butyl ether gives the corresponding tricarbonyliron complex 177

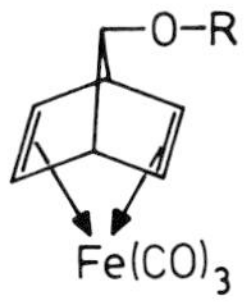

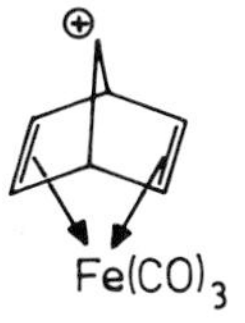

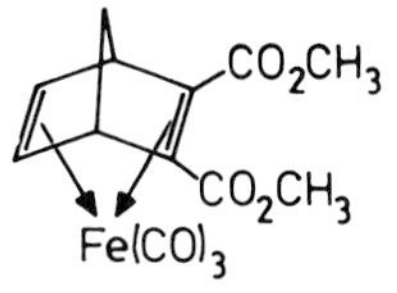

177 178 179 180

(R = CO-C_6H_5). Other 7-substituted norbornadiene-tricarbonyl-iron derivatives which have been similarly prepared include 177 (R = H, CO-CH_3, C(CH_3)$_3$, and SO_2-C_6H_4-*p*-CH_3). Hydrolysis of the (7-*p*-toluenesulfonyloxynorbornadiene)tricarbonyliron (177, R = SO_2-C_6H_4-*p*-CH_3) in 80 % aqueous acetone proceeds at least 10^6 times more slowly than that of the uncomplexed 7-*p*-toluenesulfonyloxynorbornadiene *(215)*. This suggests that the (7-norbornadienyl)tricarbonyliron cation (178) is much less stable than the uncomplexed 7-norbornadienyl cation. This is a very unusual example of a carbonium ion which is destabilized by attachment of a transition metal atom. Oxidation of (7-norbornadienol)tricarbonyliron (177, R = H) with the pyridine-(sulfur trioxide) complex in dimethyl sulfoxide containing triethylamine gives the orange crystalline tricarbonyliron complex 179 of the unstable 7-norbornadienone *(269,271)*. Photolysis of various norbornadiene complexes regenerates the free norbornadiene ligand *(270)*. In the case of the norbornadienone derivative 179 the unstable free norbornadienone was detected by its Diels-Alder reaction with 1,3-diphenylisobenzofuran to give 9,10-diphenylanthracene after dehydration *(270)*.

The dark yellow crystalline complex 180 has been obtained by heating the Diels-Alder adduct of cyclopentadiene and dimethyl acetylenedicarboxylate with either enneacarbonyl-diiron or dodecacarbonyltriiron in boiling benzene *(318)*.

Several reactions of (norbornadiene)tricarbonyliron (174) have been investigated. Treatment of 174 with fluoro-

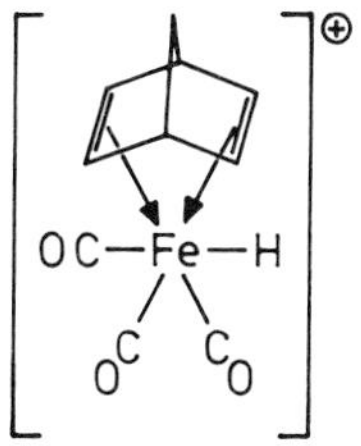

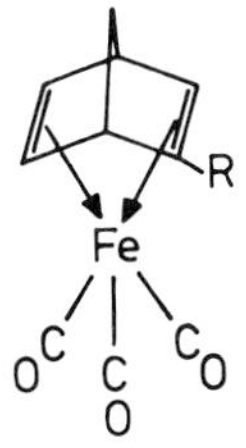

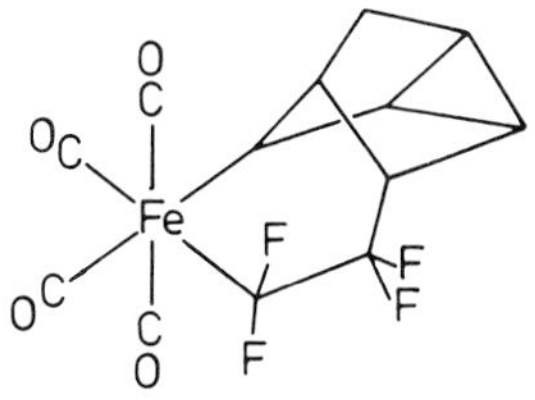

181 182 183

sulfonic acid in liquid sulfur dioxide results in protonation at the iron atom to give the hydride cation 181. This cation was identified by its ^{1}H-NMR spectrum including a resonance at τ 17.3 assigned to the proton bonded to iron *(151)*. Hydrolysis of the solutions of protonated 174 leads to recovery of unchanged 174. Treatment of (norbornadiene)tricarbonyliron (174) with equivalent amounts of dichloromethyl methyl ether and titanium tetrachloride in dichloromethane for 30 min at 0°C and subsequent hydrolysis results in formylation to give the aldehyde 182 (R = CO-H) as an orange liquid *(178)*. Reduction of this aldehyde 182 (R = CO-H) with 1:4 $LiAlH_4/AlCl_3$ gives the 2-methylnorbornadiene complex 182 (R = CH_3). The same methyl derivative is also obtained by 1:4 $LiAlH_4/AlCl_3$ reduction of the methyl ester 182 (R = CO_2-CH_3). Attempted Friedel-Crafts acylation of (norbornadiene)-tricarbonyliron (174) with acetyl chloride and various Lewis acids results in decomposition *(178)*. Ultraviolet irradiation of 174 with tetrafluoroethylene gives only a white crystalline tetracarbonyliron derivative formulated as the nortricyclyl derivative 183 on the basis of its ^{1}H- and ^{19}F-NMR spectra *(52)*.

Several tricarbonyliron complexes of bicyclo[2.2.2]-octatriene derivatives have been prepared. Treatment of bis-(trifluoromethyl)tetramethylbicyclo[2.2.2]octatriene with pentacarbonyliron in boiling ethylcyclohexane gives mainly the corresponding tricarbonyliron complex 184a in which the

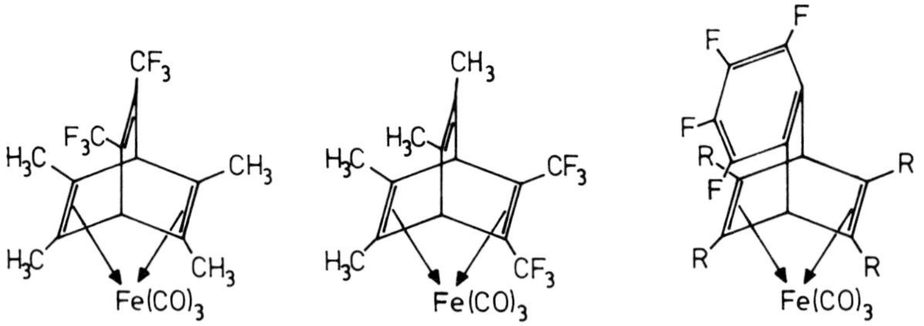

184a 184b 185

methylated carbon-carbon double bonds are coordinated to the iron atom *(240)*. However, in addition a small amount (10 - 25 % of the total) of the other isomer (184b) is also observed in the ^{1}H- and ^{19}F-NMR spectra *(240)*. An extensive series of tetrafluorobenzobicyclo[2.2.2]octatriene-tricarbonyliron derivatives of the type 185 (R = H or CH_3) have been prepared containing anywhere between zero and six methyl substituents *(346,373)*. The structure of 185 (R = H) has been confirmed by X-ray crystallography *(218)*.

Some carbonyliron derivatives containing the bicyclo-

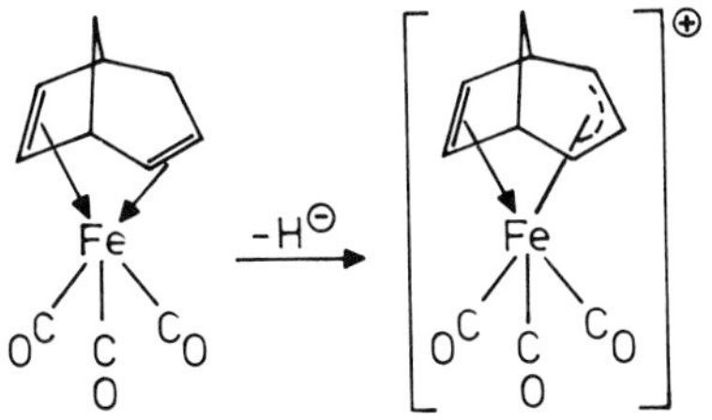

186 187 [7]

[3.2.1]octadiene system have been prepared *(295)*. Reaction of bicyclo[3.2.1]octadiene with pentacarbonyliron in methylcyclohexane solution gives the corresponding pale yellow crystalline tricarbonyliron complex 186. Reaction of 186 with triphenylmethyl tetrafluoroborate (eq. [7]) results in hydride abstraction to give the stable yellow tetrafluoroborate salt of the (bicyclo[3.2.1]octadienyl)tricarbonyliron cation 187. The structure of this cation has been confirmed by X-ray crystallography *(295)*.

Some bicyclo[3.2.2]nonadiene carbonyliron derivatives have been investigated *(139)*. Treatment of bicyclo[3.2.2]-nona-2,6,8-trien-4-ol (188) with enneacarbonyldiiron gives the three isomeric tricarbonyliron complexes 189, 190, and 191 in 8 %, 10 %, and 52 % yields, respectively, as indicated in equation [8]. Protonation of 190 with tetrafluoro-

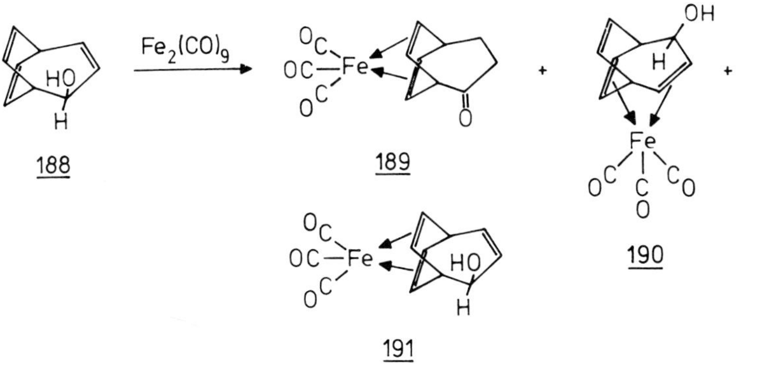

188 189 190 191 [8]

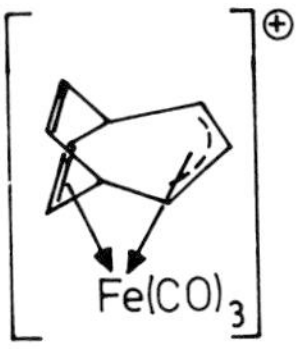

192

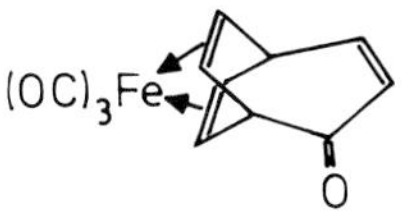

193

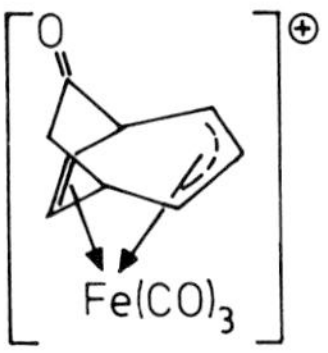

194

boric acid in acetic anhydride gives the yellow tetrafluoroborate of the (bicyclo[3.2.2]nonadienyl)tricarbonyliron cation 192. A similar protonation of 191 also gives 192 apparently through a 1,2 carbon shift (Wagner-Meerwein process). Reaction of bicyclo[3.2.2]nonatrienone with $Fe_2(CO)_9$ gives the tricarbonyliron complex 193 which forms the cation 194 upon protonation with concentrated sulfuric acid.

Some similar reactions have also been studied with 6,7-benzobicyclo[3.2.2]nonadiene derivatives *(142)*. The tricarbonyliron complex 195 can be prepared in quantitative yield by treatment of the corresponding ketone with enneacarbonyldiiron in benzene solution. Complex 196 was similarly made from the free benzobicyclononadienyl methyl ether and

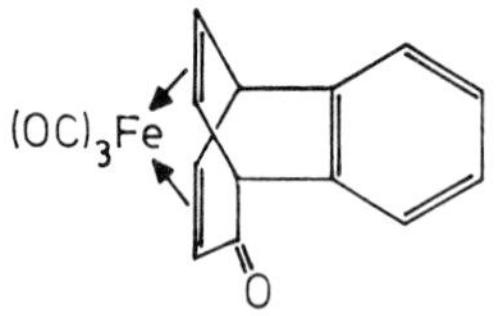

195

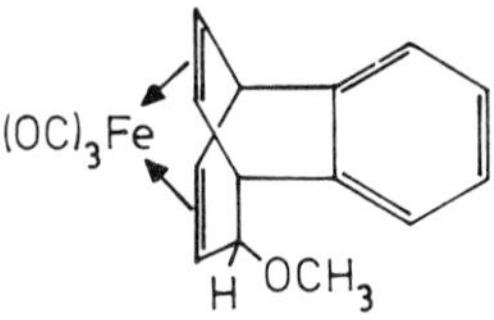

196

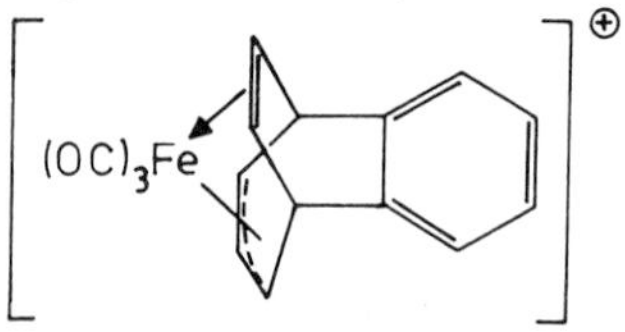

197

$Fe_2(CO)_9$. Protonation of 196 with tetrafluoroboric acid in acetic anhydride resulted in elimination of methanol to give the cation 197 isolated as its stable tetrafluoroborate salt.

Reaction of tricyclo[4.3.1.$0.^{1,6}$]deca-2,4-diene with dodecacarbonyltriiron in boiling benzene for 6 h gives a tricarbonyliron complex *(49)* shown by X-ray crystallography *(30)* to have the expected structure 198. The yellow-orange crystal-

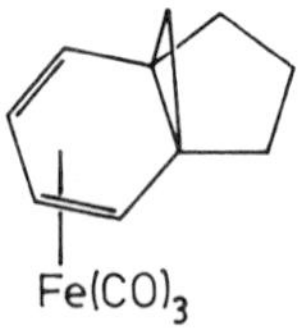

198

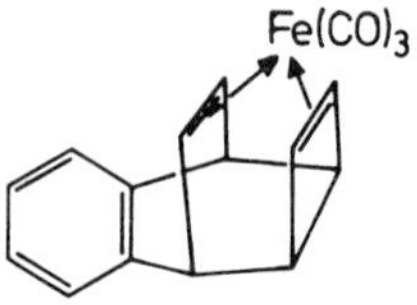

199

line complex 199 has been obtained from *anti*-7,8-benzotricyclo[4.2.2.$0^{2,5}$]deca-3,7,9-triene by reaction with enneacarbonyldiiron in hexane/benzene at 55°C *(298)*.

Several tricarbonyliron complexes of [4.4.3]propellane ethers and imides have been prepared *(7)* including the two stereoisomeric propellatetraene complexes 200a and 200b. By reactions of either 200a or 200b with only one equivalent of Ce(IV) ($Ce(NH_4)_2(NO_3)_6$) one of the two complexed tricarbonyliron groups can be selectively removed to give ~90 % yields

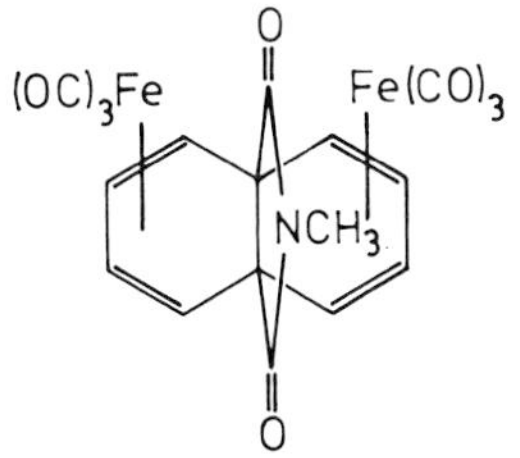

200a

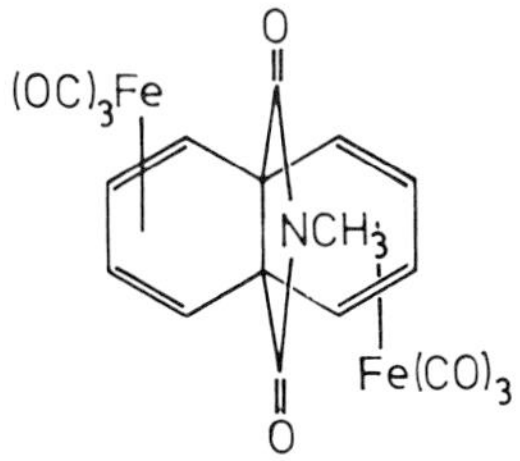

200b

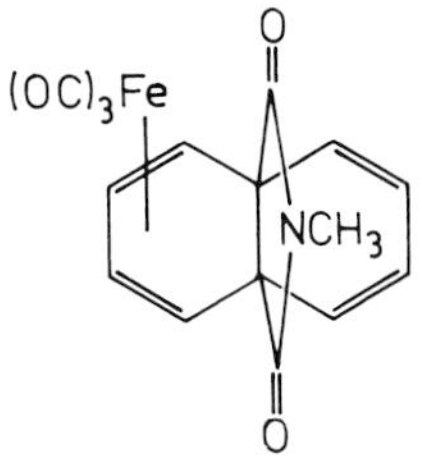

201a

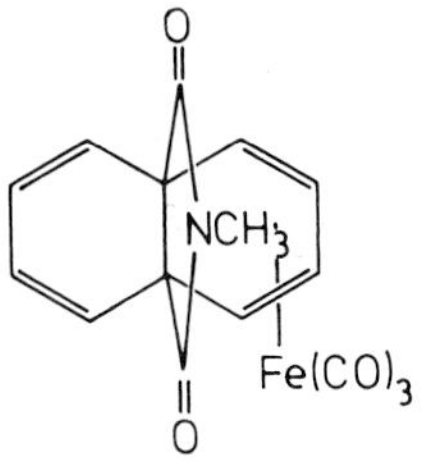

201b

of the stereoisomers 201a and 201b, respectively. Similar mononuclear tricarbonyliron complexes have been prepared from propellatriene and propelladiene *N*-methylimides and ethers *(7)*. Protection of the 1,3-cyclohexadiene ring in the azapropellane 202 and its stereoisomer by complexation with a

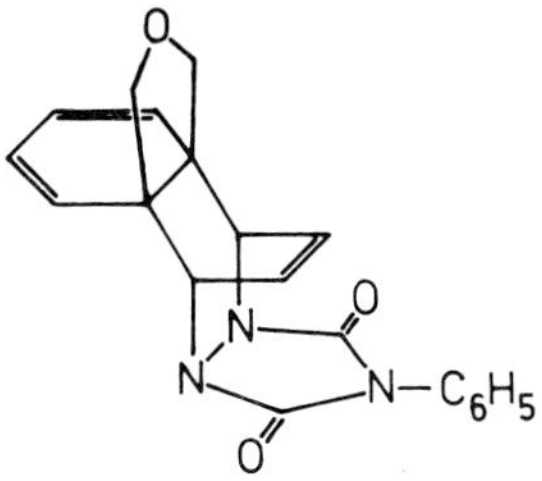

202

tricarbonyliron group has been employed prior to hydrogenation of the isolated carbon-carbon double bond *(258)*.

I. DIENE-TRICARBONYLIRON COMPLEXES FROM CYCLOPROPANE AND CYCLOBUTANE DERIVATIVES

Sometimes diene-tricarbonyliron complexes can be prepared by reactions involving the opening of cyclopropane rings. An example of such a reaction has already been given in which

1,3,3-trimethylcyclopropene reacts with dodecacarbonyltriiron to give the vinylketene derivative 69 *(241)*. The reaction of enneacarbonyldiiron with dimethyl methylenecyclopropane-2,3-

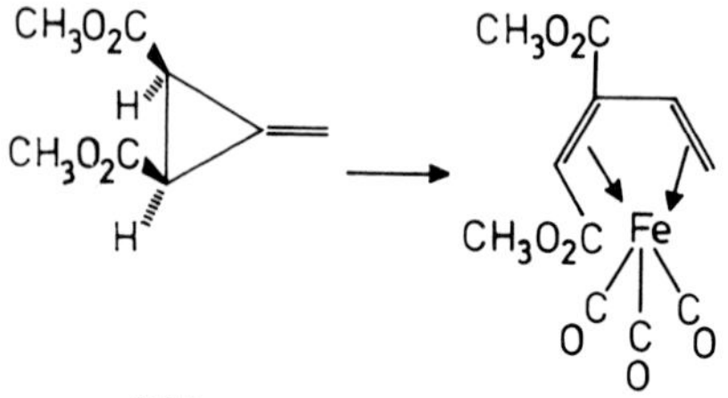

203a 204a

[9]

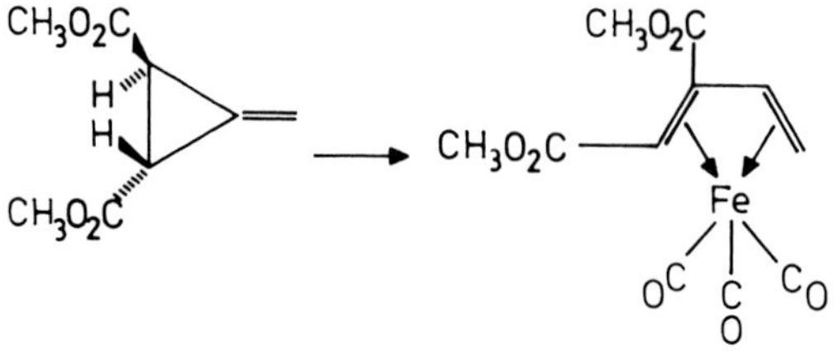

203b 204b

dicarboxylate (eq. [9]) gives ultimately a diene-tricarbonyliron derivative through an olefin-tetracarbonyliron complex *(395)*. This reaction is stereospecific since the *cis*-isomer 203a of the methylenecyclopropane derivative gives exclusively the *anti*-complex 204a, and the *trans*-isomer 203b of the methylenecyclopropane derivative gives exclusively the *syn*-complex 204b. The major product from the reaction of enneacarbonyldiiron with either *cis*- or *trans*-2,3-bis(hydroxymethyl)methylenecyclopropane is the 1,4-diene-tricarbonyliron

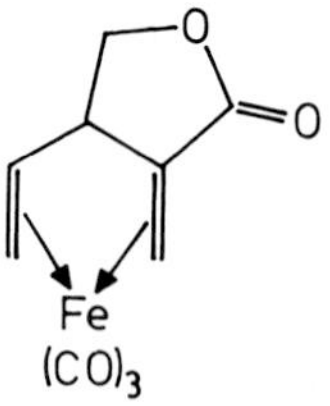

205

complex 205 *(184)*. This complex appears to be the only example of a 1,4-diene-tricarbonyliron complex in which the coordinated carbon-carbon double bonds are not incorporated into a ring system.

Some reactions of cyclopropylethylene derivatives with carbonyliron reagents give diene-tricarbonyliron complexes *(349,350)*. Treatment of the 1-aryl-1-cyclopropylethylenes

206 (R = H, Cl, OCH_3) with pentacarbonyliron in boiling di-

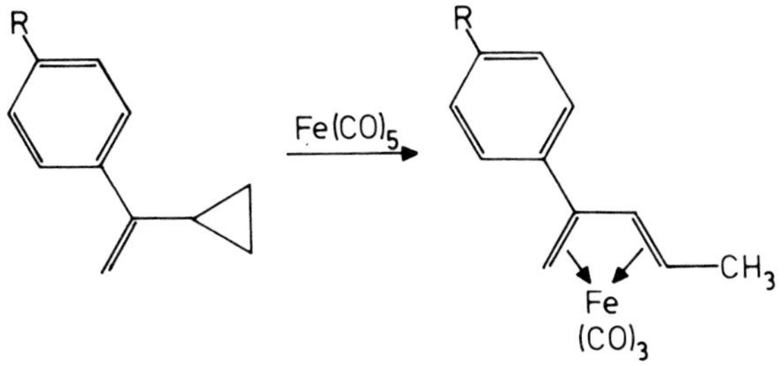

206 207 [10]

n-butyl ether gives the corresponding 2-arylpenta-1,3-diene-tricarbonyliron complexes 207 according to equation [10]. However, ultraviolet irradiation of 206 with $Fe(CO)_5$ or $Fe_2(CO)_9$ at room temperature gives the cyclohexenones 208 ap-

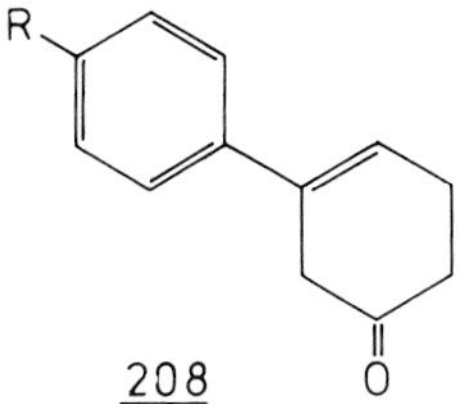

208

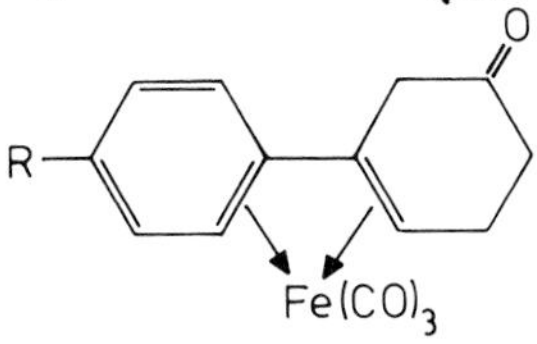

209

parently through unstable tricarbonyliron derivatives, possibly 209 *(379)*. Reaction of 1,1-dicyclopropylethylene (210) with pentacarbonyliron in boiling ethylcyclohexane results in successive opening of the two cyclopropane rings to give first the yellow liquid 211 and then the yellow crystalline 212

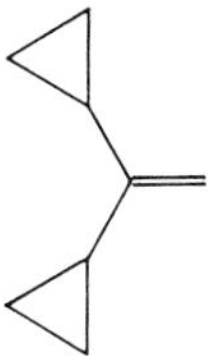

210

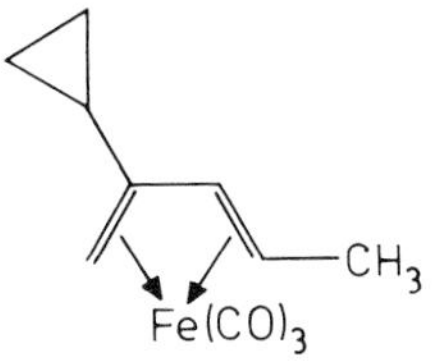

211

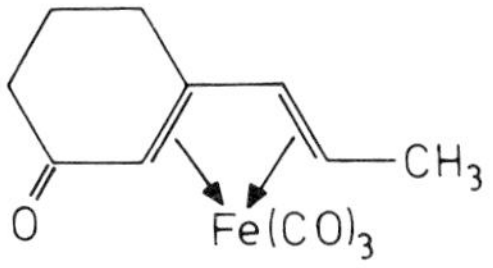

212

(34,350). The opening of the second ring is accompanied by carbonyl insertion *(34)*. However, further studies *(351)* show

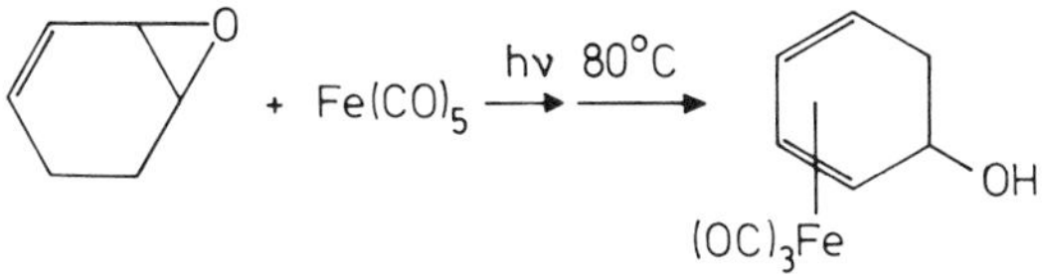

213 214 [11]

that 211 is not an intermediate in the formation of 212. Ultraviolet irradiation of the bicyclic vinyloxirane 213 with pentacarbonyliron (eq. [11]) followed by heating in boiling benzene results ultimately in the formation of (*endo*-hydroxycyclohexadiene)tricarbonyliron (214) *(19)*.

Several reactions of polycyclic hydrocarbons containing vinylcyclopropane units with iron carbonyls give various diene-tricarbonyliron derivatives. Reaction of homosemibullvalene (215) with enneacarbonyldiiron in boiling hexane gives a complex mixture of products *(20)*. One of these products is the yellow liquid (bicyclo[4.2.1]nonadiene)tricarbonyliron (216), formed by opening the cyclopropane ring. Reaction of bullvalene (217) with enneacarbonyldiiron gives the bis(tri-

215

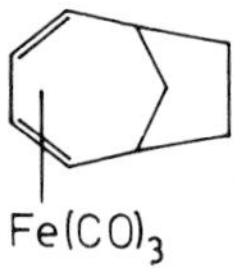

216

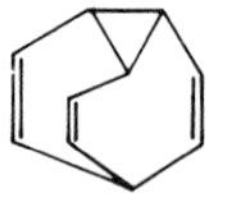

217

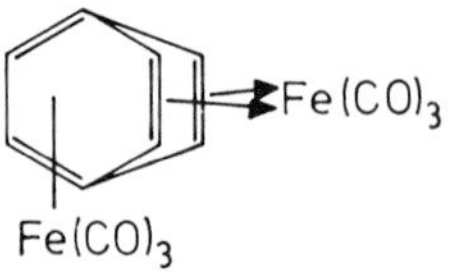

218

carbonyliron) complex 218, which rearranges upon heating to (9,10-dihydronaphthalene)bis(tricarbonyliron) *(359,361)*. Another product obtained from bullvalene and carbonyliron reagents, 219 *(14)*, undergoes isomerization upon heating to

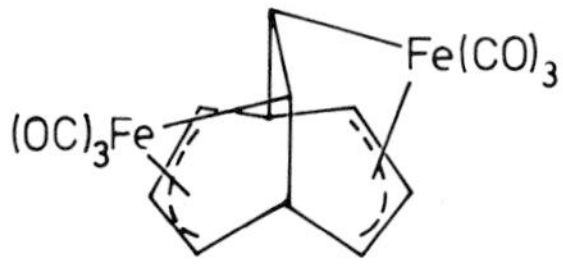

219

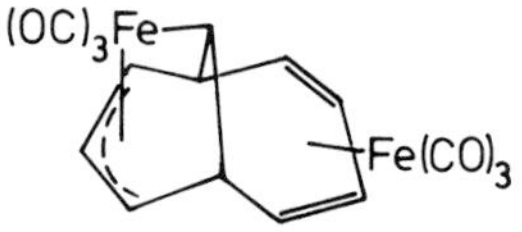

220

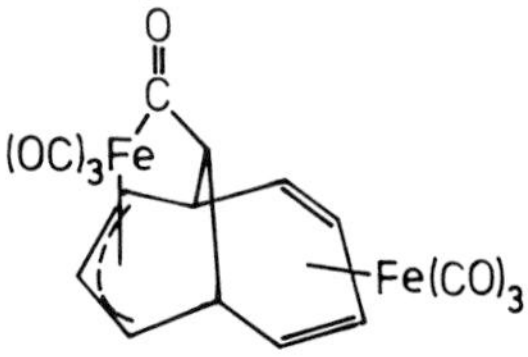

221

120°C to give the complex 220 containing a diene-tricarbonyliron unit *(15)*. Compound 220 reacts with carbon monoxide at room temperature and atmospheric pressure with carbonyl insertion into the iron-carbon σ-bond to give 221 *(383)*. The structure of 221 has been determined by X-ray crystallography *(383)*. For further details on the rather complex reaction between bullvalene and carbonyliron complexes see the chapter

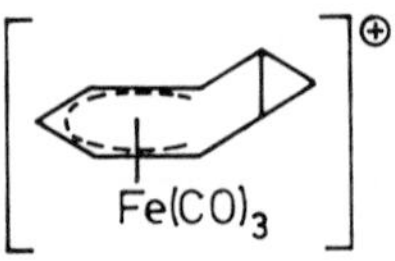

222

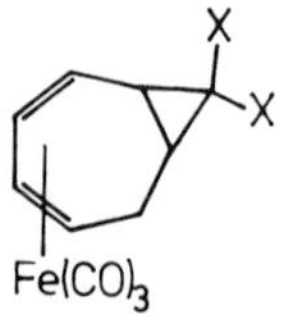

223

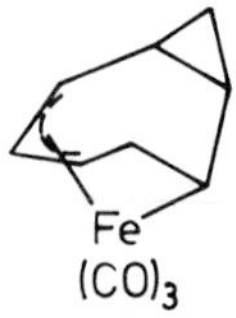

224

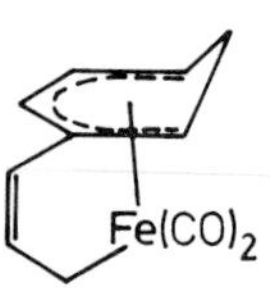

225

on η^3-allylic derivatives.

Protonation of (cyclooctatetraene)tricarbonyliron (172) readily leads to the bicyclic cation 222. Sodium borohydride reduction of 222 gives the diene-tricarbonyliron derivative 223 (X = H) containing a fused three-membered ring *(17,21)* along with the isomeric η^3-allyl-σ-acyl derivative 224. Pyrolysis of 223 (X = H) at 120°C results in rearrangement to give (bicyclo[4.2.0]octadiene)tricarbonyliron (168). A similar pyrolysis of 224 at 60°C gives (1,3,5-cyclooctatriene)-tricarbonyliron (167). The *gem*-dibromo-derivative 223 (X = Br) has been prepared by treatment of the ligand with dodecacarbonyltriiron in boiling heptane *(224)*. This compound has been characterized by X-ray crystallography *(366)*. Reaction of 223 (X = Br) with methyllithium at -65°C gives a bright yellow-orange crystalline product shown by X-ray crystallography to have the novel structure 225 formed by an unusual rearrangement *(223,224)*.

The spirane 226 reacts with enneacarbonyldiiron to give the pentacarbonyldiiron complex $[C_5H_4C_6H_6CO]Fe_2(CO)_5$ *(301)*. X-ray crystallography on this complex *(91)* indicates the structure 227.

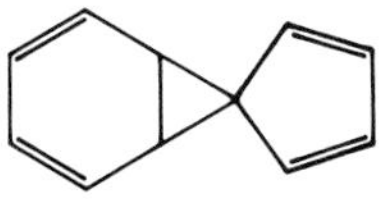

226

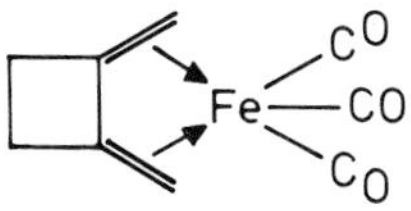

228

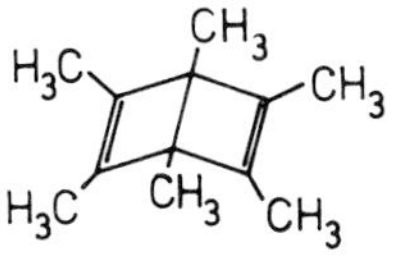

229

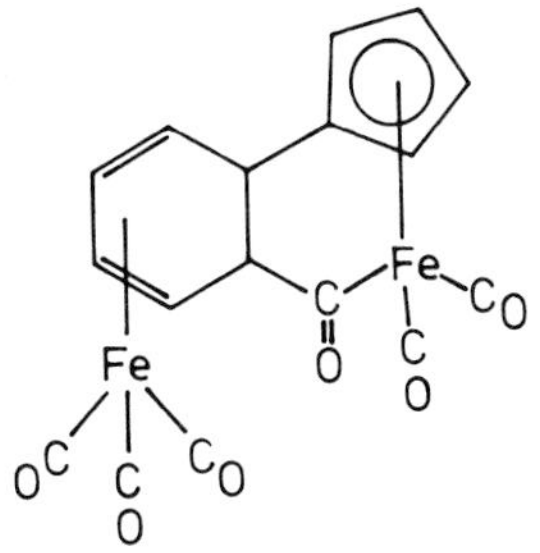

227

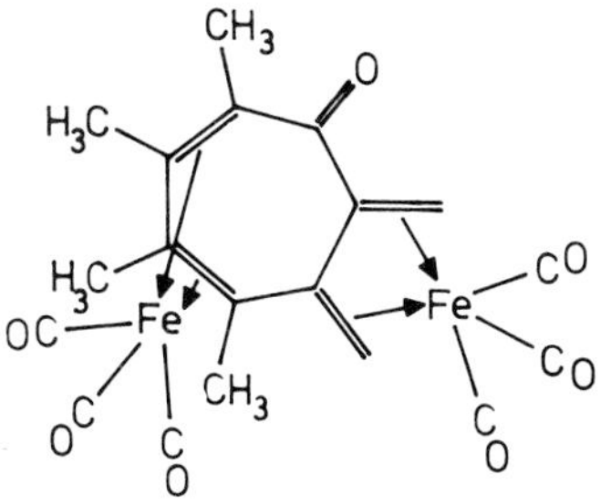

230

Several diene-tricarbonyliron complexes have been prepared from cyclobutane derivatives. 1,2-Dimethylenecyclobutane reacts with carbonyliron reagents to give the corresponding yellow-orange liquid diene-tricarbonyliron derivative $C_6H_8Fe(CO)_3$ 228 *(173,248)* in addition to yellow crystalline $C_{12}H_{16}Fe(CO)_2$ formulated as a substituted bis(η^3-allyl)dicarbonyliron derivative *(248)*. Hexamethylbicyclo[2.2.0]hexadiene (229) reacts with $Fe_2(CO)_9$ to give a low yield of a brick red solid formulated as the dimethylenecycloheptadienone derivative 230 *(164)*.

J. *DIENE-TRICARBONYLIRON COMPLEXES FROM HETEROCYCLIC SYSTEMS*

Several examples are known where heterocyclic dienes can bond to tricarbonyliron groups through the carbon-carbon double bonds in the heterocyclic ring. Phosphole derivatives (231, E = P) may form carbonyliron derivatives in which either the pair of carbon-carbon double bonds or the lone pair on the phosphorus atom are bonded to carbonyliron residues *(56,57)*. Reaction of pentaphenylphosphole (231, E = P, R = C_6H_5) with dodecacarbonyltriiron in boiling isooctane gives the yellow

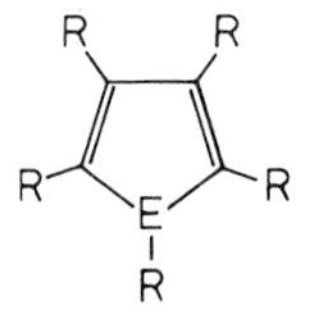

231

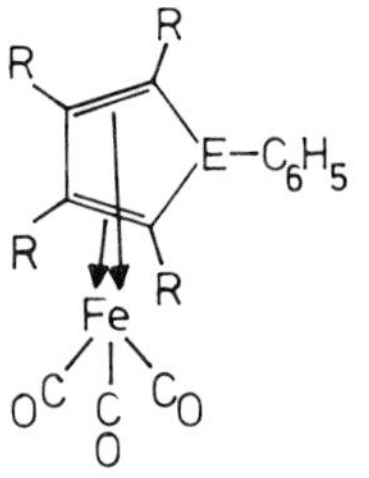

232

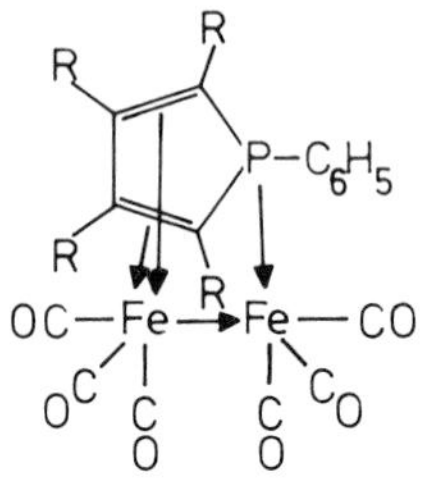

233

crystalline diene-tricarbonyliron derivative 232 (E = P) in addition to the hexacarbonyldiiron derivative 233 in which both the lone pair on phosphorus and the pair of carbon-carbon double bonds are bonded to tricarbonyliron units, and the tetracarbonyliron derivative $[(C_6H_5)_4C_4PC_6H_5]Fe(CO)_4$ involving only phosphorus coordination. A similar reaction of pentaphenylphosphole (231, E = P, R = C_6H_5) with pentacarbonyliron under more vigorous conditions (140°C) gives only the tetracarbonyliron derivative in nearly quantitative yield. However, the reaction of pentaphenylarsole (231, E = As, R = C_6H_5) with pentacarbonyliron under similar vigorous conditions gives only the diene-tricarbonyliron derivative 232 (E = As) with no evidence for the formation of derivatives containing arsenic-iron bonds.

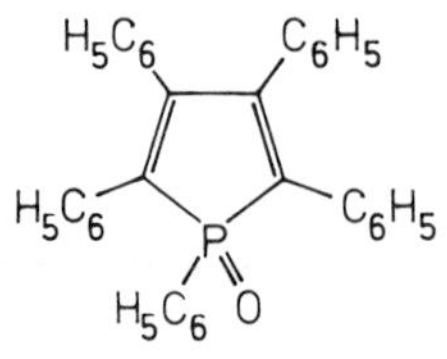

234

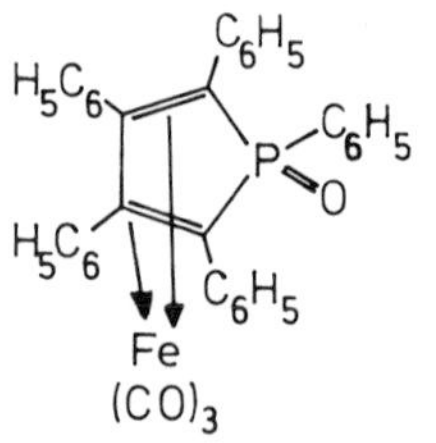

235

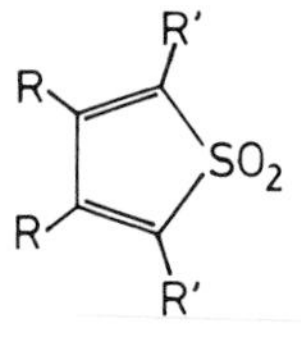

236

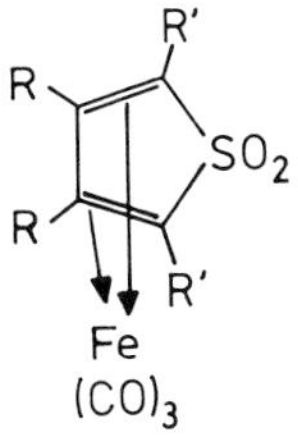

237

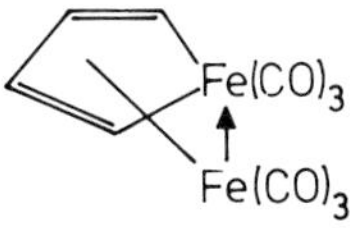

238

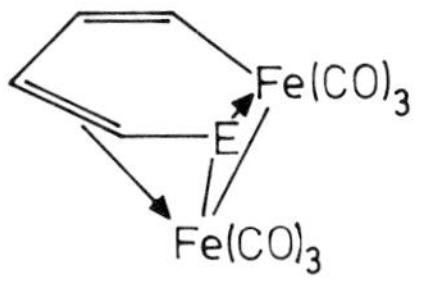

239

Oxidation of pentaphenylphosphole to the corresponding oxide (234) removes the lone pair from the phosphorus atom leaving only the phosphole diene system for possible coordination to a metal. Thus, reaction of pentaphenylphosphole oxide (234) with pentacarbonyliron gives only the diene-tricarbonyliron complex 235 in nearly quantitative yield *(56, 57)*. Similarly, ultraviolet irradiation or heating *(387)* the thiophene *S,S*-dioxides 236 (R = R' = C_6H_5; R = H, R' = CH_3) with pentacarbonyliron in benzene solution gives the corresponding diene-tricarbonyliron derivatives 237. The unsubstituted (thiophene dioxide)tricarbonyliron (237, R = R' = H) was prepared by ultraviolet irradiation of 3,4-dibromotetrahydrothiophene dioxide with excess pentacarbonyliron at room temperature since thiophene dioxide itself (236, R = R' = H) is too unstable to use as a reagent *(85)*. The reaction of thiophene with dodecacarbonyltriiron does not give the corresponding tricarbonyliron complex *(233)*. Instead, desulfurization occurs to give (tricarbonylferrole)tricarbonyliron (238) identical to a compound obtained from acetylene and carbonyliron complexes. Reaction of tellurophene with dodecacarbonyltriiron in boiling benzene gives in addition to 238 and $Fe_3(CO)_9Te_2$ the red air-sensitive sublimable C_4H_4Te-$Fe_2(CO)_6$ suggested to have structure 239 (E = Te) *(228)*. A sulfur analogue (239, E = S) has been mentioned in a footnote but has not been described in detail *(16)*.

Tricarbonyliron units have also been complexed with the carbon-carbon double bonds in a pyrone ring. Thus, ultraviolet irradiation of α-pyrone with pentacarbonyliron in diethyl ether gives the yellow crystalline tricarbonyliron complex

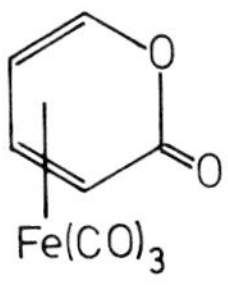

240

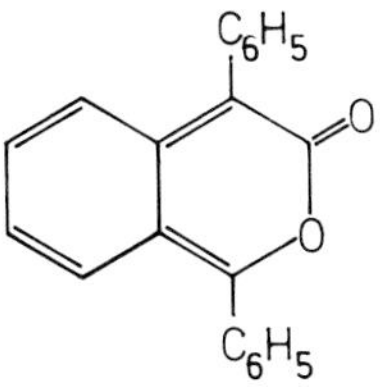

241

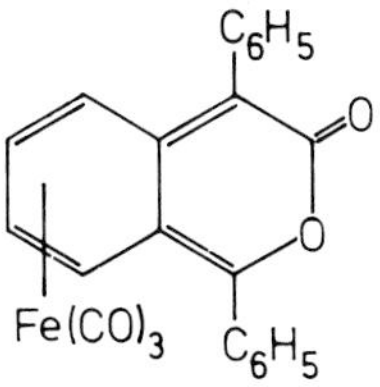

242

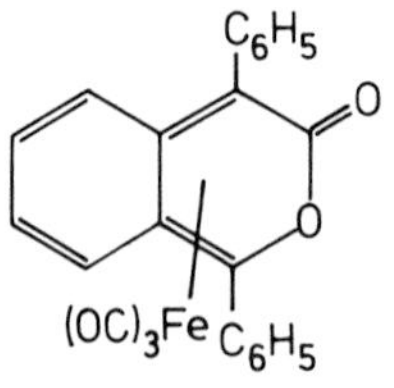

243

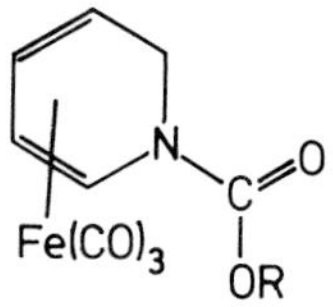

244

240 in addition to (cyclobutadiene)tricarbonyliron formed by decarboxylation *(347)*. Reaction of the bicyclic α-pyrone 241 with dodecacarbonyltriiron in boiling toluene gives two isomeric tricarbonyliron derivatives *(211)*. One is formulated as 242 in which the tricarbonyliron group is bonded to the 1,3-cyclohexadiene ring. The second is formulated as 243 in which the tricarbonyliron group is bonded to the α-pyrone ring. These isomers 242 and 243 readily undergo interconversion upon heating in xylene solution with isomer 242 predominating in the resulting equilibrium mixture.

Some tricarbonyliron derivatives of dihydropyridines have been prepared. Reactions of the *N*-alkoxycarbonyl-1,2- and -1,4-dihydropyridines with enneacarbonyldiiron in benzene at room temperature give the corresponding *N*-alkoxycarbonyl-1,2-dihydropyridine-tricarbonyliron complexes 244 (R = CH_3, C_2H_5) *(6)*. In these compounds the dihydropyridine ring is coordinated to the iron through its double bonds rather than through its nitrogen atom.

Some other reactions of carbonyliron reagents with unsaturated nitrogen heterocycles have been investigated. The reaction of *N*-phenyl-2-oxa-3-azabicyclo[2.2.2]oct-5-ene (245)

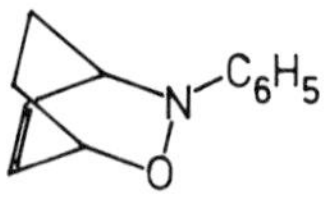

245

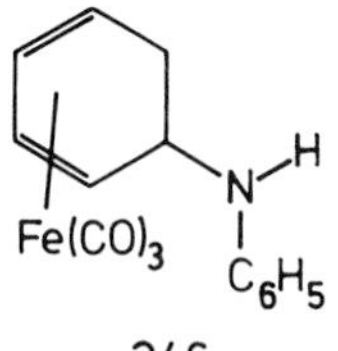

246

with enneacarbonyldiiron in benzene at 40°C for 20 min gives (*endo*-5-anilinocyclohexa-1,3-diene)tricarbonyliron (246) as the major product in addition to several iron-free derivatives *(29)*. The chemistry of tricarbonyliron complexes of azepine *(163)*, and some of its substitution products such as *N*-carbethoxyazepine *(163,177,330)* and diazepines *(123,368)* is discussed in detail elsewhere in this book.

K. DIENE-TRICARBONYLIRON COMPLEXES FROM VINYLSILICON AND VINYLBORON DERIVATIVES

Reactions of 1,4-pentadienes with carbonyliron complexes always lead to isomerization to produce an iron complex containing a 1,3-pentadiene system. However, if the central sp^3 carbon atom in 1,4-pentadiene is replaced by a silicon atom, then this type of isomerization cannot occur because of the inability of silicon to form stable double bonds. Reaction of dimethyldivinylsilane with dodecacarbonyltriiron in boiling benzene for 12 h gives the liquid air-sensitive (dimethyldivinylsilane)tricarbonyliron (247) *(167)*.

More stable compounds are obtained if the divinylsilane system is part of a silacyclopentadiene ring. The 2,5-diphenylsilacyclopentadiene-tricarbonyliron derivatives 248 (R" = H, R = R' = CH_3, C_2H_5, or n-C_4H_9; R = CH_3, R' = C_2H_5, C_6H_5) can easily be obtained as yellow crystalline solids by reaction of the corresponding free silacyclopentadiene with any

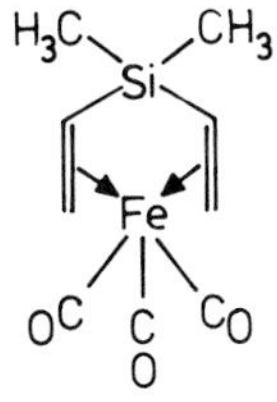
247

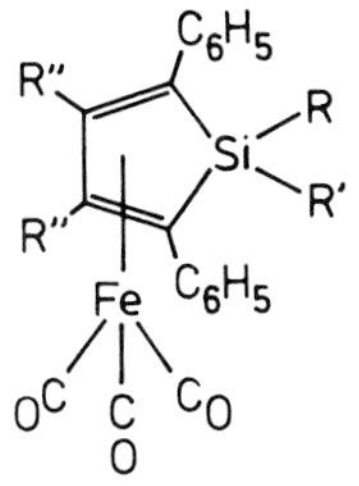
248

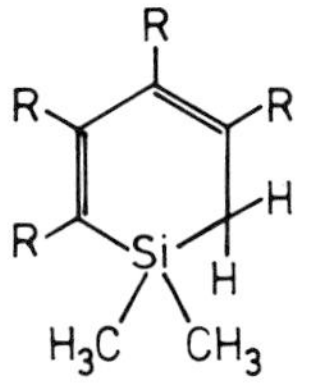
249

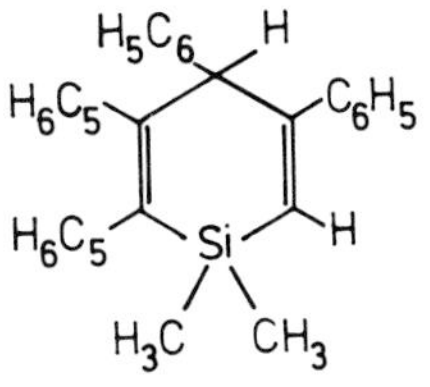
250

of the three carbonyliron complexes under appropriate conditions *(67,155)*. Furthermore, the tricarbonyliron complex 248 (R" = H, R = R' = CH_3) can also be obtained by debromination of the corresponding dibromosilacyclopentene derivatives with enneacarbonyldiiron in benzene at 50°C *(68)*. Tetraphenylsilacyclopentadiene-tricarbonyliron derivatives (248, R" = C_6H_5) can be obtained by treatment of the free tetraphenylsilacyclopentadiene with pentacarbonyliron under relatively vigorous conditions *(66,344)*. Reaction of 1,1-dimethyl-1-silacyclohexa-2,4-diene (249, R = H) with pentacarbonyliron in benzene at 150°C gives the corresponding tricarbonyliron complex as a distillable liquid *(156)*. However, 1,1-dimethyl-2,3,4,5-tetraphenyl-1-silacyclohexa-2,4-diene (249, R = C_6H_5) does not react with pentacarbonyliron under comparable conditions, presumably because of steric hindrance towards complex formation arising from the phenyl groups *(156)*. Reaction of 249 (R = C_6H_5) with pentacarbonyliron under still more vigorous conditions (190°C/120 h) results in isomerization to 250 without formation of a tricarbonyliron derivative *(156)*. Reaction of the tricarbonyliron complex of the silacyclohexadiene derivative 249 (R = H) with triphenylmethyl hexafluorophosphate in dichloromethane at room temperature results in hydride abstraction to give the corresponding silacyclohexadienyl-tricarbonyliron complex 251 as a stable yellow crystalline solid *(157)*.

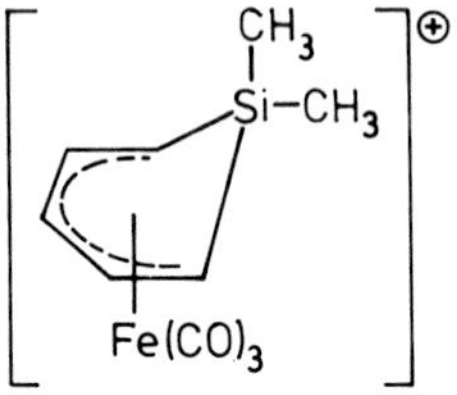

251

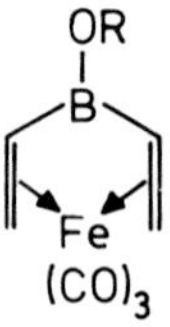

252

Reactions of alkoxydivinylboranes with enneacarbonyldiiron in benzene at 35°C followed by ultraviolet irradiation of the resulting tetracarbonyliron derivative in diethyl ether at 10°C give the unstable orange liquid tricarbonyliron complexes $[ROB(CH{=}CH_2)_2]Fe(CO)_3$ (252, R = n-C_4H_9 and $C_6H_5CH_2$). The products 252 are isoelectronic with the *cis*-pentadienyl-tricarbonyliron cations discussed above and have similar features in the complexed vinyl regions of their ^1H-NMR spectra *(203)*.

Divinylborane-tricarbonyliron derivatives are more stable if the divinylborane unit is part of a heterocyclic ring system *(205,206)*. Thus, reaction of pentaphenylborole with enneacarbonyldiiron in warm toluene gives the corresponding tricarbonyliron complex 253 as a stable yellow crystalline

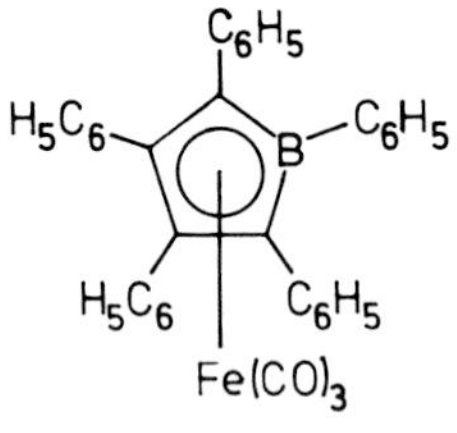

253

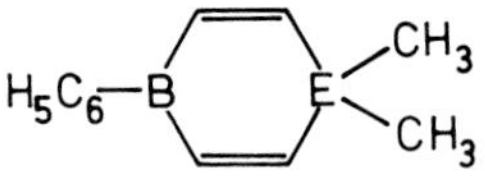

254

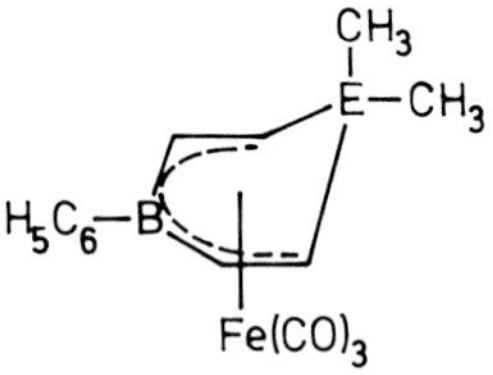

255

solid *(206)*. Similarly, the boracyclohexadienes 254 (E = C or Si) react with pentacarbonyliron to give the corresponding tricarbonyliron complexes 255 (E = C or Si) *(205)*. The dihydroborepin 256 reacts with pentacarbonyliron under relatively mild conditions to give the corresponding tricarbonyliron complex 257. Heating 257 first causes double bond migration to give 258 followed by ring contraction to give a mixture of the borole-tricarbonyliron derivatives 259 (R = C_2H_5, $CH{=}CH_2$) *(206)*. In general, the sp^2 boron in the ring systems

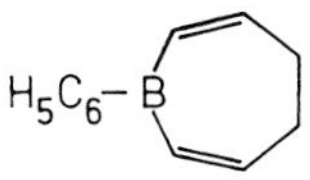

256

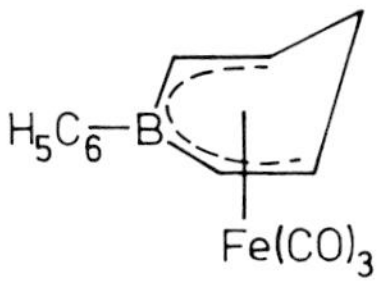

257

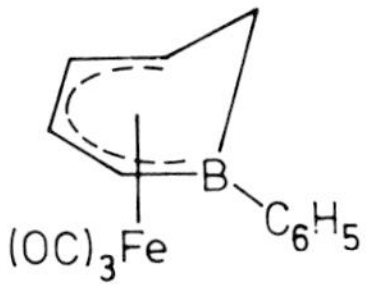

258

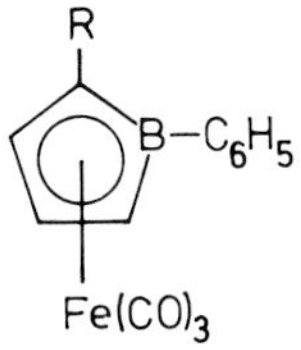

259

253, 255, 257, 258, and 259 participates in the delocalization so that these derivatives may be regarded as the indicated boradienyl complexes rather than true diene complexes.

L. *REACTIONS OF CARBONYLIRON COMPLEXES WITH PERFLUORODIENES*

The strong electron withdrawing properties of fluorine atoms are expected to reduce significantly the electron density in the π-orbital system of the 1,3-diene available for donation to a transition metal. Present information suggests that in order to form a 1,3-diene-tricarbonyliron complex, the two carbon-carbon double bonds must be part of a ring system, specifically 1,3-cyclohexadiene or 1,3-cycloheptadiene, in which the fluorinated carbon-carbon double bonds are rigidly held in the favourable *cisoid* conformation for bonding to a single metal atom.

The first tricarbonyliron complex of a perfluorodiene to be prepared was the perfluorocyclohexadiene derivative C_6F_8-$Fe(CO)_3$. This volatile air-stable pale yellow crystalline solid was first prepared by Wilkinson and co-workers *(210,386)* by reaction of perfluorocyclohexadiene (either the 1,3- or the 1,4-isomer) with dodecacarbonyltriiron in a sealed tube at 120°C. An X-ray crystallographic study *(87,88)* of this complex indicates bond distances and conformations suggesting the structure 260 with localized ring-iron bonding. The cor-

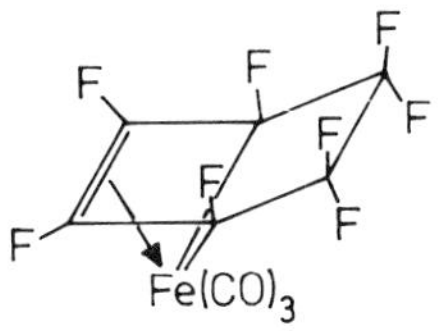

260

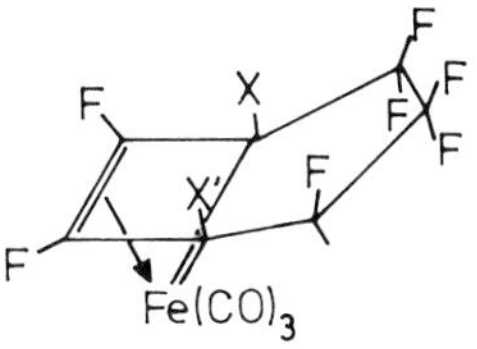

261

responding polyfluorocycloheptadiene complexes 261 (X = X' = H or F; X = H, X' = F) have been obtained similarly from the corresponding polyfluorocyclohepta-1,3-diene and dodecacarbonyltriiron as yellow volatile solids *(132)*. An X-ray crystal structure of the perfluorocycloheptadiene complex *(131)* was consistent with structure 261 (X = X' = F).

Relatively little is known about the chemical reactions of these perfluoro-1,3-diene-tricarbonyliron complexes. However, the perfluorocyclohexa-1,3-diene derivative 260 has been shown to add fluoride to give the stable yellow (η^3-perfluorocyclohexenyl)tricarbonyliron anion (262) isolated as its tetramethylammonium salt *(329)*.

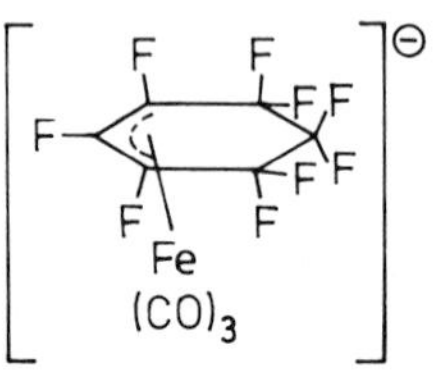

262

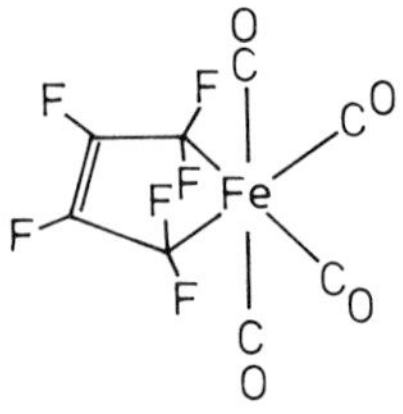

263

Other perfluoro-1,3-dienes form tetracarbonyliron complexes upon reactions with carbonyliron reagents. Thus, perfluorobutadiene reacts with pentacarbonyliron to give the complex 263 in which only the two terminal carbon atoms of the 1,3-diene chain are bonded to the iron atom *(209)*. Perfluorocyclopentadiene *(26)* and perfluorobicyclo[2.2.0]hexadiene *(97)* form only mono- and binuclear olefin-tetracarbonyliron complexes involving individual coordination of one or both of their carbon-carbon double bonds to tetracarbonyliron units.

M. CARBONYL SUBSTITUTION REACTIONS ON DIENE-TRICARBONYLIRON COMPLEXES

A ligand with very similar π-acceptor properties to the carbonyl group is the trifluorophosphine ligand. Numerous tris(trifluorophosphine)iron complexes of 1,3-dienes have been prepared *(262)*. Thus, ultraviolet irradiations of $Fe(PF_3)_5$ with butadiene *(262)*, isoprene *(262)*, *cis-* and *trans-*1,3-pentadienes *(262)*, 2,3-dimethylbutadiene *(262)*, *trans,-trans-* and *cis,trans-*2,4-hexadienes *(262)*, 2,4-dimethylpenta-1,3-diene *(262)*, methyl sorbate *(262)*, cyclopentadiene *(261)*, 1,3-cyclohexadiene *(262)*, 1,3-cycloheptadiene *(262)*, 1,3-cyclooctadiene *(262)*, and norbornadiene *(262)* give the corresponding (diene)$Fe(PF_3)_3$ complexes as air-stable sublimable yellow crystals. The corresponding 1,3-hexadiene complex (C_2H_5-CH=CH-CH=CH_2)$Fe(PF_3)_3$ can be obtained by ultraviolet irradiation of $Fe(PF_3)_5$ with allyl chloride in diethyl ether at -10°C followed by air oxidation of the unstable volatile dark green intermediate in pentane solution *(263)*. The 1,5-cyclooctadiene complex (1,5-C_8H_{12})$Fe(PF_3)_3$ can be obtained by

reaction of bis(1,5-cyclooctadiene)iron with phosphorus trifluoride at -30°C *(281)*.

Mixtures of (diene)Fe(PF$_3$)$_x$(CO)$_{3-x}$ derivatives can be
obtained by irradiating with excess phosphorus trifluoride
in hexane solution the tricarbonyliron complexes of butadiene
(384), isoprene *(75)*, *cis-* and *trans*-1,3-pentadiene *(75)*, 2,3-
dimethylbutadiene *(75)*, *trans,trans*-2,4-hexadiene *(75)*, 2,4-
dimethylpentadiene *(75)*, and 1,3-cyclohexadiene *(385)*. For a
given 1,3-diene the individual (diene)Fe(PF$_3$)$_x$(CO)$_{3-x}$ derivatives can be separated by preparative vapour phase chromatography at 60 - 118°C on copper columns several meters long
packed with 15 % Dow Corning DC-702 silicone oil on Chroma-
sorb FB.

An extensive ^{19}F-NMR study has been made on these substituted (diene)Fe(PF$_3$)$_x$(CO)$_{3-x}$ derivatives *(76)*. The trifluorophosphine ligand exhibits a strong preference for the
apical position over either of the two available basal sites
of the square pyramid coordination polyhedron. In the bis-
(trifluorophosphine) derivatives of prochiral 1,3-dienes the
trifluorophosphine ligand exhibits a secondary preference for
the basal position *trans* to the methyl group on the 1,3-diene
ligand. Intramolecular exchange of the trifluorophosphine ligands occurs in the bis- and tris(trifluorophosphine) complexes of both prochiral and symmetric dienes. The limiting
NMR spectra are generally well-developed at -100°C.

(Cyclopentadiene)tris(trifluorophosphine)iron, (η^4-C$_5$H$_6$)-
Fe(PF$_3$)$_3$ (264), is a useful source of cyclopentadienyliron
derivatives containing trifluorophosphine ligands *(261)*. Thus,

Fe(PF$_3$)$_3$

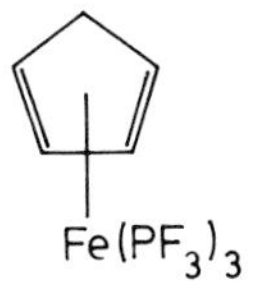

264

the reaction of 264 with triethylamine in diethyl ether gives
an essentially quantitative yield of the orange volatile liquid hydride (η^5-C$_5$H$_5$)Fe(PF$_3$)$_2$H.

Some reactions of (diene)Fe(CO)$_3$ derivatives with other
trivalent phosphorus ligands have also been investigated.
Such reactions can proceed through either carbonyl substitution to give a (diene)Fe(CO)$_2$L derivative, or through diene
substitution to give a *trans*-L$_2$Fe(CO)$_3$ derivative as indicated in equation [12]. Such reactions with triphenylphos-

$$(\text{diene})\text{Fe(CO)}_3 + \text{L} \underset{k_2}{\overset{k_1}{\rightleftarrows}} \begin{array}{l} (\text{diene})\text{Fe(CO)}_2\text{L} + \text{CO} \\ \textit{trans}\text{-L}_2\text{Fe(CO)}_3 + \text{diene} \end{array} \quad [12]$$

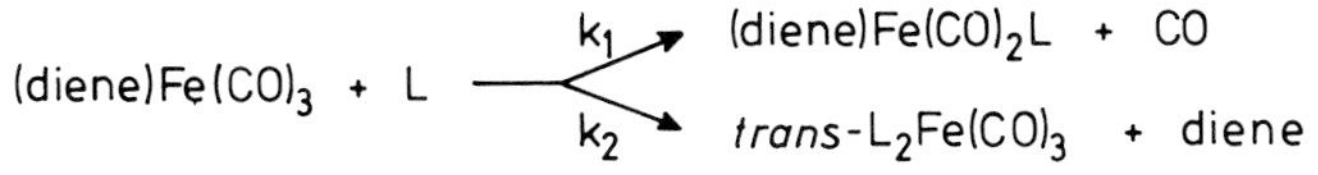

phine have been investigated for the tricarbonyliron complexes of butadiene *(83,289,340)*, 1,3-cyclohexadiene *(83,228)*, 1,3-cycloheptadiene *(83,228)*, cycloheptatriene *(83,228,289,340)*, 1,3-cyclooctadiene *(228)*, 1,5-cyclooctadiene *(100)*, and bicyclo[4.2.0]octadiene *(289)*. Thermal reactions with triphenylphosphine generally result in displacement of the 1,3-diene to give *trans*-$[(C_6H_5)_3P]_2Fe(CO)_3$ whereas photochemical reactions at room temperature result in successive replacement of carbonyl groups to give (diene)$Fe(CO)_2P(C_6H_5)_3$ and finally (diene)$Fe(CO)[P(C_6H_5)_3]_2$ derivatives *(83)*. The kinetics of some of these reactions have been investigated in some detail *(228)*. The reaction of (1,3-cyclooctadiene)tricarbonyliron with excess triphenylphosphine at 60°C in *n*-heptane proceeds smoothly by a second order process to give *trans*-$[(C_6H_5)_3P]_2$-$Fe(CO)_3$ with a rate constant (k_2 in eq. [12] above) of $3.15 \cdot 10^{-3}$ l $mole^{-1} s^{-1}$, $\Delta H^{\ddagger} = 20.76$ kcal $mole^{-1}$, and $\Delta S^{\ddagger} = -8.0$ cal deg^{-1} $mole^{-1}$. However, the tricarbonyliron complexes of 1,3-cyclohexadiene and 1,3-cycloheptadiene do not react with triphenylphosphine under such conditions. This suggests that the iron-olefin bond is weaker in (1,3-cyclooctadiene)-tricarbonyliron than in the tricarbonyliron complexes of 1,3-cycloheptadiene and 1,3-cyclohexadiene. Reaction of triphenylphosphine with (1,3-cyclohexadiene)tricarbonyliron at 154°C in decalin results in replacement of a carbonyl group rather than the cyclohexadiene to give $(C_6H_8)Fe(CO)_2P(C_6H_5)_3$ by a carbonyl-dissociative mechanism. (Cycloheptatriene)tricarbonyliron was unreactive towards triphenylphosphine at 60°C but reacted at 154°C by both first and second order processes.

Some reactions of (diene)$Fe(CO)_3$ derivatives with trivalent alkoxyphosphorus ligands have also been investigated. Thermal reactions of (butadiene)tricarbonyliron with the trivalent alkoxyphosphorus compounds C_6H_5-$P(OR)_2$ ($R = C_2H_5$, *n*-C_3H_7, and *n*-C_4H_9), C_2H_5O-CH=CH-$P(OC_2H_5)_2$, and $(C_2H_5O)_3P$ results in displacement of one carbonyl group to form the corresponding $(C_4H_6)Fe(CO)_2L$ derivative rather than in displacement of the butadiene ligand to form the corresponding L_2-$Fe(CO)_3$ derivative *(321)*. Ultraviolet irradiation of the tricarbonyliron complexes of 1,3-cyclohexadiene and 1,3-cycloheptadiene with the polycyclic phosphite 4-ethyl-1-phospha-2,6,7-trioxa-bicyclo[2.2.2]octane results in formation of products in which either one or two of the carbonyl groups have been substituted by the phosphite ligand *(397)*. The different substitution products can be separated by chromatography. Substituted cyclohexadienyl- and cycloheptadienyl(carbonyl)-(phosphite)iron salts can be obtained by reactions of these substitution products with triphenylmethyl tetrafluoroborate. The rates of ligand scrambling processes in these complexes have been determined by a study of the temperature dependence of their ^{31}P-NMR spectra *(397)*.

Radicals have been detected by electron spin resonance when the 1,3-cyclohexadiene complex $(C_6H_8)Fe(CO)_2P(C_6H_5)_3$ is irradiated with γ-rays from a cobalt-60 source *(9)*.

Reaction of the (diene)$Fe(CO)_3$ complexes (diene = butadiene, 1,3-cyclohexadiene, and cyclooctatetraene) with sodium bis(trimethylsilylamide) in benzene at 120°C results in replacement of one of the carbonyl oxygen atoms with a nitrogen atom to give the corresponding anionic cyano derivatives $[(diene)Fe(CO)_2CN]^-$. These anions can be isolated directly from the reaction mixtures as their crystalline sodium salts. These products are yellow in the case of the butadiene and cyclohexadiene derivatives and red-violet in the case of the cyclooctatetraene complex *(31)*. The anions are rather unstable in solution. Furthermore, they could not be prepared by treatment of the (diene)$Fe(CO)_3$ derivatives with potassium cyanide in ethanol *(31)*.

Ultraviolet irradiation of (diene)$Fe(CO)_3$ complexes with excess 1,3-diene can give the corresponding $(diene)_2FeCO$ derivative in many cases (*e.g.* diene = butadiene and 1,3-cyclohexadiene). This will be discussed in detail later in this chapter.

N. GENERAL THEORETICAL, SPECTROSCOPIC, AND PHYSICAL STUDIES ON DIENE-TRICARBONYLIRON COMPLEXES

Diene-tricarbonyliron complexes have been the subject of some theoretical studies *(62,145)*. In the most recent treatment *(145)* the energy ordering, symmetry, and extent in space of the valence molecular orbitals for a range of geometries of $M(CO)_5$, $M(CO)_4$, and $M(CO)_3$ fragments have been analysed in detail. This analysis provides a rationalization as to why $Cr(CO)_4$ fragments prefer to bond to chelating non-conjugated dienes whereas $Fe(CO)_3$ fragments prefer to bond to conjugated dienes.

The tricarbonyliron complexes of butadiene and 1,3-cyclohexadiene have been included in a study of the intensities of the ν(CO) bands in a wide range of metal carbonyl derivatives *(326)*. Substitution of two carbonyl groups in pentacarbonyliron by a conjugated diene had no significant effect on the specific intensities of the ν(CO) bands from the three remaining carbonyl groups.

The NMR spectra of diene-tricarbonyliron complexes have been investigated extensively. The coupling constants in the ^1H-NMR spectra of a wide range of diene-tricarbonyliron derivatives *(111,194,196)* have been analysed in terms of the metal-olefin bonding and the geometry of the substituents on the coordinated diene. The carbon-hydrogen coupling constants obtained from the ^{13}C-NMR spectra of the tricarbonyliron complexes of butadiene and methyl octadecadienoate *(345)* suggest that all four diene carbon atoms in these complexes function as sp^2 hybrids. The ^{13}C-NMR spectra of the tricarbonyliron complexes of butadiene *(264)*, 1,3-cyclohexadiene *(264)*, cycloheptatriene *(264)*, hepta-3,5-dien-2-ol *(259)*, hepta-3,5-dien-2-one *(259)*, and 1-methoxycyclohexa-1,3-diene *(268)* show only one carbonyl resonance at temperatures above 0°C but two resonances of relative intensity 2:1 at around -80°C. This indicates that such diene-tricarbonyliron derivatives are fluxional molecules in which the ligands can readily change their positions around the formally five-coordinate iron.

Diene-tricarbonyliron derivatives have been included in several extensive mass spectral studies of metal carbonyl derivatives by both electron impact *(112,114,294)* and chemical ionization methods *(217)*. Transfer of acyl substituents from the coordinated 1,3-diene to the iron has been observed *(294)*. Molecular ions have been observed for the tricarbonyliron complexes of heptafulvene, tropone, and formylcycloheptatriene in mass spectra obtained by methane chemical ionization but not if the spectra were obtained by electron impact *(217)*.

The Mössbauer spectra of diene-tricarbonyliron complexes of 2-methoxyhexa-3,5-diene, *allo*-ocimene, 2-hydroxyhexa-3,5-diene, butadiene, 7-acetoxynorbornadiene, 1-phenylbutadiene, and hexa-2,4-dienoic acid, and the dienyl-tricarbonyliron cations derived from 1,5-dimethylpentadienyl, 1-methylpentadienyl, cyclohexadienyl, and cycloheptadienyl have been investigated *(96)*. For the diene-tricarbonyliron derivatives the isomer shifts are in the range -0.218 to -0.178 mm s^{-1}, and for the dienyl-tricarbonyliron cations the isomer shifts are in the range -0.126 to -0.103 mm s^{-1} relative to a copper-cobalt-57 source. The quadrupole splittings are similar in both types of compounds and occur in the range 1.46 to 2.01 mm s^{-1}.

The photoelectron spectra of the tricarbonyliron complexes of butadiene and *trans*-1,3-pentadiene have been measured as well as the photoelectron spectra of numerous organic compounds and several cyclobutadiene-tricarbonyliron derivatives *(135)*. The first ionization potentials of the two diene-iron complexes are 9.73 and 9.58 volts, respectively. The ionization potential of (1,3-cyclohexadiene)tricarbonyliron (92) was found by mass spectrometry *(406)* to be 8.0 ± 0.2 volts.

A radical anion has been prepared by the electrochemical reduction of (1,3-cyclohexadiene)tricarbonyliron *(133)*. This radical anion was characterized by its EPR spectrum. A similar electrochemical reduction of (butadiene)tricarbonyliron was found to lead to decomposition.

The dipole moments of the tricarbonyliron complexes of butadiene, methyl 1,3-pentadienoate, and diethyl muconate were found to be 2.26, 2.75, and 2.68 D, respectively *(279)*.

III. BIS(DIENE)-MONOCARBONYLIRON DERIVATIVES AND RELATED COMPOUNDS

Diene-tricarbonyliron derivatives have been known since 1930. However, the first bis(diene)monocarbonyliron derivatives were prepared only in 1970 *(81,251)*. Several methods are now known for the preparation of $(diene)_2FeCO$ derivatives.

In many cases the low temperature ultraviolet irradiation of pentacarbonyliron or a diene-tricarbonyliron derivative with excess of the 1,3-diene can be used to prepare the corresponding $(diene)_2FeCO$ complex *(69,251,252,255)*. In this way $(diene)_2FeCO$ derivatives of butadiene, isoprene, 2,3-dimethylbutadiene, 2,4-hexadiene, methyl sorbate, diethyl muconate, and 1,3-cyclohexadiene have been prepared. Similar attempts to prepare $(diene)_2FeCO$ complexes of 1,3- and 1,5-cyclooctadiene were unsuccessful *(69)*.

Bis(butadiene)carbonyliron has also been prepared using iron atoms condensed from iron vapour *(253,405)*. Thus, co-condensation of iron atoms with butadiene at -196°C followed by warming the mixture to -80°C and adding carbon monoxide gives bis(butadiene)carbonyliron. A similar experiment but adding phosphorus trifluoride rather than carbon monoxide gives the corresponding trifluorophosphine complex $(C_4H_6)_2$-$Fe(PF_3)$.

Some $(diene)_2FeCO$ derivatives can be prepared by the reductive carbonylation of mixtures of ferric chloride and the 1,3-diene *(81,82)*. Thus, treatment of a mixture of the 1,3-diene and ferric chloride with isopropylmagnesium chloride followed by treatment with carbon monoxide at 1 - 10 atm pressure has been used to prepare the $(diene)_2FeCO$ derivatives of butadiene *(81)*, isoprene *(82)*, and 1,3-pentadiene *(82)*. Mixed (diene)(diene')FeCO derivatives (265; R = R' =

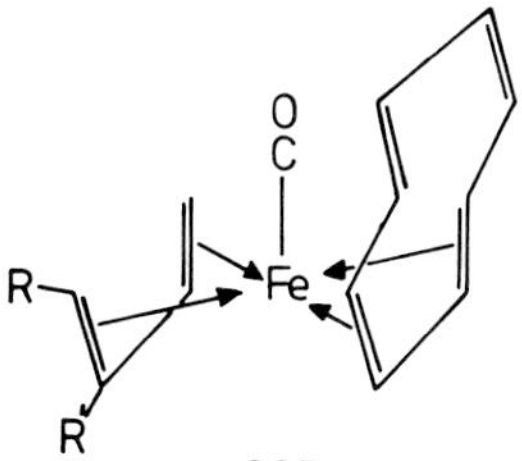

265

H; R = H, R' = CH_3; R = CH_3, R' = H) in which cyclooctatetraene functions as one of the dienes are obtained by a similar reductive carbonylation of a mixture of ferric chloride in the presence of both the 1,3-diene and cyclooctatetraene *(82)*.

Some (diene)(diene')FeCO derivatives are accessible from $[(dienyl)Fe(CO)_3]^+$ cations *(13,227)*. Cations of the type 266

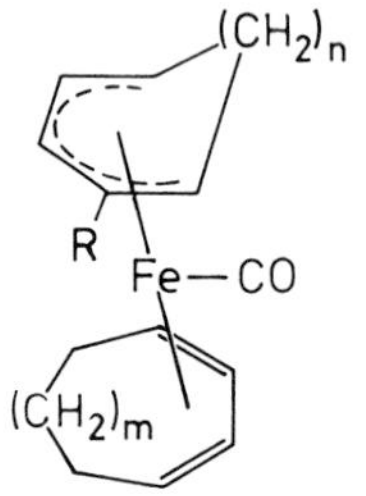

266

	n	m	R	R'
a	1	2	H	H
b	1	3	H	H
c	2	0	H	CH_3
d	2	2	H	H
e	2	3	H	H
f	3	2	H	H
g	1	2	OCH_3	H

have been prepared by ultraviolet irradiation of a $[(dienyl)Fe(CO)_3]^+$ cation with the corresponding 1,3-diene in dichloromethane solution. Similar ultraviolet irradiations of the $[(dienyl)Fe(CO)_3]^+$ cations with the eight-membered ring olefins 1,3-cyclooctadiene, 1,5-cyclooctadiene, and cyclooctatetraene do not give products of the type 266 but instead result in disproportionation of the $[(dienyl)Fe(CO)_3]^+$ cation. Reaction of the cation $[(\eta^5\text{-}C_6H_7)(\eta^4\text{-}C_6H_8)Fe(CO)]^+$ (266a) with sodium borohydride gives bis(1,3-cyclohexadiene)carbonyliron *(13)*. A similar reduction of the cation $[(\eta^5\text{-}C_7H_9)(\eta^4\text{-}C_6H_8)Fe(CO)]^+$ (266d) with sodium borohydride gives the 1,4-cycloheptadiene complex 267. This is an interesting example of a stable 1,4-diene complex of iron *(227)*. Heating a solution of 267 in heptane at 60°C results in isomerization to a mixture of 268 and 269. Further heating of 268 to 90°C gives an equilibrium mixture containing 10 % of 268 and 90 % of 269 *(227,230)*.

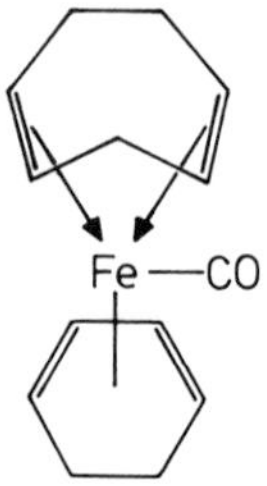

267

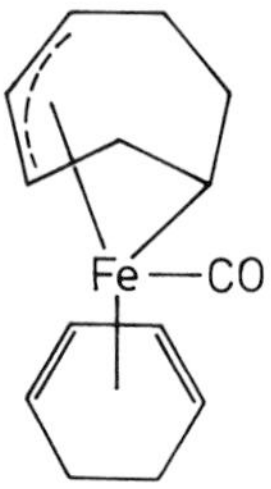

268

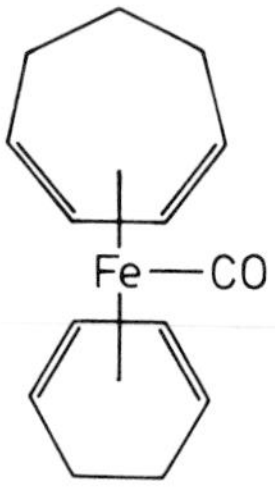

269

Some unsymmetrical (diene)(diene')FeCO derivatives can best be prepared by ligand exchange reactions. Bis(diethyl muconate)carbonyliron (270) is particularly effective for

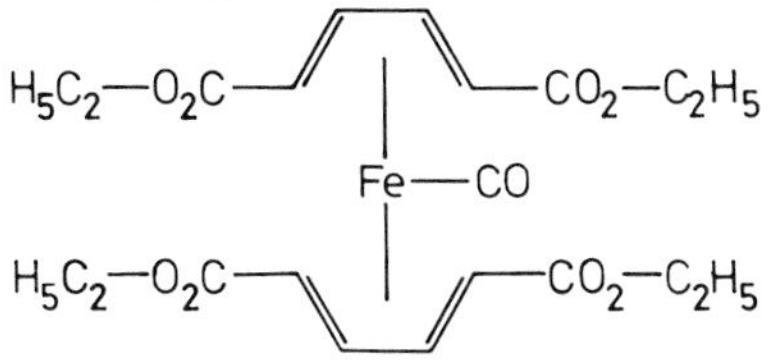

270

this purpose *(191,192)*. Thus, treatment of 270 with butadiene, isoprene, *cis*- and *trans*-1,3-pentadiene, and 2,3-dimethylbutadiene at room temperature results in a facile reaction to give the corresponding mixed complexes ($C_2H_5O_2C$-CH=CH-CH=CH-CO_2-C_2H_5)(diene)FeCO. The NMR spectra of the derivatives from the symmetrical 1,3-dienes, butadiene and 2,3-dimethylbutadiene, indicate symmetrical structures. The 1,4-cycloheptadiene ligand in 267 is also easily displaced. For example, reaction of 267 with cyclooctatetraene under mild conditions gives (1,3-cyclohexadiene)(cyclooctatetraene)carbonyliron, (C_6H_8)(C_8H_8)FeCO *(13)*.

Some reactions of $(diene)_2FeCO$ derivatives with trivalent phosphorus ligands have been investigated. The photoreaction of bis(butadiene)carbonyliron with trimethyl phosphite leads to different substitution products depending on the wavelength *(70)*. Ultraviolet irradiation (λ = 254 nm) of bis(butadiene)-carbonyliron with trimethyl phosphite results predominantly in replacement of the carbonyl group to give $(C_4H_6)_2FeP(OCH_3)_3$, whereas at λ > 400 nm displacement of one of the butadiene ligands occurs to give $(C_4H_6)Fe(CO)[P(OCH_3)_3]_2$ (for a detailed discussion see the UV chapter). Compounds of the type $(diene)Fe(CO)(PR_3)_2$ can exist as two isomers since the single carbonyl group can occupy either the axial (271a) or a basal

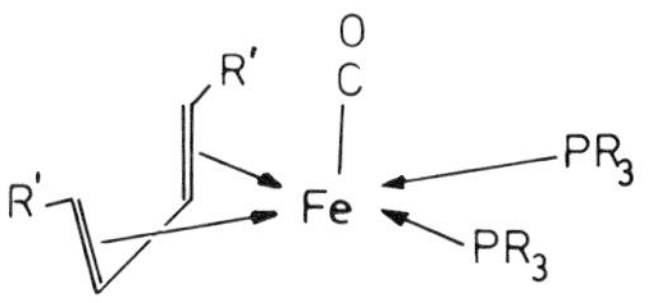

271 a

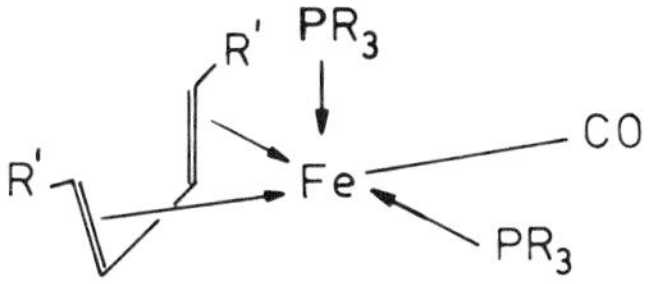

271 b

(271b) position of the square pyramid. The reaction of bis-(diethyl muconate)carbonyliron (270) with trimethyl phosphite proceeds easily upon stirring at room temperature to give the pure CO-axial isomer 271a (R = OCH_3, R' = CO_2-C_2H_5) *(152)*. On the other hand, a similar treatment of 270 with triphenylphosphine gives the pure CO-basal isomer 271b (R = C_6H_5, R' = CO_2-

C_2H_5). Reaction of 270 with tri-*n*-butylphosphine gives a mixture of both isomers 271a and 271b (R = n-C_4H_9, R' = CO_2-C_2H_5). The temperature dependence of the infrared spectrum in the ν(CO) region indicates a temperature-dependent equilibrium between these two isomers. The isomers 271a and 271b can be differentiated on the basis of their NMR spectra since 271a has a plane of symmetry and 271b does not *(152)*. Reaction of bis(butadiene)carbonyliron with the bidentate ligand $(C_6H_5)_2$-P-CH_2-CH_2-P$(C_6H_5)_2$ (abbreviated as diphos) gives a product of stoichiometry (C_4H_6)Fe(CO)(diphos) shown by its ^{31}P-NMR spectrum at low temperature (-63°C) to be a 4:1 mixture of symmetrical and unsymmetrical isomers. ^{31}P-NMR spectra at higher temperatures indicate rapid interchange of these two isomers in a fluxional process *(376)*. Bis(diethyl muconate)carbonyliron (270) reacts with 2,2'-bipyridyl at room temperature to give a (diene)Fe(CO)(bipy) derivative shown by its NMR spectra to have the low symmetry structure 272 (R' = CO_2-C_2H_5)

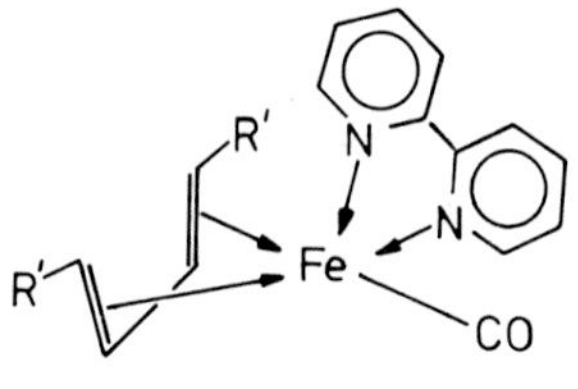

272

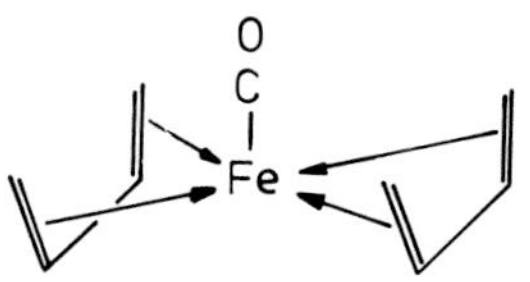

273

(193) confirmed by an X-ray structure determination.

The bis(diene)carbonyliron complexes under appropriate conditions can be catalytically active for the oligomerization of butadiene *(69,81,82,251)*. Thus, heating butadiene with a mixture of bis(butadiene)carbonyliron and triphenylphosphine in catalytic quantities results in its dimerization to give a mixture of vinylcyclohexene and 1,5-cyclooctadiene. In the absence of added triphenylphosphine or other donor ligands, the catalytic activity of bis(butadiene)carbonyliron is less and butadiene trimers are obtained.

The structure of bis(butadiene)carbonyliron has been found by X-ray crystallography *(117,401)* to be 273 in which the two butadiene ligands occupy the four basal positions of a square pyramid and the single carbonyl group occupies the axial position. An analogous structure has been found *(266)* for bis(1,3-cyclohexadiene)carbonyliron. In this latter compound the two 1,3-cyclohexadiene-rings have slightly different geometries leading to a chiral molecule with no plane of symmetry. The structure 265 (R = R' = H) for (butadiene)-(cyclooctatetraene)carbonyliron has been confirmed by X-ray crystallography *(28)*.

The vibrational spectrum of bis(butadiene)carbonyliron

(273) has been analysed *(115)*. The internal modes of the butadiene ligands are very similar to those of the butadiene ligand in (butadiene)tricarbonyliron and thus indicate very little interaction between the two butadiene ligands in 273. The dipole moments of $(diene)_2FeCO$ derivatives of butadiene, 2,3-dimethylbutadiene, 1,3-cyclohexadiene, methyl sorbate, and diethyl muconate *(279)* were found to be 2.15, 2.98, 2.05, 1.78, and 2.65 D, respectively. From the data on the butadiene complex a tentative value of 2 D for the moment from the iron atom to the carbonyl group was suggested.

IV. OTHER TYPES OF DIENE-IRON COMPLEXES

Substitution reactions of $(diene)_2FeCO$ with trivalent phosphorus ligands can give carbonyl-free complexes of the type $(diene)_2FePR_3$ (*e.g.*, diene = butadiene, R = OCH_3, *(70)*) as discussed above. Such complexes can also be synthesized directly from iron(III) acetylacetonate by electrochemical methods *(352,353)*. Thus, the cathodic reduction of iron(III) acetylacetonate in the presence of triphenylphosphine and excess diene in methanolic lithium chloride at -15°C gives the corresponding $(diene)_2FeP(C_6H_5)_3$ (diene = butadiene and isoprene) as orange crystals which are stable to air for short periods. The square pyramidal structures 274 with apical triphenylphosphine ligands (R = H and CH_3) are consistent

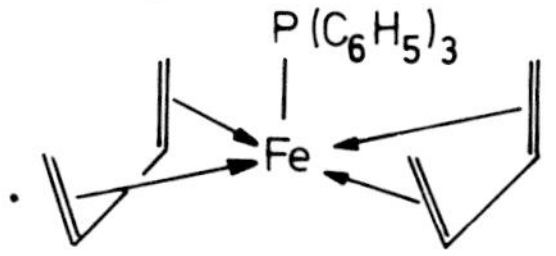

274

with the observed ^{1}H-, ^{31}P-, and ^{13}C-NMR spectra *(353)*.

A closely related $(diene)_2FeL$ derivative can be prepared from the dinitrogen complex $H_2(N_2)Fe[(C_6H_5)_2PC_2H_5]_3$ *(165,254)*.

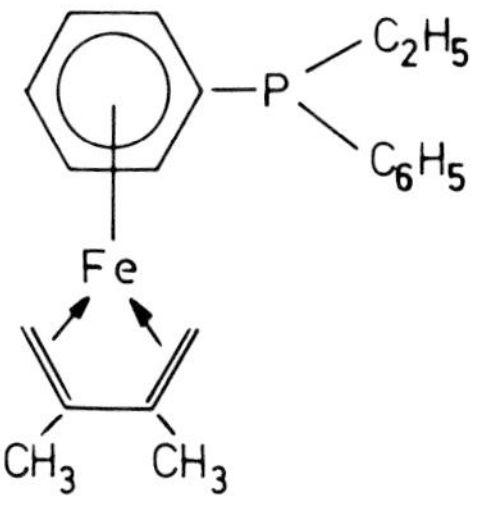

275

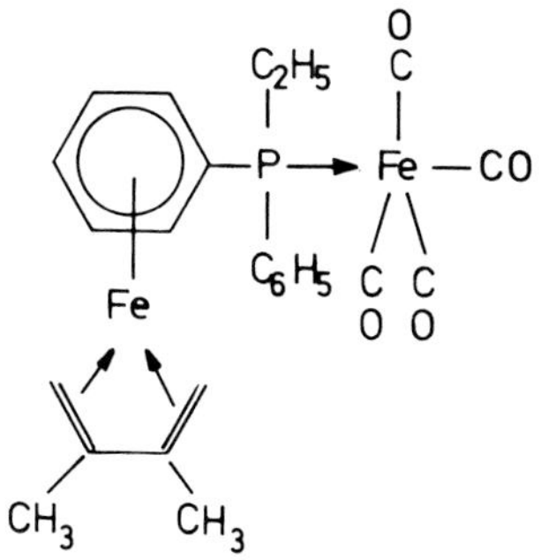

276

Ultraviolet irradiation of this dinitrogen complex with 2,3-dimethylbutadiene results in the displacement of the hydride ligand, the dinitrogen ligand, and two of the coordinated phosphines to give the complex 275 as an orange solid *(254)*. The structure 275 is unusual since the arylphosphine is bonded to iron only by η^6-coordination of one of its phenyl rings. The presence of the uncomplexed phosphorus atom in 275 is indicated by the formation of the corresponding tetracarbonyliron complex 276 upon photolysis with pentacarbonyliron at -40°C in diethyl ether solution *(254)*. Ultraviolet irradiation of 275 with excess 2,3-dimethylbutadiene results in migration of the iron from the phenyl ring to the trivalent phosphorus atom of the arylphosphine to give the $(diene)_2FeL$ derivative 277 *(165)*.

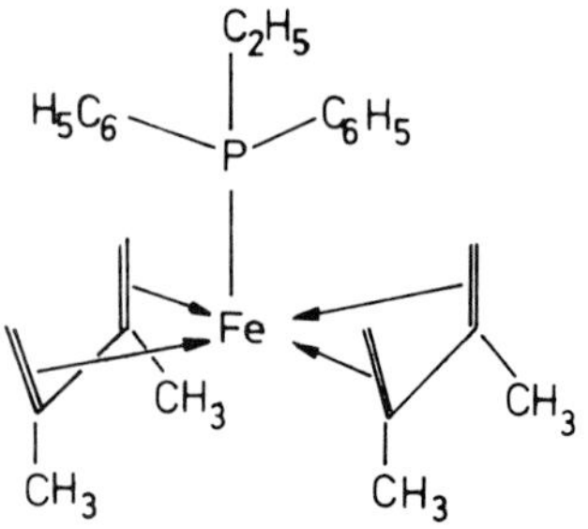

277

Reduction of the trimethylphosphine complex $[(CH_3)_3P]_2FeCl_2$ with sodium amalgam in the presence of excess trimethylphosphine and butadiene gives the complex $(C_4H_6)Fe[P(CH_3)_3]_3$ *(338)*. This complex is related to (butadiene)tricarbonyliron by complete substitution of its three carbonyl groups with trimethylphosphine ligands.

Some diene-iron complexes containing only hydrocarbon ligands have also been prepared. The first of these complexes to be reported was (benzene)(1,3-cyclohexadiene)iron, $(\eta^6-C_6H_6)(\eta^4-C_6H_8)Fe$ (278) which is obtained by treatment of a

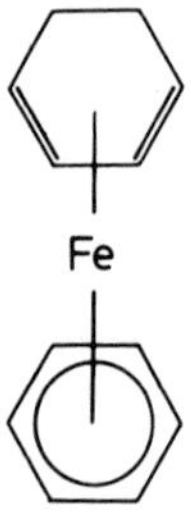

278

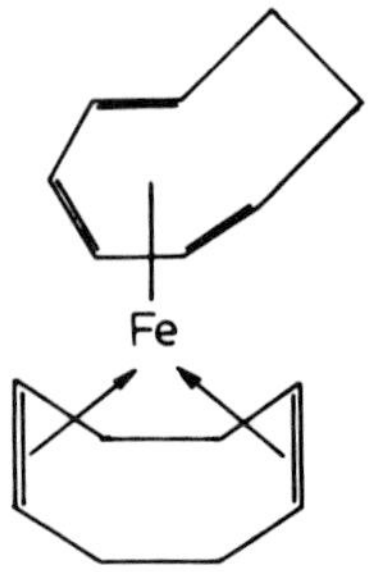

279

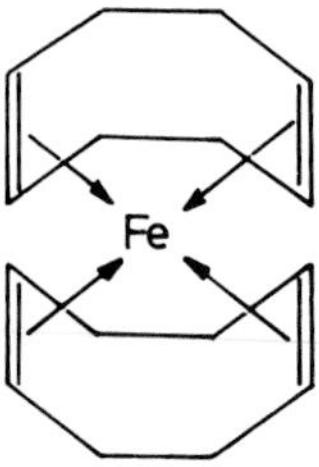

280

mixture of ferric chloride and 1,3-cyclohexadiene with excess isopropylmagnesium bromide *(160)*, preferably assisted by ultraviolet irradiation, followed by methanolysis. A similar reduction of ferric chloride with isopropylmagnesium bromide in the presence of a mixture of 1,5-cyclooctadiene and 1,3,5-cyclooctatriene gives (1,3,5-cyclooctatriene)(1,5-cyclooctadiene)iron (279) as an air-sensitive red solid *(162)*. A similar reduction of a mixture of ferric chloride with excess cycloheptatriene and 1,3-cycloheptadiene using isopropylmagnesium bromide was once believed *(161)* to give a similar mixed olefin complex (cycloheptatriene)(1,3-cycloheptadiene)-iron, $(\eta^6\text{-}C_7H_8)(\eta^4\text{-}C_7H_{10})Fe$, but this product has subsequently been shown *(303)* to be bis(cycloheptadienyl)iron, $(\eta^5\text{-}C_7H_9)_2Fe$.

The iron atom in the pure hydrocarbon complexes 278 and 279 has the favoured 18-electron noble gas configuration. A $(\text{diene})_2Fe$ complex in which the iron atom has only a 16-electron configuration has also been prepared *(281,372)* using free iron atoms. Thus, condensing iron vapour into a solution of excess 1,5-cyclooctadiene in methylcyclohexane at -120°C gives brown crystalline bis(1,5-cyclooctadiene)iron, $(C_8H_{12})_2Fe$ (280). The easy displacement of the 1,5-cyclooctadiene ligands in $(C_8H_{12})_2Fe$ (280) makes this iron compound a useful intermediate for the preparation of other iron(O) derivatives under mild conditions *(77,281)*. Reaction of 280 with cyclooctatetraene at -30°C results in almost quantitative removal of the 1,5-cyclooctadiene ligands to give bis-(cyclooctatetraene)iron *(281)*. Reaction of 280 with excess trifluorophosphine or *t*-butyl isocyanide at -78°C results in displacement of only one of the 1,5-cyclooctadiene rings to give $(C_8H_{12})FeL_3$ (L = PF_3 and $(CH_3)_3C\text{-}NC$, respectively) *(77)*. Reaction of 280 with excess trimethyl phosphite gives (1,3-cyclooctadiene)tris(trimethyl phosphite)iron. Both the ^1H- and ^{13}C-NMR spectra of this complex indicate rearrangement of the eight-membered ring ligand. Reaction of 280 with $(C_6H_5)_2P\text{-}CH_2\text{-}CH_2\text{-}P(C_6H_5)_2$ (diphos) under argon does not give a stable product. However, reaction of 280 with diphos under nitrogen gives the red crystalline dinitrogen complex $(\text{diphos})_2FeN_2$ *(77)*. Reaction of this dinitrogen complex with carbon monoxide results in displacement of the coordinated dinitrogen to give the monocarbonyl $(\text{diphos})_2Fe(CO)$ *(77)*.

Some (butadiene)cyclopentadienyliron complexes have been prepared by photochemical reactions *(322,323)*. Ultraviolet irradiation of $(C_6H_5)_3EFe(CO)_2(\eta^5\text{-}C_5H_5)$ (E = Ge, Sn, and Pb) with butadiene in benzene solution at room temperature results in the displacement of both carbonyl groups to give $(C_6H_5)_3EFe(\eta^4\text{-}C_4H_6)(\eta^5\text{-}C_5H_5)$ (281, E = Ge, Sn, and Pb).

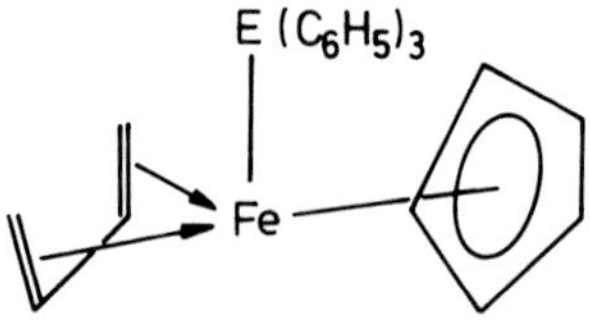

281

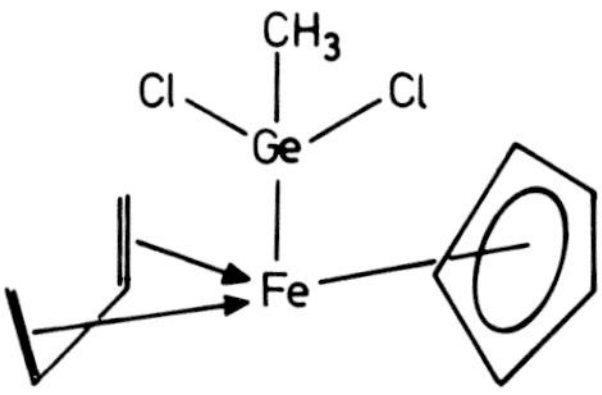

282

A similar ultraviolet irradiation of $CH_3GeCl_2Fe(CO)_2(\eta^5$-$C_5H_5)$ with butadiene proceeds analogously to give CH_3GeCl_2-$Fe(\eta^4$-$C_4H_6)(\eta^5$-$C_5H_5)$. The structure 282 for this complex has been confirmed by X-ray crystallography, which also shows the shortest known iron-germanium bond length *(10)*. Ultraviolet irradiation of $(CH_3)_3EFe(CO)_2(\eta^5$-$C_5H_5)$ (E = Sn, Pb) with butadiene proceeded differently *(159)* to give the binuclear derivatives $[(CH_3)_3EFe(\eta^5$-$C_5H_5)]_2$-μ(1,2-η:3,4-η-C_4H_6) (E = Sn, Pb) in which the butadiene double bonds are individually coordinated to different iron atoms.

(Cyclopentadienyl)(diene)iron complexes of the type $[(\eta^5$-$C_5H_5)Fe(CO)_2(\eta^2$-diene$)]^+$ (diene = butadiene *(159)*, cyclopentadiene *(189)*, and 1,3-cyclohexadiene *(159)*) have been prepared. However, these complexes only use one of the two double bonds of the diene for coordination. They are therefore discussed in greater detail in the chapter on iron complexes of monoolefins.

Reaction of the organosulfur-carbonyliron complex $[CH_3S$-$Fe(CO)_3]_2$ with norbornadiene in boiling benzene for 70 h results in the replacement of two carbonyl groups with a norbornadiene ligand to give dark brown crystalline $(CH_3S)_2Fe_2$-$(CO)_4(C_7H_8)$ apparently with structure 283 *(243)*. Analogous

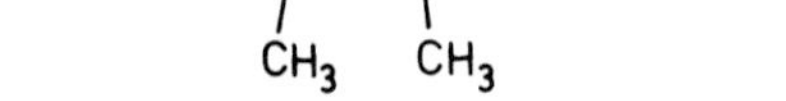

283

treatment of $[CH_3SFe(CO)_3]_2$ under the same conditions with 1,5-cyclooctadiene, 1,3,5-cyclooctatriene, cyclooctatetraene, cycloheptatriene, or butadiene did not give any corresponding olefin-iron complexes *(243)*. The mass spectrum of 283 first exhibits stepwise loss of carbonyl groups from the molecular ion to give $[(C_7H_8)Fe_2(SCH_3)_2]^+$ followed by loss of the two methyl groups to give $[(C_7H_8)Fe_2S_2]^+$ before fragmentation of the complexed norbornadiene ligand occurs *(244)*.

Reactions of the hydride $HFe(CO)_3NO$ with the 1,3-dienes butadiene and isoprene give the red distillable liquid η^3-allylic derivatives 284 (R = H, CH_3) rather than diene-carbonyliron derivatives *(84)*. Reactions of mixtures of NaFe-$(CO)_3NO$ and methyl iodide with the acyclic 1,3-dienes butadiene, isoprene, and 2,3-dimethylbutadiene followed by treatment of the crude product with triphenylphosphine in boiling benzene give a mixture of the red η^3-allylic derivatives 285 (R = R' = H; R = H, R' = CH_3; R = R' = CH_3, respectively) and the yellow dicarbonyl-diene-triphenylphosphine-iron complexes 286 (R = R' = H; R = H, R' = CH_3; R = R' = CH_3, res-

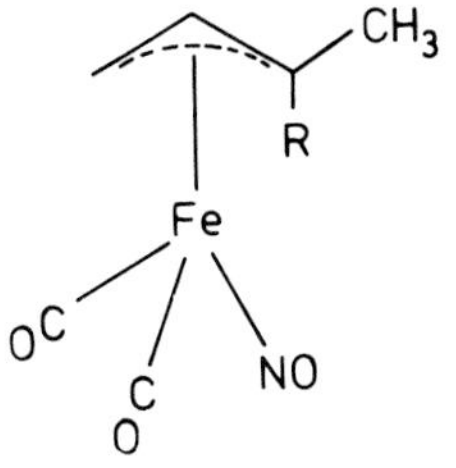

284

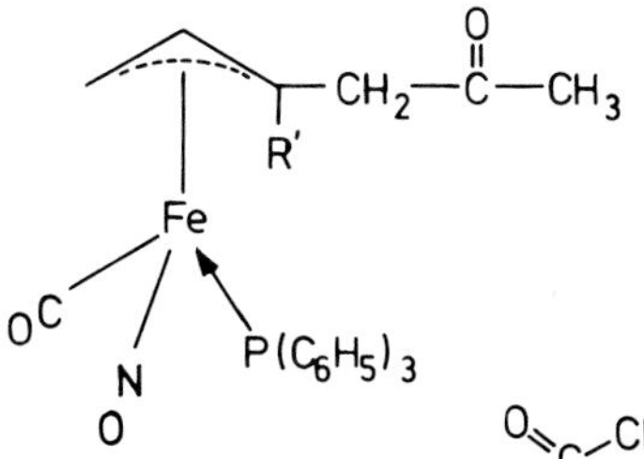

285

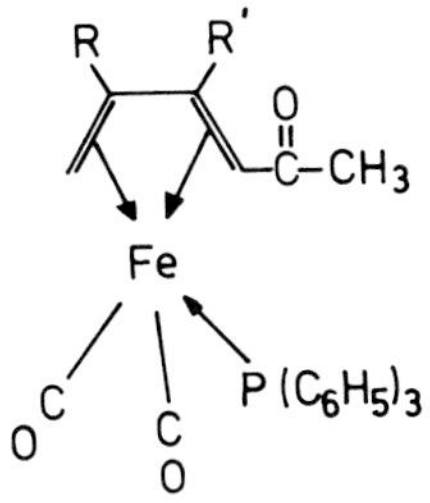

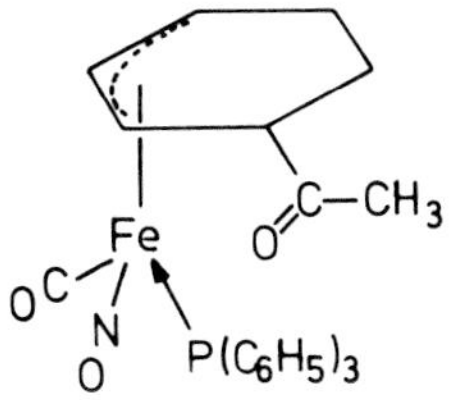

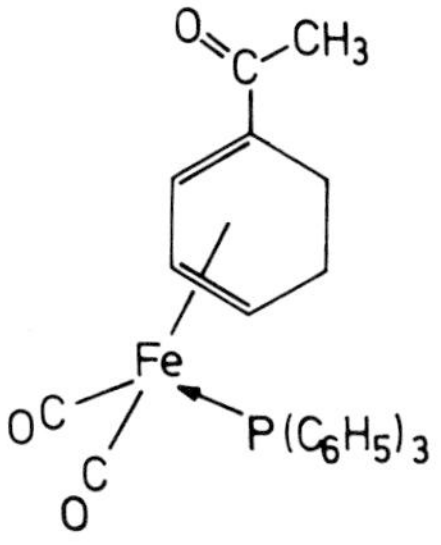

286 287 288

pectively) *(84)*. A similar reaction of a mixture of NaFe-$(CO)_3NO$ and methyl iodide with 1,3-cyclohexadiene followed by triphenylphosphine treatment gives the unstable red η^3-allylic derivative 287 and a low yield of the stable yellow substituted cyclohexadiene complex 288 *(84)*.

Acknowledgement

A fellowship of the Max-Planck-Gesellschaft during the time this article was written at the Institut für Strahlenchemie im Max-Planck-Institut für Kohlenforschung (Mülheim a.d. Ruhr, Germany) is gratefully acknowledged.

REFERENCES

1. Alper, H., and Edward, J.T., *J. Organometal. Chem.*, *14*, 411 (1968).
2. Alper, H., LePort, P.C., and Wolfe, S., *J. Amer. Chem. Soc.*, *91*, 7553 (1969).
3. Alper, H., and Edward, J.T., *J. Organometal. Chem.*, *16*, 342 (1969).
4. Alper, H., and Keung, E.C.-H., *J. Amer. Chem. Soc.*, *94*, 2144 (1972).
5. Alper, H., and Huang, C.-C., *J. Organometal. Chem.*, *50*, 213 (1973).
6. Alper, H., *J. Organometal. Chem.*, *96*, 95 (1975).
7. Amith, C., and Ginsburg, D., *Tetrahedron*, *30*, 3415 (1974).
8. Anderson, M., Clague, A.D.H., Blaauw, L.P., and Couperus, P.A., *J. Organometal. Chem.*, *56*, 307 (1973).
9. Anderson, O.P., and Symons, M.C.R., *Inorg. Chem.*, *12*, 1932 (1973).
10. Andrianov, V.G., Martynov, V.P., Anisimov, K.N., Kolobova, N.E., and Skripkin, V.V., *J. Chem. Soc. D, Chem. Commun.*, *1970*, 1252.
11. Anisimov, K.N., Magomedov, G.K., Kolobova, N.E., and Trufanov, A.G., *Izv. Akad. Nauk SSSR, Ser. Khim.*, *1970*, 2533; *Bull. Acad. Sci. USSR, Div. Chem. Ser.*, *1970*, 2379.
12. Arnet, J.E., and Pettit, R., *J. Amer. Chem. Soc.*, *83*, 2954 (1961).
13. Ashley-Smith, J., Howe, D.V., Johnson, B.F.G., Lewis, J., and Ryder, I.E., *J. Organometal. Chem.*, *82*, 257 (1974).
14. Aumann, R., *Angew. Chem.*, *83*, 176 (1971); *Angew. Chem. Int. Ed. Engl.*, *10*, 189 (1971).
15. Aumann, R., *Angew. Chem.*, *83*, 177 (1971); *Angew. Chem. Int. Ed. Engl.*, *10*, 190 (1971).
16. Aumann, R., *Angew. Chem.*, *83*, 583 (1971); *Angew. Chem. Int. Ed. Engl.*, *10*, 560 (1971).
17. Aumann, R., *Angew. Chem.*, *85*, 628 (1973); *Angew. Chem. Int. Ed. Engl.*, *12*, 574 (1973).
18. Aumann, R., *J. Organometal. Chem.*, *47*, C 29 (1973).
19. Aumann, R., Fröhlich, K., and Ring, H., *Angew. Chem.*, *86*, 309 (1974); *Angew. Chem. Int. Ed. Engl.*, *13*, 275 (1974).
20. Aumann, R., *J. Organometal. Chem.*, *77*, C 33 (1974).
21. Aumann, R., *J. Organometal. Chem.*, *78*, C 31 (1974).
22. Aumann, R., *Chem. Ber.*, *109*, 168 (1976).
23. Aumann, R., and Knecht, J., *Chem. Ber.*, *109*, 174 (1976).
24. Bailey, N.A., and Mason, R., *Acta Crystallogr.*, *21*,

652 (1966).
25. Bamford, C.H., and Finch, C.A., *Z. Naturforsch., B 17*, 500 (1962).
26. Banks, R.E., Harrison, T., Haszeldine, R.N., Lever, A.B.P., Smith, T.F., and Walton, J.B., *Chem. Commun., 1965*, 30.
27. Banthorpe, D.V., Fitton, H., and Lewis, J., *J. Chem. Soc. Perkin Trans. I, 1973*, 2051.
28. Bassi, I.W., and Scordamaglia, R., *J. Organometal. Chem., 37*, 353 (1972).
29. Becker, Y., Eisenstadt, A., and Shvo, Y., *Tetrahedron Lett., 1972*, 3183.
30. Beddoes, R.L., Lindley, P.F., and Mills, O.S., *Angew. Chem., 82*, 293 (1970); *Angew. Chem. Int. Ed. Engl., 9*, 304 (1970).
31. Behrens, H., and Moll, M., *Z. anorg. allgem. Chem., 416*, 193 (1975).
32. Ben-Shoshan, R., and Pettit, R., *J. Amer. Chem. Soc., 89*, 2231 (1967).
33. Ben-Shoshan, R., and Pettit, R., *Chem. Commun., 1968*, 247.
34. Ben-Shoshan, R., and Sarel, S., *J. Chem. Soc. D, Chem. Commun., 1969*, 883.
35. Berger, M., and Manuel, T.A., *J. Polym. Sci. Part A-1, 4*, 1509 (1966).
36. Birch, A.J., Cross, P.E., Lewis, J., and White, D.A., *Chem. Ind. (London), 1964*, 838.
37. Birch, A.J., Fitton, H., McPartlin, M., and Mason, R., *Chem. Commun., 1968*, 531.
38. Birch, A.J., Cross, P.E., Lewis, J., White, D.A., and Wild, S.B., *J. Chem. Soc. A, 1968*, 332.
39. Birch, A.J., and Haas, M., *Tetrahedron Lett., 1968*, 3705.
40. Birch, A.J., and Haas, M.A., *J. Chem. Soc. C, 1971*, 2465.
41. Birch, A.J., Chamberlain, K.B., Haas, M.A., and Thompson, D.J., *J. Chem. Soc. Perkin Trans. I, 1973*, 1882.
42. Birch, A.J., and Williamson, D.H., *J. Chem. Soc. Perkin Trans. I, 1973*, 1892.
43. Birch, A.J., Chamberlain, K.B., and Thompson, D.J., *J. Chem. Soc. Perkin Trans. I, 1973*, 1900.
44. Birch, A.J., and Jenkins, I.D., *Tetrahedron Lett., 1975*, 119.
45. Birch, A.J., Jenkins, I.D., and Liepa, A.J., *Tetrahedron Lett., 1975*, 1723.
46. Birch, A.J., and Pearson, A.J., *Tetrahedron Lett., 1975*, 2379.
47. Bird, C.W., Cookson, R.C., and Hudec, J., *Chem. Ind. (London), 1960*, 20.

48. Bird, C.W., Colinese, D.L., Cookson, R.C., Hudec, J., and Williams, R.O., *Tetrahedron Lett.*, *1961*, 373.
49. Bleck, W.E., Grimme, W., Günther, H., and Vogel, E., *Angew. Chem.*, *82*, 292 (1970); *Angew. Chem. Int. Ed. Engl.*, *9*, 303 (1970).
50. Bonazza, B.R., and Lillya, C.P., *J. Amer. Chem. Soc.*, *96*, 2298 (1974).
51. Bond, A., Green, M., Lewis, B., and Lowrie, S.F.W., *J. Chem. Soc. D, Chem. Commun.*, *1971*, 1230.
52. Bond, A., Lewis, B., and Green, M., *J. Chem. Soc. Dalton Trans.*, *1975*, 1109.
53. Bonner, T.G., Holder, K.A., and Powell, P., *J. Organometal. Chem.*, *77*, C 37 (1974).
54. Boston, J.L., Sharp, D.W.A., and Wilkinson, G., *J. Chem. Soc.*, *1962*, 3488.
55. Bottrill, M., Goddard, R., Green, M., Hughes, R.P., Lloyd, M.K., Lewis, B., and Woodward, P., *J. Chem. Soc. Chem. Commun.*, *1975*, 253.
56. Braye, E.H., and Hübel, W., *Chem. Ind. (London)*, *1959*, 1250.
57. Braye, E.H., Hübel, W., and Caplier, I., *J. Amer. Chem. Soc.*, *83*, 4406 (1961).
58. Brookhart, M., and Harris, D.L., *J. Organometal. Chem.*, *42*, 441 (1972).
59. Brookhart, M., Lippman, N.M., and Reardon, E.J., Jr., *J. Organometal. Chem.*, *54*, 247 (1973).
60. Brookhart, M., and Harris, D.L., *Inorg. Chem.*, *13*, 1540 (1974).
61. Brookhart, M., Dedmond, R.E., and Lewis, B.F., *J. Organometal. Chem.*, *72*, 239 (1974).
62. Brown, D.A., *J. Inorg. Nucl. Chem.*, *13*, 212 (1960).
63. Brune, H.A., and Schwab, W., *Tetrahedron*, *26*, 1357 (1970).
64. Brune, H.A., Horlbeck, G., and Schwab, W., *Tetrahedron*, *28*, 4455 (1972).
65. Brune, H.A., Horlbeck, G., and Müller, P., *Z. Naturforsch.*, *B 27*, 911 (1972).
66. Brunet, J.C., Resiboil, B., and Bertrand, J., *Bull. Soc. Chim. Fr.*, *1969*, 3424.
67. Brunet, J.-C., and Demey, N., *Ann. Chim.*, *8*, 123 (1973).
68. Brunet, J.C., Bertrand, J., and Lesenne, C., *J. Organometal. Chem.*, *71*, C 8 (1974).
69. Buchkremer, J., Dissertation, Ruhr-Universität Bochum, 1973.
70. Buchkremer, J., Grevels, F.-W., Jaenicke, O., Kirsch, P., Knoesel, R., Koerner von Gustorf, E.A., and Shields, J., *7th Int. Conf. Photochemistry*, Jerusalem, Israel, September 1973, Abstracts, p. 61.

71. Burrill, J.W., Bonazza, B.R., Garrett, D.W., and Lillya, C.P., *J. Organometal. Chem.*, *104*, C 37 (1976).
72. Burt, R., Cooke, M., and Green, M., *J. Chem. Soc. A*, *1970*, 2981.
73. Burton, R., Green, M.L.H., Abel, E.W., and Wilkinson, G., *Chem. Ind. (London)*, *1958*, 1592.
74. Burton, R., Pratt, L., and Wilkinson, G., *J. Chem. Soc.*, *1961*, 594.
75. Busch, M.A., and Clark, R.J., *Inorg. Chem.*, *14*, 219 (1975).
76. Busch, M.A., and Clark, R.J., *Inorg. Chem.*, *14*, 226 (1975).
77. Cable, R.A., Green, M., MacKenzie, R.E., Timms, P.L., and Turney, T.W., *J. Chem. Soc. Chem. Commun.*, *1976*, 270.
78. Cais, M., and Feldkimel, M., *Tetrahedron Lett.*, *1961*, 444.
79. Cais, M., and Maoz, N., *Israel J. Chem.*, *2*, 239 (1964).
80. Cais, M., and Maoz, N., *J. Organometal. Chem.*, *5*, 370 (1966).
81. Carbonaro, A., and Greco, A., *J. Organometal. Chem.*, *25*, 477 (1970).
82. Carbonaro, A., and Cambisi, F., *J. Organometal. Chem.*, *44*, 171 (1972).
83. Chaudhari, F.M., and Pauson, P.L., *J. Organometal. Chem.*, *5*, 73 (1966).
84. Chaudhari, F.M., Knox, G.R., and Pauson, P.L., *J. Chem. Soc. C*, *1967*, 2255.
85. Chow, Y.L., Fossey, J., and Perry, R.A., *J. Chem. Soc. Chem. Commun.*, *1972*, 501.
86. Churchill, M.R., and Mason, R., *8th Int. Conf. Coord. Chem.*, Vienna, September 1964, Abstracts, p. 249.
87. Churchill, M.R., and Mason, R., *Proc. Chem. Soc.*, *1964*, 226.
88. Churchill, M.R., and Mason, R., *Proc. Roy. Soc. (London)*, *A 301*, 433 (1967).
89. Churchill, M.R., Wormald, J., Young, D.A.T., and Kaesz, H.D., *J. Amer. Chem. Soc.*, *91*, 7201 (1969).
90. Churchill, M.R., and Wormald, J., *Inorg. Chem.*, *9*, 2430 (1970).
91. Churchill, M.R., and Chang, S.W.-Y., *Inorg. Chem.*, *14*, 1680 (1975).
92. Clack, D.W., Monshi, M., and Maguire, L.A.P., *J. Organometal. Chem.*, *107*, C 40 (1976).
93. Clinton, N.A., and Lillya, C.P., *Chem. Commun.*, *1968*, 579.
94. Clinton, N.A., and Lillya, C.P., *J. Amer. Chem. Soc.*, *92*, 3058 (1970).

95. Clinton, N.A., and Lillya, C.P., *J. Amer. Chem., Soc.*, *92*, 3065 (1970).
96. Collins, R.L., and Pettit, R., *J. Amer. Chem. Soc.*, *85*, 2332 (1963).
97. Cook, D.J., Green, M., Mayne, N., and Stone, F.G.A., *J. Chem. Soc. A*, *1968*, 1771.
98. Cope, A.C., Haven, A.C., Jr., Ramp, F.L., and Trumbull, E.R., *J. Amer. Chem. Soc.*, *74*, 4867 (1952).
99. Cotton, F.A., and Leto, J.R., *Chem. Ind. (London)*, *1958*, 1592.
100. Cotton, F.A., Deeming, A.J., Josty, P.L., Ullah, S.S., Domingos, A.J.P., Johnson, B.F.G., and Lewis, J., *J. Amer. Chem. Soc.*, *93*, 4624 (1971).
101. Cotton, F.A., and Deganello, G., *J. Amer. Chem. Soc.*, *94*, 2142 (1972).
102. Cotton, F.A., and Deganello, G., *J. Organometal., Chem.*, *38*, 147 (1972).
103. Cotton, F.A., and Deganello, G., *J. Amer. Chem. Soc.*, *95*, 396 (1973).
104. Cotton, F.A., and Troup, J.M., *J. Amer. Chem. Soc.*, *95*, 3798 (1973).
105. Cotton, F.A., Day, V.W., Frenz, B.A., Hardcastle, K.I., and Troup, J.M., *J. Amer. Chem. Soc.*, *95*, 4522 (1973).
106. Cotton, F.A., Frenz, B.A., and Troup, J.M., *J. Organometal. Chem.*, *61*, 337 (1973).
107. Cotton, F.A., and Troup, J.M., *J. Organometal. Chem.*, *77*, 360 (1974).
108. Cotton, F.A., Day, V.W., and Hardcastle, K.I., *J. Organometal. Chem.*, *92*, 369 (1975).
109. Courtot, P., and Clément, J.C., *Bull. Soc. Chim. Fr.*, *1973*, 2121.
110. Cowles, R.J.H., Johnson, B.F.G., Lewis, J., and Parkins, A.W., *J. Chem. Soc. Dalton Trans.*, *1972*, 1768.
111. Crews, P., *J. Amer. Chem. Soc.*, *95*, 636 (1973).
112. Cross, P.E., Haas, M.A., and Wilson, J.M., in R. Bonnett, and J.G. Davis (Eds.), *Some New Physical Methods in Structural Chemistry*, United Trade Press, London 1967, pp. 90-97.
113. Dauben, H.J., Jr., and Bertelli, D.J., *J. Amer. Chem. Soc.*, *83*, 497 (1961).
114. Dauben, W.G., and Lorber, M.E., *Org. Mass Spectr.*, *3*, 211 (1970).
115. Davidson, G., and Duce, D.A., *J. Organometal. Chem.*, *44*, 365 (1972).
116. Davidson, J.L., Green, M., Stone, F.G.A., and Welch, A.J., *J. Chem. Soc. Chem. Commun.*, *1975*, 286.
117. Davis, R.E., Cupper, G.L., and Simpson, H.D., *Amer.*

Crystal. Assoc., Summer Meeting 1970, Ottawa, Abstracts, p. 80.

118. Davis, R., *Inorg. Chem., 14,* 1735 (1975).

119. Davis, R., Green, M., and Hughes, R.P., *J. Chem. Soc. Chem. Commun., 1975,* 405.

120. Davison, A., Green, M.L.H., and Wilkinson, G., *J. Chem. Soc., 1961,* 3172.

121. Davison, J.B., and Bellama, J.M., *Inorg. Chim. Acta, 14,* 263 (1975).

122. De Cian, A., L'Huillier, P.M., and Weiss, R., *Bull. Soc. Chim. Fr., 1973,* 451.

123. De Cian, A., L'Huillier, P.M., and Weiss, R., *Bull. Soc. Chim. Fr., 1973,* 457.

124. Deeming, A.J., Ullah, S.S., Domingos, A.J.P., Johnson, B.F.G., and Lewis, J., *J. Chem. Soc. Dalton Trans., 1974,* 2093.

125. DePuy, C.H., Greene, R.N., and Schroer, T.E., *Chem. Commun., 1968,* 1225.

126. DePuy, C.H., Kobal, V.M., and Gibson, D.H., *J. Organometal. Chem., 13,* 266 (1968).

127. DePuy, C.H., Jones, T., and Parton, R.L., *J. Amer. Chem. Soc., 96,* 5602 (1974).

128. Dickens, B., and Lipscomb, W.N., *J. Amer. Chem. Soc., 83,* 4862 (1961).

129. Dickson, R.S., and Yawney, D.B.W., *Aust. J. Chem., 20,* 77 (1967).

130. Dickson, R.S., and Yawney, D.B.W., *Inorg. Nucl. Chem. Lett., 3,* 209 (1967).

131. Dodman, P., and Hamor, T.A., *J. Chem. Soc. Dalton Trans., 1974,* 1010.

132. Dodman, P., and Tatlow, J.C., *J. Organometal. Chem., 67,* 87 (1974).

133. Dessy, R.E., Stary, F.E., King, R.B., and Waldrop, M., *J. Amer. Chem. Soc., 88,* 471 (1966).

134. Dessy, R.E., and Pohl, R.L., *J. Amer. Chem. Soc., 90,* 1995 (1968).

135. Dewar, M.J.S., and Worley, S.D., *J. Chem. Phys., 50,* 654 (1969).

136. Ecke, G.G., U.S. Patent 3,126,401 (1964).

137. Edwards, R., Howell, J.A.S., Johnson, B.F.G., and Lewis, J., *J. Chem. Soc. Dalton Trans., 1974,* 2105.

138. Eilbracht, P., *Chem. Ber., 109,* 1429 (1976).

139. Eisenstadt, A., and Winstein, S., *Tetrahedron Lett., 1970,* 4603.

140. Eisenstadt, A., and Winstein, S., *Tetrahedron Lett., 1971,* 613.

141. Eisenstadt, A., Scharf, G., and Fuchs, B., *Tetrahedron Lett., 1971,* 679.

142. Eisenstadt, A., *J. Organometal. Chem.*, *38*, C 32 (1972).
143. Eisenstadt, A., *J. Organometal. Chem.*, *97*, 443 (1975).
144. Eiss, R., *Inorg. Chem.*, *9*, 1650 (1970).
145. Elian, M., and Hoffmann, R., *Inorg. Chem.*, *14*, 1058 (1975).
146. Emerson, G.F., and Pettit, R., *J. Amer. Chem. Soc.*, *84*, 4591 (1962).
147. Emerson, G.F., Mahler, J.E., and Pettit, R., *Chem. Ind. (London) 1964*, 836.
148. Emerson, G.F., Mahler, J.E., Kochhar, R., and Pettit, R., *J. Org. Chem.*, *29*, 3620 (1964).
149. Emerson, G.F., Watts, L., and Pettit, R., *J. Amer. Chem. Soc.*, *87*, 131 (1965).
150. Evans, J., Howe, D.V., Johnson, B.F.G., and Lewis, J., *J. Organometal. Chem.*, *61*, C 48 (1973).
151. Falkowski, D.R., Hunt, D.F., Lillya, C.P., and Rausch, M.D., *J. Amer. Chem. Soc.*, *89*, 6387 (1967).
152. Feldhoff, U., Ingenieurarbeit, Universität Essen Gesamthochschule, 1975.
153. Filbey, A.H., Wollensak, J.C., and Keblys, K.A., *American Chemical Society Meeting 1960*, New York City, Abstracts, p. 54P.
154. Filbey, A.H., U.S. Patent 3,178,463 (1965).
155. Fink, W., *Helv. Chim. Acta*, *57*, 167 (1974).
156. Fink, W., *Helv. Chim. Acta*, *58*, 1205 (1975).
157. Fink, W., *Helv. Chim. Acta*, *59*, 276 (1976).
158. Fischer, E.O., and Fischer, R.D., *Angew. Chem.*, *72*, 130 (1960).
159. Fischer, E.O., and Fichtel, K., *Chem. Ber.*, *95*, 2063 (1962).
160. Fischer, E.O., and Müller, J., *Z. Naturforsch.*, *B 17*, 776 (1962).
161. Fischer, E.O., and Müller, J., *J. Organometal. Chem.*, *1*, 89 (1963).
162. Fischer, E.O., and Müller, J., *Z. Naturforsch.*, *B 18*, 413 (1963).
163. Fischer, E.O., and Rühle, H., *Z. anorg. allgem. Chem.*, *341*, 137 (1965).
164. Fischer, E.O., Kreiter, C.G., and Berngruber, W., *J. Organometal. Chem.*, *12*, P 39 (1968).
165. Fischler, I., and Koerner von Gustorf, E.A., *Z. Naturforsch.*, *B 30*, 291 (1975).
166. Fischler, I., Grevels, F.-W., and Wakatsuki, Y., VIth IUPAC Symp. Photochem., Aix-en-Provence, July 1976, Contributed Papers.
167. Fitch, J.W., and Herbold, H.E., *Inorg. Chem.*, *9*, 1926 (1970).
168. Foreman, M.I., *J. Organometal. Chem.*, *39*, 161 (1972).

169. Franck-Neumann, M., and Martina, D., *Tetrahedron Lett.*, *1975*, 1759.

170. Frankel, E.N., Jones, E.P., and Glass, C.A., *J. Amer. Oil Chem. Soc.*, *41*, 392 (1964).

171. Frankel, E.N., Emken, E.A., Peters, H.M., Davison, V.L., and Butterfield, R.O., *J. Org. Chem.*, *29*, 3292 (1964).

172. Frankel, E.N., Emken, E.A., and Davison, V.L., *J. Org. Chem.*, *30*, 2739 (1965).

173. Gajewski, J.J., and Shih, C.N., *Tetrahedron Lett.*, *1973*, 3959.

174. Gibson, D.H., and Vonnahme, R.L., *J. Amer. Chem. Soc.*, *94*, 5090 (1972).

175. Gibson, D.H., and Vonnahme, R.L., *J. Chem. Soc. Chem. Commun.*, *1972*, 1021.

176. Gibson, D.H., and Vonnahme, R.L., *J. Organometal. Chem.*, *70*, C 33 (1974).

177. Gill, G.B., Gourlay, N., Johnson, A.W., and Mahendran, M., *J. Chem. Soc. D, Chem. Commun.*, *1969*, 631.

178. Graf, R.E., and Lillya, C.P., *J. Chem. Soc. Chem. Commun.*, *1973*, 271.

179. Graf, R.E., and Lillya, C.P., *J. Organometal. Chem.*, *47*, 413 (1973).

180. Grant, G.F., and Pauson, P.L., *J. Organometal. Chem.*, *9*, 553 (1967).

181. Greaves, E.O., Knox, G.R., and Pauson, P.L., *J. Chem. Soc. D, Chem. Commun.*, *1969*, 1124.

182. Greaves, E.O., Knox, G.R., Pauson, P.L., Toma, S., Sim, G.A., and Woodhouse, D.I., *J. Chem. Soc. Chem. Commun.*, *1974*, 257.

183. Green, M., and Lewis, B., *J. Chem. Soc. Chem. Commun.*, *1973*, 114.

184. Green, M., Hughes, R.P., and Welch, A.J., *J. Chem. Soc. Chem. Commun.*, *1975*, 487.

185. Green, M., Lewis, B., Daly, J.J., and Sanz, F., *J. Chem. Soc. Dalton Trans.*, *1975*, 1118.

186. Green, M., and Lewis, B., *J. Chem. Soc. Dalton Trans.*, *1975*, 1137.

187. Green, M.L.H., Pratt, L., and Wilkinson, G., *J. Chem. Soc.*, *1959*, 3753.

188. Green, M.L.H., Pratt, L., and Wilkinson, G., *J. Chem. Soc.*, *1960*, 989.

189. Green, M.L.H., and Nagy, P.L.I., *Z. Naturforsch.*, *B 18*, 162 (1963).

190. Greene, R.N., DePuy, C.H., and Schroer, T.E., *J. Chem. Soc. C*, *1971*, 3115.

191. Grevels, F.-W., Schneider, K., and Feldhoff, U., unpublished results, 1975.

192. Grevels, F.-W., Feldhoff, U., and Schneider, K.,

VII th Int. Conf. Organometal. Chem., Venice, Italy 1975, Abstracts, No. 192.

193. Grevels, F.-W., Frühauf, H.W., de Paoli, M.A., and Krüger, C., unpublished results, 1976.

194. Günther, H., and Wenzl, R., *Angew. Chem.*, *81*, 919 (1969); *Angew. Chem. Int. Ed. Engl.*, *8*, 900 (1969).

195. Günther, H., Wenzl, R., and Klose, H., *J. Chem. Soc. D, Chem. Commun.*, *1970*, 605.

196. Gutowsky, H.S., and Jonas, J., *Inorg. Chem.*, *4*, 430 (1965).

197. Haas, M.A., and Wilson, J.M., *J. Chem. Soc. B*, *1968*, 104.

198. Hagihara, N., *Ann. N. Y. Acad. Sci.*, *125*, 98 (1965).

199. Hallam, B.F., and Pauson, P.L., *J. Chem. Soc.*, *1958*, 642.

200. Hallam, B.F., and Pauson, P.L., *J. Chem. Soc.*, *1958*, 646.

201. Hardy, A.D.U., and Sim, G.A., *J. Chem. Soc. Dalton Trans.*, *1972*, 2305.

202. Hashmi, M.A., Munro, J.D., Pauson, P.L., and Williamson, J.M., *J. Chem. Soc. A*, *1967*, 240.

203. Herberich, G.E., and Müller, H., *Angew. Chem.*, *83*, 1020 (1971); *Angew. Chem. Int. Ed. Engl.*, *10*, 937 (1971).

204. Herberich, G.E., and Müller, H., *Chem. Ber.*, *104*, 2781 (1971).

205. Herberich, G.E., *et al.*, *Chem. Ber.*, in press.

206. Herberich, G.E., private communication, 1976.

207. Hill, A.E., and Hoffmann, H.M.R., *J. Chem. Soc. Chem. Commun.*, *1972*, 574.

208. Hine, K.E., Johnson, B.F.G., and Lewis, J., *J. Chem. Soc. Chem. Commun.*, *1975*, 81.

209. Hitchcock, P.B., and Mason, R., *Chem. Commun.*, *1967*, 242.

210. Hoehn, H.H., Pratt, L., Watterson, K.F., and Wilkinson, G., *J. Chem. Soc.*, *1961*, 2738.

211. Holland, J.M., and Jones, D.W., *Chem. Commun.*, *1967*, 946.

212. Howell, J.A.S., Johnson, B.F.G., Josty, P.L., and Lewis, J., *J. Organometal. Chem.*, *39*, 329 (1972).

213. Hübel, W., and Braye, E.H., *J. Inorg. Nucl. Chem.*, *10*, 250 (1959).

214. Hübel, K.W., and Weiss, E.L., Brit. Patent 913,763 (1962).

215. Hunt, D.F., Lillya, C.P., and Rausch, M.D., *J. Amer. Chem. Soc.*, *90*, 2561 (1968).

216. Hunt, D.F., Farrant, G.C., and Rodeheaver, G.T., *J. Organometal. Chem.*, *38*, 349 (1972).

217. Hunt, D.F., Russell, J.W., and Torian, R.L., *J. Organo-*

metal. Chem., 43, 175 (1972).

218. Hursthouse, M.B., Massey, A.G., Tomlinson, A.J., and Urch, D.S., *J. Organometal. Chem., 21,* P 51 (1970).

219. Ihrman, K.G., and Coffield, T.H., U.S. Patent 3,164,621 (1965).

220. Immirzi, A., *J. Organometal. Chem., 76,* 65 (1974).

221. Impastato, F.J., and Ihrman, K.G., *J. Amer. Chem. Soc., 83,* 3726 (1961).

222. Ireland, R.E., Brown, G.G., Jr., Stanford, R.H., Jr., and McKenzie, T.C., *J. Org. Chem., 39,* 51 (1974).

223. Janse Van Vuuren, P., Fletterick, R.J., Meinwald, J., and Hughes, R.E., *J. Chem. Soc. D, Chem. Commun., 1970,* 883.

224. Janse Van Vuuren, P., Fletterick, R.J., Meinwald, J., and Hughes, R.E., *J. Amer. Chem. Soc., 93,* 4394 (1971).

225. John, G.R., Kane-Maguire, L.A.P., and Eaborn, C., *J. Chem. Soc. Chem. Commun., 1975,* 481.

226. Johnson, B.F.G., Lewis, J., McArdle, P., and Randall, G.L.P., *J. Chem. Soc. Dalton Trans., 1972,* 456.

227. Johnson, B.F.G., Lewis, J., Matheson, T.W., Ryder, I.E., and Twigg, M.V., *J. Chem. Soc. Chem. Commun., 1974,* 269.

228. Johnson, B.F.G., Lewis, J., and Twigg, M.V., *J. Chem. Soc. Dalton Trans., 1974,* 2546.

229. Johnson, B.F.G., Lewis, J., and Thompson, D.J., *Tetrahedron Lett., 1974,* 3789.

230. Johnson, B.F.G., Lewis, J., and Twigg, M.V., *J. Chem. Soc. Dalton Trans., 1976,* 421.

231. Jones, E.R.H., Wailes, P.C., and Whiting, M.C., *J. Chem. Soc., 1955,* 4021.

232. Joshi, K.K., *J. Chem. Soc. A, 1966,* 594.

233. Kaesz, H.D., King, R.B., Manuel, T.A., Nichols, L.D., and Stone, F.G.A., *J. Amer. Chem. Soc., 82,* 4749 (1960).

234. Kalck, P., and Poilblanc, R., *C.R. Acad. Sci., Ser. C, 274,* 66 (1972).

235. Kane-Maguire, L.A.P., *J. Chem. Soc. A, 1971,* 1602.

236. Kane-Maguire, L.A.P., and Mansfield, C.A., *J. Chem. Soc. Chem. Commun., 1973,* 540.

237. Keller, C.E., Emerson, G.F., and Pettit, R., *J. Amer. Chem. Soc., 87,* 1388 (1965).

238. Kiji, J., and Iwamoto, M., *Bull. Chem. Soc. Jap., 41,* 1483 (1968).

239. King, R.B., Manuel, T.A., and Stone, F.G.A., *J. Inorg. Nucl. Chem., 16,* 233 (1961).

240. King, R.B., *J. Amer. Chem. Soc., 84,* 4705 (1962).

241. King, R.B., *Inorg. Chem., 2,* 642 (1963).

242. King, R.B., *Inorg. Chem., 2,* 807 (1963).

243. King, R.B., and Bisnette, M.B., *Inorg. Chem., 4,* 1663 (1965).

244. King, R.B., *J. Amer. Chem. Soc.*, *90*, 1429 (1968).
245. King, R.B., and Efraty, A., *J. Amer. Chem. Soc.*, *93*, 4950 (1971).
246. King, R.B., and Efraty, A., *J. Amer. Chem. Soc.*, *94*, 3773 (1972).
247. King, R.B., and Eavenson, C.W., *J. Organometal. Chem.*, *42*, C 95 (1972).
248. King, R.B., and Harmon, C.A., *J. Amer. Chem. Soc.*, *98*, 2409 (1976).
249. Kochhar, R.K., and Pettit, R., *J. Organometal. Chem.*, *6*, 272 (1966).
250. Koerner von Gustorf, E., and Hogan, J.C., *Tetrahedron Lett.*, *1968*, 3191.
251. Koerner von Gustorf, E., Buchkremer, J., Pfajfer, Z., and Grevels, F.-W., *Angew. Chem.*, *83*, 249 (1971); *Angew. Chem. Int. Ed. Engl.*, *10*, 260 (1971).
252. Koerner von Gustorf, E., Pfajfer, Z., and Grevels, F.-W., *Z. Naturforsch.*, *B 26*, 66 (1971).
253. Koerner von Gustorf, E., Jaenicke, O., and Polansky, O.E., *Angew. Chem.*, *84*, 547 (1972); *Angew. Chem. Int. Ed. Engl.*, *12*, 532 (1972).
254. Koerner von Gustorf, E., Fischler, I., Leitich, J., and Dreeskamp, H., *Angew. Chem.*, *84*, 1143 (1972); *Angew. Chem. Int. Ed. Engl.*, *12*, 1088 (1972).
255. Koerner von Gustorf, E., Buchkremer, J., Pfajfer, Z., and Grevels, F.-W., Ger. Patent 2,105,627 (1972).
256. Kolshorn, H., Meier, H., and Müller, E., *Tetrahedron Lett.*, *1971*, 1469.
257. Koptyug, V.A., Berezina, R.N., and Shubin, V.G., *Tetrahedron Lett.*, *1968*, 673.
258. Korat, M., and Ginsburg, D., *Tetrahedron*, *29*, 2373 (1973).
259. Kreiter, C.G., Stüber, S., and Wackerle, L., *J. Organometal. Chem.*, *66*, C 49 (1974).
260. Krespan, C.G., *J. Org. Chem.*, *40*, 261 (1975).
261. Kruck, T., and Knoll, L., *Chem. Ber.*, *105*, 3783 (1972).
262. Kruck, T., Knoll, L., and Laufenberg, J., *Chem. Ber.*, *106*, 697 (1973).
263. Kruck, T., and Knoll, L., *Z. Naturforsch.*, *B 28*, 34 (1973).
264. Kruczynski, L., and Takats, J., *J. Amer. Chem. Soc.*, *96*, 932 (1974).
265. Krüger, C., *J. Organometal. Chem.*, *22*, 697 (1970).
266. Krüger, C., and Tsay, Y.-H., *Angew. Chem.*, *83*, 250 (1971); *Angew. Chem. Int. Ed. Engl.*, *10*, 261 (1971).
267. Kuhn, D.E., and Lillya, C.P., *J. Amer. Chem. Soc.*, *94*, 1682 (1972).

268. Lallemand, J.-Y., Laszlo, P., Muzette, C., and Stockis, A., *J. Organometal. Chem.*, *91*, 71 (1975).
269. Landesberg, J.M., and Sieczkowski, J., *J. Amer. Chem. Soc.*, *90*, 1655 (1968).
270. Landesberg, J.M., and Sieczkowski, J., *J. Amer. Chem. Soc.*, *91*, 2120 (1969).
271. Landesberg, J.M., and Sieczkowski, J., *J. Amer. Chem. Soc.*, *93*, 972 (1971).
272. Landesberg, J.M., and Katz, L., *J. Organometal. Chem.*, *33*, C 15 (1971).
273. Landesberg, J.M., Katz, L., and Olsen, C., *J. Org. Chem.*, *37*, 930 (1972).
274. Leto, J.R., and Cotton, F.A., *J. Amer. Chem. Soc.*, *81*, 2970 (1959).
275. Lewis, J., and Parkins, A.W., *Chem. Commun.*, *1968*, 1194.
276. Lillya, C.P., and Sahatjian, R.A., *J. Organometal. Chem.*, *25*, C 67 (1970).
277. Lillya, C.P., and Sahatjian, R.A., *J. Organometal. Chem.*, *32*, 371 (1971).
278. Lombardo, L., Wege, D., and Wilkinson, S.P., *Aust. J. Chem.*, *27*, 143 (1974).
279. Lumbroso, H., and Bertin, D.M., *J. Organometal. Chem.*, *108*, 111 (1976).
280. Lundquist, R.T., and Cais, M., *J. Org. Chem.*, *27*, 1167 (1962).
281. Mackenzie, R., and Timms, P.L., *J. Chem. Soc. Chem. Commun.*, *1974*, 650.
282. Maglio, G., Musco, A., Palumbo, R., and Sirigu, A., *J. Chem. Soc. D, Chem. Commun.*, *1971*, 100.
283. Maglio, G., Musco, A., and Palumbo, R., *J. Organometal. Chem.*, *32*, 127 (1971).
284. Maglio, G., and Palumbo, R., *J. Organometal. Chem.*, *76*, 367 (1974).
285. Mahler, J.E., and Pettit, R., *J. Amer. Chem. Soc.*, *85*, 3955 (1963).
286. Mahler, J.E., Gibson, D.H., and Pettit, R., *J. Amer. Chem. Soc.*, *85*, 3959 (1963).
287. Mansfield, C.A., Al-Kathumi, K.M., and Kane-Maguire, L.A.P., *J. Organometal. Chem.*, *71*, C 11 (1974).
288. Mantzaris, J., and Weissberger, E., *Tetrahedron Lett.*, *1972*, 2815.
289. Manuel, T.A., and Stone, F.G.A., *J. Amer. Chem. Soc.*, *82*, 366 (1960).
290. Manuel, T.A., and Stone, F.G.A., *J. Amer. Chem. Soc.*, *82*, 6240 (1960).
291. Manuel, T.A., Stafford, S.L., and Stone, F.G.A., *J. Amer. Chem. Soc.*, *83*, 3597 (1961).
292. Manuel, T.A., *Inorg. Chem.*, *3*, 510 (1964).

293. Manuel, T.A., *Inorg. Chem., 3,* 1794 (1964).
294. Maoz, N., Mandelbaum, A., and Cais, M., *Tetrahedron Lett., 1965,* 2087.
295. Margulis, T.N., Schiff, L., and Rosenblum, M., *J. Amer. Chem. Soc., 87,* 3269 (1965).
296. McArdle, P., and Sherlock, H., *J. Organometal. Chem., 52,* C 29 (1973).
297. McFarlane, W., Pratt, L., and Wilkinson, G., *J. Chem. Soc., 1963,* 2162.
298. Menachem, Y., and Eisenstadt, A., *J. Organometal. Chem., 33,* C 29 (1971).
299. Mills, O.S., and Robinson, G., *Proc. Chem. Soc., 1960,* 421.
300. Mills, O.S., and Robinson, G., *Acta Crystallogr., 16,* 758 (1963).
301. Moriarty, R.M., Chen, K.-N., Churchill, M.R., and Chang, S.W.-Y., *J. Amer. Chem. Soc., 96,* 3661 (1974).
302. Müller, H., and Herberich, G.E., *Chem. Ber., 104,* 2772 (1971).
303. Müller, J., and Mertschenk, B., *Chem. Ber., 105,* 3346 (1972).
304. Müller, J., Herberich, G.E., and Müller, H., *J. Organometal. Chem., 55,* 165 (1973).
305. Murdoch, H.D., and Weiss, E., *Helv. Chim. Acta, 45,* 1156 (1962).
306. Murdoch, H.D., and Weiss, E., *Helv. Chim. Acta, 46,* 1588 (1963).
307. Musco, A., Paiaro, G., and Palumbo, R., *Chim. Ind. (Milan), 50,* 669 (1968).
308. Musco, A., Palumbo, R., and Paiaro, G., *Inorg. Chim. Acta, 5,* 157 (1971).
309. Nakamura, A., and Hagihara, N., *Mem. Inst. Sci. Ind. Res. Osaka, 17,* 187 (1960).
310. Nakamura, A., and Tsutsui, M., *J. Med. Chem., 6,* 796 (1963).
311. Nakamura, A., Kim, P.-J., and Hagihara, N., *Bull. Chem. Soc. Jap., 37,* 292 (1964).
312. Nakamura, A., Kim, P.-J., and Hagihara, N., *J. Organometal. Chem., 3,* 7 (1965).
313. Nakamura, A., *Bull. Chem. Soc. Jap., 39,* 543 (1966).
314. Nametkin, N.S., Nekhaev, A.I., Tyurin, V.D., and Dontsova, V.N., *Izv. Akad. Nauk SSSR, Ser. Khim., 1973,* 959; *Bull. Acad. Sci. USSR, Div. Chem. Ser., 1973,* 931.
315. Nametkin, N.S., Tyurin, V.D., Nazur-Al'-Laddavi, M., and Gromasheva, N.A., *Izv. Akad. Nauk SSSR, Ser. Khim., 1973,* 2170; *Bull. Acad. Sci. USSR, Div. Chem. Ser., 1973,* 2130.
316. Nametkin, N.S., Gubin, S.P., Ivanov, V.I., and Tyurin,

V.D., *Izv. Akad. Nauk SSSR, Ser. Khim., 1974,* 486; *Bull. Chem. Soc. USSR, Div. Chem. Ser., 1974,* 458.
317. Nametkin, N.S., Nekhaev, A.I., and Tyurin, V.D., *Izv. Akad. Nauk SSSR, Ser. Khim., 1974,* 890; *Bull. Acad. Sci. USSR, Div. Chem. Ser., 1974,* 852.
318. Nametkin, N.S., Tyurin, V.D., Nekhaev, A.I., Ivanov, V.I., and Bayaouova, F.S., *J. Organometal. Chem., 107,* 377 (1976).
319. Nelson, S.M., and Sloan, M., *J. Chem. Soc. Chem. Commun., 1972,* 745.
320. Nelson, S.M., Regan, C.M., and Sloan, M., *J. Organometal. Chem., 96,* 383 (1975).
321. Nesmeyanov, A.N., Anisimov, K.N., and Kolobova, N.E., *Izv. Akad. Nauk SSSR, Ser. Khim., 1962,* 722; *Bull. Acad. Sci. USSR, Div. Chem. Ser., 1962,* 669.
322. Nesmeyanov, A.N., Kolobova, N.E., Anisimov, K.N., and Skripkin, V.V., *Izv. Akad. Nauk SSSR, Ser. Khim., 1969,* 2859; *Bull. Acad. Sci. USSR, Div. Chem. Ser., 1969,* 2698.
323. Nesmeyanov, A.N., Kolobova, N.E., Skripkin, V.V., Anisimov, K.N., and Fedorov, L.A., *Dokl. Akad. Nauk SSSR, 195,* 368 (1970); *Doklady Chemistry, 195,* 816 (1970).
324. Nesmeyanov, A.N., Anisimov, K.N., and Magomedov, G.K., *Izv. Akad. Nauk SSSR, Ser. Khim., 1970,* 715; *Bull. Acad. Sci. USSR, Div. Chem. Ser., 1970,* 676.
325. Nesmeyanov, A.N., Anisimov, K.N., and Magomedov, G.K., *Izv. Akad. Nauk SSSR, Ser. Khim., 1970,* 959; *Bull. Acad. Sci. USSR, Div. Chem. Ser., 1970,* 916.
326. Noack, K., *Helv. Chim. Acta, 45,* 1847 (1962).
327. Noyori, R., Makino, S., and Takaya, H., *J. Amer. Chem. Soc., 93,* 1272 (1971).
328. Öfele, K., and Dotzauer, E., *J. Organometal. Chem., 42,* C 87 (1972).
329. Parshall, G.W., and Wilkinson, G., *J. Chem. Soc., 1962,* 1132.
330. Paul, I.C., Johnson, S.M., Paquette, L.A., Barrett, J.H., and Haluska, R.J., *J. Amer. Chem. Soc., 90,* 5023 (1968).
331. Pelter, A., Gould, K.J., and Kane-Maguire, L.A.P., *J. Chem. Soc. Chem. Commun., 1974,* 1029.
332. Pettit, R., *J. Amer. Chem. Soc., 81,* 1266 (1959).
333. Pettit, R., Emerson, G.F., and Mahler, J., *J. Chem. Educ., 40,* 175 (1963).
334. Pettit, R., and Emerson, G.F., *Advan. Organometal. Chem., 1,* 1 (1964).
335. Pettit, R., *Ann. N. Y. Acad. Sci., 125,* 89 (1965).
336. Piottukh-Peletskii, V.N., Berezina, R.N., Rezbukhin, A.I., and Shubin, V.G., *Izv. Akad. Nauk SSSR, Ser.*

Khim., *1973*, 2083; *Bull. Acad. Sci. USSR, Div. Chem. Ser.*, *1973*, 2027.

337. Piper, T.S., Cotton, F.A., and Wilkinson, G., *J. Inorg. Nucl. Chem.*, *1*, 165 (1955).

338. Rathke, J.W., and Muetterties, E.L., *J. Amer. Chem. Soc.*, *97*, 3272 (1975).

339. Rausch, M.D., and Schrauzer, G.N., *Chem. Ind. (London)*, *1959*, 957.

340. Reckziegel, A., and Bigorgne, M., *J. Organometal. Chem.*, *3*, 341 (1965).

341. Reger, D.L., and Gabrielli, A., *J. Amer. Chem. Soc.*, *97*, 4421 (1975).

342. Reihlen, H., Gruhl, A., v. Hessling, G., and Pfrengle, O., *Justus Liebigs Ann. Chem.*, *482*, 161 (1930).

343. Reppe, W., and Vetter, H., *Justus Liebigs Ann. Chem.*, *582*, 133 (1953).

344. Resibois, B., and Brunet, J.C., *Ann. Chim.*, *5*, 199 (1970).

345. Retcofsky, H.L., Frankel, E.N., and Gutowsky, H.S., *J. Amer. Chem. Soc.*, *88*, 2710 (1966).

346. Roe, D.M., and Massey, A.G., *J. Organometal. Chem.*, *17*, 429 (1969).

347. Rosenblum, M., and Gatsonis, C., *J. Amer. Chem. Soc.*, *89*, 5074 (1967).

348. Roth, W.R., and Meier, J.D., *Tetrahedron Lett.*, *1967*, 2053.

349. Sarel, S., Ben-Shoshan, R., and Kirson, B., *J. Amer. Chem. Soc.*, *87*, 2517 (1965).

350. Sarel, S., Ben-Shoshan, R., and Kirson, B., *Israel J. Chem.*, *10*, 787 (1972).

351. Sarel, S., Felzenstein, A., Victor, R., and Yovell, J., *J. Chem. Soc. Chem. Commun.*, *1974*, 1025.

352. Schäfer, W., Kerrinnes, H.-J., and Langbein, U., *Z. anorg. allgem. Chem.*, *406*, 101 (1974).

353. Schäfer, W., Zschunke, A., Kerrinnes, H.-J., and Langbein, U., *Z. Anorg. Allgem. Chem.*, *406*, 105 (1974).

354. Schiavon, G., Paradisi, C., and Boanini, C., *Inorg. Chim. Acta*, *14*, L 5 (1975).

355. Scholes, G., Graham, C.R., and Brookhart, M., *J. Amer. Chem. Soc.*, *96*, 5665 (1974).

356. Schrauzer, G.N., *Chem. Ind. (London)*, *1958*, 1403.

357. Schrauzer, G.N., *Chem. Ind. (London)*, *1958*, 1404.

358. Schrauzer, G.N., *J. Amer. Chem. Soc.*, *81*, 5307 (1959).

359. Schrauzer, G.N., Glockner, P., and Merenyi, R., *Angew. Chem.*, *76*, 498 (1964); *Angew. Chem. Int. Ed. Engl.*, *3*, 509 (1964).

360. Schrauzer, G.N., and Kratel, G., *J. Organometal. Chem.*,

2, 336 (1964).

361. Schrauzer, G.N., Glockner, P., Reid, K.I.G., and Paul, I.C., *J. Amer. Chem. Soc.*, *92*, 4479 (1970).

362. Shubin, V.G., Berezina, R.N., Derendyaev, B.G., and Koptyug, V.A., *Izv. Akad. Nauk SSSR, Ser. Khim.*, *1970*, 2747; *Bull. Acad. Sci. USSR, Div. Chem. Ser.*, *1970*, 2583.

363. Shubin, V.G., Berezina, R.N., and Piottukh-Peletski, V.N., *J. Organometal. Chem.*, *54*, 239 (1973).

364. Shvo, Y., and Hazum, E., *J. Chem. Soc. Chem. Commun.*, *1974*, 336.

365. Shvo, Y., and Hazum, E., *J. Chem. Soc. Chem. Commun.*, *1975*, 829.

366. Skarstad, P., Janse Van Vuuren, P., Meinwald, J., and Hughes, R.E., *J. Chem. Soc. Perkin Trans. II*, *1975*, 88.

367. Sorensen, T.S., and Jablonski, C.R., *J. Organometal. Chem.*, *25*, C 62 (1970).

368. Streith, J., and Cassal, J.M., *Angew. Chem.*, *80*, 117 (1968); *Angew. Chem. Int. Ed. Engl.*, *7*, 129 (1968).

369. Stüber, S., and Ugi, I., *Synthesis*, *1974*, 437.

370. Tayim, H.A., Bouldoukian, A., and Kharboush, M., *Inorg. Nucl. Chem. Lett.*, *8*, 231 (1972).

371. Thompson, D.J., *J. Organometal. Chem.*, *108*, 381 (1976).

372. Timms, P.L., *Angew. Chem.*, *87*, 295 (1975); *Angew. Chem. Int. Ed. Engl.*, *14*, 273 (1975).

373. Tomlinson, A.J., and Massey, A.G., *J. Organometal. Chem.*, *8*, 321 (1967).

374. Treichel, P.M., and Shubkin, R.L., *J. Organometal. Chem.*, *5*, 488 (1966).

375. Treichel, P.M., and Shubkin, R.L., *Inorg. Chem.*, *6*, 1328 (1967).

376. Ungermann, C.B., and Caulton, K.G., *J. Organometal. Chem.*, *94*, C 9 (1975).

377. Veltman, P.L., U.S. Patent 2,409,167 (1946).

378. Victor, R., Ben-Shoshan, R., and Sarel, S., *J. Chem. Soc. D, Chem. Commun.*, *1970*, 1680.

379. Victor, R., Ben-Shoshan, R., and Sarel, S., *Tetrahedron Lett.*, *1970*, 4253.

380. Victor, R., Ben-Shoshan, R., and Sarel, S., *Tetrahedron Lett.*, *1970*, 4257.

381. Victor, R., Ben-Shoshan, R., and Sarel, S., *J. Org. Chem.*, *37*, 1930 (1972).

382. Victor, R., and Ben-Shoshan, R., *J. Chem. Soc. Chem. Commun.*, *1974*, 93.

383. Wang, A.H.-J., Paul, I.C., and Aumann, R., *J. Organometal. Chem.*, *69*, 301 (1974).

384. Warren, J.D., and Clark, R.J., *Inorg. Chem.*, *9*, 373 (1970).

385. Warren, J.D., Busch, M.A., and Clark, R.J., *Chem.*, *11*, 452 (1972).
386. Watterson, K.F., and Wilkinson, G., *Chem. Ind. (London)*, *1959*, 991.
387. Weiss, E., and Hübel, W., *J. Inorg. Nucl. Chem.*, *11*, 42 (1959).
388. Weiss, E., Merényi, R.G., and Hübel, W., *Chem. Ind. (London)*, *1960*, 407.
389. Weiss, E., and Hübel, W., *Angew. Chem.*, *73*, 298 (1961).
390. Weiss, E., Merényi, R., and Hübel, W., *Chem. Ber.*, *95*, 1170 (1962).
391. Weiss, E., and Hübel, W., *Chem. Ber.*, *95*, 1179 (1962).
392. Weiss, E., and Hübel, W., *Chem. Ber.*, *95*, 1186 (1962).
393. Whitesides, T.H., and Arhart, R.W., *J. Amer. Chem. Soc.*, *93*, 5296 (1971).
394. Whitesides, T.H., and Neilan, J.P., *J. Amer. Chem. Soc.*, *95*, 5811 (1973).
395. Whitesides, T.H., and Slaven, R.W., *J. Organometal. Chem.*, *67*, 99 (1974).
396. Whitesides, T.H., and Arhart, R.W., *Inorg. Chem.*, *14*, 209 (1975).
397. Whitesides, T.H., and Budnik, R.A., *Inorg. Chem.*, *14*, 664 (1975).
398. Whitesides, T.H., and Neilan, J.P., *J. Amer. Chem. Soc.*, *97*, 907 (1975).
399. Whitesides, T.H., and Shelly, J., *J. Organometal. Chem.*, *92*, 215 (1975).
400. Whitesides, T.H., and Neilan, J.P., *J. Amer. Chem. Soc.*, *98*, 63 (1976).
401. Whiting, D.A., *Cryst. Struct. Comm.*, *1*, 379 (1972).
402. Whitlock, H.W., Jr., and Chuah, Y.N., *J. Amer. Chem. Soc.*, *86*, 5030 (1964).
403. Whitlock, H.W., Jr., and Chuah, Y.N., *J. Amer. Chem. Soc.*, *87*, 3605 (1965).
404. Whitlock, H.W., Jr., Reich, C.R., and Markezich, R.L., *J. Amer. Chem. Soc.*, *92*, 6665 (1970).
405. Williams-Smith, D.L., Wolf, L.R., and Skell, P.S., *J. Amer. Chem. Soc.*, *94*, 4042 (1972).
406. Winters, R.E., and Kiser, R.W., *J. Chem. Phys.*, *69*, 3198 (1965).
407. Young, D.A.T., Holmes, J.R., and Kaesz, H.D., *J. Amer. Chem. Soc.*, *91*, 6968 (1969).
408. Zakharkin, L.I., Kovredov, A.I., Orlova, L.V., and Fedorov, L.A., *Izv. Akad. Nauk SSSR, Ser. Khim.*, *1969*, 2343; *Bull. Acad. Sci. USSR, Div. Chem. Ser.*, *1969*, 2204.
409. Zakharkin, L.I., Kazantsev, A.V., and Litovchenko, L.E., *Izv. Akad. Nauk SSSR, Ser. Khim.*, *1971*, 2050; *Bull.*

Acad. Sci. USSR, Div. Chem. Ser., 1971, 1932.

410. Zakharkin, L.I., Orlova, L.V., Kovredov, A.I., Fedorov, L.A., and Lokshin, B.V., *J. Organometal. Chem., 27,* 95 (1971).

411. Zakharova, I.A., Salyn, Ya.V., Garbouzova, I.A., Aleksanyan, V.T., and Prianichnicova, M.A., *J. Organometal. Chem., 102,* 227 (1975).

412. Grevels, F.-W., Feldhoff, U., Leitich, J., and Krüger, C., *J. Organometal. Chem., 118,* 79 (1976).

Stabilizing of Unstable Species with Carbonyliron

By JOSEPH M. LANDESBERG

Department of Chemistry, Adelphi University
Garden City, New York 11530, U.S.A.

TABLE OF CONTENTS

ISBN 0-12-417101-X (v.1)

I. INTRODUCTION

In recent years organic molecules of unusually high instability have been synthesized by incorporating a transition metal into the system *(107,120)*. This review will deal with those highly elusive, unstable organic molecules that have been successfully stabilized and isolated as their carbonyliron complexes. Stabilized iron complexes of reactive intermediates (*e.g.* carbenes) are also included. Pertinent preparations, physical and chemical properties of these systems will be discussed. Omitted from this review are any comparative or speculative discussions of analogous complexes of transition metals other than iron.

The thrust in this review is to the variety of organic ligands complexed with iron. Little detailed analysis will be devoted to bonding since the diversity of the ligands is such that generalities are not appropriate and specifics demand too much space. Detailed presentations which do concern themselves with bonding properties have appeared, and the reader is referred to these monographs *(71,90)*; inherent in these discussions are the probabilities that molecular orbital combinations exist between ligand and metal which lead to an overall lowering of energy for the system. Further aspects of structure and bonding will be dealt with in the appropriate section.

II. STABILIZED OLEFINS

A. CYCLOBUTADIENE

A stabilized complex of unsubstituted cyclobutadiene was first isolated in 1965 *(55)*. Treatment of *cis*-3,4-dichlorocyclobutene with excess enneacarbonyldiiron produced (cyclobutadiene)tricarbonyliron (1) in 40 % yield (eq. [1]).

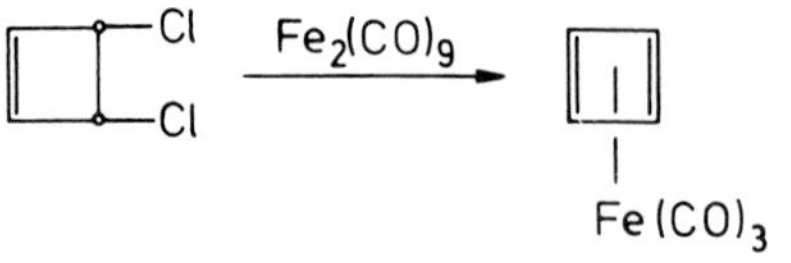

[1]

1

The synthesis is noteworthy here for several reasons. While compounds such as (tetraphenylcyclobutadiene)tricarbonyliron were already known *(45)*, complex 1 was the first unsubstituted derivative isolated and the first cyclobutadiene complex obtained by methods other than by acetylene oligomerization. The dehalogenation sequence (eq. [1]) has found general use for the preparation of other cyclobutadiene derivatives, *e.g.* (benzocyclobutadiene)tricarbonyliron (2) *(55)*,

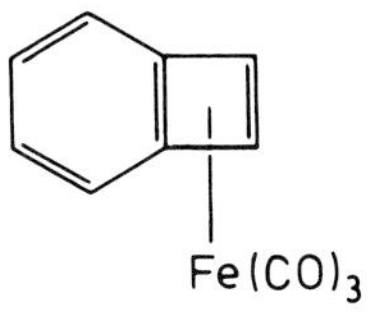

2

and for the synthesis of other unstable molecules as their carbonyliron complexes *(vide infra)*.

Ultraviolet irradiation of α-pyrone in the presence of pentacarbonyliron yields 1 and (α-pyrone)tricarbonyliron *(118)*. This preparation demonstrates the versatility of techniques for direct synthesis from suitable starting materials. Further aspects of cyclobutadiene chemistry is covered in a particular chapter of this book.

B. o-QUINODIMETHANES

o-Quinodimethanes have a comparatively young history. Experimental evidence verifies these species as kinetically unstable, reactive intermediates *(74)*. Stabilization with tricarbonyliron of the *seco* isomer of benzocyclobutene of the appropriate configuration, namely (*o*-quinodimethane)tricarbonyliron (3), was accomplished as shown in eq. [2] *(119)*.

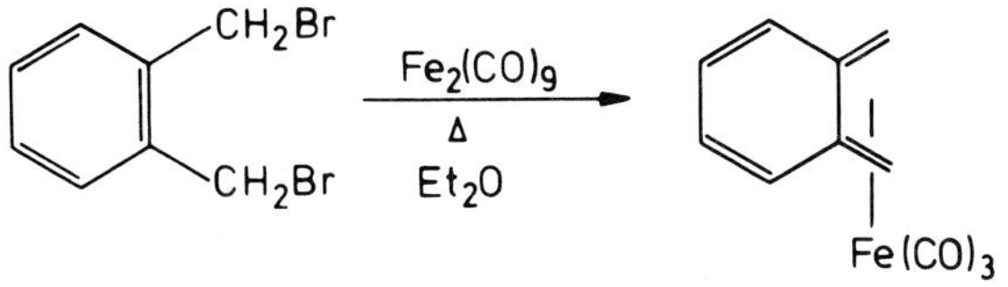

[2]

3

The similarity of the NMR spectrum of 3 to that of (butadiene)tricarbonyliron suggests that the $Fe(CO)_3$ group in η^4-bonded to C(8), C(5), C(6), and C(7) of the ligand. Similarly, (2,2-dimethylisoindene)tricarbonyliron was prepared from 2,2-dimethyl-1,3-dibromoindane *(119)*.

Compound 3 is highly stable. At 400°C the complex

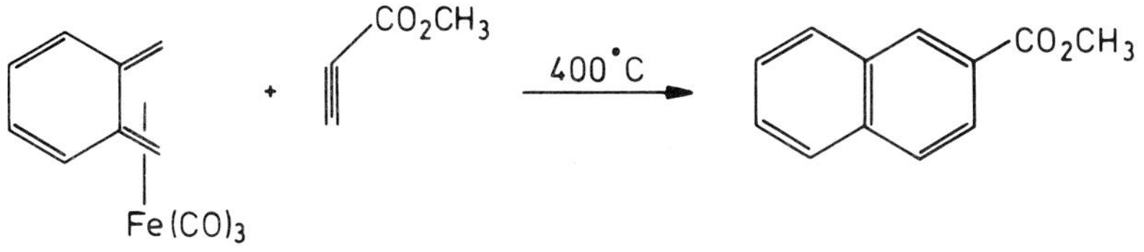

[3]

3

pyrolyzes to benzocyclobutene; if the pyrolysis is carried out in the presence of methyl acetylenecarboxylate, a naphthalene derivative is obtained (eq. [3]) *(113)*. Oxidation with Ce(IV) salts gives high yields of β-indanone *(113)*.

C. *TRIMETHYLENEMETHANES*

Trimethylenemethane is a theoretically important intermediate since the central carbon atom of this molecule attains the maximum π-bond order possible for any carbon atom. It is a short-lived, highly reactive intermediate *(33)* which has a triplet ground electronic state *(49)*. The system is stable at -185°C and has an observable ESR spectrum at that temperature *(48)*. Theoretical treatments *(56,100)* predict that little π-energy would be lost in donation to a transition metal, and since there is an unfilled nonbonding orbital, as with cyclobutadiene, trimethylenemethane should be an excellent acceptor in transition metal complexes. These predictions are realized in the complex (trimethylenemethane)tricarbonyliron (4) (from

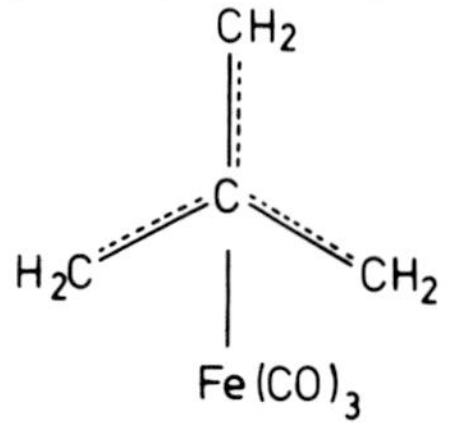

4

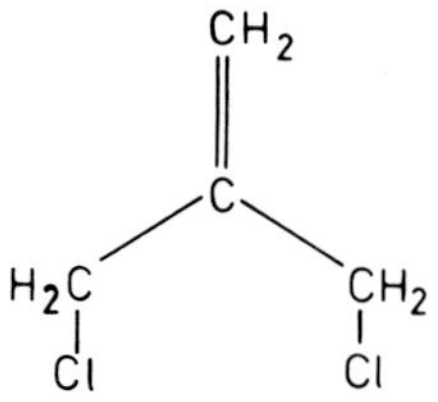

5

the interaction of 3-chloro-(2-chloromethyl)propene (5) with $Fe_2(CO)_9$ in ether at room temperature); the molecule is stable (b.p. 53-55°C/16 Torr, m.p. 28.4-29.6°C) *(56)*. This complex shows a sharp singlet at τ 8.00 in the ^{1}H-NMR spectrum and indicates the equivalency of all hydrogens. The lack of proton broadening at -60°C indicates no facile valence tautomerism commonly found in some carbonyliron complexes. The observed radial distribution curve of 4 shows that the three Fe-methylene carbon atom distances are almost identical *(1)*. In addition an electron diffraction study of gaseous 4 shows the molecular structure to be C_{3v} with a staggered arrangement of the ligands *(2)*.

Reaction of 6 (R = H, C_6H_5) with $Fe_2(CO)_9$ afforded complex 7 which, on heating under reflux, followed by fractional distillation, gave 4 (R = H, 14 - 20 %) or 8 (R = C_6H_5, 32 %) *(52,53)*. The X-ray structure *(28)* of 8 shows the Fe atom located directly beneath the central carbon atom of the trimethylenemethane residue and π-bonded to all four carbons. The carbon skeleton of the organic ligand is nonplanar with the central carbon atom 0.3 Å out of the plane of the three

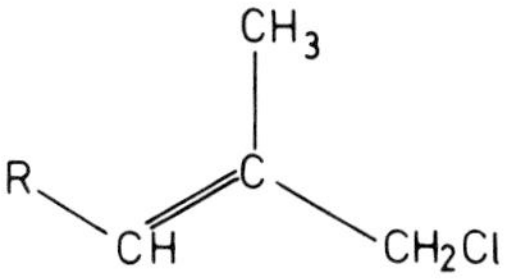

6

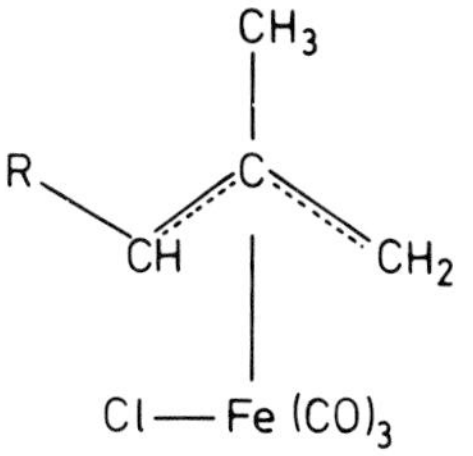

7

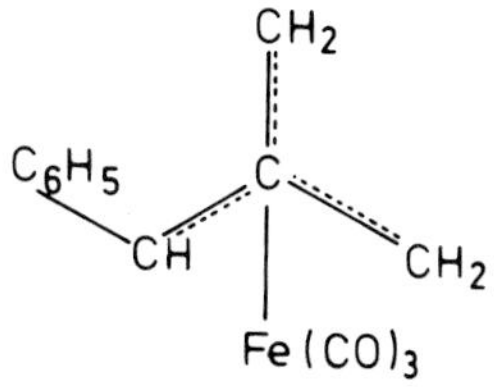

8

methylene carbons, away from iron: the distance Fe-C(central) is 1.93 Å, Fe-C(outer) is 2.10 - 2.16 Å. The phenyl residue is at an angle to the outer carbons and has little conjugation with the trimethylenemethane system. As with 4, complex 8 assumes a staggered arrangement of the ligands.

Other routes to trimethylenemethane-iron complexes have been found *(137)*. Methylene cyclopropane derivatives 9 (R = R' = C_6H_5; R = H, R' = C_6H_5; R = CH_3, R' = C_6H_5) react with

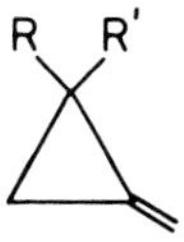

9

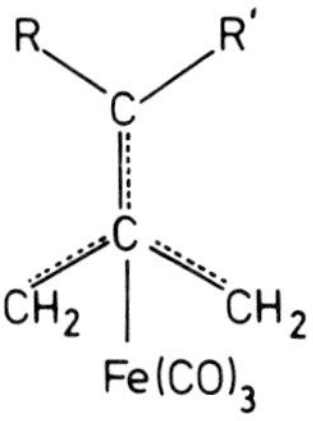

10

$Fe_2(CO)_9$ in refluxing benzene to give complexes 10 in yields of 40 - 60 % *(108)*; several vinyl derivatives of 10 (R = H, R' = $CH{=}CH_2$) have been synthesized *(9,10)*. Also, 5 reacts with sodium tetracarbonylferrate to give 4 (32 %) *(130)*.

Chemically, 4 decomposes with Ce(IV) salts; the organic ligand is trapped by tetracyanoethylene as adduct 11 *(130)*. Photochemical decomposition of 4 or 8 in tetrafluoroethylene

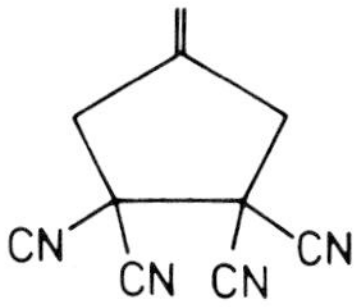

11

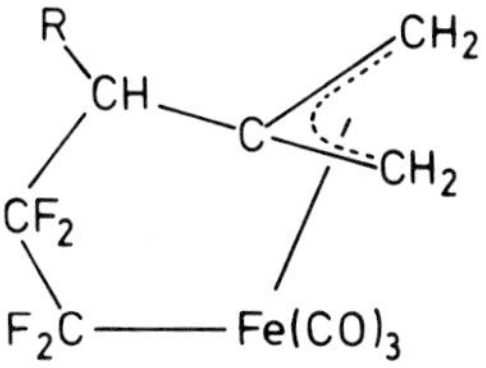

12

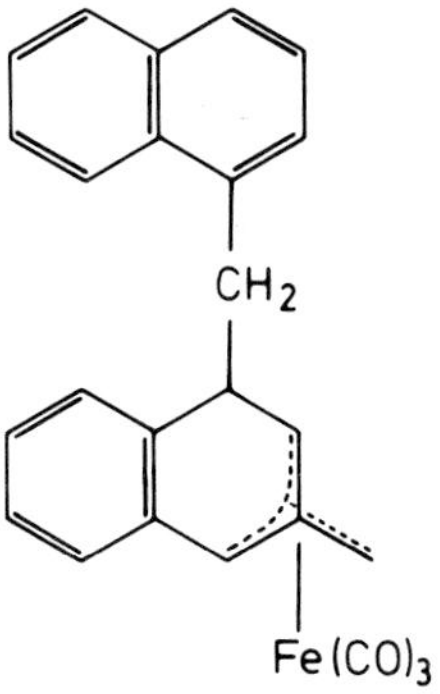

13

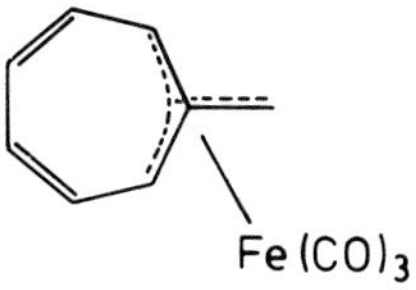

14

gives 12 (R = H or C_6H_5) *(14)*. In the solvents pentane, cyclopentene or cyclopentadiene, photolysis of 4 gives over sixteen products *(37)*. In concentrated sulfuric acid protonation of 4 leads to 7 (R = H) *(53,56)*.

The trimethylenemethane entity has been incorporated into rings; complexes 13 *(96)* and 14 *(51,87)* *(vide infra)* are recent examples.

D. *FULVENES*

1. Pentafulvenes

Many examples of pentafulvenes (15) complexed to iron are known wherein alkyl or aryl substituents are present on the exocyclic double bond; these are prepared by reacting compounds 15 with the appropriate carbonyliron reagents *(86,135)*.

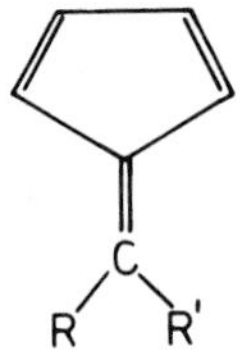

15

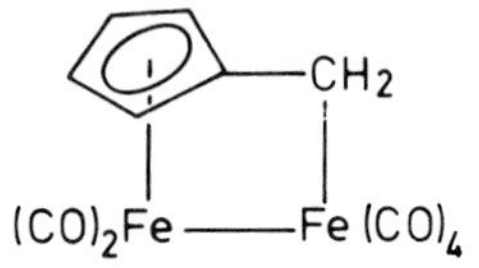

16

The only reported complex of unsubstituted pentafulvene is 16. This results (along with as many as nine other products) when $Fe_3(CO)_{12}$ is treated with acetylene in petroleum ether at pressures of 14 - 16 atm *(133)*. This structure has been confirmed by X-ray crystallography *(102)*. Triphenylphosphine does not liberate free 15 (R = R' = H) but instead displaces the ligand CO *(133)*.

2. Heptafulvenes

Stable complexes of the reactive, conjugated, nonbenzenoid hydrocarbon, heptafulvene *(47)*, have been prepared. One route is shown in eq. [4] *(116)*. Compound 17 dimerizes

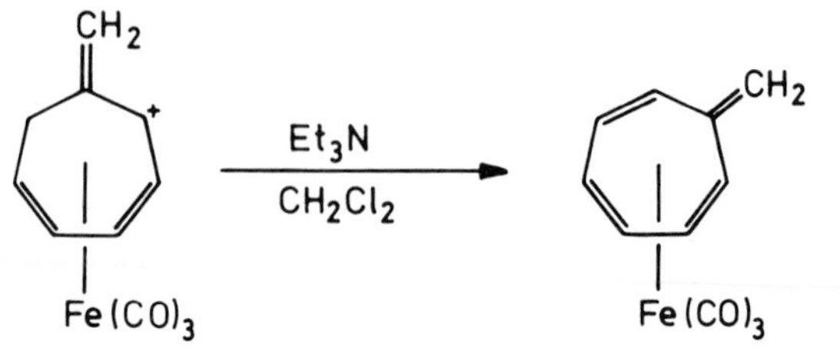

[4]

17

slowly at room temperature, is unaffected by acetic acid or

triethylamine, but is protonated by strong acids *(116)*.

Compound 14 formally is a complex of heptafulvene stabilized by a trimethylenemethane-type ligand. Complex 14 was prepared in diethyl ether from 7-hydroxymethylcycloheptatriene and a large excess of $Fe_2(CO)_9$ *(51,87)*. Decomposition of 14 takes place in refluxing xylene; in the presence of dimethyl acetylenedicarboxylate, the adduct (after dehydrogenation) 1,2-di(methoxycarbonyl)azulene is isolated *(51,87)*. This adduct is the same one obtained directly from heptafulvene and dimethyl acetylenedicarboxylate *(47)*.

A salt of heptafulvene has also been prepared (eq. [5]) *(50)*. In solution complex 18 shows properties intermediate

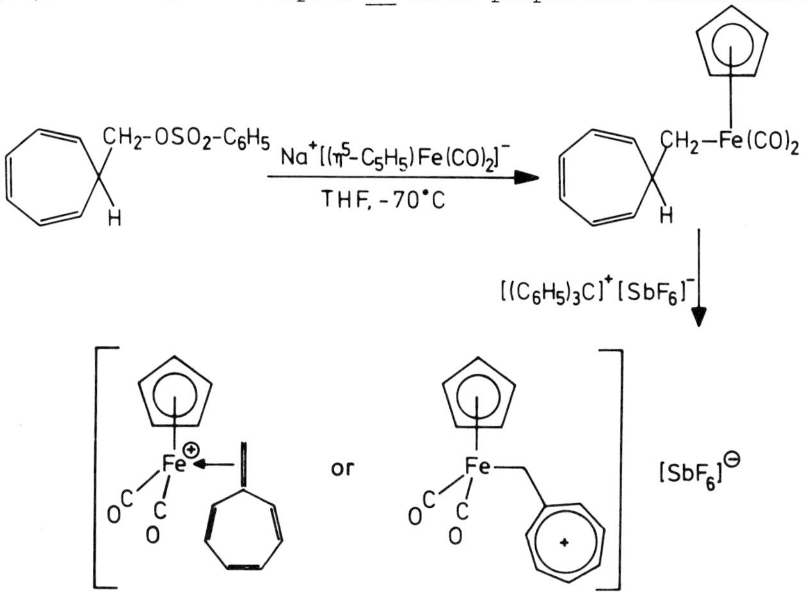

[5]

18a 18b

between those expected for 18a and 18b. However, a single crystal X-ray diffraction study established the structure as 18b *(29)*. The conversion of (tropone)tricarbonyliron into complexes of type 19 has also been accomplished through use

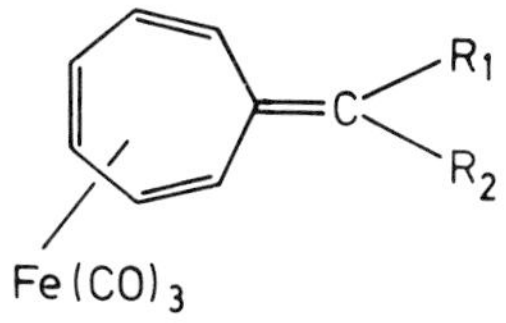

R_1	H	H	C_6H_5	CH_3
R_2	C_6H_5	$C_6H_5CH_2$	C_6H_5	CH_3

19

of the appropriate Grignard reagent followed by dehydration

(82).

E. CYCLOPENTADIENE AND DERIVATIVES

The π-complex 20, (η^4-cyclopentadiene)tricarbonyliron has been prepared directly through the reaction of cyclopentadiene and $Fe_2(CO)_9$ *(92)*. This complex is characterized by an infrared band at 2805 cm^{-1} attributed to the C-H stretching vibration. Decomposition occurs at 140°C to afford the dimeric complex $[(\eta^5\text{-}C_5H_5)Fe(CO)_2]_2$. However, reaction of cyclopentadiene with $Fe(CO)_5$ only leads to the dimer *(36,75,114)*; complex 20 has been suggested as an intermediate *(36)*.

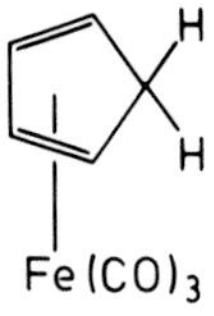

20

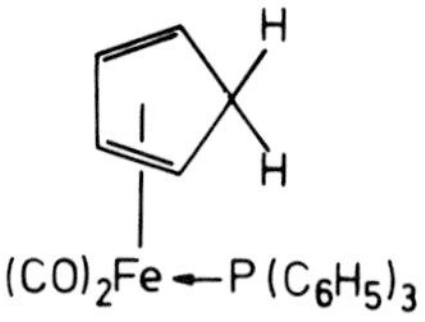

21

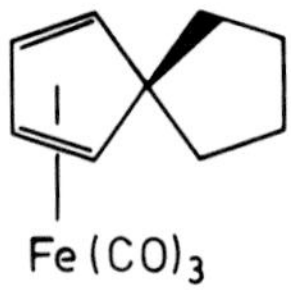

22

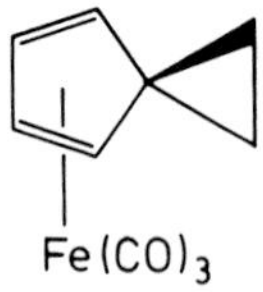

23

Reduction of the cation $[(\eta^5\text{-}C_5H_5)Fe(CO)_2P(C_6H_5)_3]^+$ with sodium borohydride gives the air sensitive complex 21 *(36)*. It is stable under nitrogen or *in vacuo* and survives for a short time in CH_2Cl_2 solution. In refluxing xylene complex 21 rapidly produces ferrocene and the dimer $[(\eta^5\text{-}C_5H_5)\text{-}Fe(CO)_2]_2$. Regeneration of the cation $[(\eta^5\text{-}C_5H_5)Fe(CO)_2PPh_3]^+$ takes place either with $[(C_6H_5)_3C]^+[BF_4]^-$ or with CCl_4 *(36)*.

Other complexes resulting when substituted cyclopentadienes are treated with $Fe_2(CO)_9$ include 22 *(66)*, 23 *(66)*, 24 *(42)*, and 25 *(54)*. Complex 24 rearranges on standing to the pentafulvene complex 26 *(42)*; however, little other chemistry is known for these species.

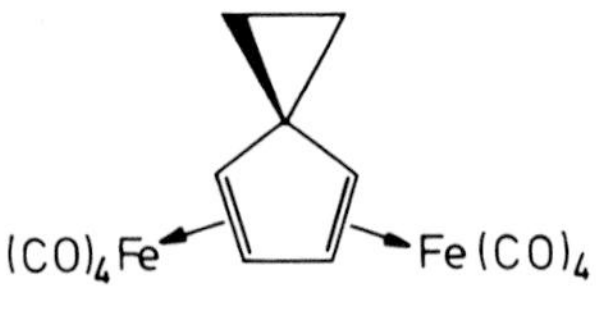

24

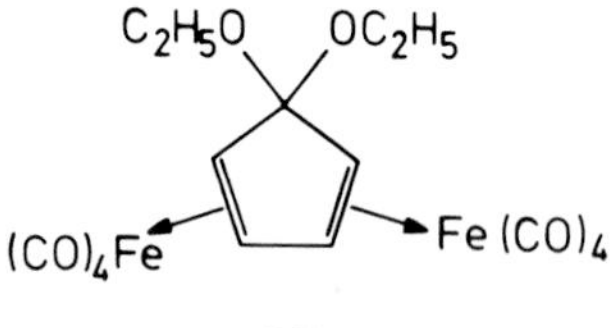

25

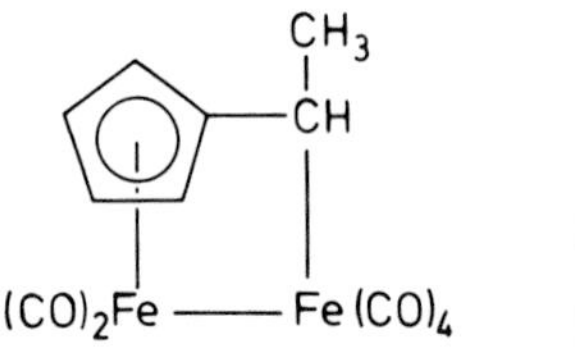

26

F. 7-NORBORNADIENONE

The simplest stabilized derivative of the highly elusive 7-norbornadienone (27) is (7-norbornadienone)tricarbonyliron (28) *(97,98)*. The air-stable complex 28 is prepared by oxidation of (7-norbornadienol)tricarbonyliron with the pyridine-sulfur trioxide complex in dimethylsulfoxide containing tri-

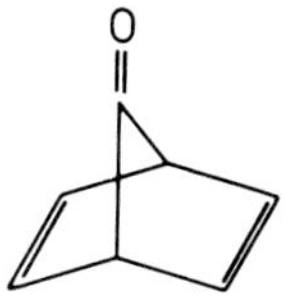

27

ethylamine (eq. [6]). The complex shows infrared absorptions

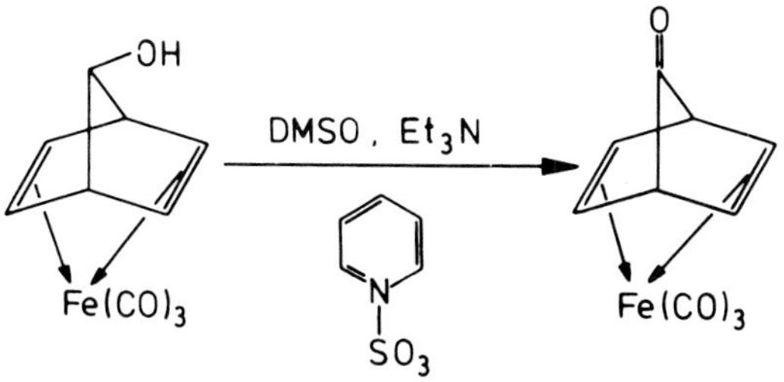

[6]

28

for a strained ketone (1860 (w) and 1780 (s) cm^{-1}) and has a simple ^{1}H-NMR spectrum (τ 6.35, quintet, J = 2.5 Hz, 2 H; τ 6.98, triplet, J = 2.5 Hz, 4 H). Chemically the ketone functional group preferentially reacts with sodium borohydride, alkyl lithium reagents, Grignard reagents, substituted hydrazines, phosphonium ylides, and sulphonium ylides. Decomplexation takes place photolytically, thermally, and with Ce(IV) salts; benzene formation and CO evolution result. Trapping experiments implicate 27 as an intermediate in these cases.

G. "FROZEN" TAUTOMERS

A number of recently prepared complexes may best be described as "frozen" tautomers: the least stable tautomeric form of an otherwise highly stable system. These include complexes 29 (R = H) *(41,43,44)*, 30 (R = H) *(41)*, 31 *(11,12)*, 32 *(57)*, 33 *(25)*, and 34 *(127)*. Enols 29 (R = H) and 30 (R = H) were prepared from the corresponding benzoates or acetates by treatment with methyl lithium in ether followed by acidi-

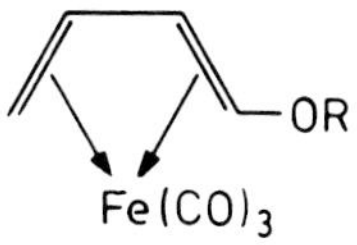

29

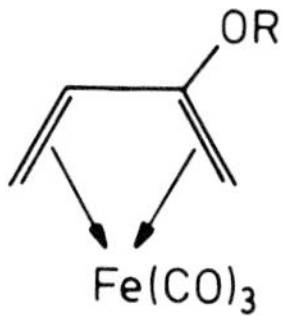

30

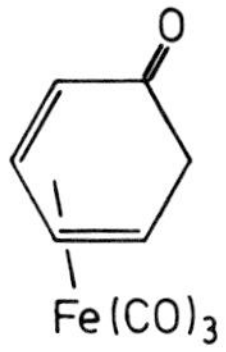

31

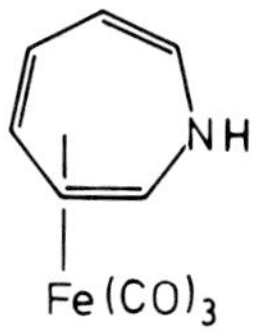

32

R = H, CH_3-CO, C_6H_5-CO

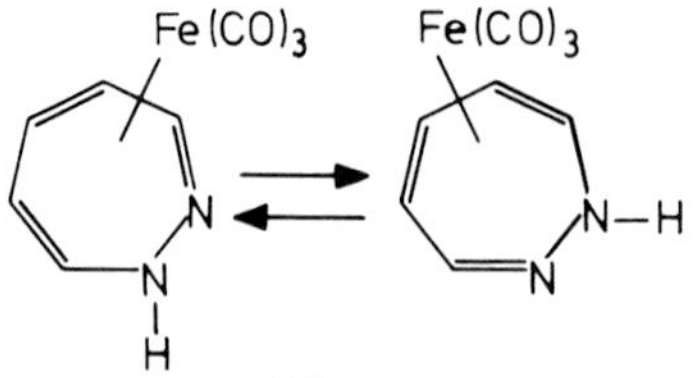

33

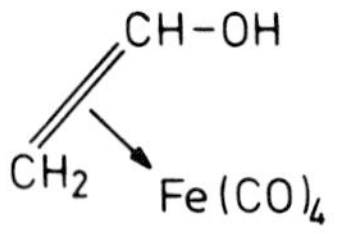

34

fication; alternatively, acid or base hydrolysis of the benzoates also gave the enol complexes. Both enol complexes are crystalline, air-sensitive solids which are stable in solution; benzoylation regenerates starting material *(41,43)*.

The solid tricarbonyliron adduct of a tautomer of phenol (31) shows a band in the infrared spectrum for the ketone at $\tilde{\nu} = 1665\ cm^{-1}$. The ketone undergoes reduction ($NaBH_4$) and 2,4 dinitrophenylhydrazon formation. Pyridine brings about decomplexation and the formation of phenol *(11,12)*.

Alkaline hydrolysis of (*N*-carbethoxyazepine)tricarbonyliron yielded 32. This is an orange-red, air-sensitive solid showing a free -NH group by IR and NMR spectroscopy. Tricarbonyliron stabilization prevents tautomerism to the more stable 2-H-azatropylidene system *(57)*.

While the unsubstituted molecule 1(1H),2-diazepine has thus far escaped synthesis, the system [1(1H),2-diazepine]-tricarbonyliron (33) has been realized *(25)* by alkaline hydrolysis of the *N*-carbethoxy precursor. The yellow, air-stable solid shows a free -NH by IR spectroscopy and exhibits fluxional behaviour by NMR spectroscopy; the latter is expected from the tautomeric behaviour of the diazepine system.

The stabilized vinyl alcohol 34 was obtained through acid catalysed hydrolysis of the precursor trimethylsilyl ether at -80°C. Acetaldehyde is obtained when 34 is treated with $(C_6H_5O)_3P$ *(127)*.

III. REACTIVE INTERMEDIATES

A. *NITRENES*

Nitrenes have resulted from metal carbonyl-catalysed decomposition of several precursors; significantly, in some cases, these reactive intermediates may be trapped and stabilized as a ligand. The R-N̈: group involving singlet nitrogen is described as being bonded to the metal by a σ-donating bond from nitrogen to metal, and a π-back donating bond from metal *d*-orbitals to empty *p*-orbitals on the nitrogen *(88)*.

Azidobenzene and $Fe_2(CO)_9$ react rapidly at room temperature in benzene to form $(C_6H_5N)_2Fe_2(CO)_6$; this complex is assigned structure 35 on the basis of the NMR, IR, Mössbauer,

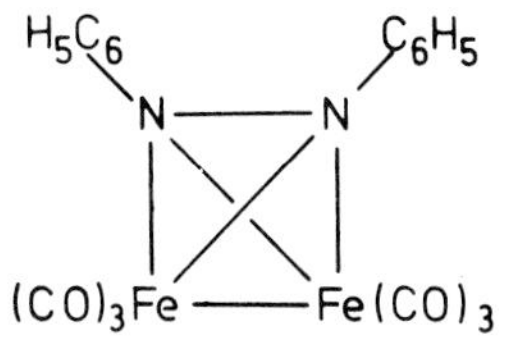

35

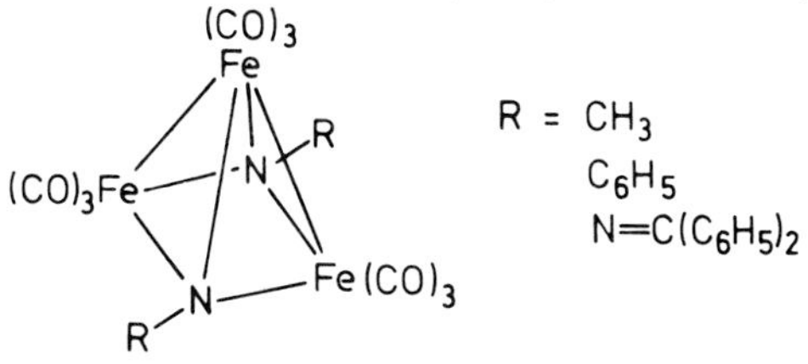

36

and mass spectral evidence and is believed to arise from a nitrene intermediate *(40)*. *ortho*-Substituted phenyl azides and *peri*-substituted α-naphthyl azides also react with Fe_2-$(CO)_9$ to give complexes whose structures are similar to 35 *(24)*. On the other hand, methyl azide reacts with $Fe_2(CO)_9$ to give a number of products, among which is 36 (R = CH_3), formally a stabilized nitrene *(40)*. The same complex is obtained from nitromethane and $Fe_2(CO)_9$ *(40)*. A closely related compound 36 (R = $N{=}C(C_6H_5)_2$) was prepared from $Fe(CO)_5$ and diphenyldiazomethane *(4)*. Both the crystal structure *(5,46)* and the Mössbauer spectra *(67)* of these two compounds 36 (R = CH_3) and 36 (R = $N{=}C(C_6H_5)_2$) confirm the similarities and indicate the presence of one unique and two equivalent iron atoms in each molecule. The phenyl analogue 36 (R = C_6H_5) is believed to result from reaction of nitrobenzene with $Fe_3(CO)_{12}$ in a methanol-benzene solvent; the Mössbauer spectrum of 36 (R = C_6H_5) is similar to that of 36 (R = CH_3) *(99)*. Decomposition of 36 (R = C_6H_5) in hydroxylic solvents gives aniline.

Analogously, trimethylsilyl azide decomposes in the presence of $Fe_2(CO)_9$ to form the trimethylsilylnitrene complex $(CH_3)_3SiN[Fe_3(CO)_{10}]$ *(95)*.

B. CARBENES

Singlet carbenes can formally be compared with the carbonyl ligand. There is presently considerable evidence for the existence of carbene complexes comparable with the terminal carbonyl arrangement *(32)*. (Phenylmethoxycarbene)-tetracarbonyliron (37), an air-sensitive, diamagnetic solid, has been prepared by a photochemical exchange reaction (eq. [7]) *(61)*. Complex 37 was not obtained from the reaction of the acylmetallate salt $Li[Fe(CO)_4C(C_6H_5)O]$ with $[(CH_3)_3O]^+$-$[BF_4]^-$; instead (phenylmethylene)octacarbonyldiiron (38) resulted. Complex 38 reportedly exists in two isomeric configurations (eq. [8]) and is the first carbonyliron compound whose configuration is influenced by temperature and solvent

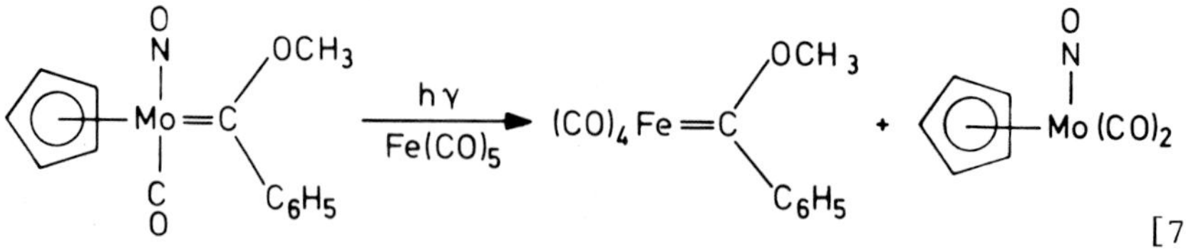

[7]

37

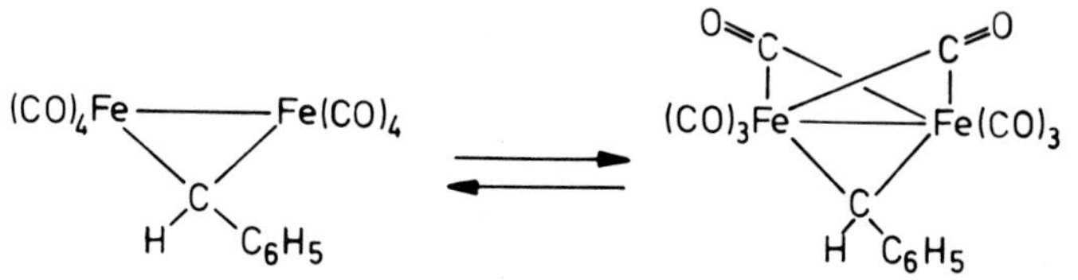

[8]

38

polarity *(60,62)*. In addition, the lithium salt is oxidized by trityl chloride to di(μ-benzoyl)hexacarbonyldiiron (39); the X-ray structure is compatible with a description as a phenylferroxycarbene *(59)*. Reacting $[(C_6H_5)_3P]Fe(CO)_4$ with CH_3Li and following with $[Et_3O]^+[BF_4]^-$ gives a carbene complex directly, $[(C_6H_5)_3P](CO)_3FeC(OC_2H_5)CH_3$ *(34)*.

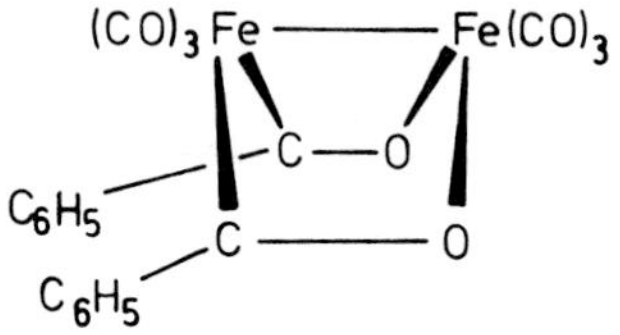

39

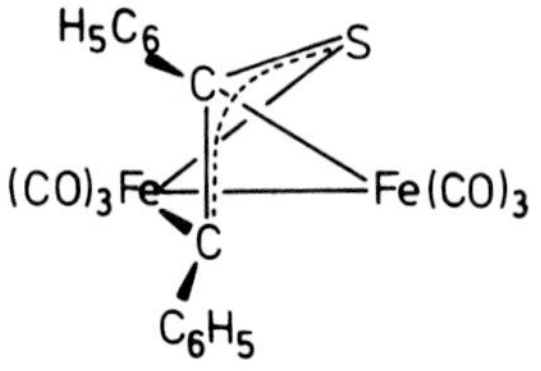

41

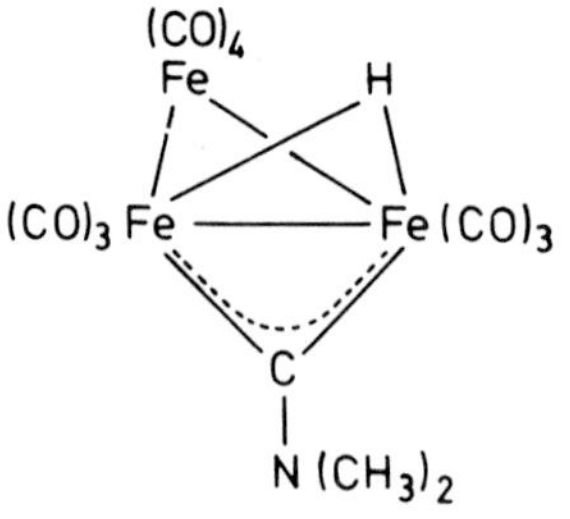

40

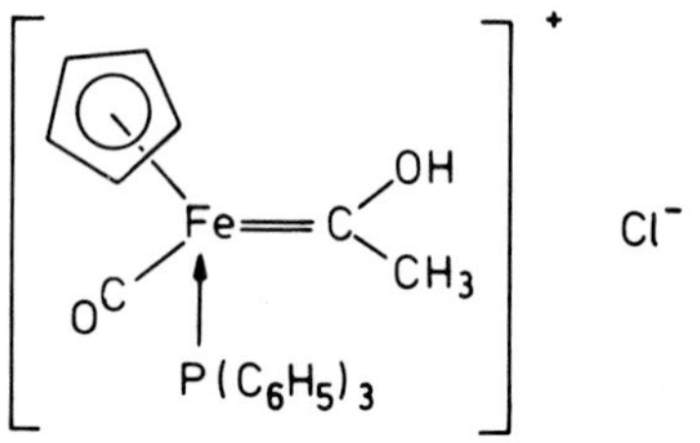

42

Other carbene complexes are known *(35,63,64,65,72,136)*. Further examples include the methylaminocarbene complex 40 *(68)*, the thioketocarbene complex 41 *(101,123,124)*, the methylenecarbene complex $[(\eta^5\text{-}C_5H_5)Fe(CO)_2CH_2)]^+$ *(83)*, and

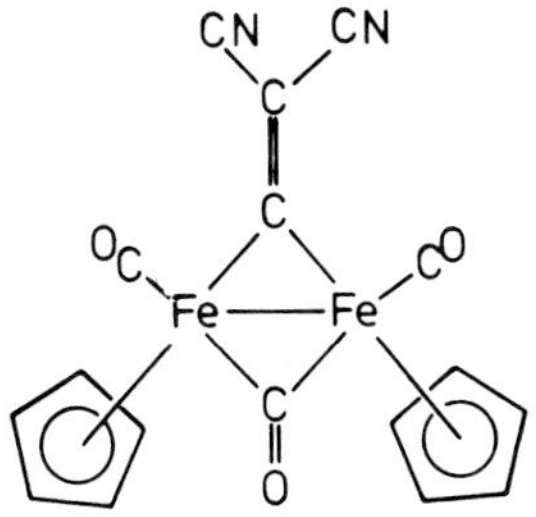

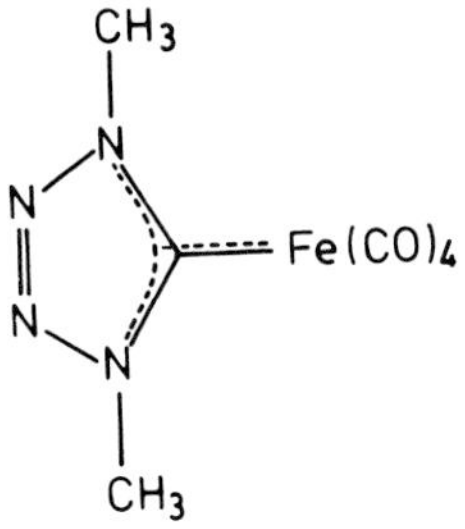

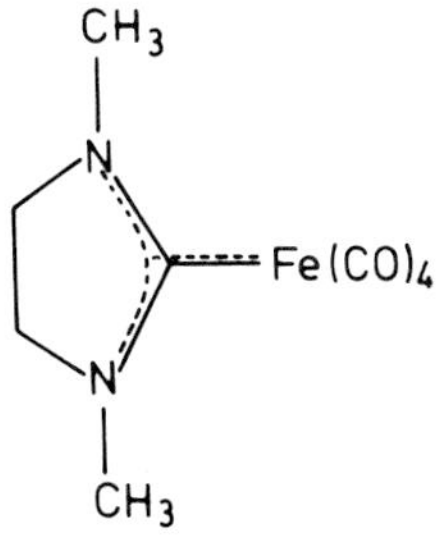

43 44 45

the hydroxycarbene complex 42 *(70)*. A complex of diphenylvinylidene has been prepared (eq. [9]) *(103)* as well as a complex of dicyanomethylenecarbene, 43 *(91)*. The heterocyclic

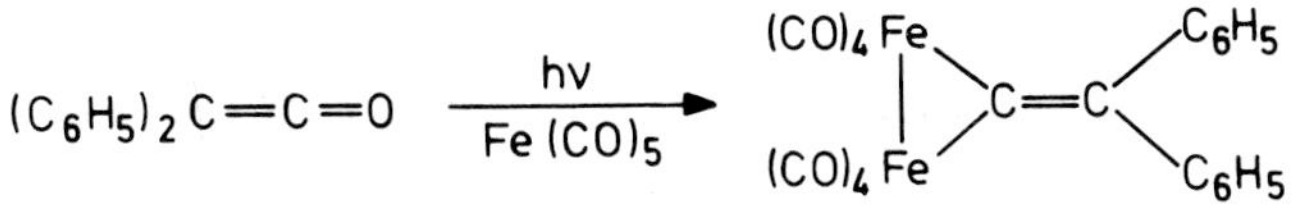

[9]

carbene complexes 44 *(109)* and 45 *(26)* have also been isolated.

C. *BENZYNE*

An air-stable, yellow complex has been prepared whose properties (mass spectral and IR) suggest a stabilized, σ-bonded benzyne; this is shown in eq. [10] *(117)*. The complex

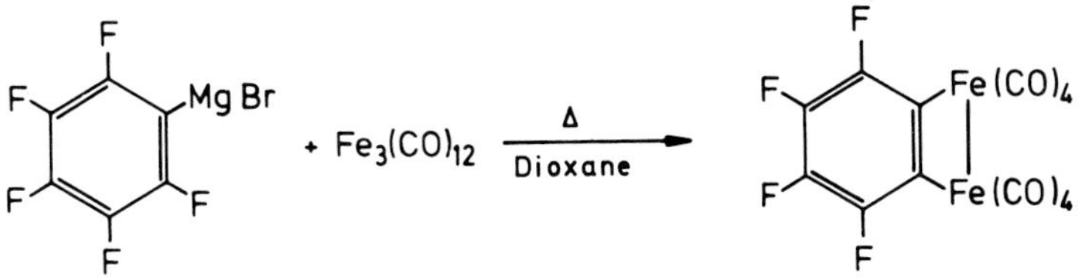

[10]

forms polymeric material upon pyrolysis in a sealed tube and releases CO. The structure, as formulated, has been confirmed by spectral analysis *(8)*.

IV. RELATED STABILIZED SYSTEMS

A. *CUMULENES*

The reaction of 1,4-dibromo- or substituted 1,4-dibromobut-2-yne with $Fe_3(CO)_{12}$ or $Na_2Fe(CO)_4$ gives products formulated as $(RR'C{=}C{=}C{=}CRR')Fe_2(CO)_6$ (R = R' = H; R = H, R' = CH_3; R = R' = CH_3; R = R' = C_6H_5) *(84,105,106,110)*. This formulation has been confirmed by both mass spectrometry *(89,106,*

110) and X-ray crystallographic analysis *(84,89)*. The preliminary structural analysis of the complex 46 (R-R = 2,2'-

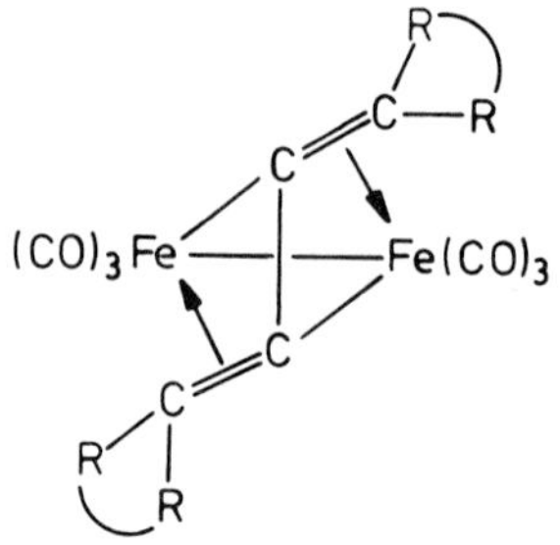

46

biphenyl) shows that the carbon backbone is nonlinear and that each iron atom is attached to the hydrocarbon by both a σ (η^1) and a π (η^2) bond.

In the case of the unsubstituted butatriene, ($H_2C{=}C{=}C{=}CH_2$)$Fe_2(CO)_6$, the complex is an air-stable, orange solid (m.p. 69 - 70°C) which thermally decomposes at 200 - 230°C *(105)*. This behaviour is in marked contrast to the uncomplexed butatriene which was reported to polymerize violently near 0°C *(125)*.

An air-stable complex of the unknown hexapentaene, ($H_2C{=}C{=}C{=}C{=}C{=}CH_2$)$Fe_3(CO)_{7-8}$, has also been reported *(104)*. The exact formulation of this complex is in doubt, however.

B. KETENIMINES

Direct interaction of substituted ketenimines with $Fe_3(CO)_{12}$ gives complexes of the type ($R_2C{=}C{=}N{-}R'$)$Fe_2(CO)_6$ *(111)*.

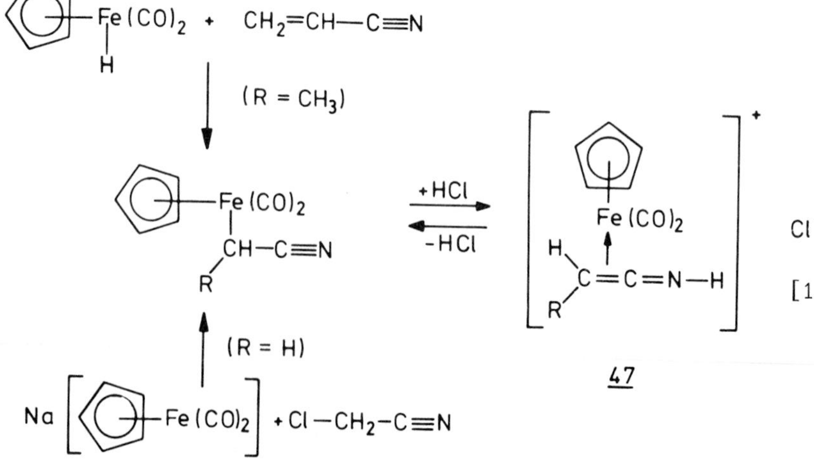

47

[11]

No examples of unsubstituted ketenimines of this type are
known. However, cationic complexes (47; R = H, CH₃) are formed from 1-cyanoalkyl complexes by reversible protonation
with dry HCl as in eq. [11] (3). Structures containing the
ketenimine η^2-bonded to the iron system are proposed for the
cations.

C. THIOCARBONYL

The ligand CS has recently been reported in cations of
the type $[(\eta^5\text{-}C_5H_5)Fe(CO)_2(CS)]^+$ (21,22,23). A typical preparative reaction is shown for a salt of this cation in eq.
[12] (21). Complex 48 shows typically high stretching fre-

$(\eta^5\text{—}C_5H_5)\,Fe\,(CO)_2$

$$\xrightarrow[\text{acetone}]{NH_4\,PF_6 \quad in} \quad \left[(\eta^5\text{-}C_5H_5)\,Fe\,(CO)_2 \atop \overset{|}{CS} \right]^+ PF_6^-$$

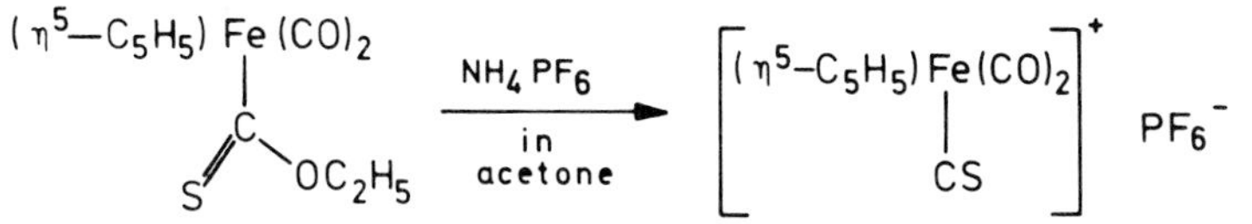

[12]

48

quencies for the CO ligands ($\tilde{\nu}(CO)$ = 2093 and 2064 cm^{-1}); in
addition, a band is observed for the CS ligand ($\tilde{\nu}(CS)$ = 1348
cm^{-1}). The shift of $\tilde{\nu}(CO)$ to higher frequencies suggests a
decrease in the back donation from the metal orbitals to the
carbonyl groups. The inference is that the CS group is a good
π-bonding ligand (22). Chemical studies show that $[(\eta^5\text{-}C_5H_5)$-
$Fe(CO)_3]^+$ and $[(\eta^5\text{-}C_5H_5)Fe(CO)_2(CS)]^+$ react similarly with
nucleophiles containing oxygen and nitrogen as donor atoms;
in the case of the CS analogue, it is the CS which undergoes
nucleophilic attack rather than the CO (23).

D. NITROSOBENZENE

Ultraviolet irradiation of a solution containing nitrobenzene and Fe(CO)₅ leads to bis[(nitrosobenzene)tricarbonyl-
iron] (49, R = C₆H₅), an air-stable, crystalline complex of

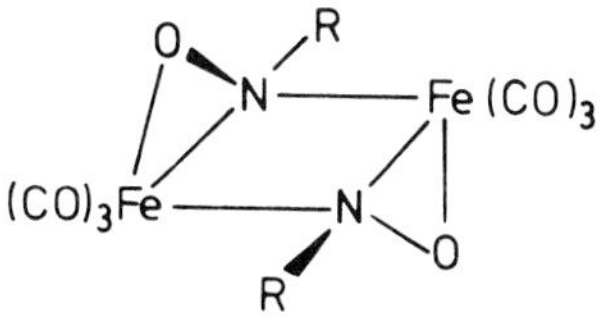

49

nitrosobenzene (93,94). Crystallographic data (7) on an
analogue (49, R = 2-methyl-3-chlorobenzene) reveals that the
Fe-N-Fe-N ring is planar and has edges of length 2.00 Å, a
similarity present in other N-bridged binuclear tricarbonyl-

iron complexes. The Fe-Fe distances is 3.13 Å, indicating no metal-metal bond; the Fe-N-Fe angle is 100°, 25 - 30° larger than corresponding angles in related systems possessing a formal Fe-Fe bond.

Treatment of 49 (R = C_6H_5) with triphenylphosphine gives the crystalline monomeric species $(C_6H_5NO)Fe(CO)_2P(C_6H_5)_3$ *(93)*.

E. COMPOUNDS DERIVED FROM ACETYLENES

Carbonyliron complexes react with alkynes and give a variety of products dependent upon the medium *(16)*. In general, a cyclic olefin forms from the combination of two or three acetylene molecules with either a carbonyl group or an iron atom in the ring; details of these reactions are presented elsewhere. Certain of these compounds are relevant to our discussion, however, and are presented below.

1. Cyclopentadienone *(58)*

Pentacarbonyliron reacts with acetylene under pressure to give the yellow solid (η^4-cyclopentadienone)tricarbonyliron (50, m.p. 114-116°C) *(69,132,134)*. Complexing of the

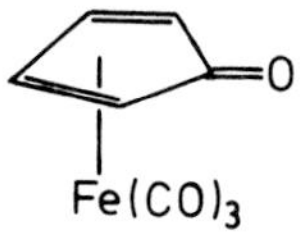

50

cyclopentadienone seems to increase the polarity of the carbonyl group. This is evidenced by a ν(C=O) at 1634 cm^{-1} and the ready reactivity with electron acceptor molecules. Thus, a 1:1 adduct is formed with either HCl or HI; hydroquinone reacts to form a complex bis[(η^4-cyclopentadienone)tricarbonyliron]hydroquinone *(132,134)*. An X-ray structural analysis of an analogue, $[\eta^4\text{-}C_5(CF_3)_4O]Fe(CO)_3$, has been done *(15)*. The ring is nonplanar with a bending angle of 20°; metal-sp^2-carbon distances are 1.99 to 2.12 Å. These distances, as well as the equivalent carbon-carbon bond lengths in the 1,3-diene portion of the molecule, suggest that both σ and π bonding exists between the metal and olefin.

2. Ferroles and Ferraindenes

Many examples of ferroles and ferraindenes are known whereby the system is stabilized by bonding to a second $Fe(CO)_3$ moiety. (In both these systems, the iron atom in the ring is electron deficient.) These systems result from suitable acetylenic substrates *(13,19,122,133)*, from heterocyclic materials *(85)*, from aromatic complexes *(129)*, or from olefins *(20,128)*. Typical examples are 51 and 52.

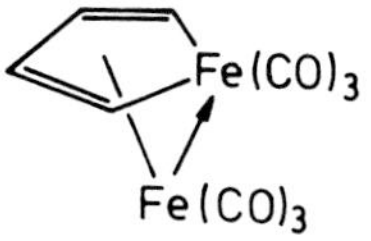

51

a ferrole complex *(85)*

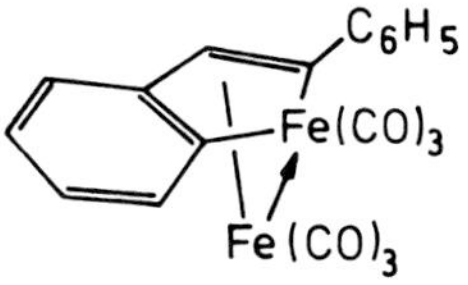

52

a ferraindene complex *(19)*

3. Penta-Coordinated Carbon

A new type of polynuclear metal complex (plus the "usual" complexes) of formula $Fe_5(CO)_{15}C$ (in 0.5 %) resulted from the treatment of either 1-phenylpropyne or 1-pentyne with $Fe_3(CO)_{12}$ *(17)*. The back crystalline carbonyliron carbide 53

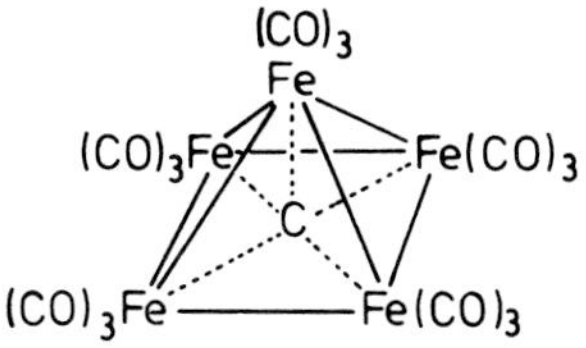

53

is diamagnetic with no bridging carbonyls. X-ray analysis shows an approximate equilateral pyramid of iron atoms with three terminal carbonyls attached to each iron. The unique structural feature is the presence of a penta-coordinated carbon atom located slightly below the center of the basal plane of iron atoms; the carbon atom is equidistant from each of the five atoms. This carbon with its four valence electrons allows each iron to attain a "closed shell" electronic structure in the ground state.

V. ADDENDA

The first transition metal formyl complex has been prepared and characterized *(31)*; the route is shown in equation [13].

$$Na_2Fe(CO)_4 + CH_3-\overset{O}{\overset{\|}{C}}-O-\overset{O}{\overset{\|}{C}}-H \xrightarrow[\text{THF, Ar}]{24\,^\circ C} \left[(CO)_4Fe-\overset{O}{\overset{\|}{C}}-H\right]^{\ominus} \qquad [13]$$

Thiete when treated thermally with $Fe_2(CO)_9$ (scheme [14]) or photochemically with $Fe(CO)_5$ yields the dimeric complex 54 which can be converted to the stable red dicarbonyliron triphenylphosphine complex 55 *(126)*. X-ray analysis has established the ligand in 55 as the planar thioacrolein, a

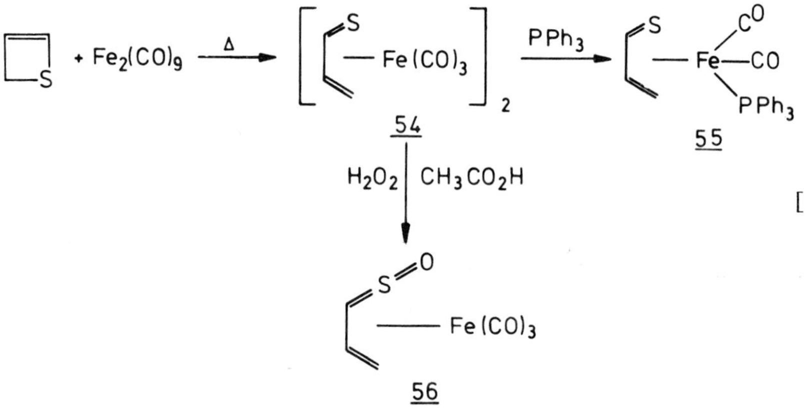

[14]

hitherto unknown thioaldehyde; the iron atom lies above the plane. Oxidation of 54 yields the yellow S-oxide complex 56, which may be considered as a derivative of the unknown vinylsulfine (CH_2=CH-CH=S=O).

The stabilized vinyl ketene was prepared according to eq. [15] *(76)*.

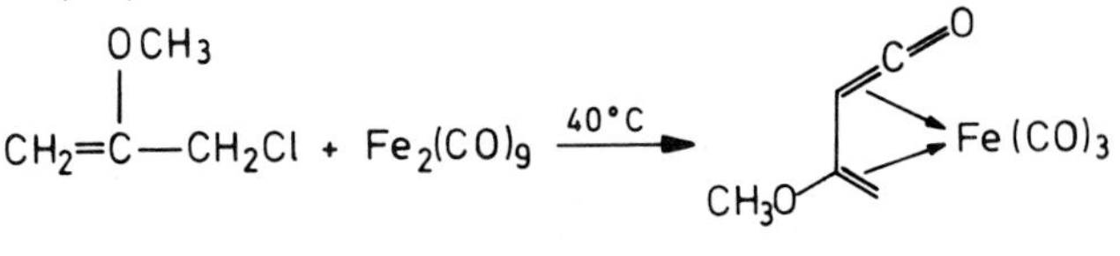

[15]

Thiophene-1,1-dioxide and *all-cis*-cyclononatetraene are two unstable compounds that can exist for short periods of time at low temperature. In the case of *all-cis*-cyclononatetraene, it is unstable relative to its ring-closed isomer, *cis*-8,9-dihydroindene ($t_{1/2}$ = 50 min, at 23°C) *(115)*. However, the complex (1-4-η-*all-cis*-cyclononatetraene)tricarbonyliron (57) is stable for days at room temperature and undergoes

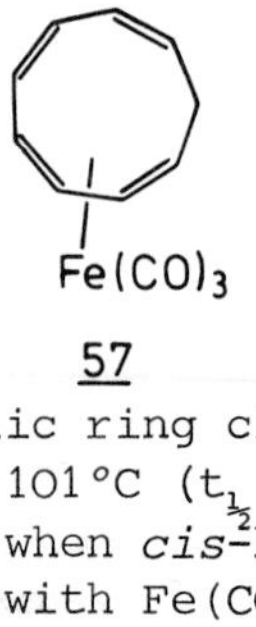

57

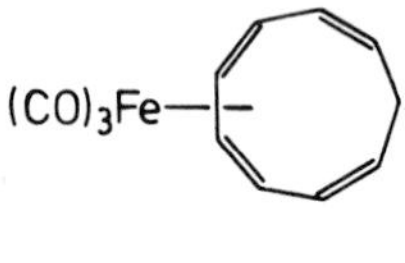

58

electrocyclic ring closure to (*cis*-8,9-dihydroindene)tricarbonyliron at 101°C ($t_{1/2}$ = 48 min, at 101°C) *(115)*. Complex 57 is trapped when *cis*-bicyclo[6.1.0]nonatriene is treated photochemically with $Fe(CO)_5$ *(115)* or thermally with either (benzylideneacetone)tricarbonyliron *(121)* or $Fe_2(CO)_9$ *(38,115)*. The isomer 58 has been identified also *(38)*.

Thiophene-1,1-dioxide readily dimerizes. On the other hand, (thiophene-1,1-dioxide)tricarbonyliron is a stable, high melting solid. The complex is obtained by treating a 0.02 *M* benzene solution of thiophene-1,1-dioxide photochemically with $Fe(CO)_5$; a 60 % yield results *(27)*.

Pentalene is another elusive compound long sought by organic chemists. Stable complexes of this ring system have been prepared with the structure established as 59 (R = H *(131)*, CH_3 *(39)*, $N(CH_3)_2$ *(80)*, C_6H_5 *(81)*).

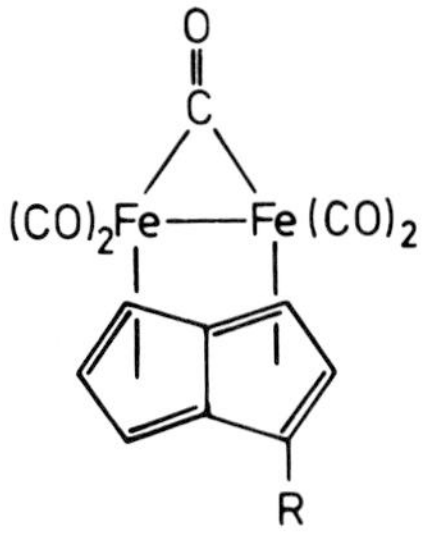

59

Two polynuclear carbidocarbonyl anions of iron have been isolated as the salts $[(CH_3)_4N]_2[Fe_5C(CO)_{14}]$ *(78)* and $[(CH_3)_4N]_2[Fe_6C(CO)_{16}]$ *(30)*. These divalent anions represent further examples of "stabilized" carbon with structures analogous to 53 *(17)*.

Finally, several new ferraboranes *(73,77)* have been reported. One of them, $(B_4H_8)Fe(CO)_3$ *(73)*, is a complex which has structural and bonding characteristics common to $(\eta^4$-$C_4H_4)Fe(CO)_3$ (1) and $C[Fe(CO)_3]_5$ (53), *i.e.* the iron atom occupies the apical position of a square-based pyramid with boron atoms in the basal plane.

REFERENCES

1. Almenningen, A., Haaland, A., and Wahl, K., *Chem. Commun., 1968,* 1027.
2. Almenningen, A., Haaland, A., and Wahl, K., *Acta Chem. Scand., 23,* 1145 (1969).
3. Ariyaratne, J.K.P., and Green, M.L.H., *J. Chem. Soc., 1963,* 2976.
4. Bagga, M.M., Baikie, P.E., Mills, O.S., and Pauson, P.L., *Chem. Commun., 1967,* 1106.
5. Baikie, P.E., and Mills, O.S., *Chem. Commun., 1967,* 1228.
6. Bailey, N.A., and Mason, R., *Acta Crystallogr., 21,* 652 (1966).
7. Barrow, M.J., and Mills, O.S., *Angew. Chem., 81,* 898 (1969); *Angew. Chem. Int. Ed. Engl., 8,* 879 (1969).
8. Bennett, M.J., Graham, W.A.G., Stewart, R.P., Jr., and Tuggle, R.M., *Inorg. Chem., 12,* 2944 (1973).
9. Billups, W.E., Lin, L.-P., and Gansow, O.A., *Angew. Chem., 84,* 684 (1972); *Angew. Chem. Int. Ed. Engl., 11,* 637 (1972).
10. Billups, W.E., Lin, L.-P., and Baker, B.A., *J. Organometal. Chem., 61,* C 55 (1973).
11. Birch, A.J., Cross, P.E., Lewis, J., and White, D.A., *Chem. Ind. (London), 1964,* 838.
12. Birch, A.J., Cross, P.E., Lewis, J., White, D.A., and Wild, S.B., *J. Chem. Soc. A, 1968,* 332.
13. Bird, C.W., Briggs, E.M., and Hudec, J., *J. Chem. Soc. C, 1967,* 1862.
14. Bond, A., Green, M., Lewis, B., and Lowrie, S.F.W., *J. Chem. Soc. D, Chem. Commun., 1971,* 1230.
15. Boston, J.L., Sharp, D.W.A., and Wilkinson, G., *J. Chem. Soc., 1962,* 3488.
16. Bowden, F.L., and Lever, A.B.P., *Organometal. Chem. Rev., 3,* 227 (1968).
17. Braye, E.H., Dahl, L.F., Hübel, W., and Wampler, D.L., *J. Amer. Chem. Soc., 84,* 4633 (1962).
18. Braye, E.H., and Hübel, W., *J. Organometal. Chem., 3,* 25 (1965).
19. Braye, E.H., and Hübel, W., *J. Organometal. Chem., 3,* 38 (1965).
20. Bruce, M.I., and Kuc, T.A., *J. Organometal. Chem., 22,* C 1 (1970).
21. Busetto, L., and Angelici, R.J., *J. Amer. Chem. Soc., 90,* 3283 (1968).
22. Busetto, L., Belluco, U., and Angelici, R.J., *J. Organometal. Chem., 18,* 213 (1969).
23. Busetto, L., Graziani, M., and Belluco, U., *Inorg.*

Chem., *10*, 78 (1971).
24. Campbell, C.D., and Rees, C.W., *J. Chem. Soc. D, Chem. Commun.*, *1969*, 537.
25. Carty, A.J., Hobson, R.F., Patel, H.A., and Snieckus, V., *J. Amer. Chem. Soc.*, *95*, 6835 (1973).
26. Cetinkaya, B., Dixneuf, P., and Lappert, M.F., *J. Chem. Soc. Chem. Commun.*, *1973*, 206.
27. Chow, Y.L., Fossey, J., and Perry, R.A., *J. Chem. Soc. Chem. Commun.*, *1972*, 501.
28. Churchill, M.R., and Gold, K., *Chem. Commun.*, *1968*, 693.
29. Churchill, M.R., and Fennessey, J.P., *J. Chem. Soc. D, Chem. Commun.*, *1970*, 1056.
30. Churchill, M.R., Wormald, J., Knight, J., and Mays, M.J., *J. Amer. Chem. Soc.*, *93*, 3073 (1971).
31. Collman, J.P., and Winter, S.R., *J. Amer. Chem. Soc.*, *95*, 4089 (1973).
32. Cotton, F.A., and Lukehart, C.M., *Progr. Inorg. Chem.*, *16*, 487 (1972).
33. Crawford, R.J., and Cameron, D.M., *J. Amer. Chem. Soc.*, *88*, 2589 (1966).
34. Darensbourg, D.J., and Darensbourg, M.Y., *Inorg. Chem.*, *9*, 1691 (1970).
35. Daub, J., Erhardt, U., Kappler, J., and Trautz, V., *J. Organometal. Chem.*, *69*, 423 (1974).
36. Davison, A., Green, M.L.H., and Wilkinson, G., *J. Chem. Soc.*, *1961*, 3172.
37. Day, A.C., and Powell, J.T., *Chem. Commun.*, *1968*, 1241.
38. Deganello, G., Maltz, H., and Kozarich, J., *J. Organometal. Chem.*, *60*, 323 (1973).
39. Deganello, G., and Toniolo, L., *J. Organometal. Chem.*, *74*, 255 (1974).
40. Dekker, M., and Knox, G.R., *Chem. Commun.*, *1967*, 1234.
41. DePuy, C.H., Greene, R.N., and Schroer, T.E., *Chem. Commun.*, *1968*, 1225.
42. DePuy, C.H., Kobal, V.M., and Gibson, D.H., *J. Organometal. Chem.*, *13*, 266 (1968).
43. DePuy, C.H., and Jablonski, C.R., *Tetrahedron Lett.*, *1969*, 3989.
44. DePuy, C.H., Jones, T., and Parton, R.L., *J. Amer. Chem. Soc.*, *96*, 5602 (1974).
45. Dodge, R.P., and Schomaker, V., *Acta Crystallogr.*, *18*, 614 (1965).
46. Doedens, R.J., *Inorg. Chem.*, *8*, 570 (1969).
47. von E. Doering, W., and Wiley, D.W., *Tetrahedron*, *11*, 183 (1960).
48. Dowd, P., *J. Amer. Chem. Soc.*, *88*, 2587 (1966).
49. Dowd, P., Gold, A., and Sachdev, K., *J. Amer. Chem.*

Soc., *90*, 2715 (1968).

50. Ehntholt, D.J., Emerson, G.F., and Kerber, R.C., *J. Amer. Chem. Soc.*, *91*, 7547 (1969).
51. Ehntholt, D.J., and Kerber, R.C., *J. Chem. Soc. D, Chem. Commun.*, *1970*, 1451.
52. Ehrlich, K., and Emerson, G.F., *J. Chem. Soc. D, Chem. Commun.*, *1969*, 59.
53. Ehrlich, K., and Emerson, G.F., *J. Amer. Chem. Soc.*, *94*, 2464 (1972).
54. Eisenstadt, A., Scharf, G., and Fuchs, B., *Tetrahedron Lett.*, *1971*, 679.
55. Emerson, G.F., Watts, L., and Pettit, R., *J. Amer. Chem. Soc.*, *87*, 131 (1965).
56. Emerson, G.F., Ehrlich, K., Giering, W.P., and Lauterbur, P.C., *J. Amer. Chem. Soc.*, *88*, 3172 (1966).
57. Fischer, E.O., and Rühle, H., *Z. anorg. allgem. Chem.*, *341*, 137 (1965).
58. Fischer, E.O., and Werner, H., *Metal π-Complexes, Vol. 1*, Elsevier, Amsterdam 1966, p. 49.
59. Fischer, E.O., Kiener, V., Bunbury, D.St.P., Frank, E., Lindley, P.F., and Mills, O.S., *Chem. Commun.*, *1968*, 1378.
60. Fischer, E.O., Kiener, V., and Fischer, R.D., *J. Organometal. Chem.*, *16*, P 60 (1969).
61. Fischer, E.O., and Beck, H.-J., *Angew. Chem.*, *82*, 44 (1970); *Angew. Chem. Int. Ed. Engl.*, *9*, 72 (1970).
62. Fischer, E.O., and Kiener, V., *J. Organometal. Chem.*, *23*, 215 (1970).
63. Fischer, E.O., and Kiener, V., *J. Organometal. Chem.*, *27*, C 56 (1971).
64. Fischer, E.O., Winkler, E., Huttner, G., and Regler, D., *Angew. Chem.*, *84*, 214 (1972); *Angew. Chem. Int. Ed. Engl.*, *11*, 238 (1972).
65. Fischer, E.O., Beck, H.-J., Kreiter, C.G., Lynch, J., Müller, J., and Winkler, E., *Chem. Ber.*, *105*, 162 (1972).
66. Grant, G.F., and Pauson, P.L., *J. Organometal. Chem.*, *9*, 553 (1967).
67. Greatrex, R., private communication.
68. Greatrex, R., Greenwood, N.N., Rhee, I., Ryang, M., and Tsutsumi, S., *J. Chem. Soc. D, Chem. Commun.*, *1970*, 1193.
69. Green, M.L.H., Pratt, L., and Wilkinson, G., *J. Chem. Soc.*, *1960*, 989.
70. Green, M.L.H., and Hurley, C.R., *J. Organometal. Chem.*, *10*, 188 (1967).
71. Green, M.L.H., in G.E. Coates, M.L.H. Green and K. Wade (Eds.), *Organometallic Compounds, Vol. 2, The*

Transition Elements, 3rd Ed., Methuen, London 1968.
72. Green, M.L.H., Mitchard, L.C., and Swanwick, M.G., *J. Chem. Soc. A, 1971*, 794.
73. Greenwood, N.N., Savory, C.G., Grimes, R.N., Sneddon, L.G., Davison, A., and Wreford, S.S., *J. Chem. Soc. Chem. Commun., 1974*, 718.
74. Grellman, K.H., Palmowski, J., and Quinkert, G., *Angew. Chem., 83*, 209 (1971); *Angew. Chem. Int. Ed. Engl., 10*, 196 (1971).
75. Hallam, B.F., Mills, O.S., and Pauson, P.L., *J. Inorg. Nucl. Chem., 1*, 313 (1955).
76. Hill, A.E., and Hoffmann, H.M.R., *J. Chem. Soc. Chem. Commun., 1972*, 574.
77. Hollander, O., Clayton, W.R., and Shore, S.G., *J. Chem. Soc. Chem. Commun., 1974*, 604.
78. Hsieh, A.T.T., and Mays, M.J., *J. Organometal. Chem., 37*, C 53 (1972).
79. Hübel, W., Braye, E.H., Clauss, A., Weiss, E., Krüerke, U., Brown, D.A., King, G.S.D., and Hoogzand, C., *J. Inorg. Nucl. Chem., 9*, 204 (1959).
80. Hunt, D.F., and Russell, J.W., *J. Amer. Chem. Soc., 94*, 7198 (1972).
81. Hunt, D.F., and Russell, J.W., *J. Organometal. Chem., 46*, C 22 (1972).
82. Johnson, B.F.G., Lewis, J., McArdle, P., and Randall, G.L.P., *J. Chem. Soc. D, Chem. Commun., 1971*, 177.
83. Jolly, P.W., and Pettit, R., *J. Amer. Chem. Soc., 88*, 5044 (1966).
84. Joshi, K.K., *J. Chem. Soc. A, 1966*, 594; see ref. 13 and refs. therein.
85. Kaesz, H.D., King, R.B., Manuel, T.A., Nichols, L.D., and Stone, F.G.A., *J. Amer. Chem. Soc., 82*, 4749 (1960).
86. Kerber, R.C., and Ehntholt, D.J., *Synthesis, 2*, 449 (1970).
87. Kerber, R.C., and Ehntholt, D.J., *J. Amer. Chem. Soc., 95*, 2927 (1973).
88. Kilner, M., *Advan. Organometal. Chem., 10*, 129 (1972).
89. King, R.B., *J. Amer. Chem. Soc., 88*, 2075 (1966); see ref. 18 therein.
90. King, R.B., *Transition-Metal Organometallic Chemistry*, Academic Press, New York 1969.
91. King, R.B., and Saran, M.S., *J. Amer. Chem. Soc., 94*, 1784 (1972).
92. Kochhar, R.K., and Pettit, R., *J. Organometal. Chem., 6*, 272 (1966).
93. Koerner von Gustorf, E., and Jun, M.J., *Z. Naturforsch. B 20*, 521 (1965).

94. Koerner von Gustorf, E., Henry, M.C., Sacher, R.E., and DiPietro, C., *Z. Naturforsch.*, *B 21*, 1152 (1966).
95. Koerner von Gustorf, E., and Wagner, R., *Angew. Chem.*, *83*, 968 (1971); *Angew. Chem. Int. Ed. Engl.*, *10*, 910 (1971).
96. Kritskaya, I.I., Zol'nikova, G.P., Leshcheva, I.F., Ustynyuk, Yu.A., and Nesmeyanov, A.N., *J. Organometal. Chem.*, *30*, 103 (1971).
97. Landesberg, J.M., and Sieczkowski, J., *J. Amer. Chem. Soc.*, *90*, 1655 (1968).
98. Landesberg, J.M., and Sieczkowski, J., *J. Amer. Chem. Soc.*, *93*, 972 (1971).
99. Landesberg, J.M., Katz, L., and Olsen, C., *J. Org. Chem.*, *37*, 930 (1972).
100. Longuet-Higgins, H.C., and Orgel, L.E., *J. Chem. Soc.*, *1956*, 1969.
101. Mente, P.G., and Rees, C.W., *J. Chem. Soc. Chem. Commun.*, *1972*, 418.
102. Meunier-Piret, J., Piret, P., and von Meerssche, M., *Acta Crystallogr.*, *19*, 85 (1965).
103. Mills, O.S., and Redhouse, A.D., *J. Chem. Soc. A.*, *1968*, 1282.
104. Nakamura, A., *Bull. Chem. Soc. Jap.*, *38*, 1868 (1965).
105. Nakamura, A., Kim, P.-J., and Hagihara, N., *J. Organometal. Chem.*, *3*, 7 (1965).
106. Nakamura, A., Kim, P.-J., and Hagihara, N., *J. Organometal. Chem.*, *6*, 420 (1966).
107. Nakamura, A., *Kagaku No Ryoiki*, *89*, 285 (1970).
108. Noyori, R., Nishimura, T., and Takaya, H., *J. Chem. Soc. D, Chem. Commun.*, *1969*, 89.
109. Öfele, K., and Kreiter, C.G., *Chem. Ber.*, *105*, 529 (1972).
110. Otsuka, S., Nakamura, A., and Yoshida, T., *Bull. Chem. Soc. Jap.*, *40*, 1266 (1967).
111. Otsuka, S., Nakamura, A., and Yoshida, T., *J. Organometal. Chem.*, *7*, 339 (1967).
112. Otsuka, S., Yoshida, T., and Nakamura, A., *Inorg. Chem.*, *8*, 2514 (1969).
113. Pettit, R., private communication.
114. Piper, T.S., Cotton, F.A., and Wilkinson, G., *J. Inorg. Nucl. Chem.*, *1*, 165 (1955).
115. Reardon, J.E., Jr., and Brookhart, M., *J. Amer. Chem. Soc.*, *95*, 4311 (1973).
116. Rodeheaver, G.T., Farrant, G.C., and Hunt, D.F., *J. Organometal. Chem.*, *30*, C 22 (1971).
117. Roe, D.M., and Massey, A.G., *J. Organometal. Chem.*, *23*, 547 (1970).
118. Rosenblum, M., and Gatsonis, C., *J. Amer. Chem. Soc.*,

89, 5074 (1967).

119. Roth, W.R., and Meier, J.D., *Tetrahedron Lett.*, *1967*, 2053.
120. Schmid, G., *Chem. Unserer Zeit*, *8*, 26 (1974).
121. Scholes, G., Graham, C.R., and Brookhart, M., *J. Amer. Chem. Soc.*, *96*, 5665 (1974).
122. Schrauzer, G.N., *J. Amer. Chem. Soc.*, *81*, 5307 (1959).
123. Schrauzer, G.N., Rabinowitz, H.N., Frank, J.A.K., and Paul, I.C., *J. Amer. Chem. Soc.*, *92*, 212 (1970).
124. Schrauzer, G.N., and Kisch, H., *J. Amer. Chem. Soc.*, *95*, 2501 (1973).
125. Schubert, W.M., Liddicoet, T.H., and Lanka, W.A., *J. Amer. Chem. Soc.*, *76*, 1929 (1954).
126. Takahashi, K., Iwanami, M., Tsai, A., Chang, P.L., Harlow, R.L., Harris, L.E., Mc Caskie, J.E., Pfluger, C.E., and Dittmer, D.C., *J. Amer. Chem. Soc.*, *95*, 6113 (1973).
127. Thyret, H., *Angew. Chem.*, *84*, 581 (1972); *Angew. Chem. Int. Ed. Engl.*, *11*, 520 (1972).
128. Victor, R., Ben-Shoshan, R., and Sarel, S., *J. Chem. Soc. D, Chem. Commun.*, *1971*, 1241.
129. Victor, R., and Ben-Shoshan, R., *J. Chem. Soc. Chem. Commun.*, *1974*, 93.
130. Ward, J.S., and Pettit, R., *J. Chem. Soc. D, Chem. Commun.*, *1970*, 1419.
131. Weidemüller, E., and Hafner, K., *Angew. Chem.*, *85*, 958 (1973); *Angew. Chem. Int. Ed. Engl.*, *12*, 925 (1973).
132. Weiss, E., Merényi, R.G., and Hübel, W., *Chem. Ind. (London)*, *1960*, 407.
133. Weiss, E., Hübel, W., and Merényi, R., *Chem. Ber.*, *95*, 1155 (1962).
134. Weiss, E., Merényi, R., and Hübel, W., *Chem. Ber.*, *95*, 1170 (1962).
135. Weiss, E., and Hübel, W., *Chem. Ber.*, *95*, 1186 (1962).
136. Yamamoto, Y., Aoki, K., and Yamazaki, H., *J. Amer. Chem. Soc.*, *96*, 2647 (1974).
137. Yasuda, N., Kai, Y., Yasuoka, N., and Kasai, N., *J. Chem. Soc. Chem. Commun.*, *1972*, 157.

INDEX

A

B

C

D

G

H

I

K

L

M

P

Q

R

S

T

V

W, X, Z

A
B
C 8
D 9
E 0
F 1
G 2
H 3
I 4
J 5